AF718696

Beyond Equilibrium Thermodynamics

Beyond Equilibrium Thermodynamics

Hans Christian Öttinger

WILEY-INTERSCIENCE

A JOHN WILEY & SONS, INC., PUBLICATION

Copyright © 2005 by John Wiley & Sons, Inc. All rights reserved.

Published by John Wiley & Sons, Inc., Hoboken, New Jersey.
Published simultaneously in Canada.

No part of this publication may be reproduced, stored in a retrieval system or transmitted in any form or by any means, electronic, mechanical, photocopying, recording, scanning or otherwise, except as permitted under Section 107 or 108 of the 1976 United States Copyright Act, without either the prior written permission of the Publisher, or authorization through payment of the appropriate per-copy fee to the Copyright Clearance Center, Inc., 222 Rosewood Drive, Danvers, MA 01923, (978) 750-8400, fax (978) 646-8600, or on the web at www.copyright.com. Requests to the Publisher for permission should be addressed to the Permissions Department, John Wiley & Sons, Inc., 111 River Street, Hoboken, NJ 07030, (201) 748-6011, fax (201) 748-6008.

Limit of Liability/Disclaimer of Warranty: While the publisher and author have used their best efforts in preparing this book, they make no representation or warranties with respect to the accuracy or completeness of the contents of this book and specifically disclaim any implied warranties of merchantability or fitness for a particular purpose. No warranty may be created or extended by sales representatives or written sales materials. The advice and strategies contained herein may not be suitable for your situation. You should consult with a professional where appropriate. Neither the publisher nor author shall be liable for any loss of profit or any other commercial damages, including but not limited to special, incidental, consequential, or other damages.

For general information on our other products and services please contact our Customer Care Department within the U.S. at 877-762-2974, outside the U.S. at 317-572-3993 or fax 317-572-4002.

Wiley also publishes its books in a variety of electronic formats. Some content that appears in print, however, may not be available in electronic format.

Library of Congress Cataloging-in-Publication Data is available.

ISBN 0-471-66658-0

Printed in the United States of America

10 9 8 7 6 5 4 3 2 1

To Ricarda

Contents

Part I Phenomenological Approach

Part II Statistical Approach

Preface

The Credo of Coarse-Graining

"Everything should be made as simple as possible, but not simpler."

Albert Einstein[1]

"It's too early to write such a book," was my standard reply to colleagues who asked me when I would finally write a book on modern nonequilibrium thermodynamics. I'm afraid that this answer is still true, even today, on September 23, 2004. Therefore, what you have in your hands should not be considered as a regular or final book. Rather it is the beginning of a project, the maybe premature birth of a textbook–encyclopedia–research guide *chimera*, a vivid vision to be materialized in forthcoming decades. The title of the project, "Beyond Equilibrium Thermodynamics," is meant to indicate that its horizon extends far beyond the well-established regimes of equilibrium and linear irreversible thermodynamics; I hope that the book-like object in your hands will actually help to shape a new paradigm of "beyond-equilibrium thermodynamics," useful to solve many existing problems and to identify even more interesting new ones, providing powerful methods and a unified language encompassing and clarifying all the competing previous formulations of nonequilibrium thermodynamics.

In my attempt at painting a balanced picture of beyond-equilibrium thermodynamics, several new problems have already been identified and solved. For example, the treatment of redundant variables in Section 1.2.1, the thermodynamically admissible

[1] According to Reader's Digest (October 1977).

incorporation of diffusion into the Navier-Stokes equations in Section 2.2.5, the formulation of linear response theory within the unified framework in Section 6.2.3, the thermodynamically admissible kinetic theory of gases in Section 7.4.3, the thermodynamically guided analysis of molecular dynamics simulations of entangled polymers in Section 8.4.6, and the geometric derivation of Poisson brackets for configurational distribution functions in Appendix B.4 are all original contributions, published for the first time. Some of the earlier developments stimulated by this project have been published in separate papers.

Over the past four decades, the phenomenological theory of irreversible processes has matured from a collection of linear constitutive laws and phenomenological balance equations into a robust, systematic, and practical methodology to treat nonequilibrium systems. However, the methods of statistical mechanics, in particular well-founded simulation techniques, need to be further developed into a powerful tool for engineering applications. Among the basic problems awaiting further clarification, boundary conditions and open systems are prominent examples, and a unified approach to nonequilibrium systems should cover rare jumps in addition to the continuous dynamics of slow variables.

Nonequilibrium thermodynamics is the key to *understanding* time-dependent phenomena. In order to appreciate such a bold statement one should realize that there are different concepts of "understanding." On the one hand, one can argue that a deep understanding of nature should be achieved through a consistent theory of all fundamental particles and interactions. On the other hand, understanding certain phenomena is more often regarded as the art of distilling the essence of the phenomena, that is, describing them in as simple terms as possible, with as few variables as absolutely necessary. For example, clearly an "understanding of turbulence" has nothing to do with fundamental particles; what is one really looking for in view of the fact that the Navier-Stokes equation can probably reproduce all the observed phenomena of turbulence? After sketching four different approaches to the scientific study of turbulence, Kadanoff states that,[2] "In each case, we are trying to isolate elements of the flow that are open to prediction, replication, and comparison among different systems." The process of understanding can be regarded as the search for the coarsest possible description of a phenomenon, and nonequilibrium thermodynamics, as the theory of coarse-graining, is hence a cornerstone for understanding the essence of time-dependent phenomena. It should therefore be useful to present a unified theory of coarse-grained systems and the procedure of coarse-graining as the road to understanding through simplicity. Thermodynamics provides the road, and one should not be disappointed that one still needs the vehicle of creative ideas for progressing along that road.

While linear irreversible thermodynamics was fully developed in the mid-20th century, a unified formulation of nonlinear nonequilibrium thermodynamics is clearly needed. Most complex fluids processed in various industries exhibit highly nonlinear material behavior, and progress in the understanding of soft condensed matter is linked

[2] Kadanoff, Physics Today 54/8 (2001) 34.

to new developments in thermodynamics. As "*non*linear *non*equilibrium" sounds too negative, we here refer to the unified new approach to nonequilibrium phenomena with an emphasis on the nonlinear regime as "beyond-equilibrium thermodynamics." This new term sounds more positive and nicely emphasizes the important message that constructive tools are now available for systems far away from equilibrium. These tools should be useful to many scientists and engineers, and my goal hence is to address the curiosity of theoretical physicists as much as the needs of chemical engineers.

Among the classical subjects of nonequilibrium thermodynamics, balance equations, hydrodynamics, the local-equilibrium assumption, the linear thermodynamics of irreversible processes, the Onsager-Casimir relationships, projection-operator techniques, Boltzmann's kinetic equation, linear response theory and the fluctuation-dissipation theorems of the first and second kind are covered extensively in the unified approach followed here. The Chapman-Enskog method and Grad's moment expansion, as important tools to solve equations, are discussed, too. On the other hand, classical topics like the BBGKY hierarchy lose their importance in the new approach to beyond-equilibrium thermodynamics and are hence only mentioned in passing.

Zürich and Rafz
September 2004

HANS CHRISTIAN ÖTTINGER

Acknowledgments

I am indebted to three colleagues who volunteered to read and criticize almost the entire manuscript form three different perspectives for their invaluable help: Rob Jongschaap as a nonequilibrium thermodynamicist, Markus Hütter as a physicist, and Manuel Laso as a chemical engineer. My own perspective on the field of nonequilibrium thermodynamics has been strongly influenced by the pioneering work of Miroslav Grmela, who has always been a most inspiring teacher for me. In addition to Miroslav Grmela and Rob Jongschaap, Antony Beris, Brian Edwards, Alexander Gorban, and Iliya Karlin with their expertise in nonequilibrium thermodynamics opened my eyes and widened my horizon in innumerable enjoyable discussions. On specific topics, I benefited from interactions with Martin Kröger, Vlasis Mavrantzas, Wolfgang Muschik, Pep Español, Juan José de Pablo, Yannis Kevrekidis, David Keffer, Daniel Lhuillier, Norman Wagner, Patrick Ilg, Jan van Meerveld, Clarisse Luap, and Dave Venerus.

hco

SYMBOLS AND NOTATION

Latin Symbols

A, B, C	Observables; functionals of independent variables – 1.2, 2, 4–6, 7.2, 7.4, 8.4, B, C, D
A', B'	Transformed observables – 1.2
$\hat{A}, \hat{B}$	Modified observables – 7.4
$A_{\text{red}}, B_{\text{red}}$	Reduced observables – B
$\bar{A}, \bar{B}$	Coefficients in Chapman-Enskog solution – 7.3
A^{LIT}	Antisymmetric transformation matrix – 3.1
A	Drift velocity – 8.3
A_k	Hilbert space operators – D
A_c	Area of cross section – 1.2, 3.1
A	Lie algebra – B
a	Scale factor – 5.5
a	Lattice spacing – 6.4
a	Contribution to drift velocity – 8.3
a_T	Shift factor for time-temperature superposition – 4.1
B, B'	Amplitude of noise – 1.2, 2.2, 2.3, 3.1, 6.4, 8.3
B_k	Bernoulli numbers – 4.3
b	Finite extensibility parameter – 4.2, 8.3
b	Impact parameter – 7.1, 7.2
$\hat{b}$	Dimensionless impact parameter – 7.1, 7.3
$b_{jk}, \bar{b}_{jk}$	Components of body tensor – B
C	Covariance matrix – 8.3
C, C_j	Closed sets of states – 8.2
C^j_{kl}	Coefficients of Lie algebra – B
c	Speed of light – 5, 8.1
c, c'	Constants – 1.2, 2.1, 6.4, B, C
c_j	Constants – B
c^j	Coordinates associated with Casimirs – B
$c^{\text{Maxw}}, c^{\text{hs}}$	Numerical constants – 7.3, 7.4
$\boldsymbol{c}, \boldsymbol{c}'$	Conformation tensors – 2.3, 4–5, 6.4, 8
$\boldsymbol{c}_k$	Conformation tensor of chain k – 8.2, 8.4
$\boldsymbol{c}, \boldsymbol{c}^*$	Conformation tensor and dual – B
$\overline{\boldsymbol{c}}$	Trace part of conformation tensor – 5.1, 5.2
$\mathring{\boldsymbol{c}}$	Traceless part of conformation tensor – 5.1, 5.2, 7.4, 8.4
$c_{\mu\nu}, C_{\mu\nu}$	Relativistic generalizations of conformation tensor – 5
$\overline{c}_{\mu\nu}, \mathring{c}_{\mu\nu}$	Relativistic generalizations of $\overline{\boldsymbol{c}}$, $\mathring{\boldsymbol{c}}$ – 5.2, 5.4
c_1, c_2, c_3	Eigenvalues of conformation tensor – 4.2
c_j	Concentration variables – 2.1, 2.2, 3.2, 7.1
$\hat{c}_V, \hat{c}_p$	Heat capacities per unit mass – 2.2, 3.1, 7.1, 7.3
C_M, D_M	Factors in the friction matrix M – 2.3, 3.1, 3.2
C'_M, D'_M	Transformed factors in the friction matrix M – 3.1

D	Diffusion matrix – 8.3
D, D'	Diffusion coefficient, and related quantity – 2, 3, 6.2, 7.1
$\boldsymbol{D}, \boldsymbol{D}'$	Symmetric tensors describing diffusion – 4.2, 4.3
D_{tube}	Curvelinear diffusion coefficient – 4.3
$D_{\text{E}}, D_{\text{GK}}, D_{\text{HE}}$	Expressions for diffusion coefficient – 8.4
D	Dirac structure – B
D_{c}	Constrained Dirac structure – B
D_k	Space of k-forms – B
$\boldsymbol{d}$	Dummy vector variable – C
d	Number of degrees of freedom – 1.2
d	Diameter of a sphere – 7, 8.4
d_P	Period of Markov chain – 8.2
dA	Surface element – 2.1
$đQ$	Heat – 1.2, A
$đW$	Work – 1.2, A
E, E'	Energy
$\tilde{E}$	Energy – 6.3
E_0	Atomistic energy – II, 6, 7.2, 7.5, 8.4
E_j	Energies – A
E_{e}	Energy of environment – 1.2
E_{internal}	Internal energy – 6.1
$\boldsymbol{E}$	Electric field – 3.1, 3.2
e	Elementary charge [$1.60217653(14) \times 10^{-19}$C]
e	Energy density – C
$\boldsymbol{F}$	Force – 3.1
$\boldsymbol{F}_j$	Force on particle j – 6.1, 6.2, 8.4
F	Helmholtz free energy – 1.2, A
F	Scalar variable related to bulk viscous pressure – 5.5
F	Hilbert space operator – D
$F, F_{jk}, F_{jk}^{(n)}$	Probability for reaching target state – 8.2
$F, \bar{F}$	Transformation rules – B
F	Real-valued functions – B
f	Function, space transformation – 2.2, 6.1, B, C
f_1, f_2	Functions, space transformations – B
f	Single-particle distribution function – 7
f, f_x	Probability density – 1.2, 4, 6, 8.3, 8.4
f^π	Atomistic single-particle distribution function – 7.2, 7.5
f^π	Atomistic single-segment distribution function – 8.4
$f_{\rho,\boldsymbol{v},T}^{\text{Maxw}}$	Maxwellian momentum distribution – 7
$f_{\text{t}}, f_{\text{r}}$	Probability densities for total and relative momentum – 7.4
$f_j, f^{(j)}$	Terms in expansion of f – 7.3, 7.4
$\hat{f}^{(1)}$	Ansatz for linearized distribution function – 7.2, 7.4
f_2	Two-particle distribution function – 7.5
f_2^π	Atomistic two-particle distribution function – 7.5

f_c	Configurational Helmholtz free energy density – 4.2, 8.2
$\overline{f}$	Number of degrees of freedom – 8.4, A
$\overline{f}, \mathring{f}, f^w_{\mu\nu}$	Factors in dissipative bracket – 5.4
G	Gibbs free energy – A
G	Shear relaxation modulus – 4
G'	Storage modulus – 4.1
G''	Loss modulus – 4.1
G^*	Complex modulus – 4.1
G	Time-ordered exponential solution – 6.1, 6.2, D
G_N	Newton's gravitational constant [$6.6742(10) \times 10^{-11}\mathrm{m}^3/(\mathrm{kg\,s}^2)$]
$\mathcal{G}, \tilde{\mathcal{G}}$	Lie group – 6.1, B
G, G^*	Lie algebra and dual – B
$\boldsymbol{g}, \tilde{\boldsymbol{g}}$	Force per unit mass – 2.1, 2.2, 6.2, 7.5
$\boldsymbol{g}, \boldsymbol{g}_1, \boldsymbol{g}_2$	Tensor functions – 4.2
g, g_1, g_2	Functions – 8, B
$\bar{g}_1, \bar{g}_2$	Averaged functions – 8.4
g	Negative of determinant of metric tensor – 5.4, 5.5
$g, \bar{g}$	Elements of Lie group – 6.1, B
$\mathsf{g}, \bar{\mathsf{g}}, \bar{\bar{\mathsf{g}}}$	Elements of Lie algebra – B
$\mathsf{g}^*, \bar{\mathsf{g}}^*$	Elements of dual Lie algebra – B
g_j	Coefficients in expansion of $\boldsymbol{g}$ – 4.2
$g_{\mu\nu}$	Metric tensor – 5.4, 5.5
$g^{\mu\nu}$	Inverse of metric tensor – 5.4, 5.5
H	Heaviside step function – 4.2
H	Hubble's parameter – 5.5
H	Enthalpy – A
$H, \bar{H}$	Spring constants – 4.3, 8.3, 8.4, C
H	H function – II
H	Hamiltonian – B, D
H_red	Reduced Hamiltonian – B
H'	Hamiltonian – 8.4
H_0	Initial value in Hamiltonian – 8.4
H_bend	Spring constant for bending – 8.2
H_B	Boltzmann's H function – 7.1, 7.2
h, h_0	Height – 7.1
h, h'	Effective spring constant – 4.2, 8.2
$\tilde{h}$	Parameter modifying effective spring constant – 8.2
$\hbar$	Planck's constant over 2π [$1.05457168(18) \times 10^{-34}$ Js]
h^*	Hydrodynamic interaction parameter – 4.3
$\hat{h}$	Auxiliary function – 5.5
$h_{\mu\nu}$	Spatial projector – 5.4, 5.5
I	Action integral – 5.3, 5.4
I_G	Gravitational action – 5.4
I_M	Matter action – 5.4
I_j	Invariants of a tensor – 4.2

I^{el}	Electric current – 3.2
$\boldsymbol{I}_\phi, \bar{\boldsymbol{I}}_\phi$	Auxiliary integrals in collision processes – 7.2, 7.4
$\hat{I}$	Interaction impulse integral – 7.4
$\boldsymbol{I}_{\mathrm{CB}}$	Fourth-rank tensor in the Curtiss-Bird model – 4.3
i	Index for components or summations – 1.2, 2.3, 7.4, B
$i = \sqrt{-1}$	Imaginary unit – 4.1, 6, 7.5, D
i, j	Labels; states – A
$\boldsymbol{i}$	Electric current density – 3.1, 3.2
i_{v}	Insertion operator – B
J	Global entropy – 5.3, 5.4
J	Rate of change of microstate – II
$J, \tilde{J}$	Momentum map, associated map – B
$J^{\mathrm{LIT}}, J^{\mathrm{LIT}\prime}$	Fluxes of linear irreversible thermodynamics – 3, 4.1, 5.1
J^{diff}	Flux associated with diffusion of configurational variables – 4.2
$J^{\mathrm{relax}}, J_j^{\mathrm{relax}}$	Fluxes associated with relaxation of configurational variables – 4.2, 5.1
J^{q}	Heat production rate in Peltier effect – 3.2
j, k, l	Indices for components or summations – 1.2, 2–4, 5.1, 5.2, 6, 7.2–7.5, 8, B, C, D
$\bar{j}, \bar{k}$	Indices for components or summations – B
j_0	Particular value of index j – 4.3
$\boldsymbol{j}_k^{\mathrm{diff}}$	Diffusion flow of component k – 2, 3
$\boldsymbol{j}^{\mathrm{q}}$	Internal energy flux – 2, 3, 7.3, 7.5
$\boldsymbol{j}^{\mathrm{s}}$	Non-convective entropy flux – 3.2, 5.1, 5.2
j^x	Flux of a molecular property – 7.1
$\boldsymbol{K}$	Elliptic integral – 7.1
K	Set of labels – 6.3, 6.4
K_j	Modified Bessel function of order j – 5.5
k	Dimensionless parameter – 5.5
k	Argument of elliptic integral – 7.1
k_{B}	Boltzmann's constant [$1.3806505(24) \times 10^{-23}$ J/K]
L, L'	Poisson matrix or operator – 1.2, 2.2, 3.1, 4, 5.1, 5.5, II, 6, 7, 8.1, 8.3, B, D
$\tilde{L}$	Poisson matrix – 6.3, 7.2
L^{LIT}	Phenomenological matrix – 3.1, 3.2
$L_{\mathrm{inst}}^{\mathrm{LIT}}, L_{\mathrm{mem}}^{\mathrm{LIT}}$	Generalizations of L^{LIT} with memory – 4.1
L_0	Cosymplectic matrix – 1.2, II, 6.1, 6.3, 7.2, 8.4
L^{cm}	Lagrangian of classical mechanics – 1.2
L_{g}	Characteristic length scale of a geometry – 1.2, 2.1, 3.1, 7.1
L_{e}	Entanglement length scale – 4.3, 8.4
L_{K}	Kuhn length – 4.3
L_{m}	Size of a monomer – 4.3
$\mathcal{L}$	Liouville operator – 6, 7.2, 7.5, 8.4, D
$\mathcal{L}_0, \mathcal{L}_1$	Contributions to Liouville operator – 6.2

$\hat{\mathcal{L}}$	Operator related to $\mathcal{L}$ – 6.1
$\mathcal{L}_j^l$	Differential operator – B
l'	Index for components or summations – 6.1
l_c	Cutoff length scale of hydrodynamics – 6.1
l_{md}	Mean distance between particles – 7.1
l_{mfp}	Mean free path – 7.1, 8.4
M, M'	Friction matrix – 1–8, D
M', M''	Contributions to friction matrix – 6.4
M	Number of phases – A
$\tilde{M}$	Friction matrix – 6.3
$\bar{M}$	Coefficient in friction matrix – 7.4, 8.4
M_μ	Momentum density four-vector – 5.4
$\boldsymbol{M}$	Momentum density – 1.1, 2–4, 5.1, 5.2, 6, 7.1, 7.3, B, C
$\tilde{\boldsymbol{M}}$	Diffusion mechanism – 4.2
M_e	Entanglement molecular weight – 4.3
M_m	Molecular weight of a monomer – 4.3
M_p	Molecular weight of a polymer – 4.2, 4.3
M_L, M_T	Multiplostat inertial parameters – 8.4
m	Mass of a particle – 1.2, 2, 3, 4.2, 5.4, 5.5, 6, 7, 8.2, 8.4
m_j	Mass of particle j – 6.1, 6.2, 7.1, 8.4
m_r	Reduced mass – 7.1
$m, \bar{m}$	Index for components or summations – 5.1, B
N	Number of particles, beads, or states – 1.2, 2, 4.3, II, 6–8, C
N_j	Number of particles – 7.1, 8.2
N, N_j	Mole numbers – A
N_p	Number of polymer chains – 8.2, 8.4
N_A	Avogadro's number [$6.0221415(10) \times 10^{23}$ mol^{-1}]
n	Number of constraints – 1.2
n	Number of components – 8.3, B
n	Index for random variables – 8.2
n	Number density of particles – 5.4, 5.5, 7.1, 8.4
n_p	Number density of polymers – 4.2, 4.3, 5.1, 8
n_ϕ	Exponent in power-law potential – 7.1, 7.3
$\boldsymbol{n}$	Normal unit vector – 2.1
P	Momentum variable – 1.2, 3.1, 8.4
P	Pressure – A
P_i	Intensive state variables – A
P_i^e	External mechanical forces – A
P, P_{jk}	Probability, transition probabilities – 8.2
P_l, P_r	Stochastic matrices – 8.2
P^{mov}, P_{jk}^{mov}	Stochastic matrix for Monte Carlo moves – 8.2
P^{qe}	Quasi-equilibrium projector – II
P^{th}	Thermodynamic projector – II
$\boldsymbol{P}$	Transverse projector – 4.3
$\boldsymbol{P}$	Total momentum – 6.1

$\boldsymbol{P}, \boldsymbol{P}'$	Relative momenta – 7.1, 7.2, 7.4
$\mathcal{P}, \tilde{\mathcal{P}}$	Projection operators – 6, 7.2, D
$\mathcal{P}$	Manifold – B
$\mathcal{P}_{\mathrm{red}}$	Reduced manifold – B
p	Pressure – 1.2, 2, 3.1, 5, 7.1, 7.3, 8.2, 8.4
p_0	Local-equilibrium pressure – 5.5
$p_{\mathrm{L}}, p_{\mathrm{T}}$	Multiplostat generalized momenta – 8.4
$\tilde{p}$	Probability – 8.2
p	Point – B
$\boldsymbol{p}, \boldsymbol{p}', \boldsymbol{p}''$	Momentum vectors – 1.2, 7
$\boldsymbol{p}_j, \boldsymbol{p}'_j$	Momentum vectors – 1.2, II, 6–8
$\boldsymbol{p}_{\mathrm{t}}$	Total momentum – 7
Q	Position variable – 1.2, 8.4
Q_0	Maximum spring extension – 8.3
Q^{G}	Transformation matrix in GENERIC – 1.2
Q^{LIT}	Transformation matrix in linear irreversible thermodynamics – 3.1
Q^{Rouse}	Transformation matrix for Rouse modes – 4.3
$\boldsymbol{Q}_{rr}, \boldsymbol{Q}_{pp}$	Transformation matrices in canonical transformations – 6.1
$\boldsymbol{Q}_{rp}, \boldsymbol{Q}_{pr}$	Transformation matrices in canonical transformations – 6.1
$\boldsymbol{Q}$	Relative momentum – 7
$\boldsymbol{Q}, \boldsymbol{Q}', \bar{\boldsymbol{Q}}$	Connector vectors – 4.3, 6.4, 8
$\boldsymbol{Q}_j, \boldsymbol{Q}'_j$	Connector vectors – 4.3, 8.4
$\boldsymbol{Q}^{\mathrm{DE}}$	Deformation measure of Doi-Edwards model – 4.3
$\mathcal{Q}, \mathcal{Q}', \tilde{\mathcal{Q}}$	Projection operators – 6, 7.2, 7.5, 8.4, D
$Q^l, q^l, \bar{q}^l$	Additional variables in a body field – B
$\boldsymbol{q}, \boldsymbol{q}'$	Momentum vectors – 7.1, 7.3
q, q_0	Generalized coordinates – 1.2, 8.4
q	Number of arms in pompon model – 4.2
q	Energy density – 5.4, 5.5
$\hat{q}, \check{q}$	Functions in energy density – 5.5
$q_{\mathrm{L}}, q_{\mathrm{T}}$	Multiplostat scaling variables – 8.4
$\boldsymbol{R}$	Relative position vector – 7
R	Distance – 7, 8.4
R, R_{jk}	Potential matrix – 8.2
$R_1, \boldsymbol{R}_2, \boldsymbol{R}_3$	Relaxation mechanisms – 4.2, 8.2
$\hat{\boldsymbol{R}}_2, \hat{\boldsymbol{R}}_3$	Tensors associated with $\boldsymbol{R}_2$ and $\boldsymbol{R}_3$ – 4.2, 8.2
$R_{\mu\nu}$	Ricci tensor – 5.4, 5.5
R_{kl}	Matrix representation of a group – 6.1
$R_{\mathrm{g}}, R^*_{\mathrm{g}}$	Representations of Lie algebra – B
Re	Reynolds number – 2.1
r	Radius, distance – 7.1, 8.2, 8.4
r	Ratio of terms in asymptotic series – 4.3
r	Number of components in a mixture – A
r_0, r_2	Auxiliary variables – 5.1, 5.2
$\boldsymbol{r}, \boldsymbol{r}', \boldsymbol{r}''$	Position vectors

$\boldsymbol{r}_j, \boldsymbol{r}'_j$	Position vectors – 1.2, II, 6–8, C
S, S'	Entropy
$\tilde{S}$	Entropy – 6.3
$\hat{S}$	Entropy as a function of Lagrange multipliers – 6.1
$S^{\mathrm{mc}}, S^{\mathrm{c}}$	Microcanonical and canonical entropies – 6.1
S^{ideal}	Entropy of ideal gas – 8.2
S^{μ}	Entropy current four-vector – 5.2
S_{e}	Entropy of environment – 1.2
$S^{(\phi)}$	Entropy associated with entropic potential – 4.3
$\boldsymbol{S}$	Orientation tensor – 4.2
$\boldsymbol{S}'$	Deviation from equilibrium orientation tensor – 4.2
$\boldsymbol{S}^{\mathrm{creation}}$	Term in the Doi-Edwards model without independent alignment approximation – 4.3
$\boldsymbol{S}^{\mathrm{drift}}$	Term in the Doi-Edwards model without independent alignment approximation – 4.3
$\mathcal{S}^{g}$	Substitution operators on phase space functions – 6.1
s	Entropy density – 1.2, 2, 4, 5, 7.4, C
s	Number of Legendre-transformed variables – A
s_0	Local-equilibrium contribution to entropy density – 5.1
$\hat{s}$	Entropy per unit mass – 2.2, 2.3
$\hat{s}$	Scalar variable – 4.2
$s, \bar{s}, s^*$	Scalar variables – B
$\hat{s}_{\mathrm{c}}$	Configurational entropy per unit mass – 4.2, 8.2
$\check{s}$	Entropy per particle – 5.5, A
$\check{s}_0$	Local-equilibrium contribution to entropy per particle – 5.5
$s^{(\phi)}$	Entropic spring potential – 4.2, 8.2, 8.3
s^{mix}	Entropy of mixing per unit volume – 2.1
T	Temperature
T^*	Temperature – 5
$\hat{T}$	Formal nonequilibrium temperature – 1.2, 2.2
T_0	Local-equilibrium temperature – 5.5
T^r	Temperature of reservoir – A
$T^{\mu\nu}$	Energy-momentum tensor – 5
$T\mathcal{P}$	Tangent bundle – B
$T_p\mathcal{P}$	Tangent space – B
$T^*\mathcal{P}$	Cotangent bundle – B
$T^*_p\mathcal{P}$	Cotangent space – B
T^m_l	Vector space of tensor fields – B
t, t', t''	Time – 1–8, B, D
$\bar{t}$	Time – B
t	Number of extensive variables – A
t_j	Discrete times – 8.3
$\mathsf{t}, \bar{\mathsf{t}}$	Tensor – B
U	Internal energy – A
U_I, V_I	Functions in action – 5.4

u	Integration variable – 6.2, 7.1, 7.2, D
$\boldsymbol{u}, \boldsymbol{u}', \widehat{\boldsymbol{u}}$	Unit vectors – 4.2, 4.3, 8.4
u^{μ}	Dimensionless velocity four-vector – 5
V	Volume – 1.2, 2, 6.1, 7.1, 8, A, C
V^{μ}	Four-vector field – 5.4
V_{f}	Characteristic flow velocity – 2.1
V, V^*	Vector space and dual – B
v	Velocity – 1.2, 3.1, 7.1
$\breve{v}$	Volume per particle – A
$\boldsymbol{v}$	Velocity – 1.2, 2, 3.1, 4, 5.1, 5.2, 6.1, 7, 8.4, B
$\boldsymbol{v}_{\xi}$	Velocity field – B
$\boldsymbol{v}_j$	Velocity of particle j – 8.4
$\boldsymbol{v}_{\mathrm{m}}$	Velocity of material element characterizing fluxes – 2.2, 3.1
v_{rel}	Relative velocity – 7
$\mathsf{v}, \bar{\mathsf{v}}, \bar{\bar{\mathsf{v}}}$	Vectors – B
$\mathsf{v}^*, \bar{\mathsf{v}}^*$	Dual vectors – B
v_H	Hamiltonian vector field – B
$\mathsf{v}_H^{\mathrm{c}}$	Constrained Hamiltonian vector field – B
$W, \boldsymbol{W}, \boldsymbol{W}'$	Wiener process – 1.2, 6.4, 8.3, 8.4
$\boldsymbol{w}$	Vector field – 2.3, 5, 7.4
w_{μ}	Relativistic four-vector – 5
w	Transition probabilities – 7.1, 7.2
X	Stochastic process – 8.3
X_n	Random variables – 8.2
X_i	Extensive state variables – A
$X^{\mathrm{LIT}}, X^{\mathrm{LIT}'}$	Forces of linear irreversible thermodynamics – 3, 4.1, 5.1
X^{diff}	Force associated with diffusion of configurational variables – 4.2
$X^{\mathrm{relax}}, X_j^{\mathrm{relax}}$	Forces associated with relaxation of configurational variables – 4.2, 5.1
$\boldsymbol{x}$	Vector – 4.2
x, x'	State variables
$\tilde{x}$	State variables – 6.1, 6.4
$\bar{x}$	Extra state variables – B
x_{ref}	Reference set of state variables – 6.1
x^{μ}	Time and space coordinates – 5.2, 5.4
x, y, z	Directions in space – 4, 7.4
$\bar{x}, \bar{y}, \bar{z}$	Auxiliary variables – 4.2, 7.4
$\dot{x}^{\mathrm{f}}$	Rapidly fluctuating part of $\dot{x}$ – 3.2, 8.4
$x^{\{j\}}$	Hydrodynamic fields and their derivatives – 7.3
x_k^{m}	Mole fractions of the components in a mixture – 2.1
x_{p}	Molecular property – 7.1
y	State variables – 6.3, 6.4, 7.4
$\tilde{y}$	State variables – 6.4
Z	Number of entanglements per polymer chain – 4.2, 4.3, 8.4
$Z, \hat{Z}, \tilde{Z}$	Partition function – 6, 7.2, 8.2, A, D

$\mathcal{Z}$	Auxiliary operators – 6.1
z, z'	Phase space coordinates – 1.2, II, 6, 7.2, 7.5, 8.4
z_0	Initial values of phase space coordinates – 6.1
$z^{\mathrm{H}}, z^{\mathrm{L}}$	Microstates – II
$\bar{z}_x, \bar{z}_x^{\mathrm{im}}, \bar{z}_x^{\mathrm{qe}}$	Representative microstates – II
z, z'	Inverse temperature variable – 5.5

Greek Symbols

α	Parameter – 1.2, 2.2, 5.2, 6.2, C
α	Giesekus parameter – 4.2
α	Normalization factor – 8.2, 8.4
α'	Auxiliary function – 5.5
α_j	Relaxation coefficients – 4.2
α_j	Prefactors – 4.3, 8.2
$\alpha_1, \alpha_2, \alpha'_1, \alpha'_2$	Coefficients in energy and entropy – 5.1
α_s	Adiabatic thermal expansivity – 2.2
$\Gamma^{\nu}_{\mu\mu'}, \bar{\Gamma}^{j}_{kl}$	Christoffel symbols – 5.4, 5.5, B
γ	Velocity-dependent relativistic factor – 5.2
γ	Auxiliary function – 5.5
γ	Total shear – 4.1, 4.3
$\dot{\gamma}$	Shear rate – 2.1, 4, 7.4, 8.3, 8.4
$\dot{\gamma}_0$	Shear rate amplitude – 4.1
γ_{jk}	Components of metric tensor – B
$\dot{\boldsymbol{\gamma}}$	Symmetrized velocity gradient tensor – 2–4, 8.2
Δ	Dimensionless energy difference – 8.2
δ	Dirac's δ function – 2.2, 4.1, 4.3, 6, 7.2, 7.5, 8.4, B, C
$\delta_{jk}, \hat{\delta}_{jk}$	Kronecker's δ (and related quantity) – 2.3, 3.2, 4.2, 5.1, 5.2, 6.1, 7.4, 8.2
ϵ	Internal energy density – 1–5, 6.2, 8.2
$\bar{\epsilon}, \bar{\epsilon}_k$	Small expansion parameters – 6.2
ϵ_{mfp}	Dimensionless mean free path – 7.3
ϵ_0	Local-equilibrium contribution to internal energy density – 5.1
ϵ_{c}	Configurational contribution to energy density – 4.2, 8.2
$\hat{\epsilon}_{\mathrm{c}}$	Configurational energy per unit mass – 4.2
$\dot{\epsilon}$	Elongational rate – 4.1
ϵ_{CB}	Link tension coefficient – 4.3
ϵ_{LJ}	Lennard-Jones energy scale – 8.2, 8.4
ε	Time-reversal behavior of observables – 3.2, 8.4
ζ	Friction coefficient – 1.2, 3.1, 4.3, 8.3, 8.4
η	Viscosity – 2–4, 5.1, 7, 8.4
η_j, η_∞	Viscosity contributions – 4.1
η_{s}	Solvent viscosity – 4.2, 4.3
η_{p}	Polymer contribution to viscosity – 8.3
η^+	Stress growth coefficient – 4.1

$\eta_{\rm E}^{+}$	Stress growth coefficient in elongational flow – 4.1
η^{-}	Stress decay coefficient – 4.1
$\eta^{*} = \eta' - i\eta''$	Complex viscosity – 4.1
$\eta_{\mu\nu}$	Minkowski tensor – 5.2, 5.4
$\hat{\eta}_{\mu\nu}$	Spatial projector – 5.2, 5.4
$\theta, \theta', \theta_0$	Angles – 4.3, 7, 8.2
κ	Dilatational or bulk viscosity – 2.1, 2.2, 5.1, 5.5
$\hat{\kappa}$	Combination of κ and η – 2, 3, 4.2
$\boldsymbol{\kappa}$	Transposed velocity gradient tensor – 2.3, 3.1, 4, 5.1, 5.2, 7.4, 8, B
Λ	Scalar configurational variable – 4.2
Λ	Cosmological constant – 5.4
λ, λ_j	Lagrange multipliers; factors – 1.2, 3.1, 6, 7.1, 7.2, 8, B, D
λ_j^A, λ_j^B	Lagrange multipliers – B
λ, λ'	Auxiliary variables – 4.3, 5.4, 7.5, A, C
λ_j	Auxiliary quantities – 4.2, 4.3, 7.4
$\lambda_E, \lambda_N, \boldsymbol{\lambda}_M$	Lagrange multipliers – 1.2, 6.1, 6.2, 8.2, 8.4
$\boldsymbol{\lambda}_c, \boldsymbol{\lambda}_c'$	Conformational Lagrange multiplier – 8.2, 8.4
$\lambda^{\rm q}$	Thermal conductivity – 2, 3, 5.1, 7
$\boldsymbol{\lambda}^{\rm q}$	Thermal conductivity tensor – 4.2
μ	Chemical potential per unit mass – 1.2, 2, 3, 4.3
μ	Chemical potential per mole – A
μ^{*}	Chemical potential per unit mass – 5.1, 5.2
μ'	Chemical potential per particle – 2.1
μ, ν, μ', ν'	Indices with values 0, 1, 2, 3 in relativity – 5
ξ	Slip parameter – 2.3, 4.2, 7.4, 8.2
ξ	Ratio of spring constants – 8.4
$\xi^j, \bar{\xi}^j, \xi_j^*, \bar{\xi}_j^*$	Local coordinates – B
Π, Π'	Slow or constrained phase space function – 1.2, II, 6, 7.5, 8.4, B
$\boldsymbol{\Pi}, \boldsymbol{\Pi}'$	Slow tensorial phase space function – 8.4
$\hat{\Pi}, \Pi_{\boldsymbol{r}}, \boldsymbol{\Pi}_{\boldsymbol{r}}$	Slow phase space functions – 6.1
$\tilde{\Pi}_y$	Atomistic probability density – 6.3
Π	Bulk viscous pressure – 5.5
π	3.1415926...
π	Permutation – B
$\pi_j, \bar{\pi}_j$	Probability distribution – 8.2
$\pi_j^{(0)}$	Initial probability distribution – 8.2
π_j^{∞}	Limit probability distribution – 8.2
$\boldsymbol{\pi}$	Pressure tensor – 2.1, 2.2, 5.1, 5.2, 8.4
$\boldsymbol{\pi}_c$	Contribution to pressure tensor – 4.2, 8.2
$\boldsymbol{\pi}^{(S)}, \boldsymbol{\pi}^{(E)}$	Contributions to pressure tensor – 4.2, 4.3
$\boldsymbol{\pi}_n$	Contact force $[\boldsymbol{\pi}_n = \boldsymbol{n} \cdot \boldsymbol{\pi}]$ – 2.1
$\pi^{\mu\nu}$	Relativistic generalization of pressure tensor – 5.2, 5.4
ρ	Mass density – 1–4, 5.1, 5.2, 6.2, 7, 8, C
ρ	Probability density in phase space – II, 6.1

$\bar{\rho}$	Density matrix – D
$\rho_x, \rho'_{x'}$	Probability densities in phase space – II, 6, 8
ρ_f	Probability density in phase space – 7.2
$\rho_{1,x}, \rho_{2,y}, \hat{\rho}_{2,y}$	Probability densities in phase space – 6.4
$\tilde{\rho}_f$	Probability density in phase space – 6.3
ρ	Energy density in rest frame – 5.4, 5.5
ρ_0, ρ_1	Contributions to energy density – 5.5
$\rho^{\rm el}$	Electric charge density – 3.1, 3.2
Σ	Total entropy production rate – 3.1, C
σ	Local entropy production rate – 2, 3, 5.2, 5.5
σ, σ'	Labels along an entangled polymer chain – 4.3, 8.4
$\sigma, \sigma', \sigma_0$	Cross section for collisions – 5.5, 7
$\sigma^{\rm el}$	Electric conductivity – 3.1, 3.2
$\sigma_{\rm LJ}$	Lennard-Jones length scale – 8.2, 8.4
$\sigma^{*(j)}$	Derivatives of energy – 5.1, 5.2
$\boldsymbol{\sigma}^*$	Tensor related to $\sigma^{*(j)}$ – 5.1, 5.2
$\hat{\boldsymbol{\sigma}}^*, \sigma^{*\mu\nu}$	Relativistic generalization of $\boldsymbol{\sigma}^*$ – 5.2, 5.4
$\boldsymbol{\tau}$	Deviatoric pressure tensor – 2, 3.1, 4.1, 7.3, 7.5
$\boldsymbol{\tau}^{(S)}$	Contribution to deviatoric pressure tensor – 4.2, 4.3
τ	Characteristic time scale – 3.2, 4, 6, 7.2, 7.4, 8, D
$\tau_{\rm b}$	Orientational relaxation time – 4.2
$\tau_{\rm s}$	Relaxation time for backbone stretch – 4.2
$\tau_{\rm s}$	Separating time scale – 8.4
$\tau_{\rm Rouse}$	Rouse time scale – 4.3, 8.4
$\tau_{\rm rept}$	Reptation time scale – 4.3, 8.4
$\tau_{\rm LJ}$	Lennard-Jones time scale – 8.4
τ_*	Characteristic time scale – 5.5
τ_j, τ_∞	Characteristic time scales – 4, 5, 6.4, 8
τ_Ψ	Mean relaxation time scale – 4.1
$\boldsymbol{\Phi}$	Coupling tensor in relaxation mechanism – 4.2
$\boldsymbol{\Phi}_{tt'}$	Solution of linear differential equation – 8.3
$\boldsymbol{\Phi}^l_j$	Infinitesimal generator – B
$\tilde{\phi}$	Coupling in diffusion mechanisms – 4.2
ϕ	Interaction potential – 4.3, 7, 8.2
ϕ_{jk}	Interaction potentials – 7.2
$\phi_{\rm eff}$	Effective potential – 7.1
$\phi_{\rm tot}$	Total potential – 8.2, 8.4
$\phi_{\rm L}$	Barostat potential – 8.4
$\phi^{(S)}$	Entropic potential – 4.3, 8.3, C
$\phi^{(\rm e)}$	External potential – 1.2, 7
$\phi^{\rm el}$	Electric potential – 3.1, 3.2
$\phi^{*(j)}$	Derivatives of energy – 5.1, 5.2
$\boldsymbol{\phi}^*$	Tensor related to $\phi^{*(j)}$ – 5.1, 5.2
$\hat{\boldsymbol{\phi}}^*, \phi^{*\mu\nu}$	Relativistic generalization of $\boldsymbol{\phi}^*$ – 5.2

$\hat{\phi}$	Parameter in power-law potential – 7
φ	Potential per unit mass – 2.1, 3.1
$\varphi, \varphi', \varphi_j$	Angles – 4.3, 7.1, 7.3, 8.2
Ψ_1, Ψ_2	Normal stress coefficients – 4, 7.4, 8.4
Ω	Massieu function – 1.2
Ω	Volume of phase space – 6, A
Ω	Two-form – B
$\bar{\Omega}$	Symplectic operator – B
Ω, Ω'	Solid angle – 7
$\boldsymbol{\omega}$	Vorticity tensor – 7.4
ω, ω'	Angular frequency – 4.1
$\omega, \bar{\omega}, \bar{\bar{\omega}}$	Differential forms – B

Mathematical Symbols and Notation

$\mathbb{R}$	Real numbers
$d\cdot/d\cdot$	Derivative [also $\dot{x}$ for time-derivative dx/dt]
$\delta\cdot/\delta\cdot$	Functional derivative
$\partial\cdot/\partial\cdot$	Partial derivative
$D\cdot/D\cdot$	Substantial or material derivative
$\partial_j\cdot = \partial\cdot/\partial x^j$	Partial derivatives
$D\cdot, D_*\cdot$	Mean forward and backward time-derivatives
$\partial\cdot$	Boundary
$\Delta\cdot$	Increment in a variable
$d\cdot$	Differential
$đ\cdot$	Small amount of a quantity; not differential of a state variable
$d^n\cdot$	Integration in n dimensions
$\cdot^*$	Conjugate of a complex number
$\cdot^\dagger$	Adjoint of an operator
$\cdot^T$	Transpose of a matrix
$\cdot^f$	Space-transformed fields
$\cdot^g$	Transformed by action of a group element
$\cdot_t$	Time-dependent quantity
$\cdot_j, \cdot_{jk}$	Subscript for labeling different particles, components, or contributions
$\cdot_{\mathrm{eq}}$	Equilibrium quantity
$\cdot_{\mathrm{f}}$	Quantity in comoving reference frame
$\cdot$	General contraction, matrix multiplication
$\bullet$	Contraction, as $\cdot$, but an additional integration is implied
$:$	Double contraction [*e.g.*, $\boldsymbol{\pi} : \dot{\boldsymbol{\gamma}} = \pi_{jk} : \dot{\gamma}_{kj}$]
$\{\cdot, \cdot\}$	Poisson bracket
$[\cdot, \cdot]$	Dissipative bracket
$\underline{\cdot}$ or $\underline{\underline{\cdot}}$	Underlines for explicit distinction between matrices and operators
$\mathbf{1}, 1$	Unit tensor
$\wedge$	Wedge product

$\otimes$	Tensor product
$\oplus$	Direct sum
$\oplus, \ominus, \bigcirc$	Dressed Onsager-Casimir symmetry
$\mathcal{OC}_{\text{bare}}$	Bare Onsager-Casimir symmetry operator
$\mathcal{OC}_{\text{dressed}}$	Dressed Onsager-Casimir symmetry operator
$\|\| \cdot \|\|$	Norm
$\langle \cdot \rangle$	Expectation of a random variable – 1.2, 6.4, 8
$\langle \cdot \rangle$	Equilibrium ensemble average – 7.1
$\langle \cdot \rangle_x$	Nonequilibrium ensemble average – 3.2, 6, 8
$\langle \cdot \rangle_t$	Time-dependent nonequilibrium ensemble average – 6.2
$\langle \cdot \rangle_{\boldsymbol{r}}$	Conditional average given a particle at $\boldsymbol{r}$ – 8.4
$\langle \cdot \rangle_\sigma$	Average over orientations at a particular position σ along a reptating polymer chain – 4.3
$\langle\langle \cdot \rangle\rangle$	Average over both orientations and positions for a reptating polymer chain – 4.3
$\langle \cdot \,; \cdot \rangle_x$	Canonical correlation – D

1

Introduction

The goal of this book is to present and illustrate a unified approach to beyond-equilibrium systems. We mostly use the term "beyond-equilibrium" instead of "nonequilibrium" to indicate that we want to go far beyond equilibrium, and the combination "nonlinear nonequilibrium," although perfectly appropriate, sounds too negative. The emphasis of this introduction is to summarize the framework of beyond-equilibrium thermodynamics. Whereas the fundamental equations are strikingly simple and natural, they are also quite challenging. First of all, one needs to get used to the compact and abstract notation in the context of various examples with different sets of variables, ranging from discrete variables to a variety of tensorial fields. Then one has to develop a feeling for the generality and implications of the proposed framework. Finally, the statistical foundations of the proposed framework provide both deeper insight into the nature of the thermodynamic approach and a powerful tool for treating systems of practical interest. Although the presentation of the framework of beyond-equilibrium thermodynamics in this introductory chapter is rather exhaustive, its full appreciation requires an attentive reading of the subsequent chapters. It is highly recommended to return to Chapter 1 after every other chapter to recognize the depth and power of the formalism. Before we describe the unified framework, however, it is worthwhile to position the field of beyond-equilibrium thermodynamics in several ways so that one knows what one can and what one cannot expect from this framework.

1.1 TO BE EXPECTED, OR NOT TO BE EXPECTED

Before getting into any mathematical details, the reader should realize that there are a number of limitations of any general framework for describing beyond-equilibrium dynamics. Although it is similarly general in its abstract formulation, beyond-equilibrium thermodynamics is intrinsically less rigorous than equilibrium thermodynamics, as explained in the first subsection. Like equilibrium thermodynamics, the beyond-equilibrium formalism cannot be derived in a rigorous way; rather, it is based on experience, elegance, plausibility, and statistical mechanics, which all motivate the fundamental equations for beyond-equilibrium systems. In describing complex fluids, one has to go beyond the balance equations of hydrodynamics, and structural variables play a central role for developing realistic theories. We finally offer some guidance on how readers with special interests can benefit from this book.

1.1.1 Thermodynamics and Rigor

Thermodynamics deals with coarse-grained descriptions of systems. Actually, the majority of degrees of freedom is usually eliminated. How rigorous can such coarse-grained descriptions be? In equilibrium thermodynamics, we have the convenient situation that we deal with systems that do not change in time, even over very or infinitely long periods of time. As elaborated by Callen,[1] there is thus a very close relationship between the variables of equilibrium thermodynamics and the symmetries of a system. The most natural candidates for variables that do not change in time are conserved quantities. According to Noether's theorem,[2] such conserved quantities result directly from the continuous symmetries of the system. For example, symmetry under time translation implies the conservation of energy, and the conservation of electric charge is a consequence of the gauge symmetry of Maxwell's equations, expressing some freedom in the choice of the scalar and vector potentials for electromagnetic fields.[3] Another possible reason for variables to become relevant to the description of time independent states lies in the observation that certain dynamic modes in the limit of long wavelengths acquire a vanishing frequency. This observation can be traced back to the Goldstone theorem, which asserts that any system with broken symmetry (and with certain weak restrictions on the atomic interactions) has a spectrum of excitations for which the frequency approaches zero as the wavelength becomes infinitely large. For a crystalline solid, for example, there exist infinitely many equally possible positions, slightly displaced by an arbitrary fraction of a lattice constant, which are related through the broken symmetry of translations. The fact that these positions are equally possible leads to a zero-frequency mode and to the occurrence of the volume as a state variable in equilibrium thermodynamics.

[1] See Chapter 21 of Callen, *Thermodynamics* (Wiley, 1985).
[2] See Section 12-7 of Goldstein, *Classical Mechanics* (Addison-Wesley, 1980).
[3] See, for example, p. 25 and p. 266 of Marsden & Ratiu, *Mechanics and Symmetry* (Springer, 1999).

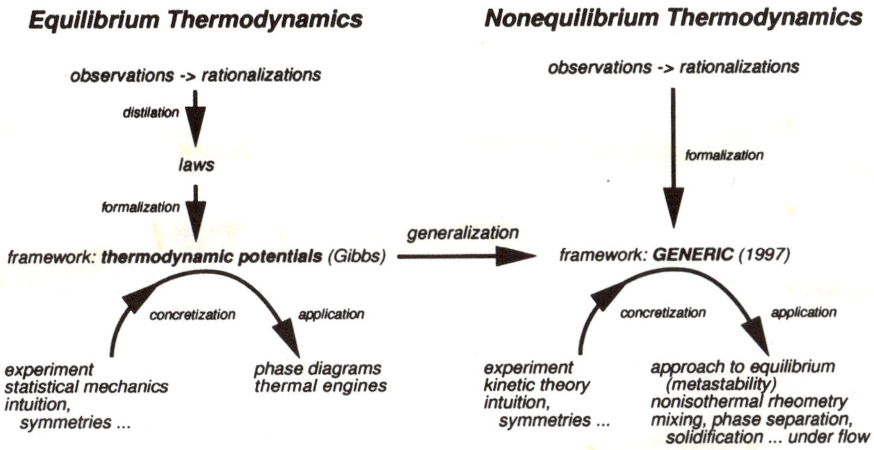

Fig. 1.1 A comparison of the foundations, formulations and applications of equilibrium and nonequilibrium thermodynamics. [Reprinted with permission from H. C. Öttinger, Appl. Rheol. 9 (1999) 17. Copyright (1999) by Applied Rheology.]

The fundamental role of symmetries in specifying equilibrium systems puts equilibrium thermodynamics on a quite rigorous footing. Beyond-equilibrium thermodynamics is in a fundamental sense less rigorous than its equilibrium counterpart. The relevant variables beyond equilibrium are found by separating fast and slow variables and keeping only the slowest ones. By neglecting the fast variables one should expect to introduce errors determined by the ratio of short to long time scales, which is zero in equilibrium thermodynamics and small but finite in beyond-equilibrium thermodynamics. In that sense, beyond-equilibrium thermodynamics is less rigorous. For example, in a detailed investigation of the validity of the Onsager-Casimir symmetry relations (see Section 3.2.1), it turned out that violations of at least second order in the time-scale ratio should be expected.[4] Even in small systems, for example, those used in computer simulations, where the time-scale ratio is not very large, beyond-equilibrium thermodynamics is remarkably successful.[5]

The diagram in Figure 1.1 illustrates further why nonequilibrium thermodynamics is usually considered as less well-founded than equilibrium thermodynamics. At equilibrium, we have fundamental laws summarizing the collected experience of centuries, and Gibbs' formalism can be developed elegantly from these fundamental laws. Away from equilibrium, there is no such set of fundamental laws from which

[4] Geigenmüller et al., Physica A 119 (1983) 53.

[5] The success of thermodynamics may be cemented by the fact that deviations from a macroscopic description are not only rare but also unsystematic, as pointed out by Tribus & Evans, Appl. Mech. Rev. 16 (1963) 765, "Only repeatable experiments (...) are published and accepted as valid data. That is, we are protected by the editors of our journals from ever finding contradictions with the macroscopic predictions ...."

a formalism could be derived, and the GENERIC framework adopted here needs to be extracted from many individual observations. However, whereas the nonexistence of well-established fundamental laws of nonequilibrium thermodynamics may be a problem for the acceptance of a framework, it is unproblematic in practical applications. At and beyond equilibrium, only the final framework is needed for all the applications, without any direct reference to the fundamental laws. A posteriori, a framework of nonequilibrium thermodynamics could be used to formulate fundamental laws, such as the occurrence of irreversible dynamics as "the rest" that cannot be incorporated into a mechanistic-reversible dynamics with an underlying geometric structure. Such a procedure, however, might confuse rather than clarify the situation.

1.1.2 Formulating Versus Deriving Irreversibility

Most scientists view the fundamental equations governing the evolution of natural processes as time reversible; however, many systems display not only reversible but also irreversible dynamical phenomena. The obvious question, then, is as follows: can one *derive* irreversible behavior from reversible equations? Barring a positive answer to this formidable question, one must ask this: is it possible to *formulate* proper equations for describing irreversible phenomena? These questions provide the motivation for the study of beyond-equilibrium thermodynamics.

The answer to the first question is, to date, only partially positive. Enormous effort has been expended toward *deriving* irreversible behavior, generally centered around the famous Boltzmann equation. Still, one does not yet possess a definitive picture of the evolution of reversible dynamics into irreversible phenomena. This book focuses on the second question, that of *formulating* proper equations for describing systems displaying both reversible and irreversible dynamical behavior. Instead of pursuing a heretofore unreached goal of obvious academic interest, we explore the substantial progress made in the formulation of beyond-equilibrium thermodynamics, which is of immense practical importance. Indeed, the situation regarding "derivation" versus "formulation" is exactly the same as in *equilibrium* thermodynamics, as subsequently explained.

Attempts at *deriving* equilibrium thermodynamics from Newton's or Hamilton's equations of motion have led to many significant developments; e.g., ergodic theory was established as a new branch of the mathematical sciences. However, this ambitious goal has not yet been rigorously achieved. Nevertheless, there does exist a fully satisfactory, elegant, and powerful *formulation* of equilibrium thermodynamics, which is enormously useful in all branches of science and engineering. All of the properties of a thermodynamic system at equilibrium can be obtained from a single potential function through application of multivariate calculus. Moreover, the pioneering work of Gibbs has shown that this potential can be evaluated from the partition function, expressed in microscopic terms. Thus, although a strict *derivation* of equilibrium thermodynamics from "first principles" has not been accomplished, one can still *formulate* equilibrium thermodynamics, and one can derive the thermodynamic properties of particular systems from microscopic principles. Indeed, the formulation of equilibrium thermodynamics is one of the high points of physics, in terms of both conceptual beauty and practical utility.

The ultimate goal of beyond-equilibrium thermodynamics is thus to understand how the dynamics of the microscopic constituents of a particular system affect both the reversible and irreversible bulk properties under dynamic conditions, irrespective of how far the system is from a state of global equilibrium. The idea of *coarse-graining*, or averaging over microscopic states, is clearly crucial to understanding how conservative atomic or molecular motions turn into irreversible macroscopic behavior. Nevertheless, it is equally important to understand exactly what is happening in the macroscopic response. Similarly, equilibrium thermodynamics was developed before statistical mechanics, and this eventually lead to a mutually beneficial infusion of knowledge into each subject from the other. The advantages of a general framework for formulating macroscopic behavior will be discussed in Section 1.2.6 after we have introduced and motivated GENERIC as a specific framework.

1.1.3 Beyond Balance Equations

In any textbook on nonequilibrium thermodynamics, one of the first chapters will be entitled "balance equations" or "conservation laws" (see also Section 2.1 below). Once the balance equations expressing local conservation laws are closed by Newton's and Fourier's expressions for the occurring fluxes, that is, for the viscous stress tensor and the heat-flux vector, the equations of *hydrodynamics* are the final outcome of the development. The assumption of a local equilibrium plays an important role in this development, for example, because it allows us to introduce a temperature field. Although the equations of hydrodynamics describe a wide and universal class of real fluids, there is an increasing number of *complex fluids* of theoretical and practical interest. Also for complex fluids, the balance equations for mass, momentum, and energy play an important role, and again the balance equations imply time-evolution equations for the position-dependent mass density ρ, the momentum density $\boldsymbol{M}$ (or the velocity), and the internal energy density ϵ (or the temperature). A closed set of time-evolution equations for complex fluids can then be obtained in two ways:

- Continue to work only with the independent variables $\rho, \boldsymbol{M}, \epsilon$ and express the fluxes as functionals of these variables. In particular, the current fluxes for complex fluids can depend on the entire history of the independent variables. Such memory effects are obviously important for describing viscoelasticity in terms of the hydrodynamic variables.

- Introduce structural variables characterizing the state of a complex fluid in addition to the hydrodynamic variables $\rho, \boldsymbol{M}, \epsilon$. The structural variables need to be chosen such that they contain all the information about the state of the fluid that is necessary for determining the stresses and the heat flux at any given time, without any memory effects.

In the first approach, closely associated with rational thermodynamics and the theory of "simple fluids with fading memory,"[6] the material information is fully contained in the form of the functionals; in the second approach, that information is contained in the relevant structural variables and their time-evolution equations. These alternative possibilities are also referred to as the choice between small and large state spaces.[7] In rheology, one often chooses the stress tensor itself or closely related configuration tensors as additional structural variables. This leads to famous differential constitutive equations, such as the Oldroyd or upper convected Maxwell model. The idea of functionals, on the other hand, is most conveniently realized in terms of integral constitutive equations.

Having the choice between small or large state spaces, between functionals or structural state variables, we are faced with this question: which approach is preferable? The author's answer to that question is very clear and firm. Large state spaces including structural variables are definitely preferable. As known from equilibrium thermodynamics, the choice of suitable state variables is a key step in describing a given thermodynamic system, and physical insight can be directly translated into the proper choice of variables. For a deep understanding of flow behavior, we need to reveal the structure of a complex fluid, that is, we need to find the relevant additional structural state variables. The popular goal of understanding structure-properties relationships can hardly be achieved in terms of abstract functionals. In a fully atomistic description of an arbitrarily complex fluid, the stress and heat flux can clearly be determined from the current microstate of the system. The art of thermodynamics is to find a strongly reduced number of coarse-grained, slow state variables so that the fluxes can still be determined from the values of the current state variables. When using only the hydrodynamic variables, one has no information about the internal structure of the fluid, and this defect is compensated by the abstract and indirect idea of memory effects in terms of functionals. After these remarks, it should not be surprising that atomistic expressions for the building blocks of beyond-equilibrium thermodynamics, which are analogous to Gibbs' expressions for thermodynamic potentials in terms of partition functions, were developed in the context of large state spaces,[8] but not in the context of small state spaces.

The extra variables associated with the structure of a complex fluid can be classified as *order parameters* or *internal variables*. Order parameters describe some long-range order connected to some broken symmetry in a material, such as the director field of a nematic liquid crystal, whereas internal variables describe some slowly relaxing microstructural processes, such as the conformational relaxation of macromolecules. It is a characteristic of internal variables that they are measurable but not controllable, that is, they cannot be adjusted to prescribed values through a direct action of surface or body forces.[9]

[6] Coleman & Noll, Arch. Rational Mech. Anal. 13 (1963) 167.

[7] Muschik et al., *Concepts of Continuum Thermodynamics* (Kielce University, 1996).

[8] Öttinger, Phys. Rev. E 57 (1998) 1416.

[9] See Sections 4.1 and 4.7 of Maugin, *Thermomechanics of Nonlinear Irreversible Behaviors* (World Scientific, 1999).

When we consider a general list of state variables, containing the hydrodynamic fields and additional structural variables, then we are immediately faced with the following questions: although the form of the time-evolution equations for the hydrodynamic variables is fixed by the balance equations, what is the general form of the time-evolution equations for the additional structural variables? Are there any general restrictions or guidelines for constructing such additional equations? The framework of beyond-equilibrium thermodynamics presented in this book gives a complete and constructive answer to such questions. One of the key themes of beyond-equilibrium thermodynamics is the structure of the time-evolution equations arising in addition to the balance equations.

1.1.4 Guide Through the Book

To read this book on beyond-equilibrium thermodynamics, a firm background in equilibrium thermodynamics and statistical mechanics (for Part II only) is required. For the reader's convenience, a brief summary of the structure of equilibrium thermodynamics, including some comments on statistical mechanics, is provided in Appendix A. This appendix should be considered as a reminder rather than as a tutorial.

Chapter 1 plays a very special role, and it should be read at least twice, at the beginning and at the end. To emphasize its significance and to illustrate the plan behind the entire book, I use the analogy of driving lessons. The first reading of Chapter 1 corresponds to studying the traffic regulations, even the entire Road Traffic Act; this may seem abstract and boring, but it nevertheless is an important first step. At the end of Part I, you can drive in Rome, and the horn may appear to be as essential as the steering wheel; looking at the traffic regulations from a completely different perspective, you should then discover deeper values in them. In Part II, you can actually learn how engines work, so that you can choose your favorite car not just by color and fashionable design and you can maintain it properly to get optimum long-term performance. Once you have driven tractors, trucks, excavators, and fork-lift trucks, you should have another look at the traffic regulations, and you might enjoy the fact that such simple rules can regulate the entire world of traffic. Even experienced accident-free drivers may benefit from reading Chapter 1, in particular, when they started driving in the early days. With this book, I want to contribute to safer driving in general, and I hope to educate some really good drivers for public transportation, in particular, for ambulances and fire trucks. I would be happy if I could encourage some young people to go rallying in deserts and jungles for the mere fun of adventure or to become formula-one pilots and to work at the forefront of beyond-equilibrium thermodynamics.

Part I is a self-contained introduction to the phenomenological description of irreversible processes in and beyond the linear regime. Chapters 2 and 3 cover the classical linear regime, whereas Chapters 4 and 5 illustrate that the unified theory of beyond-equilibrium thermodynamics works equally well in the world of strongly nonlinear material behavior. Readers not interested in the theory of fluctuations can skip Sections 1.2.5 and 2.2.4. Chapter 5 on relativistic hydrodynamics is not required to understand the rest of the book; however, Section 5.1 is nevertheless recommended for understanding certain sets of variables, in particular, the independent flux variables of extended irreversible thermodynamics (EIT), a framework discussed in Section 5.1.6 as a special case of our unified approach.

Part II contains an introduction to the statistical foundations of beyond-equilibrium thermodynamics, which can be studied independently from the phenomenological first part, provided that a reasonably sound understanding of the GENERIC framework reviewed in the introductory Section 1.2 has been acquired. The examples of simulations in Chapter 8 require a certain familiarity with the coarse-grained models of polymeric fluids presented in Chapter 4. If one is merely interested in the simulation techniques to solve practical problems, the short course on the kinetic theory of gases (Chapter 7) can be omitted.

Feeling guilty about an important omission from this book, I would like to recommend some references for further reading. GENERIC is focused on isolated systems because the equations for those are the most structured ones, but one is often interested in open systems or driven subsystems because they can be handled more conveniently. Because we often deal with local field theories, the governing equations for which are independent of the boundary conditions, we should clearly be able to learn something about the thermodynamic admissibility of such evolution equations without paying any attention to the boundary conditions. For example, the field equations of hydrodynamics can be formulated without worrying about boundary conditions so that the difference between driven and isolated systems should not be important. In other situations, it might be necessary to include the environment explicitly. In contrast to what we do in the following chapters, however, one could be more careful with the boundary conditions for treating open systems. Stokes-Dirac structures[10] might provide the proper framework for a systematic discussion of boundary conditions. Another option is the splitting of GENERIC systems into a driven subsystem of interest and an environment, possibly interacting at any point in space rather than only through the boundary of the subsystem. Such a splitting has actually been performed in the context of specific examples,[11] and a relationship between GENERIC and a most powerful and elegant formalism for driven subsystems, Rob Jongschaap's "matrix model,"[12] has been established.

Finally, I would like to emphasize the *importance of exercises*. Even more than equilibrium thermodynamics, beyond-equilibrium thermodynamics can be grasped only by doing. It is therefore recommended to work through all the exercises (completely worked-out solutions are given in Appendix F). In particular, when exercises appear in the middle of a section rather than at the end, they should be recognized as signals: they offer help to understand some mathematical details or further material for appreciating concepts, or they are simply there to make the reader pause and think. All the exercises are numbered consecutively to emphasize that they run like a thread through the book and are the key to a deep understanding of thermodynamics.

[10] van der Schaft & Maschke, J. Geom. Phys. 42 (2002) 166.
[11] Edwards, Öttinger & Jongschaap, J. Non-Equilib. Thermodyn. 22 (1997) 356.
[12] Jongschaap, Rep. Prog. Phys. 53 (1990) 1; J. Non-Newtonian Fluid Mech. 96 (2001) 63.

1.2 GENERIC FRAMEWORK

In this introductory section, we summarize and motivate the fundamental equations that represent the unified viewpoint taken in this book for looking at various kinds of nonequilibrium phenomena. The equations are presented in two equivalent formulations: an operator representation and a bracket representation. Of course, the vast implications of these equations cannot be appreciated at first sight. Nevertheless, it should be useful to get a first impression of the compact and elegant framework on which the entire building of beyond-equilibrium thermodynamics presented in this book rests. Many details and implications are elaborated throughout this book, and this section also serves as a general reference for all the later chapters.

It is not necessary to understand all the details of this rather abstract and challenging section before one continues with the more concrete chapters of phenomenological and statistical beyond-equilibrium thermodynamics. On the contrary, the goal of the book is to clarify the ideas and the significance of the framework presented in this section for a coherent approach to the field. It might hence be useful to return repeatedly to Section 1.2 at various stages of reading the later chapters.

1.2.1 Fundamental Equations

We try to motivate a framework for beyond-equilibrium thermodynamics by a sequence of steps that are comparable to steps of similar relevance in equilibrium thermodynamics.[13] The first step to equilibrium thermodynamics is the identification of "work" and "heat" as different but closely related quantities. "Work" can be understood in mechanical terms, and the rest is "heat." One then realizes that these vague concepts are related to forms of exchanging energy between thermodynamic systems, $dE = đW + đQ$. Although an energy E can be assigned to any thermodynamic system, there is no "work" or "heat" content of the system but only changes of energy by doing work on the system ($đW$) or by transferring heat to it ($đQ$). Note that the energy, as a conserved quantity, can only be changed in a system through some form of exchange with the environment, which is the first law of equilibrium thermodynamics. The next step in the development of equilibrium thermodynamics is based on the idea that the contribution $đW$ to energy exchange consists of well-understood mechanical terms like $-pdV$ for changing the volume under a certain pressure, or μdN for adding particles at a certain chemical potential. In that sense, the term $đW$ is "under mechanistic control," where electric or magnetic contributions can also be included into the "work" term. In a final step, one needs to resolve $đQ$ by associating it with a quantity that can be accumulated in a system, just like E, V or N, but unlike "work" or "heat." The second law of equilibrium thermodynamics allows us to introduce the entropy and the absolute temperature, and to write $đQ = TdS$, so that we arrive at Gibbs' fundamental equation of thermodynamics in differential form. The fundamental form $dE = -pdV + \mu dN + TdS$, on which the entire framework of

[13] For a summary of a structured approach to equilibrium thermodynamics, see Appendix A.

Table 1.1 Steps Toward Frameworks of Thermodynamics at and Beyond Equilibrium

Equilibrium	Beyond Equilibrium
"work" + "heat"	"reversible" + "irreversible"
$dE = đW + đQ$	$\frac{dx}{dt} = \dot{x}_{\text{reversible}} + \dot{x}_{\text{irreversible}}$
$dE = -pdV + \mu dN + đQ$	$\frac{dx}{dt} = L \cdot \frac{\delta E}{\delta x} + \dot{x}_{\text{irreversible}}$
$dE = -pdV + \mu dN + TdS$	$\frac{dx}{dt} = L \cdot \frac{\delta E}{\delta x} + M \cdot \frac{\delta S}{\delta x}$

thermodynamics can be built, expresses a geometric structure in the space of all thermodynamic variables. In particular, it naturally leads to Legendre transformations between thermodynamic potentials, depending on which variables one chooses as the independent ones.

The above steps toward the formulation of equilibrium thermodynamics are summarized in the left column of Table 1.1. We now propose a similar step-wise approach to beyond-equilibrium thermodynamics. While at equilibrium we need to consider the exchange of extensive quantities between a system and its environment for expressing the idea of mechanistic control, in the following steps, we consider an isolated system. Beyond equilibrium, the distinction between "reversible" and "irreversible" contributions to time evolution is the fundamental starting-point. If systems are not at equilibrium, time evolution to equilibrium is a most natural thing to look at. Again, "reversible" should be understood in mechanistic terms, and the rest is "irreversible." To arrive at a fundamental equation, we need to introduce the variables x for describing a system of interest; the time derivative dx/dt is then assumed to be the sum of a "reversible" and an "irreversible" contribution. Because the "reversible" contribution should be "under mechanistic control," this term should have a well-understood structure. Such a structure is indeed well-established in mechanics; it is associated with Hamiltonian dynamical systems (see Section 1.2.2). To obtain a time evolution $\dot{x}_{\text{reversible}}$ from a Hamiltonian, which we can think of as the energy function of the system, we need to associate a vector field with any Hamiltonian. For this task, various geometric structures have been introduced in classical mechanics, namely symplectic, Poisson and Dirac structures. These structures provide suitable linear operators that translate energy gradients into vector fields $\dot{x}_{\text{reversible}}$. The mathematically interested reader can find an extensive discussion of these geometric structures in Appendix B. To understand the framework of beyond-equilibrium thermodynamics, however, it is sufficient to accept that there exists a well-established geometric structure behind terms that are "under mechanistic control" and that it allows us to express reversible time evolution in terms of energy gradients (as will be illustrated by simple examples in the following subsection). Because symplectic structures are too restrictive, leaving no room for additional contributions that are not "under mechanistic control," we choose more general Poisson structures for translating energy gradients into "reversible" time evolution. More precisely, Poisson structures, in contrast to symplectic structures, allow us to introduce a nontrivial entropy function that is conserved under any Hamiltonian dynamics. The even more general Dirac structures introduced in Appendix B may be useful for a unified description of isolated and driven systems.[14] At this point, we are left with the final step of inventing the analogue of TdS by rationalizing $\dot{x}_{\text{irreversible}}$. It is most natural to assume that entropy gradients drive "irreversible" time evolution in a similar way as energy gradients drive "reversible" motions and that we need to introduce a further linear operator that translates entropy gradients into vector fields $\dot{x}_{\text{irreversible}}$. Motivated by these steps, which are summarized in the right column of Table 1.1, we arrive at a general time-evolution equation for beyond-equilibrium systems, known as GENERIC (general equation for the nonequilibrium reversible-irreversible coupling),[15]

$$\frac{dx}{dt} = L(x) \cdot \frac{\delta E(x)}{\delta x} + M(x) \cdot \frac{\delta S(x)}{\delta x}, \tag{1.1}$$

where x represents a set of independent variables required for a complete description of the nonequilibrium system, the real-valued functionals $E(x)$ and $S(x)$ are the total energy and entropy expressed in terms of the state variables x, and $L(x)$ and $M(x)$ are the Poisson and friction matrices representing geometric structures and dissipative material properties in terms of linear operators. The two contributions to the time evolution of x generated by the total energy $E(x)$ and the entropy $S(x)$ in (1.1) are the reversible and irreversible contributions, respectively. Because x typically contains position-dependent fields, such as the local mass, momentum and energy densities of hydrodynamics, the state variables are usually labeled by continuous (position) labels in addition to discrete ones. A matrix multiplication, which can alternatively be considered as the application of a linear operator, hence implies not only summations over discrete indices but also integrations over continuous labels, and $\delta/\delta x$ typically implies functional rather than partial derivatives. The rather compact notation and the physical significance of all the GENERIC building blocks should become entirely clear in the subsequent applications. The concept of functional derivatives is explained and illustrated in Appendix C.

A crucial part of GENERIC modeling is the choice of the state variables x or, in other words, the definition of a suitable system to describe certain phenomena of interest. There is no universal set of nonequilibrium variables. Whereas the hydrodynamic fields (mass, momentum, and internal energy density) are naturally associated with conserved quantities, for complex fluids, additional structural variables are needed, and their proper choice depends on the particular kind of fluid. If an improper set of variables is employed to describe certain phenomena, even the most sophisticated choice of the GENERIC building blocks E, S, L, M cannot lead to a successful model.

Note that the concept of the entropy of nonequilibrium states, $S(x)$, is implicitly defined by its role in (1.1). Entropy is thus associated only with sets of variables for which a closed set of well-structured time-evolution equations can be formulated; a

[14] Jongschaap & Öttinger, J. Non-Newtonian Fluid Mech. 120 (2004) 3.

[15] Grmela & Öttinger, Phys. Rev. E 56 (1997) 6620; Öttinger & Grmela, Phys. Rev. E 56 (1997) 6633.

Fig. 1.2 Has this messy room a large or small entropy? [Answer: It depends (see text for details).]

more general concept of nonequilibrium entropy for variables with potentially non-autonomous time evolution has been supported in the literature.[16] At equilibrium, the entropy is introduced to take care of "the rest;" it describes the (ex)change of energy not associated with mechanistically controllable variables. Beyond equilibrium, the entropy is introduced to describe the dynamics not associated with mechanistically controllable time evolution; it generates the irreversible dynamics. It is the functional form of the entropy on a given level of description that is relevant for generating irreversible dynamics. Particular functional dependences on certain types of variables in applications serve to enhance the understanding of the concept of entropy, in the same sense as expressions like $(1/2)m\boldsymbol{v}^2$ are important for understanding energy beyond its role of generating reversible time evolution. It may be useful to note further that, if the list of variables x consists of densities of extensive quantities, the entropy function or functional $S(x)$ must be a concave function of x. The concavity of entropy, among many other general properties, is well-known not only at equilibrium but whenever entropy is considered as a functional of probability distributions.[17]

To elucidate the fundamental concept of entropy, we discuss the entropy of the room in Figure 1.2. Parents tend to consider such a room as a mess, and, if they are

[16] Garrido, Goldstein & Lebowitz, Phys. Rev. Lett. 92 (2004) 050602.
[17] Wehrl, Rev. Mod. Phys. 50 (1978) 221.

scientists, they would claim that it is in a state of (too) high entropy. Children would argue that what we see is not at all a mess; rather, everything is perfectly well arranged for what they are playing. This everlasting discussion results from missing the fact that, without introducing an appropriate set of variables for describing it, the room as such is not really a system to which an entropy could be assigned. Parents use the "toy surface density field" (measured in kg/m^2) as the only relevant variable to define the system "p-room." Children implicitly use all the positions and orientations of all their toys as variables for the system "c-room." The almost uniform toy surface density in Figure 1.2 illustrates a high entropy state of the p-room, whereas it can have zero entropy in the c-room.[18] The entropy depends on the variables used in the system of interest, thus actually establishing it as a thermodynamic system. The entropy hence depends on the perspective. Although the object room as such does not possess an entropy, the systems p-room and c-room have meaningful entropies, which happen to be as different as parents from children. For further clarification, note that the situation is different for the total mass of the toys in the room, on which the parents after integrating the toy surface density and the children after adding the masses of all their toys would agree without any arguments. This situation should be reflected on deeply before entering any discussion about entropy.[19] ✱

Exercise 1 Concavity of Entropy

Discuss the concavity of the entropy density s of an ideal gas of particles of mass m,

$$s(\rho, \epsilon) = \frac{1}{2} \frac{k_B}{m} \rho \ln \left(c \frac{\epsilon^3}{\rho^5} \right), \tag{1.2}$$

as a function of the mass density ρ and the energy density ϵ, where k_B is Boltzmann's constant and c is a further constant. Repeat the discussion of the entropy density s as a function of the mass density ρ and the temperature T,

$$s(\rho, T) = \frac{1}{2} \frac{k_B}{m} \rho \ln \left(c' \frac{T^3}{\rho^2} \right), \tag{1.3}$$

where c' is another constant.

After elaborating the significance of entropy, and assuming familiarity with the concept of energy, we discuss some properties that give us further insight into the matrices $L(x)$ and $M(x)$. In the GENERIC framework, the time-evolution equation (1.1) is supplemented by the complementary degeneracy requirements

$$L(x) \cdot \frac{\delta S(x)}{\delta x} = 0, \tag{1.4}$$

[18] In this discussion, we neglect the temperature- and pressure-dependent entropy that each of the toys possesses because it is made out of atoms.

[19] This example of a messy room is dedicated with love to Simon, Selina, Benjamin, and Alessandra.

and

$$M(x) \cdot \frac{\delta E(x)}{\delta x} = 0. \tag{1.5}$$

The requirement (1.4) expresses the reversible nature of the L-contribution to the dynamics; the functional form of the entropy is such that it cannot be affected by the operator generating the reversible dynamics. The requirement (1.5) expresses the conservation of the total energy of an isolated system by the irreversible contribution to the dynamics; all dissipated mechanical energy is converted into internal energy.[20] Furthermore, it is required that the Poisson matrix $L(x)$ be antisymmetric and the friction matrix $M(x)$ symmetric[21] and positive semidefinite. Both the complementary degeneracy requirements (1.4) and (1.5) and the symmetry properties are very important for the formulation of proper L- and M-matrices when modeling concrete beyond-equilibrium situations.

At this point, we have not really incorporated the powerful geometric structure behind Hamiltonian dynamics, as discussed in connection with the possibility of a reversible-irreversible separation. To do so, we consider a bracket associated with the antisymmetric matrix $L(x)$,

$$\{A, B\} = \frac{\delta A(x)}{\delta x} \cdot L(x) \cdot \frac{\delta B(x)}{\delta x}, \tag{1.6}$$

which is bilinear in the observables or functionals A, B, that inherits the antisymmetry of $L(x)$,

$$\{A, B\} = -\{B, A\}, \tag{1.7}$$

and, as a consequence of the derivatives occurring in definition (1.6), satisfies Leibniz' rule

$$\{AB, C\} = A\{B, C\} + B\{A, C\}, \tag{1.8}$$

for arbitrary functionals A, B, and C. In addition, the bracket (1.6) is required to satisfy the Jacobi identity,

$$\{A, \{B, C\}\} + \{B, \{C, A\}\} + \{C, \{A, B\}\} = 0, \tag{1.9}$$

for arbitrary A, B, and C. Contrary to (1.7) and (1.8), the Jacobi identity does not follow from the previously specified properties of $L(x)$ and the bracket definition (1.6). Motivated by the discussion of what "under mechanistic control" means for the formulation of reversible dynamics, it is imposed as a further important general property in beyond-equilibrium thermodynamics.[22]

In the mathematical literature, a Poisson bracket, or Poisson structure, is defined through linearity in both arguments, the antisymmetry (1.7), Leibniz' rule (1.8), and

[20] The details behind this statement will be explained after (1.20).

[21] For a more detailed discussion of the Onsager-Casimir symmetry of M, see Section 3.2.1; for a very detailed level of description, M might not possess any well-defined symmetry properties, as is elaborated in Section 7.2.4.

[22] Beris & Edwards, *Thermodynamics of Flowing Systems* (Oxford, 1994); Grmela & Öttinger, Phys. Rev. E 56 (1997) 6620.

the Jacobi identity (1.9), as elaborated in Appendix B. By including the Jacobi identity into the GENERIC framework, we have thus fully implemented the idea that the reversible contribution to time evolution should be well structured, namely, in the form of a Hamiltonian dynamical system in the geometrical setting of Poisson manifolds.

To get a deeper understanding of the Jacobi identity and to realize its relation to time-structure invariance, we define time dependent operators A_t as the solutions of the purely reversible equation

$$\frac{dA_t}{dt} = \{A_t, E\}, \tag{1.10}$$

with the initial condition $A_0 = A$ for all functionals A, and we consider the time evolution of $C = \{A, B\}$ for arbitrary functionals A and B. The assumed time-structure invariance of reversible dynamics implies that the observable that we obtain by evolving the bracket $C = \{A, B\}$ in time can be obtained equivalently by forming the bracket of A_t and B_t,[23]

$$C_t = \{A_t, B_t\}, \tag{1.11}$$

with the same Poisson bracket $\{\cdot, \cdot\}$ for all times. By evaluating dC_t/dt in two different ways, we then find

$$\{\{A_t, B_t\}, E\} = \{\{A_t, E\}, B_t\} + \{A_t, \{B_t, E\}\}. \tag{1.12}$$

The left-hand side of this equation is obtained directly from (1.10), whereas the right-hand side of this equation is obtained from the product rule and subsequent application of (1.10) to A_t and B_t. By using the antisymmetry of the Poisson bracket, (1.12) can be written as

$$\{E, \{A_t, B_t\}\} + \{A_t, \{B_t, E\}\} + \{B_t, \{E, A_t\}\} = 0. \tag{1.13}$$

A comparison with (1.9) shows that the Jacobi identity expresses the consistency of the time evolution and the bracket represented by the commutativity of the diagram in Figure 1.3 for arbitrary energy functionals generating the time evolution.

Exercise 2 Jacobi Identity for Constant Poisson Matrices

Show that, if the Poisson matrix L is independent of x, the antisymmetry (1.7) implies the Jacobi identity (1.9) (in other words, show that the second-order-derivative terms of the functionals A, B, C, arising from the nested brackets in the Jacobi identity, always cancel).

The result of Exercise 2 allows us to rewrite the Jacobi identity (1.9) directly as a condition for the Poisson matrix $L = L(x)$. In taking the derivative of an inner bracket in the Jacobi identity, according to Exercise 2, only the L-matrix needs to be differentiated, and all terms are trilinear in the derivatives of the arbitrary functionals

[23] See Proposition 10.3.1 of Marsden & Ratiu, *Mechanics and Symmetry* (Springer, 1999).

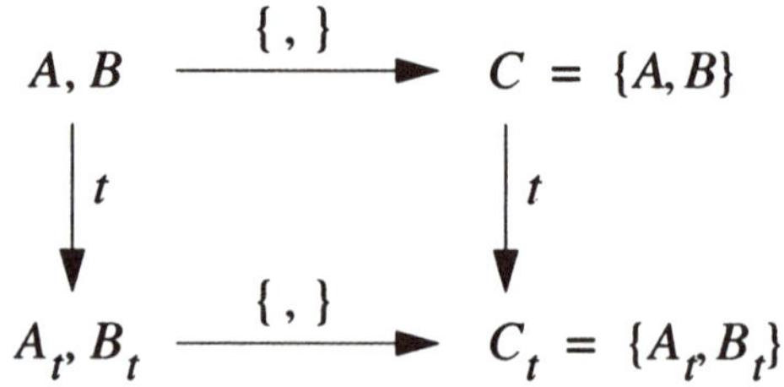

Fig. 1.3 Time-structure invariance expressed by the Jacobi identity.

A, B, and C. The expression obtained by setting the coefficients of the trilinear form equal to zero,[24]

$$\sum_l \left(L_{il} \frac{\partial L_{jk}}{\partial x_l} + L_{jl} \frac{\partial L_{ki}}{\partial x_l} + L_{kl} \frac{\partial L_{ij}}{\partial x_l} \right) = 0, \tag{1.14}$$

however, is badly singular for field theories including continuous indices i, j, k, l, such as hydrodynamics, so that condition (1.14) should be used for discrete systems only.

When done by explicit calculations, checking the Jacobi identity is by far the most laborious step in applying the GENERIC modeling approach to concrete problems. Because this step involves excessive bookkeeping and a rather limited set of manipulations needs to be performed over and over again, this is an ideal problem for symbolic mathematical computations. A symbolic code that can perform the check of the Jacobi identity for a huge class of brackets is available in the literature.[25] Another option to avoid lengthy calculations is the construction of Poisson brackets by symmetries and reduction (see Appendix B).

In the same way as an antisymmetric Poisson bracket is associated with $L(x)$, there is a dissipative bracket associated with $M(x)$,

$$[A, B] = \frac{\delta A(x)}{\delta x} \cdot M(x) \cdot \frac{\delta B(x)}{\delta x}, \tag{1.15}$$

with the symmetry property[26]

$$[A, B] = [B, A]. \tag{1.16}$$

The positive semidefinite nature of $M(x)$ is equivalent to

$$[A, A] \geq 0, \tag{1.17}$$

for all functionals A.

[24] See p. 341 of Marsden & Ratiu, *Mechanics and Symmetry* (Springer, 1999).
[25] Kröger, Hütter & Öttinger, Comput. Phys. Commun. 137 (2001) 325.
[26] See p. 14 for qualifying remarks on this symmetry property.

By writing the time evolution of an arbitrary functional A as

$$\frac{dA}{dt} = \frac{\delta A(x)}{\delta x} \cdot \frac{dx}{dt}, \tag{1.18}$$

the operator version (1.1) of GENERIC can be rewritten in the equivalent bracket form

$$\frac{dA}{dt} = \{A, E\} + [A, S]. \tag{1.19}$$

With the bracket definitions, the antisymmetry of the Poisson bracket, the symmetry of the dissipative bracket, and the degeneracy requirement (1.5), we obtain

$$\frac{dE}{dt} = \{E, E\} + [S, E] = 0. \tag{1.20}$$

Therefore, (1.5) can be considered as a strong formulation of energy conservation by assuming $[A, E] = 0$ for all A rather than for $A = S$ only. Similarly, (1.4) can be considered as a strong formulation of entropy conservation under reversible dynamics, which is the hallmark of reversibility. It leads to

$$\frac{dS}{dt} = [S, S] \geq 0. \tag{1.21}$$

The positive semidefinite nature of $M(x)$, or (1.17), is a strong formulation of the second law of thermodynamics because we would need (1.17) only for $A = S$ in order to make sure that entropy is a nondecreasing function of time.

Note that the derivation of (1.20) and (1.21) shows that the degeneracy requirements (1.4) and (1.5) are actually applied in the form

$$\frac{\delta S(x)}{\delta x} \cdot L(x) = 0, \qquad \frac{\delta E(x)}{\delta x} \cdot M(x) = 0. \tag{1.22}$$

When the symmetry of $M(x)$ cannot be taken for granted, then the second part of (1.22) is the physically relevant version of the degeneracy requirement (1.5).

We have so far assumed that x represents a set of independent variables allowing us a complete description of a nonequilibrium system. We note in passing that a redundancy among the variables x can be handled in an elegant way. If a redundancy of the form

$$A(x) = 0 \tag{1.23}$$

for some function or functional A holds even for the atomistically evaluated, fluctuating version of the relevant variables x, then the statistical expressions for $L(x)$ and $M(x)$[27] imply that $\delta A/\delta x$ is in the null spaces of the Poisson and friction matrices,

$$L(x) \cdot \frac{\delta A(x)}{\delta x} = \frac{\delta A(x)}{\delta x} \cdot L(x) = 0, \qquad M(x) \cdot \frac{\delta A(x)}{\delta x} = \frac{\delta A(x)}{\delta x} \cdot M(x) = 0. \tag{1.24}$$

[27] More specifically, the argument can be based on formulas (6.67) and (6.73) derived in Part II of this book.

The compatibility of the matrices $L(x)$ and $M(x)$ with a redundancy of the form (1.23) can thus be checked in a straightforward manner. The relationships in (1.24) conveniently imply that there is no need to evaluate the functional derivatives of energy and entropy in the basic time-evolution equation (1.1) and in the degeneracy requirements (1.4), (1.5) under the constraint (1.23) [see Appendix C.3]. The degeneracies in (1.24) imply that the constraints on the gradients of energy and entropy are irrelevant in applying the GENERIC approach. One only needs to avoid double counting of effects in the generators, energy and entropy. For example, if interactions with an electric field are important and the electric charge density can be determined from the mass densities of some components, then the electric energy is most naturally expressed in terms of the charge density, whereas the kinetic energy or short-range contributions to the internal energy are more conveniently formulated in terms of mass densities. Other options, however, are possible and lead to the same time-evolution equations.

While entropy is a primary concept of the GENERIC approach to beyond-equilibrium systems, temperature is not. Actually, temperature does not occur in any of the fundamental equations. If one wishes to define the concept of temperature in the general context, one might end up with a definition like

$$\hat{T}_k(x) = \frac{\delta E(x)}{\delta x_k} \left[\frac{\delta S(x)}{\delta x_k} \right]^{-1}, \tag{1.25}$$

where a particular variable x_k needs to be selected for the differentiations, while all other variables are kept constant. Typically, this temperature depends on all the nonequilibrium variables and, even more importantly, on the choice of x_k. In physical terms, the precise form of exchange or interaction between a potential device to measure temperature and the system of interest needs to be specified.[28] Different thermometers might measure different beyond-equilibrium temperatures associated with different sets of variables and different values of k. Although temperature might be much simpler to measure than entropy even in beyond-equilibrium situations, entropy is the clearly more fundamental concept. The lack of a fundamental concept of beyond-equilibrium temperature is also the reason why, in contrast to equilibrium thermodynamics, it is preferable to work with the two separate generators energy and entropy rather than a single generator because, simply for dimensional reasons, their combination into a single free energy would require a temperature.

Actually, the use of two separate generators for the reversible and irreversible dynamics, together with the symmetric degeneracy requirements, is a key feature of the GENERIC structure. This innovation is of crucial importance when treating systems without local equilibrium states, for example, systems described by Boltzmann's kinetic equation. The example of Boltzmann's equation furthermore shows that, despite the linear appearance of the fundamental equations (1.1)–(1.5), the formalism is not limited to the linear response regime. Rather, the almost unrestricted functional dependence of $L(x)$ and $M(x)$ on x allows for the description of systems

[28] Casas-Vázquez & Jou, Rep. Prog. Phys. 66 (2003) 1937.

far from equilibrium (it is the fully nonlinear Boltzmann equation that is treated in Chapter 7, not only its linearization). While beyond-equilibrium dynamics is usually expressed in terms of a single generator,[29] the double-generator idea, which leaves more flexibility in the choice of variables, allows us to formulate the mutual degeneracy conditions.

Several advantages of the double-generator over the single-generator framework have been identified. In addition to being applicable to systems without a local equilibrium temperature, such as Boltzmann's kinetic equation, it should be noted that "whereas as far as macroscopic systems are concerned in all the investigated systems so far a complete agreement between the one and two-generator formalisms is found, for microscopic systems [considered on the level of configurational distribution functions] small but significant differences exist with the existing evidence (by comparison to other theories) in favor of the two-generator formalism."[30] Most importantly, an atomistic derivation of the double-generator GENERIC framework with explicit coarse-graining rules is available (see Chapter 6 for classical systems and Appendix D for quantum systems). The relationships between GENERIC and several other frameworks of nonequilibrium thermodynamics have been discussed in great detail.[31]

GENERIC is not merely a new formalism; it offers an entirely new way of approaching beyond-equilibrium systems. Instead of a direct formulation of time-evolution equations by the introduction of certain plausible terms, all modeling is done through the basic building blocks of GENERIC, that is, through the generators $E(x)$, $S(x)$, and the matrices $L(x)$, and $M(x)$ or the corresponding brackets. All physical insight has to be expressed in terms of these four building blocks, and even approximations should be made on the level of building blocks. While $E(x)$ and $S(x)$ are familiar from equilibrium thermodynamics, the matrix $L(x)$ is usually determined by symmetries and geometric considerations.[32] The dynamic material information enters only in the friction matrix $M(x)$. Once the physics of a given system has been expressed in the building blocks $E(x)$, $S(x)$, $L(x)$, and $M(x)$ and all the required properties have been checked, the fundamental equation (1.1) consistently produces all the terms in the desired time-evolution equations for a beyond-equilibrium system. Among other things, such an approach guarantees that the second law of thermodynamics cannot be violated.

Exercise 3 Two Systems Exchanging Heat and Volume

Develop the GENERIC time-evolution equations describing two subsystems exchanging heat

[29] Beris & Edwards, *Thermodynamics of Flowing Systems* (Oxford, 1994).

[30] Beris, J. Non-Newtonian Fluid Mech. 96 (2001) 119.

[31] See Edwards, Öttinger & Jongschaap, J. Non-Equilib. Thermodyn. 22 (1997) 356; Edwards, J. Non-Equilib. Thermodyn. 23 (1998) 301; Edwards, Beris & Öttinger, J. Non-Equilib. Thermodyn. 23 (1998) 334; Muschik et al., Physica A 285 (2000) 448; Jou & Casas-Vázquez, J. Non-Newtonian Fluid Mech. 96 (2001) 77; Lebon et al., J. Non-Equilib. Thermodyn. 28 (2003) 1; Grmela et al., J. Non-Equilib. Thermodyn. 28 (2003) 23; Lhuillier et al., J. Non-Equilib. Thermodyn. 28 (2003) 51; Pasquali & Scriven, J. Non-Newtonian Fluid Mech. 120 (2004) 101.

[32] See Section 2.3.1 and Appendix B.

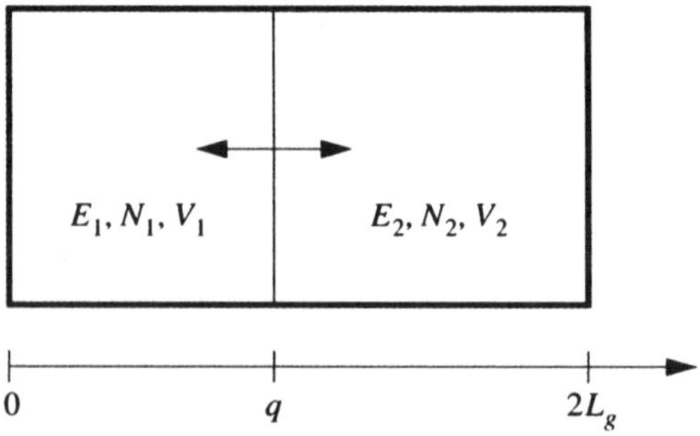

Fig. 1.4 Variables for two subsystems exchanging heat and volume.

and volume through a separating wall of mass m (see Figure 1.4). Use $x = (q, v, E_1, E_2)$, where v is the velocity of the wall. Assume that the subsystems contain the same number of particles ($N_1 = N_2 = N$), and that their equilibrium thermodynamics is given by the entropy function $S(E, V, N)$. The motion of the wall is assumed to be frictionless.

Exercise 4 Jacobi Check for a Discrete System
For properly chosen mass and size of the wall, the Poisson bracket of Exercise 3 is

$$\{A, B\} = \frac{\partial A}{\partial q}\frac{\partial B}{\partial v} - \frac{\partial A}{\partial v}\frac{\partial B}{\partial q} - p_1\left(\frac{\partial A}{\partial E_1}\frac{\partial B}{\partial v} - \frac{\partial A}{\partial v}\frac{\partial B}{\partial E_1}\right) + p_2\left(\frac{\partial A}{\partial E_2}\frac{\partial B}{\partial v} - \frac{\partial A}{\partial v}\frac{\partial B}{\partial E_2}\right),$$

where the functions A and B depend on $x = (q, v, E_1, E_2)$, whereas the functions p_1 and p_2 depend on (q, E_1, E_2) only. Verify the Jacobi identity (1.9).

1.2.2 Reversible and Irreversible Ancestors

To motivate both the use of the energy as the generator of reversible dynamics and the occurrence of the Poisson matrix L, we consider classical mechanics as a prototype example. We recall that, in the context of classical mechanics of N particles, the set of state variables is $z = (\boldsymbol{r}_1, \boldsymbol{p}_1, \ldots, \boldsymbol{r}_N, \boldsymbol{p}_N)$, where $\boldsymbol{r}_j$ and $\boldsymbol{p}_j$ are the position and momentum vectors of the jth particle. The time evolution of z is governed by Hamilton's equations of motion, which can be represented in a block-diagonal form with the following block for the jth particle

$$\frac{d}{dt}\begin{pmatrix} \boldsymbol{r}_j \\ \boldsymbol{p}_j \end{pmatrix} = L_0 \cdot \begin{pmatrix} \frac{\partial E}{\partial \boldsymbol{r}_j} \\ \frac{\partial E}{\partial \boldsymbol{p}_j} \end{pmatrix}, \tag{1.26}$$

where

$$L_0 = \begin{pmatrix} 0 & 1 \\ -1 & 0 \end{pmatrix}. \tag{1.27}$$

The matrix L_0, which is also known as the cosymplectic matrix [see remark after (B.66)], represents the kinematics of z. The general Poisson bracket (1.6) associated with the matrix L_0 in (1.27) is written out as

$$\{A, B\} = \sum_{j=1}^{N} \left(\frac{\partial A}{\partial \boldsymbol{r}_j} \cdot \frac{\partial B}{\partial \boldsymbol{p}_j} - \frac{\partial A}{\partial \boldsymbol{p}_j} \cdot \frac{\partial B}{\partial \boldsymbol{r}_j} \right), \tag{1.28}$$

which is the famous Poisson bracket of classical mechanics.

The Poisson structure used to describe reversible dynamics is thus an abstract generalization of the concepts of classical mechanics. The energy E is the most natural generator of time translations for reversible systems, and the antisymmetry and Jacobi identity are the abstract properties used in generalizing the usual Poisson bracket of classical mechanics to capture the essence of reversibility. This Poisson bracket allows us to formulate Hamiltonian dynamical systems.

Classical mechanics not only provides an important example of a symplectic structure but also is the prototype of all finite-dimensional symplectic structures. According to Darboux' Theorem (see p. 424), for any symplectic manifold, one can find canonical coordinates for which Hamilton's equations of motion take the canonical form (1.26) and (1.27). For Poisson manifolds, Darboux' Theorem implies the existence of canonical coordinates such that, in a subblock, one has the cosymplectic matrix (1.27) and all other entries are zero, thus reflecting the possibility of degeneracy [see (B.66)]. One can actually think of the Jacobi identity as the criterion for the possibility to introduce canonical coordinates. To establish the Jacobi identity, it is hence very useful to construct canonical coordinates (an explicit transformation to canonical coordinates is carried out in Exercise 9). This is also the reason why the Lagrangian formulation of classical mechanics is useful to introduce symplectic structures, as is illustrated for models with constraints in Exercise 7.

Exercise 5 Full Cosymplectic Matrix

The entries of the cosymplectic matrix (1.27) multiply three-dimensional column vectors in (1.26). What would the more detailed 6×6 cosymplectic matrix in (1.26) look like? Describe the full cosymplectic matrix for the N-particle system.

Exercise 6 Hamilton's and Newton's Equations of Motion

Write out Hamilton's equations of motion for a single particle of mass m moving in a potential $\phi^{(e)}$. Derive Newton's equation of motion for the particle.

Exercise 7 Lagrange's Treatment of Constraints

If the configuration space for N interacting particles is constrained by n independent equations of the form $\Pi_j(\boldsymbol{r}_1, \ldots, \boldsymbol{r}_N) = 0$, $j = 1 \ldots n$,[33] there are two common strategies to

[33] Constraints that can be expressed in terms of algebraic equations are said to be *holonomic*. Common nonholonomic constraints are given by inequalities or nonintegrable differential equations relating small changes of configurations. According to p. 15–16 of Goldstein, *Classical Mechanics* (Addison-Wesley, 1980), "there is no general way of attacking nonholonomic examples. True, if the constraint is nonintegrable, the differential equations of constraint can be introduced into the problem along with the differential equations of motion, and the dependent equations eliminated, in effect, by the method of Lagrange mul-

find the equations of motion. First, one can introduce $\sum_{j=1}^{n} \lambda_j \Pi_j(\boldsymbol{r}_1, \ldots, \boldsymbol{r}_N)$ as a potential of constraint forces and determine the unknown Lagrange multipliers λ_j along with the particle trajectories from Newton's equation of motion and the constraint conditions. Second, one can introduce coordinates $q_1 \ldots q_d$, $d = 3N - n$, to parametrize the constrained configuration space and formulate equations of motion by means of Hamilton's principle, that is, by looking for stationary points of the action obtained as the time integral of the Lagrangian $L^{\mathrm{cm}}(q_1 \ldots q_d, \dot{q}_1 \ldots \dot{q}_d)$ [given by the kinetic minus potential energy]. Using the second approach, show by Legendre transformation that the equations of motion, given by the second-order Euler-Lagrange equations,

$$\frac{d}{dt}\frac{\partial L^{\mathrm{cm}}}{\partial \dot{q}_j} - \frac{\partial L^{\mathrm{cm}}}{\partial q_j} = 0, \qquad j = 1 \ldots d, \tag{1.29}$$

which express Newton's equation of motion in the constrained space, can be written in a canonical Hamiltonian form (thus establishing a symplectic structure and a canonical Poisson bracket in the constrained space).

Exercise 8 From Commutators to Poisson Brackets

In the limit of small Planck constant (over 2π), $\hbar \to 0$, one can obtain the classical Poisson bracket from the commutator of quantum mechanics according to[34]

$$\{A, B\} = \frac{1}{i\hbar}[A, B]_{\mathrm{c}} = \frac{1}{i\hbar}(AB - BA), \tag{1.30}$$

where the observables A, B in quantum mechanics are linear operators or matrices. Use (1.30) to derive Leibniz' rule (1.8) and the Jacobi identity (1.9).

Whereas the reversible contribution to GENERIC can be motivated by classical mechanics, the irreversible contribution is motivated by the time dependent Ginzburg-Landau equation used in describing critical dynamics.[35] If $F(x, T)$ is the Helmholtz free energy, $F(x, T) = E(x) - TS(x)$, where T is a constant parameter that can be identified with the equilibrium temperature of a final time independent state defined by a vanishing gradient of $F(x, T)$. According to Ginzburg and Landau, the irreversible approach to the equilibrium state is governed by the equation

$$\frac{dx}{dt} = -\frac{1}{T} M \cdot \frac{\delta F(x, T)}{\delta x} = M \cdot \frac{\delta S(x)}{\delta x}, \tag{1.31}$$

tipliers. ... However, the more vicious case of nonholonomic constraint must be tackled individually, and consequently in the development of the more formal aspects of classical mechanics it is almost invariably assumed that any constraint, if present, is holonomic." The intrinsic difficulty of nonholonomic constraints is further underlined by the fact that constraints need to be holonomic in order to preserve the Jacobi identity, as shown by van der Schaft & Maschke, Rep. Math. Phys. 34 (1994) 225.

[34] See p. 401 of Goldstein, *Classical Mechanics* (Addison-Wesley, 1980).

[35] See, for example, Section 5.10.1 of Kawasaki & Gunton, Critical Dynamics (1978). A complex version of the time dependent Ginzburg-Landau equation provides a universal description of the slow modulation of oscillations in a spatially extended system near a Hopf bifurcation and has found many applications, *e.g.*, for chemical oscillations, thermal convection in binary fluids, and multimode lasers. According to Cross & Hohenberg, Rev. Mod. Phys. 65 (1993) 851, the name of this equation derives from the formal similarity with the theory of superconductivity developed by Ginzburg & Landau, Zhur. Eksp. Teor. Fiz. 20 (1950) 1064, although the latter does not deal with dynamics.

which is another special case of the GENERIC (1.1). A system tries to minimize its free energy, or to maximize its entropy, by following the gradients of these generators; however, the system does not develop directly in the direction of the gradients, but frictional effects are rather taken into account through the matrix M. In any case, there is a nonnegative component of the irreversible dynamics along the entropy gradient.

Because of the special role played by the final equilibrium temperature in $F(x, T)$, the above argument can only be of a heuristic nature. The Ginzburg-Landau equation is intended for isothermal systems, whereas GENERIC is formulated for isolated systems with conserved total energy (for open systems, the environment needs to be included in the description).

The need for the matrices L and M in the GENERIC approach can be further motivated by a formal argument. Assuming that the state variables $x = (x^j)$ form a contravariant vector and the energy E as well as the entropy S are scalars, the energy and entropy gradients $\partial E/\partial x^j$ and $\partial S/\partial x^j$ are covariant vectors. Because the time derivative of $x = (x^j)$ is a contravariant vector, the matrices L and M are needed to transform covariant into contravariant vectors or, in other words, to raise the indices on the components of the gradients. A much deeper version of this argument in terms of vector fields and differential forms can be found in Appendix B.

1.2.3 Equilibrium Thermodynamics of Stationary States

After relating the purely reversible and purely irreversible limits of GENERIC to famous special cases, we now consider the stationary limit solutions for large times, which should correspond to equilibrium states. To establish such a connection we first note that, in addition to the degeneracy requirements for the energy and entropy, there are degeneracy requirements for the total mass or, equivalently, for the total particle number

$$N(x) = \frac{1}{m} \int \rho(\boldsymbol{r}) d^3 r, \tag{1.32}$$

where m is the mass per particle. The following degeneracy conditions express the fundamental nature of the conservation of mass or particle number,

$$L(x) \cdot \frac{\delta N(x)}{\delta x} = M(x) \cdot \frac{\delta N(x)}{\delta x} = 0. \tag{1.33}$$

The most natural way of constructing thermodynamic equilibrium states is by maximizing the entropy of an isolated system under the constraints of given total energy and particle number. We hence define the function

$$\Omega(x; \lambda_E, \lambda_N) = S(x) - \lambda_E E(x) - \lambda_N N(x), \tag{1.34}$$

where the intensive parameters λ_E and λ_N play the role of Lagrange multipliers, and we define the thermodynamic states $x_{\text{eq}}(\lambda_E, \lambda_N)$ by

$$\left. \frac{\delta \Omega(x; \lambda_E, \lambda_N)}{\delta x} \right|_{x = x_{\text{eq}}(\lambda_E, \lambda_N)} = 0. \tag{1.35}$$

In the rest of this subsection we show that, on the one hand, the manifold of states $x_{\text{eq}}(\lambda_E, \lambda_N)$ carries the structure of equilibrium thermodynamics and, on the other hand, $x_{\text{eq}}(\lambda_E, \lambda_N)$ represents the set of time independent solutions of GENERIC.

We assume that $\Omega(x; \lambda_E, \lambda_N)$ is a concave function of x so that there is a unique maximum of Ω solving (1.35). The energy and the particle number of the thermodynamic states

$$E_{\text{eq}} = E(x_{\text{eq}}) \tag{1.36}$$

and

$$N_{\text{eq}} = N(x_{\text{eq}}) \tag{1.37}$$

can be varied and controlled by the Lagrange multipliers λ_E and λ_N.

We next define the function

$$\Omega_{\text{eq}}(\lambda_E, \lambda_N, V) = \Omega(x_{\text{eq}}(\lambda_E, \lambda_N); \lambda_E, \lambda_N), \tag{1.38}$$

where a dependence on the volume of the isolated system has been indicated explicitly. Note that the functionals $S(x)$, $E(x)$, and $N(x)$ depend on the system volume, even the choice of x, and hence also $x_{\text{eq}}(\lambda_E, \lambda_N)$. Because λ_E and λ_N are intensive, the extensive quantity Ω_{eq} must actually be proportional to V,

$$\frac{\partial \Omega_{\text{eq}}(\lambda_E, \lambda_N, V)}{\partial V} = \frac{\Omega_{\text{eq}}(\lambda_E, \lambda_N, V)}{V}. \tag{1.39}$$

From the definition (1.38) and equations (1.34) and (1.35), we obtain

$$\frac{\partial \Omega_{\text{eq}}(\lambda_E, \lambda_N, V)}{\partial \lambda_E} = -E(x_{\text{eq}}(\lambda_E, \lambda_N)) = -E_{\text{eq}}, \tag{1.40}$$

$$\frac{\partial \Omega_{\text{eq}}(\lambda_E, \lambda_N, V)}{\partial \lambda_N} = -N(x_{\text{eq}}(\lambda_E, \lambda_N)) = -N_{\text{eq}}. \tag{1.41}$$

Equations (1.39), (1.40), and (1.41) imply that Ω_{eq} is the Massieu function belonging to V and the conjugates of E_{eq} and N_{eq} as independent variables (see the remarks on Massieu functions at the end of Section A.4 in Appendix A). We can hence identify Ω_{eq} as the Massieu function

$$\Omega_{\text{eq}} = \frac{pV}{T}, \tag{1.42}$$

and we recognize the Lagrange multipliers as

$$\lambda_E = \frac{1}{T}, \tag{1.43}$$

$$\lambda_N = -\frac{\mu}{T}. \tag{1.44}$$

Finally, we obtain the entropy expression

$$S_{\text{eq}}(\lambda_E, \lambda_N, V) = S_{\text{eq}}(1/T, -\mu/T, V) = S(x_{\text{eq}}(\lambda_E, \lambda_N)), \tag{1.45}$$

giving the equilibrium entropy of the thermodynamic states in terms of the nonequilibrium entropy generating irreversible dynamics in GENERIC.

After the states $x_{\rm eq}$ are shown to carry the structure of equilibrium thermodynamics, how are these states related to the stationary solutions of GENERIC? If we rewrite the fundamental equation (1.1) by means of the degeneracy conditions as

$$\frac{dx}{dt} = \left(M - \frac{1}{\lambda_E} L \right) \cdot \frac{\delta \Omega}{\delta x}, \tag{1.46}$$

we immediately realize that the states $x_{\rm eq}$ defined by (1.35) are stationary solutions of this equation. By means of the conservation laws and degeneracy conditions, we further find that Ω is a Lyapunov function, that is, a nondecreasing function of time,

$$\frac{d\Omega}{dt} = \frac{\delta S}{\delta x} \cdot M \cdot \frac{\delta S}{\delta x} = \frac{\delta \Omega}{\delta x} \cdot M \cdot \frac{\delta \Omega}{\delta x} \geq 0. \tag{1.47}$$

The maximum of Ω for given energy and particle number should eventually be reached provided that there are no additional conserved quantities that would allow us only a maximization of Ω under further constraints and, at the same time, would introduce further variables into equilibrium thermodynamics. Equation (1.46) then suggests that the only stationary states are the states $x_{\rm eq}$, which carry the structure of equilibrium thermodynamics.

We have thus established that the structure of equilibrium thermodynamics is contained in the stationary solutions of GENERIC. Of course, any acceptable framework of beyond-equilibrium thermodynamics should reproduce equilibrium thermodynamics as a special case.

1.2.4 Transformation of Variables

Because the choice of variables for beyond-equilibrium systems may be a matter of taste or convenience, it is important to relate the GENERIC building blocks for different sets of independent state variables. We hence consider a one-to-one transformation $x \to x'$. For the scalar generators, energy and entropy, we expect the simple transformation behavior

$$E'(x') = E(x(x')), \tag{1.48}$$

$$S'(x') = S(x(x')). \tag{1.49}$$

By transforming the fundamental time-evolution equation (1.1) and using (1.48) and (1.49), we obtain the following natural matrix transformation laws for the Poisson and friction matrices,

$$L'(x') = \frac{\delta x'}{\delta x} \cdot L(x(x')) \cdot \frac{\delta x'}{\delta x}, \tag{1.50}$$

$$M'(x') = \frac{\delta x'}{\delta x} \cdot M(x(x')) \cdot \frac{\delta x'}{\delta x}, \tag{1.51}$$

where L' and M' are the Poisson operators and the friction matrix in the coordinates x', $\delta x'/\delta x$ is the transformation matrix (here considered as a function of x'), and L and M are to be contracted from both sides with the x in the denominator of the transformation matrix.

If the values of functionals depend only on the beyond-equilibrium state, but not on the choice of variables for describing the state,

$$A'(x') = A(x(x')), \quad B'(x') = B(x(x')), \tag{1.52}$$

then the transformation laws (1.50) and (1.51) imply

$$\frac{\delta A'(x')}{\delta x'} \cdot L'(x') \cdot \frac{\delta B'(x')}{\delta x'} = \frac{\delta A(x)}{\delta x} \cdot L(x) \cdot \frac{\delta B(x)}{\delta x}\bigg|_{x=x(x')} \tag{1.53}$$

and

$$\frac{\delta A'(x')}{\delta x'} \cdot M'(x') \cdot \frac{\delta B'(x')}{\delta x'} = \frac{\delta A(x)}{\delta x} \cdot M(x) \cdot \frac{\delta B(x)}{\delta x}\bigg|_{x=x(x')} . \tag{1.54}$$

In other words, the Poisson bracket (1.6) and the dissipative bracket (1.15) are independent of the particular choice of state variables.

The knowledge of the transformation laws of the GENERIC building blocks is very useful because, in many cases, it may be convenient to use different sets of variables in developing the reversible and irreversible contributions to the time evolution, so that a transformation is required before the results can be added. Moreover, transformations can be very useful also for simplifying an explicit check of the Jacobi identity because, according to (1.53), a sufficiently regular one-to-one transformation preserves the bracket and hence the Jacobi identity (intuitively, a transformation of variables should not affect the time-structure invariance).

Exercise 9 Transformed Poisson Matrix of a Discrete System

Construct the Poisson matrix for the discrete system of Exercise 3 for the variables $x' = (q, v, S_1, S_2)$ instead of $x = (q, v, E_1, E_2)$ with $S_j = S(E_j, V_j, N_j)$. How can the result be interpreted? Give a set of canonical coordinates for this system.

Exercise 10 Transformed Time Evolution

Show that the transformation rules (1.48)–(1.51) lead to the correct GENERIC time-evolution equation for x'.

Exercise 11 Degeneracy of Transformed Building Blocks

Show that the transformed building blocks (1.48)–(1.51) fulfill the degeneracy requirements (1.4) and (1.5).

1.2.5 Fluctuations

The description of fluctuations requires a certain background in stochastic differential equations and their relationship to diffusion equations. In the phenomenological part of this book, the ideas of the present section are needed only for understanding fluctuating hydrodynamics (see Section 2.2.4).

The time-evolution equation (1.1) is deterministic, but fluctuations can be added very naturally. A heuristic approach based on the GENERIC formulation of the Liouville equation and a Boltzmann-type modification of the entropy was given in one of the original papers on GENERIC,[36] and a statistical derivation of the noise terms was obtained by projection-operator techniques.[37] Fluctuations can be included conveniently by formulating the following diffusion equation or Fokker-Planck equation for the time evolution of the configurational distribution function, $f(x,t)$,

$$\frac{\partial f(x,t)}{\partial t} = -\frac{\delta}{\delta x} \cdot \left[\left(L \cdot \frac{\delta E}{\delta x} + M \cdot \frac{\delta S}{\delta x} \right) f(x,t) \right] + k_{\mathrm{B}} \frac{\delta}{\delta x} \cdot \left[M \cdot \frac{\delta}{\delta x} f(x,t) \right] . \tag{1.55}$$

The Fokker-Planck equation (1.55) can equivalently be rewritten as a stochastic differential equation,[38]

$$dx = L \cdot \frac{\delta E}{\delta x} dt + M \cdot \frac{\delta S}{\delta x} dt + k_{\mathrm{B}} \frac{\delta}{\delta x} \cdot M dt + B \cdot dW_t, \tag{1.56}$$

where the amplitude of the noise $B = B(x)$ is a matrix with

$$B(x) \cdot B(x)^T = 2 k_{\mathrm{B}} M(x), \tag{1.57}$$

k_{B} is Boltzmann's constant, and W_t is a multicomponent Wiener process, that is, a Gaussian process with first and second moments given by

$$\langle W_t \rangle = 0, \quad \langle W_t W_{t'}^T \rangle = \min(t, t') \, 1 \tag{1.58}$$

or, more conveniently but less rigorously, by

$$\left\langle \frac{dW_t}{dt} \right\rangle = 0, \quad \left\langle \frac{dW_t}{dt} \frac{dW_{t'}^T}{dt'} \right\rangle = \delta(t - t') \, 1. \tag{1.59}$$

The fluctuations can be eliminated from (1.56) to obtain (1.1) by going to the limit $k_{\mathrm{B}} \to 0$ (without changing the GENERIC building blocks E, S, L, M), so that k_{B} may be regarded as a small parameter controlling the importance of fluctuations. Equation (1.57) relates the amplitude of the fluctuations, B, to the friction matrix, M, and is hence known as a *fluctuation-dissipation theorem.*

The occurrence of the fluctuation-term involving the divergence of M in (1.56) is a consequence of using the Itô interpretation for the stochastic differential equation associated with the Fokker-Planck equation (1.55), in which the friction matrix M is positioned between the derivatives of the second-order term rather than behind the

[36] Grmela & Öttinger, Phys. Rev. E 56 (1997) 6620.

[37] Öttinger, Phys. Rev. E 57 (1998) 1416.

[38] Gardiner, *Handbook of Stochastic Methods* (Springer, 1990); Öttinger, *Stochastic Processes in Polymeric Fluids* (Springer, 1996).

derivatives. Although this position is most natural from a physical point of view because it leads to the occurrence of the Boltzmann distribution at equilibrium (see also p. 251), none of the familiar concepts of stochastic integration, associated with the names of Itô and Stratonovich, corresponds to this position. In applications one should avoid the numerical evaluation of the divergence of M; a promising approach for that purpose has been proposed.[39]

The transformation laws of the preceding subsection can be supplemented by the rule

$$B'(x') = \frac{\delta x'}{\delta x} \cdot B(x(x')), \tag{1.60}$$

which, in view of (1.57), is consistent with the transformation law (1.51) for the friction matrix. The multiplication in (1.60) is only from the left. Actually, one can additionally multiply B by an arbitrary orthogonal matrix from the right without changing the friction matrix (1.57). In the fluctuating GENERIC (1.56), such a multiplication by an orthogonal matrix can be compensated by a redefinition of the Wiener process, where the redefined process has exactly the same statistical properties as the original one.

The degeneracy requirement (1.5) for the friction matrix can be rewritten as an analogous condition for the noise amplitude,

$$\frac{\delta E(x)}{\delta x} \cdot B(x) = 0, \tag{1.61}$$

which is important for proving the strict conservation of energy in the presence of fluctuations. Indeed, stochastic calculus implies

$$\begin{aligned} dE &= \frac{\delta E}{\delta x} \cdot dx + \frac{\delta^2 E}{\delta x \delta x} : k_{\mathrm{B}} M dt = k_{\mathrm{B}} \left[\frac{\delta E}{\delta x} \cdot \left(\frac{\delta}{\delta x} \cdot M \right) + \frac{\delta^2 E}{\delta x \delta x} : M \right] dt \\ &= k_{\mathrm{B}} \frac{\delta}{\delta x} \cdot \left(M \cdot \frac{\delta E}{\delta x} \right) dt = 0. \end{aligned} \tag{1.62}$$

Exercise 12 Diffusing Particle

Formulate the GENERIC equations with noise for the one-dimensional diffusion of a particle of mass m in terms of the position Q, the momentum P, and the entropy of the environment S_{e}.

Exercise 13 Transformed GENERIC With Fluctuations

Consider the one-to-one transformation $x \to x'$. According to Itô calculus,[40] the transformed version of the stochastic GENERIC (1.56) is given by

$$dx' = \frac{\delta x'}{\delta x} \cdot \left(L \cdot \frac{\delta E}{\delta x} dt + M \cdot \frac{\delta S}{\delta x} dt + k_{\mathrm{B}} \frac{\delta}{\delta x} \cdot M dt + B \cdot dW_t \right) + k_{\mathrm{B}} \frac{\delta^2 x'}{\delta x \delta x} : M dt, \tag{1.63}$$

[39] Hütter & Öttinger, J. Chem. Soc. Faraday Trans. 94 (1998) 1403.

[40] See p. 98 of Öttinger, *Stochastic Processes in Polymeric Fluids* (Springer, 1996).

where the last term is a nontrivial fluctuation effect. Rewrite this equation in terms of the transformed quantities E', S', L', M', and B' of (1.48)–(1.51) and (1.60). Show that the form of (1.56) is not invariant for the above transformation laws. Show that the form of GENERIC with fluctuations becomes invariant under transformations if we modify the transformation law for the entropy,

$$S'(x') = S(x(x')) + k_{\rm B} \ln \left| \det \frac{\delta x(x')}{\delta x'} \right| . \tag{1.64}$$

1.2.6 Benefits of a Framework

The sceptical reader might and actually should ask whether the reformulation of various time-evolution equations in the form (1.1) is beneficial. Do we gain any additional insight, or is this, at best, just an esthetically nice but useless exercise? Of course, the proper answer to these questions is not only important for motivating the reader but also for revealing the full possible significance of the GENERIC framework. We hence wish to spend some time on answering these questions by an analogy.

Consider the equilibrium thermodynamics of a gas, to be described by the independent variables T (absolute temperature), V (volume), and N (number of particles). Note that, as explained in Appendix A, identifying the proper variables required for a complete description of the system under consideration is always the first fundamental problem of thermodynamics.[41] If the caloric equation of state for the energy $E = E(T, V, N)$ and the thermodynamic equations of state for the entropy $S = S(T, V, N)$, pressure $p = p(T, V, N)$, and chemical potential $\mu = \mu(T, V, N)$ are known, one has a complete description of the system. Is there any additional insight to be gained from reformulating these equations of state, by recognizing a deeper structure underlying these equations? Any reader familiar with the most basic ideas of equilibrium thermodynamics (see Appendix A) will immediately give a strongly affirmative answer to this question: Yes, there is a lot to be gained; the essence of thermodynamics actually is the identification of the unifying structure behind the various equations of state (cf. the situation sketched in Figure 1.1).

More precisely, by specifying the proper thermodynamic potential for the variables T, V, N, which is the Helmholtz free energy $F(T, V, N)$, all the above-mentioned equations of state can be obtained by taking partial derivatives [see (A.23) in Appendix A, where Callen uses the symbols U and P instead of E and p, respectively]. All information about the particular system is contained in a single thermodynamic potential F. From this structure of thermodynamics, certain consistency conditions (Maxwell relations) arise for the various equations of state (conditions such as "if E is independent of V then p must be linear in T" are certainly not immediately obvious and much more fundamental than even the most successful empirical or microscopically derived equations of state). We moreover realize that, once consistent expressions for $E(T, V, N)$ and $p(T, V, N)$ are given, the expressions for S and μ can

[41] Unfortunately, we are taught, or we are accustomed to seeing, this task as a trivial choice between well-known lists of appropriate state variables for gases, mixtures, magnetic systems etc.

be determined. By recognizing the fundamental role of thermodynamic potentials, we obtain a much more compact and transparent description of the system. We can reproduce all the equations of state, and we automatically satisfy all the consistency equations for these equations.

The formulation of equilibrium thermodynamics in terms of thermodynamic potentials incorporates, and is actually guided by and based on, the celebrated fundamental laws of thermodynamics. The proper formulation of the equations is synonymous with the deepest insight into thermodynamics. Listing all the equations of state is good enough for *understanding a particular system*; however, *understanding thermodynamics* means identifying the structure of the equations. This successful identification of the structure behind thermodynamics is also very important when calculating thermodynamic properties from a more microscopic approach: no separate (and possibly inconsistent) calculations for each equation of state are required, but only a single calculation of a thermodynamic potential via the partition functions of statistical mechanics is needed; a consistent set of equations of state is then derived from the thermodynamic potential. This is a major advantage in solving engineering problems, such as designing thermal engines or developing new materials with a knowledge of their mixing behavior (phase diagrams).

As should be clear now by analogy, the proposed general time-evolution equation (1.1) provides a deep understanding of beyond-equilibrium thermodynamics in the above spirit. Only two "potentials" or generators (E and S) and two "matrices" or linear operators (L and M) need to be determined to specify the full time-evolution equations. We refer to the GENERIC (1.1) together with the degeneracy requirements (1.4), (1.5) and (1.6)–(1.9), (1.15)–(1.17) as the GENERIC structure. The GENERIC time evolution is proposed for beyond-equilibrium systems that, if not driven by external forces, approach thermodynamic equilibrium.

Writing well-known successful equations for beyond-equilibrium systems in the form (1.1) is the essential step for validating the proposed structure of beyond-equilibrium thermodynamics. The separate but similar occurrence of the two generators E and S in the reversible and irreversible terms of the GENERIC (1.1)—which, together with the mutual degeneracy requirements, is the key feature of the proposed formalism—leads to a deeper and more complete understanding of the role of entropy than can be gained in equilibrium thermodynamics.

A number of consistency conditions are implied by the proposed form of the equations (1.1)–(1.5) (including the Maxwell relations of equilibrium thermodynamics and the Onsager-Casimir relations of linear irreversible thermodynamics). As in equilibrium thermodynamics, the predictive power of the beyond-equilibrium formalism is based on these general consistency conditions extracted from many observations. Because there exists vast experience with the form of the energy E and the entropy S in equilibrium thermodynamics and because the Poisson matrix L can be constructed in a rather straightforward manner (we will illustrate this in Section 2.3.1 and Appendix B), the formulation of dynamic equations for beyond-equilibrium systems can focus on the determination of the friction matrix M. In many situations, calculations based on more microscopic models need to provide only an expression for M and not for all the terms in the time-evolution equations.

In summary, a general formalism for beyond-equilibrium systems should provide the following (cf. Figure 1.1): (1) a unified framework for empirical (macroscopic and mesoscopic) modeling; (2) consistency conditions as the most fundamental predictions; (3) recipes for going from more microscopic to more macroscopic levels of description. It is very important for the reader to think critically about the above motivation and to keep in mind the fundamental importance of our goal throughout the phenomenological and statistical parts of this book.

The question "Why a Thermodynamic Framework?" has also been asked by G. Gallavotti in Section 5.3 of his recent book.[42] His detailed answer contains the following statement: "It is well known that for each class of problems the formalism in which they are formulated often has a clarifying and unifying role: the emergence of a 'good' formalism is often successive to the solution of important problems in the field." He also emphasized that "as with all formalisms it has interest only because it provides a natural frame (...) in which the discussion of the most important questions and applications can be situated."

1.2.7 Historical Context

A milestone in the theory of nonequilibrium thermodynamics was reached in 1962, when Sybren Ruurds de Groot and Peter Mazur published the textbook *Non-Equilibrium Thermodynamics*, a seminal work that was to define the field of irreversible processes as it stood in the early 1960s. That theory of irreversible phenomena had been assembled over the preceding 30 years through a vast collection of phenomenological observations made over the past three centuries.[43]

The first constitutive assumption for a nonequilibrium flux being proportional to a generalized force was made in 1687 by Isaac Newton in Volume II of his *Philosophiae Naturalis Principia Mathematica* where, in the final Section IX of the volume, he formulated the following hypothesis: "*Resistentiam, quae oritur ex defectu lubricitatis partium fluidi, caeteris paribus, proportionalem esse velocitati, qua partes fluidi separantur ab invicem.*" The "defect of lubrication" in a fluid, or the tenacity of the fluid elements ("tenacitas partium fluidi"), leads to a "resistance" proportional to the (relative) velocity with which different "parts of a fluid" are separated from each other (also the expression *velocitas partium relativa* occurs in the text).

In addition to Newton's law of viscosity, Fourier's law of heat conduction (1822), Ohm's law of electric conduction (1826), Fick's law of diffusion (1855), and other experimental observations, including electrokinetic (F.F. Reuss, 1809), thermoelectric (W. Thomson, or Lord Kelvin, 1854) and other cross effects, were all united into a phenomenological theory in the mid-20th century, with the aim of establishing some fundamental laws and principles that could be used to describe a wide vari-

[42] Gallavotti, *Statistical Mechanics: A Short Treatise* (Springer, 1999).

[43] The description of the historical background of "linear irreversible thermodynamics" in this section is based on Section 1.1 of de Groot & Mazur, *Non-Equilibrium Thermodynamics* (Dover, 1984) and Chapter 1 of Kreuzer, *Nonequilibrium Thermodynamics & Statistical Foundations* (Oxford, 1981); more detailed references can be found in these books.

Fig. 1.5 Sir Isaac Newton, 1643–1727.

ety of transport processes occurring in all materials. The possible linear couplings between the various thermodynamic forces and fluxes are restricted by Curie's symmetry principle (1894), which states that, in an isotropic system, fluxes and forces of different tensorial character do not couple. Key to the formulation of the unified phenomenological theory were the works of J. L. F. Bertrand (1887), L. Natanson (1896), P. Duhem (1911), G. Jaumann (1911, 1918), E. Lohr (1916, 1924), Th. De Donder (1927), and C. Eckart (1940), who recognized the importance of entropy production for irreversible processes and tried to obtain expressions for the local entropy production in nonuniform systems by combining the second law of thermodynamics with the macroscopic balance equations expressing the conservation of mass, momentum and energy. Finally, J. Meixner (1941–1943) and, somewhat later, I. Prigogine (1947) set up the consistent phenomenological theory of irreversible processes presented in the famous textbook of de Groot and Mazur (which had a noteworthy precursor[44]).

The book of de Groot and Mazur defines the state of the art of linear irreversible thermodynamics in the 1960s beautifully and comprehensively. However, its serious limitation should become clear from the following sentences, which can be found even in the introduction to the 1984 edition of the book:[45] "Non-equilibrium thermodynamics, in its present form, is mainly restricted to the study of such linear phenomena. Very little of a sufficiently general nature is known outside this linear domain. This is not a very serious restriction however, since even in rather extreme physical situations,

[44] de Groot, *Thermodynamics of Irreversible Processes* (North-Holland, 1951).
[45] de Groot & Mazur, *Non-Equilibrium Thermodynamics* (Dover, 1984).

transport processes, for example, are still described by linear laws." While de Groot and Mazur clarified in the preface to the second edition of their book that "linear" refers to the fact that the "*thermodynamic fluxes are linear functions of the gradients of the thermodynamic-state variables*," nowadays, nonlinear material behavior is the key to many technological applications; thus, there is a clear need to go beyond linearity. The purpose of the present book is to present a unified approach to nonlinear irreversible thermodynamics, the existence, necessity, and usefulness of which has currently not been fully realized by large fractions of the science and engineering communities.

Important further developments toward a more general beyond-equilibrium formalism were initiated by M. Grmela in the summer of 1983 when he introduced the idea of dissipative brackets to formulate irreversible dynamics at a research conference in the mathematical sciences on fluids and plasmas in Boulder, Colorado.[46] In the following year, three papers on this idea appeared in *Physics Letters A*.[47] In the subsequent years, the idea of combining noncanonical Poisson brackets for describing reversible dynamics, which have their roots in the work of A. Clebsch (1859) on the Hamiltonian formulation of the Euler equations for inviscid fluids,[48] and dissipative brackets for describing irreversible dynamics was further developed by M. Grmela and several people stimulated by his work. A decade after the proposal of dissipative brackets, the development was summarized, clarified and further advanced in a comprehensive textbook by A. Beris and B. Edwards.[49] A more general two-generator formalism, known by the acronym GENERIC, was proposed in equivalent bracket and operator formulations in 1997.[50] While the GENERIC two-generator approach, on which the unified approach to beyond-equilibrium phenomena in this book is based, was proposed as a general framework in 1997, the same idea was employed in the context of plasma kinetic equations already in one of the 1984 papers on dissipative brackets by Kaufman.[51] Unfortunately, A. Kaufman was discouraged from using the idea as a general tool when he failed in generalizing an equation for collisional diffusion of a magnetized pure electron plasma from cylindrical symmetry to asymmetric geometry.[52]

The pioneering work of Ludwig Boltzmann (1872) marked the beginning of nonequilibrium *statistical mechanics* in a most impressive and influential way.[53] His kinetic equation for rarefied gases provides deep insight into the problem of how irreversibility can arise from reversible equations and sets the standard for generalizations even today. The connection between Boltzmann's atomistic approach and the phenomenological hydrodynamic approach was established through the solution

[46] Grmela, Contemporary Mathematics 28 (1984) 125.

[47] Kaufman, Phys. Lett. A 100 (1984) 419; Morrison, Phys. Lett. A 100 (1984) 423; Grmela, Phys. Lett. A 102 (1984) 355.

[48] Clebsch, J. Reine Angew. Math. 56 (1859) 1.

[49] Beris & Edwards, *Thermodynamics of Flowing Systems* (Oxford, 1994).

[50] Grmela & Öttinger, Phys. Rev. E 56 (1997) 6620; Öttinger & Grmela, Phys. Rev. E 56 (1997) 6633.

[51] Kaufman, Phys. Lett. A 100 (1984) 419.

[52] Kaufman, private communication (March 2000).

[53] Boltzmann, Wien. Ber. 66 (1872) 275.

Fig. 1.6 Ludwig Boltzmann, 1844–1906.

of S. Chapman (1916) and D. Enskog (1917) and, much later, through the moment method of H. Grad (1949), thus providing atomistic expressions for transport coefficients.[54] The microscopic arguments in favor of symmetry relationships in cross effects, such as the electrokinetic and thermoelectric effects, offered by L. Onsager (1931) and H. B. G. Casimir (1945), constitute a further highlight of nonequilibrium statistical physics. Many fundamental developments in kinetic theory and various applications in different fields, often anticipating later approaches, resulted from the fruitful and inspiring work of J. G. Kirkwood in the years 1946–1959, with a series of 14 articles bearing the title "Statistical Mechanical Theory of Transport Processes" as a backbone.[55]

The fluctuation-dissipation theorem plays an important role in describing nonequilibrium systems. The theorem started its career with A. Einstein's work (1905) on Brownian motion, and his famous relation between diffusion coefficient and mobility is a special case of it. Another special case was obtained by H. Nyquist (1928) in his study of noise in resistors, and somewhat later the theorem was given in a generalized form in the work of H. B. Callen and his colleagues (1951, 1952). R. Kubo distinguished between fluctuation-dissipation theorems of the first and second kind. The fluctuation-dissipation theorem of the first kind relates the linear response of a system to an external perturbation and a two-time correlation function of the system

[54] The references to the original works mentioned in the present remarks on the history of nonequilibrium statistical mechanics can be found in the book by Kreuzer, *Nonequilibrium Thermodynamics & Statistical Foundations* (Oxford, 1981).

[55] Kirkwood, *Collected Works, Topics in Statistical Mechanics* (Gordon & Breach, 1967).

in the absence of external forces; this kind of theorem is closely related to the famous expressions for transport coefficients developed by M. S. Green (1952) and R. Kubo (1957). The fluctuation-dissipation theorem of the second kind constitutes a relationship between the frictional and the random forces in a system. The separation of frictional and random forces is a very complex problem of statistical mechanics, and the projection-operator formalism originating in the work of R. Zwanzig[56] and H. Mori[57] is an important tool for this purpose. For systems far from equilibrium, B. Robertson[58] developed the projection-operator technique for producing equations going far beyond the theory of linear irreversible processes. These techniques can be used to derive the GENERIC equations of beyond-equilibrium thermodynamics.[59]

In connection with the above-mentioned works, Nobel Prizes in Chemistry have been awarded to Lars Onsager (1968) for the discovery of the reciprocity relations and to Ilya Prigogine (1977) for his contributions to nonequilibrium thermodynamics.

[56] Zwanzig, Phys. Rev. 124 (1961) 983.
[57] Mori, Prog. Theor. Phys. 33 (1965) 423; 34 (1965) 399.
[58] Robertson, Phys. Rev. 144 (1966) 151.
[59] Öttinger, Phys. Rev. E 57 (1998) 1416; 62 (2000) 4720.

Part I

Phenomenological Approach

One of the early chapters in any textbook on nonequilibrium thermodynamics is dedicated to hydrodynamics. The Navier-Stokes equations of hydrodynamics can be explained nicely to beginners, they describe a variety of the most complicated phenomena despite their simple appearance, they express a beautiful universality rooted in balance equations, and they are of great importance in many engineering applications. Therefore, we can naturally use hydrodynamics for a first illustration of how the GENERIC framework of beyond-equilibrium thermodynamics works in concrete detail. As a next traditional highlight, we consider the well-established framework of linear irreversible thermodynamics, and we show how it fits into the form and thinking of our more general approach capable of handling nonlinear material properties. Up to that point, GENERIC has nothing to offer except maybe some interesting new perspectives on old problems that have been solved by linear irreversible thermodynamics half a century ago. From the presentation of these classical topics, one should with consummate ease develop a feeling for how the GENERIC formalism actually works, which steps are needed in concrete modeling, and how they can be carried out.

The power of the fully nonlinear framework becomes immediately clear when we study complex fluids. Nonlinear material behavior is the rule rather than the exception for the fluids processed by the polymer, food, and other industries. We hence dedicate an entire chapter to the purely phenomenological thermodynamic approach to the description of the viscoelastic and nonlinear flow behavior of complex fluids. After we address a topic of highest practical relevance, we answer a question of fundamental importance: can the GENERIC framework of beyond-equilibrium thermodynamics be applied to relativistic problems? A minimal theory of relativistic hydrodynamics arises naturally, and even gravity can be included. As an application, we formulate and study the GENERIC equations for bulk viscous cosmology.

2

Hydrodynamics

In a systematic approach to beyond-equilibrium thermodynamics, one can divide a large nonequilibrium system, say with a nonuniform mass density or temperature profile, into smaller systems for which a local density and temperature is defined (see Figure 2.1). The idea of a local temperature and temperature variations from point to point in space and time, which is most familiar from everyday experience, implicitly assumes the possibility of having local equilibrium systems. Even in flowing systems, such as the gulf stream, our intuition tells us that the temperature of a small co-moving volume element should be a meaningful concept. In a gas, say a room filled with air, maybe even in the presence of a draft in the room, we could consider little cubic volume elements, say of 1mm in size. On the one hand, such cubes contain a large number of molecules (some 3×10^{16} molecules), and they are large compared with the typical mean free path between collisions of particles (some tenths of a μm), so that thermodynamic concepts can be applied safely in a co-moving frame. On the other hand, such cubes are sufficiently small so that one obtains a reasonable spatial resolution of a temperature field in a room.

With the *local equilibrium assumption*, concepts like temperature, internal energy density, and entropy density can be defined as fields throughout a nonequilibrium system. Instead of expressing the local equilibrium thermodynamics of a small volume element V in an extensive entropy function $S(E, V, N)$ (see Section A.3 of Appendix A), we prefer to consider the densities of entropy, internal energy, and mass:

$$s = \frac{S}{V}, \quad \epsilon = \frac{E}{V}, \quad \rho = \frac{mN}{V}, \tag{2.1}$$

where m is the mass of a particle. The function $s(\rho, \epsilon)$, or alternatively the function $\epsilon(\rho, s)$, contains all of the thermodynamic information. The fact that the entropy

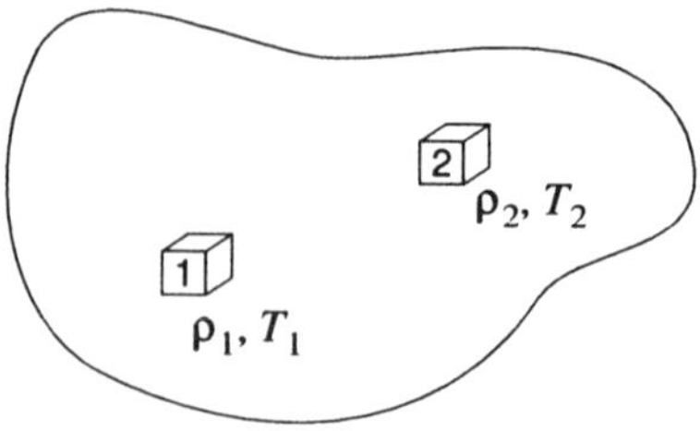

Fig. 2.1 A global nonequilibrium system with local equilibrium systems 1 and 2.

density $s(\rho, \epsilon)$ as a function of two arguments contains the same information as $S(E, V, N)$, except for the absolute system size, results from the extensive nature of all the variables involved (or, in other words, from the Gibbs-Duhem equation; see Section A.3 of Appendix A).

Exercise 14 Equilibrium Thermodynamics in Terms of Densities
Derive the following relationships:

$$\frac{\partial \epsilon(\rho, s)}{\partial s} = T(\rho, s), \quad \frac{\partial \epsilon(\rho, s)}{\partial \rho} = \mu(\rho, s), \tag{2.2}$$

$$p = \mu\rho + Ts - \epsilon, \tag{2.3}$$

$$\frac{\partial s(\rho, \epsilon)}{\partial \epsilon} = \frac{1}{T(\rho, \epsilon)}, \quad \frac{\partial s(\rho, \epsilon)}{\partial \rho} = -\frac{\mu(\rho, \epsilon)}{T(\rho, \epsilon)}, \tag{2.4}$$

where T is the temperature, μ is the chemical potential per unit mass, and p is the pressure. Discuss these relationships for

$$s(\rho, \epsilon) = \frac{1}{2} \frac{k_B}{m} \rho \ln \left(c \frac{\epsilon^3}{\rho^5} \right), \tag{2.5}$$

where k_B is Boltzmann's constant, m is the mass of a particle, and c is a constant.

2.1 BALANCE EQUATIONS

Equilibrium thermodynamics is based on two fundamental laws: the first law of thermodynamics or the law of conservation of energy and the second law of thermodynamics or entropy law. A systematic description of beyond-equilibrium processes must also be built upon these laws. Because we wish to describe systems with properties that depend continuously on position and time, we need a local formulation of conservation laws. The thermodynamics and motion of any fluid is then described by the equations of conservation of mass, momentum, and energy. Accordingly, we choose the following fields as independent variables for classical hydrodynamics: the mass density $\rho(\boldsymbol{r})$, the momentum density $\boldsymbol{M}(\boldsymbol{r})$, and the internal energy density $\epsilon(\boldsymbol{r})$. The momentum density field is closely related to the velocity field $\boldsymbol{v}(\boldsymbol{r})$, namely

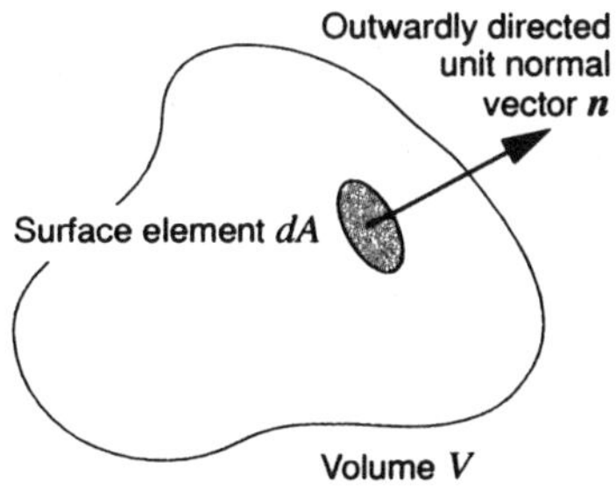

Fig. 2.2 Arbitrary volume, fixed in space, over which mass, momentum, and energy balances are made.

by

$$\boldsymbol{M}(\boldsymbol{r}) = \rho(\boldsymbol{r})\,\boldsymbol{v}(\boldsymbol{r}). \tag{2.6}$$

Because the balance equations for mass, momentum, and energy play an important role in beyond-equilibrium thermodynamics, it is reasonable to discuss their usual derivation briefly.[1]

In the derivation of local balance equations, we consider an arbitrary fixed region in space with volume V and surface ∂V, as shown in Figure 2.2. On every surface element dA, there is an outwardly directed normal unit vector $\boldsymbol{n}$. We imagine that this fixed region is in the midst of a fluid flow field and that the fluid moves across the boundaries of the region. We now want to apply the laws of conservation of mass, momentum, and energy to the fluid contained within this fixed region.

2.1.1 Mass

Suppose that at the infinitesimal surface element dA the fluid is crossing the surface of V with a velocity $\boldsymbol{v}$. Then the local volume rate of flow of fluid across the area dA is $\boldsymbol{n}\cdot\boldsymbol{v}dA$, where $\boldsymbol{n}\cdot\boldsymbol{v}$ is the component of the velocity in the normal direction. If the flow is outward, then $\boldsymbol{n}\cdot\boldsymbol{v}dA$ is positive, whereas if the flow is inward $\boldsymbol{n}\cdot\boldsymbol{v}dA$ is negative. The local mass rate of flow is then $\boldsymbol{n}\cdot\boldsymbol{v}\rho dA$. Note that $\boldsymbol{v}\rho$ is the mass flux, that is, the mass crossing a perpendicular surface per unit area and per unit time. For this argument, following standard practice in continuum mechanics, we have not taken into account that the fluid consists of small particles and that these particles do not all move with the same velocity $\boldsymbol{v}$. One would need a kinetic theory to identify the particles passing through dA and to form proper averages.

According to the law of conservation of mass, the total mass of fluid within V will increase only because of a net influx of fluid across the bounding surface ∂V. Mathematically, this is stated as

$$\frac{d}{dt}\int_V \rho d^3r = -\int_{\partial V} \boldsymbol{n}\cdot\boldsymbol{v}\rho dA, \tag{2.7}$$

[1] See, for example, § 1.1 of Bird, Armstrong & Hassager, *DPL 1, Fluid Mechanics* (Wiley, 1987).

where the left-hand side expresses the rate of increase of mass within V and the right-hand side expresses the rate of addition of mass across the surface ∂V. When Gauss's divergence theorem is used, the surface integral can be transformed into a volume integral,

$$\frac{d}{dt}\int_V \rho d^3r = -\int_V \frac{\partial}{\partial \boldsymbol{r}} \cdot (\boldsymbol{v}\rho) d^3r. \tag{2.8}$$

This equation may be rearranged by bringing the time derivative inside the integral. This is permissible because the volume V is fixed, and we obtain

$$\int_V \left[\frac{\partial \rho}{\partial t} + \frac{\partial}{\partial \boldsymbol{r}} \cdot (\boldsymbol{v}\rho)\right] d^3r = 0. \tag{2.9}$$

We now have an integral over an arbitrary volume in space, and this integral is equal to zero. Because of the arbitrariness of the volume, the integrand itself must be equal to zero. We thus obtain the following time-evolution equation for the mass density $\rho = \rho(\boldsymbol{r}, t)$,

$$\frac{\partial \rho}{\partial t} = -\frac{\partial}{\partial \boldsymbol{r}} \cdot (\boldsymbol{v}\rho), \tag{2.10}$$

which is called the *equation of continuity*. If the fluid has a constant density, the continuity equation (2.10) simplifies to

$$\frac{\partial}{\partial \boldsymbol{r}} \cdot \boldsymbol{v} = 0, \tag{2.11}$$

which is the special case of the continuity equation for an incompressible fluid.

If we consider a two-component system, we similarly have the two continuity equations

$$\frac{\partial \rho_1}{\partial t} = -\frac{\partial}{\partial \boldsymbol{r}} \cdot (\boldsymbol{v}_1\rho_1), \tag{2.12}$$

$$\frac{\partial \rho_2}{\partial t} = -\frac{\partial}{\partial \boldsymbol{r}} \cdot (\boldsymbol{v}_2\rho_2), \tag{2.13}$$

where ρ_1, ρ_2 and $\boldsymbol{v}_1, \boldsymbol{v}_2$ are the mass densities and the velocities of the two components, and chemical reactions have been excluded (chemical reactions would lead to source and sink terms in the mass balance equations for the individual components). For molecularly miscible components, it is most natural to introduce the total mass density and the center-of-mass or "barycentric" velocity,

$$\rho = \rho_1 + \rho_2, \tag{2.14}$$

$$\boldsymbol{v} = \frac{\rho_1}{\rho}\boldsymbol{v}_1 + \frac{\rho_2}{\rho}\boldsymbol{v}_2. \tag{2.15}$$

Note that the addition of the momenta in (2.15) to get a properly weighted average velocity guarantees that the continuity equation (2.10) remains valid even for two-component systems, as can be seen by adding (2.12) and (2.13). Moreover, the definition (2.15) leads to a natural momentum balance equation for mixtures. A nonuniform composition can be taken into account through the respective concentration variables

$$c_1 = \frac{\rho_1}{\rho}, \quad c_2 = \frac{\rho_2}{\rho}, \tag{2.16}$$

with $c_1 + c_2 = 1$, and a possible relative motion between the components can be described by the diffusion flows

$$\boldsymbol{j}_1^{\text{diff}} = \rho_1(\boldsymbol{v}_1 - \boldsymbol{v}), \quad \boldsymbol{j}_2^{\text{diff}} = \rho_2(\boldsymbol{v}_2 - \boldsymbol{v}), \tag{2.17}$$

with $\boldsymbol{j}_1^{\text{diff}} + \boldsymbol{j}_2^{\text{diff}} = 0$. If we introduce the "substantial derivative" or "material derivative"

$$\frac{D}{Dt} = \frac{\partial}{\partial t} + \boldsymbol{v} \cdot \frac{\partial}{\partial \boldsymbol{r}}, \tag{2.18}$$

then the time evolution of the concentration is given by the diffusion equation

$$\rho \frac{Dc_1}{Dt} = -\frac{\partial}{\partial \boldsymbol{r}} \cdot \boldsymbol{j}_1^{\text{diff}}. \tag{2.19}$$

While the mass density can be determined together with the velocity field when we consider the continuity equation (2.10) together with the momentum balance equation, the calculation of the composition from (2.19) requires additional material information about the diffusion flow $\boldsymbol{j}_1^{\text{diff}}$.

The generalization of the equations of this subsection to mixtures of more than two components is straightforward. For n components, one has n concentration variables c_k restricted by the constraint $\sum_{k=1}^{n} c_k = 1$ and n diffusion flows $\boldsymbol{j}_k^{\text{diff}}$ restricted by the constraint $\sum_{k=1}^{n} \boldsymbol{j}_k^{\text{diff}} = 0$.

Exercise 15 Time Evolution of Concentration
Derive equation (2.19).

2.1.2 Momentum

As pointed out above, the local volume rate of flow of fluid across the surface element dA is $\boldsymbol{n} \cdot \boldsymbol{v} dA$. If this is multiplied with the momentum per unit volume of the fluid, we then get $\boldsymbol{n} \cdot \boldsymbol{v} \boldsymbol{M} dA$; this is the rate at which momentum is carried across the surface element dA because the fluid itself flows across dA. The quantity $\boldsymbol{v}\boldsymbol{M} = \rho \boldsymbol{v}\boldsymbol{v}$ is the momentum flux, that is, the momentum crossing the surface per unit area and per unit time, associated with the bulk flow of fluid. This transport associated with bulk flow is sometimes referred to as "convective transport." One should note the parallelism between the convective transport of mass and momentum, but the tensorial order of the quantities involved is different. In the preceding subsection, the entity being transported is the mass (a scalar) and the mass flux is a vector ($\boldsymbol{v}\rho$). Here, the entity being transported is the momentum (a vector), and the momentum flux is a tensor ($\boldsymbol{v}\boldsymbol{M}$).

In addition to momentum transport by flow, there will also be momentum transferred by virtue of the molecular motions and interactions within the fluid. This additional momentum flux will be designated by the symbol $\boldsymbol{\pi}$, again a second-order tensor. We use the convention that the jk-component π_{jk} of this tensor represents the flux of positive k-momentum in the positive j-direction, associated with molecular motions with respect to the bulk velocity and with interactions. The rate of momentum flow across the surface element dA with orientation $\boldsymbol{n}$ is then $\boldsymbol{n} \cdot \boldsymbol{\pi} dA$. As a consequence of the conservation of angular momentum, it is found throughout this book that $\boldsymbol{\pi}$ is a symmetric tensor. The existence of the tensor $\boldsymbol{\pi}$ and its symmetry are expressed in two fundamental theorems by Cauchy.[2]

We are now ready to write down the law of conservation of momentum. According to this law, the total momentum of the fluid within V will increase because of a net influx of momentum across the bounding surface—both by bulk flow and by molecular motions and interactions. In mathematical terms this becomes

$$\frac{d}{dt}\int_V \boldsymbol{M} d^3r = -\int_{\partial V} \boldsymbol{n} \cdot (\boldsymbol{v}\boldsymbol{M} + \boldsymbol{\pi})\, dA, \tag{2.20}$$

where the left-hand side expresses the rate of increase of momentum within V and the right-hand side expresses the rate of addition of momentum across the surface ∂V by bulk flow and by molecular motions and interactions. When Gauss's divergence theorem is used, the surface integral can be transformed into a volume integral,

$$\frac{d}{dt}\int_V \boldsymbol{M} d^3r = -\int_V \frac{\partial}{\partial \boldsymbol{r}} \cdot (\boldsymbol{v}\boldsymbol{M} + \boldsymbol{\pi})\, d^3r. \tag{2.21}$$

Because the volume V is arbitrary, the integral signs may then be removed to obtain the time-evolution equation for the momentum density $\boldsymbol{M} = \boldsymbol{M}(\boldsymbol{r}, t)$,

$$\frac{\partial \boldsymbol{M}}{\partial t} = -\frac{\partial}{\partial \boldsymbol{r}} \cdot (\boldsymbol{v}\boldsymbol{M}) - \frac{\partial}{\partial \boldsymbol{r}} \cdot \boldsymbol{\pi}. \tag{2.22}$$

Exercise 16 External Forces in Momentum Balance

Incorporate the effect of an external force field, such as gravity, into the integral and local momentum balance equations (2.20) and (2.22).

In terms of the substantial derivative (2.18) and by virtue of the mass balance (2.10), the momentum balance equation (2.22) can be rewritten as

$$\rho \frac{D\boldsymbol{v}}{Dt} = -\frac{\partial}{\partial \boldsymbol{r}} \cdot \boldsymbol{\pi}. \tag{2.23}$$

In other words, if we follow the material elements by taking the time derivative D/Dt, we obtain Newton's second law, so that we can think of (2.22) and (2.23) as the *equation of motion* for the fluid. From an atomistic point of view, the application of Newton's second law to a material element is not really justified because, even if we follow the flow of a material element and keep its mass constant, there is an exchange of atoms or molecules between the material element and its environment. Based on this observation, it has been argued by Brenner[3] that one should carefully distinguish

[2] See Section 16 of Truesdell & Noll, *Non-Linear Field Theories of Mechanics* (Springer, 1992).

[3] Brenner, Physica A (2005) in press.

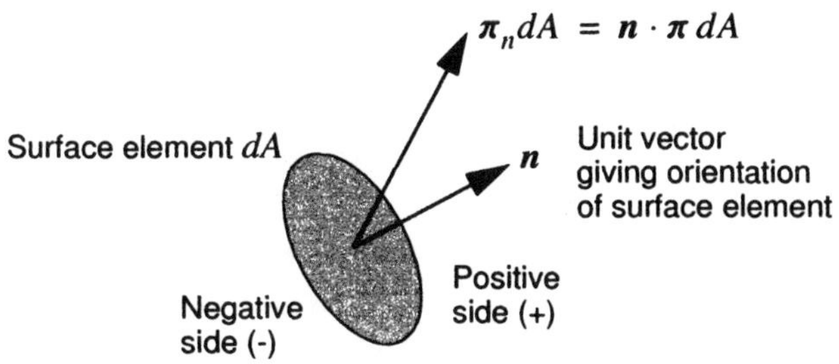

Fig. 2.3 Element of surface dA across which a force $\pi_n dA$ is transmitted.

between the velocity in the equation of motion (2.23) and the velocity characterizing the mass flux in the equation of continuity (2.10). Equation (2.23) suggests an alternative interpretation of the tensor π and its components. We could have used a somewhat different physical statement leading to (2.20): the total momentum of the fluid within the volume V will increase because of a net influx of momentum across the bounding surfaces by bulk flow, and because of the external forces acting on the fluid element—both the surface force exerted by the surrounding fluid and a possible body force exerted, for example, by gravity (see Exercise 16). We then expect a surface term of the form $-\int \pi_n dA$, where $\pi_n dA$ is a vector describing the force exerted by the fluid on the negative side of dA on the fluid on the positive side of dA (see Figure 2.3). Comparison of the above integral with the corresponding term in (2.20) shows that $\pi_n = n \cdot \pi$. That is, the force $\pi_n dA$ corresponding to an orientation n of dA can be obtained from the tensor π. When this interpretation is used, it is more natural to refer to π as the "stress tensor" or, in view of the sign conventions used here,[4] as the "pressure tensor." The component π_{jk} is the force per unit area acting in the positive k-direction on a surface perpendicular to the j-direction, the force being exerted by the negative material on the positive material. Physically speaking, the concept of force per unit area arises naturally only for short-range interactions between particles on the different sides of a surface element, which can be regarded as contact forces.

If one uses this viewpoint, then the integral $\int n \cdot \pi dA$ in (2.20) can be reinterpreted as "force of the fluid outside V acting on the fluid inside V across ∂V." For some purposes, it is useful to think of π as a momentum flux, whereas in other situations the concept of stress is more natural. We shall use both interpretations, and the terms "momentum flux tensor" and "pressure tensor" will be used interchangeably.

[4] We have adopted the sign convention of the textbooks Bird, Stewart & Lightfoot, *Transport Phenomena* (Wiley, 1960) and Bird, Armstrong & Hassager, *DPL 1, Fluid Mechanics* (Wiley, 1987); to remind the reader that this convention is different from that of many other authors, we refer to π as the "pressure tensor," indicating that a pure isotropic pressure p corresponds to $\pi = p\mathbf{1}$.

2.1.3 Energy

After establishing the local form of conservation laws in the previous two subsections, we now formulate directly the balance equations resulting from the flow of kinetic and internal energy, $\boldsymbol{n} \cdot \boldsymbol{v}(\rho \boldsymbol{v}^2/2)dA$ and $\boldsymbol{n} \cdot \boldsymbol{v}\epsilon dA$, respectively. The balance equation for the kinetic energy actually follows from the continuity equation (2.10) and the equation of motion (2.23),

$$\begin{aligned} \frac{\partial}{\partial t}\left(\frac{1}{2}\rho \boldsymbol{v}^2\right) + \frac{\partial}{\partial \boldsymbol{r}} \cdot \left(\boldsymbol{v}\frac{1}{2}\rho \boldsymbol{v}^2\right) &= \frac{1}{2}\boldsymbol{v}^2 \left[\frac{\partial \rho}{\partial t} + \frac{\partial}{\partial \boldsymbol{r}} \cdot (\boldsymbol{v}\rho)\right] + \rho \boldsymbol{v} \cdot \frac{D\boldsymbol{v}}{Dt} \\ &= -\boldsymbol{v} \cdot \left(\frac{\partial}{\partial \boldsymbol{r}} \cdot \boldsymbol{\pi}\right). \end{aligned} \tag{2.24}$$

Equation (2.24) can be rearranged as

$$\frac{\partial}{\partial t}\left(\frac{1}{2}\rho \boldsymbol{v}^2\right) = -\frac{\partial}{\partial \boldsymbol{r}} \cdot \left(\boldsymbol{v}\frac{1}{2}\rho \boldsymbol{v}^2 + \boldsymbol{\pi} \cdot \boldsymbol{v}\right) + \boldsymbol{\pi} : \left(\frac{\partial}{\partial \boldsymbol{r}}\boldsymbol{v}\right)^T. \tag{2.25}$$

We can recognize convective and conductive flux contributions of the divergence form in (2.25) and a source term of kinetic energy (see Exercise 16) which, in the absence of external forces, must be accompanied by a compensating source term of internal energy of the same magnitude and opposite sign required by energy conservation. We hence obtain the internal energy equation

$$\frac{\partial \epsilon}{\partial t} = -\frac{\partial}{\partial \boldsymbol{r}} \cdot (\boldsymbol{v}\,\epsilon + \boldsymbol{j}^{\mathrm{q}}) - \boldsymbol{\pi} : \left(\frac{\partial}{\partial \boldsymbol{r}}\boldsymbol{v}\right)^T, \tag{2.26}$$

where $\boldsymbol{j}^{\mathrm{q}}$ represents the conductive flow of internal energy.

In the presence of external forces, the corresponding changes of potential energy are compensated by changes of kinetic energy without affecting the internal energy balance (2.26). For two-component systems, an interesting effect arises in the internal energy balance when the external forces couple differently to the components.

If the external force per unit volume on the components can be written as $\rho_1 \boldsymbol{g}_1$, $\rho_2 \boldsymbol{g}_2$, where $\boldsymbol{g}_1$, $\boldsymbol{g}_2$ are the forces per unit mass and can hence be interpreted as accelerations, the momentum equation (2.23) becomes

$$\rho \frac{D\boldsymbol{v}}{Dt} = -\frac{\partial}{\partial \boldsymbol{r}} \cdot \boldsymbol{\pi} + \rho_1 \boldsymbol{g}_1 + \rho_2 \boldsymbol{g}_2. \tag{2.27}$$

For gravitational forces, we have $\boldsymbol{g}_1 = \boldsymbol{g}_2 = \boldsymbol{g}$, and we recover the result of Exercise 16. By using (2.27) in (2.24), the kinetic energy balance is generalized to

$$\frac{\partial}{\partial t}\left(\frac{1}{2}\rho \boldsymbol{v}^2\right) = -\frac{\partial}{\partial \boldsymbol{r}} \cdot \left(\boldsymbol{v}\frac{1}{2}\rho \boldsymbol{v}^2 + \boldsymbol{\pi} \cdot \boldsymbol{v}\right) + \boldsymbol{\pi} : \left(\frac{\partial}{\partial \boldsymbol{r}}\boldsymbol{v}\right)^T + \boldsymbol{v} \cdot (\rho_1 \boldsymbol{g}_1 + \rho_2 \boldsymbol{g}_2). \tag{2.28}$$

If the external forces can be derived from time independent potentials,

$$\boldsymbol{g}_1 = -\frac{\partial}{\partial \boldsymbol{r}}\varphi_1, \quad \boldsymbol{g}_2 = -\frac{\partial}{\partial \boldsymbol{r}}\varphi_2, \tag{2.29}$$

then we obtain the following equation for the potential energy density $\rho\varphi = \rho_1\varphi_1 + \rho_2\varphi_2$:

$$\begin{aligned}\frac{\partial}{\partial t}(\rho\varphi) &= \varphi_1\frac{\partial\rho_1}{\partial t} + \varphi_2\frac{\partial\rho_2}{\partial t} = -\varphi_1\frac{\partial}{\partial \boldsymbol{r}}\cdot(\boldsymbol{v}_1\rho_1) - \varphi_2\frac{\partial}{\partial \boldsymbol{r}}\cdot(\boldsymbol{v}_2\rho_2) \\ &= -\varphi_1\frac{\partial}{\partial \boldsymbol{r}}\cdot(\boldsymbol{v}\rho_1 + \boldsymbol{j}_1^{\text{diff}}) - \varphi_2\frac{\partial}{\partial \boldsymbol{r}}\cdot(\boldsymbol{v}\rho_2 + \boldsymbol{j}_2^{\text{diff}}) \\ &= -\frac{\partial}{\partial \boldsymbol{r}}\cdot(\boldsymbol{v}\rho\varphi + \boldsymbol{j}_1^{\text{diff}}\varphi_1 + \boldsymbol{j}_2^{\text{diff}}\varphi_2) - \boldsymbol{v}\cdot(\rho_1\boldsymbol{g}_1 + \rho_2\boldsymbol{g}_2) \\ &\quad -\boldsymbol{j}_1^{\text{diff}}\cdot\boldsymbol{g}_1 - \boldsymbol{j}_2^{\text{diff}}\cdot\boldsymbol{g}_2. \end{aligned} \tag{2.30}$$

Whereas the term $\boldsymbol{v}\cdot(\rho_1\boldsymbol{g}_1 + \rho_2\boldsymbol{g}_2)$ describes the exchange between kinetic and potential energies, the term in the last line of (2.30) must be compensated by internal energy, so that we have the following generalization of (2.26) in the presence of external forces [all source terms must cancel in the sum of (2.28), (2.30), and (2.31)],

$$\frac{\partial\epsilon}{\partial t} = -\frac{\partial}{\partial \boldsymbol{r}}\cdot(\boldsymbol{v}\,\epsilon + \boldsymbol{j}^{\text{q}}) - \boldsymbol{\pi} : \left(\frac{\partial}{\partial \boldsymbol{r}}\boldsymbol{v}\right)^T + \boldsymbol{j}_1^{\text{diff}}\cdot\boldsymbol{g}_1 + \boldsymbol{j}_2^{\text{diff}}\cdot\boldsymbol{g}_2. \tag{2.31}$$

For gravitational forces, the sum of the last two terms in (2.31) vanishes. If the external forces, however, couple differently to the two components of the mixture, there is an additional source of internal energy.

2.1.4 Entropy

We now have the complete set of time-evolution equations for the hydrodynamic fields, namely, the continuity equation (2.10) for the mass density, the equation of motion (2.22) for the momentum density or velocity, the internal energy equation (2.26), and, for two-component systems, the additional diffusion equation (2.19). However, before we can think about solving these equations subject to given boundary conditions, we need to express the unknown fluxes $\boldsymbol{\pi}$, $\boldsymbol{j}^{\text{q}}$, and $\boldsymbol{j}_1^{\text{diff}} = -\boldsymbol{j}_2^{\text{diff}}$ in terms of the hydrodynamic fields.

Before we make assumptions for the fluxes, it is useful to consider the entropy balance equation. One should note that this is a redundant equation because, due to the local equilibrium assumption

$$s(\boldsymbol{r}) = s(\rho(\boldsymbol{r}), \epsilon(\boldsymbol{r})), \text{ or } \quad s(\boldsymbol{r}) = s(\rho(\boldsymbol{r}), \epsilon(\boldsymbol{r}), c_1(\boldsymbol{r})), \tag{2.32}$$

the entropy balance is fully determined by the balances of mass and internal energy. Therefore, what do we gain from looking at the entropy balance? From the source term in the entropy balance equation, we obtain an explicit expression for the local entropy production rate, and we can postulate a generalized version of the second law of thermodynamics by imposing that the entropy production rate must be nonnegative. Such a formulation of the second law will restrict the possible assumptions for the fluxes $\boldsymbol{\pi}$, $\boldsymbol{j}^{\text{q}}$, and $\boldsymbol{j}_1^{\text{diff}} = -\boldsymbol{j}_2^{\text{diff}}$ in the next subsection.

For a single-component fluid, we obtain from the first part of (2.32) and the results of Exercise 14:

$$\frac{\partial s}{\partial t} = -\frac{\mu}{T}\frac{\partial \rho}{\partial t} + \frac{1}{T}\frac{\partial \epsilon}{\partial t} \tag{2.33}$$

and hence

$$\begin{aligned}
\frac{\partial s}{\partial t} &= \frac{\mu}{T}\frac{\partial}{\partial \boldsymbol{r}} \cdot (\boldsymbol{v}\rho) - \frac{1}{T}\frac{\partial}{\partial \boldsymbol{r}} \cdot (\boldsymbol{v}\epsilon + \boldsymbol{j}^{\mathrm{q}}) - \frac{1}{T}\boldsymbol{\pi} : \left(\frac{\partial}{\partial \boldsymbol{r}}\boldsymbol{v}\right)^T \\
&= \left(\frac{\mu}{T}\rho - \frac{1}{T}\epsilon\right)\frac{\partial}{\partial \boldsymbol{r}} \cdot \boldsymbol{v} - \boldsymbol{v} \cdot \frac{\partial}{\partial \boldsymbol{r}}s - \frac{1}{T}\frac{\partial}{\partial \boldsymbol{r}} \cdot \boldsymbol{j}^{\mathrm{q}} - \frac{1}{T}\boldsymbol{\pi} : \left(\frac{\partial}{\partial \boldsymbol{r}}\boldsymbol{v}\right)^T \\
&= -\frac{\partial}{\partial \boldsymbol{r}} \cdot \left(\boldsymbol{v}s + \frac{\boldsymbol{j}^{\mathrm{q}}}{T}\right) + \boldsymbol{j}^{\mathrm{q}} \cdot \frac{\partial}{\partial \boldsymbol{r}}\frac{1}{T} - \frac{1}{T}\boldsymbol{\tau} : \left(\frac{\partial}{\partial \boldsymbol{r}}\boldsymbol{v}\right)^T,
\end{aligned} \tag{2.34}$$

where, in the last step, the expression (2.3) for the local-equilibrium pressure was used. The special role of this hydrostatic pressure, to be exempt from the total pressure tensor in

$$\boldsymbol{\tau} = \boldsymbol{\pi} - p\mathbf{1}, \tag{2.35}$$

results from the necessity to use (2.3) in the derivation; this is related to the fact that hydrostatic pressure effects are accounted for in the internal energy.

The entropy flux in (2.34) is due to convection and conduction, where the conductive entropy flux differs from the heat flux only by a factor of absolute temperature, as one would anticipate from equilibrium thermodynamics. By considering the entropy production rate and imposing a local version of the second law,

$$\sigma = \boldsymbol{j}^{\mathrm{q}} \cdot \frac{\partial}{\partial \boldsymbol{r}}\frac{1}{T} - \frac{1}{T}\boldsymbol{\tau} : \left(\frac{\partial}{\partial \boldsymbol{r}}\boldsymbol{v}\right)^T \geq 0, \tag{2.36}$$

we obtain restrictions for the allowed choices of the fluxes $\boldsymbol{j}^{\mathrm{q}}$ and $\boldsymbol{\tau}$.

Exercise 17 Entropy Production for Two-Component Systems
Generalize the entropy balance equation (2.34) to the case of two-component systems.

2.1.5 Expressions for Fluxes

We next turn our attention to the task of specifying expressions for the fluxes $\boldsymbol{\pi}$ and $\boldsymbol{j}^{\mathrm{q}}$ for closing the equation of motion (2.22) for the momentum density and the internal energy equation (2.26). In formulating such constitutive equations for $\boldsymbol{\pi}$ and $\boldsymbol{j}^{\mathrm{q}}$ in terms of the hydrodynamic fields, we need to make sure that the entropy production rate (2.36) is nonnegative. The easiest and most natural way to achieve that goal is by looking for quadratic terms. For example, for a heat flux proportional to the temperature gradient,

$$\boldsymbol{j}^{\mathrm{q}} = \lambda^{\mathrm{q}}T^2\frac{\partial}{\partial \boldsymbol{r}}\frac{1}{T} = -\lambda^{\mathrm{q}}\frac{\partial T}{\partial \boldsymbol{r}}, \tag{2.37}$$

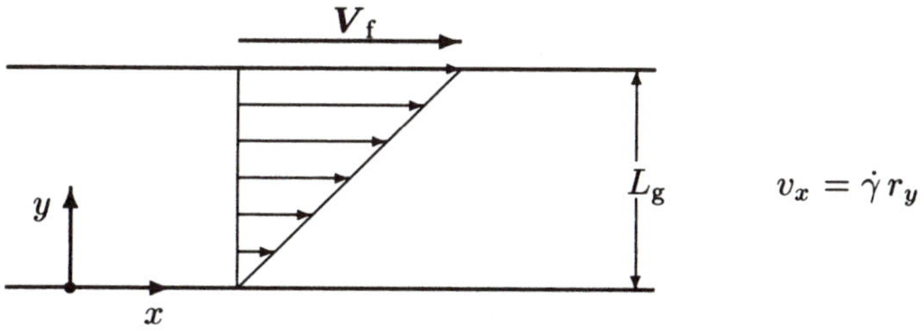

Fig. 2.4 Homogeneous simple shear flow between parallel plates with shear rate $\dot{\gamma} = |\boldsymbol{V}_{\mathrm{f}}|/L_{\mathrm{g}}$.

the first term in (2.36) is a quadratic expression and hence guaranteed to be nonnegative, provided that the thermal conductivity λ^{q} is a nonnegative material parameter. Equation (2.37) is Fourier's famous law of heat conduction.

In the same spirit, the pressure tensor $\boldsymbol{\tau}$ related to $\boldsymbol{\pi}$ by (2.35) should be taken to be proportional to the velocity gradients. At this point, however, one should keep in mind that, for fluids without internal momentum degrees of freedom, $\boldsymbol{\tau}$ is a symmetric tensor and that the trace of a symmetric tensor and its traceless part can play independent roles because they transform separately under rotation. We hence introduce Newton's expression for the pressure tensor in terms of the viscosity η and the dilatational viscosity κ, also referred to as the bulk viscosity, in the following form:

$$\boldsymbol{\tau} = -\eta \left[\frac{\partial}{\partial \boldsymbol{r}} \boldsymbol{v} + \left(\frac{\partial}{\partial \boldsymbol{r}} \boldsymbol{v} \right)^T - \frac{2}{3} \frac{\partial}{\partial \boldsymbol{r}} \cdot \boldsymbol{v} \mathbf{1} \right] - \kappa \frac{\partial}{\partial \boldsymbol{r}} \cdot \boldsymbol{v} \mathbf{1}. \tag{2.38}$$

For the homogeneous shear flow between parallel plates illustrated in Figure 2.4 with

$$\frac{\partial}{\partial \boldsymbol{r}} \boldsymbol{v} = \begin{pmatrix} 0 & 0 & 0 \\ \dot{\gamma} & 0 & 0 \\ 0 & 0 & 0 \end{pmatrix}, \tag{2.39}$$

Newton's expression (2.38) implies

$$\boldsymbol{\tau} = - \begin{pmatrix} 0 & \eta\dot{\gamma} & 0 \\ \eta\dot{\gamma} & 0 & 0 \\ 0 & 0 & 0 \end{pmatrix}. \tag{2.40}$$

The shear force per unit area on the upper plate is proportional to its velocity and inversely proportional to the distance between the plates, where the prefactor is given by the viscosity. The scalar nature of the transport coefficients λ^{q}, η, and κ is a consequence of the implicit assumption of an isotropic system.

Exercise 18 Newton's Constitutive Assumption and the Second Law

Discuss the sign of the entropy production rate (2.36) for Newton's constitutive assumption (2.38).

Together with the relationship (2.35) and the constitutive equations (2.37) and (2.38), the hydrodynamic equations (2.10), (2.22), and (2.26) form a closed set of equations that can be solved for given initial and boundary conditions. These equations are known as Navier-Stokes equations. In particular, the equation of motion, or the momentum balance equation, which can be rewritten as an equation for the velocity field,

$$\rho \frac{\partial \boldsymbol{v}}{\partial t} = -\rho \boldsymbol{v} \cdot \frac{\partial}{\partial \boldsymbol{r}} \boldsymbol{v} + \frac{\partial}{\partial \boldsymbol{r}} \cdot \eta \frac{\partial}{\partial \boldsymbol{r}} \boldsymbol{v} + \frac{\partial}{\partial \boldsymbol{r}} \left(\kappa + \frac{1}{3} \eta \right) \frac{\partial}{\partial \boldsymbol{r}} \cdot \boldsymbol{v} - \frac{\partial p}{\partial \boldsymbol{r}}, \tag{2.41}$$

is known as the Navier-Stokes equation. For an incompressible Newtonian fluid, we obtain the momentum balance

$$\rho \frac{\partial \boldsymbol{v}}{\partial t} = -\rho \boldsymbol{v} \cdot \frac{\partial}{\partial \boldsymbol{r}} \boldsymbol{v} + \frac{\partial}{\partial \boldsymbol{r}} \cdot \eta \frac{\partial}{\partial \boldsymbol{r}} \boldsymbol{v} - \frac{\partial p}{\partial \boldsymbol{r}}, \tag{2.42}$$

and the mass balance reads

$$\frac{\partial}{\partial \boldsymbol{r}} \cdot \boldsymbol{v} = 0, \tag{2.43}$$

where the constraint (2.43) on the velocity field determines the pressure in (2.42) up to an irrelevant additive constant.

The Navier-Stokes equations are the basis for the entire field of fluid mechanics,[5] which is far beyond the scope of this book. Even for isothermal flows of incompressible fluids, fully described by (2.42) and (2.43), there is a wide range of phenomena, many of them not fully understood yet (for example, turbulence[6]). The origin of these interesting phenomena is, of course, the nonlinear occurrence of velocity in the equation of motion (2.42). While Newton's constitutive equation is linear in the velocity, the nonlinearity results from the convective or inertial term. For large flow velocities, the nonlinearity becomes increasingly important. A dimensionless measure of the importance of nonlinearity can be obtained by comparing the nonlinear inertial term to the viscous term, provided that a characteristic flow velocity, V_{f}, and a characteristic length scale of the flow geometry, L_{g}, are given,

$$\mathrm{Re} = \frac{\rho V_{\mathrm{f}} L_{\mathrm{g}}}{\eta}. \tag{2.44}$$

When the Reynolds number Re becomes large, interesting effects due to nonlinearities can arise.

The irreversible contributions to the dynamics are those involving the pressure tensor $\boldsymbol{\tau}$ and the heat flux $\boldsymbol{j}^{\mathrm{q}}$ in (2.22) and (2.26); all the other terms in (2.10), (2.22), and (2.26) constitute reversible contributions. In other words, the reversible contributions are obtained by setting all the dynamic material properties or transport coefficients equal to zero, $\eta = \kappa = \lambda^{\mathrm{q}} = 0$. Note that the reversible part of the

[5] Landau & Lifshitz, *Vol. 6, Fluid Mechanics* (Pergamon, 1987).
[6] Frisch, *Turbulence* (Cambridge, 1995).

entropy balance equation (2.34) is identical in form to the continuity equation (2.10), thus expressing the conservation of entropy under reversible dynamics.

Exercise 19 Diffusion Flow for Ideal Mixtures
Show that the diffusion flow in (2.19) for ideal mixtures can be written as Fick's law,

$$\boldsymbol{j}_1^{\text{diff}} = -\rho D \frac{\partial c_1}{\partial \boldsymbol{r}}, \tag{2.45}$$

with a nonnegative diffusion coefficient D. Note that (2.19) then takes the form of the familiar diffusion equation

$$\frac{\partial c_1}{\partial t} = -\boldsymbol{v} \cdot \frac{\partial c_1}{\partial \boldsymbol{r}} + \frac{1}{\rho} \frac{\partial}{\partial \boldsymbol{r}} \cdot \rho D \frac{\partial c_1}{\partial \boldsymbol{r}}, \tag{2.46}$$

where the diffusion coefficient is placed between the two derivatives with respect to $\boldsymbol{r}$.

2.2 GENERIC FORMULATION

The discussion of Section 2.1 shows that the hydrodynamic fields provide a most natural system for modeling all fluids for which no additional variables are needed to describe changes in the structural state of the fluid. For the purpose of illustrating the GENERIC approach to beyond-equilibrium thermodynamics, we consider two slightly different choices of the independent variables $x = (\rho(\boldsymbol{r}), \boldsymbol{M}(\boldsymbol{r}), \epsilon(\boldsymbol{r}))$ and $x' = (\rho(\boldsymbol{r}), \boldsymbol{M}(\boldsymbol{r}), s(\boldsymbol{r}))$. The alternative possibility of using temperature as an independent variable instead of the internal energy or entropy density will be considered in an exercise (see Exercise 30). Of course, one could also use the velocity instead of the momentum density [see (7.150)] or, for example, the entropy per unit mass instead of the entropy per unit volume (see Exercise 26) as independent variables for describing a state.

For the hydrodynamic fields, the number of components of the state variables x or x' is infinite. More precisely, the components of x or x' are labeled by a discrete index taking five values—say 1 for the mass density, 2, 3, 4 for the components of the momentum density, and 5 for the internal energy or entropy density—and the continuous position label $\boldsymbol{r}$. Any contraction thus requires a summation over the five values of the discrete index and an integration over the position label. For the discrete part, we use a column-vector and matrix notation, typically appearing to be in terms of three-dimensional objects associated with ρ, $\boldsymbol{M}$, and ϵ or s because the momentum components are combined into a sub-block. For example, $L = L(\boldsymbol{r}, \boldsymbol{r}')$ is a 3×3-matrix with a tensorial subblock (or a 5×5-matrix) labeled by two position vectors, and

$$L \cdot \frac{\delta E}{\delta x} = \int L(\boldsymbol{r}, \boldsymbol{r}') \cdot \begin{pmatrix} \frac{\delta E}{\delta \rho(\boldsymbol{r}')} \\ \frac{\delta E}{\delta \boldsymbol{M}(\boldsymbol{r}')} \\ \frac{\delta E}{\delta \epsilon(\boldsymbol{r}')} \end{pmatrix} d^3 r' \tag{2.47}$$

contains a three-dimensional matrix multiplication and an integration over the second position label; the outcome of the operation is a column vector with three (or five) components labeled by $\boldsymbol{r}$. The components of $\delta E / \delta x$ are functional derivatives (see

Appendix C). The local nature of hydrodynamics will actually allow us to simplify the notation and to avoid the integration over $\boldsymbol{r}'$.

2.2.1 Energy and Entropy

We first formulate the energy E as a functional of the hydrodynamic fields $x = (\rho(\boldsymbol{r}), \boldsymbol{M}(\boldsymbol{r}), \epsilon(\boldsymbol{r}))$. The total energy can naturally be obtained by adding the kinetic and internal energy densities and integrating the sum over the entire volume of the flowing system,

$$E = \int \left[\frac{1}{2} \frac{\boldsymbol{M}(\boldsymbol{r})^2}{\rho(\boldsymbol{r})} + \epsilon(\boldsymbol{r}) \right] d^3r. \tag{2.48}$$

By taking functional derivatives with respect to the hydrodynamic fields we obtain, after using the definition (2.6) of the velocity,

$$\frac{\delta E}{\delta x} = \begin{pmatrix} \frac{\delta}{\delta \rho(\boldsymbol{r})} \\ \frac{\delta}{\delta \boldsymbol{M}(\boldsymbol{r})} \\ \frac{\delta}{\delta \epsilon(\boldsymbol{r})} \end{pmatrix} E(\rho, \boldsymbol{M}, \epsilon) = \begin{pmatrix} -\frac{1}{2}\boldsymbol{v}(\boldsymbol{r})^2 \\ \boldsymbol{v}(\boldsymbol{r}) \\ 1 \end{pmatrix}. \tag{2.49}$$

To obtain the energy E' as a functional of the hydrodynamic fields $x' = (\rho(\boldsymbol{r}), \boldsymbol{M}(\boldsymbol{r}), s(\boldsymbol{r}))$ we need to express the internal energy density $\epsilon(\boldsymbol{r})$ in terms of the independent variables $\rho(\boldsymbol{r})$ and $s(\boldsymbol{r})$. With the help of the local-equilibrium assumption, this can be done by evaluating the given thermodynamic relationship $\epsilon = \epsilon(\rho, s)$ at every position $\boldsymbol{r}$,

$$E' = \int \left[\frac{1}{2} \frac{\boldsymbol{M}(\boldsymbol{r})^2}{\rho(\boldsymbol{r})} + \epsilon(\rho(\boldsymbol{r}), s(\boldsymbol{r})) \right] d^3r. \tag{2.50}$$

The relationship between (2.48) and (2.50) is an example of the transformation law (1.48). By taking functional derivatives with respect to the hydrodynamic fields, we obtain, after using the results of Exercise 14 and once more the definition (2.6) of the velocity,

$$\frac{\delta E'}{\delta x'} = \begin{pmatrix} \frac{\delta}{\delta \rho(\boldsymbol{r})} \\ \frac{\delta}{\delta \boldsymbol{M}(\boldsymbol{r})} \\ \frac{\delta}{\delta s(\boldsymbol{r})} \end{pmatrix} E'(\rho, \boldsymbol{M}, s) = \begin{pmatrix} \mu(\boldsymbol{r}) - \frac{1}{2}\boldsymbol{v}(\boldsymbol{r})^2 \\ \boldsymbol{v}(\boldsymbol{r}) \\ T(\boldsymbol{r}) \end{pmatrix}. \tag{2.51}$$

External forces, such as gravity (characterized by a constant acceleration vector $\boldsymbol{g}$), can be incorporated by simply adding the corresponding potential energy, that is $-\int \rho(\boldsymbol{r})\, \boldsymbol{g} \cdot \boldsymbol{r}\, d^3r$ for gravity, to the total energy (2.48) or (2.50).

Exercise 20 Energy of Two-Component System
How does the energy (2.48) change for the two-component system of Section 2.1.1, when described by the variables $x = (\rho(\boldsymbol{r}), \boldsymbol{M}(\boldsymbol{r}), \epsilon(\boldsymbol{r}), c_1(\boldsymbol{r}))$?

Whereas, for the energy, the relationship $\epsilon = \epsilon(\rho, s)$ expressing the local equilibrium thermodynamics is needed only when working with the variables x', the situation is reversed for the entropy: The inverted thermodynamic relationship $s = s(\rho, \epsilon)$ is needed only when working with the variables x so that, in any case, only a single thermodynamic relationship is needed in specifying the two generators of hydrodynamics. In such situations one expects that single-generator formalisms can be used without any problems instead of the more general double-generator GENERIC.

The explicit expressions for the entropy are

$$S = \int s(\rho(\boldsymbol{r}), \epsilon(\boldsymbol{r})) d^3 r, \tag{2.52}$$

and

$$S' = \int s(\boldsymbol{r}) d^3 r, \tag{2.53}$$

where the relationship between (2.52) and (2.53) is an example of the transformation law (1.49). By taking functional derivatives with respect to the hydrodynamic fields we obtain, after using the results of Exercise 14,

$$\frac{\delta S}{\delta x} = \begin{pmatrix} \frac{\delta}{\delta \rho(\boldsymbol{r})} \\ \frac{\delta}{\delta \boldsymbol{M}(\boldsymbol{r})} \\ \frac{\delta}{\delta \epsilon(\boldsymbol{r})} \end{pmatrix} S(\rho, \epsilon) = \begin{pmatrix} -\frac{\mu(\boldsymbol{r})}{T(\boldsymbol{r})} \\ 0 \\ \frac{1}{T(\boldsymbol{r})} \end{pmatrix}, \tag{2.54}$$

and

$$\frac{\delta S'}{\delta x'} = \begin{pmatrix} \frac{\delta}{\delta \rho(\boldsymbol{r})} \\ \frac{\delta}{\delta \boldsymbol{M}(\boldsymbol{r})} \\ \frac{\delta}{\delta s(\boldsymbol{r})} \end{pmatrix} S'(\rho, s) = \begin{pmatrix} 0 \\ 0 \\ 1 \end{pmatrix}. \tag{2.55}$$

Exercise 21 Entropy of Two-Component System

How does the entropy (2.52) change for the two-component system of Section 2.1.1, when described by the variables $x = (\rho(\boldsymbol{r}), \boldsymbol{M}(\boldsymbol{r}), \epsilon(\boldsymbol{r}), c_1(\boldsymbol{r}))$? [Hint: Consider an ideal mixture.]

Exercise 22 Abstract Temperature in Hydrodynamics

Discuss the abstract definition of temperature (1.25) for hydrodynamics both for the variables $x = (\rho(\boldsymbol{r}), \boldsymbol{M}(\boldsymbol{r}), \epsilon(\boldsymbol{r}))$ and for $x' = (\rho(\boldsymbol{r}), \boldsymbol{M}(\boldsymbol{r}), s(\boldsymbol{r}))$.

2.2.2 Poisson Matrix

We next specify the Poisson matrix, which, as explained in the introduction to this section, is a 3×3-matrix with vector and tensor subblocks and two additional position labels. The simpler form of the Poisson matrix is obtained in terms of the variables x' rather than x because, according to the degeneracy requirement (1.4), the entropy density is the density of a conserved quantity under the reversible dynamics described by L'. As can be verified by writing out the corresponding time-evolution equations, the explicit form of the matrix L' for hydrodynamics is

$$L'(\boldsymbol{r}, \boldsymbol{r}') = \begin{pmatrix} 0 & \rho(\boldsymbol{r}')\frac{\partial \delta}{\partial \boldsymbol{r}'} & 0 \\ -\frac{\partial \delta}{\partial \boldsymbol{r}}\rho(\boldsymbol{r}) & \boldsymbol{M}(\boldsymbol{r}')\frac{\partial \delta}{\partial \boldsymbol{r}'} - \frac{\partial \delta}{\partial \boldsymbol{r}}\boldsymbol{M}(\boldsymbol{r}) & -\frac{\partial \delta}{\partial \boldsymbol{r}}s(\boldsymbol{r}) \\ 0 & s(\boldsymbol{r}')\frac{\partial \delta}{\partial \boldsymbol{r}'} & 0 \end{pmatrix}, \tag{2.56}$$

where $\delta = \delta(\boldsymbol{r} - \boldsymbol{r}')$ is Dirac's δ-function of the argument $\boldsymbol{r} - \boldsymbol{r}'$. First of all, it should be noted that generalized functions occur as entries in the matrix L'; this is most natural for every local field theory. The occurrence of derivatives of the δ-function is a consequence of the underlying conservation laws and balance equations for all the variables in x'. All the nonzero entries are in the row and column associated with the momentum density or velocity variable, where the antisymmetry of L' is obvious. The velocity-dominated structure of L' is a characteristic of modeling convection phenomena as the only reversible effect. Keeping in mind that the second entry of the energy gradients (2.51) is the velocity, one can see that the entries in the second column of (2.56) express the convection of mass, momentum, and entropy. More precisely, by multiplying the right side of (2.51), using $\boldsymbol{r}'$ instead of $\boldsymbol{r}$ in that equation, with the right side of (2.56) and integrating over $\boldsymbol{r}'$ as indicated in (2.47), one can verify explicitly that all the reversible contributions to the time-evolution equations of classical hydrodynamics are obtained, that is, (2.10) and (2.22) with $\boldsymbol{\pi} = p\boldsymbol{1}$, and (2.34) with $\boldsymbol{j}^{\mathrm{q}} = \boldsymbol{\tau} = 0$.

Exercise 23 Euler Equations

Verify that the reversible parts of the time-evolution equations of classical hydrodynamics, known as the Euler equations, are obtained by inserting (2.51) and (2.56) into (2.47).

There is an ambiguity associated with the notation of (2.56). When we translate the Poisson matrix L' directly by (1.6) into a Poisson bracket, we obtain

$$\begin{aligned}
\{A, B\} &= \int \left[\frac{\delta B}{\delta \rho(\boldsymbol{r})} \frac{\partial}{\partial \boldsymbol{r}} \cdot \left(\frac{\delta A}{\delta \boldsymbol{M}(\boldsymbol{r})} \rho(\boldsymbol{r}) \right) - \frac{\delta A}{\delta \rho(\boldsymbol{r})} \frac{\partial}{\partial \boldsymbol{r}} \cdot \left(\frac{\delta B}{\delta \boldsymbol{M}(\boldsymbol{r})} \rho(\boldsymbol{r}) \right) \right] d^3 r \\
&+ \int \left[\frac{\delta B}{\delta \boldsymbol{M}(\boldsymbol{r})} \frac{\partial}{\partial \boldsymbol{r}} : \left(\frac{\delta A}{\delta \boldsymbol{M}(\boldsymbol{r})} \boldsymbol{M}(\boldsymbol{r}) \right) \right. \\
&\qquad \left. - \frac{\delta A}{\delta \boldsymbol{M}(\boldsymbol{r})} \frac{\partial}{\partial \boldsymbol{r}} : \left(\frac{\delta B}{\delta \boldsymbol{M}(\boldsymbol{r})} \boldsymbol{M}(\boldsymbol{r}) \right) \right] d^3 r \\
&+ \int \left[\frac{\delta B}{\delta s(\boldsymbol{r})} \frac{\partial}{\partial \boldsymbol{r}} \cdot \left(\frac{\delta A}{\delta \boldsymbol{M}(\boldsymbol{r})} s(\boldsymbol{r}) \right) - \frac{\delta A}{\delta s(\boldsymbol{r})} \frac{\partial}{\partial \boldsymbol{r}} \cdot \left(\frac{\delta B}{\delta \boldsymbol{M}(\boldsymbol{r})} s(\boldsymbol{r}) \right) \right] d^3 r .
\end{aligned} \tag{2.57}$$

However, the derivatives of the δ-function with respect to $\boldsymbol{r}'$ could be replaced by derivatives with respect to $\boldsymbol{r}$, and vice versa, at the expense of a minus sign. Starting from this alternative form of the Poisson matrix we obtain the bracket

$$\begin{aligned}
\{A, B\} &= -\int \rho(\boldsymbol{r}) \left(\frac{\delta A}{\delta \boldsymbol{M}(\boldsymbol{r})} \cdot \frac{\partial}{\partial \boldsymbol{r}} \frac{\delta B}{\delta \rho(\boldsymbol{r})} - \frac{\delta B}{\delta \boldsymbol{M}(\boldsymbol{r})} \cdot \frac{\partial}{\partial \boldsymbol{r}} \frac{\delta A}{\delta \rho(\boldsymbol{r})} \right) d^3 r \\
&- \int \boldsymbol{M}(\boldsymbol{r}) \cdot \left(\frac{\delta A}{\delta \boldsymbol{M}(\boldsymbol{r})} \cdot \frac{\partial}{\partial \boldsymbol{r}} \frac{\delta B}{\delta \boldsymbol{M}(\boldsymbol{r})} - \frac{\delta B}{\delta \boldsymbol{M}(\boldsymbol{r})} \cdot \frac{\partial}{\partial \boldsymbol{r}} \frac{\delta A}{\delta \boldsymbol{M}(\boldsymbol{r})} \right) d^3 r \\
&- \int s(\boldsymbol{r}) \left(\frac{\delta A}{\delta \boldsymbol{M}(\boldsymbol{r})} \cdot \frac{\partial}{\partial \boldsymbol{r}} \frac{\delta B}{\delta s(\boldsymbol{r})} - \frac{\delta B}{\delta \boldsymbol{M}(\boldsymbol{r})} \cdot \frac{\partial}{\partial \boldsymbol{r}} \frac{\delta A}{\delta s(\boldsymbol{r})} \right) d^3 r .
\end{aligned} \tag{2.58}$$

As an integration by parts shows, the brackets (2.57) and (2.58) differ only by boundary terms. Although boundary terms have a well-defined meaning in a Poisson bracket, the δ-function in (2.56) is ambiguous whenever $\boldsymbol{r}$ or $\boldsymbol{r}'$ lies on the boundary. Therefore, Poisson brackets are clearly helpful when dealing explicitly with boundary conditions. However, because we are interested in the local field equations within an isolated system, there is no need to pay special attention to the boundary conditions, and we can consider (2.57) and (2.58) as equivalent. When Beris and Edwards[7] derived the Poisson bracket of hydrodynamics from a variational principle of classical mechanics via a Lagrangian description of the fluid in Sections 5.1–5.3 of their book, they arrived at the version (2.57). Their derivation implies that the bracket (2.57) fulfills the Jacobi identity. A direct check will be discussed in a more general context in Section 2.3.1. By Lie-Poisson reduction, which is a general approach based on the geometric structures of Hamiltonian dynamics, one obtains the version (2.58) [see (B.113) and (B.124) in Appendix B].

Passing from the variables x' to x can either be done by means of the transformation formula (1.50) or by reproducing the reversible parts of the time-evolution equations (2.10), (2.22), and (2.26) as before. The explicit form of the matrix L for hydrodynamics is

$$L(\boldsymbol{r},\boldsymbol{r}') = \begin{pmatrix} 0 & \rho(\boldsymbol{r}')\frac{\partial\delta}{\partial\boldsymbol{r}'} & 0 \\ -\frac{\partial\delta}{\partial\boldsymbol{r}}\rho(\boldsymbol{r}) & \boldsymbol{M}(\boldsymbol{r}')\frac{\partial\delta}{\partial\boldsymbol{r}'} - \frac{\partial\delta}{\partial\boldsymbol{r}}\boldsymbol{M}(\boldsymbol{r}) & -\frac{\partial\delta}{\partial\boldsymbol{r}}\epsilon(\boldsymbol{r}) - \frac{\partial\delta}{\partial\boldsymbol{r}}p(\boldsymbol{r}') \\ 0 & \epsilon(\boldsymbol{r}')\frac{\partial\delta}{\partial\boldsymbol{r}'} + p(\boldsymbol{r})\frac{\partial\delta}{\partial\boldsymbol{r}'} & 0 \end{pmatrix}, \tag{2.59}$$

where, again, $\delta = \delta(\boldsymbol{r}-\boldsymbol{r}')$. Note that the internal energy is not a conserved quantity; changes in volume, as described the divergence of the velocity field, lead to a change in the internal energy determined by the local pressure.

When checked in terms of the variables x, the degeneracy requirement (1.4) leads to the relationship (2.3), which is a deep result for the pressure in local-equilibrium thermodynamics. We thus realize that the degeneracy requirement (1.4) can lead us to important physical insights. In more complicated situations, even more important insights should be expected, such as expressions for pressure tensors.

Exercise 24 Poisson Bracket with Pressure Terms

Write out the Poisson bracket associated with the Poisson matrix (2.59).

The actual calculations in writing out the time-evolution equation and the degeneracy requirements can be simplified by observing that, as mentioned above, multiplication with the Poisson operator $L(\boldsymbol{r},\boldsymbol{r}')$ implies both a matrix multiplication and an integration over $\boldsymbol{r}'$. After integrating by parts and ignoring boundary terms, we can perform the integration over $\boldsymbol{r}'$ because δ-functions occur in all the matrix elements of $L(\boldsymbol{r},\boldsymbol{r}')$. Instead of a matrix with generalized functions as entries, we then obtain an equivalent matrix, the entries of which are differential operators (the derivatives

[7] Beris & Edwards, *Thermodynamics of Flowing Systems* (Oxford, 1994).

act on all terms to the right of them[8]):

$$L(\boldsymbol{r}) = -\begin{pmatrix} 0 & \frac{\partial}{\partial \boldsymbol{r}}\rho(\boldsymbol{r}) & 0 \\ \rho(\boldsymbol{r})\frac{\partial}{\partial \boldsymbol{r}} & \left[\frac{\partial}{\partial \boldsymbol{r}}\boldsymbol{M}(\boldsymbol{r}) + \boldsymbol{M}(\boldsymbol{r})\frac{\partial}{\partial \boldsymbol{r}}\right]^T & \epsilon(\boldsymbol{r})\frac{\partial}{\partial \boldsymbol{r}} + \frac{\partial}{\partial \boldsymbol{r}}p(\boldsymbol{r}) \\ 0 & \frac{\partial}{\partial \boldsymbol{r}}\epsilon(\boldsymbol{r}) + p(\boldsymbol{r})\frac{\partial}{\partial \boldsymbol{r}} & 0 \end{pmatrix}. \tag{2.60}$$

We will repeatedly use this kind of simplified notation throughout the book. The reader is urged to verify how directly (2.60) can be applied to (2.49) in order to reproduce the reversible parts of the equation of motion by a three-dimensional matrix multiplication without additional integrations. The antisymmetry of L is less obvious in (2.60) than in (2.59) because integrations by parts are involved for some of the entries, where we once more neglect the arising boundary terms.

Exercise 25 Poisson Operator With a Scalar Density
Write the analogue of the Poisson operator (2.60) for the variables $\rho(\boldsymbol{r})$, $\boldsymbol{M}(\boldsymbol{r})$, $s(\boldsymbol{r})$.

Exercise 26 Poisson Operator With a Scalar
Write the analogue of the Poisson operator (2.60) for the variables $\rho(\boldsymbol{r})$, $\boldsymbol{M}(\boldsymbol{r})$, $s(\boldsymbol{r})/\rho(\boldsymbol{r})$ (note that the entropy per unit mass, as a ratio of two scalar densities, is not affected by changes of volume and hence a scalar rather than a scalar density).

Exercise 27 Poisson Operator of Two-Component System
How does the Poisson operator (2.60) change for the two-component system of Section 2.1.1, when described by the variables $x = (\rho(\boldsymbol{r}), \boldsymbol{M}(\boldsymbol{r}), \epsilon(\boldsymbol{r}), c_1(\boldsymbol{r}))$?

2.2.3 Friction Matrix

For the variables $x' = (\rho(\boldsymbol{r}), \boldsymbol{M}(\boldsymbol{r}), s(\boldsymbol{r}))$, the entropy gradient (2.55) implies that the entry M'_{33} of the friction matrix $M'(\boldsymbol{r}, \boldsymbol{r}')$ is determined by the entropy production rate. According to the expression (2.36) with (2.37) and (2.38), we have

$$M'_{33} = \left[\frac{\eta}{2T}\dot{\gamma} : \dot{\gamma} + \frac{\hat{\kappa}}{4T}(\mathrm{tr}\dot{\gamma})^2 + \frac{1}{T(\boldsymbol{r})T(\boldsymbol{r}')}\frac{\partial}{\partial \boldsymbol{r}} \cdot \frac{\partial}{\partial \boldsymbol{r}'}\lambda^{\mathrm{q}}T^2\right]\delta(\boldsymbol{r} - \boldsymbol{r}'), \tag{2.61}$$

where the tensor $\dot{\gamma}$ is the symmetrized velocity gradient tensor,

$$\dot{\gamma}(\boldsymbol{r}) = \frac{\partial}{\partial \boldsymbol{r}}\boldsymbol{v}(\boldsymbol{r}) + \left[\frac{\partial}{\partial \boldsymbol{r}}\boldsymbol{v}(\boldsymbol{r})\right]^T, \tag{2.62}$$

[8] One should note the following careful distinction made in our notation. When we write $\frac{\partial f}{\partial \boldsymbol{r}}$ for any function f, the derivative applies to f only, whereas writing $\frac{\partial}{\partial \boldsymbol{r}} f$ would imply the derivation of all functions to the right of $\frac{\partial}{\partial \boldsymbol{r}}$, including the functions resulting from a vector multiplied from the right.

and the transport coefficient $\hat{\kappa}$, occasionally referred to as the second viscosity, is a combination of the dilatational viscosity κ and the viscosity η,

$$\hat{\kappa} = \kappa - \frac{2}{3}\eta. \tag{2.63}$$

In (2.61), the position argument $\boldsymbol{r}$ or $\boldsymbol{r}'$ is indicated only where the difference matters (that is, not directly in front of the δ-function). The entry M'_{33} determines the entire friction matrix; we first apply the degeneracy

$$M' \cdot \frac{\delta E'}{\delta x'} = \int M(\boldsymbol{r}, \boldsymbol{r}') \cdot \begin{pmatrix} \mu(\boldsymbol{r}') - \frac{1}{2}\boldsymbol{v}(\boldsymbol{r}')^2 \\ \boldsymbol{v}(\boldsymbol{r}') \\ T(\boldsymbol{r}') \end{pmatrix} d^3r' = 0 \tag{2.64}$$

to get M'_{32}, then the symmetry of the friction matrix to get M'_{23}, and finally again the degeneracy (2.64) to get M'_{22}. The final result for $M' = M'(\boldsymbol{r}, \boldsymbol{r}')$ is

$$M' = \begin{pmatrix} 0 & 0 & 0 \\ 0 & \left(\frac{\partial}{\partial \boldsymbol{r}'}\frac{\partial}{\partial \boldsymbol{r}} + \mathbf{1}\frac{\partial}{\partial \boldsymbol{r}'} \cdot \frac{\partial}{\partial \boldsymbol{r}}\right)\eta T\delta + \frac{\partial}{\partial \boldsymbol{r}}\frac{\partial}{\partial \boldsymbol{r}'}\hat{\kappa}T\delta & \frac{\partial}{\partial \boldsymbol{r}} \cdot \eta\dot{\gamma}\delta + \frac{\partial}{\partial \boldsymbol{r}}\frac{\hat{\kappa}}{2}\mathrm{tr}\dot{\gamma}\delta \\ 0 & \frac{\partial}{\partial \boldsymbol{r}'} \cdot \eta\dot{\gamma}\delta + \frac{\partial}{\partial \boldsymbol{r}'}\frac{\hat{\kappa}}{2}\mathrm{tr}\dot{\gamma}\delta & M'_{33} \end{pmatrix}, \tag{2.65}$$

where $\delta = \delta(\boldsymbol{r} - \boldsymbol{r}')$. Because the continuity equation is unaffected by irreversible effects, the row and the column associated with the mass density in the friction matrix are zero (this point will be reconsidered in Section 2.2.5). Although M'_{33} was chosen to get the proper irreversible terms in the entropy balance equation, it can be verified that also the expected irreversible terms in the momentum balance equation (2.22) are obtained.

The matrix M' in (2.65) has all the properties formulated in the introduction. It is symmetric and can be verified to be positive semidefinite (see the subsequent Section 2.2.4) and, by construction, $\delta E'/\delta x'$ lies in the null space of M'. We just realized that these properties are very important for obtaining a unique expression for the matrix M' of classical hydrodynamics. The matrix elements of M' for classical hydrodynamics are proportional to η, $\hat{\kappa}$, and λ^{q}, so that its relation to dissipative material properties and irreversible dynamics should be obvious. This observation finally justifies the term "friction matrix" for M'. The friction matrix M for the variables $x = (\rho(\boldsymbol{r}), \boldsymbol{M}(\boldsymbol{r}), \epsilon(\boldsymbol{r}))$ is obtained by multiplying the last column of (2.65) by $T(\boldsymbol{r}')$ and the last row by $T(\boldsymbol{r})$; because the entries of (2.65) are generalized functions and not operators, the additional factors of temperature are not to be differentiated. Also the entries of $M = M(\boldsymbol{r}, \boldsymbol{r}')$ are generalized functions, that is, the derivatives act only on the functions in the matrix elements. As previously observed for the Poisson matrix, we can simplify notation and calculations by introducing a matrix $M(\boldsymbol{r})$,

$$M(\boldsymbol{r}) = \begin{pmatrix} 0 & 0 & 0 \\ 0 & M_{22} & M_{23} \\ 0 & M_{32} & M_{33} \end{pmatrix}, \tag{2.66}$$

the entries of which are differential operators, that is, the derivatives act on all terms to the right of them,

$$M_{22} = -\left(\frac{\partial}{\partial \boldsymbol{r}}\eta T \frac{\partial}{\partial \boldsymbol{r}} + \mathbf{1}\frac{\partial}{\partial \boldsymbol{r}} \cdot \eta T \frac{\partial}{\partial \boldsymbol{r}}\right)^T - \frac{\partial}{\partial \boldsymbol{r}}\hat{\kappa}T\frac{\partial}{\partial \boldsymbol{r}}, \tag{2.67}$$

$$M_{23} = \frac{\partial}{\partial \boldsymbol{r}} \cdot \eta T \dot{\boldsymbol{\gamma}} + \frac{\partial}{\partial \boldsymbol{r}}\frac{\hat{\kappa}T}{2}\mathrm{tr}\dot{\boldsymbol{\gamma}}, \tag{2.68}$$

$$M_{32} = -\eta T \dot{\boldsymbol{\gamma}} \cdot \frac{\partial}{\partial \boldsymbol{r}} - \frac{\hat{\kappa}T}{2}\mathrm{tr}\dot{\boldsymbol{\gamma}}\frac{\partial}{\partial \boldsymbol{r}}, \tag{2.69}$$

and

$$M_{33} = \frac{\eta T}{2}\dot{\boldsymbol{\gamma}} : \dot{\boldsymbol{\gamma}} + \frac{\hat{\kappa}T}{4}(\mathrm{tr}\dot{\boldsymbol{\gamma}})^2 - \frac{\partial}{\partial \boldsymbol{r}} \cdot \lambda^{\mathrm{q}} T^2 \frac{\partial}{\partial \boldsymbol{r}}. \tag{2.70}$$

As discussed for $L(\boldsymbol{r})$, multiplying with $M(\boldsymbol{r})$ does not imply any integrations over position labels. To make such remarks absolutely clear, we introduce a more precise notation that can be used whenever careful distinctions are necessary. We use one or two underlines for a matrix, depending on whether the matrix has one or two position arguments. For example, we write $\underline{\underline{M}}'$ for the friction matrix of (2.65) and $\underline{M}$ for (2.66). The entries of $\underline{\underline{M}}'$ are (generalized) functions and ordering is unimportant, whereas the entries of $\underline{M}$ are operators so that one needs to pay attention to ordering. We further need to resolve the symbol "$\cdot$" so far used in various ways depending on the set of variables x. In the more precise notation, we use the dot "$\cdot$" when only a matrix multiplication occurs, and we use the fat dot "$\bullet$" if an additional integration is required. For example, the total entropy production (1.21) with (1.15) can be written as

$$\frac{dS}{dt} = \frac{\delta S}{\delta x} \bullet \underline{\underline{M}} \bullet \frac{\delta S}{\delta x} = \frac{\delta S}{\delta x} \bullet \underline{M} \cdot \frac{\delta S}{\delta x}, \tag{2.71}$$

where the operator $\underline{M}$ eliminates one of the integrations. The other integration remains to be performed. One could further write

$$\frac{dS}{dt} = \int \sigma(\boldsymbol{r}) d^3 r \quad \text{with} \quad \sigma = \frac{\delta S}{\delta x} \cdot \underline{M} \cdot \frac{\delta S}{\delta x}, \tag{2.72}$$

where $\sigma = \sigma(\boldsymbol{r})$ is the local entropy production rate. Below, we occasionally refer to the use of underlines and dots of different sizes as the "more careful notation."

Exercise 28 More Careful Notation

Express the reversible contribution to the time evolution in (2.47) in the more careful notation. How does the notation change when we use the Poisson operator of (2.60)?

Exercise 29 Friction Operator of Two-Component System

How does the friction operator (2.66) change for the two-component system of Section 2.1.1, when described by the variables $x = (\rho(\boldsymbol{r}), \boldsymbol{M}(\boldsymbol{r}), \epsilon(\boldsymbol{r}), c_1(\boldsymbol{r}))$? [Hint: Use the result of Exercise 19.]

Exercise 30 Temperature as an Independent Variable

Derive the GENERIC building blocks of hydrodynamics for the independent variables $x'' = (\rho(\boldsymbol{r}), \boldsymbol{M}(\boldsymbol{r}), T(\boldsymbol{r}))$.

Exercise 31 Abstract Temperature in Hydrodynamics

Discuss the abstract definition of temperature (1.25) for hydrodynamics for the variables $x'' = (\rho(\boldsymbol{r}), \boldsymbol{M}(\boldsymbol{r}), T(\boldsymbol{r}))$ of Exercise 30 (see also Exercise 22).

Exercise 32 Temperature Equation

Derive the following equation for the temperature for hydrodynamics,

$$\frac{\partial T}{\partial t} = -\boldsymbol{v} \cdot \frac{\partial T}{\partial \boldsymbol{r}} - \frac{1}{\alpha_s} \frac{\partial}{\partial \boldsymbol{r}} \cdot \boldsymbol{v} + \frac{1}{\rho \hat{c}_V} \left[\frac{\eta}{2} \dot{\gamma} : \dot{\gamma} + \frac{\hat{\kappa}}{4} (\mathrm{tr}\dot{\gamma})^2 + \frac{\partial}{\partial \boldsymbol{r}} \cdot \lambda^{\mathrm{q}} \frac{\partial T}{\partial \boldsymbol{r}} \right], \tag{2.73}$$

where α_s is the adiabatic thermal expansivity and $\hat{c}_V$ is the heat capacity per unit mass.

2.2.4 Fluctuating Hydrodynamics

The positive semidefinite nature of the matrix $M(\boldsymbol{r}, \boldsymbol{r}')$ can be shown by writing it in the form

$$\int B(\boldsymbol{r}, \boldsymbol{r}'') \cdot B(\boldsymbol{r}', \boldsymbol{r}'')^T d^3 r'' = 2 k_{\mathrm{B}} M(\boldsymbol{r}, \boldsymbol{r}'). \tag{2.74}$$

This identity can be verified explicitly for the matrix $B = B(\boldsymbol{r}, \boldsymbol{r}')$ given by

$$\begin{pmatrix} 0 & 0 & 0 \\ \frac{\partial}{\partial r_j} \sqrt{2\kappa k_{\mathrm{B}} T}\,\delta & \left(\delta_{jk} \frac{\partial}{\partial r_l} + \delta_{jl} \frac{\partial}{\partial r_k} - \frac{2}{3} \delta_{kl} \frac{\partial}{\partial r_j} \right) \sqrt{\eta k_{\mathrm{B}} T}\,\delta & 0 \\ \mathrm{tr}\dot{\gamma} \sqrt{\frac{\kappa k_{\mathrm{B}} T}{2}}\,\delta & (\dot{\gamma}_{kl} - \delta_{kl} \mathrm{tr}\dot{\gamma}/3) \sqrt{\eta k_{\mathrm{B}} T}\,\delta & \frac{\partial}{\partial r_k} \sqrt{2 \lambda^{\mathrm{q}} k_{\mathrm{B}} T}\,\delta \end{pmatrix}, \tag{2.75}$$

where, again, $\delta = \delta(\boldsymbol{r} - \boldsymbol{r}')$. In the middle row, the subscript j labels the rows corresponding to the components of the momentum vector. The indices k and l in the middle column correspond to an index pair to be summed over if $B(\boldsymbol{r}, \boldsymbol{r}')$ is multiplied with another matrix or column vector from the right. This discussion of the indices shows that, despite its first appearance due to the block notation, (2.75) is not a square matrix. The number of rows corresponds to the number of hydrodynamic fields, whereas the number of columns corresponds to the relevant fluctuations or transport mechanisms in the system.

According to the fluctuation-dissipation theorem, the matrix (2.75) characterizes the fluctuations to be added to the equations of hydrodynamics as described in (1.56); the resulting stochastic partial differential equations[9] coincide with those occurring in the standard theory of fluctuating hydrodynamics where, however, $T = T(\boldsymbol{r})$ here is the local-equilibrium temperature defined in (2.2) (see Chapter IX of Landau

[9] The mathematical theory of stochastic partial differential equations has made enormous progress in recent years; see, for example, Da Prato & J. Zabczyk, *SDEs in Infinite Dimensions* (Cambridge, 1992) or Kipnis & Landim, *Scaling Limits of Interacting Particle Systems* (Springer, 1999).

and Lifshitz[10]). The columns in the matrix (2.75) correspond to the fluctuations associated with different transport processes. The first two columns correspond to fluctuating dilatational and shear stresses, respectively, which have an impact on both the momentum and the internal energy equations. The third column represents a fluctuating heat flux vector that affects only the energy equation.

The matrix (2.75) differs slightly from an alternative version given before.[11] The advantage of the present expression is that it exists for all nonnegative values of the transport coefficients η, κ, and λ^{q}, whereas the previous version required a nonnegative $\hat{\kappa}$; the representation (2.74) hence proves the positive semidefinite nature of the friction matrix of hydrodynamics for all physical values of the transport coefficients.

As a possible application of fluctuating hydrodynamics, we mention the calculation of the fluctuating force on a Brownian particle suspended in a fluid. The force exerted by the fluid on the particle can be calculated by integrating the proper component of the total pressure tensor over the entire surface of the particle. An explicit calculation for spherical particles suspended in an incompressible fluid was performed by Zwanzig.[12] In principle, one would have to solve the Navier-Stokes equation for an incompressible fluid to obtain the velocity and pressure fields associated with the fluctuating stresses subject to stick boundary conditions on the surface of the sphere. However, by means of a theorem due to Faxen, the force can alternatively and more conveniently be evaluated from the unperturbed velocity field in the absence of the sphere, but in the presence of fluctuating stresses, by averaging over the surface of a sphere. Zwanzig's calculation confirms the well-known amplitude of the white-noise force associated with the friction coefficient of Stokes' law according to the fluctuation-dissipation theorem, which is a nice consistency check.

When fluctuating forces and torques on Brownian particles of arbitrary shape are computed from fluctuating hydrodynamics, a fluctuation-dissipation theorem can be established even when retardation effects are taken into account. Furthermore, the symmetry of the 6×6 friction matrix associated with the translational and rotational motions of the Brownian particles can be established without appeal to microscopic arguments. The corresponding calculations by Hauge and Martin-Löf[13] exhibit some interesting mathematical aspects. First, the calculation of fluctuating forces and torques from the velocity and pressure fields of an incompressible fluid sticking to the boundary of the particle can be based on an appropriate Green's identity rather than Faxen's theorem. Second, there is a possibility of choosing between distinct definitions of fluctuating stresses at the boundary of the particle; a careful consideration of boundary conditions is indispensable in applying GENERIC with fluctuations.

[10] Landau & Lifshitz, *Vol. 9, Statistical Physics 2* (Pergamon, 1992).
[11] See (32) of Öttinger & Grmela, Phys. Rev. E 56 (1997) 6633.
[12] Zwanzig, J. Res. Natl. Bur. Std. (U.S.) 68B (1964) 143.
[13] Hauge & Martin-Löf, J. Stat. Phys. 7 (1973) 259.

2.2.5 Something is Missing

We have pointed out before that the concept of a hydrodynamic velocity field is not as unambiguous as usually assumed. From an atomistic perspective, one needs to keep in mind that not all atoms or molecules in a material element of a fluid move with the same velocity and that the set of individual atoms or molecules contained in a given macroscopic material element hence changes in time, even when the mass of the material element is fixed. The far-reaching consequences of this observation have been elaborated in detail in Brenner's systematic analysis.[14] We here discuss this problem from an entirely macroscopic perspective within the GENERIC framework for the case of a single-component fluid. We offer a mathematical and a physical argument indicating that something is missing in the usual formulation of hydrodynamics.

From a mathematical perspective, it is interesting to note that we have two fundamentally different expressions for the velocity

$$\boldsymbol{v}(\boldsymbol{r}) = \frac{\boldsymbol{M}(\boldsymbol{r})}{\rho(\boldsymbol{r})} \quad \text{and} \quad \boldsymbol{v}(\boldsymbol{r}) = \frac{\delta E(\rho, \boldsymbol{M}, \epsilon)}{\delta \boldsymbol{M}(\boldsymbol{r})}, \tag{2.76}$$

where the latter expression follows from (2.49). We have thus imposed a self-dual nature on velocity, or, in other words, we have assumed the existence of a natural equivalence of vectors and covectors. The experience with tangent and cotangent bundles in the Lagrangian and Hamiltonian formulations of classical mechanics (see Appendix B) suggests that the existence of a dual pair of velocities could be considered in hydrodynamics. Note that the first expression in (2.76) is related to the momentum density $\boldsymbol{M}$ for which we formulate a time-evolution equation; it should hence be directly related to the particle momentum occurring in Newton's equation of motion. The second velocity in (2.76) determines the flux of quantities in the balance equations and describes the velocity of a material element of fluid. However, as pointed out before, the composition of this material element changes in time, and the constituent particles with fixed identity have a distribution of velocities.

We next turn to a physical argument that is much more compelling than the above mathematical consideration. In the GENERIC formulation of hydrodynamics, we have assumed that the row and the column associated with the mass density in the friction matrix (2.66) are identically zero. Although this assumption leads to the widely accepted standard equations of hydrodynamics, it imposes an unnecessary restriction. From the perspective of statistical mechanics, which will be presented in the second part of this book, this assumption excludes the possibility of density fluctuations and is hence unacceptable. A more general form of the friction matrix (2.66) is not just a logical possibility; rather, it is a physical necessity. More concretely, we need to incorporate the fundamental fact that particles perform diffusive motions. The most natural way of doing that is by adding a contribution to the friction matrix

[14] Brenner, Physica A (2005) in press.

that accounts for diffusion,

$$M^{\mathrm{diff}} = -\begin{pmatrix} \frac{\partial}{\partial \boldsymbol{r}} D' \frac{\partial}{\partial \boldsymbol{r}} & \frac{\partial}{\partial \boldsymbol{r}} D' \boldsymbol{v} \frac{\partial}{\partial \boldsymbol{r}} & \frac{\partial}{\partial \boldsymbol{r}} D' \alpha \frac{\partial}{\partial \boldsymbol{r}} \\ \frac{\partial}{\partial \boldsymbol{r}} D' \boldsymbol{v} \frac{\partial}{\partial \boldsymbol{r}} & \frac{\partial}{\partial \boldsymbol{r}} D' \boldsymbol{v}\boldsymbol{v} \frac{\partial}{\partial \boldsymbol{r}} & \frac{\partial}{\partial \boldsymbol{r}} D' \boldsymbol{v}\alpha \frac{\partial}{\partial \boldsymbol{r}} \\ \frac{\partial}{\partial \boldsymbol{r}} D' \alpha \frac{\partial}{\partial \boldsymbol{r}} & \frac{\partial}{\partial \boldsymbol{r}} D' \alpha \boldsymbol{v} \frac{\partial}{\partial \boldsymbol{r}} & \frac{\partial}{\partial \boldsymbol{r}} D' \alpha^2 \frac{\partial}{\partial \boldsymbol{r}} \end{pmatrix}$$

$$= -\frac{\partial}{\partial \boldsymbol{r}} D' \begin{pmatrix} 1 & \boldsymbol{v} & \alpha \\ \boldsymbol{v} & \boldsymbol{v}\boldsymbol{v} & \boldsymbol{v}\alpha \\ \alpha & \alpha\boldsymbol{v} & \alpha^2 \end{pmatrix} \frac{\partial}{\partial \boldsymbol{r}}, \tag{2.77}$$

where D' is a positive dynamic material property, α is an arbitrary function, and a contraction of the two derivative operators is implied for all the entries. Whereas M_{11}^{diff} defines the diffusion mechanism, the other entries are determined naturally by the symmetry and the degeneracy (1.5) of the friction matrix, where the degeneracy leaves the function α undetermined, as can be checked by multiplying (2.77) by the energy gradient (2.49). Equation (2.77) provides a most natural phenomenological extension of hydrodynamics that complies with the physical necessity of nonzero entries in the density row and column of the friction matrix. In the following, we discuss some implications of this modification.

Equation (2.77), when taken into account in the fundamental time-evolution equation (1.1), leads to additional terms in the mass, momentum, and internal energy balance equations (2.10), (2.22), and (2.26), which can be rewritten as

$$\frac{\partial \rho}{\partial t} = -\frac{\partial}{\partial \boldsymbol{r}} \cdot (\boldsymbol{v}_{\mathrm{m}} \rho), \tag{2.78}$$

$$\frac{\partial \boldsymbol{M}}{\partial t} = -\frac{\partial}{\partial \boldsymbol{r}} \cdot (\boldsymbol{v}_{\mathrm{m}} \boldsymbol{M}) - \frac{\partial}{\partial \boldsymbol{r}} \cdot \boldsymbol{\pi}, \tag{2.79}$$

and

$$\frac{\partial \epsilon}{\partial t} = -\frac{\partial}{\partial \boldsymbol{r}} \cdot [\boldsymbol{v}_{\mathrm{m}} \epsilon + (\boldsymbol{v}_{\mathrm{m}} - \boldsymbol{v})(\rho\alpha - \epsilon) + \boldsymbol{j}^{\mathrm{q}}] - \boldsymbol{\pi} : \left(\frac{\partial}{\partial \boldsymbol{r}} \boldsymbol{v}\right)^T, \tag{2.80}$$

with

$$\boldsymbol{v}_{\mathrm{m}} = \boldsymbol{v} + \frac{D'}{\rho} \left[\frac{\partial}{\partial \boldsymbol{r}} \frac{\partial s(\rho, \epsilon)}{\partial \rho} + \alpha \frac{\partial}{\partial \boldsymbol{r}} \frac{\partial s(\rho, \epsilon)}{\partial \epsilon} \right] = \boldsymbol{v} - \frac{D'}{\rho} \frac{\partial}{\partial \boldsymbol{r}} \frac{\mu}{T} + \alpha \frac{D'}{\rho} \frac{\partial}{\partial \boldsymbol{r}} \frac{1}{T}. \tag{2.81}$$

By means of the results of Exercise 14, this expression for $\boldsymbol{v}_{\mathrm{m}}$ can be rewritten as

$$\boldsymbol{v}_{\mathrm{m}} = \boldsymbol{v} - \frac{D'}{\rho^2} \frac{\partial}{\partial \boldsymbol{r}} \frac{p}{T} + (\rho\alpha - \epsilon) \frac{D'}{\rho^2} \frac{\partial}{\partial \boldsymbol{r}} \frac{1}{T}. \tag{2.82}$$

Particularly simple expressions result for $\rho\alpha = \epsilon$ (p/T drives the velocity difference) and for $\rho\alpha = \epsilon + p$ (p drives the velocity difference). For the latter choice, $\rho\alpha = \epsilon + p$, we fully reproduce the equations of Brenner.

The velocity of a material element characterizing the flow of mass and momentum, $\boldsymbol{v}_{\mathrm{m}}$, thus turns out to be different from the velocity $\boldsymbol{v} = \boldsymbol{M}/\rho$ occurring in Newton's equation of motion. In our approach, the velocity $\boldsymbol{v}$, referred to as the Lagrangian velocity by Brenner, occurs as a basic independent variable, whereas $\boldsymbol{v}_{\rm m}$ is a dependent or auxiliary variable given by the formula (2.81).

It is illuminating to consider the modified entropy balance equation (2.34), which becomes

$$\begin{aligned} \frac{\partial s}{\partial t} &= -\frac{\partial}{\partial \boldsymbol{r}} \cdot \left[\boldsymbol{v}_{\rm m} s + \frac{\boldsymbol{j}^{\rm q}}{T} - \frac{p + \epsilon - \rho\alpha}{T} (\boldsymbol{v}_{\rm m} - \boldsymbol{v}) \right] \\ &+ \frac{\rho^2 (\boldsymbol{v}_{\rm m} - \boldsymbol{v})^2}{D'} + \boldsymbol{j}^{\rm q} \cdot \frac{\partial}{\partial \boldsymbol{r}} \frac{1}{T} - \frac{1}{T} \boldsymbol{\tau} : \left(\frac{\partial}{\partial \boldsymbol{r}} \boldsymbol{v} \right)^T . \end{aligned} \tag{2.83}$$

There occurs a natural quadratic entropy production term associated with diffusion, so that the entropy production remains positive semidefinite if we keep Fourier's law (2.37) for the heat flux and Newton's expression (2.38) for the pressure tensor. Equation (2.83) shows that the added diffusion effect is on the same footing as the dissipative effects in conventional hydrodynamics. The entropy flux is particularly simple for $\rho\alpha = \epsilon + p$ because, otherwise, there occurs a subtle flux term associated with the velocity difference $\boldsymbol{v}_{\rm m} - \boldsymbol{v}$.

To understand the difference between $\boldsymbol{v}_{\rm m}$ and $\boldsymbol{v}$, we evaluate (2.82) with the entropy density (2.5) of a monatomic ideal gas. We then obtain

$$\rho(\boldsymbol{v} - \boldsymbol{v}_{\rm m}) = D \left(\frac{\partial \rho}{\partial \boldsymbol{r}} + \frac{\rho\alpha - \epsilon}{p} \frac{\rho}{T} \frac{\partial T}{\partial \boldsymbol{r}} \right), \tag{2.84}$$

with the diffusion coefficient

$$D = \frac{k_{\rm B}}{m} \frac{D'}{\rho}. \tag{2.85}$$

As a general feature, both density and temperature gradients contribute to the velocity difference. A much more detailed discussion of the modified theory of hydrodynamics has been offered by Brenner.[15]

2.3 ON CONSTRUCTING GENERIC BUILDING BLOCKS

The energy and entropy of hydrodynamics have a straightforward and transparent form; however, a number of lessons can be learned from the Poisson and friction matrices. These matrices are the GENERIC building blocks with which we have no previous experience from equilibrium thermodynamics at all; hence, we need to look at them once more and more carefully. In most of the examples of this book, the choice of the energy and entropy will appear perfectly natural, due to the local-equilibrium assumption, due to a familiar functional form such as Boltzmann's entropy, or due to a previous understanding of concepts such as entropic springs. So far, we have seen the possibility of choosing particular Poisson and friction matrices for hydrodynamics; next, we need to recognize their natural structure.

[15] Brenner, Physica A (2005) in press.

2.3.1 Poisson Matrices

If convection is the only physically relevant reversible phenomenon, the only nonzero entries in the Poisson operator are in the row and in the column associated with the momentum density. We have moreover seen that there are significant differences for the matrix elements associated with scalars and scalar densities. Such a difference is natural because, for a velocity field with a nonvanishing divergence, the corresponding volume changes have an influence on scalar densities but not on scalars. Similar differences can be observed for vectors and tensors and their densities. In this subsection, we use such observations to formulate a large class of useful Poisson operators in a pedestrian's way,[16] so that the Jacobi identity does not need to be checked in many practical applications. We start with the momentum density as the only variable; we then keep adding more and more variables of well-defined tensorial types. The result of the following exercise and the technique for obtaining it are the basis for the rest of this subsection.

Exercise 33 Jacobi Identity for Momentum Density
Verify the Jacobi identity for the bracket

$$\{A, B\}_M = -\int M_k \left(\frac{\delta A}{\delta M_j} \frac{\partial}{\partial r_j} \frac{\delta B}{\delta M_k} - \frac{\delta B}{\delta M_j} \frac{\partial}{\partial r_j} \frac{\delta A}{\delta M_k} \right) d^3 r, \tag{2.86}$$

where, following Einstein's summation convention, one needs to sum from one to three over indices occurring twice [see (2.58)]. The functionals A and B depend only on the vector field $\boldsymbol{M}$.

When we add more variables and associated terms to the bracket (2.86), then, in the nested brackets of the Jacobi identity, there arise four kinds of terms: (1) "old in old," (2) "new in old," (3) "old in new," and (4) "new in new." All terms of type (1) add up to zero because we start from a Poisson bracket satisfying the Jacobi identity.

For example, if we add the mass density ρ to the list of variables, we can define the Poisson bracket for a scalar density [see (2.58)]

$$\{A, B\}_{\text{sd}} = \{A, B\}_M - \int \rho \left(\frac{\delta A}{\delta M_j} \frac{\partial}{\partial r_j} \frac{\delta B}{\delta \rho} - \frac{\delta B}{\delta M_j} \frac{\partial}{\partial r_j} \frac{\delta A}{\delta \rho} \right) d^3 r. \tag{2.87}$$

The terms of the type "new in old" vanish because only functional derivatives with respect to $\boldsymbol{M}$ occur in the old bracket (2.86). The six terms of the type "old in new" are generated by the following term through permutations of A, B, and C (see Exercise 33 for details):

$$-\int \rho \frac{\delta B}{\delta M_j} \frac{\partial}{\partial r_j} \frac{\delta C}{\delta M_k} \frac{\partial}{\partial r_k} \frac{\delta A}{\delta \rho} d^3 r. \tag{2.88}$$

[16] Deeper insights into the structure of Poisson brackets in terms of geometric structures, symmetries, and semidirect products of groups and vector spaces may be found in Appendix B.

For "new in new," we find six terms generated by

$$\int \rho \frac{\delta A}{\delta M_j} \frac{\partial}{\partial r_j} \frac{\delta B}{\delta M_k} \frac{\partial}{\partial r_k} \frac{\delta C}{\delta \rho} d^3r, \tag{2.89}$$

where, since (2.88) and (2.89) differ only by a cyclic permutation, these terms cancel each other, and (2.87) hence satisfies the Jacobi identity. Note that, for such a cancellation of "old in new" and "new in new" terms, it is crucial that ρ occurs exactly as in the additional bracket contribution in (2.87)—not even a constant numerical factor (except 0) would be allowed.

If we introduce a further scalar density variable, say the entropy density s, with an analogous bracket contribution as in (2.87), then the "new in old" terms again vanish. This happens always when a new bracket contribution depends, except through the functionals A, B, and C, only on the new variable. For the "old in new" terms only functional derivatives with respect to the original variable $\boldsymbol{M}$ matter, and they result exclusively from the original bracket (2.86) and not from the previous additions in (2.87). The extension of the original bracket (2.86) to (2.87) is hence irrelevant for checking the Jacobi identity with a further contribution, as certainly also the "new in new" terms are unaffected by previous additions. Thus, very generally, when a new variable and a corresponding bracket contribution are added, it is sufficient to add them to the original bracket (2.86), independently of any other variables and bracket terms added before, provided that the additional terms contain explicit functions of the additional variables only, as well as functional derivatives with respect to $\boldsymbol{M}$ and the additional variable only. Therefore, adding two scalar densities to the momentum density leads to a valid Poisson bracket, and we hence have explicitly checked the Jacobi identity for the Poisson bracket (2.58) of hydrodynamics. We next discuss further possible variables in a similar spirit (see Table 2.1), with special emphasis on the tensorial nature of the variables. Appendix B.4 shows that the construction of Poisson brackets is intimately related to the particle relabeling symmetry of admissible evolution equations, which can be guaranteed if the variables are associated with well-defined body tensors of contravariant or covariant nature and which then naturally leads to so-called upper or lower convected derivatives in fluid mechanics.

The Poisson bracket (2.86) for the momentum density represents the typical term leading to a lower convected derivative of a vector density. Going from a vector density to a vector field is analogous to going from a scalar density to a scalar field (see Exercise 26), so that the corresponding Poisson bracket can be obtained by transformation (see the entries in Table 2.1). For a lower convected vector field $\boldsymbol{w}$, we can hence write

$$\begin{aligned} \{A, B\}_{\text{lcv}} &= \{A, B\}_M + \int w_j \frac{\partial}{\partial r_l} \left(\frac{\delta A}{\delta w_j} \frac{\delta B}{\delta M_l} - \frac{\delta B}{\delta w_j} \frac{\delta A}{\delta M_l} \right) d^3r \\ &\quad - \int w_l \left(\frac{\delta A}{\delta w_j} \frac{\partial}{\partial r_j} \frac{\delta B}{\delta M_l} - \frac{\delta B}{\delta w_j} \frac{\partial}{\partial r_j} \frac{\delta A}{\delta M_l} \right) d^3r, \end{aligned} \tag{2.90}$$

where, despite the similarity of the entries for the Poisson operator in Table 2.1, more terms arise for $\boldsymbol{w}$ than for $\boldsymbol{M}$ because for $\boldsymbol{w}$ the entries in the momentum column of the Poisson operator are no longer in a diagonal block.

Table 2.1 Entries in Poisson Operators

Type of variable	Example	Entry in momentum column*
Scalar density	Mass density ρ [(2.60)]	$-\frac{\partial}{\partial \boldsymbol{r}}\rho(\boldsymbol{r})$ or $-\frac{\partial}{\partial r_l}\rho$
Scalar	Entropy per mass $\hat{s}$ [Exercise 26]	$-\frac{\partial \hat{s}(\boldsymbol{r})}{\partial \boldsymbol{r}}$ or $-\frac{\partial \hat{s}}{\partial r_l}$
Lower convected vector density	Momentum density $\boldsymbol{M}$ or M_j [(2.60)]	$-\frac{\partial}{\partial r_l}M_j - M_l\frac{\partial}{\partial r_j}$
Lower convected vector	Entropy conduction $\boldsymbol{w}$ or w_j [(5.17)]	$-\frac{\partial w_j}{\partial r_l} - w_l\frac{\partial}{\partial r_j}$
Lower convected tensor	Momentum conduction $\boldsymbol{c}$ or c_{jk} [(5.17)]	$-\frac{\partial c_{jk}}{\partial r_l} - c_{lk}\frac{\partial}{\partial r_j} - c_{jl}\frac{\partial}{\partial r_k}$
Upper convected tensor	Conformation tensor $\boldsymbol{c}$ or c_{jk} [(4.44)]	$-\frac{\partial c_{jk}}{\partial r_l} + (c_{ik}\delta_{jl} + c_{ji}\delta_{kl})\frac{\partial}{\partial r_i}$

*In the index notation, the index l is always to be contracted when multiplying with a vector from the right, whereas any other indices correspond to the variable considered; Einstein's summation convention is used for the index i.

Exercise 34 Jacobi Identity With a Lower Convected Vector Variable
Verify that the bracket $\{A, B\}_{\rm lcv}$ in (2.90) satisfies the Jacobi identity.

The full brackets corresponding to lower and upper convected tensor fields, as defined through Table 2.1, are given by

$$\begin{aligned}\{A, B\}_{\rm lct} = \; & \{A, B\}_M + \int c_{jk}\frac{\partial}{\partial r_l}\left(\frac{\delta A}{\delta c_{jk}}\frac{\delta B}{\delta M_l} - \frac{\delta B}{\delta c_{jk}}\frac{\delta A}{\delta M_l}\right) d^3r \\ & - \int c_{lk}\left(\frac{\delta A}{\delta c_{jk}}\frac{\partial}{\partial r_j}\frac{\delta B}{\delta M_l} - \frac{\delta B}{\delta c_{jk}}\frac{\partial}{\partial r_j}\frac{\delta A}{\delta M_l}\right) d^3r \\ & - \int c_{jl}\left(\frac{\delta A}{\delta c_{jk}}\frac{\partial}{\partial r_k}\frac{\delta B}{\delta M_l} - \frac{\delta B}{\delta c_{jk}}\frac{\partial}{\partial r_k}\frac{\delta A}{\delta M_l}\right) d^3r, \end{aligned} \tag{2.91}$$

and

$$\begin{aligned}\{A, B\}_{\rm uct} = \; & \{A, B\}_M + \int c_{jk}\frac{\partial}{\partial r_l}\left(\frac{\delta A}{\delta c_{jk}}\frac{\delta B}{\delta M_l} - \frac{\delta B}{\delta c_{jk}}\frac{\delta A}{\delta M_l}\right) d^3r \\ & + \int c_{lk}\left(\frac{\delta A}{\delta c_{jk}}\frac{\partial}{\partial r_l}\frac{\delta B}{\delta M_j} - \frac{\delta B}{\delta c_{jk}}\frac{\partial}{\partial r_l}\frac{\delta A}{\delta M_j}\right) d^3r \\ & + \int c_{jl}\left(\frac{\delta A}{\delta c_{jk}}\frac{\partial}{\partial r_l}\frac{\delta B}{\delta M_k} - \frac{\delta B}{\delta c_{jk}}\frac{\partial}{\partial r_l}\frac{\delta A}{\delta M_k}\right) d^3r. \end{aligned} \tag{2.92}$$

It should be noted that, in addition to the pure convection term present even for scalar fields (see Exercise 26), an additional term arises in Table 2.1 for each index of a vector or tensor variable, the general structure of which can be inferred from Table 2.1 or from (2.90)–(2.92). The Jacobi identity for the bracket (2.91) can be obtained as a direct generalization of Exercise 34. An explicit check of the Jacobi identity for upper convected tensor variables can be avoided by noting that the inverse of a lower convected tensor is of the upper convected type, so that the Jacobi identity can be guaranteed to hold by a transformation to inverses (this observation is directly related to the fact that the inverse of a covariant body tensor is contravariant).

As indicated before, with the possible entries of Poisson operators listed in Table 2.1, convected derivatives of vectors and tensors occur in the final time-evolution equations (the displayed entries are to be multiplied by the derivative of energy with respect to the momentum density, which is the velocity field). These convected derivatives, which take the deformation effects on the components of vectors and tensors into account, indicate the *material objectivity* or *frame indifference* of *admissible equations*.[17] The Jacobi identity, or the time-structure invariance of a model, is hence intimately related to its frame indifference.

If one uses linear combinations of the upper and lower convected derivatives, the Jacobi identity is no longer fulfilled. The so-called *Schowalter derivative* of a tensor variable $\boldsymbol{c}$,

$$\frac{\partial \boldsymbol{c}}{\partial t} + \boldsymbol{v} \cdot \frac{\partial \boldsymbol{c}}{\partial \boldsymbol{r}} - \boldsymbol{\kappa} \cdot \boldsymbol{c} - \boldsymbol{c} \cdot \boldsymbol{\kappa}^T + \frac{\xi}{2}(\boldsymbol{c} \cdot \dot{\boldsymbol{\gamma}} + \dot{\boldsymbol{\gamma}} \cdot \boldsymbol{c}), \tag{2.93}$$

where $\kappa_{jk} = \partial v_j / \partial r_k$ represents the velocity gradient tensor and ξ is the slip parameter, expresses reversible or mechanically controllable dynamics only for the parameters $\xi = 0$ and $\xi = 2$ corresponding to upper and lower convected behavior, respectively.

In this section, we have focused on such contributions to Poisson brackets that correspond to the momentum row and column in the Poisson matrix. If one introduces further nonzero entries in the Poisson matrix, more complicated convection phenomena violating the principle of material objectivity can arise, which can be interpreted as inertial effects. An explicit example of this type arises in Grad's moment method for solving Boltzmann's kinetic equation, as discussed in Section 7.4.3.

2.3.2 Friction Matrices

In an attempt to clarify the structure of friction matrices and to relate GENERIC to the single-generator bracket formalism, Edwards[18] observed that many friction matrices can be written in the factorized form

$$M = C_M \cdot D_M \cdot C_M^T, \tag{2.94}$$

[17] See, for example, Chapter 9 of Bird, Armstrong & Hassager, *DPL 1, Fluid Mechanics* (Wiley, 1987) or p. 106–107 and Section 11.3 of Lodge, *Body Tensor Fields* (Academic, 1974).

[18] Edwards, J. Non-Equilib. Thermodyn. 23 (1998) 301.

where the matrix D_M contains all the dynamic material information, whereas the matrix C_M is considered as a mechanical component. At the end of this subsection, the decomposition (2.94) will be interpreted as a diagonalization of M. A friction matrix of the form (2.94) is symmetric and positive semidefinite provided that the matrix D_M has these properties. The degeneracy requirement (1.5) is fulfilled for

$$C_M^T \cdot \frac{\delta E}{\delta x} = 0. \tag{2.95}$$

The most natural expression for the matrix B of the fluctuation-dissipation theorem (1.57) then is

$$B = C_M \cdot \sqrt{2k_B D_M}. \tag{2.96}$$

For hydrodynamics, using the variables $x = (\rho(\boldsymbol{r}), \boldsymbol{M}(\boldsymbol{r}), \epsilon(\boldsymbol{r}))$, we need to factorize (2.66). The mechanical matrix in the operator notation with only one position label is chosen as

$$C_M^T = \begin{pmatrix} 0 & -\frac{\partial}{\partial \boldsymbol{r}} & \boldsymbol{\kappa}^T \\ 0 & 0 & -\frac{\partial}{\partial \boldsymbol{r}} \end{pmatrix}. \tag{2.97}$$

When multiplied by an appropriate column vector, the first row produces a tensor and the second row a vector. The two gradient operators are introduced because we want to model two transport processes, namely, momentum and heat transport. For relaxation processes, one should choose a constant (e.g., unity) instead of a gradient operator. The velocity gradients $\boldsymbol{\kappa}$ need to be introduced to satisfy the degeneracy requirement (2.95). Taking integrations by part into account, the matrix C_M is obtained as

$$C_M = \begin{pmatrix} 0 & 0 \\ \frac{\partial}{\partial \boldsymbol{r}} & 0 \\ \boldsymbol{\kappa} & \frac{\partial}{\partial \boldsymbol{r}} \end{pmatrix}. \tag{2.98}$$

The matrix D_M has a block-diagonal form with two entries, namely, a fourth-rank tensor for the momentum transport and a second-rank tensor for the heat transport,

$$D_M = \begin{pmatrix} \eta T(\delta_{ik}\delta_{jl} + \delta_{il}\delta_{jk}) + \hat{\kappa} T \delta_{ij}\delta_{kl} & 0 \\ 0 & \lambda^{\mathrm{q}} T^2 \delta_{ik} \end{pmatrix}. \tag{2.99}$$

By multiplying (2.97)–(2.99), one indeed recovers the friction matrix (2.66). The matrices (2.98) and (2.99) hence elaborate the essential elements of the friction matrix of hydrodynamics. The construction of the square root of the fourth-rank tensor in (2.99) is not straightforward, and the corresponding amplitude of fluctuations (2.96) is more conveniently expressed in the form (2.75).

Equation (2.99) illustrates why the factorization (2.94) is important for understanding the classical theory of irreversible thermodynamics.[19] The matrix D_M contains only the transport properties, which are the phenomenological coefficients on which linear irreversible thermodynamics is focused. The mechanical effects in the friction

[19] de Groot & Mazur, *Non-Equilibrium Thermodynamics* (Dover, 1984).

matrix M, expressing the facts that, for transport processes, the gradients of the intensive variables are the driving forces of the fluxes and that the divergences of the fluxes appear in the time-evolution equations, are pulled out through C_M. For relaxation processes, we do not expect derivatives in the factor C_M.

Exercise 35 Careful Notation for Factorization

For the specific matrices C_M^T, C_M, and D_M of (2.97)–(2.99), write the general equations (2.94)–(2.96) in the more careful notation of (2.71) and (2.72).

Exercise 36 Factorization of M for Two-Component System

Formulate the factors C_M, C_M^T, and D_M for the friction matrix of a two-component system obtained in Exercise 29.

Exercise 37 Factorization of M for Modified Hydrodynamics

Formulate the factors C_M, C_M^T, and D_M for the contribution (2.77) to the friction matrix, which introduces diffusion into hydrodynamics.

The decomposition (2.94) may be thought of as a diagonalization of a symmetric friction matrix. We can hence rewrite it as

$$M_{jk} = \sum_{l=1}^{N} (C_M)_{jl} (D_M)_{ll} (C_M)_{kl}, \tag{2.100}$$

where N is the number of dissipative processes. We thus obtain an alternative interpretation of a friction matrix as the sum of contributions associated with different transport or relaxation processes ($l = 1 \ldots N$), each of which is a multiple of the tensor product of a vector with itself [the components of the vector for the lth contribution are $(C_M)_{jl}$, the weight factor is $(D_M)_{ll}$].[20] This modeling strategy will be illustrated in Section 4.2. In particular, we expect a friction matrix to be the sum of a number of contributions, each of them being symmetric and positive semidefinite, each of them satisfying the degeneracy requirement (1.5), and each of them associated with a certain transport or relaxation process.

In some cases, it may be easier to model the fluctuations in a system rather than the dissipative processes. The friction matrix is then given by the fluctuation-dissipation theorem (1.57), with the advantage that the symmetry and the positive semidefiniteness of the friction matrix are guaranteed automatically. This procedure was, for example, applied in the construction of GENERIC equations for a discrete formulation of hydrodynamics.[21]

[20] Note that the orthogonality of the column vectors of C_M may not be immediately obvious in infinite spaces; the result could be a boundary term associated with a divergence.

[21] Español, Serrano & Öttinger, Phys. Rev. Lett. 83 (1999) 4542.

3

Linear Irreversible Thermodynamics

As already mentioned in Section 1.2.7 on the historical background, "linear irreversible thermodynamics" was an important milestone in the development of beyond-equilibrium thermodynamics, as it comprises all the phenomenological observations made over three centuries in a compact and elegant form. Any more general, nonlinear framework of beyond-equilibrium thermodynamics must undoubtedly contain linear irreversible thermodynamics as a special case. In Chapter 3, we show that, indeed, the linear theory arises naturally within the GENERIC framework. In particular, a deep understanding of the famous Onsager-Casimir relations can be obtained from the GENERIC perspective on linear irreversible thermodynamics.

3.1 THERMODYNAMIC FORCES AND FLUXES

In the formulation of balance equations, the importance of the momentum and internal energy fluxes had become apparent. As a further example of a flux, the diffusion flow in a multicomponent system occurred naturally (see Exercise 17), and diffusion was found to be relevant even in single-component fluids (see Section 2.2.5). In formulating constitutive equations for the fluxes, the gradients of temperature, velocity, and chemical potential or concentration played an important role. These gradients are examples of thermodynamic or generalized forces, which are the driving forces for the fluxes. Fourier's expression for the heat flux in terms of the temperature gradient and Newton's expression for the stress tensor in terms of the velocity gradients are most impressive illustrations of the success of assuming linear relationships between thermodynamic forces and fluxes. In this section, we construct the general, systematic approach of linear irreversible thermodynamics built on the concept of force-flux pairs from the GENERIC framework. Actually, the identification of forces and fluxes is also possible and useful in the nonlinear regime.

3.1.1 Basic Concepts

Understanding the entropy production is a key step in developing a theory of irreversible processes. From the GENERIC perspective, the fundamental expression for the total entropy production rate is (1.21) or (2.71),

$$\frac{dS}{dt} = [S, S] = \frac{\delta S}{\delta x} \cdot M \cdot \frac{\delta S}{\delta x}. \tag{3.1}$$

The direct availability of a compact entropy equation is a consequence of the double-generator structure of the fundamental time-evolution equation (1.1) and the degeneracy requirement (1.4). The quadratic form of the entropy production is a consequence of using the entropy as the generator of irreversible dynamics in the GENERIC framework, and it is the key to identifying forces and fluxes to recover the well-established special case of linear irreversible thermodynamics. One should further note that there is important background information coming with (3.1) in the GENERIC framework. That is, under the local-equilibrium assumption, the generators E and S are related by a thermodynamic relationship and M is related to E by the degeneracy requirement (1.5) expressing energy conservation.

For a more intuitive understanding of the entropy equation (3.1), we decompose the friction matrix into mechanical contributions and a core contribution containing the dynamic material information (see Section 2.3.2). This decomposition is of a general and fundamental nature because it is closely related to a diagonalization of the friction matrix and because it relates the entropy gradient with respect to all the independent variables to the transport and relaxation processes. By means of (2.94), we then obtain (3.1) in the alternative, decomposed form

$$\frac{dS}{dt} = \left(C_M^T \cdot \frac{\delta S}{\delta x} \right)^T \cdot D_M \cdot C_M^T \cdot \frac{\delta S}{\delta x}. \tag{3.2}$$

Equation (3.2) is a quadratic form in the quantities

$$X^{\mathrm{LIT}} = -C_M^T \cdot \frac{\delta S}{\delta x}, \tag{3.3}$$

with coefficients that are dynamic material parameters, such as transport coefficients or relaxation times. The arbitrary choice of a minus sign in (3.3) is motivated by particular examples, in which we wish to reproduce standard expressions. If the dynamic material parameters are collected in a phenomenological matrix,

$$L^{\mathrm{LIT}} = D_M, \tag{3.4}$$

then the total entropy production can be rewritten in the form

$$\frac{dS}{dt} = X^{\mathrm{LIT}} \cdot L^{\mathrm{LIT}} \cdot X^{\mathrm{LIT}} = X^{\mathrm{LIT}} \cdot J^{\mathrm{LIT}} \tag{3.5}$$

with

$$J^{\mathrm{LIT}} = L^{\mathrm{LIT}} \cdot X^{\mathrm{LIT}}. \tag{3.6}$$

These are the fundamental equations of linear irreversible thermodynamics, as presented in Chapter IV of the book by de Groot and Mazur.[1] The quantities X^{LIT} are the *generalized* or *thermodynamic forces*, or driving forces, and the quantities J^{LIT} are the resulting *fluxes*. A number of specific examples are considered below.

Equations (3.3) and (3.4) establish the fundamental correspondence between GENERIC and the classical theory of irreversible thermodynamics, based on the idea of diagonalizing the friction matrix. For a better appreciation of equations (3.2)–(3.6), we consider the local entropy production rate σ, which is a key quantity in the development of the classical theory. To that end, we reconsider (3.2)–(3.6) in the light of the more careful notation illustrated in (2.71) and (2.72). The matrix $C_M = \underline{\underline{C_M}}$ is taken as a function of one position label only and, for transport processes, its entries typically contain derivatives. Therefore, the dots between C_M^T and $\delta S/\delta x$ in (3.2) and (3.3) do not imply integrations, and the thermodynamic forces in (3.3) typically are gradients of intensive quantities depending on position; that is, they are the natural driving forces for transport processes.

In a local theory, the matrix $D_M = \underline{\underline{D_M}}$ contains phenomenological coefficients, which may depend on $\boldsymbol{r}$, so that the gradients of intensive variables at the position $\boldsymbol{r}$ determine the fluxes at the same position $\boldsymbol{r}$. In a nonlocal theory, the matrix $D_M = \underline{\underline{D_M}}$ could be a given function of $\boldsymbol{r}$ and $\boldsymbol{r}'$, typically depending on $\boldsymbol{r} - \boldsymbol{r}'$, so that the fluxes at the position $\boldsymbol{r}$ depend on the forces in a neighborhood of $\boldsymbol{r}$. Following standard procedure, we here focus on local theories with $L^{\mathrm{LIT}} = \underline{\underline{L}}^{\mathrm{LIT}} = \underline{\underline{D_M}}$. Then, like (2.71), (3.5) should be written more carefully as

$$\frac{dS}{dt} = X^{\mathrm{LIT}} \bullet L^{\mathrm{LIT}} \cdot X^{\mathrm{LIT}} = X^{\mathrm{LIT}} \bullet J^{\mathrm{LIT}}, \tag{3.7}$$

where the fat dots imply integrations. Equation (3.7) is the correct interpretation of (3.5) in careful notation. If one wishes to avoid the fat dots and the integrations associated with them, again in careful notation, one obtains the local entropy production rather than the total one,

$$\sigma = X^{\mathrm{LIT}} \cdot L^{\mathrm{LIT}} \cdot X^{\mathrm{LIT}} = X^{\mathrm{LIT}} \cdot J^{\mathrm{LIT}}, \tag{3.8}$$

with

$$\frac{dS}{dt} = \int \sigma(\boldsymbol{r}) d^3 r. \tag{3.9}$$

Because GENERIC is designed for isolated systems, we did not consider any exchange of entropy through the boundaries. The complete neglect of boundary conditions is not expected to cause serious problems if we are interested in the local entropy production rate $\sigma(\boldsymbol{r})$ in terms of the independent fields and their derivatives at the position $\boldsymbol{r}$.

[1] de Groot & Mazur, *Non-Equilibrium Thermodynamics* (Dover, 1984).

The only remnants from boundary conditions are possible terms of a total divergence form, the possible presence of which we should consider carefully in each application.

It is often preferred to write $T\sigma$ instead of σ in (3.8). By doing so, one can avoid some extra factors of $1/T$ occurring in the definition of certain generalized forces, which result because units of energy often arise more naturally than units of entropy [see, for example, (3.12) or (3.21) below]. Although it had been emphasized before that the concept of temperature does not play a fundamental role in GENERIC, the frequently and conveniently used extra factor of T in (3.8) indicates that linear irreversible thermodynamics relies on the local-equilibrium assumption, so that a natural temperature concept is available.

In linear irreversible thermodynamics, the second law is taken into account by requiring that the phenomenological matrix L^{LIT} is positive semidefinite, which, according to (3.8), implies that the local entropy production rate σ is positive or zero. Equations (2.94) and (3.4) imply that also the friction matrix M is positive semidefinite, which is one of the fundamental postulates of GENERIC.

To illustrate the above definitions and motivate the minus sign in (3.3), we use the mechanical matrix C_M^T of hydrodynamics given in (2.97) for the variables $x = (\rho(\boldsymbol{r}), \boldsymbol{M}(\boldsymbol{r}), \epsilon(\boldsymbol{r}))$. The definition (3.3) then implies the thermodynamic forces

$$X^{\mathrm{LIT}} = \begin{pmatrix} -\frac{1}{T}\boldsymbol{\kappa}^T \\ -\frac{1}{T^2}\frac{\partial T}{\partial \boldsymbol{r}} \end{pmatrix}, \tag{3.10}$$

and the matrix D_M of (2.99) gives the phenomenological coefficients and the resulting fluxes. We thus obtain the force-flux pair

$$X_1^{\mathrm{LIT}} = -\frac{1}{T^2}\frac{\partial T}{\partial \boldsymbol{r}}, \qquad J_1^{\mathrm{LIT}} = \boldsymbol{j}^{\mathrm{q}} = -\lambda^{\mathrm{q}}\frac{\partial T}{\partial \boldsymbol{r}} \tag{3.11}$$

of temperature gradient and heat flux, related by Fourier's law, and the pair

$$X_2^{\mathrm{LIT}} = -\frac{1}{T}\boldsymbol{\kappa}^T, \qquad J_2^{\mathrm{LIT}} = \boldsymbol{\tau} = -\eta(\boldsymbol{\kappa} + \boldsymbol{\kappa}^T) - \hat{\kappa}\mathbf{1}\mathrm{tr}\boldsymbol{\kappa} \tag{3.12}$$

of velocity gradients and momentum flux with Newton's expression for the stress tensor as examples of our somewhat abstract construction of linear irreversible thermodynamics.

Because linear irreversible thermodynamics has had an enormous impact on the unified description of nonequilibrium processes, it is reassuring to see how naturally it arises from GENERIC after a natural decomposition of the friction matrix. Conversely, the enormous experience with linear irreversible thermodynamics for a large variety of systems provides valuable guidance to the formulation of friction matrices (see Section 2.3.2). To give a feeling for the versatility and usefulness of linear irreversible thermodynamics, we list a number of applications to continuous and discrete systems from the book of de Groot and Mazur,[2] with some additions from the books of Woods[3] and Kuiken:[4]

[2] de Groot & Mazur, *Non-Equilibrium Thermodynamics* (Dover, 1984).
[3] Woods, *Thermodynamics of Fluid Systems* (Clarendon, 1975).
[4] Kuiken, *Thermodynamics of Irreversible Processes* (Wiley, 1994).

- *Chemical reactions* in uniform multicomponent systems, including the law of mass action and the principle of detailed balance; this application actually shows that the linear phenomenological laws are insufficient and that the expression for the entropy production rate in terms of forces and fluxes is more general than the linear laws [de Groot & Mazur, Chapter X, §§ 1–4]

- Isothermal and nonisothermal *diffusion* in nonreacting and reacting mixtures, without and with external Coriolis and electric forces, including the Svedberg equations for the sedimentation equilibrium and the sedimentation velocity in an ultracentrifuge, a generalized Einstein relationship between the mobilities and diffusion coefficients in a multicomponent system, the related Soret (flow of matter caused by a temperature gradient) and Dufour effects (heat flow caused by a concentration gradient), the temperature and concentration distributions in a binary reacting system, sedimentation and thermal diffusion potentials, and electrophoresis [de Groot & Mazur, Chapter XI, §§ 2–8 and Chapter XIII, §§ 8–9; Kuiken, Chapter 6]

- Discussion of the relationship between the *osmotic pressure and the relative permeability* of membranes as a generalization of the semipermeable case considered in equilibrium thermodynamics [de Groot & Mazur, Chapter XV, § 6]

- Anisotropic *heat conduction* in the presence of a magnetic field, including the Righi-Leduc effect (temperature gradient giving rise to a heat flow in an orthogonal direction) [de Groot & Mazur, Chapter XI, § 1]

- The effect of irreversible phenomena on the *propagation of sound*, including the result of Stokes and Kirchhoff for the influence of the shear viscosity and the thermal conductivity on sound attenuation [de Groot & Mazur, Chapter XII, §§ 3–5]

- Various mechanoelectric or *electrokinetic effects*, including the streaming potential (pressure difference required to suppress electric current in the presence of an electric field) and electro-osmosis effects (flow of volume caused by electric field) connected through Saxén's relation, as well as the related electroosmotic pressure (pressure difference required to suppress volume flow in the presence of an electric field) and streaming current effects (electric current caused by pressure differences) [de Groot & Mazur, Chapter XV, § 7]

- *Thermoelectric effects* occurring for thermocouples and thermocells, including the Peltier and Seebeck effects (see Section 3.2.2 for details) [de Groot & Mazur, Chapter XIII, §§ 6 and 9]

- *Thermokinetic effects*, including the related thermomolecular pressure (pressure difference required to suppress a heat flux in the presence of a temperature gradient) and mechano-caloric effects (heat flow caused by pressure differences), and thermal effusion [de Groot & Mazur, Chapter XV, §§ 5 and 8]

- Relaxation of dipole orientation in a homogeneous system in an external electric field as a prototype of the relaxation phenomena describable by linear irreversible thermodynamics, including the orientational diffusion equation leading to the Debye theory of *dielectric relaxation* and the corresponding Kramers-Kronig relation for the real and imaginary parts of the electric susceptibility [de Groot & Mazur, Chapter X, § 6 and Chapter XIV, § 5]

- Development of the equations for a *polarizable medium in an electromagnetic field* with dielectric and magnetic relaxation as the only irreversible phenomena, including the modification of the chemical potential and the pressure, and a discussion of electrostriction and magnetostriction [de Groot & Mazur, Chapter XIV, §§ 1–4]

- Discussion of *magneto-plasmas*, including the attenuation of small-amplitude waves propagating through an anisotropic medium by the basic irreversible transport mechanisms and cross effects [Woods, §§ 56–57]

- Two-fluid description of *superfluids*, including the influence of cross viscosities on wave propagation and a discussion of the role of mutual friction between the two fluids for rotating superfluids [Woods, §§ 58–61]

- Quasi-linear rheological models of *viscoelastic fluids* based on scalar and tensorial internal processes and a splitting of deformations into elastic, anelastic, and viscous parts, including a discussion of the formal aspects of linear viscoelasticity [Kuiken, Chapter 7]

To illustrate further the generality and power of linear irreversible thermodynamics and to provide the basic tools for the above applications, we next identify further force-flux pairs and we study the transformation behavior of the force-flux relationships when changing the independent variables, the coordinate system, and the basic dissipative processes. However, we also need to issue some warnings concerning rash generalizations of formally appealing principles, to avoid incorrect results and to motivate the reader for a fully nonlinear framework, such as GENERIC.

Exercise 38 Force and Flux Associated With Diffusion
Show that another example of a thermodynamic force-flux pair can be obtained in connection with diffusion effects and Fick's law in two-component systems:

$$X_3^{\mathrm{LIT}} = \frac{\partial}{\partial \boldsymbol{r}} \frac{\mu_2 - \mu_1}{T}, \qquad J_3^{\mathrm{LIT}} = \dot{\boldsymbol{j}}_1^{\mathrm{diff}} = D' \frac{\partial}{\partial \boldsymbol{r}} \frac{\mu_2 - \mu_1}{T}. \tag{3.13}$$

Formulate a similar force-flux pair for diffusion in single-component fluids (based on the decomposition of the friction matrix found in Exercise 37).

Exercise 39 Nernst-Einstein Equation
By comparing the force-flux pair

$$X^{\mathrm{LIT}} = -\frac{1}{T}\frac{\partial \varphi}{\partial \boldsymbol{r}} = \frac{1}{T}\frac{\boldsymbol{F}}{m}, \qquad J^{\mathrm{LIT}} = \rho \boldsymbol{v}, \tag{3.14}$$

where φ is a potential per unit mass, $\boldsymbol{F}$ is the corresponding force on a particle of mass m, and $\boldsymbol{v}$ is the resulting drift velocity, with the force-flux pair of Exercise 38, derive the Nernst-Einstein equation

$$\zeta = \frac{k_B T}{D} \tag{3.15}$$

for the friction coefficient ζ, introduced by $\boldsymbol{F} = \zeta \boldsymbol{v}$, in terms of the diffusion coefficient D.

Exercise 40 Force and Flux for a Diffusing Particle

Find the force-flux pair for the diffusing particle considered in Exercise 12.

Exercise 41 Implications of Positive Semidefiniteness

Show that the positive semidefinite nature of L^{LIT} implies

$$L_{jj}^{\mathrm{LIT}} \geq 0, \tag{3.16}$$

and

$$L_{jj}^{\mathrm{LIT}} L_{kk}^{\mathrm{LIT}} \geq \left(\frac{L_{jk}^{\mathrm{LIT}} + L_{kj}^{\mathrm{LIT}}}{2} \right)^2; \tag{3.17}$$

the latter condition is a restriction on the magnitude of cross effects.

3.1.2 Electric Field and Current

Heat fluxes and diffusion flows are clearly driven by entropic effects. For electric currents, however, one would expect the driving force to be of energetic nature. This might appear to be a problem because the thermodynamic forces in (3.3) are defined in terms of the entropy gradient. However, it is important to note that the mechanical matrix C_M in (3.3) can introduce energetic effects through the degeneracy condition (2.95). Indeed, the velocity gradients as the thermodynamic force X_2^{LIT} for momentum flow in (3.12) can be traced back to an energetic effect in C_M; according to the local-equilibrium assumption, velocity gradients certainly cannot occur through entropic effects. In the same way, we expect electric fields to occur as thermodynamic forces through the mechanical matrix, as a consequence of energy conservation.

For the discussion of electric effects, we need to consider the electric charge density, $\rho^{\mathrm{el}}(\boldsymbol{r})$, as an additional variable. The charge density leads to an energy contribution such that

$$\frac{\delta E}{\delta \rho^{\mathrm{el}}(\boldsymbol{r})} = \phi^{\mathrm{el}}(\boldsymbol{r}), \tag{3.18}$$

where we do not need to specify whether the electric potential ϕ^{el} is due to a given external field or due to the charges in the system (in the latter case, the energy E would be a quadratic functional of the charge distribution, and each pair of interacting charges must be counted only once).

In the spirit of Section 2.3.2, we discuss the contribution of the electric resistivity or conductivity to the friction matrix separately from other contributions. According to (2.100) we need to specify a row vector of C_M^T, which we take as

$$\left(\cdots \quad \frac{\partial \phi^{\mathrm{el}}}{\partial \boldsymbol{r}} \quad -\frac{\partial}{\partial \boldsymbol{r}} \quad \cdots \right), \tag{3.19}$$

Table 3.1 Thermodynamic Force-Flux Pairs

Force [gradient of ...]	Flux	Material Property [transport coefficient]	Equation
Temperature	Heat flux	Thermal conductivity	(3.11)
Velocity	Momentum flux	Viscosity	(3.12)
Chemical potential	Diffusion flow	Diffusion coefficient	(3.13)
Electric potential	Electric current	Electric conductivity	(3.21) (3.22)

where the displayed entries are associated with the internal energy density ϵ and the electric charge density ρ^{el}, respectively. Because we here focus on the electric effects on the energy balance, all other entries in the above vector are taken as zero. The construction is such that the degeneracy condition (2.95) is satisfied,

$$\left(\cdots \quad \frac{\partial \phi^{\mathrm{el}}}{\partial \boldsymbol{r}} \quad -\frac{\partial}{\partial \boldsymbol{r}} \quad \cdots \right) \cdot \begin{pmatrix} \vdots \\ 1 \\ \phi^{\mathrm{el}} \\ \vdots \end{pmatrix} = 0. \tag{3.20}$$

If the entropy is independent of the electric charge distribution, the general definition of thermodynamic forces in (3.3) yields

$$X_4^{\mathrm{LIT}} = -\frac{\partial \phi^{\mathrm{el}}}{\partial \boldsymbol{r}} \frac{1}{T} = \frac{1}{T} \boldsymbol{E}, \tag{3.21}$$

where $\boldsymbol{E}$ is the electric field. The weight D_M in (2.100) should be taken as $\sigma^{\mathrm{el}} T$, where σ^{el} is the electric conductivity and, as for the viscosity in (2.99) with an extra factor of T. We hence obtain the flux

$$J_4^{\mathrm{LIT}} = \sigma^{\mathrm{el}} \boldsymbol{E} = \boldsymbol{i}, \tag{3.22}$$

which is the nonconvective electric current density, expressing Ohm's law. Equations (3.21) and (3.22) define the force-flux pair associated with electric currents. The thermodynamic force-flux pairs identified in Sections 3.1.1 and 3.1.2 are compiled in Table 3.1.

All the force-flux pairs in Table 3.1 are associated with transport processes. However, the concepts of Section 3.1.1 can also be applied to relaxation processes, for which forces and fluxes can be defined. Such relaxation processes occur naturally for the variables describing the internal structure of complex fluids. Specific examples of such force-flux pairs in which the forces are not of the gradient form will be discussed in Section 4.2.3.

Exercise 42 Electric Energy Dissipation
Show that the electric force-flux pair (3.21), (3.22) leads to the proper entropy production rate, so that $\sigma^{\text{el}}T$ indeed was the proper choice for the weight D_M in (2.100).

3.1.3 Transformation Behavior

In the context of linear irreversible thermodynamics, there are three different types of transformations to be understood, all of them expressing important features of the framework. The first type of transformation behavior is associated with the *choice of the independent variables* to describe the states of a beyond-equilibrium system. For different choices of the independent variables x, one can relate the corresponding building blocks E, S, L, and M. The explicit relationships between the building blocks for different sets of independent variables were given explicitly in Section 1.2.4.

The second type of transformation behavior is associated with the *choice of the coordinate system* for the real space in which the fields $x(\boldsymbol{r})$ describing the states of a system evolve. Scalars, vectors, tensors, and their densities are the well-known objects with acceptable transformation behavior when changing coordinate systems. Constructing constitutive equations for isotropic systems that are invariant under changes of coordinate systems is the idea behind the Curie principle to be explained in the subsequent Section 3.1.4.

The third type of transformation behavior, to be discussed here in more detail, is associated with the *choice of the basic dissipative processes*. According to (3.3), there is a generalized force associated with each dissipative effect. The matrix C_M is nonsquare because it translates between dissipative processes and the variables describing beyond-equilibrium states. The definition of generalized forces in (3.3) is actually independent of the choice of the variables to describe the states of a system.

Exercise 43 Generalized Forces for Different Independent Variables
Show that the definition of generalized forces in (3.3) is indeed independent of the choice of the independent variables chosen for describing a beyond-equilibrium system.

Transformations among the dissipative processes, as discussed by de Groot and Mazur,[5] are given by the equations[6]

$$C_M'^T = Q^{\text{LIT}} \cdot C_M^T \quad \text{or} \quad C_M' = C_M \cdot (Q^{\text{LIT}})^T, \tag{3.23}$$

where Q^{LIT} is a nonsingular matrix. From (3.3), the transformation of the generalized forces is then given by

$$X^{\text{LIT}'} = Q^{\text{LIT}} \cdot X^{\text{LIT}}. \tag{3.24}$$

[5] See Chapter VI, § 5 of de Groot & Mazur, *Non-Equilibrium Thermodynamics* (Dover, 1984).

[6] Note the difference compared with the alternative type of transformations among the independent variables considered in Exercise 43, where the transformation matrix acts from the opposite side.

It is then natural to assume the transformation behavior

$$D'_M = [(Q^{\mathrm{LIT}})^T]^{-1} \cdot D_M \cdot (Q^{\mathrm{LIT}})^{-1}, \tag{3.25}$$

which leaves the fiction matrix (2.94) invariant. In other words, the transformations among dissipative processes have no influence on the friction matrix, which is fully determined by the variables used for describing states. Then also the entropy production rate is invariant under such transformations. For the fluxes, we obtain from (3.4), (3.6), and (3.25)

$$J^{\mathrm{LIT}'} = [(Q^{\mathrm{LIT}})^T]^{-1} \cdot J^{\mathrm{LIT}}. \tag{3.26}$$

As a more general transformation one could consider the still linear relationship

$$X^{\mathrm{LIT}'} = Q^{\mathrm{LIT}} \cdot (X^{\mathrm{LIT}} + A^{\mathrm{LIT}} \cdot J^{\mathrm{LIT}}), \tag{3.27}$$

instead of (3.24), with an arbitrary antisymmetric matrix A^{LIT} and without changing (3.26). The transformation (3.26), (3.27) still leaves the entropy production (3.8) invariant. However, from a GENERIC perspective, the transformations (3.23), (3.25) leading to (3.24), (3.26) are more natural than (3.26), (3.27) because they leave the entire friction matrix invariant, rather than only the entropy production rate. The GENERIC building blocks are not affected by transformations among the dissipative processes, whereas the choice of independent variables dictates their appearance. For the thermodynamic forces and fluxes, the situation is reversed: According to (3.24) and (3.26), their appearance is determined by the selection of the fundamental dissipative processes, and transformations of the independent variables do not affect them, as shown in Exercise 43.

Note that the transformations Q^{LIT} are most naturally taken as orthogonal because then, according to (3.25), also functions of D_M have a simple transformation behavior (consider Taylor expansions!). For the amplitude of the noise in GENERIC with fluctuations, we then obtain from (2.96)

$$B' = B \cdot (Q^{\mathrm{LIT}})^T. \tag{3.28}$$

The transformation among dissipative processes thus corresponds exactly to the redefinition of Wiener processes describing fluctuations, which was discussed at the end of Section 1.2.5.

The transformations among dissipative processes considered here are important for identifying the fundamental, independent processes. For example, the transport of mass and of electric charge cannot be considered as independent when, say for ions, an important contribution to the mass flux automatically comes with a proportional electric current [see (3.57) and the associated footnote]. When the forces and fluxes become dependent, there are ambiguities in the choice of L^{LIT}, which, unless sufficient care is taken, could even spoil the Onsager-Casimir symmetry[7] to

[7] See Chapter VI, § 3 of de Groot & Mazur, *Non-Equilibrium Thermodynamics* (Dover, 1984).

be discussed below (see Section 3.2). From the more general GENERIC perspective, such considerations are mainly important for constructing noise terms, as only the matrix B translates between the dissipative processes and the independent variables describing beyond-equilibrium states and hence is sensitive to transformations among the dissipative processes.

3.1.4 Curie's Principle

The Curie principle can be explained best by asking this question: "what exactly do the dots in the force-flux relationship (3.6) mean?" The answer is that they do not only imply the multiplication of a matrix and a column vector, $J_j^{\rm LIT} = L_{jk}^{\rm LIT} X_k^{\rm LIT}$, because the entries $J_j^{\rm LIT}$, $L_{jk}^{\rm LIT}$, $X_k^{\rm LIT}$ themselves can be vectors and tensors to be further contracted. For example, the generalized force $X_1^{\rm LIT}$ associated with the temperature gradient is a vector and so is the generalized force $X_3^{\rm LIT}$ for diffusion or $X_4^{\rm LIT}$ for electric currents. In all cases, the corresponding fluxes are vectors, too. In the force-flux pair (3.12), $X_2^{\rm LIT}$ and $J_2^{\rm LIT}$ are both tensors. Whereas, for Newtonian liquids, the element $L_{11}^{\rm LIT}$ translating the temperature gradient into the heat flux is a scalar (or isotropic tensor), the element $L_{22}^{\rm LIT}$ translating the velocity-gradient tensor into the momentum-flux tensor is a tensor of rank four [see (2.99)], and a double contraction is involved in obtaining $J_2^{\rm LIT}$ from $X_2^{\rm LIT}$.

For the description of an isotropic material within linear irreversible thermodynamics, the entries of the phenomenological matrix $L^{\rm LIT}$ with their well-defined tensorial nature, required to make the force-flux relations independent of the choice of coordinate systems, can depend only on the scalar local-equilibrium variables. Only the unit tensor is thus available for forming tensors of rank two or four, and the possibilities of coupling forces and fluxes are hence limited severely. These remarks, known as the Curie principle, can be summarized most appropriately in the words of de Groot and Mazur [p. 5]:

> ... possible spatial symmetries of a material system may further simplify the scheme of phenomenological coefficients. Thus in an isotropic fluid a scalar phenomenon like a chemical reaction cannot be coupled to a vectorial phenomenon like heat conduction. This reduction of the scheme of phenomenological coefficients, which results from invariance of the phenomenological equations under special orthogonal transformations, goes under the name of the Curie principle, but should more appropriately be called Curie's theorem.

In the stereotypical formulation used in most textbooks on nonequilibrium thermodynamics the Curie principle asserts that *for isotropic systems the thermodynamic forces and fluxes of different tensorial character do not couple*. However, some care must be taken in interpreting such a statement as it is not immediately obvious what is meant by tensorial character. For example, a second-rank tensor may be regarded as being composed of a scalar (the trace of the tensor), an axial vector (the antisymmetric part of the tensor), and a traceless symmetric tensor, so that a scalar can be coupled with the trace of a tensor or an axial vector with the antisymmetric part of a tensor.

In applications, one has to be careful to build up the required objects of a certain tensorial type from all available quantities possessing well-defined tensorial characters, where all kinds of decompositions, contractions, and tensor products must be taken into account. A formal analysis of the influence of symmetry on the allowed couplings of thermodynamic forces and fluxes, including proofs and examples, can be found in Chapter VI of the classical book by de Groot and Mazur,[8] and a detailed discussion from a slightly different perspective is given in Appendix C of the textbook by Reichl.[9]

3.1.5 Stationary States

In Section 1.2.3, we had considered the stationary states of GENERIC, which turned out to be equilibrium states. This is a consequence of the formulation of GENERIC for isolated systems. At equilibrium, the gradients of temperature, velocity, chemical or electric potentials, and so forth are zero, so that all fluxes of linear irreversible thermodynamics vanish. We here consider the possibility of stationary states with nonvanishing fluxes. Although such fluxes must be maintained by suitable boundary conditions, we here focus on the form and the properties of the field equations describing stationary states inside the system of interest.

As a first example, we consider heat conduction in a material at rest ($\boldsymbol{v} = 0$). The temperature equation (2.73) then simplifies to

$$\frac{\partial T}{\partial t} = \frac{1}{\rho \hat{c}_V} \frac{\partial}{\partial \boldsymbol{r}} \cdot \lambda^{\mathrm{q}} \frac{\partial T}{\partial \boldsymbol{r}} = -\frac{1}{\rho \hat{c}_V} \frac{\partial}{\partial \boldsymbol{r}} \cdot \boldsymbol{j}^{\mathrm{q}}, \tag{3.29}$$

and steady states are hence described by the condition of a divergence-free flux,

$$\frac{\partial}{\partial \boldsymbol{r}} \cdot \boldsymbol{j}^{\mathrm{q}} = 0. \tag{3.30}$$

This equation determines the distribution of temperature throughout the system for given boundary conditions.

Under certain circumstances, such stationary states can be obtained by minimizing the total entropy production rate for the given boundary conditions. Although this is an appealing idea, the purpose of this section is to show also that, unfortunately, this so-called *principle of minimum entropy production is of rather limited validity* and by no means a general principle. The discussion should clarify both the basic ideas and their limitations.

For the total entropy production rate Σ, we obtain from (2.36) for $\boldsymbol{v} = 0$

$$\Sigma = \int_V \sigma d^3r = \int_V \boldsymbol{j}^{\mathrm{q}} \cdot \frac{\partial}{\partial \boldsymbol{r}} \frac{1}{T} d^3r = \int_V \frac{\lambda^{\mathrm{q}}}{T^2} \frac{\partial T}{\partial \boldsymbol{r}} \cdot \frac{\partial T}{\partial \boldsymbol{r}} d^3r. \tag{3.31}$$

In order to find the extrema of the functional Σ, we calculate the functional derivative (see Exercise 221),

$$\frac{\delta \Sigma}{\delta T} = \frac{\partial(\lambda^{\mathrm{q}}/T^2)}{\partial T} \frac{\partial T}{\partial \boldsymbol{r}} \cdot \frac{\partial T}{\partial \boldsymbol{r}} - 2 \frac{\partial}{\partial \boldsymbol{r}} \cdot \frac{\lambda^{\mathrm{q}}}{T^2} \frac{\partial T}{\partial \boldsymbol{r}}, \tag{3.32}$$

[8] de Groot & Mazur, *Non-Equilibrium Thermodynamics* (Dover, 1984).
[9] Reichl, *Modern Course in Statistical Physics* (University of Texas, 1980).

which can be rewritten in the alternative form

$$\frac{\delta\Sigma}{\delta T} = \frac{1}{T^4}\frac{\partial(\lambda^{\rm q}T^2)}{\partial T}\left(\frac{\partial T}{\partial \boldsymbol{r}}\right)^2 + \frac{2}{T^2}\frac{\partial}{\partial \boldsymbol{r}}\cdot \boldsymbol{j}^{\rm q}. \tag{3.33}$$

By comparing (3.30) and (3.33), we note that the stationary states can only correspond to states of minimum entropy production with $\delta\Sigma/\delta T = 0$ if $\lambda^{\rm q}T^2$ is independent of temperature. Although this is the combination occurring in the matrix D_M of (2.99) or hence in the phenomenological matrix $L^{\rm LIT}$ of (3.4), and one might hence be tempted to take it as a constant, nevertheless a very particular temperature dependence of the thermal conductivity is needed for the stationary states to lead to a possible minimum of the entropy production. The result of Exercise 222 can be used to prove that we actually have a minimum, provided that the temperature is bounded and that the magnitude of the temperature gradient is larger than some positive constant.

To study the stability of stationary states, we consider the time evolution of the entropy production in a quiescent material for constant $\lambda^{\rm q}T^2$. From (3.31), we obtain for deviations from stationary states

$$\begin{aligned}\frac{d\Sigma}{dt} &= \int_V \left(-4\frac{\lambda^{\rm q}}{T^3}\frac{\partial T}{\partial t}\frac{\partial T}{\partial \boldsymbol{r}}\cdot\frac{\partial T}{\partial \boldsymbol{r}} + 2\frac{\lambda^{\rm q}}{T^2}\frac{\partial T}{\partial \boldsymbol{r}}\cdot\frac{\partial^2 T}{\partial \boldsymbol{r}\partial t}\right) d^3r \\ &= -2\int_V \lambda^{\rm q}\frac{\partial T}{\partial \boldsymbol{r}}\cdot\frac{\partial}{\partial \boldsymbol{r}}\frac{\partial}{\partial t}\frac{1}{T}d^3r. \end{aligned} \tag{3.34}$$

After an integration by parts, where the boundary term vanishes because the temperature on the boundary is time independent, we obtain

$$\frac{d\Sigma}{dt} = 2\int_V \left[\frac{\partial}{\partial \boldsymbol{r}}\cdot\left(\lambda^{\rm q}\frac{\partial T}{\partial \boldsymbol{r}}\right)\right]\frac{\partial}{\partial t}\frac{1}{T}d^3r = -2\int_V \frac{\rho\hat{c}_V}{T^2}\left(\frac{\partial T}{\partial t}\right)^2 d^3r. \tag{3.35}$$

In the last step, the time-evolution equation (3.29) for the temperature in a quiescent material was used. Equation (3.35) proves that the entropy production rate can only decrease in time until a minimum is reached when $\partial T/\partial t = 0$ throughout the system, that is, for a stationary state.

In Chapter V of their book, de Groot and Mazur showed that the results of the above example can be generalized to multicomponent materials with externally imposed gradients of temperature and chemical potentials, including chemical reactions and cross effects. As long as the phenomenological parameters are constants, stationary states are states of minimum entropy production.

As a more complicated second example, we consider a system with both temperature and velocity gradients. Concrete examples could be the shear flow between parallel plates at different temperatures in relative motion or, to avoid boundary conditions at infinity, the flow between two concentric spheres at different temperatures where, for example, the outer sphere rotates. From (2.10), (2.23), and (2.73), we obtain the equations

$$0 = -\boldsymbol{v}\cdot\frac{\partial}{\partial \boldsymbol{r}}\rho - \rho\frac{\partial}{\partial \boldsymbol{r}}\cdot\boldsymbol{v}, \tag{3.36}$$

$$0 = -\rho \boldsymbol{v} \cdot \frac{\partial}{\partial \boldsymbol{r}} \boldsymbol{v} - \frac{\partial}{\partial \boldsymbol{r}} \cdot \boldsymbol{\tau} - \frac{\partial p}{\partial \boldsymbol{r}}, \tag{3.37}$$

$$0 = -\boldsymbol{v} \cdot \frac{\partial}{\partial \boldsymbol{r}} T - \frac{1}{\alpha_s} \frac{\partial}{\partial \boldsymbol{r}} \cdot \boldsymbol{v} - \frac{1}{\rho \hat{c}_V} \left(\frac{1}{2} \boldsymbol{\tau} : \dot{\boldsymbol{\gamma}} + \frac{\partial}{\partial \boldsymbol{r}} \cdot \boldsymbol{j}^{\mathrm{q}} \right), \tag{3.38}$$

for stationary states.

If we now minimize the entropy production rate obtained from (2.36),

$$\Sigma = \int_V \sigma d^3 r = \int_V \left(\frac{\lambda^{\mathrm{q}}}{T^2} \frac{\partial T}{\partial \boldsymbol{r}} \cdot \frac{\partial T}{\partial \boldsymbol{r}} - \frac{1}{2T} \boldsymbol{\tau} : \dot{\boldsymbol{\gamma}} \right) d^3 r, \tag{3.39}$$

then we see that, not surprisingly, it is impossible to produce the proper convective terms in (3.36)–(3.38). The *minimum principle can only be associated with the dissipative terms.* We hence consider only such steady states for which ρ, $\boldsymbol{v}$ and T do not change along streamlines; this is a reasonable assumption for the concrete stationary flows mentioned above. Equation (3.36) then implies that we should consider only divergence-free velocity fields, and the remaining stationarity conditions are simplified to

$$\frac{\partial}{\partial \boldsymbol{r}} \cdot \boldsymbol{\tau} + \frac{\partial p}{\partial \boldsymbol{r}} = 0, \tag{3.40}$$

and

$$\frac{\partial}{\partial \boldsymbol{r}} \cdot \boldsymbol{j}^{\mathrm{q}} - \frac{\eta}{2} \dot{\boldsymbol{\gamma}} : \dot{\boldsymbol{\gamma}} = 0. \tag{3.41}$$

We now try to obtain the conditions (3.40) and (3.41) by minimizing the entropy production rate (3.39). By evaluating the functional derivative with respect to the temperature field at fixed velocity field (see Appendix C.1), we obtain the following generalization of (3.33):

$$\frac{\delta \Sigma}{\delta T} = \frac{1}{T^4} \frac{\partial (\lambda^{\mathrm{q}} T^2)}{\partial T} \left(\frac{\partial T}{\partial \boldsymbol{r}} \right)^2 + \frac{1}{2T^2} \frac{\partial (\eta T)}{\partial T} \dot{\boldsymbol{\gamma}} : \dot{\boldsymbol{\gamma}} + \frac{2}{T^2} \left(\frac{\partial}{\partial \boldsymbol{r}} \cdot \boldsymbol{j}^{\mathrm{q}} - \frac{\eta}{2} \dot{\boldsymbol{\gamma}} : \dot{\boldsymbol{\gamma}} \right). \tag{3.42}$$

The proper stationary temperature profiles for fixed velocity fields are obtained if, in addition to the combination $\lambda^{\mathrm{q}} T^2$, also ηT is independent of temperature. Note that these are exactly the combinations occurring in the phenomenological matrix (2.99). For these particular temperature dependencies of the thermal conductivity and the viscosity, the condition of minimum entropy production leads to the correct stationary temperature profile even in the presence of viscous dissipation.

For a fixed temperature field, we wish to calculate the functional derivative of the entropy production rate (3.39) with respect to the velocity field under the constraint of incompressibility. If we assume that neither λ^{q} nor η depends on velocity, say through velocity gradients, then, by the rules of Appendix C, we obtain the constrained functional derivative

$$\frac{\delta \Sigma}{\delta \boldsymbol{v}} = 2 \left(\frac{\partial}{\partial \boldsymbol{r}} \cdot \frac{\boldsymbol{\tau}}{T} + \frac{\partial}{\partial \boldsymbol{r}} \frac{p}{T} \right), \tag{3.43}$$

where, as in (3.40), the function p is to be determined such that the solution of $\delta \Sigma / \delta \boldsymbol{v} = 0$ is divergence free. The function $-2p/T$ in (3.43) corresponds to f in (C.27) and results from the incompressibility condition.

A comparison of (3.40) and (3.43) shows that in the isothermal case the stationary states for imposed velocity fields on the boundary lead to a minimum of the entropy production. If T is a function of $\boldsymbol{r}$, however, only in exceptional cases might it be possible to compensate for additional terms in (3.43) by choosing the function p in that equation different from the pressure in (3.40). Even under isothermal conditions, the viscosity may not depend on the shear rate because otherwise unwanted extra terms arise in (3.43). In the presence of flow, minimum entropy production is thus far from being a general principle for stationary states. To make this conclusion more memorizable, we add a mocking passage from a famous critic,[10] which was based on a citation of an overly enthusiastic biologist:

> In a more poetic version, we read that IT [irreversible thermodynamics] "...sheds new light on 'the wisdom of living organisms.' Life is a constant struggle against the tendency to produce entropy by irreversible processes. But since there is no possibility of escaping the entropic doom imposed on all natural phenomena under the Second Law of Thermodynamics, living organisms choose the least evil—they produce entropy at a minimal rate by maintaining a steady state." This statement echoes the long tradition of the concept of 'least action' in physics and theology, and is particularly pleasing esthetically ... but it was found not to be true. Therefore, we have no assurance that we can use this doctrine in biological systems, or in any other systems far from equilibrium.

A modified minimum entropy production principle would also be desirable from the GENERIC perspective. The splitting into reversible and irreversible terms in the fundamental equation (1.1) is not unique. By physical intuition, the reversible contribution, the one that is "under mechanistic control," should be as large as possible, and "the rest" shall be kept "as small as possible." One interpretation of keeping the irreversible contribution, $M \cdot \delta S/\delta x$, "as small as possible" would be to minimize the entropy production rate. In a generalized version, an appropriately defined norm of $M \cdot \delta S/\delta x$ should be minimized. In view of (3.1), this norm might be based on some properly regularized or modified version of the inverse friction matrix.

Exercise 44 Problems With Minimum Entropy Production

Why was the problematic factor $1/T$ in (3.43), which can cause a discrepancy between the state of minimum entropy production and the stationary state (3.40), not mentioned in the discussion of the "principle" of minimum entropy production in the classical textbooks on irreversible thermodynamics?

3.2 ONSAGER-CASIMIR RELATIONS

In the classical theory of linear irreversible thermodynamics, the symmetry properties of the phenomenological matrix L^{LIT} play an important role. According to (2.94) and (3.4), the symmetry of L^{LIT} is intimately related to the symmetry of the GENERIC

[10] Wei, Ind. Eng. Chem. 58 (1966) 55.

friction matrix $M = C_M \cdot L^{\mathrm{LIT}} \cdot C_M^T$,

$$M^T = (C_M \cdot L^{\mathrm{LIT}} \cdot C_M^T)^T = C_M \cdot (L^{\mathrm{LIT}})^T \cdot C_M^T. \tag{3.44}$$

If L^{LIT} is symmetric or antisymmetric, then M has the same symmetry property. However, if L^{LIT} has mixed symmetry properties, as associated with the names of Onsager and Casimir, the situation is more complicated, as the discussion in this section will show.

In Section 2.3.2, it was argued that $D_M = L^{\mathrm{LIT}}$ has a diagonal or block-diagonal structure when there are no cross effects, for example, as in (2.99) for hydrodynamics. Off-diagonal elements in L^{LIT} describe cross effects among the thermodynamic force-flux pairs. The situation is different for the friction matrix M, which typically contains off-diagonal elements even in the absence of cross effects, as (2.65) for hydrodynamics. The structure of friction matrices in the presence of cross effects is considered in Exercises 46 and 47. In general, both diagonal and off-diagonal elements of the friction matrix contain diagonal and off-diagonal elements of L^{LIT}. In that respect, the situation looks more complicated for M than for L^{LIT}. However, we shall see that the arguments concerning the symmetry of M are more direct than for L^{LIT}.

3.2.1 Bare and Dressed Symmetry

The standard analysis[11] of the symmetry behavior of the phenomenological matrix L^{LIT} is built on a very important assumption, the so-called regression hypothesis for the fluctuations. Although one is interested in the matrix L^{LIT} containing the transport coefficients for a certain set of independent extensive quantities, one actually derives the symmetry properties of a different matrix describing the relaxation of these quantities to their equilibrium values after fluctuations and assumes that these two matrices have the same symmetries.[12] In the words of Reichl,[13] one "assumes that fluctuations about the equilibrium state decay, *on the average*, according to the same laws that govern the decay of macroscopic deviations from equilibrium, that is, the hydrodynamic equations." Casimir[14] pointed out that this regression hypothesis is by no means evident.

The matrix describing the relaxation of fluctuations can be related to the behavior of the corresponding two-time correlation functions of the quantities of interest under time reversal; "microreversibility" and "detailed balance" are further keywords commonly used to describe the situation. For the behavior under time reversal, it is important whether an observable is an even or odd function of the particle velocities.

[11] See Chapter IV, § 3 of de Groot & Mazur, *Non-Equilibrium Thermodynamics* (Dover, 1984) or, for a more structured discussion, see Section 3.2 of Kreuzer, *Nonequilibrium Thermodynamics & Statistical Foundations* (Oxford, 1981).

[12] Onsager, Phys. Rev. 37 (1931) 405; 38 (1931) 2265.

[13] See p. 508 of Reichl, *Modern Course in Statistical Physics* (University of Texas, 1980).

[14] Casimir, Rev. Mod. Phys. 17 (1945) 343.

If the observables corresponding to the indices j and k in the list of variables are both even or both odd functions, then one obtains the relationship

$$L_{jk}^{\mathrm{LIT}} = L_{kj}^{\mathrm{LIT}} \tag{3.45}$$

for the phenomenological coefficients describing the cross effects; if one of the observables is an even and the other one is an odd function of the particle velocities, then one finds

$$L_{jk}^{\mathrm{LIT}} = -L_{kj}^{\mathrm{LIT}}. \tag{3.46}$$

The conditions (3.45) for cross effects are known as Onsager's reciprocal relations, and the additional possibility (3.46) was established by Casimir.

Rather than reproducing the details of the time reversal argument in the usual context of the phenomenological matrix L^{LIT}, we will explain it for the GENERIC friction matrix M. In Chapter 6, we use projection-operator techniques to derive an atomistic expression for the friction matrix of the form

$$M_{jk}(x) = \frac{1}{k_{\mathrm{B}}} \int_0^{\tau} \left\langle \dot{x}_j^{\mathrm{f}}(t) \dot{x}_k^{\mathrm{f}}(0) \right\rangle_x dt, \tag{3.47}$$

where τ is an intermediate time scale separating the slow degrees of freedom x from the fast degrees of freedom, $\dot{x}^{\mathrm{f}}$ is the rapidly fluctuating part of the time derivative of the atomistic expressions for the slow variables x, and the average indicated by the pointed brackets is over an ensemble of atomistic trajectories consistent with the coarse-grained state x at $t = 0$ and evolved according to the atomistic dynamics to the time t. Equation (3.47), which is known as a Green-Kubo formula, is a more symbolic version of (6.73), but it is sufficient for our present arguments. All details regarding notation and derivation can be found in Chapter 6.

To study the symmetry properties of (3.47), we consider the backward time evolution of the atomistic trajectories. Whereas for the forward time evolution $\dot{x}_j^{\mathrm{f}}$ is evaluated later than $\dot{x}_k^{\mathrm{f}}$, the situation is reversed for the backward time evolution. When we introduce $\varepsilon(x_j) = +1$ if the variable x_j is an even function of the particle velocities and hence invariant under time reversal, and $\varepsilon(x_j) = -1$ if x_j is an odd function of the particle velocities and hence changes sign under time reversal, then we conclude

$$\left\langle \dot{x}_j^{\mathrm{f}}(t) \dot{x}_k^{\mathrm{f}}(0) \right\rangle_x = \varepsilon(x_j)\varepsilon(x_k) \left\langle \dot{x}_j^{\mathrm{f}}(0) \dot{x}_k^{\mathrm{f}}(t) \right\rangle_x, \tag{3.48}$$

where the time derivatives $\dot{x}_j^{\mathrm{f}}$ and $\dot{x}_k^{\mathrm{f}}$ pick up the signs $-\varepsilon(x_j)$ and $-\varepsilon(x_k)$ under time reversal. In (3.48) we paid attention only to the reversed difference in time but not to absolute values of time; this simplification is allowed provided that the Hamiltonian generating the atomistic dynamics is independent of time. Furthermore, whereas the average on the left-hand side of (3.48) is over an ensemble of initial values, after time reversal, the average on the right-hand side of (3.48) is over an ensemble of final values. The difference does not matter because one can assume that the slow variables x do not change on the time scale τ, whereas the fast variables are equally

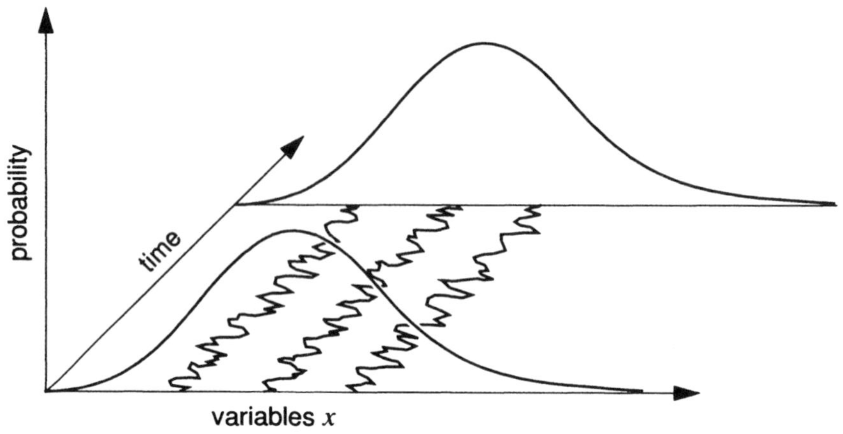

Fig. 3.1 Exchange of the roles of initial and final ensembles under time reversal.

equilibrated both at the beginning and at the end of the trajectories (see Figure 3.1).[15] We hence obtain

$$M_{jk}(x) = \varepsilon(x_j)\varepsilon(x_k)M_{kj}(x). \tag{3.49}$$

The assumption that the ensemble remains essentially unchanged on the time scale τ is crucial to obtain the property (3.49). In many cases, this assumption is justified because individual molecules have a marginal influence on the coarse-grained variables, for example, when they contribute little to averages such as the hydrodynamic fields. For very detailed levels of description, however, when individual molecules can have a considerable influence on the ensemble, the symmetry of M cannot be taken for granted. An explicit example is the friction matrix with a lack of symmetry arising for Boltzmann's kinetic equation (see Section 7.2.4).

We refer to (3.49) as the bare Onsager-Casimir symmetry of the friction matrix. For example, from $\varepsilon(\varrho) = \varepsilon(\epsilon) = +1, \varepsilon(\boldsymbol{M}) = -1$, we obtain for the bare symmetry of the friction matrix (2.66) for hydrodynamics

$$\mathcal{OC}_{\text{bare}}(M) = \begin{pmatrix} + & - & + \\ - & + & - \\ + & - & + \end{pmatrix}, \tag{3.50}$$

where the operator $\mathcal{OC}_{\text{bare}}$ lists the products $\varepsilon(x_j)\varepsilon(x_k)$ in matrix form. The subscript "bare" on the operator that identifies the Onsager-Casimir symmetry indicates that this is not our final answer yet. It should be noted that some elements of the particular friction matrix (2.66) are zero and hence are consistent with both symmetry and antisymmetry, that M_{22} and M_{33} are even functions of $\dot{\gamma}$, and that M_{23} and M_{32}

[15] Small changes of x on the time scale τ can lead to small violations of the symmetry; Geigenmüller et al., Physica A 119 (1983) 53 showed that violations of at least second-order in the ratio of fast to slow time scales should be expected.

are odd functions of $\dot{\gamma}$. Like velocity, the symmetrized velocity gradient $\dot{\gamma}$ should change sign in the time-reversal process described by the bare Onsager-Casimir symmetry. If we do not change the sign for matrix elements that are odd functions of velocity, we arrive at the "dressed" Onsager-Casimir symmetry of the friction matrix for hydrodynamics:

$$\mathcal{OC}_{\text{dressed}}(M) = \begin{pmatrix} \bigcirc & \bigcirc & \bigcirc \\ \bigcirc & \oplus & \oplus \\ \bigcirc & \oplus & \oplus \end{pmatrix}. \tag{3.51}$$

Circled signs indicate the dressed symmetry properties, whereas the open circles indicate zero entries that can be considered to be either symmetric or antisymmetric. Equation (3.51) is consistent with a fully symmetric friction matrix for hydrodynamics. Whereas the bare symmetry properties depend only on the time-reversal behavior of the variables, the dressed symmetry properties depend also on the specific form of the elements of the friction matrix. Although not obvious to many scientists, proper dressing depends on the occasion.

The bare Onsager-Casimir symmetry of the friction matrix is most fundamental because it is directly obtained from the time-reversal behavior of the independent variables of a beyond-equilibrium system, without any regression hypothesis, and without any knowledge about a particular friction matrix. However, it is the dressed Onsager-Casimir symmetry rather than the bare symmetry that matters for the entropy production associated with a certain effect because the entropy production rate of (2.72) can be rewritten as

$$\sigma = \frac{1}{2}\frac{\delta S}{\delta x} \cdot (M + M^T) \cdot \frac{\delta S}{\delta x}, \tag{3.52}$$

where the transposition is effected without changing any signs in the matrix elements of the friction matrix.

It is very important to realize that the *classical Onsager and Casimir relationships*, (3.45) and (3.46), *can be derived directly from the dressed symmetry behavior of the friction matrix* without decomposing it according to (2.94). An explicit example is considered in Exercise 46. The more rigorous approach to symmetry on the level of friction matrices is a consequence of the availability of atomistic expressions for the GENERIC building blocks (see Chapter 6).

Exercise 45 Symmetry of M for Two-Component System

Determine the bare and dressed Onsager-Casimir symmetry properties of the friction matrix for a two-component system as obtained in Exercise 29.

3.2.2 Thermoelectric Effects

As an illustrative example for the application of linear irreversible thermodynamics and for the significance of the Onsager-Casimir symmetry, we consider thermoelectric effects. As the previous discussion has shown, the study of cross effects is important

Table 3.2 Possible Experiments Probing Cross Effects

Situation	Measured Variables	Probed Coefficients
X_1^{LIT} fixed, $J_2^{\text{LIT}} = 0$	X_2^{LIT}	$L_{21}^{\text{LIT}}/L_{22}^{\text{LIT}} = -X_2^{\text{LIT}}/X_1^{\text{LIT}}$
X_2^{LIT} fixed, $X_1^{\text{LIT}} = 0$	J_1^{LIT}, J_2^{LIT}	$L_{12}^{\text{LIT}}/L_{22}^{\text{LIT}} = J_1^{\text{LIT}}/J_2^{\text{LIT}}$
X_2^{LIT} fixed, $J_1^{\text{LIT}} = 0$	X_1^{LIT}	$L_{12}^{\text{LIT}}/L_{11}^{\text{LIT}} = -X_1^{\text{LIT}}/X_2^{\text{LIT}}$
X_1^{LIT} fixed, $X_2^{\text{LIT}} = 0$	J_1^{LIT}, J_2^{LIT}	$L_{21}^{\text{LIT}}/L_{11}^{\text{LIT}} = J_2^{\text{LIT}}/J_1^{\text{LIT}}$

to appreciate the real power of symmetry considerations. Moreover, thermoelectric effects have found important applications, such as thermocouples to measure and control temperatures through electric potential differences. Finally, we can also illustrate the proper approach to discontinuous systems.

In general, when one considers two coupled phenomena, such as thermal and electric effects, the linear force-flux relationships (3.6),

$$J_1^{\text{LIT}} = L_{11}^{\text{LIT}} X_1^{\text{LIT}} + L_{12}^{\text{LIT}} X_2^{\text{LIT}}, \tag{3.53}$$

$$J_2^{\text{LIT}} = L_{21}^{\text{LIT}} X_1^{\text{LIT}} + L_{22}^{\text{LIT}} X_2^{\text{LIT}}, \tag{3.54}$$

imply that four different types of experiments can be made to study cross effects, which allow us two pairwise tests of Onsager's reciprocal relations. This situation is illustrated in Table 3.2.

As a specific example, we now consider a thermocouple consisting of two connected metal wires A and B, with an ideal capacitance C interrupting the wire A according to Figure 3.2. It is known experimentally that keeping the junctions between A and B at different temperatures T and $T + \Delta T$ will not only produce a heat current but also an electric current in the wires A and B which, in turn, will establish an electric potential difference across the capacitance, and *vice versa*.

The situation illustrated in Figure 3.2 provides a typical example of a discontinuous system with a finite number of variables of interest, such as the temperatures at the junctions and the potential difference across the capacitance. The basic equations of linear irreversible thermodynamics, for example (3.53) and (3.54), however, relate the local thermodynamic forces and fluxes. We hence need to derive the proper equations for the discontinuous system from the continuous field equations. The derivation of such equations for discontinuous systems from the continuous equations of linear irreversible thermodynamics is of general importance in many applications; in a careful and reliable approach within the framework of linear irreversible thermodynamics, one should not simply postulate linear constitutive relations with Onsager's reciprocal relations directly for a discontinuous system.

We start our analysis of the thermocouple by considering the entropy production rate (3.8) associated with the force-flux pairs (3.11), (3.13), (3.21), and (3.22),

$$\sigma = -\frac{1}{T^2}\boldsymbol{j}^{\text{q}} \cdot \frac{\partial T}{\partial \boldsymbol{r}} - \boldsymbol{j}_1^{\text{diff}} \cdot \frac{\partial}{\partial \boldsymbol{r}}\frac{\mu_1}{T} - \boldsymbol{j}_2^{\text{diff}} \cdot \frac{\partial}{\partial \boldsymbol{r}}\frac{\mu_2}{T} + \frac{1}{T}\boldsymbol{i} \cdot \boldsymbol{E}$$

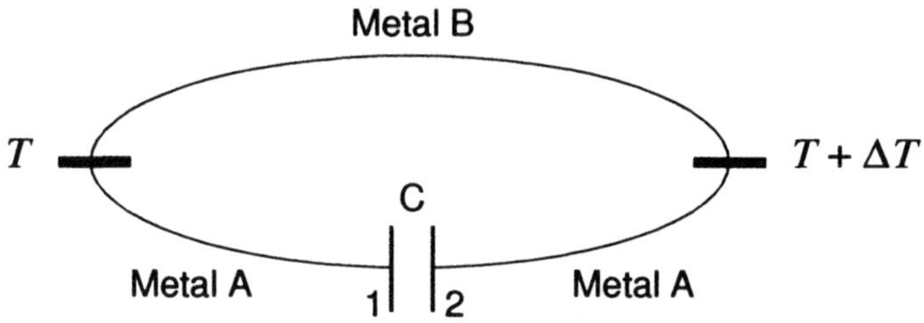

Fig. 3.2 Thermocouple.

$$= \frac{1}{T}\left[-\boldsymbol{j}^{\rm s}\cdot\frac{\partial T}{\partial \boldsymbol{r}} - \boldsymbol{j}_1^{\rm diff}\cdot\frac{\partial(\mu_1-\mu_2)}{\partial \boldsymbol{r}} + \boldsymbol{i}\cdot\boldsymbol{E}\right], \tag{3.55}$$

where $\boldsymbol{j}^{\rm s}$ is the nonconvective entropy flux (see Exercise 17),

$$\boldsymbol{j}^{\rm s} = \frac{\boldsymbol{j}^{\rm q}}{T} - \frac{\mu_1}{T}\boldsymbol{j}_1^{\rm diff} - \frac{\mu_2}{T}\boldsymbol{j}_2^{\rm diff}, \tag{3.56}$$

and the metals are considered as two-component systems, of which the first component is formed by the electrons and the second by the positive ion lattices (see also the results of Exercise 17 in the absence of viscous flow). We assume that the heavy ion lattice is fixed in space. Both the diffusion flow and the electric current density are determined entirely by the motion of the electrons so that

$$\boldsymbol{j}_1^{\rm diff} = -\frac{m}{e}\boldsymbol{i}, \tag{3.57}$$

where m is the mass of an electron and e is the elementary charge. In terms of independent force-flux pairs,[16] (3.55) becomes

$$\sigma = \frac{1}{T}\left[-\boldsymbol{j}^{\rm s}\cdot\frac{\partial T}{\partial \boldsymbol{r}} + \boldsymbol{i}\cdot\left(\boldsymbol{E} + \frac{m}{e}\frac{\partial(\mu_1-\mu_2)}{\partial \boldsymbol{r}}\right)\right]. \tag{3.58}$$

We hence write the force-flux relationships of the type (3.53) and (3.54) as

$$\boldsymbol{j}^{\rm s} = -L_{\rm ss}^{\rm LIT}\frac{1}{T}\frac{\partial T}{\partial \boldsymbol{r}} + L_{\rm se}^{\rm LIT}\frac{1}{T}\left(\boldsymbol{E} + \frac{m}{e}\frac{\partial(\mu_1-\mu_2)}{\partial \boldsymbol{r}}\right), \tag{3.59}$$

$$\boldsymbol{i} = -L_{\rm es}^{\rm LIT}\frac{1}{T}\frac{\partial T}{\partial \boldsymbol{r}} + L_{\rm ee}^{\rm LIT}\frac{1}{T}\left(\boldsymbol{E} + \frac{m}{e}\frac{\partial(\mu_1-\mu_2)}{\partial \boldsymbol{r}}\right). \tag{3.60}$$

According to (3.22), we realize that $L_{\rm ee}^{\rm LIT} = T\sigma^{\rm el}$ is given by the isothermal electric conductivity. For $\boldsymbol{i} = 0$, we can rewrite (3.59) as

$$\boldsymbol{j}^{\rm s} = -\left(L_{\rm ss}^{\rm LIT} - \frac{L_{\rm se}^{\rm LIT}L_{\rm es}^{\rm LIT}}{L_{\rm ee}^{\rm LIT}}\right)\frac{1}{T}\frac{\partial T}{\partial \boldsymbol{r}}, \tag{3.61}$$

[16] We have selected $\boldsymbol{j}^{\rm s}$ and $\boldsymbol{i}$ as the independent fluxes; for the equally possible alternative choice of $\boldsymbol{j}^{\rm s}$ and $\boldsymbol{j}_1^{\rm diff}$, the transformation formulas developed in Section 3.1.3 for different choices of the basic dissipative processes can be used to relate the respective results.

so that (3.11) suggests that

$$L_{\rm ss}^{\rm LIT} - \frac{L_{\rm se}^{\rm LIT} L_{\rm es}^{\rm LIT}}{L_{\rm ee}^{\rm LIT}} = \lambda^{\rm q} \tag{3.62}$$

is the thermal conductivity in the absence of an electric current. The phenomenological coefficients $L_{\rm se}^{\rm LIT}$ and $L_{\rm es}^{\rm LIT}$ describe the cross effects between thermal and electric phenomena, and Onsager's reciprocal relations (3.45) imply that

$$L_{\rm se}^{\rm LIT} = L_{\rm es}^{\rm LIT}. \tag{3.63}$$

We now consider the experiment corresponding to the first row in Table 3.2; that is, we keep the junctions at fixed temperatures T and $T+\Delta T$ and we observe the electric potential difference across the capacitance, $\Delta\phi^{\rm el} = \phi_1^{\rm el} - \phi_2^{\rm el}$, when no electric current flows in the wires. The need for a nonzero $\Delta\phi^{\rm el}$ proportional to ΔT to suppress the electric current is known as the *Seebeck effect*. According to (3.21), the potential difference is obtained by integrating along the wires,

$$\Delta\phi^{\rm el} = \phi_1^{\rm el} - \phi_2^{\rm el} = -\int_1^2 \frac{\partial \phi^{\rm el}}{\partial \boldsymbol{r}} \cdot d\boldsymbol{r} = \int_1^2 \boldsymbol{E} \cdot d\boldsymbol{r}, \tag{3.64}$$

where the integration can actually be performed along any line inside the metallic wires joining the two plates of the capacitance. If we use (3.60) with $\boldsymbol{i} = 0$ to eliminate $\boldsymbol{E}$, we obtain from (3.64)

$$\Delta\phi^{\rm el} = \int_1^2 \left[\frac{L_{\rm es}^{\rm LIT}}{L_{\rm ee}^{\rm LIT}} \frac{\partial T}{\partial \boldsymbol{r}} - \frac{m}{e} \frac{\partial(\mu_1 - \mu_2)}{\partial \boldsymbol{r}} \right] \cdot d\boldsymbol{r}. \tag{3.65}$$

The second term in (3.65) leads to a vanishing integral. This follows if we keep the temperature, pressure, and chemical composition of the two plates of the capacitance the same so that the chemical potentials on the two plates are identical. To evaluate the first term we assume that, at least for small ΔT, the ratio $L_{\rm es}^{\rm LIT}/L_{\rm ee}^{\rm LIT}$ is constant in each of the metals. We then obtain

$$\frac{\Delta\phi^{\rm el}}{\Delta T} = \left(\frac{L_{\rm es}^{\rm LIT}}{L_{\rm ee}^{\rm LIT}} \right)_{\rm B} - \left(\frac{L_{\rm es}^{\rm LIT}}{L_{\rm ee}^{\rm LIT}} \right)_{\rm A}. \tag{3.66}$$

Note that, for the discontinuous system, it is not the ratio of phenomenological coefficients as indicated in Table 3.2 that matters but rather the difference between those ratios for the two different metals (this essential point is ignored in some textbooks).

The Seebeck effect for thermocouples is frequently used to measure temperatures. In principle, the temperature difference ΔT is obtained directly by measuring an electric potential difference; if T is a known reference temperature, such as the freezing temperature of water under air pressure, $T + \Delta T$ is measured. In practice, one uses only one junction of the wires and measures its temperature $T + \Delta T$ through the potential difference between the metals A and B at the reference temperature T. To cover a wide range of temperatures, the nonlinear relationship between $\Delta\phi^{\rm el}$ and

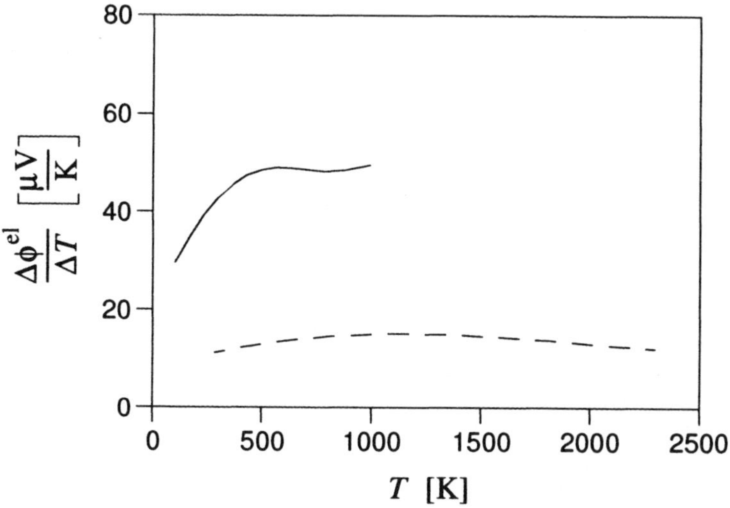

Fig. 3.3 Temperature dependence of $\Delta\phi^{\text{el}}/\Delta T$ for Iron-Constantan (Type J, continuous line) and Tungsten 5% Rhenium-Tungsten 26% Rhenium (Type C, dashed line) thermocouples.

ΔT resulting from the temperature dependence of $L_{\text{es}}^{\text{LIT}}/L_{\text{ee}}^{\text{LIT}}$ needs to be taken into account. Instead of fixing the reference temperature T, it is usually measured by another thermometer, for example, by measuring a temperature dependent resistance. The three most common thermocouple alloys for moderate temperatures are Iron-Constantan, Copper-Constantan, and Chromel-Alumel. For high temperatures, one can use Tungsten-Rhenium alloys of different composition for the wires A and B (see Figure 3.3).

What are the advantages of using a thermocouple as a thermometer when one needs another thermometer to measure the reference temperature? Most importantly, the temperature range can be extended from some 800°C to beyond 2300°C. Furthermore, the system disturbance can be kept at a minimum by using very thin wires with diameters much less than 1 mm, even down to 0.02 mm (at the expense of fragility).

We next consider the *Peltier effect*, which corresponds to the second row in Table 3.2. When the electric potential $\Delta\phi^{\text{el}}$ is fixed and both junctions are maintained at the same temperature (we can assume that the entire system is kept at this temperature), then, according to (3.59) and (3.60), there is not only an electric current density but also a corresponding entropy flux,

$$\boldsymbol{j}^{\text{s}} = \frac{L_{\text{se}}^{\text{LIT}}}{L_{\text{ee}}^{\text{LIT}}}\boldsymbol{i}. \tag{3.67}$$

As the total electric current I^{el} obtained by integrating $\boldsymbol{i}$ across the wires is constant throughout the system, the entropy flux differs in the metals A and B due to different ratios of the phenomenological coefficients in the different metals; in other words, entropy needs to be removed in one junction and added in the other one, or the junctions need to be heated or cooled in order to maintain them at a constant temperature. For the experimentally accessible amount of heat to be absorbed by the environment around the right junction of Figure 3.2 per unit time, J^{q}, we obtain

$$\frac{J^{\mathrm{q}}}{I^{\mathrm{el}}T} = \left(\frac{L_{\mathrm{se}}^{\mathrm{LIT}}}{L_{\mathrm{ee}}^{\mathrm{LIT}}}\right)_{\mathrm{A}} - \left(\frac{L_{\mathrm{se}}^{\mathrm{LIT}}}{L_{\mathrm{ee}}^{\mathrm{LIT}}}\right)_{\mathrm{B}} . \tag{3.68}$$

A comparison of (3.66) and (3.68) shows that, by a careful quantitative investigation of the Seebeck and Peltier effects, one can test Onsager's reciprocal relation (3.63). A large number of experimental tests, in particular for thermoelectric effects in metallic and electrolytic thermocouples, were collected by D. G. Miller.[17] The data from this article displayed in Table 3.3 were originally published in 1912 (Fe–Hg), in 1916 (Cu–Hg), and in 1935 (Cu–Constantan) where, for all three systems, the two different experiments for the Seebeck and Peltier effects were performed on the same specimens because minor impurities have an enormous effect on the thermoelectric properties of metals. Keeping in mind the difficulty of directly measuring the Peltier effect, the ratio in the last column of Table 3.3 is remarkably close to unity, thus indeed confirming the reciprocal relationship (3.63). In view of these and many further experimental tests for various cross effects, Miller concluded "that the experimental evidence is overwhelmingly in favor of the validity of the Onsager reciprocal relations."

For the thermocouple, the discontinuous nature of the system can be handled in a rather unproblematic way because the cross sections of the wires need not be considered explicitly. For other discontinuous systems, geometrical details enter the calculations because, for example in studying electrokinetic effects, the velocity profile for the flow of matter through a thin capillary connecting two large reservoirs needs to be determined explicitly.[18]

Exercise 46 Coupling Heat Flux and Diffusion Flow

For the coupling of the force-flux pairs (3.11) and (3.13), one can write in the spirit of (3.53), (3.54) and (3.59), (3.60),[19]

$$\begin{aligned} \boldsymbol{j}^{\mathrm{q}} &= L_{\mathrm{qq}}^{\mathrm{LIT}} \frac{\partial}{\partial \boldsymbol{r}} \frac{1}{T} + L_{\mathrm{qd}}^{\mathrm{LIT}} \frac{\partial}{\partial \boldsymbol{r}} \frac{\mu_2 - \mu_1}{T}, \\ \boldsymbol{j}^{\mathrm{diff}} &= L_{\mathrm{dq}}^{\mathrm{LIT}} \frac{\partial}{\partial \boldsymbol{r}} \frac{1}{T} + L_{\mathrm{dd}}^{\mathrm{LIT}} \frac{\partial}{\partial \boldsymbol{r}} \frac{\mu_2 - \mu_1}{T}. \end{aligned} \tag{3.69}$$

How does the dressed symmetry of the friction matrix of Exercise 45 change due to the presence of the cross couplings $L_{\mathrm{qd}}^{\mathrm{LIT}}$ and $L_{\mathrm{dq}}^{\mathrm{LIT}}$? Derive the Onsager relation $L_{\mathrm{qd}}^{\mathrm{LIT}} = L_{\mathrm{dq}}^{\mathrm{LIT}}$ from the dressed symmetry of the friction matrix.

Exercise 47 Friction Matrix for Thermoelectric Effects

Consider a two-component fluid with the electric charge density as an additional variable, $x =$

[17] Miller, Chem. Rev. 60 (1960) 15.

[18] See Chapter XV, § 4 of de Groot & Mazur, *Non-Equilibrium Thermodynamics* (Dover, 1984).

[19] See p. 49 of de Groot & Mazur, *Non-Equilibrium Thermodynamics* (Dover, 1984).

Table 3.3 Test of Onsager's Reciprocal Relation for Some Thermocouples

Couple	T [C]	$\left\|\frac{J^q}{I^{el}T}\right\|$ [10^{-6}V/K]	$\left\|\frac{\Delta\phi^{el}}{\Delta T}\right\|$ [10^{-6}V/K]	Ratio
Cu–Constantan	20	37.7	38.9	1.03
	30	40.5	41.8	1.03
	40	43.2	44.6	1.03
Cu–Hg	13.9	6.19	6.19	1.00
	56.2	7.36	7.35	1.00
	77.7	7.90	7.94	0.995
	99.7	8.49	8.54	0.994
	132.2	9.28	9.46	0.980
	184.4	10.76	10.80	0.996
Fe–Hg	18.4	16.72	16.66	1.004
	56.5	16.17	16.14	1.002
	99.6	15.57	15.42	1.010
	131.6	14.89	14.81	1.005
	182.3	13.88	13.74	1.011

$(\rho(\boldsymbol{r}), \boldsymbol{M}(\boldsymbol{r}), \epsilon(\boldsymbol{r}), c_1(\boldsymbol{r}), \rho^{el}(\boldsymbol{r}))$. In a generalization of Exercise 36, momentum transport, heat flow, diffusion, and electric conduction are described by the following phenomenological matrix of transport coefficients,

$$L^{\rm LIT} = \begin{pmatrix} \eta T(\delta_{ik}\delta_{jl} + \delta_{il}\delta_{jk}) + \hat{\kappa} T \delta_{ij}\delta_{kl} & 0 & 0 & 0 \\ 0 & L^{\rm LIT}_{\rm qq}\delta_{ik} & 0 & L^{\rm LIT}_{\rm qe}\delta_{ik} \\ 0 & 0 & D'\delta_{ik} & 0 \\ 0 & L^{\rm LIT}_{\rm eq}\delta_{ik} & 0 & L^{\rm LIT}_{\rm ee}\delta_{ik} \end{pmatrix}, \tag{3.70}$$

where thermoelectric cross effects have been taken into account. Construct the corresponding friction matrix.

Exercise 48 Psychokinetic Effects
What do Onsager's reciprocal relations tell you about people who can use their minds to bend spoons or move bodies?

3.3 PARALYZING CRITICISM

Linear irreversible thermodynamics provides a consistent methodology to ensure that a set of evolution equations describing irreversible dynamical behavior does not violate the laws of thermodynamics. Nevertheless, in the mid-1960s, the theory was coming under attack by avid critics who questioned whether anything inherently useful had been produced. An influential engineer of that period, James Wei, explicitly asked this question:[20] "Has this field produced anything other than journal papers?" His conclusion was essentially negative, and other scholars echoed it ever since. This negativism, whether justified or not, cast a pall over the subject for many years in the eyes of practically minded engineers and applied physicists, and, to date, it has not completely been lifted.

Most critics complain that the theory does not provide a sufficient framework for extending key ideas to more complicated systems:[21] "Engineers are, of course, quite familiar with linear models and used them for years before the emergence of IT [irreversible thermodynamics]." In particular, linear irreversible thermodynamics offers little guidance for processes far from equilibrium where the linear constitutive equations, (3.6), are no longer valid. Nowadays, nonlinear material behavior is the key to many technological applications, so that there is a clear need to go beyond linearity.

Moreover, specific forms for the balance equations are assumed a priori. Hence, linear irreversible thermodynamics contains no derivation of the reversible contributions to these expressions, which is a critical omission for complex materials with internal microstructure (see the discussion in Section 1.1.3).

The main new features of linear irreversible thermodynamics are the Onsager relations and the calculation of the local entropy production rate; however, the latter concept had been applied for many years in one form or another. For instance, the Stokes relations in hydrodynamics state that the viscosity and thermal conductivity parameters must be nonnegative; these were known at least a hundred years before linear irreversible thermodynamics. Furthermore, some questioned the validity of the local equilibrium hypothesis, upon which the calculation of the local entropy production rate, as well as a sensible definition of an entropic potential, depends. Thus, the sole benefit of using linear irreversible thermodynamics, in the view of many, is the reduction in the number of parameters achieved through Onsager's reciprocal relations. Onsager's relations, however, are based on the assumption that fluctuations about the equilibrium state decay according to the same laws that govern the decay of macroscopic deviations from equilibrium. Wei and Truesdell criticized this assumption, and the school of rational thermodynamics thus banished the reciprocity relations from nonequilibrium thermodynamics, along with the entropic potential function.

Chapter 3 has shown that the symmetry properties of the GENERIC friction matrix, in contrast to the closely related ones of the phenomenological matrix of linear irreversible thermodynamics, can be derived without Onsager's regression hypothesis for the fluctuations. Moreover, the GENERIC framework is not based on the assumption of a local equilibrium state and produces valuable insights for systems far from equilibrium, as further illustrated in the subsequent chapters. Today, the negativism of the mid-1960s should be fully overcome by the more general approach to beyond-equilibrium thermodynamics summarized in Section 1.2.

[20] See p. 55 of Wei, Ind. Eng. Chem. 58 (1966) 55.

[21] See p. 59 of Wei, Ind. Eng. Chem. 58 (1966) 55.

4

Complex Fluids

Hydrodynamics is a classical topic of nonequilibrium thermodynamics. When time evolution becomes important, it is most natural to consider fluids and balance equations. In this chapter, we want to dare the first step from simple to complex fluids. In the present context, a *simple fluid* is one in which the internal microstructure of the molecular or atomic constituents is essentially regular in space and time, regardless of whether the fluid is undergoing some sort of transport process. Low-molecular-weight fluids such as water classify as simple materials and can be described universally at the macroscopic level of hydrodynamic fields without regard to the motions of the individual molecular constituents.

Complex fluids offer more challenges because they have a complicated microstructure that must be accounted for through the inclusion of additional structural variables which quantify the dynamic response of the material's constituent components to the application of external forces. Such materials are driving new, innovative technologies, and examples include polymer melts, liquid crystals, fiber suspensions, colloids, emulsions, and other forms of soft condensed matter. These complex materials undergo not only nonequilibrium *transport* processes like those described above, but also have the additional complication of experiencing *relaxation* processes, where the internal microstructure rearranges itself in response to the transport processes which pass the effects of the external forces throughout the substance. The microstructure responds on time scales that are comparable with those of the transport processes, and must be described in a similar and consistent fashion. For complex fluids of practical relevance, nonlinear flow behavior is the rule rather than the exception.

We are now prepared to attack the problem of complex fluids by means of beyond-equilibrium thermodynamics. We here choose two different sets of structural variables in addition to the hydrodynamic ones. First, we use *a tensor and a scalar* as structural variables. In particular, by decoupling or "ignoring" the scalar variable, this choice includes the special case of a tensor variable alone, which has been used for some 50 years in modeling the flow behavior of complex fluids. In many cases, the pressure tensor has served as this tensorial variable. In the phenomenological approach, it is actually unnecessary to assign a physical meaning to the tensor and scalar variables, and many arguments are of a rather general type. However, it certainly helps for the physical motivation of the GENERIC building blocks to have an interpretation of the variables. Throughout this chapter, we illustrate the phenomenological procedure by specific examples.

A more detailed understanding of polymeric liquids can be achieved by using configurational distribution functions as structural variables. Polymer molecules can be oriented and stretched in flows, and we can think of the end-to-end vector as a structural variable for polymer chains. For reasons of symmetry between the chain ends, the average end-to-end vector vanishes so that we need to take deviations from the average, or fluctuations, into account. This can either be done through the second-moment tensor of the end-to-end vector, which would bring us back to using a tensor as the structural variable, or by using configurational distribution functions, that is, probability densities depending on the end-to-end vector. Similar arguments hold for other configuration vectors. *Configurational distribution functions* hence define the second level of description for complex fluids that we consider here in more detail.

Before we start modeling complex fluids, we need to discuss what kind of properties are of interest to us. For that purpose, we first present some selected basic ideas to describe non-Newtonian flow behavior. Although the science of flow behavior is a highly developed and still rapidly progressing research field of great practical importance, we here restrict ourselves to giving a flavor of its essence. In other words, the present chapter provides some basic tools to describe soft condensed matter.

4.1 BASIC RHEOLOGICAL PROPERTIES

What are the issues we want to focus on for complex fluids? For hydrodynamics, we focused on the form of the hydrodynamic equations and on the constitutive assumptions for the momentum and heat fluxes; we did not get into the wealth of properties of the solutions of these nonlinear equations, which lead to complicated phenomena such as turbulence. We certainly will not get into the even richer universe of possible flow patterns for complex fluids, such as additional viscoelastic instabilities of various types occurring at much lower flow rates than turbulence for Newtonian fluids, or complicated flow situations occurring in processing applications. Again, we focus on the structure of the equations and the fundamental constitutive behavior, including predictions of material properties in shear flow. We describe the basic features of *rheology*, which is defined as the science of deformation and flow.[1]

[1] Barnes, Hutton & Walters, *Introduction to Rheology* (Elsevier, 1989).

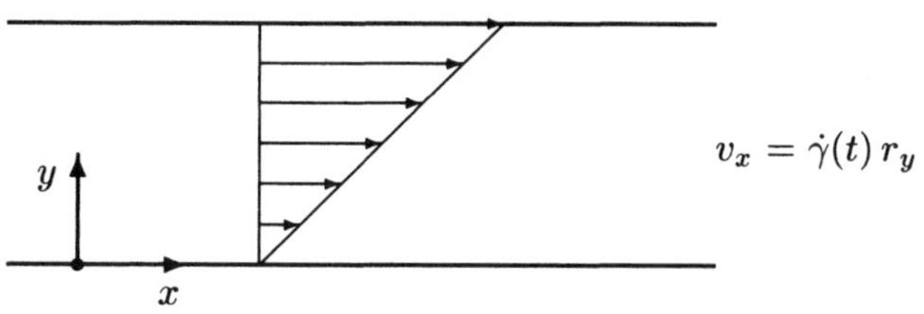

Fig. 4.1 Time dependent shear flow between parallel plates with shear rate $\dot{\gamma}(t)$.

4.1.1 Linear Viscoelasticity

Let us reconsider the homogeneous shear flow of Section 2.1.5, but now with a time dependent shear rate $\dot{\gamma}(t)$ (see Figure 4.1). For example, in *start-up of steady shear flow* one has $\dot{\gamma}(t) = 0$ for $t < 0$ and $\dot{\gamma}(t) = \dot{\gamma}$ for $t \geq 0$. For a Newtonian fluid, the shear stress $\tau_{xy}(t)$ at any time would be proportional to the instantaneous shear rate $\dot{\gamma}(t)$, that is, $\tau_{xy}(t) = 0$ for $t < 0$ and $\tau_{xy}(t) = -\eta\dot{\gamma}$ for $t \geq 0$. For a complex fluid, the measured shear stress will not jump immediately to the steady-state value at the start-up of the steady shear flow. If we assume that the shear rate $\dot{\gamma}$ is sufficiently small, then we expect an expression of the linear form

$$\tau_{xy}(t) = -\int_{-\infty}^{t} G(t-t')\dot{\gamma}(t')dt', \tag{4.1}$$

where $G(t)$ is the *shear relaxation modulus*. Equation (4.1) is the fundamental equation of linear viscoelasticity. For the stress growth in start-up of steady shear flow, we obtain from (4.1) for $t \geq 0$

$$\tau_{xy}(t) = -\dot{\gamma}\int_{0}^{t} G(t-t')dt' = -\dot{\gamma}\int_{0}^{t} G(t')dt'. \tag{4.2}$$

When, for large times t, a steady state is reached, we can identify the steady-state viscosity as

$$\eta = \int_{0}^{\infty} G(t')dt'. \tag{4.3}$$

For shorter times, we would find a stress growth as sketched in Figure 4.2. In a stress-growth experiment, one can measure the integral of $G(t')$, or *stress growth coefficient*,

$$\eta^{+}(t) = \int_{0}^{t} G(t')dt'. \tag{4.4}$$

In *cessation of steady flow*, one can measure the *stress decay coefficient*

$$\eta^{-}(t) = \int_{t}^{\infty} G(t')dt' = \eta - \eta^{+}(t). \tag{4.5}$$

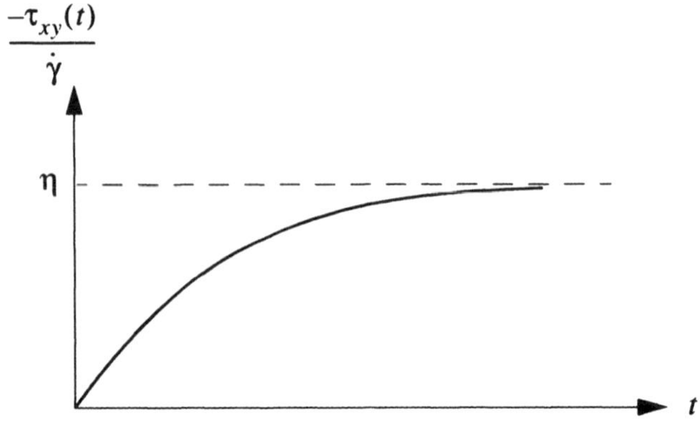

Fig. 4.2 Stress growth in start-up of steady shear flow.

If we want to measure the modulus $G(t)$ directly, we need to perform a small instantaneous shear step of size γ, described by the pulsed generalized force $\dot{\gamma}(t) = \gamma\delta(t)$. After the pulse, we find a shear stress proportional to the shear relaxation modulus

$$\tau_{xy}(t) = -\gamma \int_{-\infty}^{t} G(t-t')\delta(t')dt' = -\gamma G(t). \tag{4.6}$$

Although we have shown how the material property $G(t)$ can be measured in principle, it is quite difficult in practice to produce steps in shear strain or strain rate. It often is advantageous to perform oscillatory experiments, $\dot{\gamma}(t) = \dot{\gamma}_0 \cos\omega t$, where $\dot{\gamma}_0$ is a small amplitude and ω is the angular frequency. In oscillatory flow, one can measure the *storage modulus*

$$G'(\omega) = \omega \int_0^{\infty} G(t) \sin\omega t dt, \tag{4.7}$$

and the *loss modulus*

$$G''(\omega) = \omega \int_0^{\infty} G(t) \cos\omega t dt, \tag{4.8}$$

as is elaborated in Exercise 49. A *complex modulus* can be defined as

$$G^*(\omega) = G'(\omega) + iG''(\omega) = i\omega \int_0^{\infty} G(t)e^{-i\omega t}dt. \tag{4.9}$$

If one prefers to leave out the factor $i\omega$ in front of the integral in (4.9), one can introduce the *complex viscosity*

$$\eta^*(\omega) = \int_0^{\infty} G(t)e^{-i\omega t}dt. \tag{4.10}$$

The time integrals from 0 to ∞ in the preceding formulas can be extended to integrals from $-\infty$ to ∞ because we have $G(t) = 0$ for $t < 0$.

By inverting the Fourier transformation (4.9) we obtain

$$G(t) = \frac{1}{2\pi}\int_{-\infty}^{\infty} \frac{G^*(\omega)}{i\omega} e^{i\omega t} d\omega, \tag{4.11}$$

and, in particular,

$$G(0) = \frac{1}{2\pi}\int_{-\infty}^{\infty} \frac{G^*(\omega)}{i\omega} d\omega = \frac{1}{\pi}\int_{0}^{\infty} \frac{G''(\omega)}{\omega} d\omega. \tag{4.12}$$

Note that, at $t = 0$, $G(t)$ jumps from $G(0-) = 0$ to a finite value $G(0+)$, and, by Fourier transformation, we pick up the mean value $G(0+)/2$ so that we have

$$G(0+) = \frac{2}{\pi}\int_{0}^{\infty} \frac{G''(\omega)}{\omega} d\omega. \tag{4.13}$$

Equation (4.13) is an example of a *sum rule*. In general, sum rules express the relaxation modulus and its derivatives in the limit $t \downarrow 0$ in terms of moments of $G'(\omega)$ or $G''(\omega)$. Such sum rules are important for checking the consistency of experimental data or phenomenological models. Such consistency checks can be particularly useful when one realizes that quantities such as $G(0+)$ are amenable to equilibrium statistical mechanics.

In the start-up, cessation, step-strain, and oscillatory experiments considered so far, we specified a time dependent shear deformation. A further option is not to prescribe the time dependence of the deformation, but it may be advantageous to control the stress and to measure the resulting deformation. Again, various types of experiments are possible, for example, creep and recoil experiments.[2]

It is moreover possible to treat other types of flow than shear. For that purpose, (4.1) can be generalized to the integral constitutive equation of linear viscoelasticity by writing the full symmetric pressure tensor as

$$\boldsymbol{\tau}(t) = -\int_{-\infty}^{t} G(t-t')\dot{\boldsymbol{\gamma}}(t')dt', \tag{4.14}$$

where $\dot{\boldsymbol{\gamma}}$ is the symmetrized velocity gradient tensor introduced in (2.62). Assuming incompressibility, we have here neglected the divergence of the velocity field occurring in (2.38).

Exercise 49 Measurement of Storage and Loss Moduli
How can the storage and loss moduli be measured by small-amplitude oscillatory shear flow?

Exercise 50 Decomposition of Complex Viscosity
From (4.10), derive integral expressions for the real-valued frequency dependent viscosities $\eta'(\omega)$ and $\eta''(\omega)$ in

$$\eta^*(\omega) = \eta'(\omega) - i\eta''(\omega). \tag{4.15}$$

[2] Dealy, J. Rheol. 39 (1995) 253.

Exercise 51 Entropy Production in Oscillatory Shear Flow
Show that the average entropy production rate in oscillatory shear flow is given by

$$\sigma = \frac{\dot{\gamma}_0^2}{2T} \frac{G''(\omega)}{\omega}. \tag{4.16}$$

This expression explains why $G''(\omega)$ is referred to as the loss modulus.

Exercise 52 Simple Elongational Flow
Simple elongational flow is defined by the symmetric velocity gradient tensor

$$\boldsymbol{\kappa} = \boldsymbol{\kappa}^T = \frac{1}{2}\dot{\boldsymbol{\gamma}} = \begin{pmatrix} \dot{\epsilon} & 0 & 0 \\ 0 & -\frac{\dot{\epsilon}}{2} & 0 \\ 0 & 0 & -\frac{\dot{\epsilon}}{2} \end{pmatrix}. \tag{4.17}$$

When a sample is elongated in the x-direction with the elongational rate $\dot{\epsilon}$, incompressibility requires that the sample shrinks in the two transverse directions with half of that rate. Determine the linear viscoelastic stress growth coefficient

$$\eta_{\mathrm{E}}^{+}(t) = -\frac{\tau_{xx}(t) - \tau_{yy}(t)}{\dot{\epsilon}} \tag{4.18}$$

for start-up of steady elongational flow at time $t = 0$.

The reader may not be too excited about all the equations of this section. In fact, all the physically important information is contained in the memory integral (4.1), and all of the other equations are about the consequences of (4.1) in various types of time dependent shear flow situations. In principle, all of these equations contain equivalent information. In practice, however, different types of experiments come with different limitations. If one wants to collect information about the flow behavior of a material over the widest possible range of times or frequencies, it is important to bring together all the information available from different experiments. One hence needs to know all the different material functions of this section and how they are related.

A most convenient concept to perform the various transformations is the *relaxation spectrum*, which can be introduced through the assumption

$$G(t) = \eta_\infty \delta(t) + \sum_j \frac{\eta_j}{\tau_j} e^{-t/\tau_j} + G_0, \tag{4.19}$$

where $j = 1, 2, \ldots$ labels a finite or infinite number of relaxation times. The η_∞-term, which should be considered as the limit of $(\eta_\infty/\tau_\infty)\exp\{-t/\tau_\infty\}$ for $\tau_\infty \to 0$, describes an instantaneous Newtonian viscosity contribution, as can be seen by calculating

$$\eta^{+}(t) = \eta_\infty + \sum_j \eta_j (1 - e^{-t/\tau_j}) + G_0 t. \tag{4.20}$$

Table 4.1 Material Functions of Linear Viscoelasticity

Linear Viscoelastic Material Function [in shear]	Expression in Terms of Relaxation Spectrum	General Nomenclature*
Relaxation modulus $G(t)$	$\eta_\infty \delta(t) + \sum_j \frac{\eta_j}{\tau_j} e^{-t/\tau_j}$	Response function
Stress growth coeff. $\eta^+(t)$	$\eta_\infty + \sum_j \eta_j (1 - e^{-t/\tau_j})$	
Stress decay coeff. $\eta^-(t)$	$\sum_j \eta_j e^{-t/\tau_j}$	Relaxation function
Complex modulus $G^*(\omega)$	$i\omega\eta_\infty + \sum_j \frac{i\omega\eta_j}{1+i\omega\tau_j}$	
Storage modulus $G'(\omega)$	$\sum_j \eta_j \frac{\omega^2\tau_j}{1+\omega^2\tau_j^2}$	
Loss modulus $G''(\omega)$	$\eta_\infty\omega + \sum_j \frac{\eta_j\omega}{1+\omega^2\tau_j^2}$	
Complex viscosity $\eta^*(\omega)$	$\eta_\infty + \sum_j \frac{\eta_j}{1+i\omega\tau_j}$	Complex admittance
Dynamic viscosity $\eta'(\omega)$	$\eta_\infty + \sum_j \frac{\eta_j}{1+\omega^2\tau_j^2}$	
$\eta''(\omega)$	$\sum_j \eta_j \frac{\omega\tau_j}{1+\omega^2\tau_j^2}$	

*According to Section 3.1 of Kubo et al., *Statistical Physics II* (Springer, 1991).

The G_0-term describes a fixed modulus characteristic of rubber elasticity and hence solid-like behavior. For $G_0 = 0$, we obtain for the steady-state viscosity

$$\eta = \eta_\infty + \sum_j \eta_j. \tag{4.21}$$

In (4.20), the mode j contributes with a rubber-like term $(\eta_j/\tau_j)t$ for $t \ll \tau_j$, and with a Newtonian liquid-like term η_j for $t \gg \tau_j$.

The relationships between the various material functions of linear viscoelasticity discussed in this section in terms of the relaxation spectrum are summarized in Table 4.1. The material functions associated with stress-controlled experiments cannot be expressed directly in terms of the relaxation spectrum; rather, a retardation spectrum is needed. One should realize from Table 4.1 that all the material functions contain equivalent information, except for $\eta^-(t)$, $G'(\omega)$, and $\eta''(\omega)$, which are lacking the information about η_∞. In particular, the real and imaginary parts of the complex viscosity or, more precisely, $\eta'(\omega) - \eta_\infty$ and $\eta''(\omega)$, contain equivalent information. This equivalence is a consequence of the fact that the relaxation modulus $G(t)$ is fully

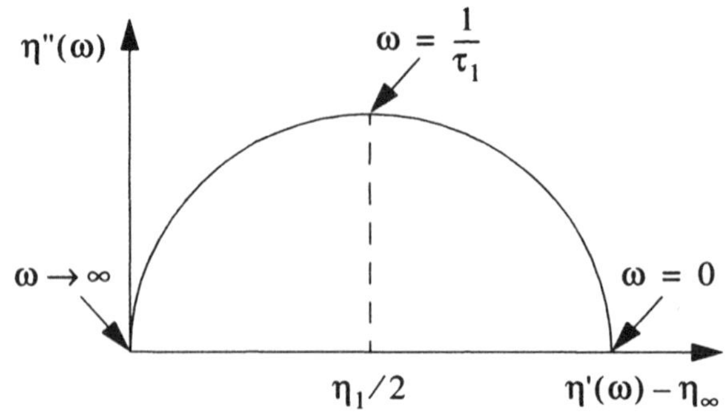

Fig. 4.3 Cole-Cole plot of the complex viscosity for single-mode relaxation.

determined by its values on the positive real axis.[3] The explicit relationship between $\eta'(\omega) - \eta_\infty$ and $\eta''(\omega)$ is an example of the *Kramers-Kronig relations*,[4]

$$\eta'(\omega) - \eta_\infty = \frac{2}{\pi} \int_\infty^\infty \frac{\omega' \eta''(\omega') - \omega \eta''(\omega)}{\omega'^2 - \omega^2} d\omega', \tag{4.22}$$

and

$$\eta''(\omega) = \frac{2\omega}{\pi} \int_0^\infty \frac{\eta'(\omega') - \eta'(\omega)}{\omega^2 - \omega'^2} d\omega'. \tag{4.23}$$

The relationship between the real and imaginary parts of the complex viscosity can be analyzed graphically by means of a *Cole-Cole plot*, in which $\eta''(\omega)$ is plotted against $\eta'(\omega) - \eta_\infty$. Such complex-plane plots are very common in the field of impedance spectroscopy for solid materials characterization and electrochemistry, where they are also known as *Nyquist plots*.[5] In Figure 4.3, a Cole-Cole plot is shown for single-mode relaxation, that is, when the sums in Table 4.1 consist of a single term only. For single-mode relaxation, the Cole-Cole plot is a semicircle. The parameters τ_1 and η_1 can be found as the inverse angular frequency at the maximum and as the diameter of the circle, respectively. If a relaxation process is dominated by a single slowest relaxation mode, but small further contributions with shorter relaxation times exist, one expects deviations from a semicircle in the Cole-Cole plot at short times or high frequencies, that is, near the origin.

An alternative to the Cole-Cole plot for the complex viscosity is obtained by plotting the loss modulus G'' versus the storage modulus G' (see Figure 4.4). For single-mode relaxation, we again find a semicircle from which the parameters τ_1 and

[3] $G(t)$ is identically zero on the negative axis; the existence of a relationship between $\eta'(\omega) - \eta_\infty$ and $\eta''(\omega)$ thus is a consequence of *causality*, that is, the stress in (4.1) cannot depend on future strain rates but is fully determined by the flow history.

[4] A guided derivation of the Kramers-Kronig relations can be found in Problem 5D.2 on p. 290–291 of Bird, Armstrong & Hassager, *DPL 1, Fluid Mechanics* (Wiley, 1987); in fact, Exercise 54 may be considered as an alternative derivation of a Kramers-Kronig relation.

[5] Macdonald, *Impedance Spectroscopy* (Wiley, 1987).

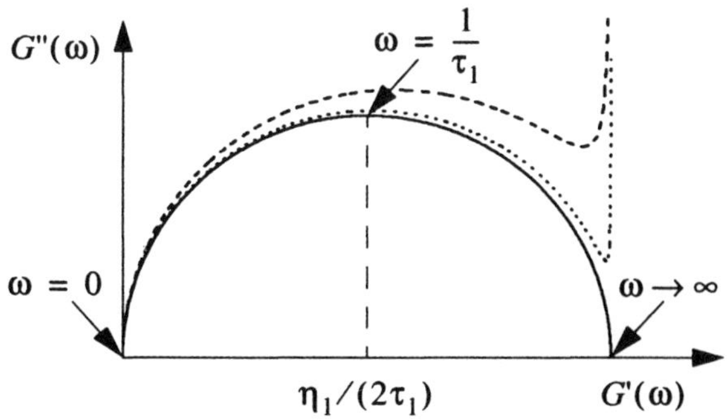

Fig. 4.4 Cole-Cole plot of the complex modulus for single-mode relaxation (*continuous line*). The *dotted* and *dashed lines* illustrate the effect of an additional, purely viscous contribution with $\eta_\infty/\eta_1 = 0.01$ and 0.05, respectively.

η_1 can be read off as indicated in the figure. To obtain a semicircle for the Cole-Cole plot of the complex modulus, however, we need to assume $\eta_\infty = 0$. Typical curves for nonzero η_∞/η_1 are included in Figure 4.4.

Cole-Cole plots are not commonly used in linear viscoelasticity. Usually, the storage modulus G' and the loss modulus G'' are plotted directly as functions of frequency. The frequency domain and the range of values of the moduli usually cover several or even many orders of magnitude, and hence the axes in plots of $G'(\omega)$ and $G''(\omega)$ are scaled logarithmically. Whereas such double-logarithmic plots can be quite misleading by wiping out subtle differences, Cole-Cole plots allow us a linear representation of measured data. However, the experimental data may be distributed in a highly nonuniform manner along a Cole-Cole plot, and it may be worthwhile to zoom into the high- or low-frequency regions. If the typical features observed for Cole-Cole plots can be related to the relaxation spectrum, and ultimately to molecular details such as the polydispersity index characterizing the width of the distribution of molecular weights, then these plots might provide very useful insight. The linear representation of data may also be useful in checking the idea of *time-temperature superposition*, that is, the idea of accounting for the effect of temperature by rescaling all relaxation times τ_j by the *same* temperature-dependent factor (see Exercise 56). The time-temperature superposition is frequently employed to extend the effective frequency range by varying the temperature, and it helps to reduce significantly the amount of experimental work required for a complete rheological characterization of viscoelastic materials.

The determination of relaxation spectra from any of the material functions in Table 4.1, which may be considered as a first step toward transformations, is much more subtle than one might expect at first sight. In mathematical terms, the determination of a spectrum is an *ill-posed problem*, that is, drastically different spectra can be used to fit the data for some linear viscoelastic material function equally good. "Spectrology" has been a heavily debated field in rheology for several decades. Two extreme strategies seem to be particularly useful, both of which resulted in commercial software written by leading research groups:

- One can choose a minimal set of modes by placing the relaxation times and selecting the weights of the corresponding modes such that a given data set can be fitted with the desired precision.[6] This approach is, for example, useful when one needs a rheological model for efficient numerical flow calculations, and experimental results for linear viscoelastic material functions should be reproduced faithfully.
- One can construct a continuous (in practice, sufficiently dense) spectrum of relaxation times, subject to certain smoothness conditions.[7] This approach is, for example, useful for theoretical investigations or for material characterization.

As the determination of spectra is an ill-posed problem, a number of approximate but direct conversion formulas between the material functions in Table 4.1 can be useful for practical purposes.[8]

Exercise 53 Material Functions in Terms of Spectra
Verify the expressions for the material functions of linear viscoelasticity given in Table 4.1.

Exercise 54 Application of Kramers-Kronig Relations
Calculate $\eta''(\omega)$ from $\eta'(\omega)$ in Table 4.1 by means of the Kramers-Kronig relation (4.23).

Exercise 55 Single-Mode Relaxation
Derive the rules to identify the parameters τ_1 and η_1 for single-mode relaxation from the Cole-Cole plots in Figures 4.3 and 4.4.

Exercise 56 Time-Temperature Superposition
Assume that the temperature dependence of all relation times is given by

$$\tau_j(T) = a_T \tau_j(T_0), \tag{4.24}$$

where the factor a_T depends also on the reference temperature T_0, and that the ratios $\eta_j(T)/\tau_j(T)$ occurring as weights in the spectrum are independent of temperature (the latter assumption is a simplification). Derive the following relationships:

$$G(t;T) = G(t/a_T;T_0), \tag{4.25}$$

$$G'(\omega;T) = G'(a_T\omega;T_0), \tag{4.26}$$

and

$$G''(\omega;T) = G''(a_T\omega;T_0). \tag{4.27}$$

[6] Winter, J. Non-Newtonian Fluid Mech. 68 (1997) 225.

[7] Honerkamp & Weese, Rheol. Acta 32 (1993) 65.

[8] See, for example, Chapter 4 of Ferry, *Viscoelastic Properties of Polymers* (Wiley, 1980), Chapter 8 of Tschoegl, *Linear Viscoelastic Behavior* (Springer, 1989), or Section 8.8 of Schwarzl, *Polymermechanik* (Springer, 1990).

How does the viscosity (4.3) depend on temperature?

Equation (4.1), together with (4.19), is a generalization of Newton's relationship (2.40) between shear stress and shear rate. In the force-flux relationship (2.38) for general flows, the momentum flux is linearly related to the velocity gradient at the same time; in view of the absence of any memory effects, such a relationship is often referred to as *Markovian*. Equation (4.1) is a *non-Markovian* linear force-flux relationship. The viscosity η_∞ in (4.19) describes the instantaneous Markovian contribution, all other terms represent memory effects. In the same spirit, the general force-flux relationship (3.6) of linear irreversible thermodynamics can be generalized to[9]

$$J^{\rm LIT}(t) = L^{\rm LIT}_{\rm inst} \cdot X^{\rm LIT}(t) + \int_{-\infty}^{t} L^{\rm LIT}_{\rm mem}(t-t') \cdot X^{\rm LIT}(t')dt', \tag{4.28}$$

and a number of time or frequency dependent material functions can be introduced and measured in different types of experiments just as for linear viscoelasticity. In the general context, one uses a terminology different from the one of rheology; the rheological equivalents of general response functions, relaxation functions, and complex admittances are indicated in Table 4.1.

While the non-Markovian generalization of force-flux relations generalizes linear irreversible thermodynamics enormously, it is still unsatisfactory to describe complex fluids. Nonlinear material behavior is a both fundamentally challenging and practically important feature of complex fluids, as will be discussed in the next subsection. Moreover, even though linear viscoelastic behavior can be described, important questions remain unanswered: *Why* can the relaxation times for complex fluids, such as polymer melts, easily be of the order of seconds or minutes? *Why* do viscosities cover the enormous range from some 10^{-3} Pas for low-molecular-weight liquids like water to 10^{6} Pas or more for the high-molecular-weight liquids commonly used in polymer processing? Just as you can easily feel the difference in viscosity between a soft drink and a milk shake when drinking with a straw, in particular when greedily trying to maximize the flow rate on a hot day, you can imagine the dramatic implications of an increase of viscosity by nine or more orders of magnitude. Pressure and energy dissipation are proportional to the viscosity. Huge and strong equipment is hence needed to generate the pressures required for the processing of such high-viscosity liquids. In addition to the major investments for processing equipment, one should keep in mind that the running costs are basically proportional to the "energy consumption" and hence to the viscosity.

There certainly are strikingly good reasons for understanding the rheological material properties of complex fluids, for going beyond the phenomenological description provided by the non-Markovian version of linear irreversible thermodynamics. The mechanisms leading to enormously large relaxation times and high viscosities need

[9] See, for example, Section 3.1.2 of Kubo et al., *Statistical Physics II* (Springer, 1991), or §64 of Woods, *Thermodynamics of Fluid Systems* (Clarendon, 1975).

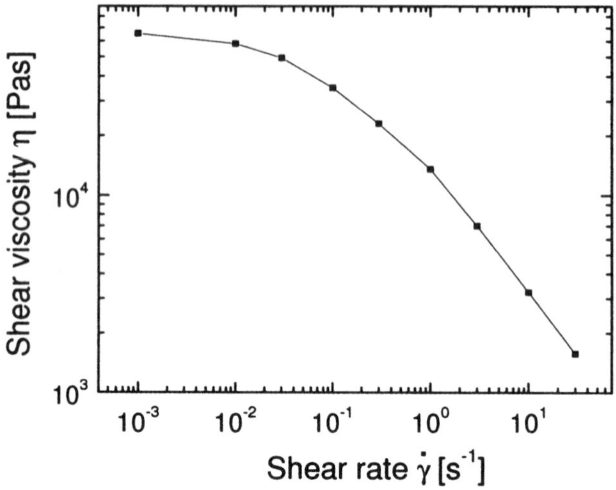

Fig. 4.5 Stress-rate-dependence of the steady shear viscosity for a low-density polyethylene melt (stabilized LDPE 1810H at 150°C). [Figure courtesy of Dr. T. Schweizer, ETH Zürich.]

to be revealed and, for that purpose, the proper structural variables must be found. These remarks are a concrete example of the general discussion of "small *versus* large state spaces" in Section 1.1.3. Non-Markovian linear irreversible thermodynamics introduces memory functionals instead of additional variables, whereas a deeper understanding of nonlinear rheological behavior or the quantitative prediction of material parameters calls for the identification of structural variables. The use of structural variables is generally advocated in the thermodynamic approach of this book, and it is illustrated in the subsequent sections of this chapter for complex fluids.

4.1.2 Nonlinear Material Behavior

The theory of linear viscoelasticity holds only for small deformations, or if the product of deformation rate and longest relaxation time is small (then, a large total deformation is irrelevant because the fluid "remembers" only a small part of the deformation). For sufficiently large and rapid deformations, most complex fluids exhibit a strongly nonlinear behavior. For example, the steady-state viscosity in shear flow depends on the shear rate, as shown in Figure 4.5 for low-density polyethylene.

The shear-rate dependence of the viscosity, referred to as *shear thinning* when the viscosity is a decreasing function of shear rate, is a pronounced effect with far-reaching consequences. By running processing equipment at high flow rates, one not only obtains higher output rates, but one can drastically reduce the viscosity and hence the energy dissipation (at least, it does not increase quadratically with shear rate). For high shear rates, one often finds a power-law decay of the viscosity with an exponent smaller than unity (an exponent larger than unity would imply the counterintuitive result that an increasing shear rate is achieved by reducing the shear stress). For typical polymeric liquids, the exponent describing the decay of the viscosity is found experimentally to be between 0.4 and 0.9.

Another important nonlinear effect is illustrated in Figure 4.6. For a Newtonian fluid, a steady shear flow between parallel plates is maintained by a constant shear stress. For a complex fluid, one typically needs to exert an additional normal stress in order to keep the plates at the prescribed distance. The intuitive explanation for the need for such a normal force is as follows: For a polymeric fluid, the molecules in the shear flow of Figure 4.6 are more stretched in the main flow direction than in transverse directions; these macromolecules hence naturally tend to retract, thus contracting the entire material in the main flow direction. The normal pressure on the plates in Figure 4.6 is needed to prevent the material from contracting in the main flow direction by pushing the plates apart. The occurrence of normal-stress differences is another characteristic of a liquid with elastic features, previously recognized in memory effects.

Going from a simple to a complex fluid, a pressure tensor of the more general form

$$\boldsymbol{\tau} = \begin{pmatrix} \tau_{xx} & \tau_{xy} & 0 \\ \tau_{xy} & \tau_{yy} & 0 \\ 0 & 0 & \tau_{zz} \end{pmatrix}, \tag{4.29}$$

instead of the Newtonian form (2.40) is expected for the shear flow (2.39). While no additional off-diagonal entries can occur for symmetry reasons, in general, all diagonal elements can be different. In addition to the viscosity,

$$\tau_{xy} = -\eta(\dot{\gamma})\dot{\gamma}, \tag{4.30}$$

one hence introduces the first and second *normal-stress coefficients* Ψ_1 and Ψ_2 by

$$\tau_{xx} - \tau_{yy} = -\Psi_1(\dot{\gamma})\dot{\gamma}^2, \tag{4.31}$$

and

$$\tau_{yy} - \tau_{zz} = -\Psi_2(\dot{\gamma})\dot{\gamma}^2. \tag{4.32}$$

These definitions take into account that we have $\tau_{xx} = \tau_{yy} = \tau_{zz}$ at equilibrium, and that, for symmetry reasons, differences between the diagonal components of the pressure tensor at small shear rates are usually quadratic in the shear rate. Therefore, the zero-shear-rate limits of Ψ_1 and Ψ_2 are finite. The rheological material properties in shear flow, η, Ψ_1, and Ψ_2, are collectively known as the *viscometric functions*.

The shear-rate dependence of the viscosity and the occurrence of normal-stress differences, which also depend on the shear rate, are typical and important nonlinear effects in viscoelastic fluids.[10] We hence need to explain these nonlinear effects in addition to the enormously high viscosities and large relaxation times. We need to

[10] Many further viscoelastic effects have been described and discussed in Bird & Curtiss, Physics Today 37/1 (1984) 36 and in Chapter 2 of Bird, Armstrong & Hassager, *DPL 1, Fluid Mechanics* (Wiley, 1987).

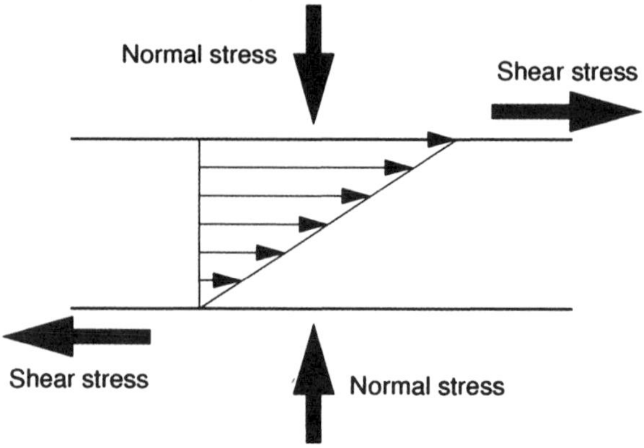

Fig. 4.6 Occurrence of normal-stress differences in shear flow of viscoelastic liquids.

understand qualitatively how viscoelasticity, shear thinning, and normal stress differences arise, and, as a next step, we need quantitative predictions for the rheological material properties, ideally in terms of the molecular architecture.

Exercise 57 Power-Law Decay of Viscosity
Estimate the exponent for the power-law decay of the viscosity at high shear rates for the polymer melt in Figure 4.5.

Exercise 58 Mean Relaxation Time from Viscometric Functions
Although the first normal-stress coefficient occurs as an intrinsically nonlinear material property, continuum mechanical considerations[11] show that Ψ_1 can be expressed in terms of the shear relaxation modulus,

$$\Psi_1 = 2 \int_0^\infty t G(t) dt. \tag{4.33}$$

Discuss the significance of $\tau_\Psi = \Psi_1/(2\eta)$ as a mean relaxation time.

4.2 TENSORS AND SCALARS AS CONFIGURATIONAL VARIABLES

We now turn to the question of how we can describe fluids with viscoelastic properties, including nonlinear material behavior, by introducing structural variables. A number of successful rheological models for complex liquids are based on a tensor as a structural variable in addition to the usual hydrodynamic variables. In particular, this structural variable may be taken as the pressure tensor. This choice is typical for many

[11] See § 9.6 on the memory-integral expansion in Bird, Armstrong & Hassager, *DPL 1, Fluid Mechanics* (Wiley, 1987), in particular Example 9.6-1, and § 6.2 on the retarded-motion expansion, in particular (6.2-6).

well-known constitutive equations, such as the Maxwell,[12] Jeffreys, Oldroyd, Gordon-Schowalter, Johnson-Segalman, Phan-Thien-Tanner, FENE-P, Giesekus, and Leonov models.[13] About half a century of work has been invested into the study of these models, resulting in a deeper understanding of many phenomena. Fully consistent nonisothermal versions of all these models are now available.[14] In general, it is more convenient and hence recommended to use a conformation tensor as the structural variable rather than to use the pressure tensor directly.

While a structural variable of tensorial nature is indispensable to obtain interesting rheological models, recent progress in understanding specific liquids clearly indicates that additional variables need to be considered. The simplest generalization is obtained by adding a scalar variable. Although that might be considered as a purely formal choice, various recent models based on deep physical insight employ such structural variables. For example, in the Doi-Ohta model of emulsions and blends,[15] the scalar describes the amount of interface per unit volume, and the tensor-like variable separately accounts for the orientation of the interface. In the pompon model for melts of branched polymers,[16] the scalar represents the stretch of a tube confining the molecular backbone, and the tensor is used to characterize the tube orientation. In transient network models of polymer melts, the scalar describes the number density of network segments, and the tensor characterizes the segment deformation. An even simpler example is provided by inhomogeneous dilute polymer solutions, where the scalar describes the polymer concentration and the tensor represents the polymer stretch and orientation.

The purpose of this section is to describe a quite general phenomenological model based on a tensor and an additional scalar as structural variables (of course, the scalar variable can be ignored in order to recover all the famous models based on a tensor as the only structural variable).[17] In order to obtain a thermodynamically admissible

[12] James Clerk Maxwell (1831–1879) appears to have been the first who developed a theory of viscoelasticity in 1867; actually, he felt that gases might be viscoelastic. Note that the pioneer of viscoelasticity is the same Maxwell who found the *Maxwell equations* for electromagnetic fields, the *Maxwell relations* guaranteeing the consistency of thermodynamic equations of state, the *Maxwell construction* for the pressure-volume relationship in first-order phase transitions, the *Maxwellian distribution* of velocities, the *Maxwell molecules* allowing closed-form results in the kinetic theory of gases, and the *Maxwell demon* causing violations of the second law by manipulating individual particles in a thermodynamic system. In 1861, Maxwell showed the world's first *color photo* during a lecture in London, namely, the picture of a tartan skirt, where he was actually lucky because the red color in the photo was produced by ultraviolet light reflect by the red parts of the kilt, passing through the red filter, and being recorded in the photographic emulsion. Remarkable is also Maxwell's unexampled devotion to his wife Katherine, whom he sent at least one detailed letter per day whenever he was traveling.

[13] See Bird, Armstrong & Hassager, *DPL 1, Fluid Mechanics* (Wiley, 1987) for a discussion of all these models.

[14] Dressler, Edwards & Öttinger, Rheol. Acta 38 (1999) 117; Wapperom & Hulsen, J. Rheol. 42 (1998) 999.

[15] Doi & Ohta, J. Chem. Phys. 95 (1991) 1242; Wagner, Öttinger & Edwards, AIChE J. 45 (1999) 1169.

[16] McLeish & Larson, J. Rheol. 42 (1998) 81; Öttinger, Rheol. Acta 40 (2001) 317.

[17] The work summarized in this section was motivated by the wish to present a balanced picture of beyond-equilibrium thermodynamics in this book. The results have previously been published in Öttinger, Rev. Mex. Fís. 48 (2002) Supl. 1, 220.

model, appropriate even under nonisothermal flow conditions, we formulate the new model within the GENERIC framework of beyond-equilibrium thermodynamics. A most general exploration of this level of description is not only interesting from a phenomenological point of view, but it also provides the proper background for thermodynamically guided atomistic simulations for systems such as unentangled polymer melts by defining all the material information of interest (see Section 8.2.3). At the end of this section, we consider a conformation-tensor model for dilute polymer solutions and the pompon model for melts of multiply branched polymers as specific examples.

4.2.1 Energy and Entropy

We choose the list of state variables as $x = (\rho(\boldsymbol{r}), \boldsymbol{M}(\boldsymbol{r}), \epsilon(\boldsymbol{r}), \boldsymbol{c}(\boldsymbol{r}), \Lambda(\boldsymbol{r}))$. For any complex fluid, it is natural and convenient to have the standard hydrodynamic fields, namely the mass density $\rho(\boldsymbol{r})$, the momentum density $\boldsymbol{M}(\boldsymbol{r})$, and the internal energy density $\epsilon(\boldsymbol{r})$ among the independent state variables. The internal energy density $\epsilon(\boldsymbol{r})$ includes all the interactions between the atoms or particles in the system, as well as the kinetic energies associated with their motions with respect to the local fluid velocity, $\boldsymbol{v}(\boldsymbol{r}) = \boldsymbol{M}(\boldsymbol{r})/\rho(\boldsymbol{r})$. The choice of the additional, structural or configurational variables, the tensor $\boldsymbol{c}(\boldsymbol{r})$ and the scalar $\Lambda(\boldsymbol{r})$, reflects insight into the behavior of specific systems, such as polymer solutions, emulsions, polymer blends, and entangled linear or branched polymers. For convenience, we here assume that the tensor $\boldsymbol{c}$ is symmetric; the generalization to nonsymmetric tensor variables is straightforward.

For the energy functional, we assume the following generalization of (2.48),

$$E(x) = \int \left[\frac{\boldsymbol{M}(\boldsymbol{r})^2}{2\rho(\boldsymbol{r})} + \epsilon(\boldsymbol{r}) + \epsilon_{\rm c}(\rho(\boldsymbol{r}), \Lambda(\boldsymbol{r}), \boldsymbol{c}(\boldsymbol{r})) \right] d^3r. \tag{4.34}$$

The additional internal energy contribution $\epsilon_{\rm c}$ describes configurational effects on the interactions between particles. For simplicity, we assume that the total entropy can be written as the sum of two contributions,

$$S(x) = \int \left[s(\rho(\boldsymbol{r}), \epsilon(\boldsymbol{r})) + \rho(\boldsymbol{r})\, \hat{s}_{\rm c}(\Lambda(\boldsymbol{r}), \boldsymbol{c}(\boldsymbol{r})) \right] d^3r, \tag{4.35}$$

where s is the entropy density typical for a low-molecular-weight, Newtonian fluid as previously encountered in (2.52), and $\hat{s}_{\rm c}$ is the configurational entropy per unit mass for a complex fluid with additional structural variables. The crucial features of (4.35) thus are

- that the entropy density is independent of the local fluid velocity,
- that there is a "local" entropy contribution s independent of the configurational aspects represented by Λ and $\boldsymbol{c}$ so that a background temperature can be defined,[18] and

[18] In many cases, one does not specify $s(\rho, \epsilon)$ explicitly because one is interested in incompressible fluids under isothermal flow conditions.

- that the extensive nature of the entropy is taken into account.

It is natural to make an analogous assumption $\epsilon_c(\rho, \Lambda, \boldsymbol{c}) = \rho\hat{\epsilon}_c(\Lambda, \boldsymbol{c})$ for the extensive configurational energy.

Exercise 59 Gradients of Energy and Entropy
Calculate the gradients of the energy (4.34) and the entropy (4.35).

4.2.2 Poisson Matrix

To formulate the Poisson matrix, we first provide the convection mechanisms for the configurational variables $\boldsymbol{c}$ and Λ. For the conformation tensor $\boldsymbol{c}$, we assume upper-convected behavior (see Section 2.3.1),

$$\dot{\boldsymbol{c}}_{\text{convect}} = -\boldsymbol{v} \cdot \frac{\partial \boldsymbol{c}}{\partial \boldsymbol{r}} + \boldsymbol{\kappa} \cdot \boldsymbol{c} + \boldsymbol{c} \cdot \boldsymbol{\kappa}^T, \tag{4.36}$$

where, as before, $\kappa_{jk} = \partial v_j / \partial r_k$ represents the velocity gradient tensor. The choice of upper-convected behavior is without loss of generality because we could equally well choose $\boldsymbol{c}^{-1}$, which possesses lower-convected behavior, as an independent variable, so that the two fundamental possibilities are covered (see Section 2.3.1 or the discussion of entrainment laws in Appendix B.3). For the scalar variable Λ, we assume a very general coupling of the velocity gradients and the configurational variables,

$$\dot{\Lambda}_{\text{convect}} = -\boldsymbol{v} \cdot \frac{\partial \Lambda}{\partial \boldsymbol{r}} + \boldsymbol{\kappa} : \boldsymbol{g}(\Lambda, \boldsymbol{c}), \tag{4.37}$$

where $\boldsymbol{g}$ is an arbitrary tensor function of Λ and $\boldsymbol{c}$. If Λ should naturally occur as a scalar density rather than the assumed scalar, such as the number density of polymer molecules in a polymer solution, then Λ/ρ should be taken as the proper scalar variable Λ. The linear occurrence of $\boldsymbol{\kappa}$ in (4.36) and (4.37) is motivated by the Lie-Poisson reduction procedure for constructing Poisson brackets (see Appendix B.3). The detailed functional form of $\boldsymbol{g}$ can be restricted by the Jacobi identity. According to the Cayley-Hamilton theorem, we can write the most general expression for $\boldsymbol{g}$ in terms of three tensors as

$$\boldsymbol{g} = g_1 \boldsymbol{c} + g_2 \boldsymbol{1} + g_3 \boldsymbol{c}^{-1}, \tag{4.38}$$

with the scalar coefficient functions

$$g_j = g_j(\Lambda, I_1, I_2, I_3) \tag{4.39}$$

depending on Λ and the three independent invariants

$$I_1 = \text{tr}\boldsymbol{c}, \tag{4.40}$$

$$I_2 = \ln \det \boldsymbol{c}, \tag{4.41}$$

$$I_3 = -\text{tr}\boldsymbol{c}^{-1}. \tag{4.42}$$

The choice of the three powers of $\boldsymbol{c}$ in (4.38) is such that the symmetry between upper- and lower-convected behavior is reflected. Moreover, to check the Jacobi identity, it is very convenient to have the following property of the chosen invariants (see Exercise 60),

$$\frac{\partial I_j}{\partial \boldsymbol{c}} = \boldsymbol{c}^{1-j}. \tag{4.43}$$

In (4.41), we have implicitly assumed that the tensor $\boldsymbol{c}$ has a positive determinant.

The Poisson matrix L, or the equivalent Poisson bracket (1.6), can now be formulated by a straightforward procedure (see Section 2.3.1); the final result is

$$\begin{aligned}
\{A, B\} = & - \int \rho \left(\frac{\delta A}{\delta M_j} \frac{\partial}{\partial r_j} \frac{\delta B}{\delta \rho} - \frac{\delta B}{\delta M_j} \frac{\partial}{\partial r_j} \frac{\delta A}{\delta \rho} \right) d^3 r \\
& - \int M_k \left(\frac{\delta A}{\delta M_j} \frac{\partial}{\partial r_j} \frac{\delta B}{\delta M_k} - \frac{\delta B}{\delta M_j} \frac{\partial}{\partial r_j} \frac{\delta A}{\delta M_k} \right) d^3 r \\
& - \int \epsilon \left(\frac{\delta A}{\delta M_j} \frac{\partial}{\partial r_j} \frac{\delta B}{\delta \epsilon} - \frac{\delta B}{\delta M_j} \frac{\partial}{\partial r_j} \frac{\delta A}{\delta \epsilon} \right) d^3 r \\
& - \int \pi_{jk}^{(S)} \left(\frac{\delta A}{\delta \epsilon} \frac{\partial}{\partial r_j} \frac{\delta B}{\delta M_k} - \frac{\delta B}{\delta \epsilon} \frac{\partial}{\partial r_j} \frac{\delta A}{\delta M_k} \right) d^3 r \\
& + \int c_{jk} \frac{\partial}{\partial r_l} \left(\frac{\delta A}{\delta c_{jk}} \frac{\delta B}{\delta M_l} - \frac{\delta B}{\delta c_{jk}} \frac{\delta A}{\delta M_l} \right) d^3 r \\
& + \int c_{lk} \left(\frac{\delta A}{\delta c_{jk}} \frac{\partial}{\partial r_l} \frac{\delta B}{\delta M_j} - \frac{\delta B}{\delta c_{jk}} \frac{\partial}{\partial r_l} \frac{\delta A}{\delta M_j} \right) d^3 r \\
& + \int c_{jl} \left(\frac{\delta A}{\delta c_{jk}} \frac{\partial}{\partial r_l} \frac{\delta B}{\delta M_k} - \frac{\delta B}{\delta c_{jk}} \frac{\partial}{\partial r_l} \frac{\delta A}{\delta M_k} \right) d^3 r \\
& - \int \Lambda \frac{\partial}{\partial r_j} \left(\frac{\delta A}{\delta M_j} \frac{\delta B}{\delta \Lambda} - \frac{\delta B}{\delta M_j} \frac{\delta A}{\delta \Lambda} \right) d^3 r \\
& + \int g_{jk} \left(\frac{\delta A}{\delta \Lambda} \frac{\partial}{\partial r_j} \frac{\delta B}{\delta M_k} - \frac{\delta B}{\delta \Lambda} \frac{\partial}{\partial r_j} \frac{\delta A}{\delta M_k} \right) d^3 r.
\end{aligned} \tag{4.44}$$

The first four integrals represent the usual Poisson bracket of hydrodynamics previously considered in Exercise 24, where a thermodynamic pressure tensor with components $\pi_{jk}^{(S)}$ is introduced instead of a scalar pressure. A pressure tensor is actually needed in the presence of the additional configurational variables to fulfill the degeneracy condition (1.4), which is found to be (see Exercise 61)

$$\boldsymbol{\pi}^{(S)} = T \left[\left(s - \rho \frac{\partial s}{\partial \rho} - \epsilon \frac{\partial s}{\partial \epsilon} \right) \mathbf{1} + \rho \left(2\boldsymbol{c} \cdot \frac{\partial \hat{s}_{\mathrm{c}}}{\partial \boldsymbol{c}} + \boldsymbol{g} \frac{\partial \hat{s}_{\mathrm{c}}}{\partial \Lambda} \right) \right], \tag{4.45}$$

with

$$\frac{1}{T} = \frac{1}{T(\rho, \epsilon)} = \frac{\partial s(\rho, \epsilon)}{\partial \epsilon}. \tag{4.46}$$

The pressure-tensor expression (4.45) is a very important first result of beyond-equilibrium thermodynamics. The integrals five through seven in (4.44) represent the upper-convected behavior (4.36) previously formulated in (2.92), whereas the last two integrals represent the convection behavior (4.37) of Λ. If we impose the requirement that the bracket defined in (4.44) fulfills the Jacobi identity (1.9), then one obtains the following set of conditions for the scalar coefficient functions in (4.38),

$$g_1 \frac{\partial g_2}{\partial \Lambda} - g_2 \frac{\partial g_1}{\partial \Lambda} = 2 \left(\frac{\partial g_1}{\partial I_2} - \frac{\partial g_2}{\partial I_1} \right), \tag{4.47}$$

$$g_1 \frac{\partial g_3}{\partial \Lambda} - g_3 \frac{\partial g_1}{\partial \Lambda} = 2 \left(\frac{\partial g_1}{\partial I_3} - \frac{\partial g_3}{\partial I_1} \right), \tag{4.48}$$

$$g_2 \frac{\partial g_3}{\partial \Lambda} - g_3 \frac{\partial g_2}{\partial \Lambda} = 2 \left(\frac{\partial g_2}{\partial I_3} - \frac{\partial g_3}{\partial I_2} \right). \tag{4.49}$$

A direct evaluation of these conditions implied by the Jacobi identity is extremely time consuming. The particular choice of $\boldsymbol{\pi}^{(S)}$ in (4.45) simplifies the calculation considerably; when the internal energy density in the list of variables is replaced by the total entropy density, $s + \rho \hat{s}_{\rm c}$, the corresponding transformation of the Poisson bracket eliminates the occurrence of $\boldsymbol{\pi}^{(S)}$ from the bracket when $\boldsymbol{\pi}^{(S)}$ is chosen according to (4.45). In general, it is recommended to check the Jacobi identity by means of symbolic mathematical computation. A useful symbolic code for that purpose exists in the literature.[19] For an alternative derivation of the Jacobi identity and a more direct physical interpretation, one can construct Λ from a scalar variable that is convected independently of $\boldsymbol{c}$ and the invariants of $\boldsymbol{c}$ (see Exercise 62).

The cyclic occurrence of the indices in (4.47)–(4.49) is remarkable. If only one of the coefficients is nonzero, say g_1, then the left-hand sides of (4.47)–(4.49) vanish, and $g_1 = g_1(\Lambda, I_1)$ cannot depend on I_2 or I_3. Only if more than one of the coefficient functions g_j are nonzero, they can depend on several of the invariants. If one of the coefficients is zero, say g_1, then the other two coefficients must be independent of the corresponding invariant I_1. These observations can be understood directly in terms of the results of Exercise 62.

Exercise 60 Derivatives of Invariants

Prove the formula (4.43) for the derivatives of the invariants of $\boldsymbol{c}$.

Exercise 61 Derivation of Pressure Tensor

Show that the pressure tensor (4.45) is a consequence of the degeneracy requirement (1.4).

Exercise 62 Jacobi Identity by Transformation

For a given scalar variable $\hat{s}$ with

$$\dot{\hat{s}}_{\rm convect} = -\boldsymbol{v} \cdot \frac{\partial \hat{s}}{\partial \boldsymbol{r}}, \tag{4.50}$$

consider the convection behavior of the variable Λ obtained by the transformation $\Lambda = \Lambda(\hat{s}, I_1, I_2, I_3)$. Determine the functions g_j introduced in (4.38) in terms of the function $\hat{s}(\Lambda, I_1, I_2, I_3)$. Verify the conditions (4.47)–(4.49) for an arbitrary choice of $\hat{s}(\Lambda, I_1, I_2, I_3)$.

[19] Kröger, Hütter & Öttinger, Comput. Phys. Commun. 137 (2001) 325.

4.2.3 Friction Matrix

There are a large number of effects that should be incorporated through the friction matrix. Of course, we must include the shear and bulk viscosity effects as well as the thermal conductivity of hydrodynamics into the friction matrix. We also consider the relaxation and diffusion effects for the configurational variables. Finally, we introduce the slip coefficient leading to Schowalter's derivative, which provides an example of odd dressed Onsager-Casimir symmetry for a contribution to the friction matrix.

The transport coefficients *viscosity and thermal conductivity* are associated with two of the standard hydrodynamic variables, $\boldsymbol{M}(\boldsymbol{r})$ and $\epsilon(\boldsymbol{r})$. We hence consider only the 2×2 block of the friction matrix associated with these components; all the other entries of the friction matrix are zero, including those associated with the other hydrodynamic variable, $\rho(\boldsymbol{r})$ [see, however, Section 2.2.5 for a natural generalization of standard hydrodynamics]. From the friction matrix (2.66) of hydrodynamics, we have

$$M^{\text{hyd}} = \begin{pmatrix} M_{22} & M_{23} \\ M_{32} & \frac{\eta T}{2}\dot{\boldsymbol{\gamma}} : \dot{\boldsymbol{\gamma}} + \frac{\hat{\kappa} T}{4}(\text{tr}\dot{\boldsymbol{\gamma}})^2 - \frac{\partial}{\partial \boldsymbol{r}} \cdot \boldsymbol{\lambda}^{\text{q}} \cdot T^2 \frac{\partial}{\partial \boldsymbol{r}} \end{pmatrix}, \tag{4.51}$$

where the expressions for M_{22}, M_{23}, and M_{32} were previously given in (2.67)–(2.69), and M_{33} has been generalized by introducing the thermal conductivity tensor $\boldsymbol{\lambda}^{\text{q}}$. As before, the transport coefficient $\hat{\kappa}$ is the combination of the viscosity η and the dilatational viscosity κ introduced in (2.63), and $\dot{\boldsymbol{\gamma}}$ is the symmetrized velocity gradient tensor (2.62). For nonnegative η, κ and a positive semidefinite tensor $\boldsymbol{\lambda}^{\text{q}}$, M^{hyd} is a symmetric and positive semidefinite matrix satisfying the degeneracy condition (1.5).

For modeling the *relaxation behavior of the configurational variables* we consider the 3×3 block of the friction matrix associated with the components $(\epsilon(\boldsymbol{r}), \boldsymbol{c}(\boldsymbol{r}), \Lambda(\boldsymbol{r}))$. As a symmetric matrix, this block has six independent entries, some of which are tensors, that are further restricted by three conditions implied by the degeneracy requirement (1.5). We therefore keep three different relaxation mechanisms. Two of them represent the independent relaxation mechanisms for $\boldsymbol{c}$ and Λ, the other one is a coupled relaxation of these two structural variables in situations where they are not fully independent. Following the modeling approach discussed around (2.100), we hence write

$$M^{\text{relax}} = M_{\Lambda}^{\text{relax}} + M_{c}^{\text{relax}} + M_{\text{coupled}}^{\text{relax}}, \tag{4.52}$$

where the independent relaxation of Λ is expressed in the product form

$$M_{\Lambda}^{\text{relax}} = \begin{pmatrix} -\frac{\partial \epsilon_c}{\partial \Lambda} \\ 0 \\ 1 \end{pmatrix} \cdot R_1 \begin{pmatrix} -\frac{\partial \epsilon_c}{\partial \Lambda} & 0 & 1 \end{pmatrix} = \begin{pmatrix} \frac{\partial \epsilon_c}{\partial \Lambda} R_1 \frac{\partial \epsilon_c}{\partial \Lambda} & 0 & -\frac{\partial \epsilon_c}{\partial \Lambda} R_1 \\ 0 & 0 & 0 \\ -R_1 \frac{\partial \epsilon_c}{\partial \Lambda} & 0 & R_1 \end{pmatrix}. \tag{4.53}$$

When R_1 is a nonnegative scalar, then $M_{\Lambda}^{\text{relax}}$ is a manifestly symmetric and positive semidefinite matrix satisfying the degeneracy condition (1.5). In an analogous way, the independent relaxation of $\boldsymbol{c}$ is expressed as

$$
\begin{aligned}
M_c^{\text{relax}} &= \begin{pmatrix} -\frac{\partial \epsilon_c}{\partial \boldsymbol{c}} \\ 1 \\ 0 \end{pmatrix} \cdot \boldsymbol{R}_2 \cdot \begin{pmatrix} -\frac{\partial \epsilon_c}{\partial \boldsymbol{c}} & 1 & 0 \end{pmatrix} \\
&= \begin{pmatrix} \frac{\partial \epsilon_c}{\partial \boldsymbol{c}} : \boldsymbol{R}_2 : \frac{\partial \epsilon_c}{\partial \boldsymbol{c}} & -\frac{\partial \epsilon_c}{\partial \boldsymbol{c}} : \boldsymbol{R}_2 & 0 \\ -\boldsymbol{R}_2 : \frac{\partial \epsilon_c}{\partial \boldsymbol{c}} & \boldsymbol{R}_2 & 0 \\ 0 & 0 & 0 \end{pmatrix},
\end{aligned} \tag{4.54}
$$

where $\boldsymbol{R}_2$ is a symmetric, positive semidefinite fourth-rank tensor; that is, $\boldsymbol{c} : \boldsymbol{R}_2 : \boldsymbol{c}' = \boldsymbol{c}' : \boldsymbol{R}_2 : \boldsymbol{c}$ and $\boldsymbol{c} : \boldsymbol{R}_2 : \boldsymbol{c} \geq 0$ for all second-rank tensors $\boldsymbol{c}$ and $\boldsymbol{c}'$. Note that the matrix multiplications in (4.54) involve double contractions whenever second-rank and fourth-rank tensors are multiplied. Finally, we describe the coupled relaxation by a second-rank coupling tensor $\boldsymbol{\Phi}$ and another symmetric, positive semidefinite fourth-rank tensor, $\boldsymbol{R}_3$,

$$
M_{\text{coupled}}^{\text{relax}} = \begin{pmatrix} -[\frac{\partial \epsilon_c}{\partial \boldsymbol{c}} + \frac{\partial \epsilon_c}{\partial \Lambda} \boldsymbol{\Phi}] \\ 1 \\ \boldsymbol{\Phi} \end{pmatrix} \cdot \boldsymbol{R}_3 \cdot \begin{pmatrix} -[\frac{\partial \epsilon_c}{\partial \boldsymbol{c}} + \frac{\partial \epsilon_c}{\partial \Lambda} \boldsymbol{\Phi}] & 1 & \boldsymbol{\Phi} \end{pmatrix}. \tag{4.55}
$$

The functional form of the various quantities (R_1, $\boldsymbol{R}_2$, $\boldsymbol{R}_3$, $\boldsymbol{\Phi}$) introduced to describe relaxation can be motivated by some plausible arguments. From the Green-Kubo-type formula for the friction matrix, previously considered in (3.47) and to be derived in Chapter 6, it is expected that the friction matrix multiplied by Boltzmann's constant is quadratic in the configurational variables, where this quadratic expression is divided by a characteristic relaxation time. The same conclusion could be reached by dimensional arguments. Further dependencies on configurational variables should be through dimensionless combinations of structural variables. Note that, if the relaxation results from a diffusion process in some configuration space, the dimensional arguments change. For example, if the tensor $\boldsymbol{c}$ is the second moment of some vector, dimensional arguments imply that the quadratic dependence of the friction matrix is changed into a linear dependence on $\boldsymbol{c}$ when relaxation occurs through noise on the connector vector. By similar arguments, we expect the coupling tensor $\boldsymbol{\Phi}$ to be a dimensionless multiple of $\Lambda\, \boldsymbol{c}^{-1}$.

The above considerations on functional dependencies should be recognized to be as nonrigorous as they are useful. Note that the arguments are not invariant under nonlinear transformations of the variables. It is hence important to find the appropriate or natural variables for a given problem before trying to apply such heuristic arguments.

The relaxation of $\boldsymbol{c}$ and Λ provides examples of force-flux pairs, which, in contrast to those for transport processes (see Table 3.1), do not involve spatial derivatives. By identifying the mechanical contribution C_M in (3.3) from (4.53), with a convenient choice of sign, and using the entropy gradient of Exercise 59, we obtain

$$
X_\Lambda^{\text{relax}} = \begin{pmatrix} -\frac{\partial \epsilon_c}{\partial \Lambda} & 0 & 1 \end{pmatrix} \cdot \begin{pmatrix} \frac{\partial s}{\partial \epsilon} \\ \rho \frac{\partial \hat{s}_c}{\partial \boldsymbol{c}} \\ \rho \frac{\partial \hat{s}_c}{\partial \Lambda} \end{pmatrix} = -\frac{1}{T} \frac{\partial f_c}{\partial \Lambda}, \tag{4.56}
$$

and similarly from (4.54),

$$X_c^{\rm relax} = \begin{pmatrix} -\frac{\partial \epsilon_c}{\partial \boldsymbol{c}} & 1 & 0 \end{pmatrix} \cdot \begin{pmatrix} \frac{\partial s}{\partial \epsilon} \\ \rho \frac{\partial \hat{s}_c}{\partial \boldsymbol{c}} \\ \rho \frac{\partial \hat{s}_c}{\partial \Lambda} \end{pmatrix} = -\frac{1}{T} \frac{\partial f_c}{\partial \boldsymbol{c}}, \tag{4.57}$$

with the *configurational Helmholtz free energy density*

$$f_c(\rho, \epsilon, \Lambda, \boldsymbol{c}) = \epsilon_c(\rho, \Lambda, \boldsymbol{c}) - T(\rho, \epsilon)\rho \hat{s}_c(\Lambda, \boldsymbol{c}). \tag{4.58}$$

The forces $X_\Lambda^{\rm relax}$ and $X_c^{\rm relax}$ are thus obtained as the conjugate partners of Λ and $\boldsymbol{c}$ by differentiating the Helmholtz free energy density. According to (3.6), the corresponding fluxes are given by the constitutive equations

$$J_\Lambda^{\rm relax} = -\frac{R_1}{T} \frac{\partial f_c}{\partial \Lambda}, \tag{4.59}$$

and

$$J_c^{\rm relax} = -\frac{\boldsymbol{R}_2}{T} : \frac{\partial f_c}{\partial \boldsymbol{c}}, \tag{4.60}$$

where (4.55) would introduce additional cross effects,

$$J_\Lambda^{\rm relax} = -\frac{R_1}{T} \frac{\partial f_c}{\partial \Lambda} - \boldsymbol{\Phi} : \frac{\boldsymbol{R}_3}{T} : \left(\frac{\partial f_c}{\partial \boldsymbol{c}} + \boldsymbol{\Phi} \frac{\partial f_c}{\partial \Lambda} \right), \tag{4.61}$$

and

$$J_c^{\rm relax} = -\frac{\boldsymbol{R}_2}{T} : \frac{\partial f_c}{\partial \boldsymbol{c}} - \frac{\boldsymbol{R}_3}{T} : \left(\frac{\partial f_c}{\partial \boldsymbol{c}} + \boldsymbol{\Phi} \frac{\partial f_c}{\partial \Lambda} \right). \tag{4.62}$$

The forces in (4.56) and (4.57) are obtained as gradients of the Helmholtz free energy, and the fluxes (4.61) and (4.62) are the relaxational contributions to $d\Lambda/dt$ and $d\boldsymbol{c}/dt$, as we shall see in (4.78) and (4.77).

For modeling *diffusion*, we again focus on the 3×3 block of the friction matrix associated with the components $(\epsilon(\boldsymbol{r}), \boldsymbol{c}(\boldsymbol{r}), \Lambda(\boldsymbol{r}))$. We assume that Λ is nonnegative. In contrast to what we did for relaxation, we now try to formulate a single fundamental diffusion mechanism of the physical objects described by the structural variables. We write

$$M^{\rm diff} = M_\Lambda^{\rm diff} + M_c^{\rm diff}, \tag{4.63}$$

with

$$M_c^{\rm diff} = \begin{pmatrix} \frac{\partial \epsilon_c}{\partial \boldsymbol{c}} \frac{\tilde{\phi}}{\rho} \frac{\partial}{\partial \boldsymbol{r}} \\ -\frac{\tilde{\phi}}{\rho} \frac{\partial}{\partial \boldsymbol{r}} \\ 0 \end{pmatrix} \cdot \frac{\Lambda \rho}{\tilde{\phi}^2} \boldsymbol{D}' \tilde{\boldsymbol{M}} \cdot \begin{pmatrix} -\frac{\partial}{\partial \boldsymbol{r}} \frac{\tilde{\phi}}{\rho} \frac{\partial \epsilon_c}{\partial \boldsymbol{c}} & \frac{\partial}{\partial \boldsymbol{r}} \frac{\tilde{\phi}}{\rho} & 0 \end{pmatrix}, \tag{4.64}$$

and

$$\begin{aligned} M_\Lambda^{\rm diff} = & \begin{pmatrix} -\frac{\partial \epsilon_c}{\partial \boldsymbol{c}} : \frac{\partial \boldsymbol{c}}{\partial \boldsymbol{r}} \frac{1}{\tilde{\phi}\rho} \frac{d\tilde{\phi}}{d\Lambda} + \frac{\partial \epsilon_c}{\partial \Lambda} \frac{1}{\rho} \frac{\partial}{\partial \boldsymbol{r}} \\ \frac{\partial \boldsymbol{c}}{\partial \boldsymbol{r}} \frac{1}{\tilde{\phi}\rho} \frac{d\tilde{\phi}}{d\Lambda} \\ -\frac{1}{\rho} \frac{\partial}{\partial \boldsymbol{r}} \end{pmatrix} \cdot \Lambda \rho \boldsymbol{D}' \\ & \cdot \begin{pmatrix} -\left(\frac{1}{\tilde{\phi}\rho} \frac{d\tilde{\phi}}{d\Lambda} \frac{\partial \boldsymbol{c}}{\partial \boldsymbol{r}} : \frac{\partial \epsilon_c}{\partial \boldsymbol{c}} + \frac{\partial}{\partial \boldsymbol{r}} \frac{1}{\rho} \frac{\partial \epsilon_c}{\partial \Lambda} \right) & \frac{1}{\tilde{\phi}\rho} \frac{d\tilde{\phi}}{d\Lambda} \frac{\partial \boldsymbol{c}}{\partial \boldsymbol{r}} & \frac{\partial}{\partial \boldsymbol{r}} \frac{1}{\rho} \end{pmatrix}. \end{aligned} \tag{4.65}$$

The positive semidefinite symmetric tensor $\boldsymbol{D}'$ describes the fundamental diffusion behavior; in all matrix products, it is contracted from both sides with the derivatives with respect to $\boldsymbol{r}$. The scalar function $\tilde{\phi} = \tilde{\phi}(\Lambda)$ can be used to introduce a proper coupling of $\boldsymbol{c}$ and Λ when the entropy terms depending on $\boldsymbol{c}$ have an additional multiplicative dependence on Λ. The most natural choices are

$$\tilde{\phi} = \frac{1}{\Lambda} \qquad \text{or} \qquad \tilde{\phi} = 1. \tag{4.66}$$

The symmetric, positive semidefinite fourth-rank tensor $\tilde{\boldsymbol{M}}$ serves for proper coupling of the fundamental diffusion mechanism to the tensor $\boldsymbol{c}$; for dimensional reasons, the most natural choice is

$$\tilde{M}_{ijkl} = \frac{1}{\Lambda^2}(c_{ik}c_{lj} + c_{jk}c_{li}). \tag{4.67}$$

The product form of the contributions to the friction matrix in (4.64) and (4.65) is chosen such that the degeneracy, symmetry, and positive semidefiniteness of the friction matrix is obvious. A strong coupling of the diffusion effects on different variables is suggested by our experience with hydrodynamics in (2.77).

So far, we have assumed that the time-evolution equation for the tensor $\boldsymbol{c}$ should contain upper-convected time derivatives; a tensor with lower-convected behavior could be obtained by using its inverse as the structural variable $\boldsymbol{c}$. However, as discussed in Section 2.3.1, mixed derivatives are incompatible with the Jacobi identity when we use Poisson operators that vanish outside the velocity row or column. This should not be surprising from a physical point of view because the slip parameter ξ occurring in mixed or *Schowalter derivatives* (2.93) should not appear in a Poisson bracket. The more natural way for such a dynamic material parameter to occur is through the friction matrix. Mixed derivatives are not expected to lead to entropy production.[20] We hence need to implement them through an antisymmetric contribution to the friction matrix.

Because we want to couple the velocity field and the tensor variable while keeping an eye on the entropy production, the relevant variables to discuss slip effects are $\boldsymbol{M}(\boldsymbol{r})$, $\epsilon(\boldsymbol{r})$, and $\boldsymbol{c}(\boldsymbol{r})$. Slip effects can be described by the following antisymmetric contribution to the M matrix:[21]

$$M^{\text{slip}} = \begin{pmatrix} 0 & -2\frac{\partial}{\partial \boldsymbol{r}} \cdot \boldsymbol{c} \cdot \frac{\partial \epsilon_c}{\partial \boldsymbol{c}} T\xi & \boldsymbol{M}^{\text{slip}}_{Mc} \\ -2T\xi \frac{\partial \epsilon_c}{\partial \boldsymbol{c}} \cdot \boldsymbol{c} \cdot \frac{\partial}{\partial \boldsymbol{r}} & 0 & \frac{T\xi}{2}(\boldsymbol{c} \cdot \dot{\boldsymbol{\gamma}} + \dot{\boldsymbol{\gamma}} \cdot \boldsymbol{c}) \\ \boldsymbol{M}^{\text{slip}}_{cM} & -\frac{T\xi}{2}(\boldsymbol{c} \cdot \dot{\boldsymbol{\gamma}} + \dot{\boldsymbol{\gamma}} \cdot \boldsymbol{c}) & 0 \end{pmatrix}, \tag{4.68}$$

with

$$\left(\boldsymbol{M}^{\text{slip}}_{Mc}\right)_{jkl} = \left\{ \frac{\partial}{\partial r_k} c_{jl} + \frac{\partial}{\partial r_l} c_{kj} + \left(\frac{\partial}{\partial \boldsymbol{r}} \cdot \boldsymbol{c}\right)_k \delta_{jl} + \left(\frac{\partial}{\partial \boldsymbol{r}} \cdot \boldsymbol{c}\right)_l \delta_{kj} \right\} \frac{T\xi}{2} \tag{4.69}$$

[20] Wapperom & Hulsen, J. Rheol. 42 (1998) 999.

[21] Dressler, Edwards & Öttinger, Rheol. Acta 38 (1999) 117.

and

$$\left(\boldsymbol{M}_{cM}^{\text{slip}}\right)_{jkl} = \frac{T\xi}{2}\left\{c_{jl}\frac{\partial}{\partial r_k} + c_{lk}\frac{\partial}{\partial r_j} + \left(\boldsymbol{c}\cdot\frac{\partial}{\partial \boldsymbol{r}}\right)_j \delta_{lk} + \left(\boldsymbol{c}\cdot\frac{\partial}{\partial \boldsymbol{r}}\right)_k \delta_{jl}\right\}. \tag{4.70}$$

At this point we need to understand how antisymmetric contributions to the friction matrix arise through the symmetry arguments of Section 3.2.1. For our current variables of interest, $\boldsymbol{M}(\boldsymbol{r})$, $\epsilon(\boldsymbol{r})$, and $\boldsymbol{c}(\boldsymbol{r})$, the velocity and hence the momentum density change signs under time reversal, whereas the internal energy density and the configurational variables should not be affected. We hence expect the following bare symmetry properties of the subblock of the friction matrix associated with $\boldsymbol{M}(\boldsymbol{r})$, $\epsilon(\boldsymbol{r})$, and $\boldsymbol{c}(\boldsymbol{r})$,

$$\mathcal{OC}_{\text{bare}}(M^{\text{slip}}) = \begin{pmatrix} + & - & - \\ - & + & + \\ - & + & + \end{pmatrix}. \tag{4.71}$$

It should be noted that in some elements of the particular friction matrix (4.68) the velocity occurs explicitly through $\dot{\gamma}$. In the notation of Section 3.2.1, we thus obtain the "dressed" Onsager-Casimir symmetry

$$\mathcal{OC}_{\text{dressed}}(M^{\text{slip}}) = \begin{pmatrix} \bigcirc & \ominus & \ominus \\ \ominus & \bigcirc & \ominus \\ \ominus & \ominus & \bigcirc \end{pmatrix}. \tag{4.72}$$

As before, circled signs indicate the dressed symmetry properties, while open circles indicate zero entries that can be considered to be either symmetric or antisymmetric.

Note that, according to (3.51), the $(\boldsymbol{M},\epsilon)$ and $(\epsilon,\boldsymbol{M})$ components of the friction matrix for hydrodynamics have the dressed symmetry $\oplus$. As a result of adding hydrodynamic and slip contributions with different dressed symmetry properties, the $(\boldsymbol{M},\epsilon)$ and $(\epsilon,\boldsymbol{M})$ components of the friction matrix display neither symmetric nor antisymmetric behavior. As discussed in Section 3.2.1, this is not at variance with the usual idea of well-defined bare Onsager-Casimir symmetry properties, which is known as an important criterion for the proper phenomenological formulation or statistical derivation of cross effects, such as the Seebeck and Peltier effects of thermoelectricity (see Section 3.2.2) or the Soret and Dufour effects of thermodiffusion.[22] However, one should remember that it is the dressed Onsager-Casimir symmetry rather than the bare symmetry that matters for the entropy production associated with a certain effect.

Exercise 63 Force-Flux Pair for Diffusion

Derive the force-flux pair associated with the diffusion mechanism for $\tilde{\phi} = 1$ and $\tilde{\boldsymbol{M}} = 0$.

[22] García-Colín & Goldstein, Rev. Mex. Fís. 44 (1998) 420.

4.2.4 Time-Evolution Equations

By inserting all the building blocks developed in the preceding subsections into the GENERIC (1.1), we obtain the final time-evolution equations for a large class of models. The continuity equation takes the familiar form,

$$\frac{\partial \rho}{\partial t} = -\frac{\partial}{\partial \boldsymbol{r}} \cdot (\boldsymbol{v}\rho), \tag{4.73}$$

whereas a number of energetic and entropic pressure-tensor contributions appear in the momentum-balance equation,

$$\frac{\partial \boldsymbol{M}}{\partial t} = -\frac{\partial}{\partial \boldsymbol{r}} \cdot (\boldsymbol{v}\boldsymbol{M}) - \frac{\partial}{\partial \boldsymbol{r}} \cdot \left(\boldsymbol{\pi}^{(S)} + \boldsymbol{\pi}^{(E)} + \xi 2\boldsymbol{c} \cdot \frac{\partial f_c}{\partial \boldsymbol{c}} - \eta \dot{\boldsymbol{\gamma}} - \frac{\hat{\kappa}}{2} \mathbf{1} \mathrm{tr} \dot{\boldsymbol{\gamma}} \right), \tag{4.74}$$

with $\boldsymbol{\pi}^{(S)}$ given by (4.45) and the following pressure tensor contribution determined by the configurational energy,

$$\boldsymbol{\pi}^{(E)} = -\left(\epsilon_c - \rho \frac{\partial \epsilon_c}{\partial \rho} \right) \mathbf{1} - \left(2\boldsymbol{c} \cdot \frac{\partial \epsilon_c}{\partial \boldsymbol{c}} + \boldsymbol{g} \frac{\partial \epsilon_c}{\partial \Lambda} \right), \tag{4.75}$$

and the configurational Helmholtz free energy density (4.58).

The equation for the nonconfigurational internal energy density shows that only the entropic, but not the energetic, pressure-tensor contribution appears in the usual form,

$$\begin{aligned}
\frac{\partial \epsilon}{\partial t} &= -\frac{\partial}{\partial \boldsymbol{r}} \cdot (\boldsymbol{v}\epsilon) - \frac{1}{2} \left(\boldsymbol{\pi}^{(S)} - \xi 2T\rho\boldsymbol{c} \cdot \frac{\partial \hat{s}_c}{\partial \boldsymbol{c}} - \eta \dot{\boldsymbol{\gamma}} - \frac{\hat{\kappa}}{2} \mathbf{1} \mathrm{tr} \dot{\boldsymbol{\gamma}} \right) : \dot{\boldsymbol{\gamma}} \\
&+ \frac{\partial}{\partial \boldsymbol{r}} \cdot \boldsymbol{\lambda}^{\mathrm{q}} \cdot \frac{\partial T}{\partial \boldsymbol{r}} + \frac{\partial \epsilon_c}{\partial \Lambda} \frac{R_1}{T} \frac{\partial f_c}{\partial \Lambda} + \frac{\partial \epsilon_c}{\partial \boldsymbol{c}} : \frac{\boldsymbol{R}_2}{T} : \frac{\partial f_c}{\partial \boldsymbol{c}} \\
&+ \left(\frac{\partial \epsilon_c}{\partial \boldsymbol{c}} + \boldsymbol{\Phi} \frac{\partial \epsilon_c}{\partial \Lambda} \right) : \frac{\boldsymbol{R}_3}{T} : \left(\frac{\partial f_c}{\partial \boldsymbol{c}} + \boldsymbol{\Phi} \frac{\partial f_c}{\partial \Lambda} \right) \\
&- \frac{\partial \epsilon_c}{\partial \boldsymbol{c}} : \frac{\tilde{\phi}}{\rho} \frac{\partial}{\partial \boldsymbol{r}} \cdot \frac{\Lambda \rho}{\tilde{\phi}^2} \tilde{\boldsymbol{M}} \boldsymbol{D}' \cdot \frac{\partial}{\partial \boldsymbol{r}} \frac{\tilde{\phi}}{\rho T} : \frac{\partial f_c}{\partial \boldsymbol{c}} \\
&- \left(\frac{\partial \epsilon_c}{\partial \Lambda} \frac{1}{\rho} \frac{\partial}{\partial \boldsymbol{r}} - \frac{\partial \epsilon_c}{\partial \boldsymbol{c}} : \frac{\partial \boldsymbol{c}}{\partial \boldsymbol{r}} \frac{1}{\tilde{\phi}\rho} \frac{d\tilde{\phi}}{d\Lambda} \right) \cdot \Lambda \rho \boldsymbol{D}' \\
&\quad \cdot \left(\frac{\partial}{\partial \boldsymbol{r}} \frac{1}{\rho T} \frac{\partial f_c}{\partial \Lambda} + \frac{1}{\tilde{\phi}\rho T} \frac{d\tilde{\phi}}{d\Lambda} \frac{\partial \boldsymbol{c}}{\partial \boldsymbol{r}} : \frac{\partial f_c}{\partial \boldsymbol{c}} \right).
\end{aligned} \tag{4.76}$$

All the relaxation and diffusion mechanisms affecting the configurational variables contribute to the generation of internal energy.

For the configurational variables, we obtain the following convection-relaxation-diffusion equations:

$$\frac{\partial \boldsymbol{c}}{\partial t} = -\boldsymbol{v} \cdot \frac{\partial \boldsymbol{c}}{\partial \boldsymbol{r}} + \boldsymbol{\kappa} \cdot \boldsymbol{c} + \boldsymbol{c} \cdot \boldsymbol{\kappa}^T - \frac{\xi}{2} (\boldsymbol{c} \cdot \dot{\boldsymbol{\gamma}} + \dot{\boldsymbol{\gamma}} \cdot \boldsymbol{c})$$

$$
\begin{aligned}
&- \frac{\boldsymbol{R}_2}{T} : \frac{\partial f_c}{\partial \boldsymbol{c}} - \frac{\boldsymbol{R}_3}{T} : \left(\frac{\partial f_c}{\partial \boldsymbol{c}} + \boldsymbol{\Phi} \frac{\partial f_c}{\partial \Lambda} \right) \\
&+ \frac{\tilde{\phi}}{\rho} \frac{\partial}{\partial \boldsymbol{r}} \cdot \frac{\Lambda \rho}{\tilde{\phi}^2} \tilde{\boldsymbol{M}} \boldsymbol{D}' \cdot \frac{\partial}{\partial \boldsymbol{r}} \frac{\tilde{\phi}}{\rho T} : \frac{\partial f_c}{\partial \boldsymbol{c}} \\
&- \frac{\partial \boldsymbol{c}}{\partial \boldsymbol{r}} \cdot \frac{\Lambda}{\tilde{\phi}} \frac{d\tilde{\phi}}{d\Lambda} \boldsymbol{D}' \cdot \left(\frac{\partial}{\partial \boldsymbol{r}} \frac{1}{\rho T} \frac{\partial f_c}{\partial \Lambda} + \frac{1}{\tilde{\phi} \rho T} \frac{d\tilde{\phi}}{d\Lambda} \frac{\partial \boldsymbol{c}}{\partial \boldsymbol{r}} : \frac{\partial f_c}{\partial \boldsymbol{c}} \right), \qquad (4.77)
\end{aligned}
$$

and

$$
\begin{aligned}
\frac{\partial \Lambda}{\partial t} &= -\boldsymbol{v} \cdot \frac{\partial \Lambda}{\partial \boldsymbol{r}} + \boldsymbol{\kappa} : \boldsymbol{g} - \frac{R_1}{T} \frac{\partial f_c}{\partial \Lambda} - \boldsymbol{\Phi} : \frac{\boldsymbol{R}_3}{T} : \left(\frac{\partial f_c}{\partial \boldsymbol{c}} + \boldsymbol{\Phi} \frac{\partial f_c}{\partial \Lambda} \right) \\
&+ \frac{1}{\rho} \frac{\partial}{\partial \boldsymbol{r}} \cdot \Lambda \rho \boldsymbol{D}' \cdot \left(\frac{\partial}{\partial \boldsymbol{r}} \frac{1}{\rho T} \frac{\partial f_c}{\partial \Lambda} + \frac{1}{\tilde{\phi} \rho T} \frac{d\tilde{\phi}}{d\Lambda} \frac{\partial \boldsymbol{c}}{\partial \boldsymbol{r}} : \frac{\partial f_c}{\partial \boldsymbol{c}} \right). \qquad (4.78)
\end{aligned}
$$

For homogeneous flows, the time-evolution equations for the configurational variables become much simpler due to the absence of spatial derivatives. For example, if we decouple the tensor from the scalar variable ($\boldsymbol{R}_3 = 0$), (4.77) reads

$$
\frac{d\boldsymbol{c}}{dt} - \boldsymbol{\kappa} \cdot \boldsymbol{c} - \boldsymbol{c} \cdot \boldsymbol{\kappa}^T + \frac{\xi}{2} (\boldsymbol{c} \cdot \dot{\boldsymbol{\gamma}} + \dot{\boldsymbol{\gamma}} \cdot \boldsymbol{c}) = -\frac{\boldsymbol{R}_2}{T} : \frac{\partial f_c}{\partial \boldsymbol{c}}, \qquad (4.79)
$$

and, if the fourth-rank tensor $\boldsymbol{R}_2$ can be inverted, we arrive at

$$
\frac{\partial f_c}{\partial \boldsymbol{c}} = -T \boldsymbol{R}_2^{-1} : \left[\frac{d\boldsymbol{c}}{dt} - \boldsymbol{\kappa} \cdot \boldsymbol{c} - \boldsymbol{c} \cdot \boldsymbol{\kappa}^T + \frac{\xi}{2} (\boldsymbol{c} \cdot \dot{\boldsymbol{\gamma}} + \dot{\boldsymbol{\gamma}} \cdot \boldsymbol{c}) \right]. \qquad (4.80)
$$

For steady homogeneous flows, this result implies that we have an explicit expression for the conformation-tensor dependence of the Helmholtz free energy in terms of the velocity gradients and the friction matrix. Moreover, because (4.74) implies that the conformation tensor leads to the pressure tensor contribution

$$
\boldsymbol{\pi}_c = -2(1 - \xi) \boldsymbol{c} \cdot \frac{\partial f_c}{\partial \boldsymbol{c}}, \qquad (4.81)
$$

we get the following equivalent formula for the pressure tensor:

$$
\boldsymbol{\pi}_c = 2(1 - \xi) T \boldsymbol{c} \cdot \boldsymbol{R}_2^{-1} : \left[\frac{d\boldsymbol{c}}{dt} - \boldsymbol{\kappa} \cdot \boldsymbol{c} - \boldsymbol{c} \cdot \boldsymbol{\kappa}^T + \frac{\xi}{2} (\boldsymbol{c} \cdot \dot{\boldsymbol{\gamma}} + \dot{\boldsymbol{\gamma}} \cdot \boldsymbol{c}) \right]. \qquad (4.82)
$$

This type of formula is known as the *Giesekus expression* for the pressure tensor. Although (4.82) seems to be more complicated than (4.81), it has an important advantage: when evaluated at equilibrium, (4.81) gives the trivial result $\boldsymbol{\pi}_c = 0$, whereas (4.82) gives the linear term in the velocity gradients

$$
\boldsymbol{\pi}_c^{\text{lin}} = -2(1 - \xi)^2 T (\boldsymbol{R}_2^{\text{eq}})^{-1} : \dot{\boldsymbol{\gamma}}. \qquad (4.83)
$$

Equation (4.83) describes the linear response in the pressure tensor to a steady homogeneous flow. The pressure tensor and hence the transport coefficients are thus given in terms of the friction matrix at equilibrium. Any relationship of this type is commonly referred to as a *fluctuation-dissipation theorem*; $\pi_c^{\rm lin}$ is a dissipative contribution due to the relaxation of the conformation tensor and, as previously indicated in (3.47), $\boldsymbol{R}_2$ as part of the friction matrix can be expressed as an equilibrium average of fluctuations.

Actually, (4.83) is quite different from (1.57), which we previously referred to as a fluctuation-dissipation theorem. Following Kubo et al.,[23] we hence distinguish between fluctuation-dissipation theorems of the first and second kind, where the first kind expresses the linear response in terms of equilibrium correlations and the second kind provides a relationship between friction and noise. The Giesekus expression for the pressure tensor may thus be considered as a nonlinear generalization of the fluctuation-dissipation theorem of the first kind.

Exercise 64 Second-Rank Instead of Fourth-Rank Tensors

In many cases, the fourth-rank tensors $\boldsymbol{R}_2$ and $\boldsymbol{R}_3$ in (4.54) and (4.55) can equivalently be replaced by the second-rank tensors $\hat{\boldsymbol{R}}_2$ and $\hat{\boldsymbol{R}}_3$,

$$(\hat{\boldsymbol{R}}_2)_{jk} = \sum_l (\boldsymbol{R}_2)_{jkll}, \qquad (\hat{\boldsymbol{R}}_3)_{jk} = \sum_l (\boldsymbol{R}_3)_{jkll}, \tag{4.84}$$

with corresponding replacements of double contractions by single contractions. Discuss the generality of this replacement.

Exercise 65 Generalized Force-Flux Pair for Diffusion

Show that, if $\epsilon_c = 0$ and if $\tilde{\phi}$ can be chosen such that $\tilde{\phi}\partial \hat{s}_c/\partial \boldsymbol{c}$ is independent of Λ, one obtains the identity

$$\frac{\partial}{\partial \boldsymbol{r}} \frac{1}{\rho T} \frac{\partial f_c}{\partial \Lambda} + \frac{1}{\tilde{\phi}\rho T} \frac{d\tilde{\phi}}{d\Lambda} \frac{\partial \boldsymbol{c}}{\partial \boldsymbol{r}} : \frac{\partial f_c}{\partial \boldsymbol{c}} = -\frac{\partial^2 \hat{s}_c}{\partial \Lambda^2} \frac{\partial \Lambda}{\partial \boldsymbol{r}}, \tag{4.85}$$

which can be used to simplify the time-evolution equations (4.76)–(4.78) considerably. Discuss the modification of the force-flux pair of Exercise 63 suggested by this result.

4.2.5 Summary of Inputs and Implications

In this section, we consider a class of thermodynamically admissible models with scalar and tensorial configurational variables. Very general assumptions for the building blocks required when modeling within the GENERIC framework of nonequilibrium thermodynamics are made, which, in particular, allow us to introduce a background temperature. In addition to the transport coefficients of standard hydrodynamics, relaxation and diffusion of the configurational variables are constructed in a very general form, and Schowalter derivatives for the tensor variables are introduced. A particular thermodynamically admissible model with scalar and tensorial configurational variables, Λ and $\boldsymbol{c}$, is defined by specifying the following quantities:

- the configurational contribution to the energy, ϵ_c, in (4.34);

[23] See Section 1.6 of Kubo et al., *Statistical Physics II* (Springer, 1991).

- the background and configurational entropies, s and $\hat{s}_c$, in (4.35);
- the three scalar coefficient functions g_j in (4.38) describing the influence of the tensor $\boldsymbol{c}$ on the convection behavior of Λ;
- the transport coefficients of hydrodynamics: the viscosity η, the dilatational viscosity κ, and the thermal conductivity tensor $\boldsymbol{\lambda}^{\mathrm{q}}$;
- the scalar R_1 and the fourth-rank tensors $\boldsymbol{R}_2$ and $\boldsymbol{R}_3$, as well as the second-rank coupling tensor $\boldsymbol{\Phi}$, in (4.53)–(4.55) to describe configurational relaxation; in many cases, $\boldsymbol{R}_2$ and $\boldsymbol{R}_3$ can equivalently be replaced by second-rank tensors;
- the diffusion tensor $\boldsymbol{D}'$ together with the coupling quantities $\tilde{\phi}$ and $\tilde{\boldsymbol{M}}$;
- the slip parameter ξ describing mixed time derivatives of the tensorial structural variable.

The GENERIC formalism imposes a number of thermodynamic consistency criteria for the model parameters:

- the functions g_j must be chosen such that the conditions (4.47)–(4.49) are fulfilled;
- the elastic pressure tensor with components $\pi_{jk}^{(S)}$ must be of the form (4.45);
- the shear and dilatational viscosities must be nonnegative, and the thermal conductivity tensor must be a positive semidefinite symmetric tensor;
- the scalar R_1 must be nonnegative and the tensors $\boldsymbol{R}_2$ and $\boldsymbol{R}_3$ must be symmetric and positive semidefinite; simple arguments for the choice of the functional form of R_1, $\boldsymbol{R}_2$, $\boldsymbol{R}_3$, and $\boldsymbol{\Phi}$ have been discussed;
- the diffusion tensor $\boldsymbol{D}'$ and the fourth-rank tensor $\tilde{\boldsymbol{M}}$ must be symmetric and positive semidefinite; particularly natural choices for $\tilde{\phi}$ and $\tilde{\boldsymbol{M}}$ have been proposed in (4.66) and (4.67).

Various generalizations of the class of models considered here are straightforward. Multimode models can be obtained by using several structural variables of the tensorial type. Models with two coupled tensor variables have also been motivated by abstract orientation and elongation tensors[24] or by the successful fitting of rheological data.[25] For example in modeling phase transitions through nucleation and growth under flow, it might be useful and natural to use a tensor variable together with several coupled scalar variables. Even without these generalizations, the class of thermodynamically admissible models introduced here cover a wide range of interesting fluids.

[24] Wilchinsky et al., J. Non-Newtonian Fluid Mech. 99 (2001) 57.
[25] Edwards et al., J. Rheol. 40 (1996) 917.

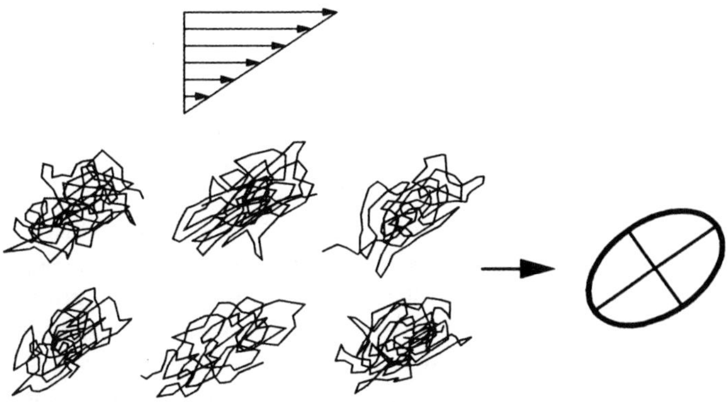

Fig. 4.7 Under flow conditions, such as the indicated shear flow, polymer molecules in solution become stretched and oriented. In a crude phenomenological model, the ensemble of possible polymer conformations is replaced by a tensor describing the overall stretch and orientation, as represented by the ellipse.

By decoupling the scalar structural variable from all the other state variables, we can reproduce a large number of successful rheological models that use only a tensor variable to describe the nature of a complex fluid (in particular, we recover all the models mentioned on p. 111). We next discuss two examples of models that make full use of the additional scalar variable (dilute polymer solutions, pompon model for melts of multiply branched polymers); a third example can be found in the literature (Doi-Ohta model for emulsions and polymer blends).[26] In all cases, we do not specify the background entropy, and the usual transport coefficients of hydrodynamics are neglected.

4.2.6 Example: Dilute Polymer Solutions

For inhomogeneous, dilute polymer solutions, we use the small mole fraction of polymer molecules as the scalar variable Λ and a conformation tensor characterizing the average stretch and orientation of the polymer molecules as the tensor variable $\boldsymbol{c}$ (see Figure 4.7). We assume that the tensor variable describes relative stretch and orientation; that is, $\boldsymbol{c}$ is normalized such that $\boldsymbol{c}_{\text{eq}} = \mathbf{1}$.

After making the fundamental decision about the structural variables, we assume that there is no configurational contribution to energy; interesting rheological behavior can result from the mere entropic effects associated with stretching and orienting the chains. A deformation of the randomly coiled equilibrium configurations leads to a reduction of the number of accessible configurations and hence to a loss of entropy.

[26] Öttinger, Rev. Mex. Fís. 48 (2002) Supl. 1, 220.

For the configurational entropy contribution, we use the expression

$$\hat{s}_c(\Lambda, \boldsymbol{c}) = \frac{k_B}{m}\Lambda\left\{-\ln\Lambda + \frac{1}{2}[\mathrm{tr}(\mathbf{1} - \boldsymbol{c}) + \ln(\det\boldsymbol{c})]\right\}, \tag{4.86}$$

where m is the mass of a solvent particle. The term involving $-\Lambda\ln\Lambda$ describes the preference for a uniform polymer distribution in space. It can be obtained from the entropy of mixing given in Exercise 19 in the limit of dilute solutions. For cases of the functional form of the entropy depending on a tensor, we have no experience yet. The simplest possibility would be a quadratic expansion around equilibrium, that is, a term proportional to $-(\mathbf{1}-\boldsymbol{c}):(\mathbf{1}-\boldsymbol{c})$. However, if we do not want to be restricted to near-equilibrium or weak-flow situations, we need a more general, nonpertubative argument. The expression in (4.86),[27] which is useful in many applications, is based on an entropy term of the functional form $-\int f\ln f$ for probability densities, where f is the centered Gaussian distribution fully characterized by the symmetric and positive-definite tensor $\boldsymbol{c}$ (see Exercise 66, and Exercise 151 for further justification).

Exercise 66 Form of Entropy Depending on Tensor
Calculate $-k_B\int f(\boldsymbol{x})\ln[f(\boldsymbol{x})/f_{eq}(\boldsymbol{x})]d^3x$ for a Gaussian probability density with covariance matrix $\boldsymbol{c}$ and $\boldsymbol{c}_{eq} = \mathbf{1}$.

Exercise 67 A Derivative of Entropy
Calculate the functional derivative $\delta S/\delta\boldsymbol{c}$ for the configurational entropy given in (4.86).

Exercise 68 Quadratic Entropy Near Equilibrium
Determine the quadratic expansion of the entropy (4.86) around $\boldsymbol{c}_{eq} = \mathbf{1}$.

The convection mechanism for Λ is independent of $\boldsymbol{c}$; that is, $g_1 = g_2 = g_3 = 0$. Equation (4.45) then implies the following expression for the pressure tensor,

$$\boldsymbol{\tau}^{(S)} = \boldsymbol{\pi}^{(S)} - T\left(s - \rho\frac{\partial s}{\partial\rho} - \epsilon\frac{\partial s}{\partial\epsilon}\right)\mathbf{1} = \Lambda\frac{\rho}{m}k_BT(\mathbf{1}-\boldsymbol{c}) = n_pk_BT(\mathbf{1}-\boldsymbol{c}), \tag{4.87}$$

where ρ/m is the number density of solvent molecules and $n_p = \Lambda\rho/m$ is the number density of polymer molecules. An additional solvent contribution to the pressure tensor can be introduced through M^{hyd} of (4.51) with the solvent viscosity $\eta = \eta_s$.

To formulate the friction matrix, we further assume that there is relaxation of only the conformation tensor ($R_1 = \boldsymbol{R}_3 = 0$). If the relaxation occurs through noise on the end-to-end vector, we should assume a linear dependence of the friction matrix on the conformation tensor

$$(\boldsymbol{R}_2)_{ijkl} = \frac{m}{\rho\Lambda k_B}\frac{1}{\tau}(\delta_{ik}c_{jl} + c_{ik}\delta_{jl}), \tag{4.88}$$

[27] Sarti & Marrucci, Chem. Eng. Sci. 28 (1973) 1053.

where τ, as a relaxation time, is a positive parameter. Assuming a diffusion mechanism with $\tilde{\phi} = 1/\Lambda$ and $\boldsymbol{D}' = (m/k_{\rm B})\boldsymbol{D}$, where the positive semidefinite symmetric tensor $\boldsymbol{D}$ has the proper dimensions for a diffusion tensor, furthermore assuming the validity of (4.67) and a vanishing slip coefficient, we obtain the following time-evolution equations for the conformation tensor and the scalar variable:

$$\frac{\partial \boldsymbol{c}}{\partial t} = -\boldsymbol{v} \cdot \frac{\partial \boldsymbol{c}}{\partial \boldsymbol{r}} + \boldsymbol{\kappa} \cdot \boldsymbol{c} + \boldsymbol{c} \cdot \boldsymbol{\kappa}^T - \frac{1}{\tau}(\boldsymbol{c} - \boldsymbol{1}) + \frac{1}{\rho \Lambda^2} \frac{\partial}{\partial \boldsymbol{r}} \cdot \rho \Lambda^2 \boldsymbol{D} \cdot \frac{\partial}{\partial \boldsymbol{r}} \boldsymbol{c} \tag{4.89}$$

and

$$\frac{\partial \Lambda}{\partial t} = -\boldsymbol{v} \cdot \frac{\partial \Lambda}{\partial \boldsymbol{r}} + \frac{1}{\rho} \frac{\partial}{\partial \boldsymbol{r}} \cdot \rho \boldsymbol{D} \cdot \frac{\partial}{\partial \boldsymbol{r}} \Lambda. \tag{4.90}$$

Slightly more transparent equations are obtained when we transform to the number density of polymers, $n_{\rm p} = \Lambda \rho / m$,

$$\frac{\partial n_{\rm p}}{\partial t} = -\frac{\partial}{\partial \boldsymbol{r}} \cdot (\boldsymbol{v} n_{\rm p}) + \frac{\partial}{\partial \boldsymbol{r}} \cdot \rho \boldsymbol{D} \cdot \frac{\partial}{\partial \boldsymbol{r}} \frac{n_{\rm p}}{\rho}, \tag{4.91}$$

and a tensor variable more directly related to the pressure tensor, $\boldsymbol{c}' = n_{\rm p} \boldsymbol{c}$

$$\frac{\partial \boldsymbol{c}'}{\partial t} = -\frac{\partial}{\partial \boldsymbol{r}} \cdot (\boldsymbol{v} \boldsymbol{c}') + \boldsymbol{\kappa} \cdot \boldsymbol{c}' + \boldsymbol{c}' \cdot \boldsymbol{\kappa}^T - \frac{1}{\tau}(\boldsymbol{c}' - n_{\rm p} \boldsymbol{1}) + \frac{\partial}{\partial \boldsymbol{r}} \cdot \rho \boldsymbol{D} \cdot \frac{\partial}{\partial \boldsymbol{r}} \frac{\boldsymbol{c}'}{\rho}. \tag{4.92}$$

For incompressible liquids under isothermal flow conditions, the expressions (4.91) and (4.92) coincide with previously given results.[28]

In homogeneous flow situations of incompressible polymer solutions, ρ, $n_{\rm p}$, and $\boldsymbol{c}'$ are independent of position, and (4.91) is automatically satisfied for time independent $n_{\rm p}$. The equation for the conformation tensor hence, after omitting spatial derivatives, becomes

$$\frac{d\boldsymbol{c}}{dt} = \boldsymbol{\kappa} \cdot \boldsymbol{c} + \boldsymbol{c} \cdot \boldsymbol{\kappa}^T - \frac{1}{\tau}(\boldsymbol{c} - \boldsymbol{1}). \tag{4.93}$$

The subsequent exercises show that our simple model of a polymer solution based on the number density and conformation tensor as structural variables predicts a single-exponential relaxation modulus and a nonvanishing first normal-stress coefficient, whereas the second normal-stress coefficient turns out to be zero. Both viscosity and normal-stress coefficients are predicted to be independent of the shear rate.

Exercise 69 Simplified Giesekus Tensor

Use the construction (4.84) for the fourth-rank tensor $\boldsymbol{R}_2$ given in (4.88) to derive a simplified version of the Giesekus expression (4.82) for the pressure tensor,

$$\boldsymbol{\pi}_c = (1 - \xi) \frac{\rho \Lambda k_{\rm B} T}{m} \tau \left[\frac{\partial \boldsymbol{c}}{\partial t} - \boldsymbol{\kappa} \cdot \boldsymbol{c} - \boldsymbol{c} \cdot \boldsymbol{\kappa}^T + \frac{\xi}{2} (\boldsymbol{c} \cdot \dot{\boldsymbol{\gamma}} + \dot{\boldsymbol{\gamma}} \cdot \boldsymbol{c}) \right]. \tag{4.94}$$

[28] See (11) and (16) of Öttinger, Rheol. Acta 31 (1992) 14, for a vanishing migration velocity, and also the careful discussion of Beris & Mavrantzas, J. Rheol. 38 (1994) 1235.

Exercise 70 Conformation Tensor in Shear Flow

Formulate the time-evolution equations for the components of the conformation tensor $\boldsymbol{c}$ in time dependent homogeneous shear flows.

Exercise 71 Linear Viscoelastic Behavior of Polymer Solutions

Calculate the shear relaxation modulus for the conformation tensor model of a dilute polymer solution.

Exercise 72 Steady Shear Properties of Polymer Solutions

Find the material properties in steady shear flow for the conformation tensor model of a dilute polymer solution.

Exercise 73 Force-Flux Pair for Relaxation of a Tensor

Calculate the force-flux pair (4.57), (4.60) for dilute polymer solutions.

A useful generalization of the configurational entropy contribution (4.86) is given by

$$\hat{s}_{\mathrm{c}}(\Lambda, \boldsymbol{c}) = \frac{k_{\mathrm{B}}}{m}\Lambda\left[-\ln\Lambda + s^{(\phi)}(\mathrm{tr}\boldsymbol{c}) + \frac{1}{2}\ln(\det\boldsymbol{c})\right], \tag{4.95}$$

where the dimensionless function $s^{(\phi)}(\mathrm{tr}\boldsymbol{c})$ can be considered as an entropic potential accounting for non-Gaussian effects. In particular, such a potential can be used to describe the finite extensibility of polymer chains, where

$$s^{(\phi)}(\mathrm{tr}\boldsymbol{c}) = \frac{1}{2}b\ln\left(\frac{b+3}{b} - \frac{1}{b}\mathrm{tr}\boldsymbol{c}\right) \tag{4.96}$$

leads to the widely used FENE-P model (FENE is the acronym for finitely extensible nonlinear elastic, and the P indicates that we employ the Peterlin approximation, that is, a closed description on the level of conformation tensors) and b is the dimensionless finite extensibility parameter proportional to the number of monomers per polymer chain and thus large compared with unity. In the limit $b \to \infty$, we recover (4.86) for infinitely extensible polymers. If we introduce an "effective spring constant,"

$$h(\mathrm{tr}\boldsymbol{c}) = -2\frac{\partial s^{(\phi)}(\mathrm{tr}\boldsymbol{c})}{\partial \mathrm{tr}\boldsymbol{c}} = \frac{b}{b+3-\mathrm{tr}\boldsymbol{c}}, \tag{4.97}$$

then we have

$$\frac{\partial \hat{s}_{\mathrm{c}}}{\partial \boldsymbol{c}} = \frac{1}{2}\frac{k_{\mathrm{B}}}{m}\Lambda[\boldsymbol{c}^{-1} - h(\mathrm{tr}\boldsymbol{c})\mathbf{1}], \tag{4.98}$$

and the time-evolution equation (4.93) for homogeneous flows is generalized to

$$\frac{d\boldsymbol{c}}{dt} = \boldsymbol{\kappa}\cdot\boldsymbol{c} + \boldsymbol{c}\cdot\boldsymbol{\kappa}^T - \frac{1}{\tau}[h(\mathrm{tr}\boldsymbol{c})\boldsymbol{c} - \mathbf{1}]. \tag{4.99}$$

Note that the normalization $\boldsymbol{c}_{\mathrm{eq}} = \mathbf{1}$ requires $h(\mathrm{tr}\boldsymbol{c}_{\mathrm{eq}}) = h(3) = 1$, which implies that τ keeps its meaning as the relaxation time for single-mode relaxation (see Exercise 71).

A generalization of the friction matrix (4.88) has been proposed by Giesekus with the idea that, in sufficiently concentrated solutions, the frictional properties are going to become anisotropic. In the present formulation, his idea can be expressed as (cf. Exercise 69)

$$\hat{\boldsymbol{R}}_2 = \frac{2m}{\rho \Lambda k_{\mathrm{B}}} \frac{1}{\tau} [(1-\alpha)\boldsymbol{c} + \alpha \boldsymbol{c}^2], \tag{4.100}$$

where the dimensionless Giesekus parameter α with $0 \leq \alpha \leq 1$ determines the strength of the anisotropy. This leads to a further generalization of (4.99):

$$\frac{d\boldsymbol{c}}{dt} = \boldsymbol{\kappa} \cdot \boldsymbol{c} + \boldsymbol{c} \cdot \boldsymbol{\kappa}^T - \frac{1}{\tau}[(1-\alpha)\boldsymbol{1} + \alpha \boldsymbol{c}] \cdot [h(\mathrm{tr}\boldsymbol{c})\boldsymbol{c} - \boldsymbol{1}]. \tag{4.101}$$

In view of (4.98), the pressure tensor (4.87) is generalized to

$$\boldsymbol{\tau}^{(S)} = n_{\mathrm{p}} k_{\mathrm{B}} T [\boldsymbol{1} - h(\mathrm{tr}\boldsymbol{c})\boldsymbol{c}], \tag{4.102}$$

independent of Giesekus' anisotropy parameter α.

Exercise 74 Traceless Derivative of Entropy

Under what conditions is $\partial \hat{s}_{\mathrm{c}} / \partial \boldsymbol{c}$ traceless for homogeneous steady-state flows of incompressible liquids?

Exercise 75 Steady Shear Properties With Finite Extensibility

Show that the solution of (4.99) for steady shear flow is of the form

$$\boldsymbol{c} = \frac{1}{h'} \begin{pmatrix} 1 + \frac{2\dot{\gamma}^2\tau^2}{h'^2} & \frac{\dot{\gamma}\tau}{h'} & 0 \\ \frac{\dot{\gamma}\tau}{h'} & 1 & 0 \\ 0 & 0 & 1 \end{pmatrix}. \tag{4.103}$$

Formulate an equation for h'. Express the viscometric functions in terms of h'.

4.2.7 Example: Pompon Model

The pompon model for melts of multiply branched polymers, developed by McLeish and Larson, is an important first step toward an understanding of industrially relevant polymers, such as the long chain branched low-density polyethylene. This pom-pom model was originally derived by using the tube model,[29] where the backbones of the molecules can "reptate" in the tubes formed by other backbones after side arms have relaxed by "deep retractions" or "breathing modes." The intuitive derivation of the pom-pom model, although based on enormous experience with the tube model for various chain architectures, raises the issue of the thermodynamic admissibility of the model, in particular, as the dynamics of the configurational variables and the pressure tensor are formulated by separate arguments.

[29] Details on the tube or reptation model, although not relevant to the present discussion, can be found in Section 4.3.2.

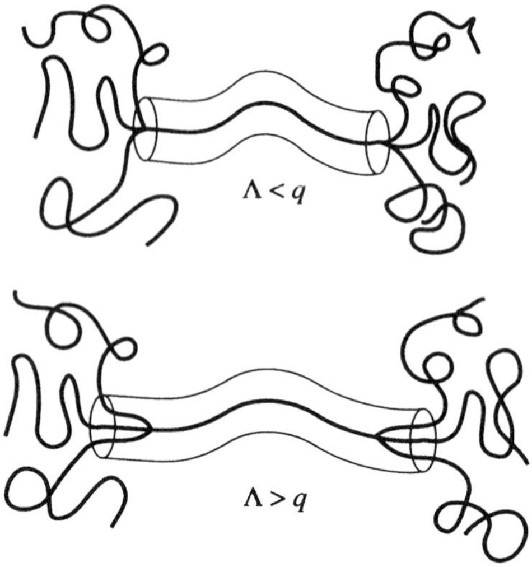

Fig. 4.8 The pompon model polymer for $q = 3$ with effective confining tube around the crossbar. [Reprinted with permission from H. C. Öttinger, Thermodynamic Admissibility of the Pompon Model for Branched Polymers, Rheol. Acta 40 (2001) 317. Copyright (2001) by Springer.]

The purpose of the present section is to express the fundamental ideas of the pom-pom model within the framework of models with a tensor and a scalar as structural variables, thus guaranteeing the thermodynamic admissibility. This thermodynamic modeling approach guarantees that only a minimum of independent input information is required, thus eliminating any unnecessary redundancies and possible inconsistencies between various assumptions. We refer to the outcome of the thermodynamic approach as the *pompon model*, thus emphasizing that the ideas of the original *pom-pom model* are reexpressed in a language with a more systematically structured grammar (and avoiding any sound-inspired associations with automatic antiaircraft guns[30]).

For the pompon model, we assume branched molecules with a "crossbar" linking two pompons of q arms each (see Figure 4.8). In that model, a backbone orientation tensor

$$\boldsymbol{S} = \frac{\boldsymbol{c}}{\mathrm{tr}\boldsymbol{c}} = \frac{\boldsymbol{c}}{I_1} \tag{4.104}$$

is constructed from a positive-definite tensor $\boldsymbol{c}$ with upper-convected behavior and normalization $\boldsymbol{c}_{\mathrm{eq}} = \mathbf{1}$. If a unit vector $\boldsymbol{u}$ describes the backbone orientation, the tensor $\boldsymbol{S}$ may be thought of as the average of the dyadic $\boldsymbol{uu}$, so that the unit trace and positive-definiteness conditions arise naturally. In the differential version of the original pom-pom model,[31] the equation for the auxiliary tensor $\boldsymbol{c}$ needs to be solved

[30] *Merriam-Webster's* (1993).

[31] McLeish & Larson, J. Rheol. 42 (1998) 81.

before $\boldsymbol{S}$ can be calculated; however, in the modified pompon model[32] presented here, a time-evolution equation directly for $\boldsymbol{S}$ is found.[33] The scalar variable Λ describes the stretch of a long-living backbone tube. For $\Lambda \leq q$, the variable Λ is equal to the dimensionless backbone stretch factor λ of the original pom-pom model of McLeish and Larson, whereas, for $\Lambda > q$, Λ accounts also for the portion of the tube into which the arms have been dragged (see Figure 4.8). For the convection mechanism, the pompon model assumes $g_2 = g_3 = 0$, so that the Jacobi identity implies $g_1 = g_1(\Lambda, I_1)$. Indeed, one chooses $g_1 = \Lambda/I_1$, that is, $\boldsymbol{g} = \Lambda\boldsymbol{S}$, and the decoupled scalar variable $\hat{s}$ of Exercise 62 can be taken as $\hat{s} = I_1/\Lambda^2$. We further assume that there is no configurational contribution to energy and an entropy contribution

$$\begin{aligned} \hat{s}_{\rm c}(\Lambda, \boldsymbol{c}) &= Z\frac{N_{\rm A}k_{\rm B}}{2M_{\rm p}}\Big\{ \ln\det(3\boldsymbol{c}) - 3\ln {\rm tr}\boldsymbol{c} \\ &- 3[\Lambda^2 - 2\ln\Lambda - 1 - (\Lambda - q)^2 H(\Lambda - q)]\Big\}, \end{aligned} \tag{4.105}$$

where $N_{\rm A}$ is Avogadro's number, $M_{\rm p}$ is the molecular weight of the pompon polymers, Z is the number of entanglements in the tube, and $H(\Lambda - q)$ is the Heaviside step function $[H(x) = 1$ for $x \geq 0$, $H(x) = 0$ for $x < 0]$. The factor in front of the curly brackets in (4.105) is chosen such that every entangled segment in the tube contributes to the entropy. The term depending on $\boldsymbol{c}$ is obtained by the replacement $\boldsymbol{c} \to 3\boldsymbol{c}/{\rm tr}\boldsymbol{c}$ in (4.86), as the entropy should depend only on the properly normalized physically relevant combination. The term $-2\ln\Lambda$ is introduced to produce a minimum at $\Lambda = 1$ and to avoid negative values of the stretch ratio Λ. Formally, this Λ dependence can be obtained by the replacement $\boldsymbol{c} \to \Lambda^2\boldsymbol{1}$ in (4.86). When $\Lambda > q$, branch-point withdrawal sets in, so that the entropy penalty for stretching beyond q is reduced; this effect is described by the quadratic term involving the Heaviside step function, which reduces the quadratic Λ-dependence of the entropy to a linear one.

Equation (4.45) gives the following expression for the configurational contribution to the pressure tensor:

$$\boldsymbol{\tau}^{(S)} = Z\frac{N_{\rm A}\rho}{M_{\rm p}}k_{\rm B}T \begin{cases} \boldsymbol{1} - 3\Lambda^2\boldsymbol{S} & \text{for } \Lambda \leq q \\ \boldsymbol{1} - 3\Lambda q\boldsymbol{S} & \text{for } \Lambda > q \end{cases}. \tag{4.106}$$

Exercise 76 Pressure Tensor for Pompon Model
Derive the pressure tensor (4.106).

For the relaxation mechanisms for Λ and $\boldsymbol{c}$, we use R_1 and $\boldsymbol{R}_2$, whereas we assume $\boldsymbol{R}_3 = 0$ for the coupled mechanism. We propose to choose $\boldsymbol{R}_2$ bilinear in $\boldsymbol{c}$ and ${\rm tr}\boldsymbol{c}$,

$$(\boldsymbol{R}_2)_{ijkl} = \frac{2M_{\rm p}}{3\rho Z N_{\rm A}k_{\rm B}}\frac{1}{\tau_{\rm b}}\Big[3c_{ik}c_{jl} + \frac{\alpha_1}{2}{\rm tr}\boldsymbol{c}(\delta_{ik}c_{jl} + c_{ik}\delta_{jl}) + \alpha_2({\rm tr}\boldsymbol{c})^2\delta_{ik}\delta_{jl}$$

[32] Öttinger, Rheol. Acta 40 (2001) 317.

[33] The thermodynamic admissibility of the integral version of the pom-pom model has been analyzed by van Meerveld, J. Non-Newtonian Fluid Mech. 108 (2002) 291.

$$+ \quad \alpha_3 (\mathrm{tr}\boldsymbol{c})^2 \delta_{ij}\delta_{kl} + \alpha_4 \mathrm{tr}\boldsymbol{c}(\delta_{ij}c_{kl} + c_{ij}\delta_{kl})\Big], \tag{4.107}$$

where τ_b is the orientational relaxation time and $\alpha_j \geq 0$. Further possible terms that would not contribute to the time-evolution equations have been omitted in (4.107). Diffusion is not considered in the pompon model, and the slip coefficient vanishes. With these assumptions, the final relaxation equation for the tensor $\boldsymbol{c}$ is

$$\begin{aligned} \frac{\partial \boldsymbol{c}}{\partial t} &= -\boldsymbol{v} \cdot \frac{\partial \boldsymbol{c}}{\partial \boldsymbol{r}} + \boldsymbol{\kappa} \cdot \boldsymbol{c} + \boldsymbol{c} \cdot \boldsymbol{\kappa}^T - \frac{1}{\tau_\mathrm{b}} \mathrm{tr}\boldsymbol{c} \left[(3\boldsymbol{S} + \alpha_1 \boldsymbol{1} + \alpha_2 \boldsymbol{S}^{-1}) \cdot \left(\boldsymbol{S} - \frac{1}{3}\boldsymbol{1} \right) \right. \\ &+ \left. (\alpha_3 \boldsymbol{1} + \alpha_4 \boldsymbol{S}) \mathrm{tr} \left(\boldsymbol{1} - \frac{1}{3}\boldsymbol{S}^{-1} \right) \right]. \end{aligned} \tag{4.108}$$

This equation can be rewritten as a closed equation for the orientation tensor $\boldsymbol{S}$ defined in (4.104),

$$\begin{aligned} \frac{\partial \boldsymbol{S}}{\partial t} &= -\boldsymbol{v} \cdot \frac{\partial \boldsymbol{S}}{\partial \boldsymbol{r}} + \boldsymbol{\kappa} \cdot \boldsymbol{S} + \boldsymbol{S} \cdot \boldsymbol{\kappa}^T - 2\boldsymbol{S}\mathrm{tr}(\boldsymbol{\kappa} \cdot \boldsymbol{S}) \\ &- \frac{1}{\tau_\mathrm{b}} \left[3(\boldsymbol{S}^2 - \boldsymbol{S}\mathrm{tr}\boldsymbol{S}^2) + \alpha_0 \left(\boldsymbol{S} - \frac{1}{3}\boldsymbol{1} \right) + \frac{\alpha_2}{3} \left(\boldsymbol{S}\mathrm{tr}\boldsymbol{S}^{-1} - \boldsymbol{S}^{-1} \right) \right], \end{aligned} \tag{4.109}$$

with

$$\alpha_0 = \alpha_1 - 3\alpha_2 - 3\alpha_3 \mathrm{tr} \left(\boldsymbol{1} - \frac{1}{3}\boldsymbol{S}^{-1} \right). \tag{4.110}$$

Exercise 77 Closed Time Evolution of Orientation Tensor

Derive the time-evolution equation (4.109) for the orientation tensor from (4.108).

The backbone stretch relaxation and branch-point withdrawal mechanisms proposed by McLeish and Larson can be summarized as

$$R_1 = \frac{\Lambda}{3Zn_\mathrm{p}k_\mathrm{B}} \frac{1}{\tau_\mathrm{s}} \begin{cases} (\Lambda + 1)^{-1} & \text{for } \Lambda \leq q \\ (\Lambda - 1/q)^{-1} & \text{for } \Lambda > q \end{cases}. \tag{4.111}$$

We then obtain from (4.78)

$$\frac{\partial \Lambda}{\partial t} = -\boldsymbol{v} \cdot \frac{\partial \Lambda}{\partial \boldsymbol{r}} + \Lambda \boldsymbol{\kappa} : \boldsymbol{S} - \frac{1}{\tau_\mathrm{s}} \begin{cases} \Lambda - 1 & \text{for } \Lambda \leq q \\ q & \text{for } \Lambda > q \end{cases}. \tag{4.112}$$

The backbone stretch and orientation time scales, τ_s and τ_b, both depend on Λ when $\Lambda > q$.[34] The entangled nature of the effective dumbbell of the pompon polymer separates these time scales (typically, $\tau_\mathrm{b} \gg \tau_\mathrm{s}$), thus allowing for the extreme differences in shear and extensional behavior observed in melts of branched polymers.

[34] McLeish & Larson, J. Rheol. 42 (1998) 81.

The pompon model reproduces the qualitative behavior of branched polymers in most respects. In particular, it predicts a shear thinning viscosity and a nonzero ratio of normal-stress coefficients,

$$\frac{\Psi_2}{\Psi_1} = -\frac{1}{2}\frac{1 - 3\alpha_2}{1 + \alpha_1 + 3\alpha_2}. \tag{4.113}$$

If we choose $\alpha_2 = 0$, thus avoiding the occurrence of the inverse of the orientation tensor in (4.109), then the experimentally observed values of Ψ_2/Ψ_1 around $-1/4$ suggest that α_1 should be close to unity.

Exercise 78 Shear Relaxation Modulus for Pompon Model
Calculate the shear relaxation modulus $G(t)$ for the pompon model.

Exercise 79 Orientation Tensor for Small Shear Rates
Derive the orientation tensor $\boldsymbol{S}$ for homogeneous steady shear flow to second order in the shear rate. Calculate the zero-shear-rate limits of the viscosity and of the normal-stress coefficients. Verify the result of Exercise 58.

Exercise 80 Viscometric Functions at High Shear Rates
Discuss the viscometric functions η, Ψ_1, and Ψ_2 predicted by the pompon model at high shear rates for $\alpha_2 = \alpha_3 = 0$.

4.3 CONFIGURATIONAL DISTRIBUTION FUNCTIONS

In many cases, the tensors used as structural variables in Section 4.2 can be interpreted as the second moments of some configurational vectors. For example, instead of describing the overall stretch and orientation of an ensemble of polymer molecules in solution by a tensor as indicated in Figure 4.7, we can describe the deformation of individual polymer molecules by their end-to-end vectors. For the further modeling steps, we introduce beads and springs in connection with the configuration vectors (see Figure 4.9), so that we can associate frictional properties with the beads and entropic forces with the springs. If we denote the connector vector between the beads of a dumbbell by $\boldsymbol{Q}$, the conformation tensor $\boldsymbol{c}$ of Section 4.2.6 can be interpreted as the second-moment tensor, that is, as the average of the dyadic $\boldsymbol{QQ}$ (after proper normalization). In a similar way, after (4.104), the backbone orientation tensor of the pompon model was interpreted as the second-moment tensor of a unit vector describing the backbone orientation.

Because the ends of a polymer chain and hence the beads of a dumbbell are equivalent, the average of $\boldsymbol{Q}$ must vanish for symmetry reasons. All the relevant information is contained in the fluctuations of $\boldsymbol{Q}$ around $\boldsymbol{Q} = 0$, and we hence use the probability density $f(\boldsymbol{Q})$ to find a configuration vector around $\boldsymbol{Q}$ as the fundamental structural variable. The use of a tensor variable $\boldsymbol{c}$ is justified after the simplifying assumption that the probability density $f(\boldsymbol{Q})$ is well-represented by its

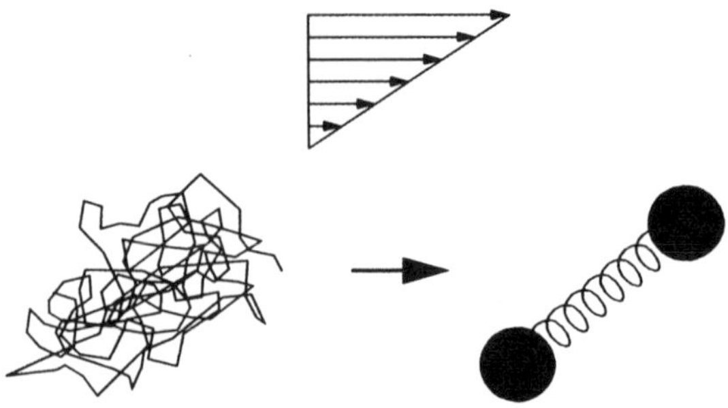

Fig. 4.9 Under flow conditions, such as the indicated shear flow, polymer molecules in solution become stretched and oriented. In a crude phenomenological model, a polymer molecule is represented by a dumbbell.

second-moment tensor (note that this connection has previously been used in Exercise 66 to obtain the functional form of the configurational entropy).

Instead of using the probability density $f(\boldsymbol{Q})$ as a structural variable, we can equivalently use a stochastic vector variable $\boldsymbol{Q}$. Then we are forced to use the GENERIC (1.56) with fluctuations because the GENERIC (1.1) without fluctuations would have the trivial deterministic solution $\boldsymbol{Q} = 0$. As pointed out before, all the relevant information is in the fluctuations of the vector variable, and one hence needs to consider a stochastic vector field governed by a stochastic differential equation. Based on a stochastic vector field, a thermodynamically consistent fluid particle model for polymeric liquids has been developed.[35] The use of a stochastic vector field as a structural variable for polymeric fluids is actually the basis of an extremely efficient simulation technique for solving complex flow problems, known as the method of Brownian configuration fields.[36] Because the correlation of the noise terms at different positions in space is irrelevant to flow calculations, an enormous variance reduction can be achieved in such flow calculations based on Brownian configuration fields when using identical noise throughout the system, that is, independent of position.[37]

4.3.1 Dumbbell Model of Polymer Solutions

According to the introductory discussion, we choose the following fields as independent variables for a dumbbell kinetic theory of a dilute polymer solution: the mass density $\rho(\boldsymbol{r})$ of solution, the momentum density $\boldsymbol{M}(\boldsymbol{r})$ of the solution, the internal energy density of solvent $\epsilon(\boldsymbol{r})$, and the configurational distribution function $f(\boldsymbol{r}, \boldsymbol{Q})$, where $\boldsymbol{Q}$ is the dumbbell configuration vector. The normalization of $f(\boldsymbol{r}, \boldsymbol{Q})$ is such

[35] Ellero, Español & Flekkøy, Phys. Rev. E 68 (2003) 041504.
[36] Hulsen, van Heel & van den Brule, J. Non-Newtonian Fluid Mech. 70 (1997) 79.
[37] Öttinger, van den Brule & Hulsen, J. Non-Newtonian Fluid Mech. 70 (1997) 255.

that

$$n_{\mathrm{p}}(\boldsymbol{r}) = \int f(\boldsymbol{r}, \boldsymbol{Q}) d^3 Q \tag{4.114}$$

is the number density of polymers. For a dilute polymer solution, the mass and momentum densities of the solvent and of the solution coincide, whereas the large polymer molecules can contribute significantly to the stress and the internal energy.

The total energy is obtained by adding the kinetic energy and the solvent and polymer potential energy contributions,

$$E = \int \left[\frac{1}{2} \frac{\boldsymbol{M}(\boldsymbol{r})^2}{\rho(\boldsymbol{r})} + \epsilon(\boldsymbol{r}) + \int \phi(Q) \, f(\boldsymbol{r}, \boldsymbol{Q}) d^3 Q \right] d^3 r, \tag{4.115}$$

where $\phi(Q)$ is the interaction potential between the beads of a dumbbell. In most cases, the spring force is entirely of entropic origin; we hence assume $\phi(Q) = 0$ from now on and only note that, within the GENERIC framework, it is naturally possible to trace back carefully the energetic and entropic origin of interactions.[38] In a non-isothermal situation, a spring potential of entropic origin, $\phi^{(S)}(\boldsymbol{r}, Q)$, would actually depend only through the solvent temperature on position. More precisely, we assume that, for potentials of purely entropic origin, the ratio $S^{(\phi)}(Q) = -\phi^{(S)}(\boldsymbol{r}, Q)/T(\boldsymbol{r})$ is independent of $\boldsymbol{r}$. Whereas true interaction potentials should be incorporated into (4.115), the classical entropic spring must be included into the entropy

$$S = \int \left\{ s(\rho(\boldsymbol{r}), \epsilon(\boldsymbol{r})) - \int f(\boldsymbol{r}, \boldsymbol{Q}) \left[\frac{\phi^{(S)}(\boldsymbol{r}, Q)}{T(\boldsymbol{r})} + k_{\mathrm{B}} \ln f(\boldsymbol{r}, \boldsymbol{Q}) \right] d^3 Q \right\} d^3 r. \tag{4.116}$$

While $s(\rho, \epsilon)$ should be interpreted as the solvent entropy, leading to the solvent temperature T as defined in (4.46), the $f \ln f$-term is clearly associated with the dumbbell translational and configurational entropy, and the entropic potential $\phi^{(S)}(\boldsymbol{r}, Q)$ accounts for effects on the much finer level of polymer segments, which have been eliminated in coarse-graining to the dumbbell level.[39] If the equilibrium probability density is given by

$$f_{\mathrm{eq}}(Q) \propto \exp\left\{ -\frac{\phi^{(S)}(\boldsymbol{r}, Q)}{k_{\mathrm{B}} T(\boldsymbol{r})} \right\} = \exp\left\{ \frac{S^{(\phi)}(Q)}{k_{\mathrm{B}}} \right\}, \tag{4.117}$$

then the configurational entropy density in (4.116) can be rewritten in the more compact and transparent form

$$-k_{\mathrm{B}} \int f(\boldsymbol{r}, \boldsymbol{Q}) \ln \frac{f(\boldsymbol{r}, \boldsymbol{Q})}{f_{\mathrm{eq}}(Q)} d^3 Q \tag{4.118}$$

(cf. Exercise 66). Such an entropy favors a number density of polymers uniform in space; to justify such a uniform polymer distribution, we assume for the rest of this subsection that the solvent density ρ is virtually independent of position.

[38] See Section III of Öttinger & Grmela, Phys. Rev. E 56 (1997) 6633.

[39] See, for example, Section I.1.1 of de Gennes, *Scaling Concepts in Polymer Physics* (Cornell, 1979).

For later use, we write down the functional derivatives of E and S with respect to the independent variables $x = (\rho(\boldsymbol{r}), \boldsymbol{M}(\boldsymbol{r}), \epsilon(\boldsymbol{r}), f(\boldsymbol{r}, \boldsymbol{Q}))$:

$$\frac{\delta E}{\delta x} = \begin{pmatrix} -\frac{1}{2}\boldsymbol{v}(\boldsymbol{r})^2 \\ \boldsymbol{v}(\boldsymbol{r}) \\ 1 \\ 0 \end{pmatrix} \tag{4.119}$$

and

$$\frac{\delta S}{\delta x} = \begin{pmatrix} -\frac{\mu(\boldsymbol{r})}{T(\boldsymbol{r})} \\ 0 \\ \frac{1}{T(\boldsymbol{r})} \\ -\frac{\phi^{(S)}(\boldsymbol{r},\boldsymbol{Q})}{T(\boldsymbol{r})} - k_{\mathrm{B}} \ln f(\boldsymbol{r}, \boldsymbol{Q}) - k_{\mathrm{B}} \end{pmatrix}. \tag{4.120}$$

To obtain the L operator, we start from (2.60) for the hydrodynamic variables,

$$L(\boldsymbol{r}, \boldsymbol{Q}, \boldsymbol{Q}') = -\begin{pmatrix} 0 & \frac{\partial}{\partial \boldsymbol{r}}\rho(\boldsymbol{r}) & 0 & 0 \\ * & \left[\frac{\partial}{\partial \boldsymbol{r}}\boldsymbol{M}(\boldsymbol{r}) + \boldsymbol{M}(\boldsymbol{r})\frac{\partial}{\partial \boldsymbol{r}}\right]^T & * & L_{24} \\ 0 & \frac{\partial}{\partial \boldsymbol{r}}\epsilon(\boldsymbol{r}) + \boldsymbol{\pi}^{(S)}(\boldsymbol{r}) \cdot \frac{\partial}{\partial \boldsymbol{r}} & 0 & 0 \\ 0 & L_{42} & 0 & 0 \end{pmatrix}, \tag{4.121}$$

where the entries replaced with $*$ can be obtained from the antisymmetry of L and, for the same reason as in (4.44), a pressure tensor $\boldsymbol{\pi}^{(S)}$ has been introduced instead of a scalar pressure. The form of L_{24} and L_{42} can be obtained because f is a scalar density in $\boldsymbol{r}$ and $\boldsymbol{Q}$ is deformed affinely with the flow field,

$$L_{42} = \frac{\partial}{\partial \boldsymbol{r}} f(\boldsymbol{r}, \boldsymbol{Q}) + \frac{\partial}{\partial \boldsymbol{Q}} f(\boldsymbol{r}, \boldsymbol{Q})\boldsymbol{Q} \cdot \frac{\partial}{\partial \boldsymbol{r}}, \tag{4.122}$$

$$L_{24} = f(\boldsymbol{r}, \boldsymbol{Q}')\frac{\partial}{\partial \boldsymbol{r}} - \frac{\partial}{\partial \boldsymbol{r}} \cdot f(\boldsymbol{r}, \boldsymbol{Q}')\boldsymbol{Q}'\frac{\partial}{\partial \boldsymbol{Q}'}. \tag{4.123}$$

The new components associated with f have additional labels $\boldsymbol{Q}$ and $\boldsymbol{Q}'$, where matrix multiplication implies an integration over $\boldsymbol{Q}'$. The Jacobi identity can be checked by an explicit calculation.[40]

As in Section 4.2, the degeneracy condition (1.4) fixes the functional form of the pressure tensor,

$$\boldsymbol{\pi}^{(S)} = T\left(s - \rho\frac{\partial s}{\partial \rho} - \epsilon\frac{\partial s}{\partial \epsilon}\right)\mathbf{1} + \int f(\boldsymbol{Q})\left[2k_{\mathrm{B}}T\mathbf{1} - \boldsymbol{Q}\frac{\partial \phi^{(S)}(Q)}{\partial \boldsymbol{Q}}\right] d^3Q. \tag{4.124}$$

The configurational contribution is known as the *Kramers expression* for the pressure tensor. The momentum transport is given by the entropic spring force and the distance over which momentum is transferred by the spring force.

[40] In Appendix B.4, the validity of the proposed Poisson operator is established by arguments based on geometry and symmetry.

For the friction matrix, we generalize the expression (2.66) used in hydrodynamics,

$$M(\boldsymbol{r},\boldsymbol{Q},\boldsymbol{Q}') = \begin{pmatrix} 0 & 0 & 0 & 0 \\ 0 & M_{22} & M_{23} & 0 \\ 0 & M_{32} & M_{33} & 0 \\ 0 & 0 & 0 & M_{44} \end{pmatrix}, \tag{4.125}$$

where the expressions for M_{22}, M_{23}, and M_{32}, in which η stands for the solvent viscosity η_s, were previously given in (2.67)–(2.69), and M_{33} can be generalized by introducing the thermal conductivity tensor as in (4.51). The entry M_{44} describes diffusion of the two beads of a dumbbell or, equivalently, of the center of the dumbbells and the connector vector,

$$M_{44} = -\frac{\partial}{\partial \boldsymbol{Q}}\cdot\frac{2T}{\zeta}f(\boldsymbol{r},\boldsymbol{Q})\frac{\partial}{\partial \boldsymbol{Q}}\delta(\boldsymbol{Q}-\boldsymbol{Q}')-\frac{\partial}{\partial \boldsymbol{r}}\cdot\frac{T}{2\zeta}f(\boldsymbol{r},\boldsymbol{Q})\frac{\partial}{\partial \boldsymbol{r}}\delta(\boldsymbol{Q}-\boldsymbol{Q}'), \tag{4.126}$$

where ζ is the bead friction coefficient. Just as for the position argument, the integration over $\boldsymbol{Q}'$ in a matrix multiplication can be omitted because it can immediately be carried out in view of the δ-functions in (4.126). With the help of the energy gradient (4.119), we realize that the degeneracy (1.5) is satisfied.

With the GENERIC building blocks of this subsection, we can write out all the time-evolution equations. Most interesting is the diffusion equation or Fokker-Planck equation for the configurational distribution function,

$$\begin{aligned}\frac{\partial f}{\partial t} &= -\frac{\partial}{\partial \boldsymbol{r}}\cdot(\boldsymbol{v}f) - \frac{\partial}{\partial \boldsymbol{Q}}\cdot\left[\left(\boldsymbol{\kappa}\cdot\boldsymbol{Q} - \frac{2}{\zeta}\frac{\partial \phi^{(S)}}{\partial \boldsymbol{Q}}\right)f\right] \\ &\quad + \frac{\partial}{\partial \boldsymbol{r}}\cdot\frac{k_B T}{2\zeta}\frac{\partial f}{\partial \boldsymbol{r}} + \frac{\partial}{\partial \boldsymbol{Q}}\cdot\frac{2k_B T}{\zeta}\frac{\partial f}{\partial \boldsymbol{Q}},\end{aligned} \tag{4.127}$$

where $\boldsymbol{\kappa}$ is the transposed velocity gradient tensor. Such diffusion equations in configuration space constitute a cornerstone of polymer kinetic theory.

By evaluating

$$\frac{d}{dt}\int f(\boldsymbol{Q})\boldsymbol{Q}\boldsymbol{Q}d^3Q = \int \frac{\partial f}{\partial t}\boldsymbol{Q}\boldsymbol{Q}d^3Q \tag{4.128}$$

for homogeneous flows by means of (4.127), we obtain after several integrations by parts

$$\begin{aligned}\frac{d}{dt}\int f(\boldsymbol{Q})\boldsymbol{Q}\boldsymbol{Q}d^3Q &= \int f(\boldsymbol{Q})\Big[\boldsymbol{\kappa}\cdot\boldsymbol{Q}\boldsymbol{Q} + \boldsymbol{Q}\boldsymbol{Q}\cdot\boldsymbol{\kappa}^T \\ &\quad + \frac{4}{\zeta}\left(k_B T\boldsymbol{1} - \boldsymbol{Q}\frac{\partial \phi^{(S)}(Q)}{\partial \boldsymbol{Q}}\right)\Big]d^3Q,\end{aligned} \tag{4.129}$$

so that the pressure tensor (4.124) can be rewritten as

$$\begin{aligned}\boldsymbol{\pi}^{(S)} &= T\left(s - \rho\frac{\partial s}{\partial \rho} - \epsilon\frac{\partial s}{\partial \epsilon} + n_p k_B\right)\boldsymbol{1} + \frac{\zeta}{4}\frac{d}{dt}\int f(\boldsymbol{Q})\boldsymbol{Q}\boldsymbol{Q}d^3Q \\ &\quad - \frac{\zeta}{4}\int f(\boldsymbol{Q})(\boldsymbol{\kappa}\cdot\boldsymbol{Q}\boldsymbol{Q} + \boldsymbol{Q}\boldsymbol{Q}\cdot\boldsymbol{\kappa}^T)d^3Q.\end{aligned} \tag{4.130}$$

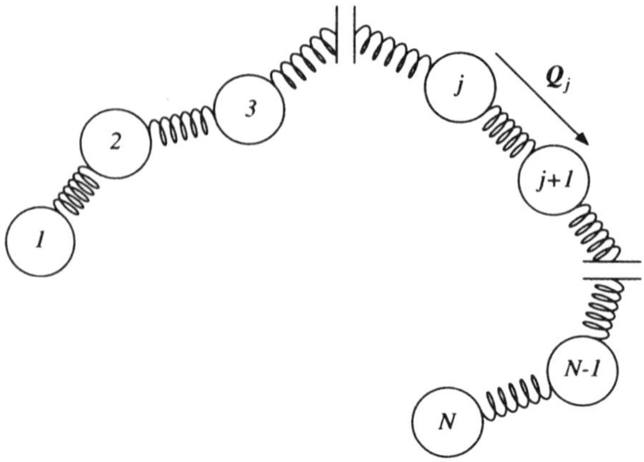

Fig. 4.10 The freely jointed bead-spring chain model with N beads and $N-1$ springs. The internal configuration of the chain is given by the connector vectors $\boldsymbol{Q}_j$ ($j = 1, 2, \ldots, N-1$).

The spring-force law has thus been eliminated in favor of a convected time derivative of the second-moment tensor. A pressure tensor of this type is usually referred to as a *Giesekus expression* [see (4.82) and (4.94)].

For a Hookean entropic potential, $\phi^{(S)} = HQ^2/2$ with the spring constant H, which, according to (4.117) corresponds to a Gaussian probability density resulting from the central limit theorem, we recover the dumbbell model of Section 4.2.6 (see Exercise 81). The second moments and the configurational distribution function provide equivalent descriptions because, for Gaussian initial conditions, the solutions of (4.91) remain strictly Gaussian at later times. A non-Hookean spring-force law is considered in Section 8.3.2, where Brownian dynamics is illustrated by calculating the viscosity function for a finitely extensible spring.

Exercise 81 Moment Equations for Hookean Dumbbells

Derive the time-evolution equations for the zeroth moment, equal to the polymer number density n_p, and the second moment tensor of the configurational distribution function from (4.127). Compare properly normalized moments with (4.91) and (4.92) and identify the parameters.

All the formulas and results can be extended from dumbbells to bead-spring chains with Hookean springs to obtain the famous *Rouse model* illustrated in Figure 4.10.[41] A detailed discussion of bead-spring chains with linear spring forces begins with the

[41] Although developed for dilute polymer solutions, the Rouse model is often and most successfully applied to melts, as long as the polymer molecules are not entangled and hence move in an effective viscous environment.

introduction of suitable eigenmodes,

$$\boldsymbol{Q}'_k = \sum_{j=1}^{N-1} Q_{jk}^{\text{Rouse}} \boldsymbol{Q}_j, \tag{4.131}$$

where Q_{jk}^{Rouse} is an orthogonal transformation matrix that decouples the motions of the springs in a chain. The modes $\boldsymbol{Q}'_k$ are independent of each other. One can hence replace a Rouse chain by $N-1$ noninteracting dumbbells with a spectrum of $N-1$ relaxation times. On the level of conformation tensors, one has a second-moment tensor for each mode in addition to the scalar concentration variable. On the level of configurational distribution functions, the Fokker-Planck equation for the full $f(\boldsymbol{r}, \boldsymbol{Q}_1, \boldsymbol{Q}_2, \ldots, \boldsymbol{Q}_{N-1})$ can be solved with the *ansatz* of a factorization into contributions from the modes. Most elegantly, the Rouse model is solved on the level of stochastic differential equations of motion for the connector vectors. Rather than elaborating all the details,[42] we note that, for chains of N beads, the relaxation of the springs is governed by a multiple of the $(N-1) \times (N-1)$ Rouse matrix,

$$-\frac{2H}{\zeta} \begin{pmatrix} 2 & -1 & & & & & \\ -1 & 2 & -1 & & & & \\ & -1 & 2 & -1 & & & \\ & & \ddots & \ddots & \ddots & & \\ & & & -1 & 2 & -1 & \\ & & & & -1 & 2 & -1 \\ & & & & & -1 & 2 \end{pmatrix}, \tag{4.132}$$

where all entries not shown in (4.132) are equal to zero. Each entry 2 describes the fact that the retracting spring force associated with a connector vector acts on each of the two beads at its ends, while an opposite force of a neighboring spring felt by either bead is taken into account through an entry -1. For a dumbbell, we simply have a single mode with the relaxation rate $4H/\zeta$, or the relaxation $\zeta/(4H)$ [see Exercise 81, which justifies the prefactor $2H/\zeta$ in (4.132)]. For a chain of N beads, Q_{jk}^{Rouse} is the diagonalizing matrix for the Rouse matrix in (4.132), and one can calculate the following set of relaxation times (see Exercise 82):

$$\tau_j = \frac{\zeta}{8H} \left(\sin \frac{j\pi}{2N} \right)^{-2}, \qquad j = 1, 2, \ldots, N-1. \tag{4.133}$$

Equation (4.133) implies that τ_j decreases with increasing j, and that τ_j divided by the longest relaxation time τ_1 is a function of j and N only. The occurrence of the combination N/j in τ_j suggests that mode j corresponds to the relaxation of subchains of N/j beads, where smaller subchains have shorter relaxation times. For

[42] A detailed discussion of bead-spring chain models with Hookean springs from different perspectives can be found in Chapter 15 of Bird, Curtiss, Armstrong & Hassager, *DPL 2, Kinetic Theory* (Wiley, 1987), or in Chapter 4 of Öttinger, *Stochastic Processes in Polymeric Fluids* (Springer, 1996).

small j and large N, more precisely, for $j \ll N$, we have $\tau_j = \tau_1/j^2$, where the longest relaxation time is proportional to N^2,

$$\tau_{\rm Rouse} = \tau_1 = \frac{\zeta N^2}{2H\pi^2}. \tag{4.134}$$

Instead of calculating the stress contributions from the $N-1$ springs of a chain, one can equivalently calculate the contributions from the $N-1$ eigenmodes of the chain. One then obtains the following multi-mode modification of the single-mode dumbbell relaxation modulus found in Exercise 71,

$$G(t) = \eta_{\rm s}\delta(t) + n_{\rm p}k_{\rm B}T\sum_{j=1}^{N-1} e^{-t/\tau_j}, \tag{4.135}$$

where the instantaneous solvent contribution has been included.

The qualitative behavior of the rheological predictions for dilute polymer solutions (see Exercises 70–72) can now be improved in various ways.[43] For example, a non-Hookean spring can be used to model the *finite extensibility* of polymer molecules; one then obtains viscometric functions that depend on the shear rate. A nonvanishing second normal-stress coefficient can be predicted for dilute solutions after introducing *hydrodynamic interactions* between the beads of a dumbbell. As a further important universal effect, the *excluded-volume interactions* between beads in good solvents must be taken into account. Quantitative in addition to qualitative understanding certainly requires bead-spring chain instead of dumbbell models. The bead-spring chain model with Hookean springs and equilibrium-averaged hydrodynamic interactions, known as the *Zimm model*, can predict the experimentally observed linear viscoelastic behavior of dilute polymer solutions reasonably well. The longest relaxation time scales as $N^{3/2}$ for the Zimm model, whereas, according to (4.134), the corresponding chain length dependence for the Rouse model is N^2. More precisely, one has the relaxation times

$$\tau_j = \alpha_j \left(\frac{N}{j}\right)^{3/2} \frac{\zeta}{4h^* H\pi^2} \qquad j = 1, 2, \ldots, N-1, \tag{4.136}$$

where each α_j is a numerical constant close to unity, h^* is the hydrodynamic interaction parameter characterizing the strength of the interactions (typical values of h^* lie between 0.1 and 0.25), the leading-order dependence of the numerical prefactors on j has been introduced through the idea that the jth relaxation time corresponds to the relaxation of a mode involving N/j beads, and the factor $4\pi^2$ has been introduced for convenience. The original numerical calculation of the relaxation times[44]

[43] For a detailed discussion of the various improvements see, for example, Chapter 13 of Bird, Curtiss, Armstrong & Hassager, *DPL 2, Kinetic Theory* (Wiley, 1987), or Chapter 4 of Öttinger, *Stochastic Processes in Polymeric Fluids* (Springer, 1996).

[44] Zimm et al., J. Chem. Phys. 24 (1956) 279.

corresponds to $\alpha_1 = 1.22$, $\alpha_2 = 1.09$, and $\alpha_3 = 1.06$, where the latter two values are already consistent with the asymptotic expression for large j, $\alpha_j = \left(1 - \frac{1}{2\pi j}\right)^{-1}$.

As a final comment on bead-spring models, intended to be useful for readers with an interest in *renormalization-group theory*, we point out that the beads are fictitious objects. In a dumbbell model, a bead roughly represents the frictional properties of half a polymer molecule, and the excluded volume associated with it. In a chain model, one makes the beads sufficiently small so that the shortest time scales relevant to the rheological properties of interest can be resolved, or the corresponding length scales in a scattering experiment. One might expect that a bead should be larger than a monomer, but such a constraint on the minimum size of a bead has no practical relevance. Indeed, the limit of infinitely many infinitely small beads, leading to a continuous chain description, may offer some mathematical advantages and is actually standard in renormalization-group calculations for polymers.

If two people use two different numbers of beads in modeling a specific polymer solution of interest, say 100 or 200 beads, it must be possible to relate the respective bead properties, for example, by matching the overall size of the polymer molecules and their longest relaxation times. The same relationship between the properties of beads differing by a factor of two in the number of monomers that they represent would also allow one to find the properties of the beads when modeling the system of interest with 50 beads only; this observation reflects the semi-group nature of the renormalization procedure. If one goes all the way to the continuous description with an infinite number of beads, one should not be surprised if the corresponding bead properties become singular, very much like the infinite bare electric charge of an electron in a point-particle picture.

What is the practical usefulness of these abstract remarks on renormalization? The most important benefit is the possibility of refining the results of perturbation theories. Perturbation theory can treat only a few hydrodynamic or excluded-volume interactions between beads; however, the universal properties of polymers are determined exactly by the fact that very many interactions occur in large molecules. This discrepancy makes a direct application of perturbation theory meaningless and calls for more sophisticated considerations. A physically motivated analysis is based on calculating the properties of larger beads from those of smaller beads, where the ratio in numbers can be chosen such that the interactions between only a few beads need to be taken into account, well within the scope of perturbation theory. By performing a large number of renormalization steps, one can infer the properties of rather large beads, effectively incorporating a large number of interactions between much smaller beads. If the chain of interest is finally represented by a moderate number of large beads, its large-scale properties can be predicted by perturbation theory. This procedure reflects the self-similar nature of polymer molecules over a wide range of length

Fig. 4.11 Probe chain in an entangled polymer melt.

scales. Although the renormalization-group approach is standard for equilibrium properties,[45] rheological properties can also be obtained.[46]

Exercise 82 Relaxation Spectrum of Rouse Model
Verify that the symmetric matrix

$$Q_{jk}^{\mathrm{Rouse}} = \sqrt{\frac{2}{N}} \sin \frac{jk\pi}{N} \tag{4.137}$$

is orthogonal and that it diagonalizes the matrix (4.132), where the eigenvalues are $1/\tau_j$ with τ_j given by (4.133).

Exercise 83 Viscometric Functions for Rouse Chains
Calculate the viscosity and the first normal-stress difference for Rouse chains.

4.3.2 Reptation Model of Polymer Melts

Concentrated polymer solutions and polymer melts are extremely complex, many-particle systems. For that reason, it is unquestionably important to make both ingenious and far-reaching assumptions in describing the dynamics of polymers in such undiluted systems. A widely and successfully applied class of molecular models for the polymer dynamics in concentrated solutions and melts relies on the notion of *reptational motion*.[47] The first detailed reptation model for the rheology of undiluted polymers was developed in a series of papers by Doi and Edwards published in 1978

[45] Freed, *Renormalization Group Theory of Macromolecules* (Wiley, 1987); Schäfer, *Excluded Volume Effects in Polymer Solutions* (Springer, 1999).
[46] Öttinger & Y. Rabin, J. Non-Newtonian Fluid Mech. 33 (1989) 53; Öttinger, Phys. Rev. A 40 (1989) 2664; 41 (1990) 4413.
[47] de Gennes, J. Chem. Phys. 55 (1971) 572.

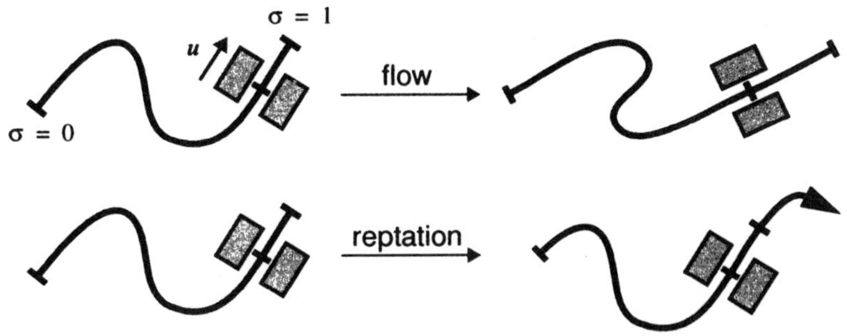

Fig. 4.12 Illustration of reptation dynamics. The curves represent a probe chain, and the rectangular bars indicate the constraining effect of the matrix material via a tube segment or anisotropic frictional properties. Top: illustration of how the flow deforms the matrix material and hence changes the orientation of the probe chain (rotation of the tangent vector $\boldsymbol{u}$). Bottom: illustration of how the probe chain reptates through the matrix material (random displacement of the position label σ).

and 1979.[48] The Doi-Edwards model is based on the assumption that each polymer in an entangled system moves ("reptates") in a tube formed by other polymers (see Figure 4.11). Several further assumptions need to be made, and several different representations of the reptating polymer are used (the "slip-link model" and the "cage model" in addition to the "tube model") in order to derive the final diffusion equation describing the dynamics of the polymers. For the pressure tensor required in deriving rheological properties, the Doi-Edwards model employs a formula from rubber elasticity. In such a situation, when combining different kinds of ingredients, the issue of thermodynamic consistency arises naturally.

From 1981 on, another series of papers on a kinetic theory for polymer melts was published by Curtiss, Bird, and coworkers.[49] In their systematic kinetic theory derivation of a diffusion equation for the polymer dynamics, Curtiss and Bird used anisotropic friction tensors to describe the hindrance of sideway motions of the polymers in concentrated systems. These anisotropic friction tensors express essentially the same physical idea of constrained motion as the tube in the Doi-Edwards model. After several other well-defined assumptions were made, the diffusion equation obtained by Curtiss and Bird has exactly the same form as in the Doi-Edwards model (however, the two models differ in the interpretation of the time constant appearing in the diffusion equation). In the Curtiss-Bird model, the pressure tensor is also derived in a systematic manner by resorting to the same kind of approximations previously made in the derivation of the diffusion equation. In the resulting expression, a dis-

[48] Doi & Edwards, J. Chem. Soc. Faraday Trans. II 74 (1978) 1789; 1802; 1818; 75 (1979) 38.

[49] Curtiss & Bird, J. Chem. Phys. 74 (1981) 2016; 2026; Bird, Saab & Curtiss, J. Phys. Chem. 86 (1982) 1102; J. Chem. Phys. 77 (1982) 4747; Saab, Bird & Curtiss, J. Chem. Phys. 77 (1982) 4758; Fan & Bird, J. Non-Newtonian Fluid Mech. 15 (1984) 341.

Table 4.2 Characteristic Molecular Weights (in g/mol) and Length Scales (in Å) in Melts of Entangled Linear Polymers*

Polymer	M_{m}	M_{e}	L_{m}	L_{K}	L_{e}
Polystyrene	104	13000	2.7	16.9	75
Polyethylene	28	830	2.8	12.3	32
Polypropylene	42	4600	2.6	11.1	56
Polymethylmethacrylate	100	10000	2.5	17	65
Polydimethylsiloxane	74	12000	3.3	10.2	74

*M_{m} is the molecular weight of the monomers; M_{e} is the entanglement molecular weight (obtained from the plateau modulus); L_{m} is a rough estimate of the size of the monomers, shown as a consistency check [obtained as $L_{\mathrm{m}} = M_{\mathrm{m}} L_{\mathrm{e}}^2 / (M_{\mathrm{e}} L_{\mathrm{K}})$]; L_{K} is the Kuhn length, that is, a measure of the local chain stiffness (according to Kuhn, the size and the contour length of a polymer molecule can be reproduced by a random walk with steps of length L_{K}; $L_{\mathrm{K}}/2$ is also known as the persistence length); L_{e} is the entanglement length scale (obtained from the number of entanglements per chain and the radius of gyration measured by small-angle neutron scattering). Data for L_{K} from Table I of Aharoni, Macromolecules 16 (1983) 1722; all data for M_{e} and L_{e} from Table 24.2 of Fetters et al., Chain Dimensions and Entanglement Spacings (1996) for melts at 140°C.

sipative term arises in addition to the pressure tensor assumed in the Doi-Edwards model.

In the reptation model, the chains reptating through an entangled environment are characterized by their local orientation. One employs a unit vector $\boldsymbol{u}$ that describes the direction of the polymer chain at the position σ within the chain, where the label σ varies from 0 to 1 in going from one chain end to the other. Speaking about chain orientation is meaningful because the polymer molecule can be represented by a smooth wormlike chain, or primitive path, after coarse-graining to a length scale that is of the order of the distance between entanglements or of the tube diameter (see Figure 4.12). Typical length scales and molecular weights for several polymers are listed in Table 4.2. Entanglement length scales (L_{e}) are large compared to typical bond lengths (1.54 Å for the C—C bond) and monomer dimensions (L_{m}). As the polymer molecules in a melt can be considered as random walks of segments between entanglements, the size of a molecule is $\sqrt{Z}\, L_{\mathrm{e}}$, where Z is the number of entanglements per chain, which can be obtained by dividing the molecular weight of the polymers by the entanglement molecular weight also given in Table 4.2.

The effects of flow and reptation on the configurations of a linear polymer chain in a melt are illustrated in Figure 4.12. These motions can be translated directly into a Fokker-Planck equation for the configurational distribution function $f = f(\boldsymbol{r}, \boldsymbol{u}, \sigma)$,

$$\frac{\partial f}{\partial t} = -\boldsymbol{v} \cdot \frac{\partial f}{\partial \boldsymbol{r}} - \frac{\partial}{\partial \boldsymbol{u}} \cdot [(\mathbf{1} - \boldsymbol{u}\boldsymbol{u}) \cdot \boldsymbol{\kappa} \cdot \boldsymbol{u} f] + \frac{1}{\tau_{\mathrm{rept}} \pi^2} \frac{\partial^2 f}{\partial \sigma^2}, \tag{4.138}$$

where, again, we need to consider equations with fluctuations or to employ a configurational distribution function as the structural variable because the averages of unit vector $\boldsymbol{u}$ and of the label σ contain only trivial information (the average of $\boldsymbol{u}$ vanishes for symmetry reasons, and the average of σ is the label $1/2$ of the chain center). The first term on the right-hand side of (4.138) describes the convection of the chains; the second term describes the rotation of the chain orientation by the flow, where $\boldsymbol{\kappa} \cdot \boldsymbol{u}$ would be the deformation by the flow and the transverse projector $\mathbf{1} - \boldsymbol{u}\boldsymbol{u}$ makes sure that changes can only be perpendicular to $\boldsymbol{u}$ so that $\boldsymbol{u}$ stays normalized; the last term describes the diffusive motion of the chains in the tube, where τ_{rept} is the reptation time scale. The Fokker-Planck equation (4.138) is to be solved subject to the boundary conditions

$$f(\boldsymbol{r}, \boldsymbol{u}, 0) = f(\boldsymbol{r}, \boldsymbol{u}, 1) = \frac{1}{4\pi}, \tag{4.139}$$

which express the fact that the orientation at the chain ends is random because the ends of a reptating molecule are free to move in arbitrary directions. The goal of the present section is to derive (4.138) by casting the ideas of the reptation model into the GENERIC framework. In doing so, we automatically obtain an expression for the pressure tensor.

Guided by the dumbbell kinetic theory of a dilute polymer solution, we choose the mass density $\rho(\boldsymbol{r})$, the momentum density $\boldsymbol{M}(\boldsymbol{r})$, the internal energy density $\epsilon(\boldsymbol{r})$, and the configurational distribution function $f(\boldsymbol{r}, \boldsymbol{u}, \sigma)$ as the full set of independent variables. However, the normalization of $f(\boldsymbol{r}, \boldsymbol{u}, \sigma)$ is now chosen as

$$\int_0^1 \int f(\boldsymbol{r}, \boldsymbol{u}, \sigma) d^2u d\sigma = 1, \tag{4.140}$$

rather than the number density of polymers in (4.114). This difference is important because, in a melt, ρ and n_{p} would be redundant variables. We assume that the energy of the melt does not change with the polymer conformation so that E is independent of the configurational distribution function,

$$E = \int \left[\frac{1}{2} \frac{\boldsymbol{M}(\boldsymbol{r})^2}{\rho(\boldsymbol{r})} + \epsilon(\boldsymbol{r}) \right] d^3r. \tag{4.141}$$

The entropy is analogous to (4.116) in the absence of entropic springs,

$$S = \int \left[s(\rho(\boldsymbol{r}), \epsilon(\boldsymbol{r})) - \rho(\boldsymbol{r}) \frac{N_{\text{A}}}{M_{\text{e}}} k_{\text{B}} \int_0^1 \int f(\boldsymbol{r}, \boldsymbol{u}, \sigma) \ln f(\boldsymbol{r}, \boldsymbol{u}, \sigma) d^2u d\sigma \right] d^3r. \tag{4.142}$$

Note that the configurational distribution function f is normalized to unity, whereas every entanglement contributes to the entropy. Therefore, the number density of entanglements, $\rho N_{\text{A}}/M_{\text{e}}$, has been included as an extra factor in the entropy expression; the introduction of this factor can be regarded as the definition of an entanglement in the context of thermodynamic modeling. For the gradients of energy and entropy, we have

$$\frac{\delta E}{\delta x} = \begin{pmatrix} -\frac{1}{2}\boldsymbol{v}(\boldsymbol{r})^2 \\ \boldsymbol{v}(\boldsymbol{r}) \\ 1 \\ 0 \end{pmatrix} \tag{4.143}$$

and

$$\frac{\delta S}{\delta x} = \begin{pmatrix} -\frac{\mu(\boldsymbol{r})}{T(\boldsymbol{r})} - \frac{N_{\mathrm{A}}}{M_{\mathrm{e}}} k_{\mathrm{B}} \int_0^1 \int f(\boldsymbol{r},\boldsymbol{u},\sigma) \ln f(\boldsymbol{r},\boldsymbol{u},\sigma) d^2u d\sigma \\ 0 \\ \frac{1}{T(\boldsymbol{r})} \\ -\rho(\boldsymbol{r}) \frac{N_{\mathrm{A}}}{M_{\mathrm{e}}} k_{\mathrm{B}} \left[\ln f(\boldsymbol{r},\boldsymbol{u},\sigma) - \int_0^1 \int f(\boldsymbol{r},\boldsymbol{u}',\sigma') \ln f(\boldsymbol{r},\boldsymbol{u}',\sigma') d^2u' d\sigma' \right] \end{pmatrix}, \tag{4.144}$$

where the normalization constraint (4.140) has been taken into account in calculating the functional derivative with respect to f (see Appendix C.3).

As a next step, the Poisson operator is taken from the dumbbell model for dilute polymer solutions, (4.121) with (4.122) and (4.123) where, except for the different normalization turning a scalar density into a scalar, the only difference is that the codeformational behavior of the connector vector $\boldsymbol{Q}$ needs to be turned into a corotational behavior of the unit vector $\boldsymbol{u}$. We therefore use

$$L_{42} = \frac{\partial f(\boldsymbol{r},\boldsymbol{u},\sigma)}{\partial \boldsymbol{r}} + \frac{\partial}{\partial \boldsymbol{u}} \cdot f(\boldsymbol{r},\boldsymbol{u},\sigma) \boldsymbol{P}(\boldsymbol{u}) \boldsymbol{u} \cdot \frac{\partial}{\partial \boldsymbol{r}} \tag{4.145}$$

and

$$L_{24} = -\frac{\partial f(\boldsymbol{r},\boldsymbol{u}',\sigma')}{\partial \boldsymbol{r}} - \frac{\partial}{\partial \boldsymbol{r}} \cdot f(\boldsymbol{r},\boldsymbol{u}',\sigma') \boldsymbol{u}' \boldsymbol{P}(\boldsymbol{u}') \cdot \frac{\partial}{\partial \boldsymbol{u}'} \tag{4.146}$$

instead of (4.122) and (4.123), where the transverse projector

$$\boldsymbol{P}(\boldsymbol{u}) = \boldsymbol{1} - \frac{\boldsymbol{u}\boldsymbol{u}}{\boldsymbol{u}^2} \tag{4.147}$$

makes sure that the length of $\boldsymbol{u}$ cannot change. Even though $\boldsymbol{u}$ is a unit vector, a normalizing factor $\boldsymbol{u}^2$ has been included in the definition of $\boldsymbol{P}(\boldsymbol{u})$. This has been done because, then, the constraint $\boldsymbol{u}^2 = 1$ needs not explicitly be taken into account in calculating derivatives with respect to $\boldsymbol{u}$. Even after the projector $\boldsymbol{P}(\boldsymbol{u})$ has been introduced, the Jacobi identity remains valid, as can be checked by an explicit calculation.[50]

Finally, the friction matrix for the Doi-Edwards model is also obtained by modifying the corresponding expression for the dumbbell model. In the operator version of (4.125), M_{44} is changed from (4.126) to

$$M_{44} = -\frac{\partial}{\partial \sigma} \frac{M_{\mathrm{e}} f}{\tau_{\mathrm{rept}} \pi^2 \rho N_{\mathrm{A}} k_{\mathrm{B}}} \frac{\partial}{\partial \sigma} \tag{4.148}$$

[50] In Appendix B.4, the validity of the proposed Poisson operator is established by arguments based on geometry and symmetry.

because we want to describe the one-dimensional diffusion in the tube rather than the three-dimensional diffusion of the beads in space.[51] Whereas the central 2×2-block of (4.125) contains the solvent viscosity, we here rather introduce a dissipative pressure tensor as derived by Curtiss and Bird by writing

$$M = \begin{pmatrix} 0 & 0 & 0 & 0 \\ 0 & -2\frac{\partial}{\partial \boldsymbol{r}} \cdot \boldsymbol{I}_{\mathrm{CB}} \cdot \frac{\partial}{\partial \boldsymbol{r}} & \frac{\partial}{\partial \boldsymbol{r}} \cdot \boldsymbol{I}_{\mathrm{CB}} : \dot{\boldsymbol{\gamma}} & 0 \\ 0 & -\dot{\boldsymbol{\gamma}} : \boldsymbol{I}_{\mathrm{CB}} \cdot \frac{\partial}{\partial \boldsymbol{r}} & \frac{1}{2}\dot{\boldsymbol{\gamma}} : \boldsymbol{I}_{\mathrm{CB}} : \dot{\boldsymbol{\gamma}} - \frac{\partial}{\partial \boldsymbol{r}} \cdot \boldsymbol{\lambda}^{\mathrm{q}} \cdot T^2 \frac{\partial}{\partial \boldsymbol{r}} & 0 \\ 0 & 0 & 0 & M_{44} \end{pmatrix}, \tag{4.149}$$

where $\boldsymbol{\lambda}^{\mathrm{q}}$ is the thermal conductivity tensor; the fourth-rank tensor $\boldsymbol{I}_{\mathrm{CB}}$ is chosen as

$$\boldsymbol{I}_{\mathrm{CB}} = \frac{3}{2}\pi^2 \epsilon_{\mathrm{CB}} \tau_{\mathrm{rept}} \rho \frac{N_{\mathrm{A}}}{M_{\mathrm{e}}} k_{\mathrm{B}} T^2 \int_0^1 \int \sigma(1-\sigma) \boldsymbol{u}\boldsymbol{u}\boldsymbol{u}\boldsymbol{u} f(\boldsymbol{u}, \sigma) d^2u d\sigma, \tag{4.150}$$

and ϵ_{CB} is the link tension coefficient. In the Doi-Edwards model, the dissipative pressure tensor contribution is neglected, that is, $\epsilon_{\mathrm{CB}} = 0$. It can be argued that ϵ_{CB} should be a small parameter of order $1/Z$.[52]

The friction matrix (4.149) is constructed such that the degeneracy condition (1.5) is satisfied. The complementary degeneracy condition, (1.4), implies the following pressure tensor expression for the Doi-Edwards model,

$$\boldsymbol{\pi}^{(S)} = T\left(s - \rho\frac{\partial s}{\partial \rho} - \epsilon\frac{\partial s}{\partial \epsilon}\right)\mathbf{1} + \rho\frac{N_{\mathrm{A}}}{M_{\mathrm{e}}} k_{\mathrm{B}} T \int_0^1 \int (\mathbf{1} - 3\boldsymbol{u}\boldsymbol{u}) f(\boldsymbol{u}, \sigma) d^2u d\sigma. \tag{4.151}$$

The additional dissipative pressure tensor contribution in the Curtiss-Bird model resulting from (4.150) is given by

$$-\frac{3}{2}\pi^2 \epsilon_{\mathrm{CB}} \rho \frac{N_{\mathrm{A}}}{M_{\mathrm{e}}} k_{\mathrm{B}} T \tau_{\mathrm{rept}} \dot{\boldsymbol{\gamma}} : \int_0^1 \int \sigma(1-\sigma) \boldsymbol{u}\boldsymbol{u}\boldsymbol{u}\boldsymbol{u} f(\boldsymbol{u}, \sigma) d^2u d\sigma. \tag{4.152}$$

It is remarkable that the pressure tensor in reptation models can be expressed in terms of $f(\boldsymbol{u}, \sigma)$, that is, in terms of the local orientation $\boldsymbol{u}$ of a single chain at the position σ. When the pressure tensor for a melt of linear polymer chains is evaluated on the atomistic level, say by Monte-Carlo simulation, then one clearly obtains contributions from interactions between different polymer chains. The answer to the question of whether the calculation of stresses requires more than single-chain information depends on the level of description. At the same time, the origin of stresses changes from energetic to entropic. On the reptation level, the information about interactions between different chains enters only indirectly through the constraints imposed by a tube or anisotropic friction. Any explicit formula for the pressure tensor depends on the interplay between the momentum density field and the structural

[51] A considerably richer structure of M_{44}, as given in (8.169), is found by detailed atomistic considerations. For the present purpose, however, the simplified expression (4.148) is sufficient.

[52] See p. 279 of Öttinger, *Stochastic Processes in Polymeric Fluids* (Springer, 1996).

variables chosen on a particular level of description of a complex fluid. It has been shown in simulations of polymer melts that, for an appropriate coarse-grained level of description, the pressure tensors of energetic and entropic origin, accounting for all pair interactions or just single-chain deformation, lead to identical results.[53]

Exercise 84 Derivation of Pressure Tensor

Show that the pressure tensor (4.151) is a consequence of the degeneracy requirement (1.4).

At this point, we have established the thermodynamic consistency of the Doi-Edwards and Curtiss-Bird reptation models. As a next step, we solve the Fokker-Planck equation (4.138) and eliminate the configurational distribution function from the list of variables. This step is useful to obtain the rheological properties predicted by reptation models; moreover, it illustrates the passage from a large to a small state space at the expense of introducing explicit memory (see Section 1.1.3). As a result, one then works with the hydrodynamic variables and balance equations only which, however, contain a momentum-flux tensor depending on deformation history. Actually, we obtain an explicit memory integral expression, or integral constitutive equation, for the pressure tensor as a functional of deformation history of a type that is well-established in continuum mechanics.

To solve (4.138), we introduce the quantity $\hat{\boldsymbol{u}} = \hat{\boldsymbol{u}}(\boldsymbol{u}', t, t')$ as the result of the deterministic time evolution of the unit vector $\boldsymbol{u}$ in the time interval from t' to t, that is, as the solution of $d\hat{\boldsymbol{u}}/dt = (\mathbf{1} - \hat{\boldsymbol{u}}\hat{\boldsymbol{u}}) \cdot \boldsymbol{\kappa} \cdot \hat{\boldsymbol{u}}$ with the initial condition $\hat{\boldsymbol{u}} = \boldsymbol{u}'$ at $t = t'$. For an inhomogeneous flow, one needs to trace the trajectory of a volume element between t' and t so that the velocity gradient $\boldsymbol{\kappa}$ can be evaluated at the proper positions along the trajectory. Finding the rotation of unit vectors by the flow field must clearly be an important step in solving the Fokker-Planck equation. We can now write the explicit solution of (4.138) as

$$f(\boldsymbol{u}, \sigma, t) = \int_{-\infty}^{t} dt' \sum_{\substack{j=1 \\ j \text{ odd}}}^{\infty} \frac{4j}{\pi \tau_{\text{rept}}} \sin(j\pi\sigma) e^{-j^2(t-t')/\tau_{\text{rept}}} \int \frac{d^2u'}{4\pi} \delta(\boldsymbol{u} - \hat{\boldsymbol{u}}(\boldsymbol{u}', t, t')). \tag{4.153}$$

To verify this expression for the solution, we observe that the second-order derivative with respect to σ produces a factor $-j^2\pi^2$, which, when divided by $\tau_{\text{rept}}\pi^2$, is canceled by the time derivative of the exponential. The further time dependence through $\hat{\boldsymbol{u}}(\boldsymbol{u}', t, t')$, when following a volume element because (4.138) contains the material time derivative, is given by the chain rule (see Exercise 85)

$$\frac{\partial}{\partial t} \delta(\boldsymbol{u} - \hat{\boldsymbol{u}}(\boldsymbol{u}', t, t')) = -\frac{\partial}{\partial \boldsymbol{u}} \cdot [(\mathbf{1} - \boldsymbol{u}\boldsymbol{u}) \cdot \boldsymbol{\kappa} \cdot \boldsymbol{u} \delta(\boldsymbol{u} - \hat{\boldsymbol{u}}(\boldsymbol{u}', t, t'))], \tag{4.154}$$

and it thus cancels the $\partial/\partial \boldsymbol{u}$ term in (4.138). After verifying that (4.153) indeed solves (4.138), we need to check the boundary conditions (4.139). For any $t - t' > 0$,

[53] Mavrantzas & Theodorou, Macromolecules 31 (1998) 6310; see Section 8.2.3 for further details.

only a finite number of terms in the sum over j are important because further terms become exponentially small. By choosing σ sufficiently close to 0 or 1, $\sin(j\pi\sigma)$ can be made arbitrarily small for this finite number of terms, so that the time integral must be fully determined by the singularity of the integrand at $t = t'$. Sufficiently close to the boundaries, we can thus assume $\widehat{\boldsymbol{u}}(\boldsymbol{u}', t, t') = \boldsymbol{u}'$, and we then obtain

$$\begin{aligned} f(\boldsymbol{u}, \sigma, t) &= \frac{1}{4\pi} \int_{-\infty}^{t} dt' \sum_{\substack{j=1 \\ j \text{ odd}}}^{\infty} \frac{4j}{\pi \tau_{\text{rept}}} \sin(j\pi\sigma) e^{-j^2(t-t')/\tau_{\text{rept}}} \\ &= \frac{1}{4\pi} \sum_{\substack{j=1 \\ j \text{ odd}}}^{\infty} \frac{4}{\pi} \frac{\sin(j\pi\sigma)}{j} = \frac{1}{4\pi}, \end{aligned} \tag{4.155}$$

where the Fourier series of the constant function 1 has been used.[54] While each term in the sum over j solves the Fokker-Planck equation (4.138) individually, the boundary conditions require the particular combination proposed in (4.153).

Exercise 85 Proof of Auxiliary Equation

Prove the auxiliary equation (4.154).

With the solution (4.153) of the Fokker-Planck equation for the reptation model, we obtain an explicit integral expression for the pressure tensor (4.151) of the Doi-Edwards model,

$$\begin{aligned} \boldsymbol{\tau}^{(S)}(t) &= \rho \frac{N_{\text{A}}}{M_{\text{e}}} k_{\text{B}} T \left\{ \mathbf{1} - 3 \int_{-\infty}^{t} dt' \left[\frac{8}{\pi^2 \tau_{\text{rept}}} \sum_{\substack{j=1 \\ j \text{ odd}}}^{\infty} e^{-j^2(t-t')/\tau_{\text{rept}}} \right] \right. \\ &\quad \times \left. \int \frac{d^2u'}{4\pi} \widehat{\boldsymbol{u}}(\boldsymbol{u}', t, t') \widehat{\boldsymbol{u}}(\boldsymbol{u}', t, t') \right\}, \end{aligned} \tag{4.156}$$

which allows us to evaluate the pressure tensor as a functional of deformation history. The factor in square brackets can be considered as as memory function, which is normalized such that its integral over t' from $-\infty$ to t is equal to unity. The integral over $\boldsymbol{u}'$ in the second line of (4.156) defines a deformation measure; it describes how the flow affects the dyadic $\boldsymbol{u}'\boldsymbol{u}'$ of an initially isotropic ensemble from time t' to t. Equation (4.156) fits into the class of factorized K-BKZ constitutive equations, which are a subclass of the Rivlin-Sawyers integral equations, where both classes are well-known in continuum mechanics.[55]

[54] See, for example, 1.442.1 of Gradshteyn & Ryzhik, *Tables* (Academic, 1980).

[55] See § 8.3 of Bird, Armstrong & Hassager, *DPL 1, Fluid Mechanics* (Wiley, 1987), in particular, Table 8.3-2.

From the integral expression for the pressure tensor, one can derive the following relaxation modulus (see Exercise 88)

$$G(t) = G(0)\frac{8}{\pi^2} \sum_{\substack{j=1 \\ j \text{ odd}}}^{\infty} \frac{1}{j^2} e^{-j^2 t/\tau_{\text{rept}}}, \tag{4.157}$$

with

$$G(0) = \frac{3}{5}\rho \frac{N_{\text{A}}}{M_{\text{e}}} k_{\text{B}} T. \tag{4.158}$$

The weight factor $1/j^2$ in (4.157) implies that the modulus is strongly dominated by a single relaxation mode, even at short times. The Cole-Cole plot for the Doi-Edwards model is thus practically indistinguishable from the one shown in Figure 4.3. Such a dominance of a single relaxation mode is never observed in experiments on entangled melts, even for a narrow distribution of molecular weights.

Experimentally, $G(0)$ is obtained as the plateau modulus after all rapid relaxation processes are completed, when the slow relaxation due to reptation begins. From the modulus $G(0)$ of an entangled network of chains, one can thus determine the entanglement molecular weight

$$M_{\text{e}} = \frac{3\rho N_{\text{A}} k_{\text{B}} T}{5G(0)}. \tag{4.159}$$

Another estimate of the entanglement molecular weight can be obtained from the viscosity and the first normal-stress coefficient in the limit of zero shear rate, as obtained from the relaxation modulus (4.157) (see Exercise 89):

$$\eta = \frac{\pi^2}{20}\rho \frac{N_{\text{A}}}{M_{\text{e}}} k_{\text{B}} T \tau_{\text{rept}} \tag{4.160}$$

and

$$\Psi_1 = \frac{\pi^4}{100}\rho \frac{N_{\text{A}}}{M_{\text{e}}} k_{\text{B}} T \tau_{\text{rept}}^2, \tag{4.161}$$

which imply

$$M_{\text{e}} = \rho N_{\text{A}} k_{\text{B}} T \frac{\Psi_1}{4\eta^2}. \tag{4.162}$$

The experimental data for M_{e} in Table 4.2 were obtained from (4.159) with a prefactor $4/5$ instead of $3/5$, which corresponds to a more rigorous version of the Doi-Edwards model [conversely, the more rigorous model predicts a 3 instead of the 4 in the denominator of (4.162)].

Because the density and the entanglement molecular weight are independent of the molecular weight of the entangled polymer molecules, the zero-shear-rate viscosity (4.160) has the same molecular-weight-dependence as the reptation time scale. This is an important prediction of the reptation model. A classical argument further leads to

$$\eta \propto \tau_{\text{rept}} \propto M_{\text{p}}^3, \tag{4.163}$$

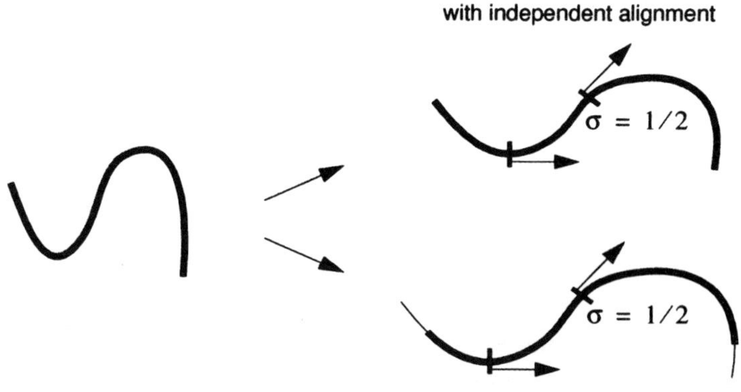

Fig. 4.13 Relaxation of a chain into the confining tube upon deformation is neglected under the independent alignment approximation; the thin lines indicate the part of the deformed tube that is no longer occupied by the probe chain.

which explains the enormously high viscosities observed for melts of highly entangled polymers. The simplest version of de Gennes argument[56] can be presented as follows. Because the total friction experienced by an entangled chain is proportional to the length of the confining tube, the curvelinear diffusion coefficient in the tube is inversely proportional to the molecular weight, $D_{\rm tube} \propto M_{\rm p}^{-1}$. The reptation or disentanglement time, which is characterized by the condition that the chain diffuses along the tube by the full contour length $ZL_{\rm e}$, that is, $(ZL_{\rm e})^2 = D_{\rm tube}\tau_{\rm rept}$, then implies $\tau_{\rm rept} \propto Z^2 M_{\rm p}$ or (4.163). Experimental observations for moderately entangled chains actually suggest an exponent even somewhat larger than 3, namely, around 3.4.

Another important prediction is made for the ratio of normal-stress coefficients at low shear rates (see Exercise 89),

$$\frac{\Psi_2}{\Psi_1} = -\frac{2}{7}. \tag{4.164}$$

For a long time, the value $-2/7$ was considered to be an artifact of the so-called *independent alignment approximation* made in the original Doi-Edwards model. This approximation is illustrated in Figure 4.13. When a melt is deformed, the tube becomes stretched, and the chain shrinks rapidly into the tube to recover its original contour length (Figure 4.13, bottom). With the independent alignment approximation, this stretching effect is neglected, and the local orientations are independently aligned by the deformation field (Figure 4.13, top). When avoiding the independent alignment approximation, Figure 4.13 shows that the orientation at the chain center is unaffected; however, the other orientational constraints are felt at σ-labels further

[56] See p. 226 of de Gennes, *Scaling Concepts in Polymer Physics* (Cornell, 1979).

away from the chain center, as determined by the tube-stretching rate accumulated from the center. Doi[57] proposed the following drift away from the center:

$$\boldsymbol{\kappa} : \int_{1/2}^{\sigma} \langle \boldsymbol{u}\boldsymbol{u} \rangle_{\sigma'} \, d\sigma', \tag{4.165}$$

with the locally averaged orientation tensor at the position σ within the chain,

$$\langle \boldsymbol{u}\boldsymbol{u} \rangle_{\sigma} = \int \boldsymbol{u}\boldsymbol{u} f(\boldsymbol{u}, \sigma) d^2 u. \tag{4.166}$$

The segments lost at the tube ends must be compensated by a creation term in the Fokker-Planck equation, as more chain segments now feel the inner tube orientations. We hence generalize (4.138) to

$$\begin{aligned} \frac{\partial f}{\partial t} &= -\boldsymbol{v} \cdot \frac{\partial f}{\partial \boldsymbol{r}} - \frac{\partial}{\partial \boldsymbol{u}} \cdot [(\mathbf{1} - \boldsymbol{u}\boldsymbol{u}) \cdot \boldsymbol{\kappa} \cdot \boldsymbol{u} f] + \frac{1}{\tau_{\text{rept}} \pi^2} \frac{\partial^2 f}{\partial \sigma^2} \\ &- \frac{\partial}{\partial \sigma} (\boldsymbol{\kappa} : \boldsymbol{S}^{\text{drift}} f) + \boldsymbol{\kappa} : \boldsymbol{S}^{\text{creation}} f, \end{aligned} \tag{4.167}$$

where the drift (4.165) corresponds to

$$\boldsymbol{S}^{\text{drift}}(\sigma) = \int_{1/2}^{\sigma} \langle \boldsymbol{u}\boldsymbol{u} \rangle_{\sigma'} \, d\sigma', \tag{4.168}$$

and Doi's original proposal was

$$\boldsymbol{S}^{\text{creation}}(\sigma) = \langle \boldsymbol{u}\boldsymbol{u} \rangle_{\sigma}. \tag{4.169}$$

The choice (4.169) is referred to as "erroneous" in the book by Doi and Edwards,[58] where they rather propose the unaveraged version

$$\boldsymbol{S}^{\text{creation}}(\boldsymbol{u}) = \boldsymbol{u}\boldsymbol{u}. \tag{4.170}$$

Because avoiding the independent alignment approximation can be regarded as a more careful treatment of the convection effect on a chain reptating in a tube, all the assumptions summarized here can be translated into modifications of the Poisson operator. One then notes that the modifications with (4.169) and (4.170) both lead

[57] Doi, J. Polym. Sci., Polym. Phys. Ed. 18 (1980) 2055.
[58] See footnote on p. 277 of Doi & Edwards, *Theory of Polymer Dynamics* (Clarendon, 1986); it should be noted that the derivation of the "rigorous constitutive equation" in the book of Doi and Edwards, which in our notation suggests the replacement of (4.169) by (4.170), is mathematically questionable because their equations (7.246)–(7.250) don't seem to follow from (7.243).
[59] Doi, J. Polym. Sci., Polym. Phys. Ed. 18 (1980) 2055.
[60] Section 7.9 of Doi & Edwards, *Theory of Polymer Dynamics* (Clarendon, 1986).
[61] Öttinger & Beris, J. Chem. Phys. 110 (1999) 6593.
[62] Section V.B of Öttinger, J. Rheol. 43 (1999) 1461.

Table 4.3 Summary of Reptation Models Avoiding the Independent Alignment Approximation

Reference	$\boldsymbol{S}^{\mathrm{drift}}$	$\boldsymbol{S}^{\mathrm{creation}}$	Ψ_2/Ψ_1	Admissiblility
Doi 1980[59]	$\int_{1/2}^{\sigma}\langle\boldsymbol{uu}\rangle_{\sigma'}d\sigma'$	$\langle\boldsymbol{uu}\rangle_{\sigma}$	$-2/7$	no
Doi/Edwards 1986[60]	$\int_{1/2}^{\sigma}\langle\boldsymbol{uu}\rangle_{\sigma'}d\sigma'$	$\boldsymbol{uu}$	$-1/7$	no
GENERIC 1999a[61]	$\int_{1/2}^{\sigma}\langle\boldsymbol{uu}\rangle_{\sigma'}d\sigma'$	$\langle\langle\boldsymbol{uu}\rangle\rangle$	$-2/7$	yes
GENERIC 1999b[62]	$\left(\sigma-\frac{1}{2}\right)\langle\langle\boldsymbol{uu}\rangle\rangle$	$\langle\langle\boldsymbol{uu}\rangle\rangle$	$-2/7$	yes

to a violation of the Jacobi identity; therefore, they violate an important criterion of thermodynamic admissibility. At this point, one could argue that the violation of the Jacobi identity is a consequence of nonholonomic constraints in restricting σ to the interval $[0, 1]$, or one may look for alternative formulations satisfying the Jacobi identity. It was shown that a globally averaged variation of (4.169) is thermodynamically admissible,[63]

$$\boldsymbol{S}^{\mathrm{creation}} = \langle\langle\boldsymbol{uu}\rangle\rangle = \int_0^1 \int \boldsymbol{uu}f(\boldsymbol{u},\sigma)d^2u d\sigma. \tag{4.171}$$

However, because the creation term (4.171) is uniform along the chain while the drift term (4.168) can have a complicated σ-dependence, a nonuniform monomer distribution along the tube can result from the combination of these two equations (one hence needs to be careful with the interpretation of (4.166) as an average at constant σ). To maintain a uniform monomer distribution, one should use (4.171) together with

$$\boldsymbol{S}^{\mathrm{drift}}(\sigma) = \left(\sigma - \frac{1}{2}\right)\langle\langle\boldsymbol{uu}\rangle\rangle\,. \tag{4.172}$$

This expression for $\boldsymbol{S}^{\mathrm{drift}}$ was constructed in an alternative manner by starting with an additional contour length variable and eliminating it in the limit of fast relaxation of the contour length.[64]

All the suggestions for avoiding the independent alignment approximation are summarized in Table 4.3. It should be noted that the thermodynamically admissible models satisfying the Jacobi identity predict $\Psi_2/\Psi_1 = -2/7 \approx -0.28$, as the original Doi-Edwards model. Experiments suggest a normal-stress ratio close to -0.25.

We finally discuss the shear-rate dependence of the viscosity. From the integral constitutive equation (4.156), we obtain the viscosity for arbitrary shear rates as a time integral. In particular, one can construct the following expansion of the viscosity in

[63] Öttinger & Beris, J. Chem. Phys. 110 (1999) 6593.

[64] See Section V.B of Öttinger, J. Rheol. 43 (1999) 1461.

terms of $\dot{\gamma}^2$ (see Exercise 91),

$$\eta(\dot{\gamma}) = 24\pi^2 \rho \frac{N_A}{M_e} k_B T \tau_{rept} \sum_{j=0}^{\infty} \frac{(-1)^j}{j+1} \frac{(2j+1)!!(2j+3)!!}{(4j+3)!!(4j+5)!!} \times \left(1 - \frac{1}{2^{4j+4}}\right) |B_{4j+4}| (2\pi^2 \tau_{rept} \dot{\gamma})^{2j}, \tag{4.173}$$

where $(2j + 1)!!$ is the product of all odd integers from 1 to $2j + 1$ and the rational numbers B_k are known as the Bernoulli numbers.[65] The asymptotic behavior of the Bernoulli numbers for even k is given by

$$|B_k| \sim 2 \frac{k!}{(2\pi)^k}. \tag{4.174}$$

The expansion (4.173) for the viscosity is a so-called *asymptotic series*, that is, it is not convergent, not even for very small values of $\tau_{rept}\dot{\gamma}$. To see this, we consider the ratio $r(j)$ of the $(j + 1)$th and the jth term in the infinite sum (4.173); for large j, one has

$$|r(j)| = (j \tau_{rept} \dot{\gamma})^2. \tag{4.175}$$

For all values of $\tau_{rept}\dot{\gamma}$, there exists a certain value j_0 such that $|r(j)| \gg 1$ for $j \geq j_0$; that is, the absolute values of the contributions to the sum (4.173) increase very rapidly with j, and the series is clearly divergent. However, for very small values of $\tau_{rept}\dot{\gamma}$, the first few terms of the series suggest that the sum converges to a well-defined value; the sum then changes almost unnoticeably with j, and only for very large values of j (more precisely, for $j\tau_{rept}\dot{\gamma} > 1$) does the sum begin to diverge. For example for $\tau_{rept}\dot{\gamma} = 0.05$, the sum of the first few terms suggests rapid convergence; the minimum contribution to the sum is reached for $j = 19$, and inspection of the sum of the first j terms for $j \leq 19$ suggests that 15 significant figures for $\eta(0.05)$ can be obtained. For $\tau_{rept}\dot{\gamma} = 0.1$, the minimum term in the sum (4.173) is reached for $j = 9$, and only seven significant figures are found before the series begins to diverge. Figure 4.14 suggests that, for $\tau_{rept}\dot{\gamma} > 0.2$, much less precision can be obtained.

The above examples indicate that there is useful information contained in the divergent series (4.173). This is so because, by increasing the number of terms in (4.173), one in fact obtains better approximations for the viscosity, however, only for sufficiently small values of $\tau_{rept}\dot{\gamma}$; for larger values of $\tau_{rept}\dot{\gamma}$, the approximation obtained by increasing the number of terms gets poorer. The larger the number of terms, the smaller is the range of shear rates for which a better approximation is achieved by taking additional terms into account. This situation is illustrated in Figure 4.14. For a convergent series, one could, for any $\tau_{rept}\dot{\gamma}$, increase the number of terms until the sum is sufficiently close to the limiting value. For an asymptotic series, this is not possible; one has to stop summation when the minimum term is reached, and there is no systematic possibility to obtain a better approximation from

[65] See 0.233.5 and 9.71 of Gradshteyn & Ryzhik, *Tables* (Academic, 1980).

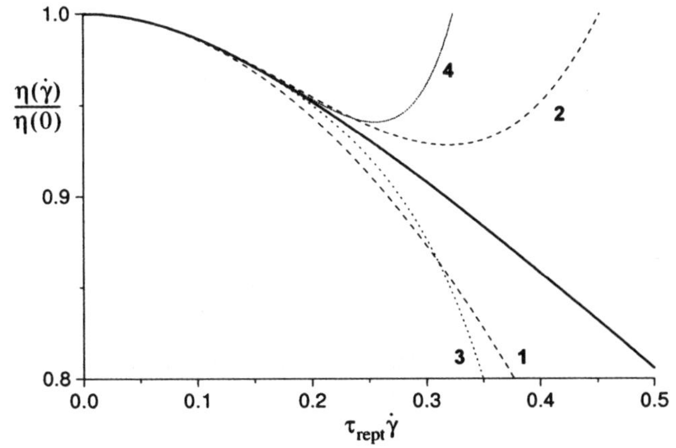

Fig. 4.14 The exact result obtained by numerical integration (thick line) and various approximations for the normalized viscosity as a function of the dimensionless shear rate; the numbers attached to the approximate curves give the power of $(\tau_{\text{rept}}\dot{\gamma})^2$ after which the low-shear-rate expansion (4.173) has been truncated to obtain the respective curve. [Reprinted with permission from H. C. Öttinger, Series Expansions for Shear-Flow Properties from the Doi-Edwards Model, J. Non-Newtonian Fluid Mech. 37 (1990) 265. Copyright (1990) by Elsevier.]

an asymptotic series. Because the terms in the sum (4.173) alternate in signs, the error resulting from truncation of the series can be estimated quite reliably by inspection of the last few terms before the series has to be truncated (see Figure 4.14).

For high shear rates, where the expansion (4.173) is of no use, the Doi-Edwards model predicts[66]

$$\eta = 1.66791\rho\frac{N_{\text{A}}}{M_{\text{e}}}k_{\text{B}}T\tau_{\text{rept}}(\tau_{\text{rept}}\dot{\gamma})^{-3/2}, \tag{4.176}$$

which suggests a too strong shear thinning. In particular, the shear stress increases with shear rate at low shear rates, goes through a maximum near $\tau_{\text{rept}}\dot{\gamma} = 1$, and decreases at high shear rates as $\dot{\gamma}^{-1/2}$. The fact that one can have the same shear stress at low and high shear rates leads to an instability of shear flow.

Exercise 86 Deformation of a Unit Vector by Shear Flow

Construct a closed-form expression for $\widehat{\boldsymbol{u}}(\boldsymbol{u}', t, t')$ for shear flow with an arbitrary time dependent shear rate.

Exercise 87 Deformation Measure in Shear Flow

Consider the deformation measure of the Doi-Edwards model,

$$\boldsymbol{Q}^{\text{DE}}(\gamma) = \int \frac{d^2u'}{4\pi}\widehat{\boldsymbol{u}}(\boldsymbol{u}', t, t')\widehat{\boldsymbol{u}}(\boldsymbol{u}', t, t') \tag{4.177}$$

[66] See (16) and (19) of Öttinger, J. Non-Newtonian Fluid Mech. 37 (1990) 265.

for a total shear deformation $\gamma = \int_{t'}^{t} \dot{\gamma}(t'')dt''$. Demonstrate the identity

$$Q_{xx}^{\rm DE}(\gamma) - Q_{yy}^{\rm DE}(\gamma) = \gamma Q_{xy}^{\rm DE}(\gamma), \tag{4.178}$$

for all γ, and, for $\gamma \ll 1$, derive the asymptotic behavior

$$Q_{xy}^{\rm DE}(\gamma) = \frac{1}{5}\gamma, \tag{4.179}$$

$$Q_{yy}^{\rm DE}(\gamma) - Q_{zz}^{\rm DE}(\gamma) = -\frac{2}{35}\gamma^2. \tag{4.180}$$

Exercise 88 Relaxation Modulus for the Doi-Edwards Model

Derive (4.157) and (4.158) for the shear relaxation modulus of the Doi-Edwards model.

Exercise 89 Zero-Shear-Rate Viscometric Functions

Calculate the zero-shear-rate viscometric functions (4.160), (4.161), and (4.164) for the Doi-Edwards model.

Exercise 90 Poisson Operator Without Independent Alignment

Write down the Poisson operator for reptation models without the independent alignment approximation.

Exercise 91 Expansion for the Viscosity

Derive the asymptotic expansion (4.173) for the viscosity.

5

Relativistic Hydrodynamics

The clearly distinguished role of time in the fundamental time-evolution equation (1.1) raises the question of whether the GENERIC structure can be consistent with the principles of relativity. If the GENERIC approach is really fundamental, it should work also for relativistic systems. In Chapter 5, we show that the principles of nonequilibrium thermodynamics and of special relativity are indeed compatible. Because the respective structures are so different, the combination of these two theories actually leads to a rather restricted form of the possible model equations or, in other words, to high predictive power.

For example, due to the intimate relation between space and time in the theory of relativity, the time derivatives on the left side of the GENERIC (1.1) must be matched by first-order space derivatives on the right side, which naturally occur as convective effects in the Poisson matrix. Higher-order space derivatives are not allowed on the right side. In particular, in a relativistic theory, the friction matrix should not contain any derivatives. This situation is quite different from the various nonrelativistic examples considered in the preceding chapters, where the friction matrices typically contain second-order derivatives associated with transport processes, and even higher-order derivatives or more nonlocal expressions for the friction matrix would be possible in nonrelativistic theories. Therefore, relativistic realizations of GENERIC are strongly restricted. Only relaxation processes are admitted in the irreversible contribution. We here show for a fluid with heat flow and viscous pressure tensor how the dissipative effects typically described by second-order derivative terms can be formulated in accordance with both GENERIC and special relativity. The purpose of this chapter is to propose a minimum setting, in which the hydrodynamics of a relativistic imperfect fluid can be described in a GENERIC form.[1] Although we are forced to introduce some faster independent variables in addition to the hydrodynamic ones to avoid the occurrence of second-order space derivatives, we do not attempt to describe the multitude of fast processes present on the level of kinetic theory (see Chapter 7).

In addition to these rather formal considerations, there are good physical reasons for being interested in relativistic hydrodynamics. The following quote from Weinberg's book on gravitation and cosmology underlines this statement:[2]

> The correct treatment of dissipative effects for relativistic fluids raises certain delicate questions of principle, which do not arise in the nonrelativistic case. For this reason, and also because dissipation plays an increasingly important role in theories of the early universe, it will be worth our while here to develop the outlines of the general theory of relativistic imperfect fluids.

After including some additional independent variables and pointing out the relationship to extended irreversible thermodynamics, we consider the equations of special relativistic hydrodynamics in the noncovariant formulation of GENERIC presented so far, where time plays a distinguished role in the fundamental equations. The final equations, of course, can be written in a covariant form. Before introducing gravity, we present a covariant formulation of the GENERIC framework itself. Then, gravity is not treated as an external field; rather, the gravitational field is part of the isolated system within a general relativistic formulation. For illustration, we discuss the application of our equations in bulk viscous cosmology, including a detailed comparison to previous theories of relativistic hydrodynamics.

For this chapter on relativistic hydrodynamics, a background on relativity is needed. Because we use both special and general relativity, the famous textbooks of Weinberg[3] and of Misner, Thorne and Wheeler[4] are used here as standard references. Only Section 5.1 can be read without any knowledge of relativity. It provides the ideas and equations to describe a complex fluid in terms of a vector and a tensor variable in addition to the hydrodynamic fields, and it can thus be considered as a natural addition to Section 4.2. In particular, extended irreversible thermodynamics is based on an additional vector and a tensor field, as is discussed in Section 5.1.6. The rest of Chapter 5 is not required to understand the subsequent chapters.

5.1 PRELUDE: A TENSOR AND A VECTOR AS VARIABLES

The description of dissipative phenomena, such as heat flow and viscous stresses, typically involves second-order derivatives. For example, the energy equation (2.31) involves the divergence of the heat flux (2.37), which is itself proportional to the gra-

[1] In the literature on relativistic hydrodynamics, one usually refers to *imperfect* and *perfect fluids*, depending on whether dissipative momentum and heat fluxes are present or not. In the usual terminology of nonrelativistic hydrodynamics, perfect fluids are described by the Euler equations.

[2] See p. 53 of Weinberg, *Gravitation and Cosmology* (Wiley, 1972).

[3] Weinberg, *Gravitation and Cosmology* (Wiley, 1972).

[4] Misner, Thorne & Wheeler, *Gravitation* (Freeman, 1973).

dient of temperature. The standard procedure to avoid higher-order derivatives is to introduce additional variables that are closely related to these derivatives. More precisely, we envision a simple dissipative relaxation mechanism by which an additional vector variable rapidly converges to the temperature gradient, while the divergence of this variable occurs through the Poisson operator in the energy or entropy balance equation. Similarly, we expect an additional tensor variable that converges to the velocity gradients. This procedure is analogous to the idea of avoiding explicit memory effects by using additional structural variables (see Section 1.1.3 and the concrete example on p. 148): additional variables can be used to make the equations local in both time and space. Because the construction of "good" variables is a key step in formulating beyond-equilibrium thermodynamic models, we try to motivate our choice carefully during the construction.[5] We develop a nonrelativistic theory for an augmented set of variables consisting of the hydrodynamic variables mass density, momentum density, and entropy density, $(\rho, \boldsymbol{M}, s)$, a further vector variable, $\boldsymbol{w}$, and an additional tensor variable, $\boldsymbol{c}$. In Section 4.2, we have discussed the modeling of complex fluids with a tensor and a scalar structural variable in addition to the hydrodynamic variables; here, we work with additional tensor and vector variables. Our approach is in the spirit of what we previously did for complex fluids, rather than being motivated by kinetic theory.

5.1.1 Energy and Entropy

For the energy functional of the fields $x = (\rho, \boldsymbol{M}, s, \boldsymbol{w}, \boldsymbol{c})$, we assume the following version of (2.48):

$$E(x) = \int \left[\frac{\boldsymbol{M}(\boldsymbol{r})^2}{2\rho(\boldsymbol{r})} + \epsilon(\boldsymbol{r}) \right] d^3r, \tag{5.1}$$

where, for an isotropic fluid, we postulate a general thermodynamic relationship for the internal energy density ϵ as a function of the independent variables in the local rest frame, which is of the form

$$\epsilon = \epsilon(\rho, s, \mathrm{tr}\boldsymbol{c}, \mathrm{tr}\boldsymbol{c}^2, \mathrm{tr}\boldsymbol{c}^3, \boldsymbol{w}^2, \boldsymbol{w} \cdot \boldsymbol{c} \cdot \boldsymbol{w}, \boldsymbol{w} \cdot \boldsymbol{c}^2 \cdot \boldsymbol{w}). \tag{5.2}$$

Rather than formally writing $\epsilon = \epsilon(\rho, s, \boldsymbol{w}, \boldsymbol{c})$ (see Section 5.4), we here use explicitly scalar arguments in the internal energy density. Following the previous work on which this section is based,[6] we use scalar invariants of the lower-convected tensor $\boldsymbol{c}$ that are different from those introduced in (4.40)–(4.42) for a upper-convected tensor,[7] and other choices could also be useful. We introduce the following notation for the

[5] Since the pioneering work of Cattaneo, Atti Sem. Mat. Fis. Univ. Modena 3 (1948) 3, the need to introduce additional variables related to heat flux has carefully been motivated in the nonrelativistic approach, for example, in Müller, Z. Physik 198 (1967) 329 or in Section 10.1 of Beris & Edwards, *Thermodynamics of Flowing Systems* (Oxford, 1994).

[6] Öttinger, Physica A 254 (1998) 433.

[7] In passing from an upper-convected tensor to its lower-convected inverse, I_1 and I_3 are exchanged, and all three invariants pick up a minus sign.

partial derivatives of ϵ with respect to the arguments listed in (5.2),

$$\mu^* = \frac{\partial \epsilon}{\partial \rho}, \tag{5.3}$$

$$T^* = \frac{\partial \epsilon}{\partial s}, \tag{5.4}$$

$$\phi^{*(1)} = 2\frac{\partial \epsilon}{\partial \mathrm{tr}\boldsymbol{c}}, \tag{5.5}$$

$$\phi^{*(2)} = 4\frac{\partial \epsilon}{\partial \mathrm{tr}\boldsymbol{c}^2}, \tag{5.6}$$

$$\phi^{*(3)} = 6\frac{\partial \epsilon}{\partial \mathrm{tr}\boldsymbol{c}^3}, \tag{5.7}$$

$$\sigma^{*(1)} = 2\frac{\partial \epsilon}{\partial \boldsymbol{w}^2} \tag{5.8}$$

$$\sigma^{*(2)} = 2\frac{\partial \epsilon}{\partial \boldsymbol{w}\cdot\boldsymbol{c}\cdot\boldsymbol{w}}, \tag{5.9}$$

$$\sigma^{*(3)} = 2\frac{\partial \epsilon}{\partial \boldsymbol{w}\cdot\boldsymbol{c}^2\cdot\boldsymbol{w}}, \tag{5.10}$$

where the superscript "*" is used as a warning to indicate that the corresponding derivatives are formed for a list of variables in which only the first two variables are densities of extensive variables, whereas all the other variables are intensive ones. With this notation, we obtain the following energy gradient:

$$\frac{\delta E(x)}{\delta x} = \begin{pmatrix} \mu^* - \frac{1}{2}\boldsymbol{v}^2 \\ \boldsymbol{v} \\ T^* \\ \boldsymbol{j}^{\mathrm{s}} \\ \frac{1}{2}\boldsymbol{\phi}^* \end{pmatrix}, \tag{5.11}$$

with

$$\boldsymbol{j}^{\mathrm{s}} = \boldsymbol{\sigma}^* \cdot \boldsymbol{w}, \tag{5.12}$$

$$\boldsymbol{\sigma}^* = \sigma^{*(1)}\mathbf{1} + \sigma^{*(2)}\boldsymbol{c} + \sigma^{*(3)}\boldsymbol{c}^2, \tag{5.13}$$

and

$$\boldsymbol{\phi}^* = \phi^{*(1)}\mathbf{1} + \phi^{*(2)}\boldsymbol{c} + \phi^{*(3)}\boldsymbol{c}^2 + \sigma^{*(2)}\boldsymbol{w}\boldsymbol{w} + \sigma^{*(3)}(\boldsymbol{w}\boldsymbol{w}\cdot\boldsymbol{c} + \boldsymbol{c}\cdot\boldsymbol{w}\boldsymbol{w}). \tag{5.14}$$

The entropy is fully determined by the independent variable s,

$$S(x) = \int s(\boldsymbol{r})d^3r. \tag{5.15}$$

The entropy gradient is thus trivial because only the entropy component of the gradient, which is unity, is different from zero.

Exercise 92 Gradient of Energy
Derive the expression (5.11)–(5.14) for the gradient of the energy.

Exercise 93 Internal Energy Density as Independent Variable
Write out the entropy density in terms of the independent variables $(\rho, \boldsymbol{M}, \epsilon, \boldsymbol{w}, \boldsymbol{c})$. How does one obtain the functions (5.3)–(5.10) from this expression for the entropy density? Calculate the functions $\phi^{*(1)}, \phi^{*(2)}, \phi^{*(3)}$ and the tensor $\boldsymbol{\phi}^*$ for a configurational entropy density contribution of the form

$$s = -\frac{1}{2} n_{\rm p} k_{\rm B} [\mathrm{tr}(\mathbf{1} - \boldsymbol{c})^{-1} + \ln \det(\mathbf{1} - \boldsymbol{c}) - 3], \tag{5.16}$$

which is obtained from (4.86) by replacing the tensor $\boldsymbol{c}$ by $(\mathbf{1} - \boldsymbol{c})^{-1}$.

5.1.2 Poisson Matrix

To construct the Poisson matrix, the following assumptions are important: $\boldsymbol{w}$ is a lower-convected vector variable, $\boldsymbol{c} - \mathbf{1}$ is a lower-convected tensor variable, and both variables are of intensive nature. These assumptions are motivated by the fact that $\boldsymbol{w}$ should be closely related to $\partial T^*/\partial \boldsymbol{r}$, where the temperature T^* is intensive and the derivative with respect to position possesses lower-convected behavior. Similarly, $\boldsymbol{c}$ is closely related to $\partial \boldsymbol{v}/\partial \boldsymbol{r}$. According to Table 2.1, we write down the Poisson matrix for $x = (\rho, \boldsymbol{M}, s, \boldsymbol{w}, \boldsymbol{c})$

$$L = -\begin{pmatrix} 0 & \frac{\partial}{\partial \boldsymbol{r}} \rho & 0 & 0 & 0 \\ \rho \frac{\partial}{\partial \boldsymbol{r}} & \left(\frac{\partial}{\partial \boldsymbol{r}} \boldsymbol{M} + \boldsymbol{M} \frac{\partial}{\partial \boldsymbol{r}}\right)^T & s \frac{\partial}{\partial \boldsymbol{r}} & -\left[\frac{\partial \boldsymbol{w}}{\partial \boldsymbol{r}}\right] + \left(\frac{\partial}{\partial \boldsymbol{r}} \boldsymbol{w}\right)^T & L_{25} \\ 0 & \frac{\partial}{\partial \boldsymbol{r}} s & 0 & \frac{\partial}{\partial \boldsymbol{r}} & 0 \\ 0 & \left[\frac{\partial \boldsymbol{w}}{\partial \boldsymbol{r}}\right]^T + \left(\boldsymbol{w} \frac{\partial}{\partial \boldsymbol{r}}\right)^T & \frac{\partial}{\partial \boldsymbol{r}} & 0 & 0 \\ 0 & L_{52} & 0 & 0 & 0 \end{pmatrix}, \tag{5.17}$$

with

$$(L_{52})_{jk,l} = \left[\frac{\partial c_{jk}}{\partial r_l}\right] + (c_{jl} - \delta_{jl}) \frac{\partial}{\partial r_k} + (c_{kl} - \delta_{kl}) \frac{\partial}{\partial r_j} \tag{5.18}$$

and

$$(L_{25})_{j,lm} = -\left[\frac{\partial c_{lm}}{\partial r_j}\right] + \frac{\partial}{\partial r_l} (c_{jm} - \delta_{jm}) + \frac{\partial}{\partial r_m} (c_{jl} - \delta_{jl}). \tag{5.19}$$

The $(4, 2)$- and $(2, 4)$-elements are the natural choices for the intensive lower-convected vector $\boldsymbol{w}$, and the $(5, 2)$- and $(2, 5)$-elements are the natural choices for the intensive lower-convected tensor $\boldsymbol{c} - \mathbf{1}$. Not so obvious is the modification in the $(4, 3)$- and $(3, 4)$-elements. The purpose of these is to introduce temperature gradients into the $\boldsymbol{w}$-equation and, at the same time, the entropy current vector into the entropy equation.

Whereas the antisymmetry of the L-operator defined in (5.17) is obvious and the degeneracy requirement (1.4) follows immediately form the trivial form of the entropy (5.15), the Jacobi identity needs to be checked after the derivatives in the $(4, 3)$- and $(3, 4)$-elements are introduced. Indeed, the Jacobi identity can be verified explicitly for the Poisson matrix in (5.17) (see Exercise 94). This identity would not be satisfied if we had assumed upper-convected behavior of $\boldsymbol{w}$, or if we had simply neglected deformational flow effects on this vector variable. As emphasized before, the Jacobi identity implies severe restrictions for the choice of L, and the natural simplicity of L largely determines the structure of the final equations.

Exercise 94 Jacobi Identity with Heat Flux Variable

Verify that the Poisson bracket associated with (5.17) satisfies the Jacobi identity.

5.1.3 Friction Matrix

In order to formulate the irreversible dynamics in a compact way, we introduce the trace and the traceless parts of the tensor $\boldsymbol{c}$,

$$\overline{\boldsymbol{c}} = \frac{1}{3}(\mathrm{tr}\boldsymbol{c})\boldsymbol{1}, \tag{5.20}$$

$$\mathring{\boldsymbol{c}} = \boldsymbol{c} - \frac{1}{3}(\mathrm{tr}\boldsymbol{c})\boldsymbol{1}, \tag{5.21}$$

and the auxiliary scalar variables

$$r_0 = \frac{1}{2}\boldsymbol{\phi}^* : \overline{\boldsymbol{c}}, \qquad r_2 = \frac{1}{2}\boldsymbol{\phi}^* : \mathring{\boldsymbol{c}}. \tag{5.22}$$

Suppressing the first two rows and columns, which are chosen to be identically zero, we can now specify the friction matrix as follows,

$$M = \begin{pmatrix} \frac{1}{T^*\tau_1}\boldsymbol{w}\cdot\boldsymbol{\sigma}^*\cdot\boldsymbol{w} + \frac{r_0}{T^*\tau_0} + \frac{r_2}{T^*\tau_2} & -\frac{1}{\tau_1}\boldsymbol{w} & -\frac{1}{\tau_0}\overline{\boldsymbol{c}} - \frac{1}{\tau_2}\mathring{\boldsymbol{c}} \\ -\frac{1}{\tau_1}\boldsymbol{w} & \frac{T^*}{\tau_1}\boldsymbol{\sigma}^{*-1} & 0 \\ -\frac{1}{\tau_0}\overline{\boldsymbol{c}} - \frac{1}{\tau_2}\mathring{\boldsymbol{c}} & 0 & \frac{T^*}{\tau_0 r_0}\overline{\boldsymbol{c}}\overline{\boldsymbol{c}} + \frac{T^*}{\tau_2 r_2}\mathring{\boldsymbol{c}}\mathring{\boldsymbol{c}} \end{pmatrix}, \tag{5.23}$$

where τ_0, τ_1, and τ_2 are characteristic relaxation times. This symmetric friction matrix satisfies the fundamental GENERIC degeneracy requirement (1.5), which expresses the conservation of the total energy by the irreversible contribution to the dynamics. For this matrix M to be positive semidefinite, we need to satisfy the following conditions:

$$T^* > 0, \quad \tau_0 > 0, \tau_1 > 0, \tau_2 > 0, \quad r_0 > 0, r_2 > 0, \tag{5.24}$$

and $\boldsymbol{\sigma}^*$ must be positive definite. This latter condition and $r_0 > 0$, $r_2 > 0$ are, for example, guaranteed for

$$\phi^{*(1)} = \phi^{*(3)} = \sigma^{*(2)} = \sigma^{*(3)} = 0 \tag{5.25}$$

and

$$\phi^{*(2)} \geq 0, \quad \sigma^{*(1)} \geq 0. \tag{5.26}$$

Equations (5.25) and (5.26) are sufficient but not necessary conditions for establishing the full GENERIC structure.[8] Furthermore, for arbitrary $\sigma^{*(3)} > 0$, a positive semidefinite friction matrix of a different functional form can be constructed for $\tau_0 = \tau_2$ by modifying the (5,5)-element of (5.23) as follows:

$$\frac{\overline{c}\,\overline{c}}{r_0} + \frac{\mathring{c}\mathring{c}}{r_2} \rightarrow \frac{(\overline{c} + \mathring{c})(\overline{c} + \mathring{c})}{r_0 + r_2} = \frac{cc}{r_0 + r_2}. \tag{5.27}$$

The possibility of choosing $\sigma^{*(3)} > 0$ is important to obtain an anisotropic thermal conductivity.

Exercise 95 Vanishing Elements of Friction Matrix

Why are the first two rows and columns in the friction matrix, not shown in (5.23), chosen to be zero?

Exercise 96 Degeneracy of the Friction Matrix

Verify the degeneracy requirement (1.5) for the friction matrix (5.23).

Exercise 97 Force-Flux-Pairs for Relaxation

Construct three natural force-flux pairs associated with the relaxation of $\boldsymbol{w}$, $\overline{c}$, and $\mathring{c}$.

Exercise 98 Positive-Semidefinite Friction Matrix

Show that the friction matrix (5.23) with (5.24) is positive semidefinite.

5.1.4 Time-Evolution Equations

By inserting all the building blocks developed in the preceding sections into the GENERIC (1.1), we obtain a set of time-evolution equations for the independent fields. The continuity equation takes the familiar form

$$\frac{\partial \rho}{\partial t} = -\frac{\partial}{\partial \boldsymbol{r}} \cdot (\boldsymbol{v}\rho), \tag{5.28}$$

and a configuration-dependent pressure tensor,

$$\begin{aligned} \boldsymbol{\pi} &= (\rho\mu^* + sT^* - \epsilon)\mathbf{1} + (\phi^{*(1)}\mathbf{1} + \phi^{*(2)}\boldsymbol{c} + \phi^{*(3)}\boldsymbol{c}^2) \cdot (\boldsymbol{c} - \mathbf{1}) \\ &+ (\sigma^{*(1)} - \sigma^{*(2)})\boldsymbol{w}\boldsymbol{w} + (\sigma^{*(2)} - \sigma^{*(3)})(\boldsymbol{w}\boldsymbol{w} \cdot \boldsymbol{c} + \boldsymbol{c} \cdot \boldsymbol{w}\boldsymbol{w}) \\ &+ \sigma^{*(3)}(\boldsymbol{w}\boldsymbol{w} \cdot \boldsymbol{c}^2 + \boldsymbol{c}^2 \cdot \boldsymbol{w}\boldsymbol{w} + \boldsymbol{c} \cdot \boldsymbol{w}\boldsymbol{w} \cdot \boldsymbol{c}), \end{aligned} \tag{5.29}$$

appears in the momentum-balance equation

$$\frac{\partial \boldsymbol{M}}{\partial t} = -\frac{\partial}{\partial \boldsymbol{r}} \cdot (\boldsymbol{v}\boldsymbol{M}) - \frac{\partial}{\partial \boldsymbol{r}} \cdot \boldsymbol{\pi}. \tag{5.30}$$

[8] One could actually try to simplify the discussion by letting the internal energy ϵ depend on a different set of scalar variables; see the remarks following (5.2).

Whereas the pressure tensor $\boldsymbol{\pi}$ occurs in the momentum balance (5.29), the stress-related contribution to the entropy production is determined by the tensor $\boldsymbol{\phi}^*$,

$$\frac{\partial s}{\partial t} = -\frac{\partial}{\partial \boldsymbol{r}} \cdot (\boldsymbol{v} s) - \frac{\partial}{\partial \boldsymbol{r}} \cdot \boldsymbol{j}^{\mathrm{s}} + \frac{\boldsymbol{j}^{\mathrm{s}} \cdot \boldsymbol{w}}{T^* \tau_1} + \frac{1}{2} \frac{\boldsymbol{\phi}^* : \overline{\boldsymbol{c}}}{T^* \tau_0} + \frac{1}{2} \frac{\boldsymbol{\phi}^* : \mathring{\boldsymbol{c}}}{T^* \tau_2}, \tag{5.31}$$

and the time evolution of the remaining generalized force variables is governed by

$$\frac{\partial \boldsymbol{w}}{\partial t} + \boldsymbol{v} \cdot \frac{\partial}{\partial \boldsymbol{r}} \boldsymbol{w} + \boldsymbol{\kappa}^T \cdot \boldsymbol{w} = -\frac{\partial}{\partial \boldsymbol{r}} T^* - \frac{1}{\tau_1} \boldsymbol{w} \tag{5.32}$$

and

$$\frac{\partial \boldsymbol{c}}{\partial t} + \boldsymbol{v} \cdot \frac{\partial}{\partial \boldsymbol{r}} \boldsymbol{c} + \boldsymbol{\kappa}^T \cdot \boldsymbol{c} + \boldsymbol{c} \cdot \boldsymbol{\kappa} = \boldsymbol{\kappa} + \boldsymbol{\kappa}^T - \frac{1}{\tau_0} \overline{\boldsymbol{c}} - \frac{1}{\tau_2} \mathring{\boldsymbol{c}}, \tag{5.33}$$

where $\boldsymbol{\kappa}$ has the usual meaning of the transposed velocity gradient tensor.

Exercise 99 Derivation of Pressure Tensor
Derive the expression (5.29) for the pressure tensor.

5.1.5 Limit of Classical Hydrodynamics

To rederive the hydrodynamic equations of Chapter 2, we assume a specific quadratic thermodynamic relation of the form (5.2),

$$\epsilon = \epsilon_0 + \frac{1}{2}(\alpha_1 \boldsymbol{w}^2 + \alpha_2 \mathrm{tr} \boldsymbol{c}^2), \tag{5.34}$$

where $\epsilon_0 = \epsilon_0(\rho, s)$, $\alpha_j = \alpha_j(\rho, s)$, and $\alpha_j \geq 0$. In the limit of very short relaxation times τ_0, τ_1, and τ_2, (5.32) and (5.33) imply a very rapid relaxation of the generalized force variables to the small values

$$\boldsymbol{w} = -\tau_1 \frac{\partial}{\partial \boldsymbol{r}} T^*, \tag{5.35}$$

$$\boldsymbol{c} = \tau_2 \left(\boldsymbol{\kappa} + \boldsymbol{\kappa}^T - \frac{2}{3} \frac{\partial}{\partial \boldsymbol{r}} \cdot \boldsymbol{v} \boldsymbol{1} \right) + \frac{2}{3} \tau_0 \frac{\partial}{\partial \boldsymbol{r}} \cdot \boldsymbol{v} \boldsymbol{1}, \tag{5.36}$$

and we obtain the following leading-order terms in the relaxation times:

$$\boldsymbol{j}^{\mathrm{s}} = -\tau_1 \alpha_1 \frac{\partial}{\partial \boldsymbol{r}} T^*, \tag{5.37}$$

$$\boldsymbol{\pi} = \left(\rho \frac{\partial \epsilon_0}{\partial \rho} + s \frac{\partial \epsilon_0}{\partial s} - \epsilon_0 \right) \boldsymbol{1} - 2\tau_2 \alpha_2 \left(\boldsymbol{\kappa} + \boldsymbol{\kappa}^T - \frac{2}{3} \frac{\partial}{\partial \boldsymbol{r}} \cdot \boldsymbol{v} \boldsymbol{1} \right) - \frac{4}{3} \tau_0 \alpha_2 \frac{\partial}{\partial \boldsymbol{r}} \cdot \boldsymbol{v} \boldsymbol{1}. \tag{5.38}$$

By comparison with (2.37) and (2.38), we can thus identify the thermal conductivity

$$\lambda^{\mathrm{q}} = \tau_1 \alpha_1 T^*, \tag{5.39}$$

the viscosity

$$\eta = 2\tau_2\alpha_2, \tag{5.40}$$

and the dilatational or bulk viscosity

$$\kappa = \frac{4}{3}\tau_0\alpha_2, \tag{5.41}$$

so that (5.28)–(5.33) reproduce classical hydrodynamics with the correct dissipation terms. All these transport coefficients need to be nonnegative. As discussed in Chapter 2, the limiting equations for the classical hydrodynamic variables, when the generalized forces (5.35), (5.36) become dependent variables, also possess the GENERIC structure.

Exercise 100 Quadratic Entropy Density
Compare the choice (5.34) for the internal energy density to the assumption

$$s = s_0 - \frac{1}{2}(\alpha_1' \boldsymbol{w}^2 + \alpha_2' \mathrm{tr}\boldsymbol{c}^2), \tag{5.42}$$

where $s_0 = s_0(\rho, \epsilon)$, $\alpha_j' = \alpha_j'(\rho, \epsilon)$, and $\boldsymbol{w}^2$, $\mathrm{tr}\boldsymbol{c}^2$ can be treated as small quantities.

Exercise 101 Force-Flux-Pairs for Relaxation, Continued
Evaluate the limit of short relaxation times for the force-flux pairs determined in Exercise 97.

5.1.6 Extended Irreversible Thermodynamics

The most detailed and thoughtful discussion of the relationship between GENERIC and *extended irreversible thermodynamics* (EIT) can be found in an article by Jou and Casas-Vázquez,[9] two of the pioneers of EIT. These authors emphasize two main features that can be considered as the hallmarks of EIT: (1) the choice of dissipative fluxes as independent variables in addition to the usual hydrodynamic fields and (2) an emphasis on the exploration of nonequilibrium equations of state in the process of developing time-evolution equations for the additional variables where, as in any approach, evolution equations and equations of state, or dynamics and thermodynamics, need to be formulated consistently. In particular, the roles of different thermodynamic potentials and Legendre transformations are elaborated in depth within EIT. The special choice of dissipative fluxes as independent variables implies that the constitutive equations in EIT do not involve gradients; this feature of EIT is the deeper reason why, in the process of developing relativistic hydrodynamics, we come across that framework.

As shown above in this section, the vector $\boldsymbol{w}$ and the tensor $\boldsymbol{c}$ are closely related to the entropy flux (5.12) and the pressure tensor (5.29), where shear and bulk viscous effects can be treated separately. By a transformation to these flux variables one

[9] Jou & Casas-Vázquez, J. Non-Newtonian Fluid Mech. 96 (2001) 77.

would obtain equivalent equations that fit into the framework of EIT with its special choice of variables.

Jou and Casas-Vázquez emphasize rightly that there is no counterpart of the Jacobi identity for the reversible dynamics in EIT. However, in view of the principle of material objectivity or frame indifference, one usually introduces convected time derivatives that, according to Appendix B.4, are intimately related to the validity of the Jacobi identity. This important identity is thus implicitly taken into account. Sometimes, when it is argued that EIT can be derived from Boltzmann's kinetic equation for dilute gases by means of Grad's moment method, the exact form of the resulting derivatives is not considered with sufficient care; as we realize in Section 7.4.3, the flow-modified derivatives obtained from Grad's moment expansion coincide neither with upper-convected nor with lower-convected derivatives, nor with any other Schowalter time derivative of the mixed type. This discrepancy is related to the fact that, in most applications of EIT, stresses result from configurational changes rather than from anisotropic momentum distributions. Therefore, kinetic theory can at best provide arguments in favor of using fluxes as additional independent variables.

Like GENERIC, EIT uses a nonequilibrium entropy expression that depends on the additional variables. However, in EIT one usually assumes a quadratic dependence of the entropy on the additional variables. Although more general entropy expressions are allowed in EIT and have actually been constructed within EIT by means of the maximum-entropy formalism,[10] the formalism does not require a particular form of the entropy function for internal consistency, in contrast to the degeneracy requirement (1.4) of GENERIC.[11] The restriction to quadratic entropies within the EIT approach is natural because the mere requirement of a nonnegative entropy production is not sufficiently restrictive beyond the second order. However, the discussion of the entropy production as calculated from the resulting equations has to be performed with great care because, beyond second order, reversible dynamics can be an unphysical source of entropy in EIT.

As an advantage of GENERIC, Jou and Casas-Vázquez mentioned that this framework provides a clear-cut expression for the beyond-equilibrium thermodynamic pressure, while the pressure expression is a subject of discussion in EIT. This is an example of how GENERIC may be useful to eliminate some of the hesitations that arise in the traditional formulation of EIT.

For further details on EIT, the reader is referred to a textbook dedicated to that approach.[12] A variety of problems have been treated within EIT, and thus many interesting ideas and results can be carried over into GENERIC, at least after modification of the entropy expression beyond the quadratic form. A comprehensive review of the

[10] See p. 82 of Jou & Casas-Vázquez, J. Non-Newtonian Fluid Mech. 96 (2001) 77 for an example, and the references given there for further details.

[11] See, for example, the discussion after the entropy expression (4.86), which leads to the anticipated form (4.87) of the pressure tensor, Exercise 102, or Exercise 181.

[12] Jou et al., *Extended Irreversible Thermodynamics* (Springer, 1996).

literature on beyond-equilibrium thermodynamics, with a strong emphasis on EIT, has been compiled by Jou et al.[13]

Exercise 102 Pressure Tensor for Quadratic Entropy Density
Show that, if we assume that the entropy density is of the quadratic type (5.42) used in EIT, GENERIC requires a particular form of the pressure tensor. How is it related to the form of the pressure tensor obtained by assuming (5.34)?

5.2 SPECIAL RELATIVISTIC HYDRODYNAMICS

In the construction of the extended theory of hydrodynamics presented in Section 5.1, first-order time and space derivatives occur in a balanced way. Based on this observation, we here substantiate the ideas and results of Section 5.1 into a set of field equations that is strictly form invariant under Lorentz transformations. Like for Maxwell's equations of electromagnetism and other relativistic theories, the invariance is proved most elegantly by grouping the variables into objects with proper transformation behavior and by writing the final field equations in a covariant form. Such a covariant formulation is obtained after the variables and GENERIC building blocks of Section 5.1 are modified in a natural way.

In the conventional procedure, the thermodynamic admissibility of relativistic hydrodynamics is implemented by formulating an entropy balance equation and simply verifying that the source term therein is always nonnegative. The classical theories obtained in this way by Eckart[14] and by Landau and Lifshitz[15] have problems concerning causality and stability;[16] therefore, most current analyses of dissipative phenomena are based on the causal second-order theory of Israel and Stewart[17] or on extended irreversible thermodynamics.[18] The purpose of this section is to develop a set of equations for imperfect fluids that are in accordance with the principles of both relativity and beyond-equilibrium thermodynamics, where our thermodynamic principles are by far more restrictive than those embodied in previous theories.

In the Israel-Stewart theory, the basic variables, which can be regarded as thermodynamic fluxes, are introduced through a straightforward decomposition of the energy-momentum tensor; for the entropy current vector, a quadratic expansion in the flux variables is then postulated, and evolution equations for the flux variables are chosen so that the divergence of the entropy current is nonnegative.[19] Within the

[13] Jou et al., Rep. Prog. Phys. 62 (1998) 1035.
[14] Eckart, Phys. Rev. 58 (1940) 919.
[15] See Chapter XV of Landau & Lifshitz, *Vol. 6, Fluid Mechanics* (Pergamon, 1987).
[16] Wodarzik, Phys. Lett. A 91 (1982) 279; Hiscock & Lindblom, Phys. Rev. D 31 (1985) 725.
[17] Israel, Ann. Phys. (N.Y.) 100 (1976) 310; Israel & Stewart, Ann. Phys. (N.Y.) 118 (1979) 341; Hiscock & Lindblom, Ann. Phys. (N.Y.) 151 (1983) 466.
[18] Jou et al., *Extended Irreversible Thermodynamics* (Springer, 1996).
[19] There are a number of possibilities for formulating such equations; see the note added in proof in Hiscock & Lindblom, Ann. Phys. (N.Y.) 151 (1983) 466.

GENERIC framework, the choice of the basic configurational variables, which can be regarded as thermodynamic forces, is strongly restricted by the Poissonian structure of the reversible dynamics, in particular, by the Jacobi identity expressing the time-structure invariance of the reversible dynamics (see Section 5.1). Furthermore, the dependence of the energy-momentum tensor and of the entropy current vector on these variables is determined by the GENERIC structure. Whereas the second law of thermodynamics is crucial to formulate the Israel-Stewart equations, in our thermodynamic approach it basically leads to stability conditions for the thermodynamic potential expressing the internal energy as a nontrivial function of the thermodynamic variables. The postulate of a quadratic expression for the entropy current in the Israel-Stewart theory, for which there is no counterpart in the new thermodynamic approach, can be regarded as the result of an expansion around equilibrium; this is a severe restriction that is not shared by the GENERIC approach. As a consequence of all these differences, the structure of the resulting equations is considerably different in general nonequilibrium situations but coincides for the linearized equations. Note that only the linearized equations are relevant to the discussion of stability and causality.

5.2.1 Notation and Variables

To formulate local equilibrium thermodynamic relationships in a relativistic theory of hydrodynamics, it is important to identify the comoving reference frame of the fluid. In particular, the mass, energy, and entropy densities, ρ_{f}, ϵ_{f} and s_{f}, play important roles, where, here and throughout this chapter, we reserve the subscript "f" for quantities evaluated in the comoving local reference frame of the fluid. There are actually various possibilities for defining the local fluid velocity. For the two most popular ones, the particle flow[20] or the momentum density[21] vanish in the local rest frame of the fluid, and we here prefer Eckart's definition associated with the particle flow.[22] In the corresponding reference frame of the fluid, the timelike Lorentz vector for the dimensionless fluid velocity, u^μ ($\mu = 0, 1, 2, 3$), becomes $u_{\mathrm{f}}^0 = 1$, $u_{\mathrm{f}}^j = 0$ ($j = 1, 2, 3$). In the fixed Cartesian laboratory reference frame, we then have

$$u^0 = \gamma, \qquad u^j = \gamma \frac{v_j}{c}, \tag{5.43}$$

[20] Eckart, Phys. Rev. 58 (1940) 919.

[21] Landau & Lifshitz, *Vol. 6, Fluid Mechanics* (Pergamon, 1987).

[22] We have tried to find another realization of GENERIC that is consistent with Landau and Lifshitz's definition of the hydrodynamic velocity. However, a fundamental thermodynamic relationship in the local rest frame involving an additional vector variable $\boldsymbol{w}$ in general seems to be incompatible with the required simple form of the pressure tensor. Notice that already in the nonrelativistic expression for the pressure tensor (5.29), there occurs the dyadic product of the additional vector variable $\boldsymbol{w}$, and this is inconsistent with Landau and Lifshitz's definition.

where c is the speed of light. The relationship between the factor γ and the velocity vector $\boldsymbol{v}$,

$$\gamma = \frac{1}{\sqrt{1 - \boldsymbol{v}^2/c^2}}, \tag{5.44}$$

can be derived from the identity $u_\mu u^\mu = -1$ implied by the choice of u_{f}^μ. Note that we use the summation convention for repeated indices throughout this chapter and that, in special relativity, *Minkowski's tensor*, $\eta_{00} = \eta^{00} = -1$, $\eta_{11} = \eta^{11} = \eta_{22} = \eta^{22} = \eta_{33} = \eta^{33} = 1$, $\eta_{\mu\nu} = \eta^{\mu\nu} = 0$ for $\mu \neq \nu$, is used for lowering or raising indices. We further introduce a spatial projector:

$$\hat{\eta}^{\mu\nu} = \eta^{\mu\nu} + u^\mu u^\nu. \tag{5.45}$$

Eventually, the fluid velocity needs to be expressed in terms of the independent variables in order to get a close set of equations.

Now we are in a position to imbed the set of variables $x = (\rho, \boldsymbol{M}, s, \boldsymbol{w}, \boldsymbol{c})$ of Section 5.1 into proper relativistic variables. This step is important to obtain covariant equations from GENERIC. For the mass density, we define

$$\rho = \gamma \rho_{\mathrm{f}}. \tag{5.46}$$

Note that ρ could actually be called the rest mass density because the rest mass is a scalar invariant proportional to the particle number, and the prefactor γ results from the FitzGerald-Lorentz contraction of volume elements in the direction of motion.

One might be inclined to define the entropy density in a similar way, but one should note that the simplicity of (5.46) is actually a consequence of Eckart's definition of the local rest frame through a vanishing particle flow. From a more general point of view, ρc is the time component of the mass current four-vector that, according to Eckart's convention, is given by $\rho_{\mathrm{f}} c u^\mu$. We hence define the entropy density in terms of the time component of the entropy current four-vector S^μ:

$$s = \frac{S^0}{c}. \tag{5.47}$$

The entropy current four-vector S^μ must satisfy a balance equation. Balance equations can be formulated conveniently in terms of the operators $\partial_\mu = \partial/\partial x^\mu$ ($\mu = 0, 1, 2, 3$), where $x^0 = ct$, t is the time, and x^j ($j = 1, 2, 3$) are the Cartesian components of the position vector $\boldsymbol{r}$. We then obtain the relativistic mass balance or equation of continuity

$$\partial_\mu(\rho_{\mathrm{f}} u^\mu) = 0 \tag{5.48}$$

and the entropy balance

$$\partial_\mu S^\mu = \sigma, \tag{5.49}$$

where σ is the entropy production rate. We expect to find constitutive equations for the space components S^j, so that S^0 or s is the only independent variable associated with entropy.

It is now natural to introduce the momentum density through certain components of the energy-momentum tensor $T^{\mu\nu}$:

$$M_j = \frac{1}{c} T^{0j}. \tag{5.50}$$

Again, all other components of $T^{\mu\nu}$, in particular the energy component, should be given by symmetry and by constitutive equations. The energy and momentum balances are combined into the four equations

$$\partial_\mu T^{\mu\nu} = 0. \tag{5.51}$$

Of course, the energy conservation must be consistent with the balances of mass, momentum, and entropy. In other words, we must expect a redundancy in the balance equations (5.48), (5.49), and (5.51).

If we want to eliminate the kinetic contribution from $T^{\mu\nu}$ to construct a tensor $\pi^{\mu\nu}$ corresponding to the pressure tensor $\boldsymbol{\pi}$ of Section 5.1, we start from

$$T^{\mu\nu} = \alpha u^\mu u^\nu + \pi^{\mu\nu}, \tag{5.52}$$

as motivated for example by (2.22). By contracting with u_μ and u_ν, we obtain

$$\alpha = u_\mu T^{\mu\nu} u_\nu - u_\mu \pi^{\mu\nu} u_\nu. \tag{5.53}$$

In the local rest frame of the fluid, T^{00} is the energy density at rest, consisting of the energy densities associated with the rest mass and with the internal energy,

$$T^{00}_{\mathrm{f}} = u_\mu T^{\mu\nu} u_\nu = \rho_{\mathrm{f}} c^2 + \epsilon_{\mathrm{f}}. \tag{5.54}$$

We thus obtain the decomposition

$$T^{\mu\nu} = (\rho_{\mathrm{f}} c^2 + \epsilon_{\mathrm{f}} - u_{\mu'} \pi^{\mu'\nu'} u_{\nu'}) u^\mu u^\nu + \pi^{\mu\nu}. \tag{5.55}$$

The convenient definition of a pressure p_{f} is related to fixing $u_\mu \pi^{\mu\nu} u_\nu$.

After imbedding the classical hydrodynamic variables ρ, $\boldsymbol{M}$, and s into proper relativistic objects, we now turn to the additional variables $\boldsymbol{w}$ and $\boldsymbol{c}$. The tensor $\boldsymbol{c}$ we imbed as follows. In the rest frame of the fluid, the components of a corresponding relativistic tensor are chosen as

$$(c_{\mu\nu}) = \begin{pmatrix} -1 & 0 \\ 0 & \boldsymbol{c}_{\mathrm{f}} \end{pmatrix}, \tag{5.56}$$

where $\boldsymbol{c}_{\mathrm{f}}$ is the matrix of spatial components. The time component of this tensor is chosen such that we have the useful property

$$c_{\mu\nu} u^\nu = u_\mu. \tag{5.57}$$

In an arbitrary frame, we obtain the following representation of $c_{\mu\nu}$ (see Exercise 103):

$$(c_{\mu\nu}) = \begin{pmatrix} \frac{\boldsymbol{v}}{c} \cdot \boldsymbol{c} \cdot \frac{\boldsymbol{v}}{c} - \frac{v^2}{c^2} - 1 & (\boldsymbol{1} - \boldsymbol{c}) \cdot \frac{\boldsymbol{v}}{c} \\ (\boldsymbol{1} - \boldsymbol{c}) \cdot \frac{\boldsymbol{v}}{c} & \boldsymbol{c} \end{pmatrix}. \tag{5.58}$$

The independent variable $\boldsymbol{c}$ determines the full tensor $c_{\mu\nu}$.

For the four-vector w_μ to be associated with the independent $\boldsymbol{w}$, we impose the requirement

$$w_\mu u^\mu = -\frac{T_\mathrm{f}^*}{c}, \tag{5.59}$$

which leads to the explicit representation (see Exercise 103)

$$(w_\mu) = \begin{pmatrix} -\frac{1}{c}\left(\frac{1}{\gamma}T_\mathrm{f}^* + \boldsymbol{v}\cdot\boldsymbol{w}\right) \\ \boldsymbol{w} \end{pmatrix}. \tag{5.60}$$

The role of T_f^* as a temperature variable will be established below. This occurrence of temperature is important to obtain the temperature gradient in (5.32). For $c_{\mu\nu}$, the velocity gradients in (5.33) arise naturally through the convection mechanism.

We have now imbedded all the independent variables $x = (\rho, \boldsymbol{M}, s, \boldsymbol{w}, \boldsymbol{c})$ into proper relativistic objects. Moreover, we can immediately specify the generators of reversible and irreversible dynamics, energy, and entropy. Because s is the entropy density, (5.15) for the entropy remains unchanged. For the energy, the theory of relativity gives E in terms of the energy-momentum tensor:

$$E(x) = \int T^{00} d^3r. \tag{5.61}$$

Before we show that we can produce a set of covariant extended hydrodynamic equations without modifying the Poisson matrix (5.17) at all and by merely working the effect of time dilation into the friction matrix (5.23) through modifying the relaxation rates, we need to derive some useful auxiliary results for the gradients of energy and entropy. Readers not interested in the mathematical details involved in the calculation of the energy gradient should notice the modification (5.64) and can then proceed directly to the final gradients (5.88) and (5.93).

Exercise 103 Components of Relativistic Extra Variables

Derive the expressions (5.58) and (5.60).

5.2.2 Gradients of Energy and Entropy

In the GENERIC formalism, we need to know the functional derivatives of the total energy (5.61), or the partial derivatives of T^{00}, with respect to the independent variables, $x = (\rho, \boldsymbol{M}, s, \boldsymbol{w}, \boldsymbol{c})$. These derivatives can be obtained by the following procedure. We first differentiate the expression

$$M_j = (T^{00} - \pi^{00})\frac{v_j}{c^2} + \frac{1}{c}\pi^{0j}, \tag{5.62}$$

obtained by combining (5.50) and (5.55) with respect to the variables x, and multiply by $\boldsymbol{v}$ to obtain an equation for $(v^2/c^2)\partial T^{00}/\partial x$. We then differentiate the expression for T^{00} implied by (5.55) and divide by γ^2 to obtain an expression for

$(1/\gamma^2)\partial T^{00}/\partial x$. By adding these two expressions, we find after some rearrangements

$$\begin{aligned}\frac{\partial T^{00}}{\partial x} &= \frac{c^2}{\gamma}\frac{\partial \rho}{\partial x} + \boldsymbol{v}\cdot\frac{\partial \boldsymbol{M}}{\partial x} + \frac{\partial \epsilon_{\rm f}}{\partial x} - \frac{\partial}{\partial x}\left(\frac{1}{\gamma}\hat{\eta}^{0\mu}\pi_{\mu\nu}u^{\nu}\right) \\ &+ (\epsilon_{\rm f} - u_\mu \pi^{\mu\nu} u_\nu)\frac{1}{\gamma}\frac{\partial \gamma}{\partial x} + \frac{1}{c}\sum_{j=1}^{3}\pi^{0j}\frac{\partial v_j}{\partial x}\,. \end{aligned} \tag{5.63}$$

The derivative of $\epsilon_{\rm f}$ can be evaluated by adopting the thermodynamic relationship (5.2) and the definitions (5.3)–(5.10) in the local rest frame of the fluid,

$$\epsilon_{\rm f} = \epsilon_{\rm f}(\rho_{\rm f}, s_{\rm f}, {\rm tr}\boldsymbol{c}_{\rm f}, {\rm tr}\boldsymbol{c}_{\rm f}^2, {\rm tr}\boldsymbol{c}_{\rm f}^3, \boldsymbol{w}_{\rm f}^2, \boldsymbol{w}_{\rm f}\cdot\boldsymbol{c}_{\rm f}\cdot\boldsymbol{w}_{\rm f}, \boldsymbol{w}_{\rm f}\cdot\boldsymbol{c}_{\rm f}^2\cdot\boldsymbol{w}_{\rm f}), \tag{5.64}$$

and the partial derivatives,

$$\mu_{\rm f}^* = \frac{\partial \epsilon_{\rm f}}{\partial \rho_{\rm f}}, \tag{5.65}$$

$$T_{\rm f}^* = \frac{\partial \epsilon_{\rm f}}{\partial s_{\rm f}}, \tag{5.66}$$

$$\phi_{\rm f}^{*(1)} = 2\frac{\partial \epsilon_{\rm f}}{\partial {\rm tr}\boldsymbol{c}_{\rm f}}, \tag{5.67}$$

$$\phi_{\rm f}^{*(2)} = 4\frac{\partial \epsilon_{\rm f}}{\partial {\rm tr}\boldsymbol{c}_{\rm f}^2}, \tag{5.68}$$

$$\phi_{\rm f}^{*(3)} = 6\frac{\partial \epsilon_{\rm f}}{\partial {\rm tr}\boldsymbol{c}_{\rm f}^3}, \tag{5.69}$$

$$\sigma_{\rm f}^{*(1)} = 2\frac{\partial \epsilon_{\rm f}}{\partial \boldsymbol{w}_{\rm f}^2} \tag{5.70}$$

$$\sigma_{\rm f}^{*(2)} = 2\frac{\partial \epsilon_{\rm f}}{\partial \boldsymbol{w}_{\rm f}\cdot\boldsymbol{c}_{\rm f}\cdot\boldsymbol{w}_{\rm f}}, \tag{5.71}$$

$$\sigma_{\rm f}^{*(3)} = 2\frac{\partial \epsilon_{\rm f}}{\partial \boldsymbol{w}_{\rm f}\cdot\boldsymbol{c}_{\rm f}^2\cdot\boldsymbol{w}_{\rm f}}. \tag{5.72}$$

If we introduce powers of $c_{\mu\nu}$ recursively by

$$c^0_{\mu\nu} = \eta_{\mu\nu}, \quad c^{n+1}_{\mu\nu} = c_{\mu\mu'}\eta^{\mu'\nu'}c^n_{\nu'\nu}, \tag{5.73}$$

then the scalars occurring in (5.64) can be evaluated in any reference frame:

$${\rm tr}\boldsymbol{c}_{\rm f}^n = \eta^{\mu\nu}c^n_{\mu\nu} - 1 \tag{5.74}$$

and

$$\boldsymbol{w}_{\rm f}\cdot\boldsymbol{c}_{\rm f}^n\cdot\boldsymbol{w}_{\rm f} = w^\mu c^n_{\mu\nu} w^\nu + \frac{T_{\rm f}^{*2}}{c^2}. \tag{5.75}$$

We thus obtain

$$\begin{aligned}\frac{\partial \epsilon_{\rm f}}{\partial x} &= \mu_{\rm f}^* \frac{\partial \rho_{\rm f}}{\partial x} + T_{\rm f}^* \frac{\partial s_{\rm f}}{\partial x} + \frac{\phi^{*\mu\nu}}{2} \frac{\partial c_{\mu\nu}}{\partial x} \\ &+ w_\mu \sigma^{*\mu\nu} \frac{\partial w_\nu}{\partial x} + (\sigma_{\rm f}^{*(1)} + \sigma_{\rm f}^{*(2)} + \sigma_{\rm f}^{*(3)}) \frac{T_{\rm f}^*}{c^2} \frac{\partial T_{\rm f}^*}{\partial x},\end{aligned} \tag{5.76}$$

with the following generalizations of (5.13) and (5.14):

$$\sigma^{*\mu\nu} = (\sigma_{\rm f}^{*(1)} c^0 + \sigma_{\rm f}^{*(2)} c^1 + \sigma_{\rm f}^{*(3)} c^2)^{\mu\nu} \tag{5.77}$$

and

$$\begin{aligned}\phi^{*\mu\nu} &= (\phi_{\rm f}^{*(1)} c^0 + \phi_{\rm f}^{*(2)} c^1 + \phi_{\rm f}^{*(3)} c^2)^{\mu\nu} + \sigma_{\rm f}^{*(2)} w^\mu w^\nu \\ &+ \sigma_{\rm f}^{*(3)} (w^\mu w_{\mu'} c^{\mu'\nu} + c^{\mu\mu'} w_{\mu'} w^\nu).\end{aligned} \tag{5.78}$$

By inserting (5.76) into (5.63) and using

$$w_\mu \sigma^{*\mu\nu} u_\nu = -(\sigma_{\rm f}^{*(1)} + \sigma_{\rm f}^{*(2)} + \sigma_{\rm f}^{*(3)}) \frac{T_{\rm f}^*}{c}, \tag{5.79}$$

we obtain

$$\begin{aligned}\frac{\partial T^{00}}{\partial x} &= \frac{c^2 + \mu_{\rm f}^*}{\gamma} \frac{\partial \rho}{\partial x} + \boldsymbol{v} \cdot \frac{\partial \boldsymbol{M}}{\partial x} + \frac{T_{\rm f}^*}{\gamma} \frac{\partial (\gamma s_{\rm f})}{\partial x} \\ &- \frac{\partial}{\partial x} \left(\frac{1}{\gamma} \hat{\eta}^{0\mu} \pi_{\mu\nu} u^\nu \right) - \frac{1}{c} w^\mu \sigma_{\mu\nu}^* \hat{\eta}^{\nu 0} \frac{\partial}{\partial x} \left(\frac{T_{\rm f}^*}{\gamma} \right) \\ &+ \frac{1}{2} \phi^{*\mu\nu} \frac{\partial c_{\mu\nu}}{\partial x} + w_\mu \left(\sigma^{*\mu\nu} - \sigma^{*\mu 0} \frac{u^\nu}{\gamma} \right) \frac{\partial w_\nu}{\partial x} \\ &+ \frac{1}{c} \sum_{j=1}^{3} (\pi^{0j} - w_\mu \sigma^{*\mu 0} w_j) \frac{\partial v_j}{\partial x} \\ &+ \left(\epsilon_{\rm f} - \rho_{\rm f} \mu_{\rm f}^* - s_{\rm f} T_{\rm f}^* - u_\mu \pi^{\mu\nu} u_\nu - \frac{T_{\rm f}^*}{c} w_\mu \sigma^{*\mu\nu} u_\nu \right) \frac{1}{\gamma} \frac{\partial \gamma}{\partial x}.\end{aligned} \tag{5.80}$$

So far, we have not made any constitutive assumptions for the fluxes $\pi^{\mu\nu}$ and S^μ. To simplify (5.80) we now assume

$$\pi^{\mu\nu} u_\nu = p_{\rm f} u^\mu - \frac{T_{\rm f}^*}{c} \sigma^{*\mu\nu} w_\nu \tag{5.81}$$

with the definition

$$p_{\rm f} = -\epsilon_{\rm f} + \rho_{\rm f} \mu_{\rm f}^* + s_{\rm f} T_{\rm f}^*. \tag{5.82}$$

In particular, this choice fixes the scalar quantity $u_\mu \pi^{\mu\nu} u_\nu$, which was left undetermined in (5.55), such that the last line of (5.80) vanishes. The constitutive assumption

(5.81) further allows us to simplify the second line of (5.80) considerably, so that we obtain

$$\begin{aligned}\frac{\partial T^{00}}{\partial x} &= \frac{c^2+\mu_{\rm f}^*}{\gamma}\frac{\partial \rho}{\partial x} + \boldsymbol{v}\cdot\frac{\partial \boldsymbol{M}}{\partial x} + \frac{T_{\rm f}^*}{\gamma}\frac{\partial}{\partial x}\left(\gamma s_{\rm f} + \frac{1}{c}\hat{\eta}^{0\mu}\sigma^*_{\mu\nu}w^\nu\right) \\ &+ \frac{1}{2}\phi^{*\mu\nu}\frac{\partial c_{\mu\nu}}{\partial x} + w_\mu\left(\sigma^{*\mu\nu} - \sigma^{*\mu 0}\frac{u^\nu}{\gamma}\right)\frac{\partial w_\nu}{\partial x} \\ &+ \frac{1}{c}\sum_{j=1}^{3}(\pi^{0j} - w_\mu\sigma^{*\mu 0}w_j)\frac{\partial v_j}{\partial x}. \end{aligned} \tag{5.83}$$

We are thus led to the identification

$$S^\mu = s_{\rm f} c u^\mu + \hat{\eta}^{\mu\nu}\sigma^*_{\nu\nu'}w^{\nu'}, \tag{5.84}$$

which is chosen such that (5.47) implies that the entropy density

$$s = \gamma s_{\rm f} + \frac{1}{c}\hat{\eta}^{0\mu}\sigma^*_{\mu\nu}w^\nu \tag{5.85}$$

occurs in (5.83). The expression (5.84) shows that w^μ describes the entropy conduction on top of the entropy convection, so that we expect an intimate relation between w^μ and the temperature gradient.

To finalize the constitutive assumptions, we next extend (5.81) into an equation for the full pressure tensor $\pi^{\mu\nu}$, where we have the freedom to add a term that vanishes upon contraction with u_ν. Guided by the nonrelativistic expression found in Exercise 99, the most natural symmetric generalization is (see Exercise 104)

$$\pi^{\mu\nu} = p_{\rm f}\eta^{\mu\nu} - \phi^{*\mu\nu} + \phi^{*\mu\nu'}c_{\nu'}{}^{\nu} + \sigma^{*\mu\nu'}w_{\nu'}w^\nu. \tag{5.86}$$

The constitutive equation (5.86) implies the auxiliary equation

$$\pi^{0j} - w_\mu\sigma^{*\mu 0}w_j = -\phi^{*0j} + \phi^{*0\mu}c_\mu{}^j = -\sum_{k=1}^{3}\left(\phi^{*0k} - \phi^{*00}\frac{v_k}{c}\right)(\delta_{jk} - c_{jk}), \tag{5.87}$$

which is useful in obtaining the final relativistic generalization of the result (5.11) for the energy gradient

$$\frac{\delta E(x)}{\delta x} = \begin{pmatrix} \frac{c^2+\mu_{\rm f}^*}{\gamma} \\ \boldsymbol{v} \\ \frac{T_{\rm f}^*}{\gamma} \\ \boldsymbol{j}^{\rm s} \\ \frac{1}{2}\hat{\boldsymbol{\phi}}^* \end{pmatrix}. \tag{5.88}$$

Here, we have introduced a generalization of (5.12):

$$(\boldsymbol{j}^{\rm s})_j = \left(\sigma^{*j\mu} - \frac{v_j}{c}\sigma^{*0\mu}\right)w_\mu \tag{5.89}$$

and

$$\hat{\phi}^{*jk} = \phi^{*jk} - \phi^{*j0}\frac{v_k}{c} - \frac{v_j}{c}\phi^{*0k} + \phi^{*00}\frac{v_j}{c}\frac{v_k}{c}. \tag{5.90}$$

If we define

$$\hat{\sigma}^{*jk} = \sigma^{*jk} - \sigma^{*j0}\frac{v_k}{c} - \frac{v_j}{c}\sigma^{*0k} + \sigma^{*00}\frac{v_j}{c}\frac{v_k}{c}, \tag{5.91}$$

then we can write

$$\boldsymbol{j}^{\mathrm{s}} = \hat{\boldsymbol{\sigma}}^{*} \cdot \left(\boldsymbol{w} - \gamma T_{\mathrm{f}}^{*}\frac{\boldsymbol{v}}{c^2}\right). \tag{5.92}$$

The matrix $\hat{\boldsymbol{\sigma}}^*$ with elements $\hat{\sigma}^{*jk}$ can be shown to be positive semidefinite if the matrix with elements $\sigma_{\mathrm{f}}^{*jk}$ is positive semidefinite.

Because the total entropy density s is among the independent variables, it is trivial to calculate the derivatives of the entropy $S(x) = \int s d^3r$:

$$\frac{\delta S(x)}{\delta x} = \begin{pmatrix} 0 \\ 0 \\ 1 \\ 0 \\ 0 \end{pmatrix}. \tag{5.93}$$

With (5.50), (5.55), and (5.86), we are now in a position to determine the velocity four-vector from $\boldsymbol{M}$ or $T^{\mu\nu}$ and the other independent variables, for which we construct explicit time evolution or covariant field equations. The component $\delta E/\delta s = T_{\mathrm{f}}^*/\gamma$ in (5.88) leads to the interpretation of the scalar quantity T_{f}^* introduced in (5.59) as a temperature.

Exercise 104 Symmetry of Relativistic Pressure Tensor
Show that the pressure tensor (5.86) is symmetric.

Exercise 105 Calculation of the Energy Gradient
Elaborate the details of the calculation of the energy gradient (5.88) from (5.83).

5.2.3 Poisson and Friction Matrices

Because the form of the Poisson matrix is dictated by the transformation behavior of the variables, we do not modify the expression (5.17) and the degeneracy condition (1.4) thus remains valid. To modify the friction matrix, we should keep in mind the effect of time dilation; that is, relaxation times should be multiplied by γ. In view of the degeneracy condition (1.5), the relativistic modifications of the energy gradient in (5.88) provide further guidance to modify the friction matrix. In particular, temperature should be divided by γ, $\boldsymbol{\sigma}^*$ and $\boldsymbol{\phi}^*$ should be changed into $\hat{\boldsymbol{\sigma}}^*$ and $\hat{\boldsymbol{\phi}}^*$, and one should modify the procedure to separate trace and traceless parts. We introduce $\overline{c}_{\mu\nu}$ and $\mathring{c}_{\mu\nu}$ such that they vanish upon contraction with the velocity four-vector:

$$\overline{c}_{\mu\nu} = \frac{1}{3}(c_{\mu'}{}^{\mu'} - 1)\hat{\eta}_{\mu\nu}, \tag{5.94}$$

$$\mathring{c}_{\mu\nu} = c_{\mu\nu} + u_\mu u_\nu - \bar{c}_{\mu\nu}, \tag{5.95}$$

and thus

$$\bar{\boldsymbol{c}} = \frac{1}{3}\mathrm{tr}\boldsymbol{c}_\mathrm{f}\left(\mathbf{1} + \gamma^2 \frac{\boldsymbol{vv}}{c^2}\right), \tag{5.96}$$

$$\mathring{\boldsymbol{c}} = \boldsymbol{c} + \gamma^2 \frac{\boldsymbol{vv}}{c^2} - \bar{\boldsymbol{c}}. \tag{5.97}$$

Again suppressing the first two rows and columns because they are identically zero, we can now specify the modified friction matrix (5.23) as follows:

$$M = \begin{pmatrix} M_{33} & -\frac{1}{\gamma\tau_1}\boldsymbol{j}^\mathrm{s}\cdot\hat{\boldsymbol{\sigma}}^{*-1} & -\frac{1}{\gamma\tau_0}\bar{\boldsymbol{c}} - \frac{1}{\gamma\tau_2}\mathring{\boldsymbol{c}} \\ -\frac{1}{\gamma\tau_1}\hat{\boldsymbol{\sigma}}^{*-1}\cdot\boldsymbol{j}^\mathrm{s} & \frac{T_\mathrm{f}^*}{\gamma^2\tau_1}\hat{\boldsymbol{\sigma}}^{*-1} & 0 \\ -\frac{1}{\gamma\tau_0}\bar{\boldsymbol{c}} - \frac{1}{\gamma\tau_2}\mathring{\boldsymbol{c}} & 0 & \frac{T_\mathrm{f}^*}{\gamma^2\tau_0 r_0}\bar{\boldsymbol{c}}\bar{\boldsymbol{c}} + \frac{T_\mathrm{f}^*}{\gamma^2\tau_2 r_2}\mathring{\boldsymbol{c}}\mathring{\boldsymbol{c}} \end{pmatrix}, \tag{5.98}$$

with

$$M_{33} = \frac{1}{T_\mathrm{f}^*\tau_1}\boldsymbol{j}^\mathrm{s}\cdot\hat{\boldsymbol{\sigma}}^{*-1}\cdot\boldsymbol{j}^\mathrm{s} + \frac{r_0}{T_\mathrm{f}^*\tau_0} + \frac{r_2}{T_\mathrm{f}^*\tau_2} \tag{5.99}$$

and

$$r_0 = \frac{1}{2}\hat{\boldsymbol{\phi}}^* : \bar{\boldsymbol{c}} = \frac{1}{2}\phi^{*\mu\nu}\bar{c}_{\mu\nu}, \tag{5.100}$$

$$r_2 = \frac{1}{2}\hat{\boldsymbol{\phi}}^* : \mathring{\boldsymbol{c}} = \frac{1}{2}\phi^{*\mu\nu}\mathring{c}_{\mu\nu}. \tag{5.101}$$

Within the GENERIC framework, (5.88), (5.93), (5.17), and (5.98) determine the time evolution of the independent fields $x = (\rho, \boldsymbol{M}, s, \boldsymbol{w}, \boldsymbol{c})$. As a next step, we show that the resulting equations can be expressed as covariant field equations.

5.2.4 Covariant Field Equations

The relativistic time-evolution equations can be expressed in the following covariant form: mass balance (5.48), energy-momentum balance (5.51), entropy balance (5.49) with the entropy production rate

$$\sigma = \frac{w_\mu \hat{\eta}^{\mu\nu}\sigma^*_{\nu\nu'}w^{\nu'}}{T_\mathrm{f}^*\tau_1} + \frac{1}{2}\frac{\phi^{*\mu\nu}\bar{c}_{\mu\nu}}{T_\mathrm{f}^*\tau_0} + \frac{1}{2}\frac{\phi^{*\mu\nu}\mathring{c}_{\mu\nu}}{T_\mathrm{f}^*\tau_2}, \tag{5.102}$$

a field equation for the four-vector variable associated with heat flow,

$$u^{\mu'}(\partial_{\mu'}w_\mu - \partial_\mu w_{\mu'}) = -\frac{1}{c\tau_1}\hat{\eta}_{\mu\mu'}w^{\mu'}, \tag{5.103}$$

and a field equation for the tensor variable associated with viscous momentum flow,

$$u^{\mu'}(\partial_{\mu'}c_{\mu\nu} - \partial_\mu c_{\mu'\nu} - \partial_\nu c_{\mu\mu'}) = -\frac{1}{c\tau_0}\bar{c}_{\mu\nu} - \frac{1}{c\tau_2}\mathring{c}_{\mu\nu}. \tag{5.104}$$

The rest of this subsection is dedicated to the derivation of these covariant equations. For the full verification of these equations, we need to take into account that GENERIC provides time-evolution equations for $\boldsymbol{M}$, $\boldsymbol{w}$, $\boldsymbol{c}$, and hence only for the spatial components of (5.51), (5.103), and (5.104). In addition, we hence need to show that the time components of (5.51), (5.103), and (5.104) follow consistently. In particular, we have to establish the consistency of the energy and the entropy balance.

We first look at the scalar equations for ρ and s. The equation for the mass density, which first arises in the usual form (5.28), can be rewritten in the covariant formulation (5.48) by using (5.43) and (5.46). The entropy balance can be rearranged by means of the identity

$$S^j = s v_j + (\boldsymbol{j}^{\mathrm{s}})_j, \tag{5.105}$$

which follows from (5.84), (5.85), and (5.89), where the entropy production rate arises as

$$\sigma = \frac{\boldsymbol{j}^{\mathrm{s}} \cdot \hat{\boldsymbol{\sigma}}^{*-1} \cdot \boldsymbol{j}^{\mathrm{s}}}{T_{\mathrm{f}}^* \tau_1} + \frac{1}{2}\frac{\hat{\boldsymbol{\phi}}^* : \overline{\boldsymbol{c}}}{T_{\mathrm{f}}^* \tau_0} + \frac{1}{2}\frac{\hat{\boldsymbol{\phi}}^* : \mathring{\boldsymbol{c}}}{T_{\mathrm{f}}^* \tau_2}. \tag{5.106}$$

The covariant formulation of the first term on the right-hand side of (5.106) is left as an exercise to the reader (see Exercise 106); the covariant expressions for r_0 and r_2 given in (5.100) and (5.101) can actually be verified in a similar way.

We now turn to the vector variable $\boldsymbol{w}$. By combining the GENERIC building blocks, we obtain the following generalization of (5.32),

$$\frac{\partial \boldsymbol{w}}{\partial t} + \boldsymbol{v} \cdot \frac{\partial}{\partial \boldsymbol{r}} \boldsymbol{w} + \boldsymbol{\kappa}^T \cdot \boldsymbol{w} = -\frac{\partial}{\partial \boldsymbol{r}} \frac{T_{\mathrm{f}}^*}{\gamma} - \frac{1}{\gamma \tau_1} \hat{\boldsymbol{\sigma}}^{*-1} \cdot \boldsymbol{j}^{\mathrm{s}}, \tag{5.107}$$

which, according to (5.92), can also be written as

$$u^\mu \partial_\mu \boldsymbol{w} + \frac{\gamma}{c} \boldsymbol{\kappa}^T \cdot \boldsymbol{w} + \gamma \frac{\partial}{\partial \boldsymbol{r}} \frac{T_{\mathrm{f}}^*}{\gamma c} = -\frac{1}{c \tau_1} \left(\boldsymbol{w} - \gamma T_{\mathrm{f}}^* \frac{\boldsymbol{v}}{c^2} \right). \tag{5.108}$$

With the auxiliary equation

$$u^\mu \partial_j w_\mu = \gamma \partial_j \frac{u^\mu w_\mu}{\gamma} - \gamma w_\mu \partial_j \frac{u^\mu}{\gamma} = -\gamma \partial_j \frac{T_{\mathrm{f}}^*}{\gamma c} - \frac{\gamma}{c} \boldsymbol{w} \cdot \partial_j \boldsymbol{v}, \tag{5.109}$$

we then obtain

$$u^\mu (\partial_\mu w_j - \partial_j w_\mu) = -\frac{1}{c \tau_1} \hat{\eta}_{j\mu} w^\mu. \tag{5.110}$$

The spatial components of the covariant equation (5.103) are thus equivalent to (5.107). To identify any further possible information in the time component of the covariant equation (5.103) we contract it with u^μ and obtain the condition

$$u^\mu u^{\mu'} (\partial_{\mu'} w_\mu - \partial_\mu w_{\mu'}) = 0, \tag{5.111}$$

which is trivially satisfied as can be seen by exchanging the summation indices μ and μ'. The time component of the field equation for w_μ is thus redundant. Note that it is important to have the factor $\hat{\eta}_{\mu\mu'}$ on the right-hand side of (5.103) to make sure that the right side vanishes on contraction with u^μ.

The procedure to verify the equation for $c_{\mu\nu}$ is very similar to the above steps for w_μ. The modified version of (5.33)

$$\frac{\partial \boldsymbol{c}}{\partial t} + \boldsymbol{v} \cdot \frac{\partial}{\partial \boldsymbol{r}} \boldsymbol{c} + \boldsymbol{\kappa}^T \cdot \boldsymbol{c} + \boldsymbol{c} \cdot \boldsymbol{\kappa} = \boldsymbol{\kappa} + \boldsymbol{\kappa}^T - \frac{1}{\gamma \tau_0} \overline{\boldsymbol{c}} - \frac{1}{\gamma \tau_2} \mathring{\boldsymbol{c}}, \tag{5.112}$$

with (5.96) and (5.97), leads to

$$u^\mu \partial_\mu \boldsymbol{c} + \frac{\gamma}{c} \left(\boldsymbol{\kappa}^T \cdot \boldsymbol{c} + \boldsymbol{c} \cdot \boldsymbol{\kappa} - \boldsymbol{\kappa} - \boldsymbol{\kappa}^T \right) = -\frac{1}{c\tau_0} \overline{\boldsymbol{c}} - \frac{1}{c\tau_2} \mathring{\boldsymbol{c}}. \tag{5.113}$$

With the auxiliary equation

$$u^\mu \partial_j c_{\mu k} = \gamma \partial_j \frac{u^\mu c_{\mu k}}{\gamma} - \gamma c_{\mu k} \partial_j \frac{u^\mu}{\gamma} = \frac{\gamma}{c} \left(\partial_j v_k - \sum_{l=1}^{3} c_{lk} \partial_j v_l \right), \tag{5.114}$$

we realize that (5.113) indeed represents the spatial components of (5.104). The time components of (5.104) are obtained by contracting with u^μ,

$$u^\mu u^{\mu'} \partial_\nu c_{\mu\mu'} = 0, \tag{5.115}$$

which follows from (5.57) (see Exercise 107). The trivially satisfied condition (5.115) restricts the construction of the trace and traceless parts $\overline{c}_{\mu\nu}$ and $\mathring{c}_{\mu\nu}$ such that they vanish on contraction with u^μ, as mentioned before (5.94) and (5.95).

Finally, the time-evolution equation for the momentum density vector field $\boldsymbol{M}$, obtained by combining the GENERIC building blocks, arises as

$$\begin{aligned} \frac{\partial \boldsymbol{M}}{\partial t} &= -\frac{\partial}{\partial \boldsymbol{r}} \cdot (\boldsymbol{v}\boldsymbol{M}) - \frac{\partial}{\partial \boldsymbol{r}} \left(\rho_{\mathrm{f}} c^2 + \rho_{\mathrm{f}} \mu_{\mathrm{f}}^* + \frac{s}{\gamma} T_{\mathrm{f}}^* + \boldsymbol{M} \cdot \boldsymbol{v} - T_{00} \right) \\ &- \frac{\partial}{\partial \boldsymbol{r}} \cdot [\boldsymbol{j}^{\mathrm{s}} \boldsymbol{w} + \hat{\boldsymbol{\phi}}^* \cdot (\boldsymbol{c} - \mathbf{1})]. \end{aligned} \tag{5.116}$$

By means of (5.50), (5.55), (5.82) and (5.86), with further use of (5.58) and (5.90), we obtain the intermediate result

$$\begin{aligned} \partial_\mu T^{\mu j} &= -\partial_k \left[\left(\frac{v_k}{c} \sigma^{*0\mu} - \sigma^{*k\mu} \right) w_\mu w_j + (\boldsymbol{j}^{\mathrm{s}})_k w_j \right] \\ &- \partial_j \left[\rho_{\mathrm{f}} c^2 + \epsilon_{\mathrm{f}} + \left(\frac{s}{\gamma} - s_{\mathrm{f}} \right) T_{\mathrm{f}}^* + \boldsymbol{M} \cdot \boldsymbol{v} - T_{00} \right]. \end{aligned} \tag{5.117}$$

The first term on the right-hand side of this equation vanishes due to (5.89). From (5.55) and (5.62), we get

$$T_{00} - \boldsymbol{M} \cdot \boldsymbol{v} = \rho_{\mathrm{f}} c^2 + \epsilon_{\mathrm{f}} - u_\mu \pi^{\mu\nu} u_\nu + \pi^{00} - \sum_{j=1}^{3} \pi^{0j} \frac{v_j}{c}. \tag{5.118}$$

By inserting (5.86) and using (5.58) and (5.60), we further obtain

$$T_{00} - \boldsymbol{M} \cdot \boldsymbol{v} = \rho_{\mathrm{f}} c^2 + \epsilon_{\mathrm{f}} + \frac{T_{\mathrm{f}}^*}{\gamma c} \hat{\eta}^{0\mu} \sigma_{\mu\nu}^* w^\nu, \tag{5.119}$$

and, after introducing (5.119) and (5.85) into (5.117), we finally conclude that (5.116) is equivalent to the relativistic momentum balance,

$$\partial_\mu T^{\mu j} = 0. \tag{5.120}$$

We still need to derive the energy balance, $\partial_\mu T^{\mu 0} = 0$, from the GENERIC equations. Because the momentum balance (5.120) is already established, we eliminate kinetic energy effects by contracting (5.51) with u_ν and, after inserting (5.55) and using the mass balance, we obtain the following condition:

$$\partial_\mu(\epsilon_{\rm f} u^\mu - \hat{\eta}^{\mu\mu'}\pi_{\mu'\nu}u^\nu) + \pi^{\mu\nu}\partial_\mu u_\nu = 0. \tag{5.121}$$

Equation (5.121) has the form of an internal energy balance equation in which we can recognize a heat flux and a source term due to deformations. We expect that we can obtain (5.121) from the thermodynamic relationship (5.64) and the field equations for the arguments of this relationship. Now, to derive (5.121) from the thermodynamic relationship, we use (5.76) and (5.79) to obtain

$$\begin{aligned} \partial_\mu \epsilon_{\rm f} u^\mu &= (\epsilon_{\rm f} - \rho_{\rm f}\mu_{\rm f}^* - s_{\rm f} T_{\rm f}^*)\partial_\mu u^\mu + \mu_{\rm f}^*\partial_\mu \rho_{\rm f} u^\mu + T_{\rm f}^*\partial_\mu s_{\rm f} u^\mu \\ &+ \frac{1}{2}\phi^{*\mu'\nu'} u^\mu \partial_\mu c_{\mu'\nu'} + w_{\mu'}\sigma^{*\mu'\nu'} u^\mu \partial_\mu w_{\nu'} - w_{\mu'}\sigma^{*\mu'\nu'} u_{\nu'} u^\mu \partial_\mu \frac{T_{\rm f}^*}{c}. \end{aligned} \tag{5.122}$$

By inserting (5.82) and (5.86), and using (5.84) for S^μ as well as the mass balance, we find

$$\begin{aligned} \partial_\mu(\epsilon_{\rm f} u^\mu - \hat{\eta}^{\mu\mu'}\pi_{\mu'\nu}u^\nu) &= -p_{\rm f}\partial_\mu u^\mu + \frac{T_{\rm f}^*}{c}\partial_\mu S^\mu + \sigma^{*\mu\nu} w_\nu \partial_\mu \frac{T_{\rm f}^*}{c} \\ &+ \frac{1}{2}\phi^{*\mu'\nu'} u^\mu \partial_\mu c_{\mu'\nu'} + w_{\mu'}\sigma^{*\mu'\nu'} u^\mu \partial_\mu w_{\nu'}. \end{aligned} \tag{5.123}$$

With the field equations for S^μ, w_μ and $c_{\mu\nu}$, that is, (5.49), (5.103), and (5.104), this expression can be simplified by means of (5.57) and (5.59):

$$\partial_\mu(\epsilon_{\rm f} u^\mu - \hat{\eta}^{\mu\mu'}\pi_{\mu'\nu}u^\nu) = -p_{\rm f}\partial_\mu u^\mu + [\phi^{*\mu\nu} - \phi^{*\mu\nu'} c_{\nu'}{}^\nu - \sigma^{*\mu\nu'} w_{\nu'} w^\nu]\partial_\mu u_\nu. \tag{5.124}$$

In view of the pressure-tensor expression (5.86), we have thus derived the internal energy balance (5.121).

At this point, we have fully established the GENERIC form of the covariant field equations for relativistic hydrodynamics presented at the beginning of this subsection, with the building blocks obtained by natural modifications of their nonrelativistic counterparts. The elegance and simplicity of the equations for the additional fields w_μ and $c_{\mu\nu}$, (5.103) and (5.104), should be noted.

Exercise 106 Covariant Form of Entropy Production Rate

Prove (5.92) and the identity

$$\boldsymbol{j}^{\rm s} \cdot \hat{\boldsymbol{\sigma}}^{*-1} \cdot \boldsymbol{j}^{\rm s} = w_\mu \hat{\eta}^{\mu\nu} \sigma^*_{\nu\nu'} w^{\nu'}. \tag{5.125}$$

Exercise 107 Verification of a Redundancy

Derive (5.115).

5.3 COVARIANT GENERIC FRAMEWORK

Although the core part of the GENERIC formalism is an explicit time-evolution equation, we have seen in the previous sections that it nevertheless can produce a set of covariant equations for relativistic hydrodynamics as a special realization. In formulating a relativistic hydrodynamic theory, one is forced to introduce additional generalized force variables, related to velocity and temperature gradients and hence eventually also to the fluxes of momentum and energy, which are known to be essential for a causal theory. Moreover, the structure of the transport equations is severely restricted by the combination of the rather complementary principles of relativity and GENERIC.

According to a general recipe, the previous set of hydrodynamic equations could be generalized to self-gravitating fluids by replacing all partial derivatives by covariant derivatives ("minimal coupling"). One can even verify that the system of equations obtained by such a replacement and by including Einstein's equation for the additional gravitational field variables possess the full GENERIC structure, as we have checked explicitly within the original noncovariant formalism of Section 1.2. However, to establish confidence in the GENERIC formalism and the resulting equations for a self-gravitating imperfect fluid in the relativistic context, we here present a covariant version of the formalism. In the subsequent section, we employ this covariant formalism to formulate a wide class of hydrodynamic equations for self-gravitating fluids.

5.3.1 Fundamental Equation

In the abstract formulation of GENERIC, equivalent Poisson operator and bracket representations have been proposed (see Section 1.2.1). We here present a covariant formulation of the GENERIC idea based on the bracket formulation, as inspired by work on reversible relativistic field theories[23] and originally proposed for a dissipative system by Kaufman.[24] In the covariant formulation, the field equations are expressed in the form

$$\{A, I\} + [A, J] = 0, \tag{5.126}$$

where I is an action integral, J is the integral of the entropy density over space and time, referred to as the global entropy by Kaufman, and A is an arbitrary functional of the fields. Note that the functionals I, J and A do not depend on the independent fields only as functions of position at a given time,[25] but rather as functions of both position

[23] Marsden et al., Ann. Phys. (N.Y.) 169 (1986) 29.

[24] Kaufman, Phys. Lett. A 109 (1985) 87.

[25] The parameter t was suppressed in the formulation of brackets in the previous chapters.

and time. Furthermore, $\{A, B\}$ is a Poisson bracket, and $[A, B]$ is a dissipative bracket with $[A, B] = [B, A]$ for arbitrary functionals A, B, and $[A, A] \geq 0$ for all A. The two contributions to the field equations generated by I and J in (5.126) are called the reversible and irreversible contributions, respectively. In the reversible situation considered in previous work,[24] the second term in (5.126) is absent.

5.3.2 Degeneracy Requirements

While (5.126) replaces the fundamental time-evolution equation (1.1) of the noncovariant GENERIC formalism, the mutual degeneracy requirements (1.4) and (1.5) of that formalism are replaced by

$$\{J, A\} = 0, \tag{5.127}$$

and

$$[I, A] = 0, \tag{5.128}$$

where, again, A is an arbitrary functional of the fields. These requirements, which are again strong formulations of the conservation of entropy by the reversible dynamics and of the conservation of energy by the irreversible dynamics, are known to be important parts of the GENERIC framework. The covariant formulation (5.126)–(5.128) retains the key features of the GENERIC structure, that is, the use of two separate generators for the reversible and irreversible dynamics, together with the complementary degeneracy requirements.

5.4 HYDRODYNAMICS IN THE PRESENCE OF GRAVITY

As a next step, we use the proposed covariant formalism to construct generally covariant equations for self-gravitating imperfect fluids. The general relativistic hydrodynamic equations proposed in this section are expected to be relevant to describing the expansion of the universe, the formation of galaxies, and the gravitational collapse of stars into neutron stars whenever one needs to go beyond the frequently used model of a dissipationless or perfect fluid. We first select the independent variables and, after determining all the building blocks of the covariant GENERIC framework, we write out all the field equations.[26]

5.4.1 Notation and Variables

To apply GENERIC, we first need to specify the list of independent variables for our thermodynamic system of interest, that is, for a self-gravitating imperfect fluid. For the basic hydrodynamic variables, we use the particle number density per unit rest

[26] Compared with Öttinger, Phys. Rev. D 60 (1999) 103507, on which Section 5.4 is based, we reshuffle factors of c in the definition of some of the variables for better consistency with the preceding sections, and we change the sign convention for the action.

Table 5.1 Corresponding Quantities

Section 5.2	Sections 5.4 and 5.5
$\rho_{\rm f}/m$	n
$s_{\rm f} + w_\mu \sigma^{*\mu\nu} u_\nu / c$	s
$u_\mu T^{\mu\nu} u_\nu = \rho_{\rm f} c^2 + \epsilon_{\rm f}$	ρ
$c_{\mu\nu} - \eta_{\mu\nu}$	$C_{\mu\nu}$
$p_{\rm f}$	p
$\rho_{\rm f} c^2 + \epsilon_{\rm f} + T_{\rm f}^* w_\mu \sigma^{*\mu\nu} u_\nu / c$	q
$\hat{\eta}_{\mu\nu} = \eta_{\mu\nu} + u_\mu u_\nu$	$h_{\mu\nu} = g_{\mu\nu} + u_\mu u_\nu$

volume, n, the total entropy density per unit rest volume, s, and a kind of momentum density four-vector, M_μ ($\mu = 0, 1, 2, 3$). In this section, we find that the four-vector M_μ turns out to be proportional to the previously considered four-velocity u_μ, where the factor of proportionality is an inertial mass density that depends not only on n and s but also on the configurational variables. Actually, the final field equations are more conveniently expressed in terms of u_μ rather than M_μ. To render a thermodynamically admissible theory possible, we further use the four-vector w_μ closely related to the temperature gradient and the symmetric tensor variable $C_{\mu\nu}$, corresponding to $c_{\mu\nu} - \eta_{\mu\nu}$ in our previous notation and thus closely related to the velocity gradient tensor, as independent variables, as elaborated in Section 5.2. A careful interpretation of all these variables is most reliably done in the light of the final field equations.

Because, in the covariant framework, we deal with only covariant objects, we here suppress the subscript "f" for quantities evaluated in the local reference frame of the fluid. For simplicity, we also omit the superscript "*", which has so far been used in this chapter as a warning that the corresponding derivatives are formed for a list of variables in which only some of the variables are densities of extensive variables, whereas the other variables are intensive ones. Furthermore, we use a notation that is as close as possible to what is usual in general relativity and cosmology. In particular, in the rest of this chapter, ρ is used for the energy density in the rest frame rather than for the mass density. For clarity and convenience, the quantities used from now on, when specialized to the situation of negligible gravitational interactions, are translated in Table 5.4.1 into the notation of Section 5.2. In particular, the previously used mass density is given by mn, where m is the rest mass per particle; in addition, one has to be careful with the identification of the covariant entropy density s, which is different from $s_{\rm f}$, and hence with the interpretation of the entropy integral J.

In addition to the extended hydrodynamic fields n, M_μ, s, w_μ, and $C_{\mu\nu}$, we use the components of the contravariant symmetric tensor representing the dual metric, $g^{\mu\nu}$, and the Christoffel symbols $\Gamma^\nu_{\mu\mu'}$ as the independent variables describing gravity. The familiar relationship between $\Gamma^\nu_{\mu\mu'}$ and the derivatives of $g^{\mu\nu}$ will eventually result from the GENERIC field equations. As usual, $g^{\mu\nu}$ is used to raise indices, whereas its inverse, the metric tensor $g_{\mu\nu}$, is used to lower indices. In accordance with our previous choice for Minkowski's tensor, the signature of the space-time metric $g_{\mu\nu}$ is (−+++), and g is the absolute value, that is, the negative, of the determinant of $g_{\mu\nu}$. Semicolons (;) denote covariant derivatives compatible with the metric $g_{\mu\nu}$, and, as before, Einstein's summation convention is used massively. The following formulas for the covariant divergence of a vector field V^μ and a symmetric tensor field $T^{\mu\nu}$[27]

$$(V^\mu)_{;\mu} = \frac{1}{\sqrt{g}}\partial_\mu\left(\sqrt{g}V^\mu\right), \tag{5.129}$$

and

$$(T^\nu{}_\mu)_{;\nu} = \frac{1}{\sqrt{g}}\partial_\nu\left(\sqrt{g}T^\nu{}_\mu\right) + \frac{1}{2}T_{\nu\nu'}\partial_\mu g^{\nu\nu'}, \tag{5.130}$$

will be useful. Equation (5.130) is a special case of the general rule for performing covariant derivatives of arbitrary mixed tensors with upper and lower indices,

$$(T^{\mu'}{}_\mu)_{;\nu} = \partial_\nu T^{\mu'}{}_\mu + \Gamma^{\mu'}_{\nu\nu'}T^{\nu'}{}_\mu - \Gamma^{\nu'}_{\mu\nu}T^{\mu'}{}_{\nu'}, \tag{5.131}$$

where one term involving Christoffel symbols appears for each index (also for tensors of higher rank).

In the context of general relativity, functional derivatives are most conveniently defined as scalars, vectors, or tensors, and not as densities (see Appendix C.1); for example,

$$\left.\frac{d}{d\lambda}\right|_{\lambda=0} A(n+\lambda\delta n) = \int \delta n \frac{\delta A}{\delta n}\sqrt{g}d^4x. \tag{5.132}$$

The only exception is made to keep the Christoffel symbols $\Gamma^\nu_{\mu\mu'}$, which do not define a third-rank tensor, as convenient variables,

$$\left.\frac{d}{d\lambda}\right|_{\lambda=0} A(\Gamma+\lambda\delta\Gamma) = \int \delta\Gamma^\nu_{\mu\mu'}\frac{\delta A}{\delta\Gamma^\nu_{\mu\mu'}}d^4x. \tag{5.133}$$

Exercise 108 Derivative of Determinant
Derive the formulas

$$\frac{1}{\sqrt{g}}\partial_\nu\sqrt{g} = \frac{1}{2}g^{\mu\mu'}\partial_\nu g_{\mu\mu'} = -\frac{1}{2}g_{\mu\mu'}\partial_\nu g^{\mu\mu'}. \tag{5.134}$$

5.4.2 Reversible Contribution

As a first step, we generalize the Poisson bracket and action integral proposed by Marsden et al. by including the variables w_μ and $C_{\mu\nu}$. Inspired by the considerations

[27] See, for example, Section 4.7 of Weinberg, *Gravitation and Cosmology* (Wiley, 1972).

in Section 2.3.1, in particular by (2.90) and (2.91), we use the following bracket:[28]

$$\begin{aligned}
\{A, B\} &= \int \sqrt{g} n \left(\frac{\delta B}{\delta M_\mu} \partial_\mu \frac{\delta A}{\delta n} - \frac{\delta A}{\delta M_\mu} \partial_\mu \frac{\delta B}{\delta n} \right) d^4x \\
&+ \int \sqrt{g} M_\nu \left(\frac{\delta B}{\delta M_\mu} \partial_\mu \frac{\delta A}{\delta M_\nu} - \frac{\delta A}{\delta M_\mu} \partial_\mu \frac{\delta B}{\delta M_\nu} \right) d^4x \\
&+ \int \sqrt{g} s \left(\frac{\delta B}{\delta M_\mu} \partial_\mu \frac{\delta A}{\delta s} - \frac{\delta A}{\delta M_\mu} \partial_\mu \frac{\delta B}{\delta s} \right) d^4x \\
&+ \int V^\nu \left(\frac{\delta A}{\delta g^{\mu\mu'}} \frac{\delta B}{\delta \Gamma^\nu_{\mu\mu'}} - \frac{\delta B}{\delta g^{\mu\mu'}} \frac{\delta A}{\delta \Gamma^\nu_{\mu\mu'}} \right) d^4x \\
&+ \int \sqrt{g} (\partial_\nu w_\mu) \left(\frac{\delta A}{\delta M_\nu} \frac{\delta B}{\delta w_\mu} - \frac{\delta B}{\delta M_\nu} \frac{\delta A}{\delta w_\mu} \right) d^4x \\
&+ \int \sqrt{g} w_\mu \left(\frac{\delta B}{\delta w_\nu} \partial_\nu \frac{\delta A}{\delta M_\mu} - \frac{\delta A}{\delta w_\nu} \partial_\nu \frac{\delta B}{\delta M_\mu} \right) d^4x \\
&+ \int \sqrt{g} \left(\frac{\delta B}{\delta w_\mu} \partial_\mu \frac{\delta A}{\delta s} - \frac{\delta A}{\delta w_\mu} \partial_\mu \frac{\delta B}{\delta s} \right) d^4x \\
&+ \int \sqrt{g} (\partial_\nu C_{\mu\mu'}) \left(\frac{\delta A}{\delta M_\nu} \frac{\delta B}{\delta C_{\mu\mu'}} - \frac{\delta B}{\delta M_\nu} \frac{\delta A}{\delta C_{\mu\mu'}} \right) d^4x \\
&+ \int \sqrt{g} C_{\mu\nu} \left(\frac{\delta B}{\delta C_{\mu\mu'}} \partial_{\mu'} \frac{\delta A}{\delta M_\nu} - \frac{\delta A}{\delta C_{\mu\mu'}} \partial_{\mu'} \frac{\delta B}{\delta M_\nu} \right) d^4x \\
&+ \int \sqrt{g} C_{\mu\nu} \left(\frac{\delta B}{\delta C_{\nu'\nu}} \partial_{\nu'} \frac{\delta A}{\delta M_\mu} - \frac{\delta A}{\delta C_{\nu'\nu}} \partial_{\nu'} \frac{\delta B}{\delta M_\mu} \right) d^4x. \qquad (5.135)
\end{aligned}$$

In this bracket expression, the partial derivatives ∂_μ could equivalently be replaced by covariant derivatives.

The Poisson bracket (5.135) is a natural generalization of the bracket considered in Exercise 94. The first three integrals in (5.135) represent the convection mechanism for the standard hydrodynamic fields. The fourth contribution involves only the variables of the gravitational field, $g^{\mu\mu'}$ and $\Gamma^\nu_{\mu\mu'}$, and an arbitrary vector field V^ν. The occurrence of V^ν is related to the freedom of arbitrary coordinate transformations in four-dimensional space, but a direct physical interpretation of V^ν, which has units of time per mass, has not yet been offered. Two further integrals express the convection mechanism for the covariant four-vector field w_μ, and a further contribution, involving w_μ and s, establishes the relation between w_μ and the temperature gradients. This coupling between w_μ and s, which corresponds to the (4, 3)- and (3, 4)-elements in

[28] Compared with Marsden et al., Ann. Phys. (N.Y.) 169 (1986) 29, we use different sign conventions for the action and in the Poisson bracket (related to the choice between push-forward and pull-back actions of space transformations on the number and entropy densities), and we avoid an artificial factor of two (necessary if one insists on a standard form of the Hamiltonian); see also Bao, Marsden & Walton, Commun. Math. Phys. 99 (1985) 319.

(5.17), is the only nonstandard contribution in the bracket (5.135). The last three integrals express the convection mechanism for the covariant four-tensor field $C_{\mu\nu}$.

The action I proposed by Marsden et al., which is the sum of a gravitational action $I_{\rm G}$ (involving $g^{\mu\nu}R_{\mu\nu}$) and a matter action $I_{\rm M}$ (involving U_I and V_I),

$$I = \int \sqrt{g}\left(\frac{-c^4}{16\pi G_{\rm N}} g^{\mu\nu}R_{\mu\nu} + \frac{1}{2}g^{\mu\nu}\frac{M_\mu M_\nu}{U_I} - V_I\right) d^4x, \tag{5.136}$$

is modified only by letting the functions U_I and V_I depend not only on the scalar quantities n and s but also on further scalar variables to be formed out of w_μ and $C_{\mu\nu}$, with the help of $g^{\mu\nu}$. For example, one could use the scalars

$$g^{\mu\nu}C_{\nu\mu}, \quad g^{\mu\nu}C_{\nu\mu'}g^{\mu'\nu'}C_{\nu'\mu}, \quad g^{\mu\nu}C_{\nu\mu'}g^{\mu'\nu'}C_{\nu'\mu''}g^{\mu''\nu''}C_{\nu''\mu}, \tag{5.137}$$

and

$$w_\mu g^{\mu\nu}w_\nu, \quad w_\mu g^{\mu\nu}C_{\nu\mu'}g^{\mu'\nu'}w_{\nu'}, \quad w_\mu g^{\mu\nu}C_{\nu\mu'}g^{\mu'\nu'}C_{\nu'\mu''}g^{\mu''\nu''}w_{\nu''}, \tag{5.138}$$

which, according to Exercise 109, are closely related to the configurational arguments in (5.64). In (5.136), $G_{\rm N}$ is Newton's gravitational constant, and the *Ricci tensor* describing space-time curvature,

$$R_{\mu\nu} = R_{\mu\nu}(\Gamma) = \partial_{\nu'}\Gamma^{\nu'}_{\mu\nu} - \partial_\nu\Gamma^{\nu'}_{\mu\nu'} + \Gamma^{\nu'}_{\mu'\nu'}\Gamma^{\mu'}_{\mu\nu} - \Gamma^{\nu'}_{\mu'\nu}\Gamma^{\mu'}_{\mu\nu'}, \tag{5.139}$$

is to be understood as a function of the Christoffel symbols alone. To calculate functional derivatives of I, it is convenient to introduce the auxiliary quantities

$$p = \frac{1}{2}U_I c^2 + V_I, \qquad q = \frac{1}{2}U_I c^2 - V_I, \tag{5.140}$$

to be interpreted below (see also Table 5.4.1).

Exercise 109 Additional Scalar Arguments in Action

Express the scalars in (5.137) and (5.138) in terms of the scalars appearing as arguments in (5.64).

Exercise 110 An Operator Identity

Derive the operator identity

$$g^{\nu'\nu}\frac{\partial}{\partial g^{\nu'\mu}} = C_{\mu\mu'}\frac{\partial}{\partial C_{\nu\mu'}} + \frac{1}{2}w_\mu\frac{\partial}{\partial w_\nu} \tag{5.141}$$

for the action on scalars such as the ones defined in (5.137) and (5.138).

5.4.3 Irreversible Contribution

In constructing the dissipative bracket, we make an *ansatz* inspired by (5.98). Only the entropy density, s, and the new variables, w_μ and $C_{\mu\nu}$, are assumed to be involved in the dissipative bracket. The following expression is manifestly symmetric and covariant, and it clearly fulfills the degeneracy requirement (5.128):

$$\begin{aligned}[A,B] &= \int \sqrt{g}\left(\frac{\delta I}{\delta w_\mu}\frac{\delta A}{\delta s} - \frac{\delta I}{\delta s}\frac{\delta A}{\delta w_\mu}\right) f^w_{\mu\nu}\left(\frac{\delta I}{\delta w_\nu}\frac{\delta B}{\delta s} - \frac{\delta I}{\delta s}\frac{\delta B}{\delta w_\nu}\right) d^4x \\ &+ \int \sqrt{g}\left(\frac{\delta I}{\delta C_{\mu\nu}}\frac{\delta A}{\delta s} - \frac{\delta I}{\delta s}\frac{\delta A}{\delta C_{\mu\nu}}\right)(\overline{f}\overline{c}_{\mu\nu}\overline{c}_{\mu'\nu'} \\ &\qquad + \mathring{f}\mathring{c}_{\mu\nu}\mathring{c}_{\mu'\nu'})\left(\frac{\delta I}{\delta C_{\mu'\nu'}}\frac{\delta B}{\delta s} - \frac{\delta I}{\delta s}\frac{\delta B}{\delta C_{\mu'\nu'}}\right) d^4x, \qquad (5.142)\end{aligned}$$

where $\overline{f}$ and $\mathring{f}$ are nonnegative, $f^w_{\mu\nu}$ is positive semidefinite, and

$$\overline{c}_{\mu\nu} = \left(1 + \frac{1}{3}C_{\mu'\nu'}h^{\mu'\nu'}\right) h_{\mu\nu}, \qquad (5.143)$$

$$\mathring{c}_{\mu\nu} = h_{\mu\mu'}C^{\mu'\nu'}h_{\nu'\nu} - \frac{1}{3}C_{\mu'\nu'}h^{\mu'\nu'}h_{\mu\nu}, \qquad (5.144)$$

are the natural generalizations of the trace and traceless parts (5.94) and (5.95) in the presence of gravitation, involving

$$h_{\mu\nu} = g_{\mu\nu} + u_\mu u_\nu \qquad (5.145)$$

as a generalization of (5.45). Expressions for the dissipative bracket more general than (5.142) can be written down easily; in particular, one can couple the indices μ, ν and μ', ν' in the second integral in a nested way, or the trace and traceless parts can be superimposed differently.

Because the entropy density is among the independent variables, we can immediately write the entropy integral J as

$$J = \int s\sqrt{g}d^4x. \qquad (5.146)$$

Whereas s is the entropy density per unit three space, the integration is not only over three space but also over time. The degeneracy requirement (5.127) is trivially fulfilled because the functional derivative of J with respect to s is unity, and the functional derivatives of J with respect to all other independent variables vanish.

5.4.4 Field Equations

After defining the brackets and the integrals I and J in the preceding subsections, we can now write out all the explicit field equations implied by the covariant form of GENERIC. By letting A in the fundamental equation (5.126) successively depend on only one of the fields in the list of variables n, M_μ, s, $g^{\mu\nu}$, $\Gamma^\nu_{\mu\mu'}$, w_μ, and $C_{\mu\nu}$, we obtain the following list of equations:

$$\frac{1}{\sqrt{g}}\partial_\mu\left(\sqrt{g}n\frac{\delta I}{\delta M_\mu}\right) = 0, \qquad (5.147)$$

$$n\partial_\mu \frac{\delta I}{\delta n} + s\partial_\mu \frac{\delta I}{\delta s} + M_\nu \partial_\mu \frac{\delta I}{\delta M_\nu} - \frac{\delta I}{\delta w_\nu}\partial_\mu w_\nu - \frac{\delta I}{\delta C_{\nu\nu'}}\partial_\mu C_{\nu\nu'}$$
$$+\frac{1}{\sqrt{g}}\partial_\nu \left[\sqrt{g}\left(\frac{\delta I}{\delta M_\nu}M_\mu + \frac{\delta I}{\delta w_\nu}w_\mu + \frac{\delta I}{\delta C_{\nu\nu'}}C_{\mu\nu'} + \frac{\delta I}{\delta C_{\nu'\nu}}C_{\nu'\mu}\right)\right] = 0, \tag{5.148}$$

$$\frac{1}{\sqrt{g}}\partial_\mu \left(\sqrt{g}s\frac{\delta I}{\delta M_\mu} + \sqrt{g}\frac{\delta I}{\delta w_\mu}\right) = \frac{\delta I}{\delta w_\mu} f^w_{\mu\nu}\frac{\delta I}{\delta w_\nu} + \overline{f}\left(\overline{c}_{\mu\nu}\frac{\delta I}{\delta C_{\mu\nu}}\right)^2 + \mathring{f}\left(\mathring{c}_{\mu\nu}\frac{\delta I}{\delta C_{\mu\nu}}\right)^2, \tag{5.149}$$

$$\frac{\delta I}{\delta \Gamma^\nu_{\mu\mu'}} = 0, \tag{5.150}$$

$$\frac{\delta I}{\delta g^{\mu\nu}} = 0, \tag{5.151}$$

$$\frac{\delta I}{\delta M_\nu}(\partial_\nu w_\mu - \partial_\mu w_\nu) + \partial_\mu \left(w_\nu \frac{\delta I}{\delta M_\nu} + \frac{\delta I}{\delta s}\right) = -\frac{\delta I}{\delta s} f^w_{\mu\nu}\frac{\delta I}{\delta w_\nu}, \tag{5.152}$$

$$\frac{\delta I}{\delta M_{\mu'}}(\partial_{\mu'}C_{\mu\nu} - \partial_\mu C_{\mu'\nu} - \partial_\nu C_{\mu\mu'}) + \partial_\mu \left(\frac{\delta I}{\delta M_{\nu'}}C_{\nu'\nu}\right) + \partial_\nu \left(C_{\mu\mu'}\frac{\delta I}{\delta M_{\mu'}}\right)$$
$$= -\frac{\delta I}{\delta s}(\overline{f}\overline{c}_{\mu\nu}\overline{c}_{\mu'\nu'} + \mathring{f}\mathring{c}_{\mu\nu}\mathring{c}_{\mu'\nu'})\frac{\delta I}{\delta C_{\mu'\nu'}}. \tag{5.153}$$

Exercise 111 General Form of Field Equations
Derive some of the equations (5.147)–(5.153).

For a further discussion of the above field equations, we proceed as follows. We first construct the expressions for the Christoffel symbols and the four-velocity from (5.150) and (5.147), respectively, so that we can safely use these expressions in all other equations. We then derive the pressure tensor and its role in the field equation and in the momentum balance equation resulting from (5.151) and (5.148). We then turn to the equations with dissipative contributions, namely, the entropy balance implied by (5.149) and the relaxation equations (5.152) and (5.153) for the configurational variables.

If we write out (5.150) more explicitly by inserting (5.136), we obtain

$$\frac{\delta \int \sqrt{g} g^{\mu''\nu''} R_{\mu''\nu''}(\Gamma) d^4x}{\delta \Gamma^\nu_{\mu\mu'}} = 0. \tag{5.154}$$

By means of (5.139), we then find

$$-\partial_\nu g^{\mu\mu'} - \Gamma^{\mu}_{\nu'\nu} g^{\nu'\mu'} - \Gamma^{\mu'}_{\nu\nu'} g^{\mu\nu'} + \left(\Gamma^{\nu'}_{\nu\nu'} - \frac{1}{\sqrt{g}}\partial_\nu \sqrt{g}\right) g^{\mu\mu'}$$
$$+ g_\nu{}^{\mu'} \left(\partial_{\nu'} g^{\mu\nu'} + g^{\mu''\nu''}\Gamma^{\mu}_{\mu''\nu''} + g^{\mu\nu'} \frac{1}{\sqrt{g}}\partial_{\nu'}\sqrt{g}\right) = 0. \tag{5.155}$$

By contracting this formula with $g_{\mu'}{}^{\nu}$, we realize that the term in the second line of (5.155) actually vanishes; this observation expresses the fact that the covariant divergence of the dual metric tensor vanishes. By further contraction of the simplified version of (5.155) with $g_{\mu\mu'}$, we obtain

$$\Gamma^{\nu'}_{\nu\nu'} = \frac{1}{\sqrt{g}}\partial_\nu \sqrt{g}, \tag{5.156}$$

where (5.134) has been used. We thus arrive at

$$-\partial_\nu g^{\mu\mu'} - \Gamma^{\mu}_{\nu'\nu} g^{\nu'\mu'} - \Gamma^{\mu'}_{\nu\nu'} g^{\mu\nu'} = 0. \tag{5.157}$$

With the rule for the derivative of the inverse of $g^{\mu\mu'}$, we alternatively get

$$\partial_\nu g_{\mu\mu'} - \Gamma^{\nu'}_{\nu\mu} g_{\nu'\mu'} - \Gamma^{\nu'}_{\mu'\nu} g_{\mu\nu'} = 0. \tag{5.158}$$

According to (5.131), (5.158) and (5.157) express the fact that the covariant derivative of the metric tensor and of its inverse vanish. By evaluating the proper combination of the identity (5.158) for different index combinations, we obtain the usual expression for the Christoffel symbols in terms of the metric:

$$\Gamma^{\nu}_{\mu\mu'} = \frac{1}{2} g^{\nu\nu'} \left[\partial_\mu g_{\nu'\mu'} + \partial_{\mu'} g_{\mu\nu'} - \partial_{\nu'} g_{\mu\mu'}\right]. \tag{5.159}$$

Equation (5.147) suggests that we introduce the velocity four-vector as

$$u^\mu = \frac{1}{c}\frac{\delta I}{\delta M_\mu} = \frac{M^\mu}{c U_I}. \tag{5.160}$$

This definition is in the spirit of Eckart's approach (see Section 5.2); that is, we have introduced the four-vector u^μ as the velocity of particle transport. More precisely, we then have the following conservation equation for the particle number:

$$\frac{1}{\sqrt{g}}\partial_\mu \left(\sqrt{g} n u^\mu\right) = \left(n u^\mu\right)_{;\mu} = 0, \tag{5.161}$$

where (5.129) has been used. In view of the constraint $g^{\mu\nu} u_\mu u_\nu = -1$, the factor U_I is given by

$$g^{\mu\nu} M_\mu M_\nu = -c^2 U_I^2. \tag{5.162}$$

This constraint is to be imposed in the final equations *after* all the functional derivatives leading to (5.147)–(5.153) have been taken. With (5.162), a number of functional derivatives can be simplified considerably,

$$\frac{\delta I}{\delta n} = \frac{\partial q}{\partial n}, \quad \frac{\delta I}{\delta s} = \frac{\partial q}{\partial s}, \quad \frac{\delta I}{\delta w_\mu} = \frac{\partial q}{\partial w_\mu}, \quad \frac{\delta I}{\delta C_{\mu\nu}} = \frac{\partial q}{\partial C_{\mu\nu}}, \tag{5.163}$$

where q has been defined in (5.140).

After establishing the connections between the metric tensor and the Christoffel symbols as well as between the velocity and momentum four-vectors, we next turn to the equations to calculate them, that is, Einstein's field equation and the equation of motion. In particular, we thus reveal the form of the energy-momentum tensor, including the pressure tensor. By inserting the action (5.136) into (5.151) and (5.148), we obtain

$$R_{\mu\nu} - \frac{1}{2} g_{\mu\nu} g^{\mu'\nu'} R_{\mu'\nu'} = \frac{8\pi G_\mathrm{N}}{c^4} \left(q u_\mu u_\nu + p h_{\mu\nu} + 2\frac{\partial q}{\partial g^{\mu\nu}} \right) \tag{5.164}$$

and

$$\begin{aligned} \frac{1}{\sqrt{g}} \partial_\nu \left[\sqrt{g} \left((p+q) u^\nu u_\mu + \frac{\partial q}{\partial w_\nu} w_\mu + 2 \frac{\partial q}{\partial C_{\nu\nu'}} C_{\mu\nu'} \right) \right] & \\ + \partial_\mu \left(n \frac{\partial q}{\partial n} + s \frac{\partial q}{\partial s} - q \right) + \frac{1}{2} \left[(p+q) u_\nu u_{\nu'} + 2 \frac{\partial q}{\partial g^{\nu\nu'}} \right] \partial_\mu g^{\nu\nu'} &= 0, \end{aligned} \tag{5.165}$$

respectively. By means of the operator identity (5.141) and the formula (5.130) for the covariant divergence of a tensor, we can rewrite (5.165) as

$$\left[(p+q) u^\nu u_\mu + \left(n \frac{\partial q}{\partial n} + s \frac{\partial q}{\partial s} - q \right) g^\nu{}_\mu + 2 g^{\nu\nu'} \frac{\partial q}{\partial g^{\nu'\mu}} \right]_{;\nu} = 0, \tag{5.166}$$

which plays the role of an equation of motion for the velocity four-vector u^μ. By comparing (5.164) and (5.166), we realize that the same energy-momentum tensor,

$$T_{\mu\nu} = q u_\mu u_\nu + p h_{\mu\nu} + 2\frac{\partial q}{\partial g^{\mu\nu}}, \tag{5.167}$$

occurs in Einstein's field equation and in the energy-momentum balance provided that

$$p = n\frac{\partial q}{\partial n} + s\frac{\partial q}{\partial s} - q, \tag{5.168}$$

where p can be interpreted as the pressure. Note that we previously introduced p and q in terms of U_I and V_I in (5.140), and thus (5.168) is a consistency condition for the choice of U_I and V_I. Although we have introduced two free functions U_I and V_I in the action (5.136), in view of this consistency condition, the covariant approach does not lead to any more freedom than the original one based on the single free function

(5.64). Note also that the energy-momentum tensor (5.167) arises from the general formula

$$\frac{\delta I_{\mathrm{M}}}{\delta g^{\mu\nu}} = \frac{1}{2} T_{\mu\nu}. \tag{5.169}$$

All the variables that remain to be considered, s, w_μ, and $C_{\mu\nu}$, involve irreversible contributions in their field equations. We start with the entropy balance (5.149), which can be rewritten as

$$\left(scu^\mu + \frac{\partial q}{\partial w_\mu} \right)_{;\mu} = \frac{\partial q}{\partial w_\mu} f^{w}_{\mu\nu} \frac{\partial q}{\partial w_\nu} + \bar{f} \left(\bar{c}_{\mu\nu} \frac{\partial q}{\partial C_{\mu\nu}} \right)^2 + \mathring{f} \left(\mathring{c}_{\mu\nu} \frac{\partial q}{\partial C_{\mu\nu}} \right)^2 . \tag{5.170}$$

From (5.152) and (5.153), we finally obtain

$$u^\nu (w_{\mu;\nu} - w_{\nu;\mu}) + \left(w_\nu u^\nu + \frac{1}{c} \frac{\partial q}{\partial s} \right)_{;\mu} = -\frac{1}{c} \frac{\partial q}{\partial s} f^{w}_{\mu\nu} \frac{\partial q}{\partial w_\nu}, \tag{5.171}$$

and

$$u^{\mu'} (C_{\mu\nu;\mu'} - C_{\mu'\nu;\mu} - C_{\mu\mu';\nu}) + \left(u^{\nu'} C_{\nu'\nu} \right)_{;\mu} + \left(C_{\mu\mu'} u^{\mu'} \right)_{;\nu} = -\frac{1}{c} \frac{\partial q}{\partial s} (\bar{f} \bar{c}_{\mu\nu} \bar{c}_{\mu'\nu'} + \mathring{f} \mathring{c}_{\mu\nu} \mathring{c}_{\mu'\nu'}) \frac{\partial q}{\partial C_{\mu'\nu'}}. \tag{5.172}$$

As in special relativity, we can introduce the conditions $C_{\mu\mu'} u^{\mu'} = 0$ and $w_\nu u^\nu = -T/c$, where we have identified the temperature T as $\partial q / \partial s$. Because the equations obtained from the covariant formulation of GENERIC in the presence of gravity have exactly the same structure as those in Section 5.2, we can easily identify all the corresponding quantities and parameters. In particular, a comparison of the energy-momentum tensor (5.167) with the previous expressions (5.55) and (5.86) gives us the equivalents of p and q listed in Table 5.4.1, and we obtain the correspondence

$$\pi_{\mu\nu} = p g_{\mu\nu} + 2 \frac{\partial q}{\partial g^{\mu\nu}}. \tag{5.173}$$

Similarly, a comparison of the entropy flux in (5.170) with the previous expression (5.84) gives us the equivalent of s in Table 5.4.1, and we obtain the correspondence

$$\sigma^{*\mu\nu} w_\nu = \frac{\partial q}{\partial w_\mu}. \tag{5.174}$$

We thus realize the full equivalence of the field equations resulting from the original and covariant frameworks.

Exercise 112 Consistency of Velocity Constraint

Show that the constraint $u^\mu u_\mu = -1$ is consistent with the evolution equations. [Hint: Verify that $u^\mu u_\mu = -1$ implies $u^\nu (u^\mu u_\mu)_{;\nu} = 0$.]

Exercise 113 Cosmological Constant
How do the field equations of this section change when we use the modified action

$$I = \int \sqrt{g} \left[\frac{-c^4}{16\pi G_{\rm N}} (g^{\mu\nu} R_{\mu\nu} - 2\Lambda) + \frac{1}{2} g^{\mu\nu} \frac{M_\mu M_\nu}{U_I} - V_I \right] d^4x, \tag{5.175}$$

with a constant Λ instead of (5.136)?

Exercise 114 Energy-Momentum Tensor From Material Action
Derive the general formula (5.169) for the energy-momentum tensor.

5.5 BULK VISCOUS COSMOLOGY

Bulk viscous cosmology has attracted much attention in studying the evolution of the universe because bulk viscosity is the only dissipative effect that can occur in an isotropic spatially homogeneous system.[29] For example, mixtures of perfect fluids[30] or the effect of particle production[31] may be described phenomenologically via a bulk viscosity, provided that the hydrodynamic description is reasonable, as is assumed here.

In this section, we simplify the general relativistic equations of the preceding section for the case of an isotropic spatially homogeneous system and, for comparison, we recall the main features of two previous approaches to irreversible relativistic thermodynamics. We then apply these theories to the case of a relativistic Boltzmann gas in a flat Robertson-Walker universe. Finally, some results of numerical integrations of the time-evolution equations are shown.

5.5.1 Relativistic Thermodynamics

As a first step, we reduce our previous list of independent variables, n, u^μ, s, $g^{\mu\nu}$, $\Gamma^\nu_{\mu\mu'}$, w_μ, and $C_{\mu\nu}$, for bulk viscous cosmology. We assume a flat Robertson-Walker universe, that is,

$$g_{00} = -1, \quad g_{11} = g_{22} = g_{33} = a(t)^2, \quad g_{\mu\nu} = 0 \text{ for } \mu \neq \nu, \tag{5.176}$$

with the scale factor $a = a(t)$. This scale factor can be related to the determinant of the metric

$$a^3 = \sqrt{g}. \tag{5.177}$$

Note that in (5.176), we have introduced a special coordinate system in which time plays a clearly distinguished role. In such an isotropic homogeneous universe, the

[29] See, for example, Zimdahl, Phys. Rev. D 53 (1996) 5483 or Coley, van den Hoogen & Maartens, Phys. Rev. D 54 (1996) 1393.
[30] Udey & Israel, Mon. Not. R. Astron. Soc. 199 (1982) 1137.
[31] Zimdahl, Phys. Rev. D 57 (1998) 2245.

bulk viscous fluid follows the overall expansion, and the velocity four-vector is fully determined:

$$u^0 = 1, \quad u^1 = u^2 = u^3 = 0. \tag{5.178}$$

Likewise, the spatial components of the four-vector w_μ vanish, and the time component contains redundant information, so that we can neglect w_μ. We expect a second-order differential equation for the scale factor $a(t)$ and the Christoffel symbols are proportional to $\dot{a}(t)$; therefore, we introduce a further variable in addition to $a(t)$,

$$H(t) = \frac{\dot{a}(t)}{a(t)}, \tag{5.179}$$

which is known as the expansion rate of the universe or *Hubble's parameter*. From (5.129), (5.177), and (5.178), we obtain the alternative expression

$$H = \frac{1}{3} c u^\mu{}_{;\mu}. \tag{5.180}$$

In an isotropically expanding universe, $C_{\mu\nu}$ must be of the form

$$C_{\mu\nu} = (F - 1) h_{\mu\nu}, \tag{5.181}$$

with a scalar variable F. At this point, we have reduced the list of independent variables for bulk viscous cosmology to the five scalar quantities n, s, a, H, and F, all of which depend on time only. The thermodynamics of the system is contained in the function $q(n, s, F^2)$, which implies the pressure $p(n, s, F^2)$ in (5.168) and the energy-momentum tensor in (5.164),

$$T_{\mu\nu} = q u_\mu u_\nu + (p + \Pi) h_{\mu\nu}, \tag{5.182}$$

with the bulk viscous pressure contribution

$$\Pi = \frac{2}{3} h^{\mu\nu} \frac{\partial q}{\partial g^{\mu\nu}} = \frac{2}{3} C_{\mu\nu} \frac{\partial q}{\partial C_{\mu\nu}} = -\frac{4}{3} F(1 - F) \frac{\partial q}{\partial F^2}. \tag{5.183}$$

Note that we obtain $T_{00} = q = \rho$ for the energy density.

As a next step, we consider the time-evolution equations for the five variables of bulk viscous cosmology. The particle number balance (5.161) becomes

$$\dot{n} = c u^\nu \partial_\nu n = -3Hn, \tag{5.184}$$

and the entropy balance (5.170) can be written as

$$\dot{s} = -3Hs + \sigma = -3Hs + \frac{2}{T\tau_0} F^2 \frac{\partial q}{\partial F^2} = -3Hs + \frac{3}{2T\tau_0} \frac{(-\Pi)F}{1 - F}, \tag{5.185}$$

where the relaxation time $\tau_0(n, s)$ must be a given function. Einstein's field equations (5.164) imply the following two equations:

$$\dot{a}^2 = \frac{8\pi G_{\rm N}}{3c^2} q a^2 \tag{5.186}$$

as the time-time component and

$$\dot{H} = -\frac{3}{2}H^2 - \frac{4\pi G_N}{c^2}(p + \Pi) \tag{5.187}$$

as the single differential equation resulting from the space-space components (see Exercise 116). With (5.186), the evolution equation for the Hubble parameter can be rewritten as

$$\dot{H} = -\frac{4\pi G_N}{c^2}(q + p + \Pi). \tag{5.188}$$

Equation (5.172), with a relaxation term as in (5.104), finally implies the following time-evolution equation for F (see Exercise 117):

$$\dot{F} = 2H(1 - F) - \frac{1}{\tau_0}F. \tag{5.189}$$

To the lowest order in the time scale τ_0, we hence obtain the physical interpretation of $F = 2H\tau_0$ as the ratio of the time scales for the relaxation of the bulk viscous stresses and for the expansion of the universe. More generally, the dimensionless expansion rate F may be regarded as the thermodynamic force variable associated with bulk viscous stresses.

Note that the above time-evolution equations imply the conservation of energy,

$$\dot{q} = \frac{\partial q}{\partial n}\dot{n} + \frac{\partial q}{\partial s}\dot{s} + 2F\frac{\partial q}{\partial F^2}\dot{F} = -3H(q + p + \Pi), \tag{5.190}$$

and $q = \rho$ can hence be used as an independent variable instead of n, s, or F. The particle number balance (5.184), with the explicit solution na^3 = constant, can be exploited to reduce the number of independent variables from five to four, for example, by eliminating a. Finally, we observe that, in view of their close similarity, (5.188) and (5.190) can be combined into the form of a conservation law,

$$\frac{d}{dt}\left(\frac{4\pi G_N}{c^2}q - \frac{3}{2}H^2\right) = 0, \tag{5.191}$$

where (5.186) shows that the value of the conserved quantity is actually zero.

Exercise 115 Relationship to Linear Irreversible Thermodynamics

Under which circumstances is Π proportional to F, as expected in linear irreversible thermodynamics?

Exercise 116 Ricci Tensor for Flat Robertson-Walker Universe

Derive the relationship

$$R_{\mu\nu} = \frac{1}{c^2}h_{\mu\nu}\left(\frac{\ddot{a}}{a} + 2\frac{\dot{a}^2}{a^2}\right) - \frac{3}{c^2}u_\mu u_\nu \frac{\ddot{a}}{a}, \tag{5.192}$$

for the metric tensor given in (5.176), and show that it implies (5.186) and (5.187).

Exercise 117 Velocity Gradients in Bulk Viscous Cosmology
Derive the relationship

$$u_{\mu;\nu} = \frac{\dot{a}}{ca} h_{\mu\nu} \tag{5.193}$$

for the velocity gradients and use it to obtain (5.189).

As pointed out before, time plays a distinguished role in the equations of cosmology. It is hence natural to derive these equations directly within the original, noncovariant GENERIC framework. We use $x = (n, H, s, F)$ as independent variables, that is, we obtain a physically relevant example of GENERIC in only four dimensions; no position dependences are involved. Equation (5.191) leads us to use

$$E(x) = \frac{4\pi G_{\rm N}}{c^2} q(n, s, F^2) - \frac{3}{2} H^2 \tag{5.194}$$

as the generator of reversible dynamics. This conserved generator E, although closely related to the energy density $q = \rho$, has not the proper dimensions of an energy, but it conveniently simplifies the subsequent equations.[32] As the generator of irreversible dynamics, we use the entropy per particle,

$$S(x) = \frac{s}{n}. \tag{5.195}$$

In view of the conservation of the total number of particles, this generator S is actually proportional to the total entropy of the universe. The gradients of the generators are given by partial derivatives in four-dimensional space:

$$\frac{\partial E(x)}{\partial x} = \begin{pmatrix} \frac{4\pi G_{\rm N}}{c^2} \frac{\partial q}{\partial n} \\ -3H \\ \frac{4\pi G_{\rm N}}{c^2} \frac{\partial q}{\partial s} \\ \frac{4\pi G_{\rm N}}{c^2} 2F \frac{\partial q}{\partial F^2} \end{pmatrix} \tag{5.196}$$

and

$$\frac{\partial S(x)}{\partial x} = \begin{pmatrix} -\frac{s}{n^2} \\ 0 \\ \frac{1}{n} \\ 0 \end{pmatrix}. \tag{5.197}$$

We further introduce the Poisson matrix

$$L(x) = \begin{pmatrix} 0 & n & 0 & 0 \\ -n & 0 & -s & \frac{2}{3}(1-F) \\ 0 & s & 0 & 0 \\ 0 & -\frac{2}{3}(1-F) & 0 & 0 \end{pmatrix} \tag{5.198}$$

[32] Note that the total energy and entropy of a homogeneous flat universe would be infinite.

Table 5.2 Sets of Variables Used in Bulk Viscous Cosmology

Theory	Development of Equations	Numerical Solutions[c]
GENERIC	$(n, H, s, F)^a$	(n, H, z, F)
Israel-Stewart, Maartens-Méndez	$(n, H, \rho, \Pi)^b$	(n, H, z, Π)

[a] The additional time-evolution equation (5.190) for $q = \rho$ and the formula (5.183) for Π are available.
[b] The expression (5.209) for the entropy production rate is available.
[c] The proportionality (5.186) between H^2 and $q = \rho$ may be used to reduce the number of time-evolution equations from four to three.

and the friction matrix

$$M(x) = \frac{n}{T\tau_0} \begin{pmatrix} 0 & 0 & 0 & 0 \\ 0 & 0 & 0 & 0 \\ 0 & 0 & 2F^2 \frac{\partial q}{\partial F^2} & -FT \\ 0 & 0 & -FT & \frac{1}{2}T^2 \left(\frac{\partial q}{\partial F^2}\right)^{-1} \end{pmatrix}. \tag{5.199}$$

Whereas the symmetries and degeneracies can be checked immediately, a check of the Jacobi identity requires some work or experience (see Exercise 118). The positive semidefiniteness of the friction matrix follows for $\partial q/\partial F^2 \geq 0$, so that (5.183) implies that Π is negative, in accordance with the expectation that bulk viscosity reduces the pressure. The resulting GENERIC time-evolution equations are (5.184), (5.185), (5.188), and (5.189).

Exercise 118 Jacobi Check for Bulk Viscous Cosmology
Verify the Jacobi identity for the Poisson bracket associated with the matrix (5.198).

For comparison, we next introduce two older theories of bulk viscosity: the widely used second-order theory of Israel and Stewart[33] and a modification proposed by Maartens and Méndez.[34] Both theories are based on the variables $x = (n, H, \rho, \Pi)$, and three functions must be given: the specific entropy[35] of the fluid in the absence of a bulk viscous pressure, $\check{s}_0(n, \rho)$, the relaxation time for bulk viscous stress, $\tau_0(n, \rho)$, and the bulk viscosity, $\kappa(n, \rho)$. In addition to the two different sets of independent variables introduced so far, for the explicit solution of the resulting equations we will use an inverse temperature variable z instead of s or ρ because we can then introduce the known temperature-dependent material properties of a relativistic Boltzmann gas most conveniently. The different sets of variables employed to formulate and solve the various theories of bulk viscous cosmology are summarized in Table 5.2.

[33] Israel, Ann. Phys. (N.Y.) 100 (1976) 310; Israel & Stewart, Ann. Phys. (N.Y.) 118 (1979) 341.
[34] Maartens & Méndez, Phys. Rev. D 55 (1997) 1937.
[35] That is, the entropy per particle.

The "fictitious" local equilibrium, described by the specific entropy $\check{s}_0 = \check{s}_0(n, \rho)$, is used to define the temperature T_0 and the pressure p_0:

$$\frac{1}{T_0(n,\rho)} = n \frac{\partial \check{s}_0(n,\rho)}{\partial \rho} \tag{5.200}$$

and

$$p_0(n,\rho) = -\rho - n^2 T_0(n,\rho) \frac{\partial \check{s}_0(n,\rho)}{\partial n}. \tag{5.201}$$

The equations for n, H, and ρ coincide with (5.184), (5.188), and (5.190), where the pressure p_0 is identified with the pressure p entering the energy-momentum tensor (5.182):

$$\dot{n} = -3Hn, \tag{5.202}$$

$$\dot{H} = -\frac{4\pi G_{\rm N}}{c^2}(\rho + p_0 + \Pi), \tag{5.203}$$

and

$$\dot{\rho} = -3H(\rho + p_0 + \Pi). \tag{5.204}$$

The formulation of an equation for Π is based on the discussion of the dissipation rate, which, in turn, is obtained from a beyond-equilibrium entropy expression. Israel and Stewart, and also Maartens and Méndez, assumed the effective specific entropy of the nonequilibrium state to be of second order in the deviations from equilibrium,

$$\check{s}(n,\rho,\Pi) = \check{s}_0(n,\rho) - \frac{\tau_0(n,\rho)}{2nT_0(n,\rho)\kappa(n,\rho)}\Pi^2. \tag{5.205}$$

Equation (5.41) shows that the ratio τ_0/κ can naturally be considered as information about the dependence of a thermodynamic potential on a beyond-equilibrium variable.

In the famous theory of Israel and Stewart, the time-evolution equation for Π is obtained by assuming that the entropy production rate resulting from the entropy expression (5.205) should be given by the expression of linear irreversible thermodynamics implied by (3.8) and (3.12) for pure bulk viscosity,

$$\sigma = n\dot{\check{s}} = \frac{\Pi^2}{T_0\kappa}. \tag{5.206}$$

This procedure, which is usually considered to guarantee thermodynamic consistency, implies the time-evolution equation (see Exercise 119)

$$\dot{\Pi} = -\frac{1}{\tau_0}(\Pi + 3\kappa H) - \frac{\Pi}{2}\frac{nT_0\kappa}{\tau_0}\frac{d}{dt}\left(\frac{\tau_0}{nT_0\kappa}\right). \tag{5.207}$$

Note that (5.206) can be used for the following reformulation of the beyond-equilibrium entropy (5.205) in terms of the entropy production rate,

$$\check{s}(n,\rho,\Pi) = \check{s}_0(n,\rho) - \frac{1}{2}\tau_0(n,\rho)\frac{\sigma(n,\rho,\Pi)}{n}. \tag{5.208}$$

In the Israel-Stewart theory, the bulk viscous pressure $|\Pi|$ is unbounded. For that reason, Maartens and Méndez proposed a generalization of the entropy production rate (5.206) of linear irreversible thermodynamics,

$$\sigma = n\dot{\check{s}} = \frac{\Pi^2}{T_0(\kappa + \tau_* \Pi)}, \tag{5.209}$$

which implies the bound

$$-\Pi \leq \frac{\kappa}{\tau_*}, \tag{5.210}$$

and the modified time-evolution equation (see Exercise 119)

$$\dot{\Pi} = -\frac{1}{\tau_0}\left(\frac{\Pi}{1 + \tau_* \Pi/\kappa} + 3\kappa H\right) - \frac{\Pi}{2}\frac{nT_0\kappa}{\tau_0}\frac{d}{dt}\left(\frac{\tau_0}{nT_0\kappa}\right). \tag{5.211}$$

The Maartens-Méndez theory introduces the extra function $\tau_*(n, \rho)$, which we assume to be of the form $\tau_*(n, \rho) = k^2\tau_0(n, \rho)$, where k^2 is a dimensionless constant. The Israel-Stewart theory is recovered in the limit $k^2 = 0$. The phenomenological coefficient k^2 remains an undetermined parameter that is supposed to be chosen such that we stay sufficiently close to equilibrium for the second-order expression (5.205) to be valid, that is, $|\Pi| \ll p_0$.

The GENERIC approach requires the functions $\tau_0(n, s)$ and $q(n, s, F^2)$, whereas the Israel-Stewart theory requires the functions $\tau_0(n, \rho)$, $\check{s}_0(n, \rho)$, and $\kappa(n, \rho)$; the Maartens-Méndez theory has an additional dimensionless parameter k^2. For a better comparison of GENERIC with the older theories, we choose the function $q(n, s, F^2)$ of three arguments such that we can identify p with p_0, that is, $p = p(n, q)$ as a function of n and q becomes independent of F^2 because p_0 plays a key role in the time-evolution equations (5.203) and (5.204). According to (5.168), this situation occurs for the following expression involving two functions of two arguments:

$$q(n, s, F^2) = \check{q}(n, \hat{q}(s/n, F^2)), \tag{5.212}$$

which implies that the quantity $\hat{q}(s/n, F^2)$ can be expressed as a function of q and n. A most natural choice is given by

$$q(n, s, F^2) = n\hat{q}(s/n, F^2), \tag{5.213}$$

which reduces the material information to two functions of two arguments, $\tau_0(n, s)$ and $\hat{q}(s/n, F^2)$. Note that T still depends on F^2 and thus does not coincide with T_0, so that a comparison between GENERIC and the older theories is not straightforward. By comparing the entropy production rate obtained from (5.183) and (5.185),

$$\sigma = \frac{9}{8}\frac{1}{T\tau_0}\frac{\Pi^2}{(1-F)^2}\left(\frac{\partial q}{\partial F^2}\right)^{-1}, \tag{5.214}$$

near equilibrium, that is, for small F, with (5.206) we find

$$\frac{\partial q(n, s, F^2)}{\partial F^2} = \frac{9}{8}\frac{\kappa}{\tau_0}. \tag{5.215}$$

Equation (5.183) then implies the bound

$$-\Pi \leq \frac{3}{8}\frac{\kappa}{\tau_0}. \tag{5.216}$$

By comparing (5.216) to the bound (5.210) for the Maartens-Méndez theory, we identify $k^2 = 8/3$ as a natural parameter.

Exercise 119 Theories of Bulk Viscous Pressure

Derive the time-evolution equations (5.207) and (5.211) for the bulk viscous pressure contribution Π.

Exercise 120 Matching Pressures

Show that the pressure p obtained from the assumption (5.212) can be expressed as a function of q and n.

5.5.2 Input from Relativistic Boltzmann Gas

Before we discuss and compare the cosmological predictions of the various theories introduced in the preceding subsections, we need to choose a number of free functions. We use the model of the relativistic Boltzmann gas to fix these functions.[36] Although the relevance of the Boltzmann gas as a meaningful model of the material contents of the universe is doubtful, we employ it here as a test case for the theories under investigation. For the classical case, we find in Chapter 7 that the corresponding equilibrium thermodynamics is that of an ideal gas [see (7.142)], and that the bulk viscosity vanishes [see remark after (7.114)]. In the relativistic case, however, the Boltzmann gas leads to a nonzero bulk viscosity so that we can employ it for the present discussion.

To express the thermodynamic quantities, as indicated in Table 5.2, it is convenient to switch from the internal energy density ρ or the entropy density s to the dimensionless inverse temperature variable

$$z = \frac{mc^2}{k_B T}, \tag{5.217}$$

where m is the mass of a gas particle. The equilibrium reference state is characterized by two equations of state: the ideal gas law,

$$p_0(n, z) = n k_B T = \frac{nmc^2}{z}, \tag{5.218}$$

and the internal energy density,

$$\rho_0(n, z) = nmc^2 \left(\frac{3}{z} + \frac{K_1(z)}{K_2(z)} \right). \tag{5.219}$$

[36] We use de Groot et al., *Relativistic Kinetic Theory* (North-Holland, 1980) as a standard reference.

The relativistic corrections to the classical equation of state for the internal energy in (5.219) are expressed in terms of the modified Bessel functions K_j of order j,[37] which play a central role in relativistic kinetic theory. The functions κ and τ_0 involve collision integrals, which can be evaluated in closed form, assuming a constant cross section σ_0 (hard spheres). The bulk viscosity turns out to be independent of n,

$$\kappa(z) = \frac{mc}{96\pi^2\sigma_0}\frac{zK_2(z)^2[(5-3\gamma)\hat{h}-3\gamma]^2}{2K_2(2z)+zK_3(2z)}, \tag{5.220}$$

with the auxiliary functions $\hat{h} = zK_3(z)/K_2(z)$ and $\gamma/(\gamma-1) = 5\hat{h} - \hat{h}^2 + z^2$, which can be interpreted as the reduced Gibbs free energy per particle and the ratio of the heat capacities, respectively.[38] The relaxation time τ_0 is related to κ by

$$\tau_0(n,z) = \frac{\kappa(z)\alpha'(z)}{p_0(n,z)} = \frac{z\kappa(z)\alpha'(z)}{nmc^2}, \tag{5.221}$$

where the function $\alpha'(z)$ is defined as

$$\alpha'(z) = \frac{(10-7\gamma)\hat{h}+z^2(5/3-\gamma)}{[(5/3-\gamma)\hat{h}-\gamma]^2}. \tag{5.222}$$

We next develop the equations for the inverse temperature parameter, z, and the variable describing bulk viscous effects, Π or F, for the hard sphere Boltzmann gas for the approaches considered in the previous subsection. For the approach of Israel and Stewart, and for the modification of Maartens and Méndez, the temperature T_0 is to be used in the definition of z. We then obtain from (5.219) with (5.202) and (5.204)

$$\dot{z} = 6H\frac{p_0+\Pi}{nmc^2}\left(\frac{6}{z^2} - 2\frac{d}{dz}\frac{K_1(z)}{K_2(z)}\right)^{-1}. \tag{5.223}$$

In view of the identity

$$\frac{\tau_0}{nT_0\kappa} = \frac{k_B\alpha'}{p_0^2}, \tag{5.224}$$

(5.211) can be rewritten as

$$\frac{d}{dt}\left(\frac{\Pi}{p_0}\right) = -\frac{1}{p_0\tau_0}\left(\frac{\Pi}{1+\tau_*\Pi/\kappa} + 3\kappa H\right) - \frac{1}{2}\frac{\Pi}{p_0}\frac{d}{dt}\ln\alpha'. \tag{5.225}$$

The entropy production (5.209) is given by

$$\sigma = \frac{nk_B\alpha'}{\tau_0}\left(\frac{\Pi}{p_0}\right)^2\left(1+\frac{\tau_*}{\kappa}\Pi\right)^{-1}, \tag{5.226}$$

[37] See, for example, Section 6.6 of Press et al., *Numerical Recipes* (Cambridge, 1992).

[38] Note, that we use the cross section σ_0 of Hiscock & Salmonson, Phys. Rev. D 43 (1991) 3249, which is 3π times larger than the one used in the textbook by de Groot et al.

and the corresponding formulas for the Israel-Stewart theory are obtained for $\tau_* = 0$. The time-evolution equations (5.223) and (5.225) replace (5.204) and (5.211) for the older theories.

Finally, we need to find an equation for z within the GENERIC approach, for which T does not coincide with T_0. We thus need to choose the function $q(n, s, F^2)$ as suggested by the relativistic Boltzmann gas, while τ_0 is still given by (5.221). For the numerical work with the inverse temperature variable z, the following specific assumption is most convenient:

$$q(n, z, F^2) = \rho_0(n, z) + F^2 \rho_1(n, z). \tag{5.227}$$

For this functional form of $q(n, z, F^2)$, the explicit relationship between the entropy density s and the inverse temperature z can be written as[39]

$$s(n, z, F^2) = -\frac{k_{\rm B}}{mc^2} \left[\int_z^\infty z' \frac{\partial \rho_0(n, z')}{\partial z'} dz' + F^2 \int_z^\infty z' \frac{\partial \rho_1(n, z')}{\partial z'} dz' \right], \tag{5.228}$$

and we further obtain (see Exercise 121 for both steps)

$$\frac{\partial q(n, s, F^2)}{\partial F^2} = -\frac{1}{z} \int_z^\infty \rho_1(n, z') dz'. \tag{5.229}$$

In other words, if $\partial q(n, s, F^2)/\partial F^2$ is known as a function of z, we can directly calculate $\rho_1(n, z) = \partial q(n, z, F^2)/\partial F^2$ by multiplication with z and subsequent differentiation. From (5.215) and (5.221), we obtain

$$\frac{\partial q(n, s, F^2)}{\partial F^2} = \frac{9}{8} \frac{nmc^2}{z \alpha'(z)} \tag{5.230}$$

and hence

$$\rho_1(n, z) = \frac{9}{8} nmc^2 \frac{d}{dz} \left(\frac{1}{\alpha'(z)} \right) \tag{5.231}$$

or the following explicit functional form of the energy density:

$$q(n, z, F^2) = nmc^2 \left[\frac{3}{z} + \frac{K_1(z)}{K_2(z)} + \frac{9}{8} F^2 \frac{d}{dz} \left(\frac{1}{\alpha'(z)} \right) \right]. \tag{5.232}$$

For this particular choice, the functional form of $q(n, s, F^2)$ turns out to be of the previously suggested type (5.213).

In analogy to the derivation of (5.223) from (5.219), we obtain from (5.232)

$$\left[\frac{6}{z^2} - 2 \frac{d}{dz} \frac{K_1(z)}{K_2(z)} - \frac{9}{4} F^2 \frac{d^2}{dz^2} \left(\frac{1}{\alpha'(z)} \right) \right] \dot{z} = 6H \frac{p_0 + \Pi}{nmc^2} + \frac{9}{2} F \dot{F} \frac{d}{dz} \left(\frac{1}{\alpha'(z)} \right), \tag{5.233}$$

[39] For the first integral in (5.228) to be finite, the heat capacity must go to zero for $T \to 0$ or $z \to \infty$; this condition is not fulfilled for an ideal gas, so that the upper limit of integration cannot really be chosen at infinity, or quantum corrections need to be taken into account. This subtlety is irrelevant to the purpose of the present analysis.

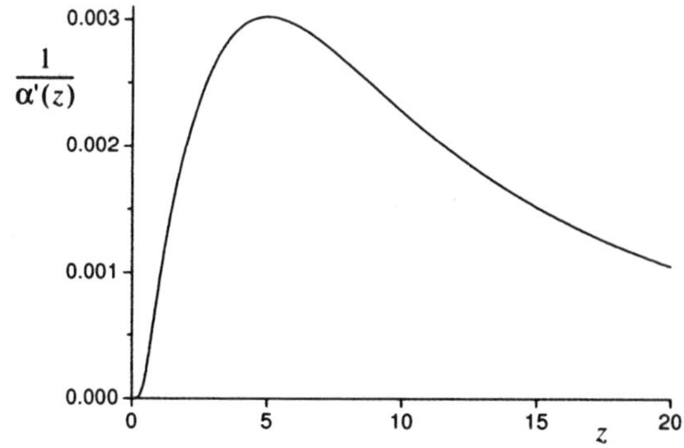

Fig. 5.1 The auxiliary function $1/\alpha'(z)$ [according to (A6) of Ilg & Öttinger, Phys. Rev. D 61 (2000) 023510].

from which $\dot{z}$ can be calculated. Together with (5.184), (5.188), and (5.189), we have all the required time-evolution equations, in which Π is given by

$$-\frac{\Pi}{p_0} = \frac{3}{2}F(1-F)\frac{1}{\alpha'(z)}, \tag{5.234}$$

as can be shown by inserting (5.230) into (5.183) and using (5.218). The resulting bound on the bulk viscous stress turns out to be very restrictive for the case of a Boltzmann gas,

$$-\frac{\Pi}{p_0} \leq \frac{3}{8}\frac{1}{\alpha'(z)}, \tag{5.235}$$

which is of the order of 10^{-3} for intermediate temperatures and vanishing in the ultra- and nonrelativistic limits (see Figure 5.1). Bulk viscosity thus is a phenomenon of mildly relativistic regimes but is almost negligible for the case of a Boltzmann gas. The same holds for the Maartens-Méndez theory for which the 3/8 in (5.235) must be replaced by $1/k^2$. Within the GENERIC framework, the entropy production is also bounded,

$$\frac{\tau_0 \sigma}{k_B n} = \frac{9}{4}\frac{F^2}{\alpha'(z)} \leq \frac{9}{4}\frac{1}{\alpha'(z)}. \tag{5.236}$$

Exercise 121 Transformation From Inverse Temperature to Entropy
Derive (5.228) and (5.229).

5.5.3 Model Predictions

All thermodynamic information is now available to integrate the various bulk viscous cosmological models presented in this section. The two parameters entering these equations are the particle mass m and the cross section σ_0. Assuming a cross section of the form $\sigma_0 = \pi[\hbar/(mc)]^2$, where $\hbar$ is Planck's constant divided by 2π, the only remaining free parameter can be taken as the particle mass m. However, we do not expect this mass to have a particular physical significance because the Boltzmann gas is not a realistic model of the cosmological fluid.

In all our numerical calculations, we choose the same initial conditions as Hiscock and Salmonson; that is, the initial scale factor is conveniently taken as $a = 1$ and the initial inverse temperature is $z = 10^{-3}$, which assures that the integration is begun in the highly relativistic high-temperature regime. The initial time multiplied by the speed of light, or $c/(2H)$, is chosen to be 10 cm, and the initial bulk stress is set equal to the Eckart value, $\Pi = -3\kappa H$. Although stated differently in their paper,[40] Hiscock and Salmonson chose an initial mass density by assigning fixed initial values of H and z, which determine $n_0 m$ by (5.186) and (5.219).[41] We choose the initial value of n according to the same procedure. All the details on the numerical integration scheme and the evaluation of modified Bessel functions can be found in the original paper on which this section is based.[42]

For the Israel-Stewart theory and the modification of Maartens and Méndez, a critical value of the mass parameter divides two qualitatively different scenarios. With the chosen parameters and initial conditions, the relaxation time $\tau_0 \propto 1/(n\sigma_0)$ is proportional to the cube of the mass. Therefore, the two terms in the transport equations (5.207) or (5.211) for Π can be weighted differently by varying the mass. Below the critical value, which turns out to be of the order 3×10^9 GeV$/c^2$ for the chosen initial conditions, all three theories predict a similar behavior of the dynamical variables.

The absolute value of the ratio of the bulk viscous pressure over the equilibrium pressure $|\Pi/p_0|$ first raises from the initial value to a maximum at intermediate temperatures, and it then decays in the nonrelativistic regime. Therefore, the scale factor a, the inverse temperature z, and the internal energy density ρ show essentially perfect fluid behavior, that is, a crossover from a radiation gas behavior $a \propto t^{1/2}$, $z \propto a$, and $\rho \propto a^{-4}$ in the ultra-relativistic regime ($z \ll 1$) to a pressureless gas $a \propto t^{2/3}$, $z \propto a^2$, and $\rho \propto a^{-3}$ in the nonrelativistic regime ($z \gg 1$) (see Figure 5.2 and Exercise 122).

The maximum value of $|\Pi/p_0|$ is, however, different in the various approaches, as well as the rate of the decay of this quantity (see Figure 5.3). The Israel-Stewart theory always predicts a higher absolute value of the bulk viscous stress than GENERIC. For small masses, the approach of Maartens and Méndez in a sense interpolates between these two theories by varying the parameter k^2, coming closer to the GENERIC bulk stress as the parameter k^2 approaches $8/3$. The maximum value of $|\Pi/p_0|$ increases with increasing mass, leading to a smaller and smaller range in which k^2 is allowed to vary without violating the bound (5.210). The prediction of the truncated Israel-Stewart theory in Figure 5.3, for which the last term in (5.207) is omitted, is significantly different from all the other results.

[40] Hiscock & Salmonson, Phys. Rev. D 43 (1991) 3249.

[41] Hiscock, private communication (June 1998).

[42] Ilg & Öttinger, Phys. Rev. D 61 (2000) 023510.

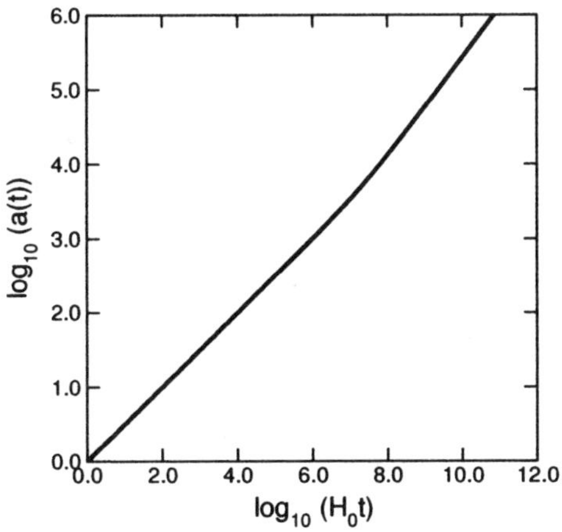

Fig. 5.2 Scale factor $a(t)$ as a function of time. There is a smooth crossover from a radiation gas behavior in the ultra-relativistic regime ($a \propto t^{1/2}$) to a pressureless gas-like behavior in the nonrelativistic regime ($a \propto t^{2/3}$). The particle mass is taken to be 5.5×10^5 GeV$/c^2$. [Reprinted with permission from P. Ilg and H. C. Öttinger, Phys. Rev. D 61 (2000) 023510. Copyright (2000) by the American Physical Society.]

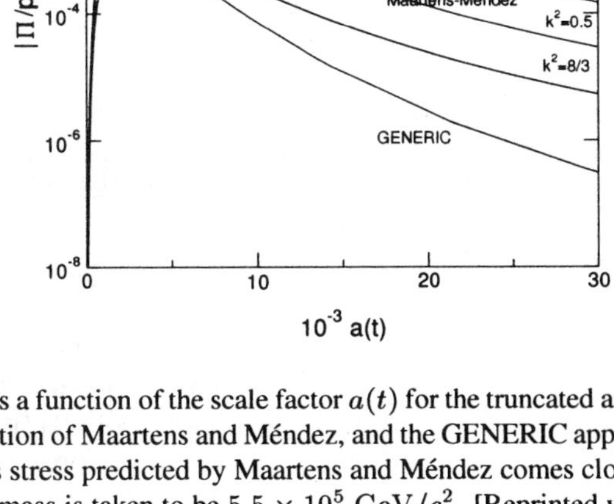

Fig. 5.3 $|\Pi/p_0|$ as a function of the scale factor $a(t)$ for the truncated and full Israel-Stewart theory, the modification of Maartens and Méndez, and the GENERIC approach. By increasing k^2, the bulk viscous stress predicted by Maartens and Méndez comes closer to the GENERIC value. The particle mass is taken to be 5.5×10^5 GeV$/c^2$. [Reprinted with permission from P. Ilg and H. C. Öttinger, Phys. Rev. D 61 (2000) 023510. Copyright (2000) by the American Physical Society.]

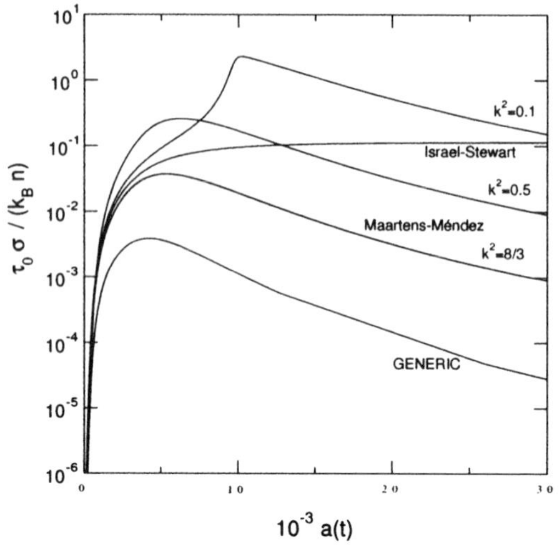

Fig. 5.4 Dimensionless entropy production per particle as a function of the scale factor $a(t)$. In the nonrelativistic regime, Israel and Stewart predicted a constant value of $\tau_0\sigma/(k_B n)$; however, in the GENERIC formalism, this quantity decreases. The modification of Maartens and Méndez predicts values of the entropy production, which approach the GENERIC values for increasing k^2. The particle mass is taken to be 5.5×10^5 GeV$/c^2$. [Reprinted with permission from P. Ilg and H. C. Öttinger, Phys. Rev. D 61 (2000) 023510. Copyright (2000) by the American Physical Society.]

Because the second law of thermodynamics is handled differently in the various theories, the entropy production is also quite different (see Figure 5.4). In the Israel-Stewart theory, the dimensionless entropy per particle generated on the relaxation time scale, $\tau_0\sigma/(k_B n)$, tends toward a positive constant in the nonrelativistic regime, whereas this quantity goes to zero in the GENERIC formalism, with a maximum located at about the same value of z where $|\Pi/p_0|$ shows a maximum. In the modification of Maartens and Méndez, (5.209) suggests that the entropy production is higher than the corresponding value for the Israel-Stewart theory. Because the bulk viscous pressure is lower than in the Israel-Stewart theory, however, the entropy production can fall below the value of the Israel-Stewart theory. Because τ_0 is proportional to the volume multiplied by the square root of the inverse temperature and goes to infinity in the nonrelativistic regime, the second term in the Israel-Stewart transport equation (5.207) dominates the relaxation term. As a result of the quadratic form of the entropy (5.205), this leads to a constant value of $\check{s}_0 - \check{s}$, which is itself related to $\tau_0\sigma/(k_B n)$ via (5.208). The similar form of $\tau_0\sigma/(k_B n)$ compared with $|\Pi/p_0|$ in the GENERIC approach is related to the bound (5.236), which is proportional to the bound on the bulk pressure (5.235).

When the mass approaches the critical value, the entropy predicted by the Maartens-Méndez theory diverges for smaller and smaller values of k^2. Above the critical value, the Israel-Stewart theory exhibits a different qualitative behavior when

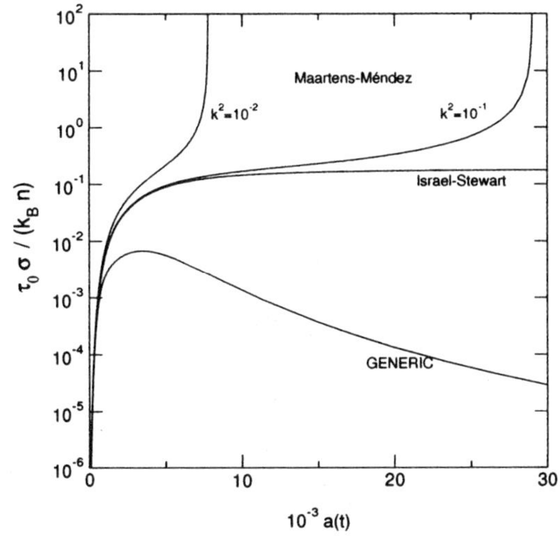

Fig. 5.5 Dimensionless entropy production per particle as a function of the scale factor $a(t)$. Whereas the entropy productions in the Israel-Stewart theory and the GENERIC formalism slowly vary with the particle mass, in the modification of Maartens and Méndez, this entropy production diverges. The particle mass is taken to be 5.5×10^6 GeV$/c^2$. [Reprinted with permission from P. Ilg and H. C. Öttinger, Phys. Rev. D 61 (2000) 023510. Copyright (2000) by the American Physical Society.]

compared with smaller values of mass. As found by Hiscock and Salmonson, the maximum value of $|\Pi/p_0|$ increases with increasing mass until, at the critical value, $|\Pi/p_0|$ no longer decays but asymptotically approaches unity (although we have found the same qualitative behavior, our quantitative results for the Israel-Stewart theory deviate from those of Hiscock and Salmonson, presumably due to numerical problems in the previous work). In this case, the bulk viscous pressure cancels the equilibrium pressure, so that the solution gets dynamically trapped in an expanding state with constant temperature (5.223), leading to a constant value of Π/p_0 (5.225) for $\tau_* = 0$. Again, the entropy production shows a plateau (see Figure 5.5), as can be seen from (5.226).

In the GENERIC formalism, the bounds on the ratio of bulk pressure over equilibrium pressure (5.235) and the entropy production (5.236) forbid a similar scenario. Rather than violating the near-equilibrium condition $|\Pi| \ll p_0$, as does the Israel-Stewart solution, GENERIC predicts small deviations from the local equilibrium state for all values of mass. In contrast to GENERIC, the modification of Maartens and Méndez is insufficient to keep the system near local equilibrium states. Figure 5.5 illustrates that the divergence of the entropy production rate (5.226) does not keep the bulk viscous stress sufficiently small, so that a singularity of the entropy production occurs for a finite scale factor.

In summary, our results do not support the often-mentioned hope that the theory of Israel and Stewart holds even far away from local equilibrium. The way in which the second law is imposed misses an upper bound on the bulk viscous stress, so that the theory may quickly move out of its range of validity. It seems that the extension proposed by Maartens and Méndez goes essentially in the direction of the GENERIC theory, provided that their phenomenological time τ_* is closely related to the relaxation time τ_0. Both theories imply a bound on the bulk viscous stress, and in many cases the bulk viscous stresses and the entropy production rates predicted by the Maartens-Méndez theory lie between the predictions of the Israel-Stewart theory and GENERIC. However, the dynamics of the Maartens-Méndez theory cannot prevent an aphysically diverging entropy. Maybe this behavior can be traced back to the quadratic form of the specific entropy, which seems to miss the upper bound on the entropy production, given by GENERIC.

Exercise 122 Perfect Fluid Cosmology

Derive the ultra-relativistic and the nonrelativistic behavior illustrated in Figure 5.2. [Hint: Use $K_1(z)/K_2(z) = z/2$ for $z \ll 1$ and $K_1(z)/K_2(z) = 1 - [3/(2z)]$ for $z \gg 1$.]

Part II

Statistical Approach

The second part of this book is devoted to the idea to "bridge scales" or connect different levels of description. To get a better feeling for this problem, one should first reflect on some key ideas behind "bridging scales." We need to specify the variables for two different levels of description, say z for the more detailed level and x for the less detailed level. Any explicit connection between these two levels requires a map $\Pi : z \mapsto x = \Pi(z)$ that associates "macrostates" x with "microstates" z. Typically, there are many microstates mapped into a given macrostate. We can hence classify the microstates according to macrostates, and we use the notation $[z]_x = \{z|\Pi(z) = x\}$ for the equivalence class of microstates corresponding to x. At this point, we need to distinguish between two fundamentally different situations.

Scenario A: There is a natural, distinguished representative of the equivalence class $[z]_x$, which we denote by $\bar{z}_x$, We assume that we have a smooth dependence of $\bar{z}_x$ on x, so that $\bar{z}_x$ defines a submanifold of representatives in the space of microstates z.

Scenario B: All the members of the equivalence class $[z]_x$ are on an equal footing so that there is no natural representative.

Both scenarios can be illustrated by the example of a fluid for which one would like to understand the hydrodynamics in terms of the interactions in an N-particle system. The coarser variables x correspond to the hydrodynamic fields. For the detailed description, we can either choose the positions and momenta of the particles as the basic variables, $z^{\mathrm{H}} = (\boldsymbol{r}_1, \boldsymbol{p}_1, \ldots, \boldsymbol{r}_N, \boldsymbol{p}_N)$, or the probability density for finding certain particle positions and momenta, $z^{\mathrm{L}} = \rho(\boldsymbol{r}_1, \boldsymbol{p}_1, \ldots, \boldsymbol{r}_N, \boldsymbol{p}_N)$. The fundamental time-evolution equations for the components of z^{H} are given by Hamilton's equations of motion, whereas the evolution of z^{L} is governed by the Liouville equation, which is a continuity equation in phase space for the probability density z^{L}.

Working with $z^{\rm H}$, one is faced with Scenario B. That is, there are many particle configurations that lead to the same hydrodynamic fields, and all these configurations are on a, maybe more or less, equal footing. The space of variables $z^{\rm L}$ is much richer than the phase space of $z^{\rm H}$ because fixed particle positions and momenta correspond to δ-function-type probability densities. There are also smooth probability densities $z^{\rm L}$, and the natural representative $\bar{z}^{\rm L}_x$ of an equivalence class could be chosen as the smoothest probability density consistent with the hydrodynamic fields x, so that we naturally arrive at Scenario A.

The example shows that both scenarios call for additional ingredients to be really meaningful. In Scenario A, we need to identify a distinguished representative $\bar{z}_x$ of the equivalence class $[z]_x$. The most natural type of criterion to identify representatives is based on some function, say $H(z)$, which is minimized in the class $[z]_x$ to find the representative $\bar{z}_x$. To ensure a unique minimum, we should postulate the existence of a convex function $H(z)$. In our example of the N-particle system, minimizing the function $H(z)$ corresponds to maximizing the smoothness of the probability density. In Scenario B, we need to specify what it means that "all microstates are on a more or less equal footing," and that is most naturally done in terms of a probability density $\rho(z)$ for weighting the importance of microstates. Typically, $\rho(z)$ is chosen "as flat as possible," which corresponds to minimizing some functional H of $\rho(z)$. This observation explains why one moves from Scenario B to Scenario A when using $\rho(z)$ instead of z as the more microscopic variable (note that we had indeed chosen $z^{\rm L} = \rho(z^{\rm H})$ for the N-particle system). The significance of the two scenarios to obtain time-evolution equations for the variables on the more macroscopic level of description is described in Figure 6.0.

In Scenario A, the quantity $-H$ can be considered as an entropy, and the set of representatives obtained by maximizing the entropy for all macrostates x is known as the *quasi-equilibrium manifold.* The time evolution of a macrostate can be obtained from the time evolution of the representative microstate by means of the chain rule

$$\frac{dx}{dt} = \left.\frac{\partial \Pi}{\partial z}\right|_{\bar{z}_x} \cdot J(\bar{z}_x),$$

where $J(z)$ represents the rate of change of the microstate z. An improved set of representatives—better adapted to the exact time evolution—can be obtained by the method of *invariant manifolds.*[1] This approach, in contrast to the quasi-equilibrium method, does not require a global classification of all microstates implied by a map $x = \Pi(z)$; it relies on projecting the small changes of a representative microstate resulting from the exact fundamental time evolution to the tangent space of the manifold of representatives, so that approximate time-evolution equations for the coordinates characterizing the representative manifold are implied. This construction requires a *thermodynamic projector*, for which an elegant unique construction has been found.[2] The *entropic scalar product* implied by $\partial^2 H(z)/\partial z \partial z$ plays a central role in the

[1] Gorban & Karlin, Physica A 190 (1992) 393; Transp. Theory Stat. Phys. 23 (1994) 559.
[2] Gorban & Karlin, Chem. Eng. Sci. 58 (2003) 4751.

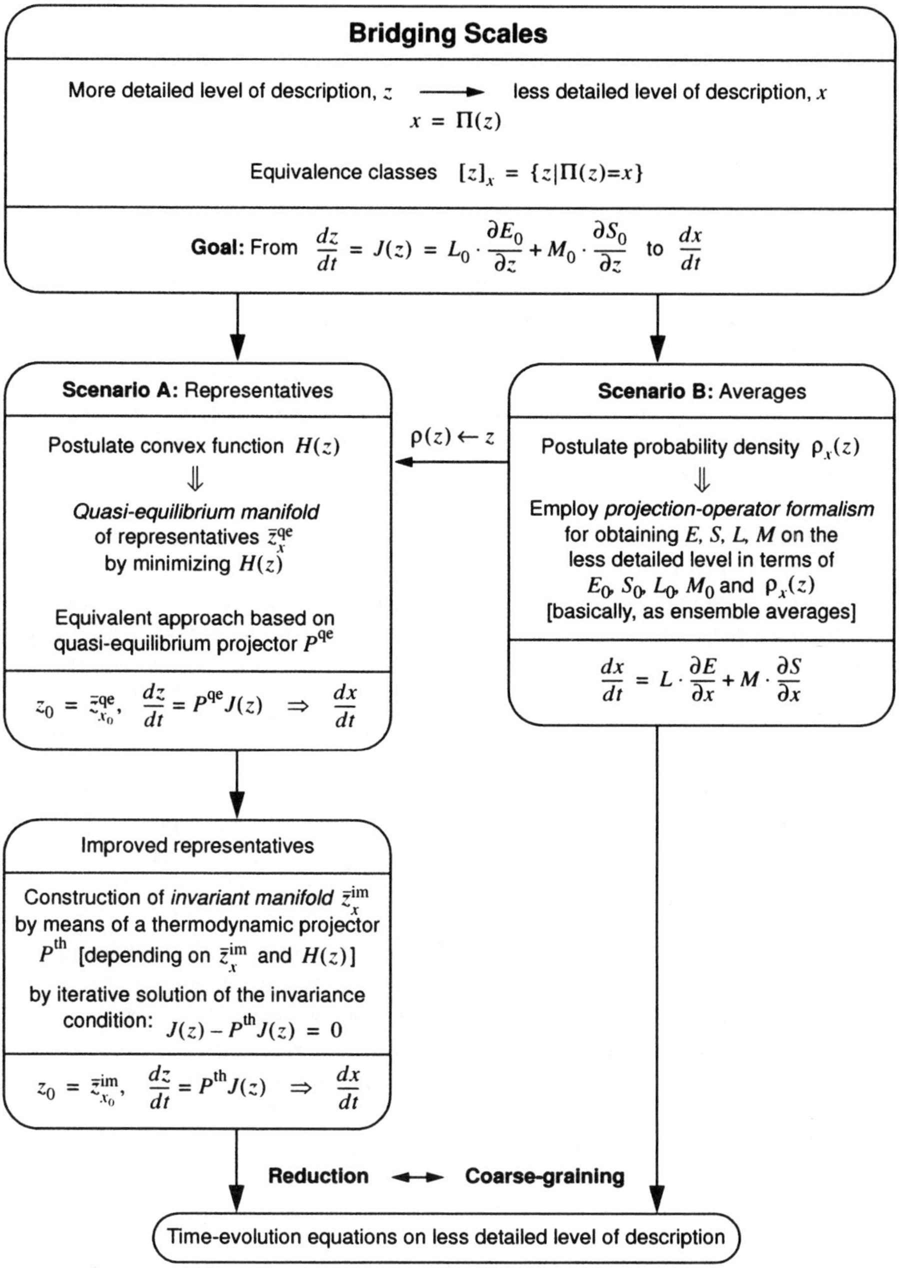

Fig. 6.0 Typical scenarios involved in bridging scales.

formulation of thermodynamic projectors. When the quasi-equilibrium approach is equivalently rewritten in terms of a projector, it turns out to be the orthogonal projection with respect to the entropic scalar product. Within the method of invariant manifolds, one iteratively improves the representative manifold by minimizing the discrepancy between the exact and the projected time evolution of representative microstates, usually starting from the quasi-equilibrium manifold. Ideally, such an iterative procedure would lead to equations with the full structure of GENERIC after each iteration step.

A deep result for the method of invariant manifolds based on thermodynamic projectors states that the entropy production rate found on the less detailed level of description equals the entropy production rate on the more detailed level of description.[3] Eliminating details of the description does not lead to an increased entropy production. In particular, one cannot derive irreversible equations from reversible ones by means of the method of invariant manifolds; we hence classify it as a *reduction technique* or as a solution technique. If we want to obtain irreversible equations from reversible ones, we need a *coarse-graining technique*. We will study in detail how coarse-grained equations can be obtained within Scenario B by means of projection-operator techniques separating slow from fast motions (see Figure 6.0).[4] In this book, we focus on coarse-graining techniques rather than reduction techniques because only with those can we cover the entire range from reversible atomistic equations to irreversible mesoscopic and macroscopic equations. Reduction techniques are here regarded as tools to solve the irreversible equations obtained by coarse-graining methods, usually in an approximate manner; for example, we will discuss Grad's moment method to solve Boltzmann's kinetic equation (see Section 7.4).

In going beyond equilibrium, the scenario for coarse-graining is very much like in equilibrium thermodynamics. If one works on one level of description only, modeling requires a suitable choice of the relevant variables and phenomenological assumptions for the building blocks of equilibrium thermodynamics (a thermodynamic potential) or beyond-equilibrium thermodynamics (the generators energy and entropy as well as the Poisson and friction matrices of GENERIC). At equilibrium, following the ideas of Willard Gibbs, thermodynamic potentials can be calculated from first principles through partition functions. The statistical approach provides unambiguous expressions for thermodynamic potentials in principle and, even more importantly, it provides the foundations to solve practical problems by means of approximations or Monte-Carlo simulations. Beyond equilibrium, the projection-operator technique is the key to first-principle expressions for the building blocks of GENERIC on coarse-grained levels. Generalizations of the more fundamental microcanonical and of the more tractable canonical approaches of equilibrium statistical mechanics have been established.

[3] Gorban & Karlin, Chem. Eng. Sci. 58 (2003) 4751.

[4] Attempts to obtain coarse-grained equations within Scenario A have been made by Gorban et al., J. Non-Newtonian Fluid Mech. 96 (2001) 203; Phys. Rev. E 63 (2001) 066124; a more general approach based on "films" has been proposed by Gorban & Karlin, arXiv:cond-mat/0308331.

Whereas, at equilibrium, it is customary to consider only two levels of description, the atomistic and the thermodynamic ones, beyond equilibrium one frequently considers more than two levels of description. For example, between Hamilton's equations of motion and the Navier-Stokes equation of hydrodynamics there is the intermediate level of Boltzmann's kinetic equation. For complex fluids, we have already seen that one has the choice between more detailed descriptions in terms of configurational distribution functions or coarser descriptions using configuration tensors or moments. In the second part of this book, we are hence concerned with connecting any two levels of description, with the goal to obtain insight into the building blocks on the coarser level from the more detailed level.

6

Projection-Operator Method

This chapter deals with the foundations of statistical mechanics or, in more inspiring words, with the *theory of coarse-graining*. We here consider the coarse-graining for classical systems only (quantum systems are discussed in Appendix D). The goal is to obtain thermodynamic information about a system from more detailed levels of description, in particular, from the atomistic level. We produce atomistic expressions for the GENERIC building blocks, where the procedure is strongly inspired by equilibrium statistical mechanics and its role in "deriving" equilibrium thermodynamics. A systematic approach is required because naive ways of coarse-graining can fail easily. For example, and by no means surprisingly, naive blob models of polymer molecules with effective interactions between blobs lack thermodynamic consistency.[1]

In the projection-operator approach, one can account for the effects of the eliminated fast variables either by memory effects, while considering linear equations for the relevant slow variables, or by suitable nonlinearities, while using a Markovian approximation, or by a combination of both.[2] To arrive at GENERIC, one needs to keep sufficiently many variables and the appropriate nonlinearities for achieving a realistic description of a system by *Markovian time-evolution equations.* We here insist on avoiding explicit memory effects and take the standpoint that memory effects always indicate the existence of unrecognized variables relevant to the definition of a proper system for understanding certain phenomena of interest. In the terminology of Grabert,[3] a full renormalization accounting for non-Markovian effects is here

[1] Akkermans & Briels, J. Chem. Phys. 114 (2001) 1020.

[2] Grabert, *Projection Operators* (Springer, 1982).

[3] See p. 4 of Grabert, *Projection Operators* (Springer, 1982).

considered to be unnecessary due to a proper choice of slow variables, that is, for sufficiently large state spaces; the significance of a renormalization accounting for fluctuation effects is discussed at the end of Section 6.3.

In our development, we assume that all conserved quantities are included in the list of relevant variables or that they can be expressed in terms of those. It would be possible to treat the conserved or infinitely slow quantities, which are the variables of equilibrium thermodynamics, separately from the other slow observables. However, we find a unified treatment more convenient.

6.1 MOTIVATION OF BASIC FORMULAS

The atomistic expressions for the GENERIC building blocks E, S, L, and M are the cornerstones for this framework of beyond-equilibrium thermodynamics. We hence produce these expressions by the standard projection-operator procedure, to be explained in full detail. Before getting into the rather formal procedure, however, we want to guess the results to emphasize how natural and robust these cornerstones are. In addition, we want to motivate the concepts that must play a fundamental role: the *ensemble* and the *projection operator*. With these concepts, and the underlying notation for classical many-particle systems, we can then write the atomistic expressions for the GENERIC building blocks. Only after guessing all the results will we independently "derive" them by the projection-operator method under the often plausible but usually not rigorously provable assumption of a separation of time scales.

Many of the results for general statistical ensembles are illustrated for generalized microcanonical and canonical ensembles. As we jump frequently between the different ensembles, the reader should try to reflect continuously on whether the equations are for general, canonical, or microcanonical ensembles; the necessary information is always provided in the text.

6.1.1 Notation of Classical Mechanics

Hamilton's formulation of classical mechanics was previously considered in Section 1.2.2, where it was recognized as a special case of GENERIC in the absence of irreversible effects. We are generally concerned with isolated systems with a large number of degrees of freedom, such as N interacting point particles in a box. The motion of such objects is described by Newton's laws or, equivalently, by Hamilton's equations of motion (see Exercise 6). In three dimensions, such a system is specified by $6N$ independent position and momentum coordinates whose motion is uniquely determined from Hamilton's equations of motion. If we set up a $6N$-dimensional phase space, whose $6N$ coordinates, $z = (\boldsymbol{r}_1, \boldsymbol{p}_1, \ldots, \boldsymbol{r}_N, \boldsymbol{p}_N)$, consist of the $3N$ momentum and $3N$ position variables of the particles with masses m_j ($j = 1, \ldots, N$), then the atomistic state of the system is given by a single point in the phase space, which moves according to Hamiltonian dynamics as the state of the system changes in time.

The starting point for the subsequent analysis of classical atomistic systems is the purely reversible time-evolution equation

$$\frac{dz}{dt} = L_0 \cdot \frac{\partial E_0(z)}{\partial z}, \tag{6.1}$$

which represents Hamilton's equations of motion for the $6N$ position coordinates and momenta of all the N particles in a system; here $E_0(z)$ is the time independent atomistic Hamiltonian, or energy function, and the antisymmetric matrix L_0, the cosymplectic matrix, does not depend on z (see Exercise 5). For an arbitrary phase-space function A, we obtain the time-evolution equation

$$\frac{dA(z)}{dt} = \frac{\partial A(z)}{\partial z} \cdot L_0 \cdot \frac{\partial E_0(z)}{\partial z} = \{A, E_0\} = i\mathcal{L}A(z), \tag{6.2}$$

where we have introduced the canonical Poisson bracket $\{.,.\}$ of classical mechanics associated with L_0 [see (1.28)], and the first-order differential operator $\mathcal{L}$ is the Liouville operator. The imaginary unit i is included into the definition as a factor in order to make the Liouville operator self-adjoint,

$$\mathcal{L}^\dagger = \mathcal{L}. \tag{6.3}$$

The adjoint operator for the standard scalar product for phase-space functions, given by the integral of the product where the complex conjugate is taken for the first factor, is defined by

$$\int A^*(z)\mathcal{L}B(z)d^{6N}z = \int B(z)[\mathcal{L}^\dagger A(z)]^* d^{6N}z. \tag{6.4}$$

For real-valued phase-space functions A and B, the complex conjugation indicated by the asterisk in (6.4) corresponds to replacing i by $-i$ in the Liouville operator. By integrating (6.2) for an arbitrary phase-space function, which we also refer to as an observable, we find that the operator $\exp\{i\mathcal{L}t\}$ propagates any observable by a finite time t.

Exercise 123 Explicit Form of Liouville Operator
Show that the Liouville operator possesses the following explicit form,

$$\mathcal{L} = i \sum_{j=1}^{N} \left(\frac{\partial E_0}{\partial \boldsymbol{r}_j} \cdot \frac{\partial}{\partial \boldsymbol{p}_j} - \frac{\partial E_0}{\partial \boldsymbol{p}_j} \cdot \frac{\partial}{\partial \boldsymbol{r}_j} \right). \tag{6.5}$$

Exercise 124 Self-Adjointness of Liouville Operator
Show that the Liouville operator is self-adjoint.

Exercise 125 Liouville Equation
Assume that a probability density in phase space, $\rho(z, 0)$, is given. Find an expression for

the time dependent probability density, $\rho(z,t)$, which results from the time evolution of the atomistic configurations. Show that $\rho(z,t)$ solves the Liouville equation

$$\frac{\partial \rho(z,t)}{\partial t} = -i\mathcal{L}\rho(z,t) = \sum_{j=1}^{N}\left[\frac{\partial E_0(z)}{\partial \boldsymbol{r}_j}\cdot\frac{\partial \rho(z,t)}{\partial \boldsymbol{p}_j} - \frac{\partial E_0(z)}{\partial \boldsymbol{p}_j}\cdot\frac{\partial \rho(z,t)}{\partial \boldsymbol{r}_j}\right]$$
$$= -\sum_{j=1}^{N}\left[\frac{\boldsymbol{p}_j}{m_j}\cdot\frac{\partial \rho(z,t)}{\partial \boldsymbol{r}_j} + \boldsymbol{F}_j\cdot\frac{\partial \rho(z,t)}{\partial \boldsymbol{p}_j}\right], \qquad (6.6)$$

where $\boldsymbol{F}_j$ is the total force on particle j. In this equation, one recognizes the velocity $\boldsymbol{p}_j/m_j$ and the force $\boldsymbol{F}_j$ as the rates of change of the position and momentum of particle j.

6.1.2 Ensembles

In establishing a relationship between two different levels of description, the variables describing states on the two levels need to be related. For simplicity, we refer to *microstates*, described by all the particle positions and momenta z, and to *macrostates*, characterized by a list of phenomenological variables x. We need atomistic expressions for the macroscopic variables, which have the form of phase-space functions $\Pi(z)$. For example, for the mass and momentum densities at the position $\boldsymbol{r}$, we have the atomistic expressions

$$\sum_{j=1}^{N} m_j\delta(\boldsymbol{r}_j - \boldsymbol{r}) \qquad (6.7)$$

and

$$\sum_{j=1}^{N} \boldsymbol{p}_j\delta(\boldsymbol{r}_j - \boldsymbol{r}), \qquad (6.8)$$

which need to be averaged over particle configurations to obtain the macroscopic mass and momentum densities as smoother functions of the position $\boldsymbol{r}$.[4] We hence need the concept of an ensemble $\rho_x(z)$, that is, a probability density on the phase space that describes how much a microstate z contributes to the properties of a given

[4] The transition from the atomistic level to hydrodynamic fields illustrates a general paradox. We start with a finite number of degrees of freedom, namely, the particle positions and momenta, and we end up with the infinite number of degrees of freedom of the hydrodynamic fields. The coarse-grained level of description thus has more degrees of freedom than the atomistic one. Of course, there is a natural way out of this paradox, which would occur for any coarse-grained description using field variables. These fields should be assumed to be sufficiently smooth, and one should not try to resolve any information below a characteristic length scale $l_{\rm c}$, which clearly should be larger than the characteristic inter-particle distance. Quite surprisingly, as discussed by Alder & Alley, Physics Today 37/1 (1984) 56, the analysis of simulation data shows that the hydrodynamic description works successfully down to lengths $l_{\rm c}$ of only a few, say five, interparticle distances. We here assume that the range of particle interactions is smaller than $l_{\rm c}$. Once one realizes that the hydrodynamic fields can be represented by a discretization on the length scale $l_{\rm c}$ and no finer details are relevant, it becomes clear that the coarse-grained hydrodynamic level of description indeed uses fewer degrees of freedom than the atomistic description, as anticipated.

macrostate x. We postulate that an ensemble is properly normalized,

$$\int \rho_x(z) d^{6N} z = 1, \tag{6.9}$$

and that by averaging their atomistic expressions one indeed obtains the corresponding macroscopic variables,

$$\int \rho_x(z) \Pi_k(z) d^{6N} z = x_k. \tag{6.10}$$

In general, the label k for the macroscopic variables represents both discrete and continuous indices. For example, k represents the continuous position label $\boldsymbol{r}$ for the mass density in (6.7), plus an additional discrete label for the three components of the momentum density in (6.8).

The key step in any statistical analysis now is the proper definition of an ensemble. The most fundamental choice is the *generalized microcanonical ensemble* for which the microstates z are partitioned into equivalence classes according to the values of the observables $\Pi_k(z)$, and all microstates are postulated to have equal weights,

$$\rho_x(z) = \frac{1}{\Omega(x)} \delta(\Pi(z) - x) = \frac{1}{\Omega(x)} \prod_k \delta(\Pi_k(z) - x_k), \tag{6.11}$$

with the volume of phase space associated with a macrostate x given by

$$\Omega(x) = \int \delta(\Pi(z) - x) d^{6N} z. \tag{6.12}$$

This construction is analogous to the microcanonical ensemble of equilibrium statistical mechanics, where the only generalization for beyond-equilibrium systems consists in the consideration of a much larger list of macroscopic variables x, for example, the position-dependent hydrodynamic fields. The normalization (6.9) is guaranteed by the definition of the volume of phase space in (6.12), and we can verify (6.10) directly,

$$\begin{aligned} \int \rho_x(z) \Pi_k(z) d^{6N} z &= \frac{1}{\Omega(x)} \int \delta(\Pi(z) - x) \Pi_k(z) d^{6N} z \\ &= \frac{1}{\Omega(x)} \int \delta(\Pi(z) - x) x_k d^{6N} z = x_k. \end{aligned} \tag{6.13}$$

Although the generalized microcanonical ensemble is most suitable for theoretical considerations and can be taken as a reference ensemble, it turns out to be rather inconvenient for practical calculations.[5] Even calculations for the simplest beyond-equilibrium systems, such as hydrodynamics, require considerable efforts in the generalized microcanonical approach,[6] and practical simulations for treating complex

[5] Even a thermodynamic potential for an ideal gas at equilibrium is much more easily obtained in the canonical than in the microcanonical ensemble; in the canonical approach, one needs to evaluate single-particle integrals, whereas in the counting of microstates states for the microcanonical ensemble all particles are coupled through energy conservation, even though noninteracting, and Stirling's formula needs to be employed.

[6] de Pablo & Öttinger, J. Non-Newtonian Fluid Mech. 96 (2001) 137.

fluids require alternative ensembles.[7] We hence consider the generalized canonical ensemble ρ_x, which is obtained by maximizing the entropy

$$S[\rho] = -k_{\rm B} \int \rho(z) \ln \rho(z) d^{6N} z, \tag{6.14}$$

where $k_{\rm B}$ is Boltzmann's constant, under the constraints (6.9) and (6.10). By carrying out the maximization of (6.14) under constraints, we obtain the following explicit expression for the relevant ensemble of the generalized canonical type[8]

$$\rho_x(z) = \frac{1}{Z(x)} \exp\left[-\sum_k \lambda_k(x) \Pi_k(z)\right], \tag{6.15}$$

where the partition function $Z(x)$ is chosen to fulfill the normalization condition for the probability density ρ_x, that is, $Z(x) = \hat{Z}(\lambda(x))$ with

$$\hat{Z}(\lambda) = \int \exp\left[-\sum_k \lambda_k \Pi_k(z)\right] d^{6N} z, \tag{6.16}$$

and the Lagrange multipliers $\lambda_k(x)$ are chosen such that the averages of the phase space functions $\Pi_k(z)$ coincide with the given values (6.10) of the beyond-equilibrium variables x_k. The $\sum$ in (6.15) and (6.16) may represent not only summations but also integrations, depending on the nature of the label k. Equations (6.10) and (6.16) imply the useful relationship

$$-\left.\frac{\delta \ln \hat{Z}(\lambda)}{\delta \lambda_k}\right|_{\lambda=\lambda(x)} = \frac{1}{\hat{Z}(\lambda)} \int \exp\left[-\sum_j \lambda_j \Pi_j(z)\right] \Pi_k(z) d^{6N} z = x_k. \tag{6.17}$$

The term *generalized canonical ensemble* emphasizes the fact that the list of variables x_k is again much more general than in the canonical ensemble of equilibrium statistical mechanics; thus (6.15) corresponds to *quasi-equilibrium states* characterized by the slow variables x_k. In view of the above remarks, it should be clear why the generalized canonical ensemble is often associated with the alternative keywords "quasi-equilibrium ensemble" or "principle of maximum entropy."

Ensembles should *not* be considered as exact representations of the true state of the atomistic system. For example, in general it is practically impossible to prepare a system that corresponds to the generalized canonical ensemble (6.15). Whereas one can carry out such a preparation at equilibrium by coupling the system to suitable reservoirs, say a heat bath for adjusting the temperature, this cannot really be done for the generalized situation with a much larger, usually position-dependent set of variables and Lagrange multipliers; even if possible in principle, such a preparation

[7] Mavrantzas & Theodorou, Macromolecules 31 (1998) 6310.
[8] Jaynes, Phys. Rev. 106 (1957) 620; Jaynes, Phys. Rev. 108 (1957) 171.

should not be a prerequisite for our developments. The generalized canonical ensemble should rather be considered as the best possible choice given that average values are available only for some experimentally accessible slow observables $\Pi_k(z)$ that we consider to be the relevant ones for understanding a certain set of phenomena of interest.[9] Strictly speaking, the generalized canonical ensemble (6.15) is not even fully consistent with the existence of conserved quantities and, whenever these are important, other mixed ensembles must be considered.

The generalized microcanonical and canonical ensembles are extreme cases. Further mixed or intermediate ensembles can be defined by assigning particular, well-defined values to a number of variables and introducing Lagrange multipliers for the remaining ones by maximizing the entropy (6.14) for given averages of those remaining variables. If the variables x are obtained as averages of $\Pi(z)$ by suitable Lagrange multipliers, and the variables $\tilde{x}$ arise from a generalized microcanonical treatment of $\tilde{\Pi}(z)$, then we can write

$$\rho_{(x,\tilde{x})}(z) = \frac{1}{Z(x,\tilde{x})} \exp\left[-\sum_k \lambda_k(x,\tilde{x})\Pi_k(z)\right] \delta(\tilde{\Pi}(z) - \tilde{x}), \tag{6.18}$$

where the normalization factor $Z(x, \tilde{x})$ is given by

$$Z(x,\tilde{x}) = \int \exp\left[-\sum_k \lambda_k(x,\tilde{x})\Pi_k(z)\right] \delta(\tilde{\Pi}(z) - \tilde{x}) d^{6N}z \tag{6.19}$$

and the Lagrange multipliers $\lambda_k(x, \tilde{x})$ are determined by

$$\int \rho_{(x,\tilde{x})}(z)\Pi_k(z) d^{6N}z = x_k. \tag{6.20}$$

How can the various ensembles be equivalent? To be equivalent to a generalized microcanonical reference ensemble, the fluctuations of the macroscopic variables in any alternative ensemble must be sufficiently small. From equilibrium thermodynamics one knows that, for large systems, sharply peaked distributions typically arise from the product of a decreasing exponential factor with a rapidly increasing number of microstates. For a more detailed discussion, we consider the probability density for finding a macrostate x' in the neighborhood of the average x in any ensemble,

$$f_x(x') = \int \delta(\Pi(z) - x')\rho_x(z) d^{6N}z. \tag{6.21}$$

For the generalized microcanonical ensemble, only the precise value x can occur, and we thus have $f_x(x') = \delta(x - x')$. For the generalized canonical ensemble, we find

$$f_x(x') = \frac{\Omega(x')}{Z(x)} \exp\left[-\sum_k \lambda_k(x) x'_k\right], \tag{6.22}$$

[9] Español, Phys. Lett. A 146 (1990) 21.

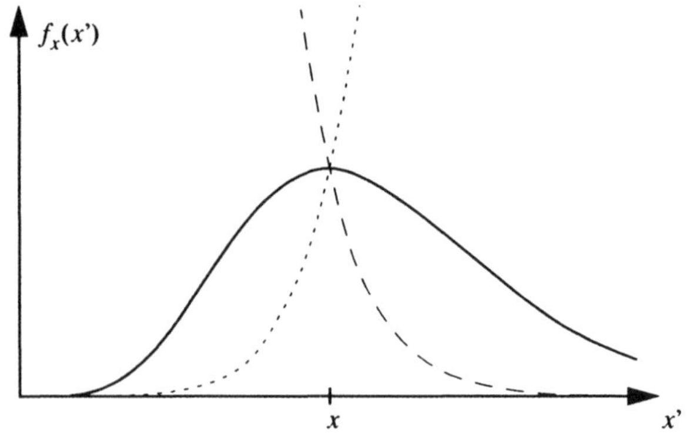

Fig. 6.1 Peaked distribution of macroscopic variables in a canonical ensemble.

where the microcanonical volume of phase space Ω associated with a macrostate and the canonical partition function Z are defined in (6.12) and (6.16), respectively. A typical example of $f_x(x')$ is illustrated in Figure 6.1. The suitably normalized factors $\Omega(x')$ and $\exp[-\sum_k \lambda_k(x)x'_k]$ in (6.22) are indicated as functions of x' by the dotted and dashed lines, respectively, whereas $f_x(x')$ is given by the thick continuous curve. The slope of the factors at $x' = x$ is typically proportional to the system size, and the relative width of the peak decreases with the square root of the system size. The example of the figure actually corresponds exactly to the distribution of the relative energy for an ideal gas of only four particles.

As a rule, for beyond-equilibrium thermodynamics, the local equilibrium or quasi-equilibrium systems are typically much smaller than the systems considered in equilibrium thermodynamics so that one should always check that the fluctuations in the chosen ensemble are indeed small because that cannot be taken for granted (see also Exercise 127). In particular, in analytical calculations, one needs to verify the conditions

$$\left|\frac{\delta^2 \ln \hat{Z}(\lambda)}{\delta\lambda_j \delta\lambda_k}\right| \ll \left|\frac{\delta \ln \hat{Z}(\lambda)}{\delta\lambda_j}\right| \left|\frac{\delta \ln \hat{Z}(\lambda)}{\delta\lambda_k}\right|, \tag{6.23}$$

where the left-hand side corresponds to the magnitude of the correlation between the fluctuations of Π_j and Π_k, and the right-hand side is given by the product of the averages of Π_j and Π_k. When large fluctuations become physically relevant, one needs to be careful with the choice of the ensemble to match the proper physical conditions.

Exercise 126 Derivatives of Generalized Canonical Ensemble

Derive the following expression for the derivatives of the generalized canonical ensemble with respect to the macroscopic variables x:

$$\frac{\delta\rho_x(z)}{\delta x_k} = -\sum_l \frac{\delta\lambda_l}{\delta x_k}[\Pi_l(z) - x_l]\rho_x(z). \tag{6.24}$$

Show that $\delta\lambda_l/\delta x_k$ is symmetric in the indices k and l.

Exercise 127 Peaked Distribution for Noninteracting Particles
Consider N noninteracting particles of mass m in a box of volume V and fix the average of the total momentum, $\boldsymbol{P}$, and the total energy, E, through corresponding Lagrange multipliers. Under what circumstances are the conditions (6.23) fulfilled? What happens if the internal energy is used as a macroscopic variable instead of the total energy?

6.1.3 Projection Operators

Once we have an ensemble that describes how individual microstates contribute to certain macrostates, we need to define a projection operator that allows us to separate the macroscopically relevant contribution of an arbitrary observable from the irrelevant one to be eliminated. It is generally assumed that the projection operator separates slow from fast degrees of freedom, and the projector is hence a key to the atomistic calculation of dynamic material properties.

Following Grabert,[10] we can define the fundamental projection operator in terms of the chosen ensemble through the following action on arbitrary observables A in a straightforward manner:

$$\mathcal{P}(x)A = \int \rho_x(z')A(z')d^{6N}z' + \sum_k (\Pi_k - x_k) \int A(z')\frac{\delta\rho_x(z')}{\delta x_k}d^{6N}z', \quad (6.25)$$

where the z-dependence of the projected observable or phase-space function $\mathcal{P}(x)A$ arises only through the z-dependence of Π_k. Note that $\mathcal{P}(x)A$ is a linear combination of the observables Π_k. The idea behind this projection operator is to get a rough approximation for the macroscopically relevant contribution to an observable by simply averaging it with the given ensemble and to represent deviations from the average linearly in terms of the deviations of the slow variables from their averages. We further introduce the complementary projector

$$\mathcal{Q}(x) = 1 - \mathcal{P}(x), \quad (6.26)$$

and we use the notation $\mathcal{P}(t) = \mathcal{P}(x(t))$, $\mathcal{Q}(t) = \mathcal{Q}(x(t))$ for the time dependent projectors. The rest of this subsection is devoted to developing a number of properties of these projection operators.

To motivate the introduction of the adjoint of the projector (6.25), we note that one often needs to calculate averages of the form

$$\int \rho_x(z)\mathcal{P}(x)A(z)d^{6N}z, \quad (6.27)$$

which can be considered as the standard scalar product in which the projection operator acts on one of the factors. If $\mathcal{P}(x)$ is considered as an operator acting on observables,

[10] Grabert, *Projection Operators* (Springer, 1982).

the adjoint operator $\mathcal{P}^\dagger(x)$ should therefore be considered as an operator acting on ensembles or probability densities. The explicit expression for the adjoint projection operator

$$\mathcal{P}^\dagger(x)B = \rho_x \int B(z')d^{6N}z' + \sum_k \frac{\delta \rho_x}{\delta x_k} \int (\Pi_k(z') - x_k)B(z')d^{6N}z', \qquad (6.28)$$

where the z-dependence of the phase-space function $\mathcal{P}^\dagger(x)B$ arises only through the z-dependence of ρ_x and its derivative, has thus the appearance of an approximate expression for an ensemble. The operator $\mathcal{P}(x)$ should be used to project on the slow part of observables, whereas $\mathcal{P}^\dagger(x)$ projects on the slow part of ensembles. While $\mathcal{P}(x)$ acts on the space of observables and $\mathcal{P}^\dagger(x)$ acts on the space of probability densities, both observables and densities are of course just phase-space functions.

Exercise 128 Fast Part of Particle Velocity

Assume that the average momentum of particle j in a system of N equal particles of mass m can be expressed in terms of the macroscopic momentum density field $\boldsymbol{M}(\boldsymbol{r})$,

$$\int \rho_x(z)\boldsymbol{p}_j d^{6N}z = \frac{1}{N}\int \boldsymbol{M}(\boldsymbol{r})d^3r. \qquad (6.29)$$

From this natural assumption, derive the formula

$$\mathcal{Q}(x)i\mathcal{L}\boldsymbol{r}_j = \frac{\boldsymbol{p}_j}{m} - \frac{1}{N}\sum_{l=1}^{N}\frac{\boldsymbol{p}_l}{m}. \qquad (6.30)$$

Exercise 129 Projection Operator Property

Show that $\mathcal{P}(x)$ is a projection operator, that is,

$$\mathcal{P}(x)[\mathcal{P}(x)A] = \mathcal{P}(x)A, \qquad (6.31)$$

for all observables A. This result can also be expressed in the more compact form $\mathcal{P}(x)\mathcal{P}(x) = \mathcal{P}(x)$.

Exercise 130 Adjoint Projection Operator

Derive the expression (6.28) for the adjoint projection operator.

Exercise 131 More on Self-Adjointness of Projectors

Although the projection operator $\mathcal{P}$ is not self-adjoint for the standard scalar product, prove the following identity for the generalized canonical ensemble and arbitrary observables A and B:

$$\int \rho_x(z)A(z)\mathcal{P}(x)B(z)d^{6N}z = \int \rho_x(z)B(z)\mathcal{P}(x)A(z)d^{6N}z. \qquad (6.32)$$

Some authors use the simplified projectors

$$\mathcal{P}(x)A = \sum_k \Pi_k \int A(z') \frac{\delta \rho_x(z')}{\delta x_k} d^{6N} z', \tag{6.33}$$

$$\mathcal{P}^\dagger(x)B = \sum_k \frac{\delta \rho_x}{\delta x_k} \int \Pi_k(z') B(z') d^{6N} z', \tag{6.34}$$

instead of (6.25) and (6.28) where, in particular, (6.34) was used in a pioneering paper on projection operators by Robertson.[11] The operators (6.25) and (6.33) differ only by terms that are independent of z. Therefore, when the Liouville operator (6.5) is applied, these constant terms are eliminated by differentiation. In other words, the difference between the projectors $\mathcal{P}$ of (6.25) and (6.33) does not matter in the combination $\mathcal{LP}$ (or $\mathcal{LQ}$);[12] indeed, this combination is discovered to be the relevant one below. The subsequent exercises show that both projectors possess essentially the same properties (except for the invariance of ρ_x under projection).

Exercise 132 Consistency of Slow Variables
Show that

$$\mathcal{P}(x)\Pi_j = \Pi_j, \tag{6.35}$$

both for the full projector (6.25) and for the simplified one (6.33).

Exercise 133 Consistency of Ensembles
Evaluate $\mathcal{P}^\dagger(x)\rho$ for any ensemble with $\int \rho(z) d^{6N} z = 1$, $\int \rho(z)\Pi_k(z) d^{6N} z = x_k$, both for the full projector (6.28) and for the simplified one (6.34).
Under the same conditions, prove the identity

$$\mathcal{P}^\dagger(x)\frac{d\rho}{dt} = \frac{d\rho_x}{dt}, \tag{6.36}$$

for both projectors, which plays a key role in the above-mentioned approach by Robertson.

Exercise 134 Generalized Projection Operator Property
With the consistency results obtained in Exercise 132, show that the property (6.31) can be generalized to

$$\mathcal{P}(x')\mathcal{P}(x) = \mathcal{P}(x) \tag{6.37}$$

for all x and x'. What is the deeper reason for the validity of this generalized projector identity?

Exercise 135 Time-Derivative of Projection Operator
Derive the identities

$$\dot{\mathcal{P}}(t) = \mathcal{P}(t)\dot{\mathcal{P}}(t) \tag{6.38}$$

and

$$\dot{\mathcal{P}}(t) = \dot{\mathcal{P}}(t)\mathcal{Q}(t) \tag{6.39}$$

[11] Robertson, Phys. Rev. 144 (1966) 151.
[12] See p. 24 of Grabert, *Projection Operators* (Springer, 1982).

for the time derivative of the projection operator $\mathcal{P}(t) = \mathcal{P}(x(t))$. [Hint: Exploit the result $\mathcal{P}(t')\mathcal{P}(t) = \mathcal{P}(t)$ of Exercise 134.]

The projection-operator formalism can be based on the following rigorous identity for the time-evolution operator associated with (6.2):

$$e^{i\mathcal{L}t} = e^{i\mathcal{L}t}\mathcal{P}(t) + \mathcal{Q}(0)G(0,t) + \int_0^t e^{i\mathcal{L}u}\mathcal{P}(u)[i\mathcal{L} - \dot{\mathcal{P}}(u)]\mathcal{Q}(u)G(u,t)du, \quad (6.40)$$

where the dot indicates differentiation with respect to time and, in the time ordered exponential,

$$G(u,t) = \text{T-exp}\left\{\int_u^t \mathcal{Q}(s)i\mathcal{L}\mathcal{Q}(s)ds\right\}, \quad (6.41)$$

the operators are ordered from left to right as time increases. The physical relevance of the formal identity (6.40) can be understood as follows. It expresses the well-known solution of the inhomogeneous linear differential equations for the fast observables, as projected out by $\mathcal{Q}(t)$, in the presence of inhomogeneities given by the known slow contributions, as projected out by $\mathcal{P}(t)$.

Because the identity (6.40) is a cornerstone of the projection-operator technique, we here present the details of the derivation just outlined. Readers not interested in such formal details can proceed to the definition (6.46).

We write

$$e^{i\mathcal{L}t} = e^{i\mathcal{L}t}\mathcal{P}(t) + \mathcal{Z}(t), \quad (6.42)$$

with the auxiliary operators

$$\mathcal{Z}(t) = e^{i\mathcal{L}t}\mathcal{Q}(t), \quad (6.43)$$

for which we obtain the differential equation

$$\begin{aligned} \dot{\mathcal{Z}}(t) &= e^{i\mathcal{L}t}i\mathcal{L}\mathcal{Q}(t) - e^{i\mathcal{L}t}\dot{\mathcal{P}}(t) \\ &= e^{i\mathcal{L}t}\mathcal{Q}(t)i\mathcal{L}\mathcal{Q}(t) + e^{i\mathcal{L}t}\mathcal{P}(t)i\mathcal{L}\mathcal{Q}(t) - e^{i\mathcal{L}t}\mathcal{P}(t)\dot{\mathcal{P}}(t)\mathcal{Q}(t) \\ &= \mathcal{Z}(t)\mathcal{Q}(t)i\mathcal{L}\mathcal{Q}(t) + e^{i\mathcal{L}t}\mathcal{P}(t)[i\mathcal{L} - \dot{\mathcal{P}}(t)]\mathcal{Q}(t), \end{aligned} \quad (6.44)$$

where $\dot{\mathcal{Q}}(t) = -\dot{\mathcal{P}}(t)$, $\mathcal{Z}(t) = \mathcal{Z}(t)\mathcal{Q}(t)$, and the results of Exercise 135 have been used. We consider (6.44) as an inhomogeneous linear differential equation for the fast quantity $\mathcal{Z}(t)$ in the presence of a given slow inhomogeneity. Solutions of the homogeneous equation are given by the operators $G(u,t)$ defined in (6.41) for any initial time u. We hence obtain the standard solution of the inhomogeneous equation (6.44),

$$\mathcal{Z}(t) = \mathcal{Z}(0)G(0,t) + \int_0^t e^{i\mathcal{L}u}\mathcal{P}(u)[i\mathcal{L} - \dot{\mathcal{P}}(u)]\mathcal{Q}(u)G(u,t)du, \quad (6.45)$$

which can be verified by taking the derivative with respect to t and by checking the initial condition for $t = 0$. By inserting (6.45) into (6.42) and using $\mathcal{Z}(0) = \mathcal{Q}(0)$, we obtain the fundamental operator identity (6.40).

According to the above arguments, operator identities such as (6.40) are entirely plausible; however, a rigorous mathematical justification is highly nontrivial and, even in the simpler situation of time independent projectors, unresolved mathematical difficulties still exist.[13] The above arguments are formal and do not address the existence of the exponentials of operators or the interpretation of the occurring operator-valued integrals.

It is inconvenient that the projection operator $\mathcal{P}(x)$ associated with the generalized microcanonical ensemble does not possess the self-adjointness property (6.32) with respect to the scalar product implied by the ensemble. There exists an alternative projection operator that fixes this problem. Instead of using a projection operator leaving the *linear* functions of Π_j invariant, we can choose a projection operator leaving *all* functions of Π_j invariant,

$$[\tilde{\mathcal{P}}A](z) = \langle A \rangle_{\Pi(z)}, \tag{6.46}$$

for an arbitrary phase space function A, where $\langle A \rangle_x$ is the generalized microcanonical average

$$\langle A \rangle_x = \int A(z) \frac{\delta(\Pi(z) - x)}{\Omega(x)} d^{6N} z. \tag{6.47}$$

This projection operator turns an arbitrary function of z into another phase space function that depends on z only through $\Pi(z)$. The operator $\tilde{\mathcal{P}}$ is independent of x. Although the projection operator $\mathcal{P}(x)$ is a perfectly natural choice for the generalized canonical ensemble, it will become clear later that it is more natural to use $\tilde{\mathcal{P}}$ rather than $\mathcal{P}(x)$ for the generalized microcanonical ensemble if one wishes to include a theory of fluctuations into beyond-equilibrium thermodynamics (see Exercise 148). The subsequent exercise shows that, among several other important and useful properties, $\tilde{\mathcal{P}}$ possesses the property (6.32).

Exercise 136 Properties of Alternative Projection Operator

Derive the following properties of the alternative projection operator (6.46) for the generalized microcanonical ensemble:

1. $\tilde{\mathcal{P}}\tilde{\mathcal{P}} = \tilde{\mathcal{P}}$;
2. $\tilde{\mathcal{P}}$ is self-adjoint;
3. $\int \rho_x(z) A(z) \tilde{\mathcal{P}} B(z) d^{6N} z = \int \rho_x(z) B(z) \tilde{\mathcal{P}} A(z) d^{6N} z$, where A and B are arbitrary observables;
4. $\tilde{\mathcal{P}} \rho_x = \rho_x$;
5. $\tilde{\mathcal{P}} \frac{d\rho_x}{dt} = \frac{d\rho_x}{dt}$.

Exercise 137 Fast Part of Particle Velocity, Continued

Under the conditions of Exercise 128, evaluate $\tilde{\mathcal{P}} i \mathcal{L} \boldsymbol{r}_j$.

[13] Lamb et al., Physica A 298 (2001) 121.

6.1.4 Atomistic Expressions for *E, S, L, M*

The purpose of this subsection is to obtain atomistic expressions for the GENERIC building blocks by plausible, heuristic arguments. We start with the energy $E(x)$. It is most natural to assume that the energy of the macrostate is given by the average of the Hamiltonian over the corresponding microstates,

$$E(x) = \int \rho_x(z) E_0(z) d^{6N} z. \tag{6.48}$$

If we introduce a more compact notation for the integral involving the probability density ρ_x as an average, $\langle \ldots \rangle_x$, we have

$$E(x) = \langle E_0 \rangle_x . \tag{6.49}$$

Equation (6.49) is a completely general relationship valid for all ensembles, and it probably is beyond any doubts. In a generalized microcanonical ensemble, where microstates are classified according to the precise values of the macroscopic variables, one can assume that, if two microstates have different energies, they cannot be in the same equivalence class of microstates because they could easily be distinguished macroscopically. The value of the atomistic energy must hence be a property of equivalence classes or, in terms of an equation,

$$E_0(z) = E(\Pi(z)), \tag{6.50}$$

which is consistent with (6.49). This is a formulation of the idea that the value of the energy should be accessible on the macroscopic level. An alternative formulation of this idea of the accessibility of energy is given by

$$\mathcal{P}(x) E_0 = E_0, \tag{6.51}$$

stating that the energy has only a macroscopic contribution and no rapidly fluctuating part. Strictly speaking, (6.51) requires that $E_0(z)$ is a linear combination of the variables $\Pi(z)$, which, in turn, implies (6.50). The simplest option for realizing (6.51) is by choosing the energy as one of the variables $\Pi_j(z)$ [see (6.35)]. In any case, one can safely assume that (6.51) is a valid approximation for any admissible ensemble.

The more general nonlinear dependence (6.50) of the energy on $\Pi(z)$ arising in the generalized microcanonical ensemble can be expressed as

$$\tilde{\mathcal{P}} E_0 = E_0. \tag{6.52}$$

This is our first evidence that it is natural to use the projection operator $\tilde{\mathcal{P}}$ of (6.46) for the generalized microcanonical ensemble.

Turning to the entropy of a macrostate, for the generalized microcanonical ensemble, we expect it to be given by the logarithm of the number of microstates z consistent with a macrostate x,

$$S(x) = k_{\rm B} \ln \left[\int \delta(\Pi(z) - x) d^{6N} z \right], \tag{6.53}$$

or, in a more compact formulation based on (6.12),

$$S(x) = k_{\mathrm{B}} \ln \Omega(x). \tag{6.54}$$

These expressions constitute the natural generalization of the Gibbs entropy formula of equilibrium thermodynamics. As usual, the logarithms in (6.53) and (6.54) mediate the transition from the product of the number of states of subsystems to the additivity of their entropies.

For the generalized canonical ensemble, we can evaluate the functional $S[\rho]$ of (6.14) for the probability density ρ_x of (6.15), thus obtaining a relationship between the entropy and the partition function

$$S(x) = k_{\mathrm{B}} \left[\ln Z(x) + \sum_k \lambda_k(x) x_k \right]. \tag{6.55}$$

Equations (6.17) and (6.55), after a cancellation of two of the three terms resulting from the differentiation, lead to

$$\frac{\delta S(x)}{\delta x_k} = k_{\mathrm{B}} \lambda_k(x), \tag{6.56}$$

which offers a nice interpretation of the Lagrange multipliers λ_k as the thermodynamic conjugates of the variables x_k, or as the entropy gradients occurring in the definition (3.3) of thermodynamic forces.

Notice that, with $S(x) = \hat{S}(\lambda(x))$ and the formula (6.17), we obtain

$$\hat{S}(\lambda) = k_{\mathrm{B}} \left[\ln \hat{Z}(\lambda) - \sum_k \lambda_k \frac{\delta \ln \hat{Z}(\lambda)}{\delta \lambda_k} \right]. \tag{6.57}$$

Equations (6.17) and (6.55)–(6.57) express the fact that the entropy $S(x)$ and the logarithm of the partition function of a generalized canonical ensemble, $k_{\mathrm{B}} \ln \hat{Z}(\lambda)$, are related by Legendre transformation. As in equilibrium statistical mechanics, for the generalized microcanonical ensemble, the logarithm of the volume of phase space gives the entropy, whereas, for the generalized canonical ensemble, the logarithm of the partition function is related to entropy by a Legendre transform. For ensembles (6.18) intermediate between microcanonical and canonical, for which some of the variables are fixed and the remaining ones are controlled by Lagrange multipliers, the entropy is

$$S(x, \tilde{x}) = k_{\mathrm{B}} \left[\ln Z(x, \tilde{x}) + \sum_k \lambda_k(x, \tilde{x}) x_k \right], \tag{6.58}$$

and there is a Legendre transformation associated with each of the occurring Lagrange multipliers (cf. Appendix A.4 for the analogous situation in equilibrium thermodynamics).

Equation (6.56) provides an interesting expression for the derivatives of the generalized canonical ensemble with respect to the macroscopic variables x occurring in the definition of the projection operators (6.25) and (6.28). The derivatives $\delta\lambda_l/\delta x_k$ in (6.24) can be expressed as second-order derivatives of the entropy, thus explaining the previously observed symmetry in the indices k and l, and all terms in the definition of the projection operator can be recognized as suitable averages.

Upon critical consideration, there is a deep problem with the application of the above entropy expressions, say (6.53) for the generalized microcanonical ensemble. If one uses the state variables x' instead of x, where we assume that the transformation is one-to-one so that all states are labeled equivalently by x or x', then one obtains

$$S'(x') = S(x(x')) + k_{\rm B} \ln \left| \det \frac{\delta x(x')}{\delta x'} \right|, \tag{6.59}$$

instead of the previously formulated transformation law (1.49). The surprising term involving the logarithm of the Jacobian indicates a problem with the proper counting of states. Whereas an overall rescaling of volume elements in the coarse-grained state space would only lead to an irrelevant constant contribution to the entropy, local nonuniformities in the parametrization of the coarse-grained state space could lead to severe ambiguities. If a proper reference parametrization, $x_{\rm ref}$ of the coarse-grained state space is known, one should actually change the entropy definition (6.53) into

$$S(x) = k_{\rm B} \left\{ \ln \left[\int \delta(\Pi(z) - x) d^{6N} z \right] - \ln \left| \det \frac{\delta x_{\rm ref}(x)}{\delta x} \right| \right\}, \tag{6.60}$$

so that one recovers the natural transformation law (1.49).

We will return to this problem of *correct counting* several times in this book. Heuristically, the problem can be seen as follows. For a proper counting of atomistic states one needs to assign a certain volume in phase space to the submanifold associated with a given thermodynamic state or, in other words, one needs to assign a certain thickness to the submanifold to obtain a nonzero measure for it. The idea of a uniform thickness, however, is not invariant under nonlinear transformations. The thickness should rather be associated with the amplitude of fluctuations. Actually, the same problem occurs in the special case of equilibrium thermodynamics. In the microcanonical approach to equilibrium thermodynamics, however, one typically assigns a thickness by specifying a fixed "energy window" while keeping the other extensive variables at fixed values, which corresponds to the above step of selecting the reference variables $x_{\rm ref}$.

In the generalized canonical approach, the problem is shifted to the observation that the mathematical form of the ensemble (6.15) is not invariant under nonlinear transformations so that there is a presumably small but nontrivial fluctuation effect. Indeed, the problem of the proper definition of the entropy and the resulting transformation behavior can be resolved by looking at GENERIC with fluctuations. Under nonlinear transformations, the fluctuations introduce systematic, that is, deterministic, effects that are exactly cancelled by the additional entropy contribution involving the Jacobian (see Exercise 13). As long as fluctuations are irrelevant, we can expect the additional entropy contribution to be equally irrelevant.

It should be noted that it is important to know the functional form of the entropy in terms of the chosen state variables x and that the entropy defined in (6.53) or

(6.60) depends on the level of coarse-graining. The more atomistic configurations are mapped into a thermodynamic state, the larger is the entropy, so that *an increasing amount of entropy occurs in any step of coarse-graining*. The amount of entropy depends on the degree of coarse-graining in the system chosen to describe certain phenomena of interest. In particular, there is no entropy on the atomistic level of description. This situation is completely different for the energy, where (6.50) shows that the concept of energy is independent of the level of description.

Exercise 138 Functional Form of Peaked Canonical Distribution
Show that (6.22) can be rewritten as

$$f_x(x') = \exp\left[\frac{S^{\mathrm{mc}}(x') - S^{\mathrm{c}}(x)}{k_{\mathrm{B}}} - \sum_k \lambda_k(x)(x'_k - x_k)\right]. \tag{6.61}$$

What can be concluded for the relationship between the microcanonical and canonical entropies S^{mc} and S^{c}? Derive the approximate formula

$$f_x(x') \propto \exp\left[\frac{1}{2k_{\mathrm{B}}}\sum_{jk}\frac{\delta^2 S(x)}{\delta x_j \delta x_k}(x'_j - x_j)(x'_k - x_k)\right]. \tag{6.62}$$

To obtain a meaningful generalized canonical ensemble, the variables must be chosen such that the entropy is a concave function, and the peaked distribution $f_x(x')$ can then be approximated by a Gaussian.

Exercise 139 Projection Operator for Generalized Canonical Ensemble
Show that the projection operator (6.25) for the generalized canonical ensemble can be rewritten in terms of averages,

$$\mathcal{P}(x)A = \langle A\rangle_x - \frac{1}{k_{\mathrm{B}}}\sum_{jk}\frac{\delta^2 S(x)}{\delta x_j \delta x_k}(\Pi_j - x_j)\,\langle(\Pi_k - x_k)A\rangle_x\,, \tag{6.63}$$

where the z-dependence of the projected observable or phase-space function $\mathcal{P}(x)A$ arises only through the z-dependence of Π_j. Verify that the matrix

$$-\frac{1}{k_{\mathrm{B}}}\frac{\delta^2 S(x)}{\delta x_j \delta x_k}$$

is the inverse of the matrix $\langle(\Pi_j - x_j)(\Pi_k - x_k)\rangle_x$.

Exercise 140 Derivatives of Generalized Microcanonical Averages
Show that, for an arbitrary observable A, one has the following relationship for averages within the generalized microcanonical ensemble:

$$\sum_k \frac{\delta}{\delta x_k}\left\langle\frac{\partial \Pi_k}{\partial z_j}A\right\rangle_x = \left\langle\frac{\partial A}{\partial z_j}\right\rangle_x - \frac{1}{k_{\mathrm{B}}}\sum_k\left\langle\frac{\partial \Pi_k}{\partial z_j}A\right\rangle_x\frac{\delta S}{\delta x_k}. \tag{6.64}$$

Exercise 141 Evolution of Ensembles
Derive the formulas

$$k_{\mathrm{B}} i\mathcal{L}\rho_x(z) = -\sum_k \frac{\delta S(x)}{\delta x_k}\rho_x(z) i\mathcal{L}\Pi_k(z) \tag{6.65}$$

for the generalized canonical ensemble and

$$k_{\mathrm{B}} i\mathcal{L}\rho_x(z) = -\sum_k \left[\frac{\delta S(x)}{\delta x_k}\rho_x(z) + k_{\mathrm{B}}\frac{\delta\rho_x(z)}{\delta x_k}\right] i\mathcal{L}\Pi_k(z) \tag{6.66}$$

for the generalized microcanonical ensemble.

Whereas the concepts of energy and entropy are familiar from equilibrium statistical mechanics, the Poisson matrix must be obtained from a new type of argument. As a first guess, one might use the transformation formula (1.50) for transforming from the Poisson matrix of classical mechanics to the level of slow variables. However, the expression

$$\frac{\partial\Pi(z)}{\partial z} \cdot L_0 \cdot \frac{\partial\Pi(z)}{\partial z}$$

is still a function of microstates z rather than the coarse-grained variables x. The coarse-graining transformation is certainly not one-to-one; thus, it cannot be inverted so that there are many possible microstates z in which the above expression for given x could be evaluated. The most natural way to proceed is by averaging these contributions and to write

$$L(x) = \left\langle \frac{\partial\Pi(z)}{\partial z} \cdot L_0 \cdot \frac{\partial\Pi(z)}{\partial z} \right\rangle_x . \tag{6.67}$$

With the general definition (1.6) of Poisson brackets, the atomistic expression (6.67) for the Poisson matrix on the coarse-grained level can be written in the following elegant and compact way in terms of the Poisson bracket of classical mechanics:

$$L_{jk}(x) = \langle\{\Pi_j, \Pi_k\}\rangle_x . \tag{6.68}$$

The expression (6.68) is so general and so elegant that it must be correct.

Equations (6.49) for the energy, (6.54) or (6.55) for the entropy, and (6.68) for the Poisson matrix are all in terms of the atomistic expressions for the slow variables, $\Pi(z)$, and of the ensemble, $\rho_x(z)$; the projection operator $\mathcal{P}(x)$ separating the slow and fast dynamic behavior is not needed, and no time dependences occur. We can hence refer to E, S, and L as static building blocks of GENERIC that do not contain any dynamic material information. These building blocks are evaluated with the distribution function determined by the macrostate at a particular time. The situation must be different for the friction matrix, which, as we have seen repeatedly in previous chapters, does contain dynamic material properties such as transport coefficients or relaxation times. In the simplest case, two-time correlations become important. One might hence expect an expression of the form

$$M_{jk}(x) = \frac{1}{k_{\rm B}} \int_0^\infty \left\langle \hat{\Pi}_k(z) e^{i\hat{\mathcal{L}}t} \hat{\Pi}_j(z) \right\rangle_x dt, \tag{6.69}$$

where the integrand is the two-time correlation function of some observable $\hat{\Pi}$ with a time difference t and some dynamics generated by a Liouville operator $\hat{\mathcal{L}}$. The time integral eliminates any explicit time dependence from the friction matrix, and the Boltzmann factor $k_{\rm B}$ with dimensions of entropy has been introduced to simplify the dimensional analysis. The fundamental time-evolution equation (1.1) then implies that $\hat{\Pi}_j$ must have the dimensions of the time derivative of Π_j. Because the friction matrix is related to fluctuations, only the rapid contribution to the time derivative should matter; hence, we expect

$$\hat{\Pi}_j(z) = \mathcal{Q}(x) i\mathcal{L}\Pi_j(z). \tag{6.70}$$

By the same argument, we consider the rapid contribution to the time evolution generated by the Liouville operator

$$\hat{\mathcal{L}} = \mathcal{Q}(x)\mathcal{L}\mathcal{Q}(x). \tag{6.71}$$

To avoid convergence problems for the integral, we choose an intermediate time scale τ rather than ∞ for the upper limit of integration in (6.69). The existence of such a time scale is crucial for the approach. The relevant slow variables singled out by the projection operator $\mathcal{P}(x)$ evolve on a time scale larger than τ and change little over the time τ, whereas the irrelevant fast variables eliminated through the projection-operator approach change on a time scale shorter than τ so that their correlations decay over the time τ, and the integral need not be extended any further. In summary, we arrive at the expression

$$M_{jk}(x) = \frac{1}{k_{\rm B}} \int_0^\tau \left\langle [\mathcal{Q}(x) i\mathcal{L}\Pi_k(z)] e^{\mathcal{Q}(x) i\mathcal{L}\mathcal{Q}(x)t} [\mathcal{Q}(x) i\mathcal{L}\Pi_j(z)] \right\rangle_x dt. \tag{6.72}$$

Because only the fast variables are evolving on the time scale τ and the slow variables can be considered as fixed parameters, the rapid time-evolution operator $\mathcal{Q}(x)\mathcal{L}\mathcal{Q}(x)$ in this equation is often replaced by the full time-evolution operator $\mathcal{L}$:

$$M_{jk}(x) = \frac{1}{k_{\rm B}} \int_0^\tau \langle [\mathcal{Q}(x) i\mathcal{L}\Pi_k(z)] e^{i\mathcal{L}t} [\mathcal{Q}(x) i\mathcal{L}\Pi_j(z)] \rangle_x dt. \tag{6.73}$$

For the interpretation of (6.73), the integrand may be considered as the time correlation between fluctuations of the thin curve, representing $\Pi(z)$, around the thick curve, representing x in Figure 6.2. More precisely, $i\mathcal{L}\Pi_j(z)$ is the time derivative of the thin curve, and the projection operator $\mathcal{Q}(x)$ indicates that only the rapid fluctuations on top of the slower changes of the thick curve should be considered. The amplitude and the relaxation time of the correlation function for the fluctuations determine the friction matrix.

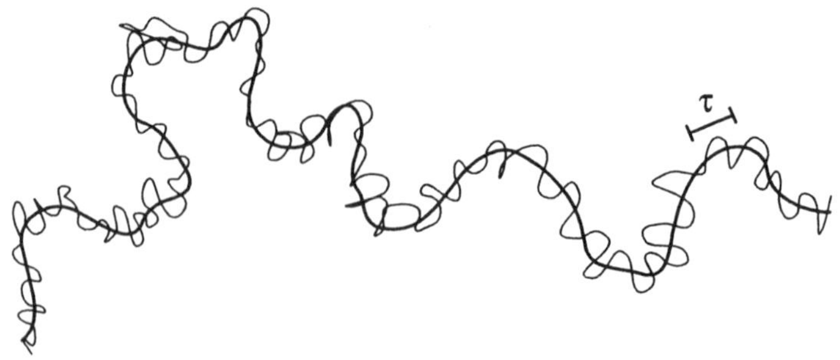

Fig. 6.2 Trajectories on two levels of description.

The arguments leading to the expression (6.72) for the friction matrix are clearly more ambiguous than for the other building blocks. Readers somewhat familiar with beyond-equilibrium thermodynamics should recognize the Green-Kubo-type nature[14] of the expression (6.72) or (6.73), and thus may gain some confidence in our heuristic arguments leading to that equation. In any case, (6.72) arises naturally in the projection-operator approach described below.

6.1.5 GENERIC Properties

In summary, we now have the atomistic expressions (6.49) for the energy, (6.54) or (6.55) for the entropy, (6.68) for the Poisson matrix, and (6.72) for the friction matrix. As a next step, we attempt to derive the various properties of the GENERIC building blocks from these expressions. Some properties can be obtained rigorously for any ensemble. Other properties can be derived only for particular ensembles, whereas small extra terms occur in other ensembles. Such extra terms can be related to the previously mentioned fluctuation effects by which the various ensembles differ, as we check explicitly.

For any ensemble, the antisymmetry of the Poisson matrix L follows immediately from the atomistic expression (6.68) and is recognized to be a consequence of the antisymmetry of the atomistic Poisson matrix. The discussion of the Onsager-Casimir symmetry properties of the friction matrix M in Section 3.2.1, including the limitations for too detailed levels of description, was based on (6.69) with the identification (6.70).

The degeneracy requirement (1.4) expressing the conservation of entropy under reversible dynamics, another important part of the GENERIC framework, is derived most convincingly within the generalized canonical ensemble. From (6.56) and (6.67), one

[14] See Section 4.6.2 of Kubo et al., *Statistical Physics II* (Springer, 1991) or Evans & Morriss, *Statistical Mechanics of Nonequilibrium Liquids* (Academic, 1990).

obtains

$$\begin{aligned}\sum_k L_{jk}(x)\frac{\delta S(x)}{\delta x_k} &= \sum_k\sum_{ll'} k_B \int \frac{\partial \Pi_j}{\partial z_l}(L_0)_{ll'}\frac{\partial \Pi_k}{\partial z_{l'}}\lambda_k(x)\rho_x(z)d^{6N}z \\ &= -\sum_{ll'} k_B \int \frac{\partial \Pi_j}{\partial z_l}(L_0)_{ll'}\frac{\partial \rho_x(z)}{\partial z_{l'}}d^{6N}z, \end{aligned} \tag{6.74}$$

where, in the second step, the explicit form (6.15) of the generalized canonical ensemble has been used. After an integration by parts and neglecting boundary terms, the antisymmetry of L_0 implies the degeneracy requirement

$$\sum_k L_{jk}(x)\frac{\delta S(x)}{\delta x_k} = 0. \tag{6.75}$$

Within the generalized microcanonical ensemble, only a heuristic argument in favor of the degeneracy requirement (1.4) based on Liouville's theorem for the reversible dynamics has been offered.[15] As an illuminating identity, we obtain from (6.64)

$$\begin{aligned}k_B\sum_k \frac{\delta}{\delta x_k}L_{jk}(x) &= k_B\sum_k\sum_{ll'}\frac{\delta}{\delta x_k}\left\langle\frac{\partial \Pi_j}{\partial z_l}(L_0)_{ll'}\frac{\partial \Pi_k}{\partial z_{l'}}\right\rangle_x \\ &= k_B\sum_k\sum_{ll'}\left\langle\frac{\partial^2 \Pi_j}{\partial z_l \partial z_{l'}}(L_0)_{ll'}\right\rangle_x - \sum_k L_{jk}(x)\frac{\delta S(x)}{\delta x_k},\end{aligned}$$

or, after using the antisymmetry of Poisson matrices,

$$k_B\frac{\delta}{\delta x}\cdot L(x) = L(x)\cdot\frac{\delta S(x)}{\delta x}. \tag{6.76}$$

Equation (6.76) should be interpreted as follows. Because the divergence of the Poisson matrix is multiplied by Boltzmann's constant, the left-hand side should be considered as a small fluctuation effect, and the degeneracy requirement (1.4) is hence obtained within such small fluctuation effects. By no means, one should argue in the reverse direction; one cannot conclude from the fundamental degeneracy requirement (1.4) that the divergence of L in (6.76) must also be zero because small fluctuation effects could be enhanced on division by k_B. A deeper discussion of this situation will be possible after (6.163) has been introduced.

The other basic degeneracy requirement of GENERIC, expressing energy conservation,

$$\sum_k M_{jk}(x)\frac{\delta E(x)}{\delta x_k} = 0, \tag{6.77}$$

can be shown for any ensemble by observing

$$\sum_k [i\mathcal{L}\Pi_k(z)]\frac{\delta E(x)}{\delta x_k} = i\mathcal{L}\mathcal{P}(x)E_0(z) = i\mathcal{L}E_0(z) = \{E_0, E_0\} = 0, \tag{6.78}$$

[15] Öttinger, Phys. Rev. E 57 (1998) 1416.

where once more the assumption (6.51), which expresses the accessibility of energy on the coarse-grained level of description, has been used. In the generalized microcanonical ensemble, the weaker condition (6.50) is sufficient to establish the degeneracy (6.77) because we have the following sequence of equalities:

$$\sum_k [i\mathcal{L}\Pi_k(z)] \frac{\delta E(x)}{\delta x_k} = \sum_k \frac{\delta E(x)}{\delta x_k} \frac{\partial \Pi_k(z)}{\partial z} \cdot L_0 \cdot \frac{\partial E_0(z)}{\partial z}$$
$$= \frac{\partial E_0(z)}{\partial z} \cdot L_0 \cdot \frac{\partial E_0(z)}{\partial z} = \{E_0, E_0\} = 0. \qquad (6.79)$$

The positive semidefinite character of the friction matrix M, which can be thought of as the second law of beyond-equilibrium thermodynamics, is not immediately obvious from (6.72). However, the fluctuation-dissipation theorem (1.57), derived below by the projection-operator method for distribution functions as coarse-grained variables (see Section 6.3.3), is meaningful if, and only if, M is positive semidefinite; a violation of this condition would lead to physically unacceptable properties of the fluctuating forces. The clear separation of time scales is crucial for the argumentation because it allows us to treat the eliminated degrees of freedom as noise having the natural physical properties of fluctuations.

The transformation behavior of the GENERIC building blocks (see Section 1.2.4) is another property that should be checked for the atomistic expressions. For the entropy, we had discussed the transformation behavior in connection with (6.59), and in particular, we have seen how to recover (1.49). The transformation behavior of the other building blocks is based on the identity

$$\rho'_{x'}(z) = \rho_x(z), \qquad (6.80)$$

which, for the generalized microcanonical ensemble, follows from the equation

$$\left|\det\left(\frac{\delta x'}{\delta x}\right)\right| \delta(\Pi'(z) - x'(x)) = \delta(\Pi(z) - x) \qquad (6.81)$$

after normalization, where the slow observables in the transformed setting are given by the phase space functions $\Pi'(z) = x'(\Pi(z))$. The transformation laws (1.48) and (1.50) for the energy (6.48) and the Poisson matrix (6.67) follow directly from (6.80). The proof of the transformation law (1.51) for the friction matrix (6.72) requires the additional property

$$\mathcal{Q}'(x') = \mathcal{Q}(x). \qquad (6.82)$$

The expression (6.25) for $\mathcal{P}(x) = 1 - \mathcal{Q}(x)$ implies that (6.82) can be rigorous only for linear transformations $x'(x)$.[16] However, the corrections for nonlinear transformations are only fluctuation effects, which we have neglected before for the transformation behavior of the entropy. Thus, in the absence of fluctuations, all the transformation laws have been rederived from the atomistic expressions for the GENERIC building blocks.

[16] For the operator $\tilde{\mathcal{Q}} = 1 - \tilde{\mathcal{P}}$ associated with the projector $\tilde{\mathcal{P}}$ of (6.46), the identity $\tilde{\mathcal{Q}}' = \tilde{\mathcal{Q}}$ holds rigorously even for nonlinear transformations.

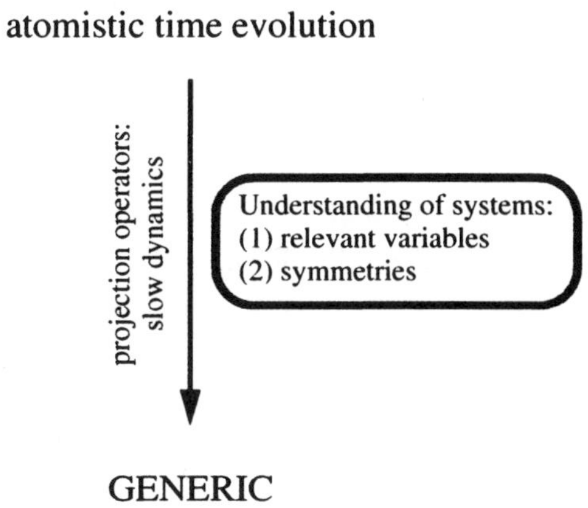

Fig. 6.3 Inputs for the successful application of statistical mechanics to specific systems.

The Jacobi identity (1.9) for the bracket associated with the Poisson matrix according to (1.6) has not yet been derived from the atomistic expression (6.68) for L. The mathematical problem is that the Poisson matrix (6.68) and hence the Poisson bracket $\{\cdot,\cdot\}$ defined in (1.6) involve an average $\langle\cdots\rangle_x$, so that the nested brackets in the Jacobi identity imply products of averages, which cannot be obtained by averaging the Jacobi identity for the canonical Poisson bracket given by L_0. Because the Jacobi identity, which expresses the time-structure invariance of the reversible dynamics, is of great importance for the predictive power of the GENERIC framework, we next consider the role of transformations and symmetries to obtain a better understanding of this identity on the coarse-grained level.

6.1.6 Symmetries

The purpose of this section is to make sure that all the symmetries in a given problem, such as isotropy or Galilean invariance, are properly expressed on any level of description. In particular, we discuss conditions under which the symmetries on one level of description are automatically inherited on a more coarse-grained level constructed by projection.

A key step in applying GENERIC to particular systems is the identification of the proper relevant variables that allow for a self-contained description of the phenomena of interest. Correspondingly, the formal projection-operator technique produces meaningful equations only if all the slow variables have been identified correctly. Whereas the importance of choosing the proper variables has been emphasized before, the main purpose of this section is to show the *importance of symmetries* and to elaborate how they can be incorporated into the coarse-graining procedure. Our understanding of particular systems crucially depends not only on the identification of the proper variables but also on the recognition and implementation of the underlying symmetries (see Figure 6.3).

Typically, the symmetry group $\mathcal{G}$ will be a Lie group.[17] We assume that the group of symmetries acts in a well-known way on the phase space, that is, the map

$$z \to z^g(z), \tag{6.83}$$

is given for all group elements $g \in \mathcal{G}$. The transformations (6.83) can be regarded as symmetries of the atomistic system if the energy,

$$E_0(z^g(z)) = E_0(z), \tag{6.84}$$

and the canonical Poisson bracket

$$\{z_j^g, z_k^g\} = (L_0)_{jk}, \tag{6.85}$$

are invariant or, in other words, if the time independent transformations (6.83) are canonical for all group elements $g \in \mathcal{G}$.

As an example of canonical transformations, we consider the point transformations

$$\boldsymbol{r}_j \to \boldsymbol{r}_j^f = f(\boldsymbol{r}_j), \tag{6.86}$$

which can be extended to canonical transformations by defining

$$\boldsymbol{p}_j \to \boldsymbol{p}_j^f = \left(\frac{\partial f(\boldsymbol{r}_j)}{\partial \boldsymbol{r}_j}\right)^{-1} \cdot \boldsymbol{p}_j, \tag{6.87}$$

where the dot represents a contraction of the components of $\boldsymbol{p}_j$ with those of $\boldsymbol{r}_j$. This example is useful to discuss coordinate transformations. In a more general context, the canonical character of extended point transformations is established in Appendix B.3 (see Exercise 215). Note that any extended point transformation (6.86), (6.87) has a unit determinant of the Jacobian,

$$\left|\det\left(\frac{\partial z^g(z)}{\partial z}\right)\right| = 1, \tag{6.88}$$

which is a general feature of canonical transformations, as can be shown by taking the determinant of both sides of (6.85).

How can we make sure now that the symmetries of a microscopic system are properly inherited by a macroscopic system? For the generalized canonical ensemble (6.15) with its linear dependence on the slow variables in the exponential to be form invariant under the transformations (6.83), we need to assume that the slow variables are constructed such that we obtain a linear relationship

$$\Pi_k(z^g(z)) = \sum_l R_{kl}(g)\Pi_l(z), \tag{6.89}$$

[17] See, for example, Hamermesh, *Group Theory* (Dover, 1989); see also Appendix B.1.

where $R_{kl}(g)$ represents the group element g. Equation (6.89) provides a sufficient condition for establishing the successful implementation of a symmetry group.

By averaging (6.89), we obtain a linear action, or a representation, of the symmetry group on the space of macroscopic variables,

$$x_k \to x_k^g(x) = \sum_l R_{kl}(g) x_l, \tag{6.90}$$

for all group elements $g \in \mathcal{G}$. If we introduce the following transformation of Lagrange multipliers,

$$\lambda_k^g(\lambda) = \sum_l \lambda_l R_{lk}^{-1}(g), \tag{6.91}$$

the generalized canonical averages of the slow variables Π_k transform according to (6.90). Moreover, we obtain the following natural transformation behavior of ensembles:

$$\rho_{x^g(x)}(z^g(z)) = \rho_x(z), \tag{6.92}$$

allowing us to study the transformation behavior of the atomistic expressions for the GENERIC building blocks, that is, the energy, the entropy, the Poisson matrix, and the friction matrix on the macroscopic level. From the atomistic expression (6.48) for the energy, we obtain

$$E(x^g(x)) = \int \rho_{x^g(x)}(z) E_0(z) d^{6N}z = \int \rho_{x^g(x)}(z^g(z)) E_0(z^g(z)) d^{6N}z, \tag{6.93}$$

where the integration variable has been changed and (6.88) has been used. With (6.84) and (6.92), we arrive at the macroscopic invariance of the energy under symmetry transformations,

$$E(x^g(x)) = E(x). \tag{6.94}$$

In the generalized canonical approach, the invariance of the entropy,

$$S(x^g(x)) = S(x), \tag{6.95}$$

is obtained by using (6.92) in (6.14) after changing the integration variables. By using (6.85), (6.89), and (6.92) in (6.67), after another change of integration variables, one obtains the invariance of the Poisson matrix under symmetry transformations,

$$R(g) \cdot L(x) \cdot R(g)^T = L(x^g), \tag{6.96}$$

or

$$\{x_j^g, x_k^g\} = L_{jk}(x^g). \tag{6.97}$$

To show the invariance of the friction matrix (6.72) under symmetry transformations, we need to establish an invariance property for the Liouville operator and for the projection operators. If we introduce a family of operators $\mathcal{S}^g$ acting on phase space functions $A(z)$ by substituting $z^g(z)$ for the argument z for every symmetry transformation $g \in \mathcal{G}$,

$$[\mathcal{S}^g A](z) = A(z^g(z)), \tag{6.98}$$

then these properties can be formulated as the following operator identities:

$$\mathcal{Q}(x)\mathcal{S}^g = \mathcal{S}^g\mathcal{Q}(x^g) \tag{6.99}$$

and

$$\mathcal{L}\mathcal{S}^g = \mathcal{S}^g\mathcal{L}. \tag{6.100}$$

With these identities and the invariance (6.84) of the atomistic energy we obtain the invariance of the friction matrix analogous to the corresponding properties for the Poisson matrix,

$$R(g) \cdot M(x) \cdot R(g)^T = M(x^g), \tag{6.101}$$

or, for the dissipative bracket,

$$[x_j^g, x_k^g] = M_{jk}(x^g). \tag{6.102}$$

In summary, we have derived the macroscopic invariance properties under symmetry transformations for the GENERIC building blocks, (6.94)–(6.96) and (6.101), from the atomistic symmetry properties (6.84) and (6.85). This derivation is based on the assumption that the slow variables admit a linear representation (6.89) of the symmetry group. For example if, for an isotropic system, we have such a representation of the group of rotations, then our derivation leads to Curie's principle on the macroscopic level.

When the linear representation (6.89) on the slow observables, with the corresponding representation (6.90) on the macroscopic variables, can be constructed for the extended space transformations (6.86), (6.87), then we obtain a representation of the group of space transformations on the macroscopic variables. According to the general idea presented in Appendix B.3 and the discussion in Appendix B.4, such a representation is exactly what is needed to construct Poisson brackets. For example, for the mass density

$$\Pi_{\boldsymbol{r}}(z) = \sum_{j=1}^{N} m_j \delta(\boldsymbol{r}_j - \boldsymbol{r}), \tag{6.103}$$

we have the transformed field

$$\Pi_{\boldsymbol{r}}^f(z) = \sum_{j=1}^{N} m_j \delta(f(\boldsymbol{r}_j) - \boldsymbol{r}), \tag{6.104}$$

and thus, with the transformation properties of δ-functions,

$$\left|\det \frac{\partial f(\boldsymbol{r})}{\partial \boldsymbol{r}}\right| \Pi_{f(\boldsymbol{r})}^f(z) = \sum_{j=1}^{N} m_j \left|\det \frac{\partial f(\boldsymbol{r})}{\partial \boldsymbol{r}}\right| \delta(f(\boldsymbol{r}_j) - f(\boldsymbol{r})) = \Pi_{\boldsymbol{r}}(z), \tag{6.105}$$

which corresponds to (B.117) for the representation of a scalar density. For the momentum density

$$\boldsymbol{\Pi}_{\boldsymbol{r}}(z) = \sum_{j=1}^{N} \boldsymbol{p}_j \delta(\boldsymbol{r}_j - \boldsymbol{r}), \tag{6.106}$$

by using (6.87) in addition to (6.86), we similarly obtain the following natural action of extended space transformations:

$$\left| \det \frac{\partial f(\boldsymbol{r})}{\partial \boldsymbol{r}} \right| \frac{\partial f(\boldsymbol{r})}{\partial \boldsymbol{r}} \cdot \boldsymbol{\Pi}^{f}_{f(\boldsymbol{r})}(z) = \boldsymbol{\Pi}_{\boldsymbol{r}}(z). \tag{6.107}$$

For each further factor of $\boldsymbol{p}_j$ there is an additional factor $\partial f(\boldsymbol{r})/\partial \boldsymbol{r}$, where the space-component index of f is contracted with that of $\boldsymbol{p}_j^f$. For a tensor of rank two, one should compare the resulting representation with (B.116).

Exercise 142 Invariance under Galilean Transformations
Do the invariance properties (6.84) and (6.85) hold for the Galilean transformations

$$\begin{pmatrix} \boldsymbol{r}_j \\ \boldsymbol{p}_j \end{pmatrix} \to \begin{pmatrix} \boldsymbol{r}_j \\ \boldsymbol{p}_j - m_j \Delta \boldsymbol{v} \end{pmatrix}? \tag{6.108}$$

Exercise 143 Linear Single Particle Transformations
Consider the linear transformation

$$\begin{pmatrix} \boldsymbol{r}_j \\ \boldsymbol{p}_j \end{pmatrix} \to \begin{pmatrix} \boldsymbol{Q}_{rr} & \boldsymbol{Q}_{rp} \\ \boldsymbol{Q}_{pr} & \boldsymbol{Q}_{pp} \end{pmatrix} \cdot \begin{pmatrix} \boldsymbol{r}_j \\ \boldsymbol{p}_j \end{pmatrix} \tag{6.109}$$

for each particle j. Under what conditions for the tensors $\boldsymbol{Q}_{rr}$, $\boldsymbol{Q}_{rp}$, $\boldsymbol{Q}_{pr}$, $\boldsymbol{Q}_{pp}$ does this transformation leave the canonical Poisson bracket invariant? Show that these conditions are fulfilled for linear extended point transformations of the type (6.86), (6.87).

Exercise 144 Properties of Substitution Operator
Prove the identities (6.99) and (6.100). Furthermore, show the analogous relation $\tilde{\mathcal{P}}\mathcal{S}^g = \mathcal{S}^g\tilde{\mathcal{P}}$ for the projection operator defined in (6.46).

6.2 DIRECT APPROACH

With ensembles and projection operators, as well as their most important properties, we now have all the ingredients to present the standard projection-operator arguments in a compact form. We first consider a direct projection on the variables of interest, which gives us equations for averages only. Fluctuations are considered in the subsequent section, where we project on distribution functions of the configurational variables of interest.

6.2.1 Exact Time-Evolution Equation

By acting with both sides of the operator identity (6.40) on $i\mathcal{L}\Pi_j$ and by averaging with respect to the initial ensemble $\rho_{x(0)}$, we obtain an exact formal time-evolution equation for $x(t)$:

$$\frac{dx_j(t)}{dt} = \int \rho_{x(t)}(z) i\mathcal{L}\Pi_j(z) d^{6N}z + \int_0^t \int \rho_{x(u)}(z) i\mathcal{L}\mathcal{Q}(u) G(u,t) i\mathcal{L}\Pi_j(z) d^{6N}z\, du. \tag{6.110}$$

In deriving (6.110), we assumed that the initial ensemble is of the generalized canonical form (6.15). This assumption should be considered to adequately define the relevant ensemble rather than to restrict the initial states.[18] The following exercises should clarify some mathematical details behind the derivation of (6.110).

Exercise 145 Occurrence of Time-Dependent Ensemble
Derive the identity

$$\int \rho_{x(0)}(z) e^{i\mathcal{L}t} \mathcal{P}(t) A(z) d^{6N}z = \int \rho_{x(t)}(z) A(z) d^{6N}z \tag{6.111}$$

for any observable A and for any type of ensemble ρ_x. Discuss the analogous property

$$\int \rho_{x(0)}(z) e^{i\mathcal{L}t} \tilde{\mathcal{P}} A(z) d^{6N}z = \int \rho_{x(t)}(z) A(z) d^{6N}z \tag{6.112}$$

for the projection operator (6.46) and the generalized microcanonical ensemble.

Exercise 146 Absence of Initial Effects
Derive the identity

$$\int \rho_x(z) \mathcal{Q}(x) A(z) d^{6N}z = 0 \tag{6.113}$$

for any observable A and for any type of ensemble ρ_x. This identity implies that the term resulting from $\mathcal{Q}(0)G(0,t)$ in (6.40) disappears in (6.110).

Exercise 147 Absence of Explicit Time Derivatives
Derive the identity

$$\int \rho_x(z) \dot{\mathcal{P}}(x) A(z) d^{6N}z = 0 \tag{6.114}$$

to show that the term resulting from $\dot{\mathcal{P}}(t)$ in (6.40) disappears in (6.110).

The first term on the right-hand side of (6.110) is referred to as reversible, and the term in the second line as irreversible. By means of (6.2), the reversible term can be written as

$$\left[\frac{dx_j(t)}{dt}\right]_{\text{rev}} = \int \rho_{x(t)}(z) \{\Pi_j, E_0\} d^{6N}z. \tag{6.115}$$

[18] See p. 20 of Grabert, *Projection Operators* (Springer, 1982).

The accessibility of the energy through the relevant variables, expressed through the formal assumption (6.51), implies

$$\frac{\partial E_0}{\partial z} = \frac{\partial \mathcal{P}(x) E_0}{\partial z} = \sum_k \frac{\partial \Pi_k}{\partial z} \frac{\delta}{\delta x_k} \int \rho_x(z') E_0(z') d^{6N} z'. \tag{6.116}$$

By using this identity in the Poisson bracket of (6.115), one obtains the final expression for the reversible contribution to the time evolution in the form

$$\left[\frac{dx_j(t)}{dt}\right]_{\text{rev}} = \sum_k L_{jk}(x(t)) \frac{\delta E(x(t))}{\delta x_k} \tag{6.117}$$

with the expected energy expression (6.48) and the Poisson matrix

$$L_{jk}(x) = \int \rho_x(z) \{\Pi_j, \Pi_k\} d^{6N} z, \tag{6.118}$$

which coincides with the expected expression (6.68). Indeed, the coarse-grained Poisson matrix is thus obtained by averaging the atomistic Poisson bracket of the relevant variables with the underlying ensemble.

After an integration by parts, to let the first Liouville operator in the second term of (6.110) act on the distribution function to its left, we obtain the following still exact result for the irreversible contribution:

$$\left[\frac{dx_j(t)}{dt}\right]_{\text{irr}} = \int_0^t \int \rho_{x(u)}(z) \sum_l \lambda_k(x(u)) [i\mathcal{L}\Pi_k(z)] \mathcal{Q}(u) G(u,t) i\mathcal{L}\Pi_j(z) d^{6N} z du, \tag{6.119}$$

where the first Liouville operator acts only on the observable in the square brackets. To obtain (6.119), we for the first time assume a generalized canonical ensemble; however, we can argue that this assumption is without loss of generality. From the results of Exercise 141, one can conclude that, for the generalized microcanonical ensemble, an extra term occurs that carries a prefactor of k_B and should hence correspond to a small correction due to fluctuations. More explicitly, the extra term (possibly among other terms) produces the contribution $k_\text{B} \frac{\delta}{\delta x} \cdot M$ on top of $M \cdot \frac{\delta S}{\delta x}$ in the GENERIC with fluctuations (1.56), and it is hence clear that such a term is relevant only in connection with fluctuations.

6.2.2 Markovian Approximation

The crucial assumption of the projection-operator formalism is that there should exist a clear separation of time scales. The relevant variables x_k evolve on a large time scale compared with some intermediate scale τ, and all other variables evolve rapidly compared with τ, as illustrated in Figure 6.2. Experience shows that this assumption works surprisingly well for many macroscopic systems of interest.[19] In view of

[19] Grabert, *Projection Operators* (Springer, 1982).

the occurrence of the projectors $\mathcal{Q}$ (projecting on the fast variables) in the integral of (6.119), the integrand should decay rapidly, and the integral is expected to be dominated by values of u between $t - \tau$ and t. All slow variables in the integral can then be evaluated at time t; a factor of entropy gradients [see (6.56)] can also be pulled out of the integral. With this Markovian approximation, we hence obtain the following approximate version of (6.119):

$$\left[\frac{dx_j(t)}{dt}\right]_{\text{irr}} = \sum_k M_{jk}(x(t)) \frac{\delta S(x(t))}{\delta x_k} \tag{6.120}$$

with the friction matrix

$$M_{jk}(x) = \frac{1}{k_{\mathrm{B}}} \int_0^{\tau} \int \rho_x(z)[i\mathcal{L}\Pi_k(z)]\mathcal{Q}(x)e^{i\mathcal{Q}(x)\mathcal{L}\mathcal{Q}(x)u}\mathcal{Q}(x)[i\mathcal{L}\Pi_j(z)]d^{6N}z\,du, \tag{6.121}$$

which, in view of the self-adjointness result of Exercise 131, coincides with (6.72). Equation (6.121), which holds for general beyond-equilibrium systems, even in coarse-grained configurations x far away from equilibrium, has the form of the Green-Kubo expressions for transport coefficients known from linear response theory.[20] The expression (6.121) appears to be symmetric in the indices j and k. However, when the adjoint of the exponential in (6.121) is taken to establish the actual symmetry properties in j and k, then the property $\mathcal{L}\rho_x(z) \neq 0$ (see Exercise 141) introduces some subtleties (see the remarks after (3.49) and the lack of symmetry of the friction matrix in Section 7.2.4).

It should be noted that, formally, the u-integration in the expression (6.121) can actually be carried out. In view of the occurrence of $\mathcal{Q}(x)$ on both sides of the exponential, $i\mathcal{Q}(x)\mathcal{L}\mathcal{Q}(x)u$ in the exponential can be replaced by $\mathcal{Q}(x)i\mathcal{L}u$, so that a total derivative with respect to u arises. When the limits of integration are introduced, the contribution from the lower limit vanishes because of the identity $\mathcal{Q}(x)\Pi_j(z) = 0$ (see Exercise 132). After using the result of Exercise 131, we then obtain

$$M_{jk}(x) = \frac{1}{k_{\mathrm{B}}} \int \rho_x(z)[\mathcal{Q}(x)i\mathcal{L}\Pi_k(z)]e^{\mathcal{Q}(x)i\mathcal{L}\tau}\Pi_j(z)d^{6N}z. \tag{6.122}$$

Although this expression for the friction matrix is more compact than (6.121), the origin of the symmetry properties of the friction matrix and their limitations become less obvious, and the well-definedness of the exponential is even less clear. Note that the time integration can only be performed when the operator $\mathcal{Q}(x)\mathcal{L}\mathcal{Q}(x)$ in the exponential of (6.121) is not replaced by $\mathcal{L}$. The compact expression (6.122) is useful, for example, in deriving the friction matrix for the Boltzmann equation in Section 7.2.4.

It is interesting to note that the time evolution in the expressions (6.121) and (6.122) for the friction matrix is represented by the projected Liouville operator $\mathcal{Q}(x)i\mathcal{L}$

[20] See, for example, Chapter 4 and Section 6.1 of Evans & Morriss, *Statistical Mechanics of Nonequilibrium Liquids* (Academic, 1990).

rather than the full operator $i\mathcal{L}$. Therefore, the time-evolution equations to be used in molecular dynamics simulations are

$$\frac{d\boldsymbol{r}_j}{dt} = \mathcal{Q}(x)i\mathcal{L}\boldsymbol{r}_j = \mathcal{Q}(x)\frac{\boldsymbol{p}_j}{m_j} = \frac{\boldsymbol{p}_j}{m_j} - \mathcal{P}(x)\frac{\boldsymbol{p}_j}{m_j} \tag{6.123}$$

and

$$\frac{d\boldsymbol{p}_j}{dt} = \mathcal{Q}(x)i\mathcal{L}\boldsymbol{p}_j = \mathcal{Q}(x)\boldsymbol{F}_j = \boldsymbol{F}_j - \mathcal{P}(x)\boldsymbol{F}_j, \tag{6.124}$$

instead of Hamilton's equations of motion. For example, (6.30) shows that the relationship between $d\boldsymbol{r}_j/dt$ and $\boldsymbol{p}_j/m_j$ differs from the usual one, unless we use a reference frame in which the total momentum vanishes. Such modifications of Hamilton's equations of motion, which we are going to use in molecular dynamics simulations, are hence suggested by the projection-operator formalism. While the dynamics (6.123), (6.124) is projected along the fast variables, neither the slow variables nor the ensemble are rigorously conserved by the modified dynamics.

6.2.3 Linear Response Theory

Near equilibrium, an alternative interpretation of the friction matrix as a linear response function can be offered. To do so, we first rewrite (6.122) as

$$M_{jk}(x) = \frac{1}{k_\mathrm{B}}\int \rho_x(z)[i\mathcal{L}\Pi_k(z)]\left[e^{\mathcal{Q}(x)i\mathcal{L}\tau} - 1\right]\Pi_j(z)d^{6N}z. \tag{6.125}$$

We then choose the variables $x = x_\mathrm{eq}$ such that they describe equilibrium states (see Section 1.2.3). We work with a fixed number of particles N and a constant Lagrange multiplier λ_E so that the equilibrium ensemble is given by

$$\rho_\mathrm{eq}(z) = \frac{1}{Z_\mathrm{eq}}e^{-\lambda_E E_0(z)/k_\mathrm{B}}, \qquad Z_\mathrm{eq} = \int e^{-\lambda_E E_0(z)/k_\mathrm{B}}d^{6N}z, \tag{6.126}$$

and we use the notation

$$\mathcal{Q}_\mathrm{eq} = \mathcal{Q}(x_\mathrm{eq}). \tag{6.127}$$

By observing that we have the identities

$$\frac{\partial\rho_\mathrm{eq}(z)}{\partial\boldsymbol{r}_j} = -\frac{\lambda_E}{k_\mathrm{B}}\rho_\mathrm{eq}(z)\frac{\partial E_0(z)}{\partial\boldsymbol{r}_j}, \qquad \frac{\partial\rho_\mathrm{eq}(z)}{\partial\boldsymbol{p}_j} = -\frac{\lambda_E}{k_\mathrm{B}}\rho_\mathrm{eq}(z)\frac{\partial E_0(z)}{\partial\boldsymbol{p}_j}, \tag{6.128}$$

the Liouville operator in (6.125), after an integration by parts, can be rewritten to obtain

$$M_{jk} = M_{jk}(x_\mathrm{eq}) = -\frac{1}{\lambda_E}\int \rho_\mathrm{eq}(z)\left\{\left(e^{\mathcal{Q}_\mathrm{eq}i\mathcal{L}\tau} - 1\right)\Pi_j(z), \Pi_k(z)\right\}d^{6N}z, \tag{6.129}$$

or, by means of (6.118),

$$\lambda_E M_{jk} - L_{jk} = -\int \rho_\mathrm{eq}(z)\left\{e^{\mathcal{Q}_\mathrm{eq}i\mathcal{L}\tau}\Pi_j(z), \Pi_k(z)\right\}d^{6N}z. \tag{6.130}$$

Note that $\lambda_E M - L$ at equilibrium is exactly the combination of GENERIC building blocks that we previously encountered in our discussion of the limit of equilibrium thermodynamics in (1.46). Whereas the previous expressions for the friction matrix involved Liouville operators and hence Poisson brackets with the energy $E_0(z)$, the expression (6.130) involves a Poisson bracket with $\Pi_k(z)$ instead of $E_0(z)$. This observation suggests to consider $\Pi_k(z)$ as a Hamiltonian or, actually, as a small perturbation of the Hamiltonian $E_0(z)$. A perturbative treatment of the effect of $\Pi_k(z)$ then involves an expression that is remarkably similar to (6.130). To see that, we write the Liouville operator as the sum of the time-evolution operator $\mathcal{L}_0$ implied by $E_0(z)$ and a perturbation,

$$\mathcal{L} = \mathcal{L}_0 + \bar{\epsilon}\mathcal{L}_1, \tag{6.131}$$

where $\bar{\epsilon}$ is a small parameter and $\mathcal{L}_1$ is defined by the Poisson bracket with Π_k. If the perturbation is switched on at $t = 0$, the solution of (6.2) to first order in $\bar{\epsilon}$ is given by

$$A(z,t) = e^{i\mathcal{L}_0 t} A(z) + \bar{\epsilon} \int_0^t e^{i\mathcal{L}_0 (t-t')} i\mathcal{L}_1 e^{i\mathcal{L}_0 t'} A(z) dt'. \tag{6.132}$$

By averaging with $\rho_{\text{eq}}(z)$ and assuming that $\rho_{\text{eq}}(z)$ is invariant under the unperturbed time evolution, that is,

$$e^{-i\mathcal{L}_0 t} \rho_{\text{eq}}(z) = \rho_{\text{eq}}(z), \tag{6.133}$$

we obtain

$$\langle A \rangle_t = \langle A \rangle_0 + \bar{\epsilon} \int_0^t \int \rho_{\text{eq}}(z) i\mathcal{L}_1 e^{i\mathcal{L}_0 t'} A(z) d^{6N}z dt', \tag{6.134}$$

or, after differentiation with respect to time,

$$\frac{d\langle A \rangle_t}{dt} = \bar{\epsilon} \int \rho_{\text{eq}}(z) i\mathcal{L}_1 e^{i\mathcal{L}_0 t} A(z) d^{6N}z = \bar{\epsilon} \int \rho_{\text{eq}}(z) \left\{ e^{i\mathcal{L}_0 t} A(z), \Pi_k(z) \right\} d^{6N}z. \tag{6.135}$$

By comparing with (6.130), we obtain the strikingly simple, elegant, and general formula

$$\frac{d\langle \Pi_j \rangle_\tau}{d\tau} = -\bar{\epsilon}(\lambda_E M_{jk} - L_{jk}). \tag{6.136}$$

The fundamental result (6.136) describes the linear response of Π_j to a small perturbation of the Hamiltonian proportional to Π_k on the time scale τ. To obtain the proper friction matrix from the linear response, the unperturbed dynamics should actually be taken as the rapidly fluctuating part of the equilibrium dynamics, $\mathcal{L}_0 = \mathcal{Q}_{\text{eq}}\mathcal{L}$. Note that τ in (6.136) corresponds to the intermediate time scale of the projection-operator formalism, so that the fast variables can explore the relevant phase space, whereas the slow variables do not change appreciably. From the perspective of the slow variables, (6.136) describes the instantaneous response. The matrix M_{jk} on the right-hand side of (6.136) describes the δ-correlated fluctuations, as discussed in Section 1.2.5. Also the expression (6.121) suggests that M_{jk} is related to temporal fluctuations. The left-hand side of (6.136) represents the immediate linear response of the system in the form of a dissipative flux. Such a relationship is often referred to as a *fluctuation-dissipation theorem* of the first kind.

For the linear response, the effects of several perturbations can be superimposed. If the perturbed Hamiltonian is of the form

$$E_0(z) + \sum_k \bar{\epsilon}_k \Pi_k(z), \tag{6.137}$$

then the fluctuation-dissipation theorem (6.136) takes the more general form

$$\frac{d\,\langle \Pi_j \rangle_\tau}{d\tau} = -\sum_k (\lambda_E M_{jk} - L_{jk})\bar{\epsilon}_k. \tag{6.138}$$

According to (1.43), the Lagrange multiplier λ_E occurring in (6.136) and (6.138) is the inverse of a constant overall equilibrium temperature.

As an example, we consider the effect of a gravitational field on a gas. According to (6.7), we have

$$\sum_k \bar{\epsilon}_k \Pi_k(z) = -\int \boldsymbol{g}\cdot\boldsymbol{r} \sum_{j=1}^{N} m\delta(\boldsymbol{r}_j - \boldsymbol{r})d^3r = -\sum_{j=1}^{N} m\boldsymbol{g}\cdot\boldsymbol{r}_j, \tag{6.139}$$

where k is actually the continuous position label $\boldsymbol{r}$. The fluctuation-dissipation theorem (6.138) then predicts the following linear response of the mass density:

$$\frac{\partial\rho(\boldsymbol{r})}{\partial\tau} = \lambda_E \int M_{11}(\boldsymbol{r},\boldsymbol{r}')\boldsymbol{g}\cdot\boldsymbol{r}'d^3r', \tag{6.140}$$

where $L_{11}(\boldsymbol{r},\boldsymbol{r}') = 0$ has been used [see (2.59)]. From (2.77) and (2.85), we obtain

$$\frac{\partial\rho(\boldsymbol{r})}{\partial\tau} = -\lambda_E \frac{\partial}{\partial\boldsymbol{r}}\cdot D'\boldsymbol{g} = -\frac{\partial}{\partial\boldsymbol{r}}\cdot\frac{D}{k_{\rm B}T}m\boldsymbol{g}\,\rho(\boldsymbol{r}), \tag{6.141}$$

in which the drift velocity is given by the gravitational force, $m\boldsymbol{g}$, times the mobility, $D/(k_{\rm B}T)$. The analogous response for the momentum density response consists of a reversible contribution in addition to the dissipative one, which is again determined by the drift velocity,

$$\frac{\partial\boldsymbol{M}(\boldsymbol{r})}{\partial\tau} = -\frac{\partial}{\partial\boldsymbol{r}}\cdot\frac{D}{k_{\rm B}T}m\boldsymbol{g}\,\boldsymbol{M}(\boldsymbol{r}) + \rho(\boldsymbol{r})\boldsymbol{g}, \tag{6.142}$$

and for the internal energy density we find

$$\frac{\partial\epsilon(\boldsymbol{r})}{\partial\tau} = -\frac{\partial}{\partial\boldsymbol{r}}\cdot\frac{D}{k_{\rm B}T}m\boldsymbol{g}\,\alpha(\boldsymbol{r})\rho(\boldsymbol{r}), \tag{6.143}$$

which allows us to study α by linear response theory.

Equation (6.141) expresses the linear response in the form of the divergence of a flux, where the flux is proportional to the gradient of the gravitational energy. The derivatives occurring in the friction matrix are thus important for the proper imbedding of the diffusion coefficient in the fluctuation-dissipation theorem (6.138). In the usual approach,[21] the fluctuation-dissipation theorem is formulated directly for the coefficients of the phenomenological Onsager matrix, which is related to the friction matrix according to (2.94) and (3.4). The advantage of (6.138) is that it provides a unified and unambiguous treatment of all transport and relaxation phenomena, including the full imbedding of transport coefficients or relaxation rates into proper equations.

6.3 PROBABILITY DENSITY APPROACH

The justification of the Markovian approximation used to derive GENERIC by the projection-operator technique depends entirely on the choice of the relevant variables. All slow variables must be taken into account in composing the appropriate system to describe certain phenomena. Because we always rely on the Markovian approximation, we should make sure that the set of relevant variables is sufficiently rich.

For example, all functions of slow variables should be expected to be slow; this has not been taken into account in the direct approach of the preceding section. As a next step, we hence do not take the variables $\Pi(z)$ as the slow variables of the projection-operator procedure but rather the probability density for finding particular values of $\Pi(z)$. This corresponds to using a much larger space of slow variables; in particular, any function of the observables $\Pi(z)$ can be reconstructed.

In the probability density approach, we have an important advantage in addition to the better justification of the Markovian approximation due to the choice of a more complete set of slow variables: We obtain a time-evolution equation for a configurational distribution function, and this equation contains the full description of fluctuations.

The results of this section cannot be generalized in a straightforward way from classical to quantum systems. The reason for this limitation to classical systems is that we need to speak about the probability for finding particular values for all the observables $\Pi(z)$, which, in general, will not commute for quantum systems. In other words, the proper coupling of quantum and statistical fluctuations is a subtle problem (see Appendix D.5 for further remarks).

6.3.1 Lifting Direct Results

Given the phase space functions $\Pi(z)$ as the relevant variables of interest, we here do not project directly on the linear space spanned by $\Pi(z)$ but rather on the space spanned by the probability densities

$$\tilde{\Pi}_y(z) = \delta(\Pi(z) - y), \tag{6.144}$$

[21] See, for example, Chapter 5 of Evans & Morriss, *Statistical Mechanics of Nonequilibrium Liquids* (Academic, 1990).

which describe the probability for finding a value y for $\Pi(z)$. That's all! Note that any function $A(\Pi(z))$ can be expressed as a linear combination of the phase space functions $\tilde{\Pi}_y(z)$,

$$A(\Pi(z)) = \int A(\Pi(z))\delta(\Pi(z) - y)d^K y = \int A(y)\tilde{\Pi}_y(z)d^K y, \tag{6.145}$$

where K indicates the set of components $\Pi_k(z)$ and the number of integrations associated with them. The richness of the set of variables in (6.144) can be further illustrated in the following way. Because the set of indices k labeling the real-valued phase space functions $\Pi_k(z)$ is K, the labels y of $\tilde{\Pi}_y(z)$ take values in $\mathbb{R}^K$.

The goal of this section is to apply the results of Section 6.2 not directly to the observables $\Pi_k(z)$; rather, we lift them to the richer level of probability densities $\tilde{\Pi}_y(z)$. This is merely a straightforward application of the previous results to a richer set of observables. However, one has to be very careful with the notation, in particular, because a close relationship between the GENERIC building blocks on the levels determined by $\tilde{\Pi}_y(z)$ and $\Pi_k(z)$ can be established. More precisely, the building blocks we obtain by using the generalized canonical approach on the level of probability densities are related to their generalized microcanonical counterparts on the direct level. We hence reserve the notation of the previous section for the direct level (for example, Π, ρ, Z, E, S, L, M), and we use a tilde over the symbols referring to the level of probability densities (for example, $\tilde{\Pi}$, $\tilde{\rho}$, $\tilde{Z}$, $\tilde{E}$, $\tilde{S}$, $\tilde{L}$, $\tilde{M}$).

The average of $\tilde{\Pi}_y(z)$ over an ensemble yields the probability density for finding the values y_k for the observables Π_k in that ensemble. Instead of writing $\tilde{x}_y$ for this average, we use the more suggestive symbol $f(y)$. The generalized canonical ensemble (6.15) takes the form

$$\tilde{\rho}_f(z) = \frac{1}{\tilde{Z}(f)} \exp\left[-\int \lambda(y)\tilde{\Pi}_y(z)d^K y\right] = \frac{e^{-\lambda(\Pi(z))}}{\tilde{Z}(f)}, \tag{6.146}$$

and the Lagrange multipliers $\lambda(y)$ are determined by (6.10),

$$f(y) = \int \tilde{\rho}_f(z)\tilde{\Pi}_y(z)d^{6N}z = \frac{e^{-\lambda(y)}}{\tilde{Z}(f)} \int \delta(\Pi(z) - y)d^{6N}z = \frac{\Omega(y)}{\tilde{Z}(f)} e^{-\lambda(y)}, \tag{6.147}$$

where $\Omega(y)$ is the generalized microcanonical volume of phase space (6.12) on the direct level. By comparing (6.146) and (6.147), we obtain the generalized canonical ensemble on the level of probability densities in a compact form, free of Lagrange multipliers:

$$\tilde{\rho}_f(z) = \frac{f(\Pi(z))}{\Omega(\Pi(z))}. \tag{6.148}$$

As an immediate consequence, we obtain the useful connection to the microcanonical ensemble $\rho_y(z)$ on the direct level, as given in (6.11),

$$\tilde{\rho}_f(z)\tilde{\Pi}_y(z) = \tilde{\rho}_f(z)\delta(\Pi(z) - y) = f(y)\rho_y(z), \tag{6.149}$$

often to be applied in connection with an auxiliary integration over y.

Exercise 148 Rich Projection Operator

Show that the projection operator on the level of probability densities is given by $\tilde{\mathcal{P}}$ defined in (6.46). Why is $\tilde{\mathcal{P}}$ independent of f?

6.3.2 GENERIC Building Blocks

After introducing the proper notation, we can now elegantly evaluate the GENERIC building blocks $\tilde{E}$, $\tilde{S}$, $\tilde{L}$, and $\tilde{M}$. From (6.48) we obtain

$$\tilde{E}(f) = \int \tilde{\rho}_f(z) E_0(z) d^{6N}z = \int f(y) E(y) d^K y, \tag{6.150}$$

where (6.50) and (6.149) with an integration over y have been used. The entropy is obtained from (6.55) as

$$\begin{aligned} \tilde{S}(f) &= k_B[\ln \tilde{Z}(f) + \int f(y)\lambda(y) d^K y] = -k_B \int f(y) \ln\left(\frac{e^{-\lambda(y)}}{\tilde{Z}(f)}\right) d^K y \\ &= -k_B \int f(y) \ln \frac{f(y)}{\Omega(y)} d^K y = \int f(y)[S(y) - k_B \ln f(y)] d^K y, \end{aligned} \tag{6.151}$$

where first (6.147) and then (6.54) for the entropy of the generalized microcanonical ensemble on the direct level have been used. For the gradients of the generators, we obtain

$$\frac{\delta \tilde{E}(f)}{\delta f(y)} = E(y) \tag{6.152}$$

and

$$\frac{\delta \tilde{S}(f)}{\delta f(y)} = S(y) - k_B \ln f(y), \tag{6.153}$$

where unimportant additional constants could be fixed by carefully accounting for the normalization condition for f (see Appendix C.3).

The Poisson matrix on the level of probability densities is obtained from (6.67),

$$\begin{aligned} \tilde{L}_{y_1 y_2}(f) &= \int \tilde{\rho}_f(z) \frac{\partial \tilde{\Pi}_{y_1}(z)}{\partial z} \cdot L_0 \cdot \frac{\partial \tilde{\Pi}_{y_2}(z)}{\partial z} d^{6N}z \\ &= \int f(y) \frac{\delta\delta(y - y_1)}{\delta y_1} \cdot L(y) \cdot \frac{\delta\delta(y - y_2)}{\delta y_2} d^K y, \end{aligned} \tag{6.154}$$

where the chain rule and again (6.149) with an integration over y have been used. If we switch from the matrix notation (with continuous indices y_1, y_2) to an operator notation, (6.154) can be simplified to

$$\tilde{L}(f) = -\frac{\delta}{\delta y} \cdot f(y) L(y) \cdot \frac{\delta}{\delta y}. \tag{6.155}$$

With the auxiliary property

$$i\mathcal{L}\tilde{\Pi}_y(z) = -\frac{\delta\delta(\Pi(z)-y)}{\delta y}\cdot i\mathcal{L}\Pi(z), \tag{6.156}$$

the evaluation of the friction matrix (6.72) becomes very similar to the one for the Poisson matrix. We obtain

$$\begin{aligned}\tilde{M}_{y_1y_2}(f) &= \frac{1}{k_B}\int_0^{\mathcal{T}}\int \tilde{\rho}_f(z)[i\mathcal{L}\tilde{\Pi}_{y_2}(z)]\tilde{\mathcal{Q}}e^{\tilde{\mathcal{Q}}i\mathcal{L}\tilde{\mathcal{Q}}t}\tilde{\mathcal{Q}}[i\mathcal{L}\tilde{\Pi}_{y_1}(z)]d^{6N}z\,dt \\ &= \int f(y)\frac{\delta\delta(y-y_2)}{\delta y_2}\cdot M(y)\cdot\frac{\delta\delta(y-y_1)}{\delta y_1}d^K y,\end{aligned} \tag{6.157}$$

with

$$M_{jk}(y) = \frac{1}{k_B}\int_0^{\mathcal{T}}\left\langle [i\mathcal{L}\Pi_k(z)]\tilde{\mathcal{Q}}e^{\tilde{\mathcal{Q}}i\mathcal{L}\tilde{\mathcal{Q}}t}\tilde{\mathcal{Q}}[i\mathcal{L}\Pi_j(z)]\right\rangle_y dt. \tag{6.158}$$

If we again switch from the matrix notation to the operator notation, (6.157) can be simplified to

$$\tilde{M}(f) = -\frac{\delta}{\delta y}\cdot f(y)M(y)\cdot\frac{\delta}{\delta y}. \tag{6.159}$$

It is important to realize that the projection operator $\tilde{\mathcal{Q}}$ in the friction matrix (6.158) does not coincide with the projection operator $\mathcal{Q}(x)$ in (6.72), even when a generalized microcanonical ensemble on the direct level is used in both equations; this is a consequence of the richness of the level of description leading to (6.158). On the level of probability densities, on which all functions of the slow variables are automatically and consistently taken into account, and which hence rests on the most solid foundations, this projection operator $\tilde{\mathcal{Q}} = 1 - \tilde{\mathcal{P}}$ is independent to the coarse-grained state [see (6.46)].

The results for $\tilde{E}$, $\tilde{S}$, $\tilde{L}$, and $\tilde{M}$ obtained here by projection-operator techniques in terms of the building blocks of the direct level have previously been proposed on the basis of intuitive arguments.[22] Also, it was noted previously[23] that, if the degeneracy

$$L(x)\cdot\frac{\delta S(x)}{\delta x} = 0 \tag{6.160}$$

is valid on the direct level, then the corresponding degeneracy on the level of distribution functions,

$$\tilde{L}(f)\cdot\frac{\delta\tilde{S}(f)}{\delta f} = 0, \tag{6.161}$$

cannot be fulfilled rigorously. With (6.153) and (6.155), we can now write out explicitly

$$\tilde{L}(f)\cdot\frac{\delta\tilde{S}(f)}{\delta f} = \frac{\delta f(y)}{\delta y}\cdot\left[k_B\frac{\delta}{\delta y}\cdot L(y) - L(y)\cdot\frac{\delta S(y)}{\delta y}\right] - f(y)\left[\frac{\delta}{\delta y}\cdot L(y)\right]\cdot\frac{\delta S(y)}{\delta y}, \tag{6.162}$$

[22] See Section III.B of Grmela & Öttinger, Phys. Rev. E 56 (1997) 6620.

[23] See p. 6630 of Grmela & Öttinger, Phys. Rev. E 56 (1997) 6620.

where some terms vanishing due to the antisymmetry of L have been omitted. The degeneracy (6.161) thus holds rigorously if we have

$$L(x) \cdot \frac{\delta S(x)}{\delta x} = k_{\rm B} \frac{\delta}{\delta x} \cdot L(x) \tag{6.163}$$

instead of (6.160). Equation (6.163) is exactly the condition (6.76), which we previously discovered to hold for the microcanonical ensemble when small fluctuation terms are not neglected.

In conclusion, we get a nicely consistent atomistic picture behind GENERIC, including noise, when we use a generalized microcanonical ensemble on the direct level in combination with a generalized canonical ensemble on the probability density level. Three important lessons can be learned:

1. If all functions of slow variables are considered to be also slow variables, then we end up with GENERIC on the level of probability densities or, in other words, with GENERIC in the presence of fluctuations or noise.

2. For a fully consistent treatment of fluctuations, instead of the original degeneracy condition (1.4) or (6.160), one must use (6.163) on the direct level with a generalized microcanonical ensemble.

3. The projection operator $\tilde{\mathcal{Q}}$ to be used in the friction matrix is independent of x and given by

   $$[\tilde{\mathcal{Q}}A](z) = A(z) - \langle A \rangle_{\Pi(z)} , \tag{6.164}$$

   where the average is formed with a generalized microcanonical ensemble.

6.3.3 Fluctuation-Dissipation Theorem

The final time-evolution equation on the level of probability densities is obtained by combining (6.152), (6.153), (6.155), and (6.159):

$$\frac{\partial f}{\partial t} = -\frac{\delta}{\delta x} \cdot \left[\left(L \cdot \frac{\delta E}{\delta x} + M \cdot \frac{\delta S}{\delta x} \right) f \right] + k_{\rm B} \frac{\delta}{\delta x} \cdot \left[M \cdot \frac{\delta}{\delta x} f \right] . \tag{6.165}$$

This Fokker-Planck equation or diffusion equation has been given before as (1.55) in Section 1.2.5 on fluctuations. The occurrence of the friction matrix M in the second-order derivative term, describing the fluctuations in a diffusion equation, is the famous fluctuation-dissipation theorem, which has been expressed in an alternative form as an equation for the amplitude of the noise in the stochastic generalization of the fundamental GENERIC in (1.57). Following the nomenclature of Kubo et al.,[24] we refer to it as a fluctuation-dissipation theorem of the second kind because it relates friction and noise in the stochastic differential equation of motion (1.56).

[24] See Section 1.6 of Kubo et al., *Statistical Physics II* (Springer, 1991).

In conclusion, we have derived the full structure of GENERIC with noise, as discussed in Section 1.2.5, by the projection-operator technique. The position of M between the derivatives of the second-order term in (6.165) is an unambiguous result of this procedure; it is closely related to the symmetry of $\tilde{M}$ on the richer level of probability densities. The projection-operator derivation of the fluctuation-dissipation theorem of the second kind is valid even far away from equilibrium.

When fluctuations are small, we expect a sharply peaked probability density of the canonical form (6.61) to be a solution of the diffusion equation (6.165),

$$f(x) \propto \exp\left[\frac{S(x)}{k_\text{B}} - \sum_k \lambda_k x_k\right], \tag{6.166}$$

where the time dependent Lagrange multipliers λ_k correspond to the proper evolution of the averages of x_k. Indeed, if we insert this *ansatz* into (6.165), we obtain

$$\frac{\partial f}{\partial t} = -\frac{\delta}{\delta x} \cdot \left[\left(L \cdot \frac{\delta E}{\delta x} + M \cdot k_\text{B}\lambda\right) f\right]. \tag{6.167}$$

For a sharply peaked distribution, the relationship between Lagrange multipliers and entropy gradients (6.56) then implies that the probability distribution follows the standard GENERIC time evolution [the continuity equation (6.167) for the probability density expresses the fundamental GENERIC (1.1) in the same way as the Liouville equation expresses Newtonian or Hamiltonian mechanics].

When fluctuations must be taken into account, a possible strategy is to derive a time-evolution equation for the averages $\langle x \rangle$ from the stochastic differential equation (1.56) for x. In nonlinear systems, more complicated averages arise, and one is faced with the problem of moment closure. The elimination of higher moments can be achieved formally by the projection-operator techniques presented in this chapter. Renormalized evolution equations for $\langle x \rangle$ that differ from the GENERIC (1.56) for x with suppressed noise (formally achieved by setting $k_B = 0$) are obtained with this method.[25] In particular, this fluctuation renormalization introduces memory effects into the equation for $\langle x \rangle$, even though the stochastic process x itself is Markovian. Small fluctuation effects can be taken into account through an expansion in k_B.[26] We here do not further consider such approximate fluctuation renormalized equations because we feel strongly that a stochastic differential equation should be treated in its own right; efficient numerical methods are available so that there is no need to rely on equations for the averages $\langle x \rangle$.[27] If one nevertheless feels the desire to work with moment equations, the closure problem should be considered on a case-to-case basis, and it should be verified that the resulting time-evolution equations are thermodynamically consistent.

[25] See Section 7.4 of Grabert, *Projection Operators* (Springer, 1982).
[26] See Section 7.5 of Grabert, *Projection Operators* (Springer, 1982).
[27] Öttinger, *Stochastic Processes in Polymeric Fluids* (Springer, 1996).

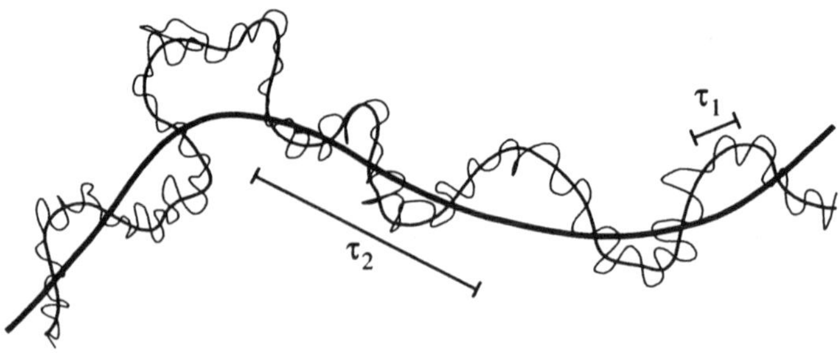

Fig. 6.4 Trajectories on three levels of description.

6.4 RELATIONSHIP BETWEEN COARSE-GRAINED LEVELS

The atomistic expressions for the building blocks E, S, L and M may be regarded as analogues of the famous expressions for the equilibrium entropy or for the Helmholtz free energy in terms of the logarithm of the partition function for beyond-equilibrium systems. The complexity of practical calculations can be reduced dramatically by starting from an intermediate level of description. For example, the viscoelastic properties of polymeric fluids are often determined from mechanical bead-rod-spring rather than truly atomistic models of polymers. Such intermediate levels play a more important role for beyond-equilibrium systems than for equilibrium systems, and they create new opportunities. This observation is related to the finite ratio of time scales separating different levels of description, which is also the origin of the more approximate nature of beyond-equilibrium thermodynamics compared with the equilibrium counterpart dealing with the uniquely determined, infinitely slow variables (see Section 1.1.1). We therefore need to formulate the rules for passing from any given level of description to a more macroscopic one.

We consider two coarse-grained levels of description, where the level with variables x and $\Pi_1(z)$, $E_1(x)$, $S_1(x)$, $L_1(x)$, $M_1(x)$ is more microscopic than the level with variables y and $\Pi_2(z)$, $E_2(y)$, $S_2(y)$, $L_2(y)$, $M_2(y)$. The respective intermediate time scales are τ_1 and τ_2, where $\tau_2 > \tau_1$ (see Figure 6.4). Both levels of coarse-graining are introduced within a generalized microcanonical ensemble because this type of ensemble is most suitable for theoretical developments. We assume that the coarsest states can be constructed from the intermediate states by a mapping Π or, equivalently for the generalized microcanonical ensemble, that $\Pi_2(z)$ can be decomposed in the form $\Pi_2(z) = \Pi(\Pi_1(z))$ (see Figure 6.5). By comparing the atomistic expressions for the building blocks of the coarse-grained levels, we can now construct transformation rules relating the GENERIC building blocks on these two levels. A direct projection of the time-evolution equation from one coarse-grained level of description to another one, both described by Fokker-Planck equations, has also been offered.[28]

[28] Español & Vázquez, Phil. Trans. R. Soc. Lond. A 360 (2002) 383.

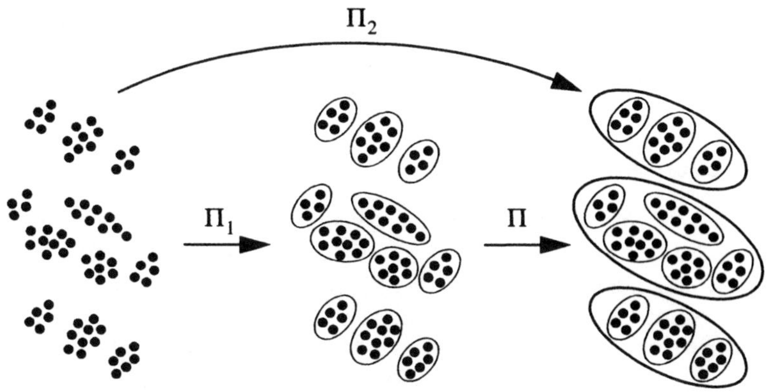

Fig. 6.5 Grouping of microstates (black circles) into macrostates (ellipses) at two different levels of coarse-graining.

6.4.1 Static Properties

Because the total energy, at least to a good approximation, must be accessible on any level of description, we should obtain $E_2(y)$ by writing $E_1(x)$ in the form

$$E_1(x) = E_2(\Pi(x)). \tag{6.168}$$

This is a direct consequence of (6.50) for the generalized microcanonical ensemble,

$$E_0(z) = E_1(\Pi_1(z)) = E_2(\Pi_2(z)) = E_2(\Pi(\Pi_1(z))). \tag{6.169}$$

For the entropy, the explicit counting of the microstates on the coarsest level according to (6.53) can be avoided by just adding the results of the previous counts for the intermediate states (see Figure 6.5 for a discrete system). In other words, we can add contributions from intermediate states with proper weight factors, and we thus obtain the following general formula for the entropy associated with the generalized microcanonical ensemble,

$$\exp\left\{\frac{S_2(y)}{k_\mathrm{B}}\right\} = \int \exp\left\{\frac{S_1(x)}{k_\mathrm{B}}\right\} \delta(\Pi(x) - y)dx. \tag{6.170}$$

In practical applications, the integral in (6.170) might be dominated by states x near the maximum of the integrand; the saddle-point approximation would then be useful for evaluating the coarse-grained entropy. Such arguments will also be important for proving the equivalence of different ensembles.

By using the chain rule for differentiating $\Pi(\Pi_1(z))$ with respect to z and doing the partial summation over intermediate states as suggested by Figure 6.5, one obtains the following transformation behavior for the Poisson matrix:

$$L_2(y) = \int \exp\left\{\frac{S_1(x) - S_2(y)}{k_\mathrm{B}}\right\} \frac{\delta\Pi(x)}{\delta x} \cdot L_1(x) \cdot \frac{\delta\Pi(x)}{\delta x} \delta(\Pi(x) - y)dx. \tag{6.171}$$

In conclusion, within the generalized microcanonical ensemble, explicit and plausible results for the transformation of the static building blocks of GENERIC between any two coarse-grained levels of description are obtained in a straightforward manner.

6.4.2 Friction Matrix

Whereas the above three relations for the successive coarse-graining of the energy, entropy, and Poisson matrix involve only static properties, the time evolution comes in when coarse-graining the friction matrix. The situation for the friction matrix can be understood best in terms of Figure 6.4. We had previously realized that the friction matrix on a coarse-grained level is determined by what is to be considered as fluctuations [see the discussion after (6.73)]. The fluctuations on the coarsest level in Figure 6.4 can be split into two contributions. First, fluctuations faster than τ_1, which have already been taken into account on the intermediate level and should be carried over as fluctuation effects to any coarser level, and second, effects that are slower than τ_1 but are newly considered as rapid fluctuations when coarse-graining to the time scale τ_2. The second contribution corresponds to additional friction associated with additional coarse-graining.

We further realize from Figure 6.4 that, if we want to construct the friction matrix on the coarsest level from the intermediate level, then we need to take into account the fluctuations on the intermediate level. We should hence describe the intermediate level with the GENERIC framework including noise.

The most fundamental starting point for the friction matrix, in particular when fluctuation effects are expected to be important, is the atomistic expression obtained on the probability density level. We hence start from the matrix $M_2(y)$ obtained from (6.158) and rewrite it by means of (6.164) in the equivalent form

$$\begin{aligned}(M_2)_{jk}(y) &= \frac{1}{k_B}\int_0^{\tau_2}\int \exp\left\{-\frac{S_2(y)}{k_B}\right\}\delta(\Pi_2(z)-y)[i\mathcal{L}\Pi_2(z)-\langle i\mathcal{L}\Pi_2\rangle_y]_k \\ &\times\ e^{\tilde{Q}i\mathcal{L}\tilde{Q}t}[i\mathcal{L}\Pi_2(z)-\langle i\mathcal{L}\Pi_2\rangle_y]_j d^{6N}z\,dt. \end{aligned} \tag{6.172}$$

By means of (6.2), (6.50), and (6.67), the exact average $\langle i\mathcal{L}\Pi_2\rangle_y$ is obtained as

$$\langle i\mathcal{L}\Pi_2\rangle_y = L_2(y)\cdot\frac{\delta E_2(y)}{\delta y}, \tag{6.173}$$

where the building blocks E_2 and L_2 are already known in terms of the properties of the intermediate level. We hence need to concentrate on $i\mathcal{L}\Pi_2(z)$, which is the rapidly fluctuating time derivative of $\Pi_2(z(t)) = \Pi[\Pi_1(z(t))]$. In the spirit of the introductory remarks, we approximate the fluctuating quantity $\Pi_1(z(t))$ by the solution $x(t)$ of the GENERIC with noise (1.56) on the intermediate level. From the Itô formula for transforming stochastic differentials,[29] we then obtain for the time

[29] See p. 98 of Öttinger, *Stochastic Processes in Polymeric Fluids* (Springer, 1996); see also Exercise 194.

derivative of $\Pi_2(z(t)) \approx \Pi(x(t))$

$$\begin{aligned} i\mathcal{L}\Pi_2 &\approx \frac{\delta \Pi}{\delta x} \cdot \left(L_1 \cdot \frac{\delta E_1}{\delta x} + M_1 \cdot \frac{\delta S_1}{\delta x} + k_B \frac{\delta}{\delta x} \cdot M_1 \right) + k_B M_1 : \frac{\delta^2 \Pi}{\delta x \delta x} \\ &+ \frac{\delta \Pi}{\delta x} \cdot B_1 \cdot \frac{dW_t}{dt}, \end{aligned} \tag{6.174}$$

where the peculiarity of stochastic calculus lies in the occurrence of the second-order derivative of Π (see also Exercise 13). By introducing the mean forward time derivative of the stochastic process $\Pi(x(t))$,[30]

$$D\Pi = \frac{\delta \Pi}{\delta x} \cdot \left(L_1 \cdot \frac{\delta E_1}{\delta x} + M_1 \cdot \frac{\delta S_1}{\delta x} + k_B \frac{\delta}{\delta x} \cdot M_1 \right) + k_B M_1 : \frac{\delta^2 \Pi}{\delta x \delta x}, \tag{6.175}$$

we can rewrite (6.174) in the more compact and transparent form

$$i\mathcal{L}\Pi_2 \approx D\Pi + \frac{\delta \Pi}{\delta x} \cdot B_1 \cdot \frac{dW_t}{dt}, \tag{6.176}$$

which reflects nicely the splitting into the intermediate scale dynamics and the rapid fluctuations in Figure 6.4.

As a word of warning, it should be pointed out that, due to the "wild nature" of white noise or Brownian motion, the time derivatives on the right-hand side of (6.176) do not exist for individual trajectories. Nevertheless, the right-hand side of (6.176) can be given a rigorous meaning within stochastic calculus and, in particular, averages can be evaluated safely. At this point, one should, however, remember that the symmetry of the friction matrix is associated with the time reversal symmetry of the correlations in (6.172). Whereas the time reversed form of the deterministic left-hand side of (6.176) is obtained in a straightforward manner, one needs to be careful with the stochastic right-hand side, to maintain the symmetry of the friction matrix. Under time reversal, the mean forward time derivative, $D\Pi$, should strictly speaking be replaced by the mean backward time derivative, $D_*\Pi$, which turns out to coincide with the mean forward derivative.[31] We hence obtain our final approximations as

$$e^{\tilde{Q}i\mathcal{L}\tilde{Q}t} i\mathcal{L}\Pi_2(z) \approx D\Pi(x(t)) + \frac{\delta \Pi(x(t))}{\delta x} \cdot B_1(x(t)) \cdot \frac{dW_t}{dt} \tag{6.177}$$

[30] See p. 80 of Nelson, *Dynamical Theories of Brownian Motion* (Princeton, 1967).

[31] See p. 95 and §13 of Nelson, *Dynamical Theories of Brownian Motion* (Princeton, 1967). The calculation of the mean forward time derivative requires an expression for $f(x)$. Contrary to Español & Vázquez, Phil. Trans. R. Soc. Lond. A 360 (2002) 383 and our previous work [Öttinger, Phys. Rev. E 57 (1998) 1416], we have assumed (6.166), or (6.22), rather than the equilibrium-motivated expression $f(x) \propto \exp[S(x)/k_B]$ in evaluating the mean backward time derivative. With the previous assumption, the mean backward time derivative turns out to be different from the forward derivative,

$$D_*\Pi = \frac{\delta \Pi}{\delta x} \cdot \left(L_1 \cdot \frac{\delta E_1}{\delta x} - M_1 \cdot \frac{\delta S_1}{\delta x} - k_B \frac{\delta}{\delta x} \cdot M_1 \right) - k_B M_1 : \frac{\delta^2 \Pi}{\delta x \delta x}.$$

and

$$i\mathcal{L}\Pi_2(z) \approx D_*\Pi(x(0)) + \frac{\delta\Pi(x(0))}{\delta x} \cdot B_1(x(0)) \cdot \frac{dW_0}{dt}, \tag{6.178}$$

where $x(t)$ is the solution of the GENERIC with noise (1.56), using the initial condition $x(0) = x = \Pi(z)$. Our goal of eliminating all dependencies on the atomistic variables z in favor of the variables x of the intermediate scale is now reached; thus, we can rewrite (6.172) as

$$\begin{aligned}(M_2)_{jk}(y) &= \frac{1}{k_B}\int_0^{\tau_2}\int \exp\left\{\frac{S_1(x) - S_2(y)}{k_B}\right\} \delta(\Pi(x) - y) \\ &\times \left\langle \left[D_*\Pi(x(0)) + \frac{\delta\Pi(x(0))}{\delta x} \cdot B_1(x(0)) \cdot \frac{dW_0}{dt} - \langle i\mathcal{L}\Pi_2\rangle_y\right]_k \right. \\ &\left. \left[D\Pi(x(t)) + \frac{\delta\Pi(x(t))}{\delta x} \cdot B_1(x(t)) \cdot \frac{dW_t}{dt} - \langle i\mathcal{L}\Pi_2\rangle_y\right]_j \right\rangle d^K x dt,\end{aligned} \tag{6.179}$$

where the pointed brackets denote an average over trajectories of the Wiener process. From the correlation of noise terms, we obtain the following contribution to the friction matrix:

$$M_2'(y) = \int \exp\left\{\frac{S_1(x) - S_2(y)}{k_B}\right\} \frac{\delta\Pi(x)}{\delta x} \cdot M_1(x) \cdot \frac{\delta\Pi(x)}{\delta x} \delta(\Pi(x) - y) d^K x, \tag{6.180}$$

which has the same form as (6.171) for the Poisson matrix and, as anticipated, corresponds to carrying over the fluctuation effects faster than τ_1. For the additional friction due to fluctuations on a time scale intermediate between τ_1 and τ_2, we obtain

$$\begin{aligned}(M_2'')_{jk}(y) &= \frac{1}{k_B}\int_0^{\tau_2}\int \exp\left\{\frac{S_1(x) - S_2(y)}{k_B}\right\} \delta(\Pi(x) - y) \\ &\times \left\langle [D_*\Pi(x(0)) - \langle i\mathcal{L}\Pi_2\rangle_y]_k [D\Pi(x(t)) - \langle i\mathcal{L}\Pi_2\rangle_y]_j \right\rangle d^K x dt.\end{aligned} \tag{6.181}$$

After we use the approximate expression (6.174) for $i\mathcal{L}\Pi_2$ to obtain (6.179) and (6.181), in these expressions, the rigorous averages $\langle i\mathcal{L}\Pi_2\rangle_y$ of (6.173) should also be replaced by the corresponding approximate averages (see Exercise 149),

$$\langle D\Pi\rangle_y = \langle D_*\Pi\rangle_y = L_2(y) \cdot \frac{\delta E_2(y)}{\delta y} + M_2'(y) \cdot \frac{\delta S_2(y)}{\delta y} + k_B \frac{\delta}{\delta y} \cdot M_2'(y). \tag{6.182}$$

The full friction matrix is given by

$$M_2(y) = M_2'(y) + M_2''(y). \tag{6.183}$$

The expressions (6.180)–(6.183) for the coarse-grained friction matrix agree with the results of a direct projection[32] between the coarse-grained levels after adapting them to a microcanonical ensemble.

As expected, the two contributions (6.180) and (6.181) to $M_2(y)$ correspond to fluctuations resulting from the fast time evolution of z compared with x and from the fast time evolution of x compared with y, respectively. If the step from Level 1 to Level 2 involves coarse-graining, then we expect $M_2'' \neq 0$. There may be reductions in the number of variables describing beyond-equilibrium states for which $M_2'' = 0$; the passage from Level 1 to Level 2 may then be considered as a "solution" or "reduction" of the Level 1 model rather than a coarse-graining procedure.

To determine M_2 by numerical simulations on the intermediate level, one generally needs Brownian dynamics simulations because $x(t)$ is obtained as the solution of a stochastic differential equation. The situation is different when one determines the friction matrix from the most microscopic level, where molecular dynamics simulations of (a projected version of) Hamilton's equations are required.

Exercise 149 Approximate Averages

Derive the approximation (6.182) to the rigorous average (6.173).

6.4.3 Generalizations

To establish the relationship between two different coarse-grained levels of description, we have so far used generalized microcanonical ensembles on both levels of description. We briefly consider two levels with $\Pi_1(z)$, $\rho_{1,x}(z)$ and $\Pi_2(z)$, $\rho_{2,y}(z)$ as the respective relevant variables and ensembles. If we want to introduce a probability density $\rho_y(x)$ for finding a value x for given y, this is most naturally done through the implicit equation

$$\rho_{2,y}(z) = \int \rho_y(x)\rho_{1,x}(z)d^K x. \tag{6.184}$$

For (6.184) to be meaningful, it must be possible to decompose $\Pi_2(z)$ in the form $\Pi_2(z) = \Pi(\Pi_1(z))$.

For arbitrary choices of $\rho_{1,x}(z)$ and $\rho_{2,y}(z)$, the defining equation (6.184) may not possess a solution, or one can find only an approximation. The most convenient and important situation arises when $\rho_{1,x}(z)$ is a generalized microcanonical ensemble,

$$\rho_{1,x}(z) = \frac{1}{\Omega_1(x)}\delta(\Pi_1(z) - x). \tag{6.185}$$

We then obtain the following explicit solution of (6.184),

$$\rho_y(x) = \Omega_1(x)\hat{\rho}_{2,y}(\Pi(x)), \tag{6.186}$$

[32] Español & Vázquez, Phil. Trans. R. Soc. Lond. A 360 (2002) 383.

where the fact that $\rho_{2,y}(z)$ depends on z only through $\Pi_2(z)$,

$$\rho_{2,y}(z) = \hat{\rho}_{2,y}(\Pi_2(z)), \tag{6.187}$$

has been used. If also $\rho_{2,y}(z)$ is of the generalized microcanonical type, we obtain

$$\rho_y(x) = \frac{\Omega_1(x)}{\Omega_2(y)}\delta(\Pi(x) - y) = \exp\left\{\frac{S_1(x) - S_2(y)}{k_\mathrm{B}}\right\}\delta(\Pi(x) - y), \tag{6.188}$$

which occurs frequently in the expressions of the previous subsections. If $\rho_{2,y}(z)$ is of the generalized canonical type, the Lagrange multipliers of this ensemble occur in $\rho_y(x)$ according to (6.186) without any change (see Exercise 150).

When (6.184) is introduced into (6.48) and (6.67), we immediately obtain

$$E_2(y) = \int \rho_y(x) E_1(x) d^K x, \tag{6.189}$$

and, after using the chain rule,

$$L_2(y) = \int \rho_y(x) \frac{\delta\Pi(x)}{\delta x} \cdot L_1(x) \cdot \frac{\delta\Pi(x)}{\delta x} d^K x, \tag{6.190}$$

which are the generalizations of (6.168) and (6.171). It is natural to use a general $\rho_y(x)$ instead of (6.188) also in the contributions (6.180) and (6.181) to the friction matrix. The explicit form of the entropy depends on the type of the example. For a generalized canonical ensemble $\rho_{2,y}(z)$, the entropy is

$$\frac{S_2(y)}{k_\mathrm{B}} = \ln\left\{\int \Omega_1(x)\exp\left[-\sum_k \lambda_k(y)\Pi_k(x)\right] d^K x\right\} + \sum_k \lambda_k(y) y_k, \tag{6.191}$$

which is a generalization of (6.55). This result for the generalized canonical ensemble can be rewritten in the elegant form

$$S_2(y) = \int \rho_y(x)[S_1(x) - k_\mathrm{B} \ln \rho_y(x)] d^K x, \tag{6.192}$$

which can be generalized to mixed ensembles. According to (6.18), we consider the situation where the variables $(y, \tilde{y})$ of the coarsest level are partially controlled by Lagrange multipliers (y) and partially fixed by δ-functions $(\tilde{y})$. Because the microcanonically treated variables $\tilde{y}$ must also be available on the intermediate level, we assume a generalized microcanonical ensemble for $(x, \tilde{x})$ with $\tilde{y} = \tilde{x}$ and $y = \pi(x)$. We then find

$$\rho_{(y,\tilde{y})}(x, \tilde{x}) = \rho_{(y,\tilde{y})}(x)\,\delta(\tilde{x} - \tilde{y}) \tag{6.193}$$

with

$$\rho_{(y,\tilde{y})}(x) = \frac{\Omega_1(x, \tilde{y}) \exp\left[-\sum_k \lambda_k(y, \tilde{y})\Pi_k(x)\right]}{\int \Omega_1(x', \tilde{y}) \exp\left[-\sum_k \lambda_k(y, \tilde{y})\Pi_k(x')\right] d^K x'} \tag{6.194}$$

and

$$\Omega_1(x,\tilde{x}) = \int \delta(\Pi_1(z) - x)\delta(\tilde{\Pi}_1(z) - \tilde{x})d^{6N}z. \tag{6.195}$$

The elegant entropy formula (6.192) can now be generalized to mixed ensembles:

$$S_2(y,\tilde{y}) = \int \rho_{(y,\tilde{y})}(x)[S_1(x,\tilde{y}) - k_B \ln \rho_{(y,\tilde{y})}(x)]d^K x. \tag{6.196}$$

The additional entropy in coarse-graining is associated with the passage from x to y; the variables $\tilde{x} = \tilde{y}$, however, are taken over without affecting entropy.

Exercise 150 Canonical Ensemble on Intermediate Level

Give the explicit form of $\rho_y(x)$ when $\rho_{1,x}(z)$ is a generalized microcanonical ensemble and $\rho_{2,y}(z)$ is of the generalized canonical type.

Exercise 151 Entropy of a Random Walk

Consider a random-walk model of polymers with the end-to-end vector $\boldsymbol{Q} = x$ and the number of steps $N = \tilde{x}$. Assume that the entropy per random walk on this (intermediate) level of description is given by Gaussian statistics,[33]

$$S_1(\boldsymbol{Q},N) = k_B\left[N\ln c - \frac{3}{2}\frac{\boldsymbol{Q}^2}{Na^2} - \frac{3}{2}\ln\left(\frac{2\pi}{3}Na^2\right)\right], \tag{6.197}$$

where the constant parameters a and c are the lattice spacing and the coordination number of the underlying lattice. For $\boldsymbol{\Pi}(x) = \boldsymbol{QQ}$, evaluate and discuss the entropy (6.196) on the configuration-tensor level.

[33] See, for example, Section I.1.1 of de Gennes, *Scaling Concepts in Polymer Physics* (Cornell, 1979).

7

Kinetic Theory of Gases

Chapter 7 deals with the kinetic theory of rarefied gases of monatomic molecules and focuses strongly on Boltzmann's kinetic equation. We first present an elementary kinetic theory approach so that the reader can develop a feeling for the mechanical model of a monatomic gas, the relevant length scales, and the phenomena that can be explained. Transport coefficients are obtained in terms of the mean free path. Also Boltzmann's kinetic equation is motivated in an elementary way, before the systematic projection-operator technique is used to derive the GENERIC building blocks for Boltzmann's equation. The Chapman-Enskog expansion and Grad's moment method for constructing approximate solutions to Boltzmann's equation are described, and a few results are presented. We also show how the equations of hydrodynamics (see Chapter 2) and extended irreversible thermodynamics (see Section 5.1.6) can be derived from Boltzmann's equation and how transport coefficients are expressed in terms of the differential cross section for particle interactions. Finally, a GENERIC version of Grad's moment expansion is presented for the first time.

The scope of this chapter is restricted to the presentation of some basic ideas, equations, predictions, and conclusions. Illustration of the abstract coarse-graining procedure of Chapter 6 is a primary goal. We by no means attempt to give a comprehensive review of the kinetic theory of gases. Much more about the underlying principles and philosophical subtleties can be found in a famous handbook article by Grad,[1] and a wealth of information on the kinetic theory of gases is contained in a classical book by Hirschfelder, Curtiss, and Bird.[2]

[1] Grad, Principles of the Kinetic Theory of Gases (1958).

[2] Hirschfelder, Curtiss & Bird, *Molecular Theory of Gases and Liquids* (Wiley, 1954).

7.1 ELEMENTARY KINETIC THEORY

This chapter begins with material from a physics text book that I used in school when I was 16. These elementary remarks should help to provide or to refresh an understanding for the properties of rarefied gases, thus providing a background for the more sophisticated developments of the subsequent sections. The mean free path is recognized as a key concept, and it leads directly to the functional forms of the transport coefficients, that is, of the diffusion coefficient, of the viscosity, and of the thermal conductivity of gases. This section concludes with a brief glance at Boltzmann's kinetic equation, which implies an increasing entropy until equilibrium is reached.

7.1.1 Model of a Rarefied Gas

A systematic kinetic theory can be developed for rarefied gases. We assume that the gas consists of N small structureless spherical particles of diameter d and mass m, which obey the laws of classical mechanics. We make the following assumptions.

1. The mean distance between particles is so large compared with their diameter that, to obtain equilibrium properties, they can be treated as point particles.
2. There are no long-range interactions; the particles of the gas interact only through short-range collisions.
3. The conservation laws for momentum and energy can be used for the elastic collisions between particles and for the reflections of particles at the walls of the container.

More precisely, it is possible to fade out one aspect of the interaction between particles, namely, nonideal behavior of the gas; the effect of collisions on the temporal evolution of the gas, however, remains finite and is described by Boltzmann's kinetic equation.[3] Such a situation can be realized in the following limit. We assume $d \to 0$ and $N \to \infty$ such that the effect of collisions between the particles of decreasing size remains important, which translates into $Nd^2 = \text{constant}$ (see Section 7.1.2). In this limit, we have $Nd^3 \to 0$, so that the volume fraction of the spherical particles in a container of fixed volume vanishes, and ideal gas behavior is thus approached. If one wants to keep the total mass constant, $Nm = \text{constant}$, then the mass density of the particles must diverge as $1/d$. This formal limit, which is useful for justifying the validity of Boltzmann's kinetic equation in a rigorous manner, is known as the *Grad limit*. In the alternative limit $Nd^2 \to 0$, at least formally, collisions become unimportant. This limit is known as the Knudsen gas, that is, a gas of particles moving independently without any collisions, except at the walls of the container.[4]

[3] See p. 214–216 of Grad, Principles of the Kinetic Theory of Gases (1958).

[4] One should consider a Knudsen gas with stochastic boundary conditions because, with specular reflections at the walls, collisions between gas particles will always dominate the properties of the gas, given enough time.

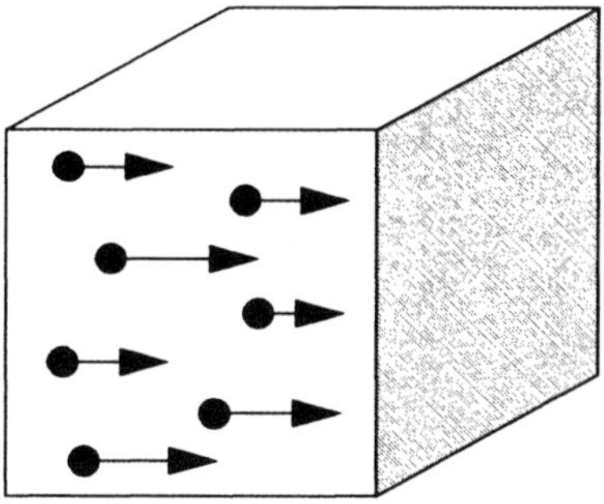

Fig. 7.1 Gas particles with a positive velocity component toward the right wall of a cubic container (particles with a negative velocity component are not shown).

To illustrate the ideal gas behavior of our hard-sphere gas, we calculate the pressure on the walls of a container (see Figure 7.1). If N_j is the number of particles that has a positive velocity component v_j toward the right wall, the fraction $v_j \Delta t / L_g$ of those particles will be reflected at the wall during the time interval Δt, where L_g denotes the size of the cubic container. The momentum transfer to the wall by a reflected particle with velocity v_j is $2mv_j$. The total momentum transferred to the wall per unit time, divided by the area of the wall, then gives the pressure

$$p = \frac{1}{L_g^2} \sum_j 2mv_j \frac{v_j}{L_g} N_j = \frac{2m}{L_g^3} \sum_j N_j v_j^2. \tag{7.1}$$

If we note that $\sum_j N_j = N/2$ because only half of the particles have a positive velocity component toward the right wall and that, on average, the square of each velocity component is one third of the square of the full velocity vector in three-dimensional space, we obtain

$$p = \frac{N}{V} m \frac{1}{3} \left\langle \boldsymbol{v}^2 \right\rangle = \frac{2}{3} \frac{N}{V} \frac{1}{2} m \left\langle \boldsymbol{v}^2 \right\rangle \tag{7.2}$$

or the *law of Boyle and Mariotte* in the form

$$pV = \frac{2}{3} E. \tag{7.3}$$

An alternative implication of (7.2) is

$$\left\langle \boldsymbol{v}^2 \right\rangle = \frac{3p}{\rho}, \tag{7.4}$$

which gives a root-mean-square velocity of 1840 m/s for hydrogen, 490 m/s for nitrogen, and 460 m/s for oxygen under normal conditions.

Exercise 152 Velocity of Hydrogen Molecules

Evaluate the mean-square velocity in (7.4) for a gas of hydrogen molecules under normal conditions.

Exercise 153 Average Speed

Show that, for an isotropic Gaussian velocity distribution around zero, the average speed is given by

$$\langle |\boldsymbol{v}| \rangle = \sqrt{\frac{8}{3\pi} \langle \boldsymbol{v}^2 \rangle} = \sqrt{\frac{8 k_{\mathrm{B}} T}{\pi m}}. \tag{7.5}$$

7.1.2 Mean Free Path

To calculate the pressure, we considered the particle-wall interactions. To understand transport phenomena in a gas, it is important to find the frequency of collisions between particles. Because we can determine the characteristic particle velocity from (7.4) and (7.5), we can equivalently determine the mean free path between collisions. We here focus on the functional form of the mean free path rather than prefactors.

Consider a spherical shell of radius r and thickness Δr around a particle. The number of particles in the shell is $4\pi r^2 \Delta r n$, where $n = N/V$ is the number density of particles. Because each particle blocks an area of πd^2 for the collision-free passage of a particle of equal size, the fraction of the solid angle covered by the particles in the shell, or the probability for a collision, is $\pi d^2 \Delta r n$. By accumulating the equal contributions from a number of shells with thickness Δr, the mean free path l_{mfp} is determined by the condition $\pi d^2 l_{\mathrm{mfp}} n \approx 1$ because this implies a unit probability for a collision. A more elaborate calculation for moving target particles gives a smaller result by $1/\sqrt{2}$:

$$l_{\mathrm{mfp}} = \frac{1}{\sqrt{2}\pi d^2 n} = \frac{l_{\mathrm{md}}^3}{\sqrt{2}\pi d^2}, \tag{7.6}$$

where $l_{\mathrm{md}} = n^{-1/3}$ is the characteristic distance between particles. The expression (7.6) for the mean free path was obtained by Maxwell and used by Boltzmann; Clausius calculated a slightly larger value for the mean free path [he found $3/4$ instead of the factor $1/\sqrt{2} \approx 0.71$ in (7.6)].[5]

For an ideal gas under normal conditions, one finds $l_{\mathrm{md}} = 30\,\text{Å}$. When we approximate the diameter of a diatomic gas molecule by four times the Bohr radius, $d \approx 2\,\text{Å}$, the mean free path turns out to be $0.15\,\mu\text{m}$. According to the first equality in (7.6), the Grad limit Nd^2 = constant in a fixed volume V corresponds to a constant mean free path at a negligible volume fraction of the particles.

Exercise 154 Collision Frequency

Estimate the frequency of collisions between gas particles for hydrogen under normal conditions.

[5] See p. 65 of Boltzmann, *Gastheorie, Teil I* (Barth, 1896).

7.1.3 Transport Coefficients

Various types of transport processes can occur in gases. In the balance equations of hydrodynamics (see Chapter 2), we have used empirical linear laws for the transport of mass, momentum, and energy. If we consider an inhomogeneous initial distribution of marked tracer particles in a gas of unmarked but otherwise identical particles, the distribution will even out and become uniform through a diffusion process. There is a mass transport of tracer particles, and Fick's law (2.45) characterizes the corresponding diffusion flow in terms of the concentration gradient and the self-diffusion coefficient.

If a fluid is stirred so that one part moves relative to another part, then friction between the moving parts will bring them to equilibrium, and the average velocity of the particles will become uniform throughout the gas. The quantity that is transported is the average momentum of the particles. The rate at which momentum is transported is determined by the viscosity introduced in (2.38) for Newtonian fluids.

If one part of a fluid is hotter than another part, the particles in the hot region will have a greater average kinetic energy than particles in the cooler region. Transport processes will equalize the average kinetic energies and hence the temperatures in the two regions. The quantity transported is the average kinetic energy, and the process that brings the gas to equilibrium is called heat conduction. The rate at which the temperatures in different parts of the gas are equalized is determined by the thermal conductivity introduced in Fourier's law (2.37).

It is possible to treat all of these transport processes in gases in a unified manner, and we here follow the treatment of Reichl.[6] The same arguments can be found already in Boltzmann's lectures.[7] Let us assume that $x_{\rm p} = x_{\rm p}(h)$ is the molecular property to be transported and that it varies in the vertical direction, as indicated by the dependence on the height coordinate h (see Figure 7.2). Let us draw an imaginary plane in the gas at $h = h_0$. When a particle crosses the plane, it transports the value of $x_{\rm p}$ it obtained in its last collision and transfers that to another particle in its next collision.

Let $x_{\rm p}(h + \Delta h)$ be the value of $x_{\rm p}$ transported from above across the plane. The distance Δh above the plane where the particle had its last collision is of the order of the mean free path $l_{\rm mfp}$, and the same is true for the distance Δh below the plane at which the next collision takes place (see Figure 7.2). The average number of particles crossing the plane per unit area and per unit time is of the order of $n\,\langle|\boldsymbol{v}|\rangle$, where $\langle|\boldsymbol{v}|\rangle$ is the average speed of the particles. For the flux of $x_{\rm p}$ transported in the positive h-direction per unit area and per unit time we hence obtain

$$j^x(h) \propto n\,\langle|\boldsymbol{v}|\rangle\,[x_{\rm p}(h_0 - \Delta h) - x_{\rm p}(h_0 + \Delta h)] \propto -n\,\langle|\boldsymbol{v}|\rangle\,l_{\rm mfp}\frac{dx_{\rm p}(h)}{dh}. \tag{7.7}$$

[6] See Section 13.B.4 of Reichl, *Modern Course in Statistical Physics* (University of Texas, 1980).

[7] See §11 of Boltzmann, *Gastheorie, Teil I* (Barth, 1896).

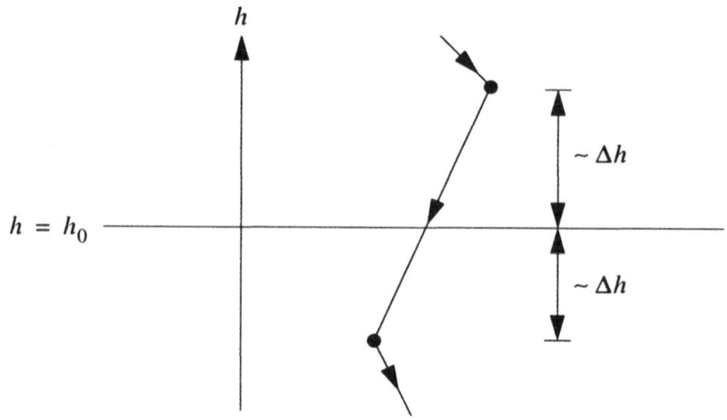

Fig. 7.2 A particle has a collision a distance Δh above the plane and transports property x_{p} to another particle a distance Δh below the plane.

The proportionality constant in (7.7) has, for example, been determined by a more detailed geometrical consideration for the diffusion coefficient;[8] the final result is

$$j^x(h) = -\frac{1}{3} n \langle |\boldsymbol{v}| \rangle l_{\mathrm{mfp}} \frac{dx_{\mathrm{p}}(h)}{dh} \tag{7.8}$$

for the flux of any molecular property x_{p}. The dimensions of the flux j^x are those of x_{p} per m^2 and s.

Let us first consider the concentration of tracer particles, $x_{\mathrm{p}}(h) = c_1(h)$. We then obtain for the number of tracer particles passing the plane per unit area and per unit time

$$j^x = -\frac{1}{3} n \langle |\boldsymbol{v}| \rangle l_{\mathrm{mfp}} \frac{dc_1}{dh}. \tag{7.9}$$

A comparison with (2.45) for the mass flux caused by diffusion results in the following expression for the *diffusion coefficient*:

$$D = \frac{1}{3} \langle |\boldsymbol{v}| \rangle l_{\mathrm{mfp}} = \frac{\langle |\boldsymbol{v}| \rangle}{3\sqrt{2}\pi d^2 n} = \frac{2}{3d^2\sqrt{m}} \frac{1}{p} \left(\frac{k_{\mathrm{B}}T}{\pi} \right)^{3/2}. \tag{7.10}$$

Quite intuitively, the diffusion coefficient increases with temperature and decreases under pressure.

If the gas in Figure 7.2 is sheared between horizontally moving plates, we choose $x_{\mathrm{p}}(h) = m \langle v_{\mathrm{horizontal}}(h) \rangle$ as the average momentum per particle. We then obtain the momentum flux

$$j^x = -\frac{1}{3} nm \langle |\boldsymbol{v}| \rangle l_{\mathrm{mfp}} \frac{d \langle v_{\mathrm{horizontal}}(h) \rangle}{dh}. \tag{7.11}$$

[8] See p. 463 of Reichl, *Modern Course in Statistical Physics* (University of Texas, 1980).

A comparison with (2.39) and (2.40) results in the following expression for the *viscosity*:

$$\eta = \frac{1}{3} nm \langle |\boldsymbol{v}| \rangle \, l_{\text{mfp}} = \frac{m \langle |\boldsymbol{v}| \rangle}{3\sqrt{2}\pi d^2} = \frac{2\sqrt{m}}{3\pi d^2} \left(\frac{k_{\text{B}}T}{\pi} \right)^{1/2}. \tag{7.12}$$

The viscosity of the gas turns out to be independent of density or pressure and increases with temperature. In 1859, when James Clerk Maxwell found this curious behavior of viscosity by using dimensional analysis and simple arguments that revealed the essential physics, he was so surprised that, together with his wife Katherine, he undertook detailed experiments in their home laboratory to confirm this "remarkable result."[9] Also, the increase of the viscosity of gases with increasing temperature has been verified experimentally. One should also note that larger particles have a shorter mean free path and hence a smaller viscosity.

If the temperature of the gas varies in the vertical direction, then we choose $x_{\text{p}}(h) = m \left\langle \boldsymbol{v}(h)^2 \right\rangle /2 = (3/2) k_{\text{B}} T(h)$, and

$$j^x = -\frac{1}{2} n \langle |\boldsymbol{v}| \rangle \, l_{\text{mfp}} k_{\text{B}} \frac{dT(h)}{dh} \tag{7.13}$$

is the heat flux. A comparison with (2.37) results in the following expression for the *thermal conductivity*:

$$\lambda^{\text{q}} = \frac{1}{2} n \langle |\boldsymbol{v}| \rangle \, l_{\text{mfp}} k_{\text{B}}. \tag{7.14}$$

The molecular expressions (7.10), (7.12), and (7.14) for the transport coefficients have been derived by very simple arguments, but they describe fairly well the observed qualitative behavior of transport coefficients in rarefied gases. If we form ratios of the transport coefficients for gases, the mean free path drops out and the ratios are given by static averages available in equilibrium thermodynamics. For example, we obtain

$$\frac{\eta}{D} = nm = \rho \tag{7.15}$$

and

$$\frac{\lambda^{\text{q}}}{\eta} = \frac{3}{2} \frac{k_{\text{B}}}{m}, \tag{7.16}$$

which is the heat capacity per unit mass at constant volume for an ideal gas. The thermal conductivity of gases thus has the same temperature dependence as the viscosity, and it is also independent of the pressure or density. However, the prefactors in the expressions (7.10), (7.12), and (7.14) should not be taken too seriously; additional factors of order unity should rather be expected in the ratios (7.15) and (7.16). Experimental results for these ratios and the predictions of a more rigorous kinetic theory are compiled in Table 7.1.3, where it should be noted that the heat capacity per unit mass at constant pressure is $\hat{c}_p = \frac{5}{2} \frac{k_{\text{B}}}{m}$ for an ideal gas.[10] The ratio in the last column of Table 7.1.3 is known as the Prandtl number.

[9] See p. xv and p. 351 of Maxwell, *Theory of Heat* (Dover, 2001).

[10] The data are taken from Table 1.2-3 of Hirschfelder, Curtiss & Bird, *Molecular Theory of Gases and Liquids* (Wiley, 1954).

Table 7.1 Universal Ratios of Transport Coefficients for Gases at 0°C

Gas	$\frac{\eta}{\rho D}$	$\frac{\eta \hat{c}_p}{\lambda^q}$
Ne	0.73	0.66
A	0.75	0.67
H_2	0.73	0.71
N_2	0.74	0.71
O_2	0.74	0.72
CO_2	0.71	0.75
CH_4	0.70	0.74
Kinetic theory	5/6	2/3
(7.15), (7.16)	1	5/3

Exercise 155 Viscosity of Nitrogen Gas
Estimate the viscosity of nitrogen gas at room temperature.

7.1.4 Boltzmann's Equation

The elementary kinetic theory considerations of the preceding subsection gave us the correct functional form of the transport coefficients for the dependence on temperature and pressure as well as on size and mass of the molecules. The prefactors, however, are not correct. To obtain quantitatively correct predictions, one needs a more rigorous kinetic theory. For rarefied gases, such a theory can be built on Boltzmann's equation. One can get not only improved results for the hard sphere gas but also can obtain the predictions resulting from a more detailed description of the interactions between molecules.

An elementary derivation of Boltzmann's kinetic equation can be based on the evolution equation for the single-particle distribution function

$$\frac{\partial f(\boldsymbol{r},\boldsymbol{p})}{\partial t} = -\left(\frac{\boldsymbol{p}}{m}\cdot\frac{\partial}{\partial \boldsymbol{r}} - \frac{\partial \phi^{(e)}(\boldsymbol{r})}{\partial \boldsymbol{r}}\cdot\frac{\partial}{\partial \boldsymbol{p}}\right) f(\boldsymbol{r},\boldsymbol{p}) + \left[\frac{\partial f(\boldsymbol{r},\boldsymbol{p})}{\partial t}\right]_{\text{coll}} . \quad (7.17)$$

The function $f(\boldsymbol{r},\boldsymbol{p})$ is the probability density for finding a particle with momentum $\boldsymbol{p}$ at the position $\boldsymbol{r}$. The first term on the right-hand side of this equation corresponds to the Liouville equation (6.6) for a single particle in an external field. The second term describes the change of the single-particle distribution function due to the occurrence of collisions with the other particles in the system.

The collision term in (7.17) consists of a gain and a loss contribution. If a particle with the given momentum $\boldsymbol{p}$ collides with another particle with any momentum $\boldsymbol{p}'$, this leads to a decrease of $f(\boldsymbol{r},\boldsymbol{p})$. If two particles with momenta $\boldsymbol{q}$ and $\boldsymbol{q}'$ collide such that one of the particles acquires the momentum $\boldsymbol{p}$, this leads to an increase of

$f(\boldsymbol{r}, \boldsymbol{p})$. For a precise formulation, we introduce the transition rate $w(\boldsymbol{q}, \boldsymbol{q}'|\boldsymbol{p}, \boldsymbol{p}')$ for a pair of particles with initial momenta $\boldsymbol{p}, \boldsymbol{p}'$ to the final momenta $\boldsymbol{q}, \boldsymbol{q}'$. For an elastic collision, $w(\boldsymbol{q}, \boldsymbol{q}'|\boldsymbol{p}, \boldsymbol{p}')$ vanishes unless momentum and energy are conserved,

$$\boldsymbol{p} + \boldsymbol{p}' = \boldsymbol{q} + \boldsymbol{q}', \tag{7.18}$$

$$\boldsymbol{p}^2 + \boldsymbol{p}'^2 = \boldsymbol{q}^2 + \boldsymbol{q}'^2. \tag{7.19}$$

The two free parameters in the choice of $\boldsymbol{q}$ and $\boldsymbol{q}'$ [six components minus four conditions (7.18), (7.19)] correspond to the solid angle into which a colliding particle is scattered. Furthermore, the transition rate $w(\boldsymbol{q}, \boldsymbol{q}'|\boldsymbol{p}, \boldsymbol{p}')$ has the following symmetry properties corresponding to time reversal and particle exchange in the two-particle collisions:

$$w(\boldsymbol{q}, \boldsymbol{q}'|\boldsymbol{p}, \boldsymbol{p}') = w(\boldsymbol{p}, \boldsymbol{p}'|\boldsymbol{q}, \boldsymbol{q}'), \tag{7.20}$$

and

$$w(\boldsymbol{q}, \boldsymbol{q}'|\boldsymbol{p}, \boldsymbol{p}') = w(\boldsymbol{q}', \boldsymbol{q}|\boldsymbol{p}', \boldsymbol{p}). \tag{7.21}$$

We can now write

$$\left[\frac{\partial f(\boldsymbol{r}, \boldsymbol{p})}{\partial t}\right]_{\text{coll}} = \int w(\boldsymbol{q}, \boldsymbol{q}'|\boldsymbol{p}, \boldsymbol{p}')[f(\boldsymbol{r}, \boldsymbol{q}) f(\boldsymbol{r}, \boldsymbol{q}') - f(\boldsymbol{r}, \boldsymbol{p}) f(\boldsymbol{r}, \boldsymbol{p}')] d^3p' d^3q d^3q', \tag{7.22}$$

and (7.17) with (7.22) is Boltzmann's famous kinetic equation. The assumption that the probability for a two-particle collision is proportional to the product $f(\boldsymbol{r}, \boldsymbol{p}) f(\boldsymbol{r}, \boldsymbol{p}')$ is Boltzmann's celebrated *Stoßzahlansatz*. It has been assumed that the collisions are local so that the same argument $\boldsymbol{r}$ appears in all single-particle distribution functions occurring in Boltzmann's equation. If $f(\boldsymbol{r}, \boldsymbol{p})$ is normalized to a number density,

$$\int f(\boldsymbol{r}, \boldsymbol{p}) d^3r d^3p = N, \tag{7.23}$$

then the transition rates for two-particle collisions must be independent of the number of particles in the system; they only depend on the impact conditions and on the interaction potential.

In the remainder of this subsection we consider two important consequences of Boltzmann's equation (7.17) and (7.22): the H theorem and the local Maxwellian distribution in velocity space. We first consider the time evolution of the functional

$$H_{\text{B}} = \int f(\boldsymbol{r}, \boldsymbol{p}) \ln f(\boldsymbol{r}, \boldsymbol{p}) d^3r d^3p. \tag{7.24}$$

We obtain

$$\frac{dH_{\text{B}}}{dt} = \int [\ln f(\boldsymbol{r}, \boldsymbol{p}) + 1] \frac{\partial f(\boldsymbol{r}, \boldsymbol{p})}{\partial t} d^3r d^3p, \tag{7.25}$$

where the constant term in the square bracket does not contribute to the integral because of the normalization (7.23) of f for all times. Actually, this observation provides also the justification for not making f dimensionless before taking the logarithm—a strictly speaking necessary factor under the logarithm would appear as a further additive constant in the square brackets. The noncollision term in (7.17) leads to total derivatives under the integral and hence to surface terms, which can be neglected for an isolated system, so that we arrive at

$$\frac{dH_{\mathrm{B}}}{dt} = \int \left[\frac{\partial f(\boldsymbol{r},\boldsymbol{p})}{\partial t} \right]_{\mathrm{coll}} \ln f(\boldsymbol{r},\boldsymbol{p}) d^3r d^3p. \tag{7.26}$$

By using (7.22) and the symmetry properties (7.20), (7.21), we obtain

$$\begin{aligned} \frac{dH_{\mathrm{B}}}{dt} &= -\frac{1}{4} \int w(\boldsymbol{q},\boldsymbol{q}'|\boldsymbol{p},\boldsymbol{p}')[f(\boldsymbol{r},\boldsymbol{q})f(\boldsymbol{r},\boldsymbol{q}') - f(\boldsymbol{r},\boldsymbol{p})f(\boldsymbol{r},\boldsymbol{p}')] \\ &\times \{\ln[f(\boldsymbol{r},\boldsymbol{q})f(\boldsymbol{r},\boldsymbol{q}')] - \ln[f(\boldsymbol{r},\boldsymbol{p})f(\boldsymbol{r},\boldsymbol{p}')]\} d^3r d^3p d^3p' d^3q d^3q', \end{aligned} \tag{7.27}$$

and hence

$$\frac{dH_{\mathrm{B}}}{dt} \leq 0. \tag{7.28}$$

The final inequality results from the fact that $w(\boldsymbol{q},\boldsymbol{q}'|\boldsymbol{p},\boldsymbol{p}')$ is nonnegative, and, in view of the monotonic behavior of the logarithm, the two remaining factors involving differences must have the same sign.

We have thus found a functional H_{B} that can only decrease or stay constant in the course of the time evolution described by Boltzmann's equation; (7.28) is *Boltzmann's H theorem*. It suggests that, as an increasing function with proper dimensions, $S = -k_{\mathrm{B}} H_{\mathrm{B}}$ plays the role of an entropy. Furthermore, Boltzmann's equation clearly describes an irreversible time evolution.

A steady state in the evolution of the single-particle distribution function can be reached when $dH_{\mathrm{B}}/dt = 0$; at this, collisions establish an equilibration in momentum space at each position $\boldsymbol{r}$. In such a situation, we obtain

$$f(\boldsymbol{r},\boldsymbol{p})f(\boldsymbol{r},\boldsymbol{p}') = f(\boldsymbol{r},\boldsymbol{q})f(\boldsymbol{r},\boldsymbol{q}'), \tag{7.29}$$

where the momenta $\boldsymbol{p}$ and $\boldsymbol{p}'$ are arbitrary and the choice of $\boldsymbol{q}$ and $\boldsymbol{q}'$ is restricted by the conservation laws (7.18), (7.19). Equation (7.29) can be rewritten as an additive conservation law:

$$\ln f(\boldsymbol{r},\boldsymbol{p}) + \ln f(\boldsymbol{r},\boldsymbol{p}') = \ln f(\boldsymbol{r},\boldsymbol{q}) + \ln f(\boldsymbol{r},\boldsymbol{q}'). \tag{7.30}$$

Because momentum and energy are the only conserved quantities restricting the choice of $\boldsymbol{q}$ and $\boldsymbol{q}'$, we conclude

$$\ln f(\boldsymbol{r},\boldsymbol{p}) = \lambda_0(\boldsymbol{r}) + \boldsymbol{\lambda}_1(\boldsymbol{r}) \cdot \boldsymbol{p} + \lambda_2(\boldsymbol{r})\boldsymbol{p}^2, \tag{7.31}$$

which implies a Gaussian distribution in momentum space. The interpretation of the factors λ_0, $\boldsymbol{\lambda}_1$, λ_2 becomes more obvious if we rewrite (7.31) as

$$f^{\mathrm{Maxw}}_{\rho,\boldsymbol{v},T}(\boldsymbol{r},\boldsymbol{p}) = \frac{\rho(\boldsymbol{r})}{m\sqrt{2\pi m k_{\mathrm{B}} T(\boldsymbol{r})}^3} \exp\left\{ -\frac{[\boldsymbol{p} - m\boldsymbol{v}(\boldsymbol{r})]^2}{2mk_{\mathrm{B}}T(\boldsymbol{r})} \right\}, \tag{7.32}$$

which is known as the *Maxwellian momentum distribution*. To get the interpretation of $\rho(\boldsymbol{r})$, $\boldsymbol{v}(\boldsymbol{r})$, and $T(\boldsymbol{r})$ as the mass density, velocity, and temperature, these fields are expressed as moments in Exercise 156. The H theorem and this Maxwellian distribution are immediate but important consequences of Boltzmann's equation.

The general role of Boltzmann's kinetic equation has been described very clearly by Grad:[11]

> The principal use of the Boltzmann equation is when there are significant changes in gas flow properties over a mean-free-path or over a collision time. There is no reason to doubt its validity in shock waves of arbitrary strength or in sound waves of arbitrary frequency; it is, indeed, the only way of studying such problems. From this viewpoint we must consider the use of the Boltzmann equation to compute the values of transport coefficients as a secondary application.

A further important use of the Boltzmann equation is to understand the origin of irreversible behavior on coarse-grained levels of description. There exists a rigorous result by Lanford based on a comparison of series expansions for hierarchies of multi-particle distribution functions,[12] showing that the Boltzmann equation is exact for a hard-sphere gas in the Grad limit, at least for an initial time interval of the order of a fraction of the collision time. This time interval is large enough for Boltzmann's H function to decrease by a strictly positive amount. Lanford's "results show unambiguously that there is no contradiction between the reversibility of molecular dynamics and the irreversibility implied by the H-Theorem."[13] The irreversibility arises in passing to the Grad limit ($d \to 0$), and *not* through the assumed factorization of multi-particle distribution functions. To understand the logical possibility of irreversibility it is important to also realize that the convergence to be assumed for the initial distributions (uniform on compact sets) is stronger than the convergence obtained at later times (pointwise almost everywhere). This last remark should also be taken as a serious warning: before a discussion about the the existence and the properties of the Boltzmann equation in a mathematically rigorous manner can occur, one should specify the appropriate space of single-particle distribution functions, including the concepts of convergence and proximity, in which one looks for solutions.

Exercise 156 Moments of Maxwellian Distribution
Evaluate the integrals

$$\int f^{\rm Maxw}_{\rho,\boldsymbol{v},T}(\boldsymbol{r},\boldsymbol{p})d^3p, \quad \int \boldsymbol{p} f^{\rm Maxw}_{\rho,\boldsymbol{v},T}(\boldsymbol{r},\boldsymbol{p})d^3p, \quad \int \frac{[\boldsymbol{p}-m\boldsymbol{v}(\boldsymbol{r})]^2}{2m} f^{\rm Maxw}_{\rho,\boldsymbol{v},T}(\boldsymbol{r},\boldsymbol{p})d^3p.$$

[11] See p. 214 of Grad, Principles of the Kinetic Theory of Gases (1958).
[12] See Chapter 7 of Lanford, Lecture Notes in Physics 38 (1975) 1.
[13] See p. 99 of Lanford, Lecture Notes in Physics 38 (1975) 1.

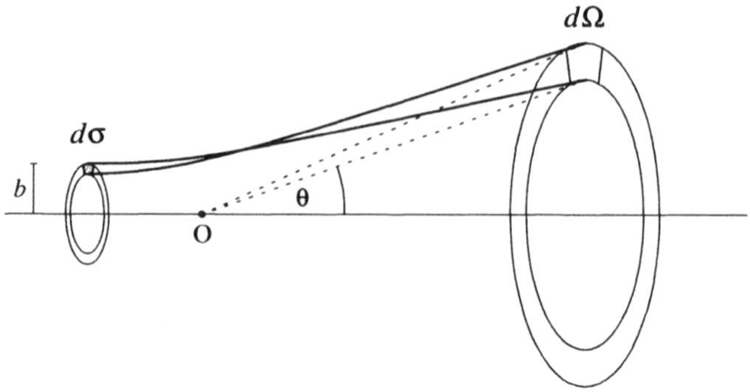

Fig. 7.3 Illustration of the differential cross section for two colliding particles in the center-of-mass frame or for a particle scattered at a fixed potential centered at the origin.

Exercise 157 More General H Functions
Show that, in the absence of external forces, a term

$$\Delta H_{\mathrm{B}} = \int [\lambda_0 + \boldsymbol{\lambda}_1 \cdot \boldsymbol{p} + \lambda_2 \boldsymbol{p}^2] f(\boldsymbol{r}, \boldsymbol{p}) d^3r d^3p$$

with arbitrary factors λ_0, $\boldsymbol{\lambda}_1$, λ_2 independent of time, position and momentum can be added to Boltzmann's H function (7.24) without affecting the rate of change (7.27).

7.1.5 Differential Cross Section for Collisions

If we wish to obtain detailed predictions for transport coefficients from Boltzmann's equation, or if we wish to treat flow problems inaccessible to a hydrodynamic description with that kinetic equation, then we need to specify the transition rates $w(\boldsymbol{q}, \boldsymbol{q}'|\boldsymbol{p}, \boldsymbol{p}')$ in the collision term (7.22) in more detail. This specification can be done in terms of the differential cross section for particle collisions (see Exercise 171 for the explicit relationship), and we thus need to consider this concept in more detail.

The primary purpose of the concept of the differential cross section is the description of scattering experiments. For an incoming beam of particles hitting a target, the differential cross section describes the angular distribution of scattered particles. A properly normalized definition can be formulated as

$$\frac{d\sigma}{d\Omega} = \frac{\text{number of scattered particles per time and per unit solid angle}}{\text{number of incident particles per time and per unit area}}. \tag{7.33}$$

Although this definition appears to be almost like an instruction for measurements, $d\sigma/d\Omega$ can also be obtained directly by theory, so that the differential cross section actually provides a beautiful interface between theory and experiment. The definition (7.33) implies that the differential cross section is the quotient formed out of an area and a solid angle, namely, the area $d\sigma$ to be hit by a particle in order to be scattered into the solid angle $d\Omega$ (see Figure 7.3). From the theoretical point of view, it is

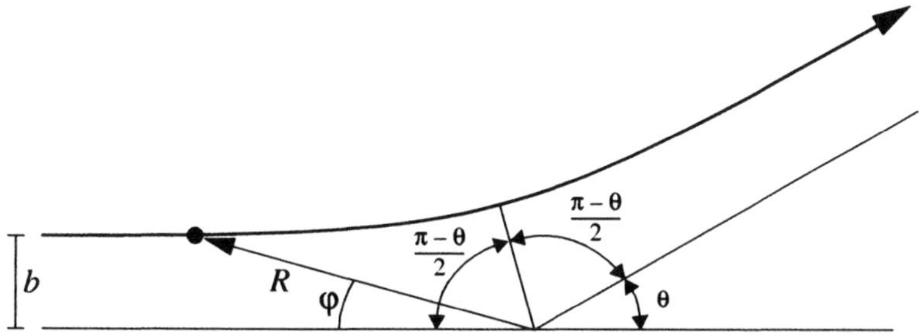

Fig. 7.4 Polar coordinates for describing a scattering process.

hence important to construct the possible trajectories for the interaction between the scattered particles and the target. If we can calculate the scattering angle θ for a given impact parameter b (see Figure 7.3), then the differential cross section is obtained as

$$\frac{d\sigma}{d\Omega} = \left| \frac{b db d\varphi}{\sin\theta d\theta d\varphi} \right| = \frac{b}{\sin\theta} \left| \frac{db}{d\theta} \right|, \tag{7.34}$$

where the azimuthal angle φ is irrelevant for central interactions. An increasing impact parameter usually leads to a decreasing scattering angle so that $db/d\theta$ is negative.

The scattering process is most transparent when it occurs at a fixed target or for the scattering of a light particle by a heavy target. However, even if the target is at rest initially, it is accelerated by the interaction with the scattered particle. In a gas, both particles of an interacting pair have equal weight, and they are typically both moving even before the collision. This situation can be handled most conveniently by considering the relative motion of the particles or, in other words, by considering the scattering process in the center-of-mass frame. The relative motion of the interacting particles is mathematically equivalent to the motion of a single particle in a fixed force field. If we start with the equations of motion for two particles with masses m_1, m_2 and interaction potential $\phi(\boldsymbol{r}_2 - \boldsymbol{r}_1)$,

$$m_1 \frac{d^2\boldsymbol{r}_1}{dt^2} = -\frac{\partial\phi(\boldsymbol{r}_2 - \boldsymbol{r}_1)}{\partial\boldsymbol{r}_1}, \qquad m_2 \frac{d^2\boldsymbol{r}_2}{dt^2} = -\frac{\partial\phi(\boldsymbol{r}_2 - \boldsymbol{r}_1)}{\partial\boldsymbol{r}_2}, \tag{7.35}$$

then we obtain the following equation for $\boldsymbol{R} = \boldsymbol{r}_2 - \boldsymbol{r}_1$:

$$\frac{d^2\boldsymbol{R}}{dt^2} = -\frac{1}{m_2}\frac{\partial\phi(\boldsymbol{r}_2 - \boldsymbol{r}_1)}{\partial\boldsymbol{r}_2} + \frac{1}{m_1}\frac{\partial\phi(\boldsymbol{r}_2 - \boldsymbol{r}_1)}{\partial\boldsymbol{r}_1} = -\left(\frac{1}{m_1} + \frac{1}{m_2}\right)\frac{\partial\phi(\boldsymbol{R})}{\partial\boldsymbol{R}} \tag{7.36}$$

or

$$\frac{m_1 m_2}{m_1 + m_2}\frac{d^2\boldsymbol{R}}{dt^2} = -\frac{\partial\phi(\boldsymbol{R})}{\partial\boldsymbol{R}}. \tag{7.37}$$

For particles with equal masses, $m_1 = m_2 = m$, the *reduced mass* $m_r = m_1 m_2/(m_1 + m_2)$ occurring in (7.37) is equal to $m/2$.

After reducing the two-particle collision problem to a one-particle problem, we next show how to reduce it even further to the one-dimensional motion of a single particle. This further simplification is obtained for any central force, that is, for any potential $\phi(R)$, depending only on the distance R between the scattered particle and the target. We use polar coordinates R and φ in the plane in which the motion of the particle with reduced mass m_r takes place (see Figure 7.4). The equations for the planar motion are obtained from the conservation of energy and angular momentum,

$$\frac{1}{2}m_r\left[\left(\frac{dR}{dt}\right)^2 + R^2\left(\frac{d\varphi}{dt}\right)^2\right] + \phi(R) = \frac{P^2}{2m_r} \tag{7.38}$$

and

$$m_r R^2 \frac{d\varphi}{dt} = bP, \tag{7.39}$$

where P is the magnitude of the initial momentum, $\boldsymbol{P} = m_r d\boldsymbol{R}/dt$, and b is the impact parameter. These are the equations of motion for the system and give a complete dynamical description of $R(t)$ and $\varphi(t)$ in terms of the interaction potential and the two parameters P and b characterizing the collision. By combining (7.38) and (7.39), we obtain

$$\frac{1}{2}m_r\left(\frac{dR}{dt}\right)^2 + \frac{P^2}{2m_r}\frac{b^2}{R^2} + \phi(R) = \frac{P^2}{2m_r}. \tag{7.40}$$

Because (7.40) contains no terms in φ, it can be regarded as a description of the one-dimensional motion of a particle in an effective potential,

$$\phi_{\text{eff}}(R) = \phi(R) + \frac{P^2}{2m_r}\frac{b^2}{R^2}, \tag{7.41}$$

which contains a centrifugal contribution in addition to the interaction potential.

According to Figure 7.4, the scattering angle θ can be obtained via $(\pi - \theta)/2$ from the position of the scattered particle at the closest approach to the target. The condition of minimum distance is given by $dR/dt = 0$, and, from (7.40), we hence obtain

$$\phi_{\text{eff}}(R_{\text{min}}) = \frac{P^2}{2m_r}. \tag{7.42}$$

Once R_{min} is determined from this condition, we integrate a combination of (7.39) and (7.40),

$$d\varphi = \frac{-bP}{R^2\sqrt{P^2 - 2m_r\phi_{\text{eff}}(R)}}dR, \tag{7.43}$$

to obtain

$$\theta = \pi - 2bP\int_{R_{\text{min}}}^{\infty}\frac{1}{R^2\sqrt{P^2 - 2m_r\phi_{\text{eff}}(R)}}dR. \tag{7.44}$$

Equation (7.44) provides the fundamental relationship $\theta = \theta(b, P)$ for a given interaction potential. If this relationship is inverted, then (7.34) gives the differential cross section in terms of $b(\theta, P)$.

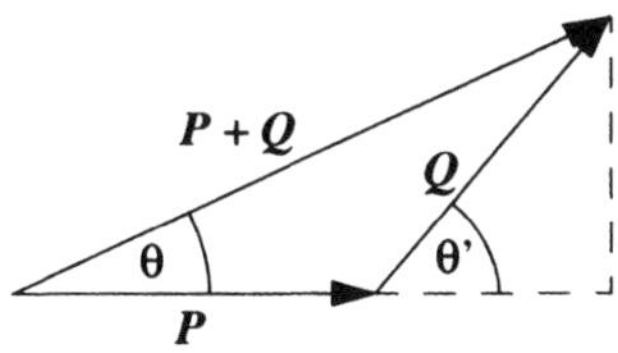

Fig. 7.5 Relationship between the scattering angles θ and θ' in the laboratory and center-of-mass frames.

We finally remark that the scattering angle and hence also the differential cross section depend on the reference frame. An example of a transformation of a differential cross section is considered in Exercise 158.

Exercise 158 Scattering Angle in Laboratory Frame

Consider the collision between two particles of equal mass in the laboratory and center-of-mass frames and denote the respective scattering angles by θ and θ'. How are the scattering angles θ and θ' for Particle 2 related when, before the collision, Particle 1 is at rest in the laboratory frame? [Hint: see Figure 7.5.]

Exercise 159 Collision for Power-Law Potential

Discuss the differential cross section, $d\sigma/d\Omega$, for the scattering of a particle with mass $m_{\rm r}$ at a fixed power-law potential

$$\phi(R) = \frac{\hat{\phi}}{R^{n_\phi - 1}}. \tag{7.45}$$

Determine the dependence of $d\sigma/d\Omega$ on $m_{\rm r}$, on $\hat{\phi}$, and on the magnitude of the initial momentum, P. Calculate the explicit dependence of the scattering angle on the impact parameter for $n_\phi = 5$.

Exercise 160 Geometric Approach to Collision of Hard Spheres

Calculate the differential cross section, $d\sigma'/d\Omega'$, in the center-of-mass frame for the collision of two hard spheres of diameter d sketched in Figure 7.6. Calculate the total cross section by integrating over all solid angles and interpret the result.

Exercise 161 A Check for Hard Spheres

Calculate $b(\theta)$ for hard spheres by means of (7.44) and compare the result with that of the geometrical consideration in Exercise 160.

7.2 PROJECTION-OPERATOR APPROACH

In this section, the power of the atomistically founded GENERIC approach to beyond-equilibrium systems developed in Chapter 6 is illustrated by coarse-graining from the full N-particle phase space to the level of single-particle distribution functions.

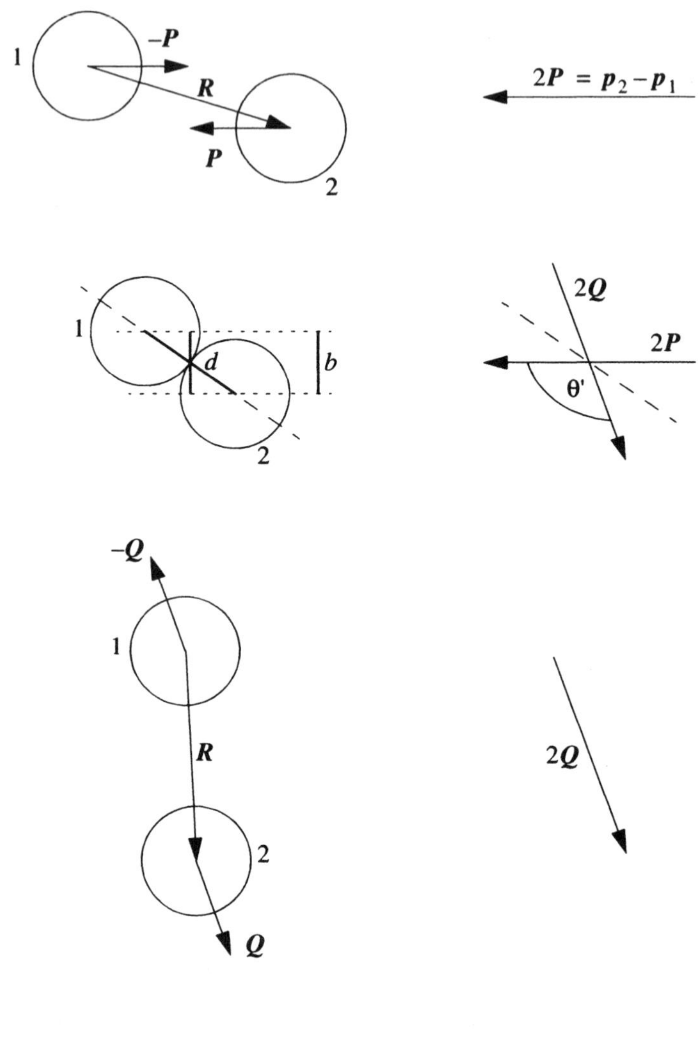

Fig. 7.6 Collision of hard spheres.

For a rarefied gas treated in the generalized canonical ensemble, the elements of GENERIC can be evaluated without any fundamental assumptions, and Boltzmann's famous kinetic equation is obtained when the final time-evolution equation is written. However, the validity of the level of single-particle distribution functions is crucial.

The derivation of Boltzmann's kinetic equation in this section is interesting not only as an illustration of the statistically founded GENERIC. It is remarkable that the derivation of the equation in this section does not require Boltzmann's famous *Stoßzahlansatz* nor any other assumption leading from a reversible to an irreversible equation. Although this might be surprising or even irritating at first sight, it should be clear, however, that the proper irreversibility has been introduced through the separation-of-time-scales assumption in the projection-operator derivation of the statistical expressions for the GENERIC building blocks. Boltzmann's equation must hence arise from a straightforward evaluation of these building blocks without any fundamental assumptions, except that the single-particle distribution function constitutes a meaningful level of description for a rarefied gas. Single-particle distribution functions should indeed provide an excellent level of description for sufficiently rarefied gases due to the fact that collisions are very rare, and thus two-particle effects are absent at almost all times. The original attempt at performing a direct projection-operator calculation[14] was strongly encouraged by work of Curtiss and Bird.[15]

For the projection-operator derivation of the famous Boltzmann equation for rarefied gases (see, for example, the articles by Bogolyubov[16] and by Lewis[17] or the book by Cercignani[18] for alternatives), we here start from Hamilton's equations of motion for an N particle system, and we use the generalized canonical approach to GENERIC summarized in Section 6.1. The potential energy consists of contributions from pairwise particle interactions, $\phi_{jk} = \phi(|\boldsymbol{r}_j - \boldsymbol{r}_k|)$ for $j \neq k$, and from external forces, $\phi^{(e)}(\boldsymbol{r}_j)$. For convenience, we set $\phi_{jk} = 0$ for $j = k$ to suppress self-interactions. The total energy is thus given by

$$E_0(z) = \sum_{j=1}^{N} \left[\frac{\boldsymbol{p}_j^2}{2m} + \phi^{(e)}(\boldsymbol{r}_j) \right] + \frac{1}{2} \sum_{j,k=1}^{N} \phi(|\boldsymbol{r}_j - \boldsymbol{r}_k|). \tag{7.46}$$

For the Boltzmann equation, the coarse-grained level of description is given by the single-particle distribution function, $f(\boldsymbol{r}, \boldsymbol{p})$, to be obtained by averaging the atomistically defined distribution function,

$$f^{\Pi}(\boldsymbol{r}, \boldsymbol{p}; z) = \sum_{j=1}^{N} \delta(\boldsymbol{r} - \boldsymbol{r}_j)\delta(\boldsymbol{p} - \boldsymbol{p}_j). \tag{7.47}$$

[14] The work summarized in this section was motivated by the wish to present a balanced picture of beyond-equilibrium thermodynamics in this book. The results have previously been published in Öttinger, J. Non-Equilib. Thermodyn. 27 (2002) 105.

[15] See Section IV.B of Curtiss & Bird, J. Chem. Phys. 111 (1999) 10362.

[16] Bogolyubov, Dynamical Theory in Statistical Physics (1962).

[17] Lewis, J. Math. Phys. 8 (1967) 1448.

[18] Cercignani, *Boltzmann Equation* (Springer, 1988).

This single-particle distribution function is normalized to the total number of particles. For more general kinetic theories, the distribution function for a larger number of particles—of course, small compared with the total number of particles in the system—can be chosen for defining the coarse-grained level of description (see Section 7.5).

7.2.1 Generalized Canonical Ensemble

Using $\lambda(\boldsymbol{r},\boldsymbol{p})$ as the Lagrange multiplier to control the average $f(\boldsymbol{r},\boldsymbol{p})$ of (7.47) in a generalized canonical ensemble, the partition function (6.16) is obtained as

$$\begin{aligned}\hat{Z}[\lambda] &= \int \exp\left\{-\int \lambda(\boldsymbol{r},\boldsymbol{p}) f^{\Pi}(\boldsymbol{r},\boldsymbol{p};z)\, d^3r\, d^3p\right\} d^{6N}z \\ &= \int \exp\left\{-\sum_{j=1}^{N} \lambda(\boldsymbol{r}_j,\boldsymbol{p}_j)\right\} d^{6N}z = \left[\int e^{-\lambda(\boldsymbol{r},\boldsymbol{p})}\, d^3r\, d^3p\right]^N . \qquad (7.48)\end{aligned}$$

A direct relationship between the single-particle distribution function $f(\boldsymbol{r},\boldsymbol{p})$ and the Lagrange multiplier $\lambda(\boldsymbol{r},\boldsymbol{p})$ is given by equation (6.17), which yields the result

$$f(\boldsymbol{r},\boldsymbol{p}) = -\frac{\delta \ln \hat{Z}[\lambda]}{\delta \lambda(\boldsymbol{r},\boldsymbol{p})} = N \frac{e^{-\lambda(\boldsymbol{r},\boldsymbol{p})}}{\int e^{-\lambda(\boldsymbol{r}',\boldsymbol{p}')}\, d^3r'\, d^3p'} . \qquad (7.49)$$

The generalized canonical ensemble (6.15) for deriving the Boltzmann equation is thus given by

$$\begin{aligned}\rho_f(z) &= \frac{1}{\hat{Z}[\lambda]} \exp\left\{-\int \lambda(\boldsymbol{r},\boldsymbol{p}) f^{\Pi}(\boldsymbol{r},\boldsymbol{p};z)\, d^3r\, d^3p\right\} \\ &= \frac{1}{\hat{Z}[\lambda]} \prod_{j=1}^{N} e^{-\lambda(\boldsymbol{r}_j,\boldsymbol{p}_j)} = \frac{1}{N^N} \prod_{j=1}^{N} f(\boldsymbol{r}_j,\boldsymbol{p}_j), \qquad (7.50)\end{aligned}$$

that is, by the properly normalized product of single-particle distribution functions.

Exercise 162 Canonical Ensemble and Liouville Equation
Show that, in the absence of pair collisions, the canonical ensemble $\rho_f(z)$ of (7.50) satisfies the Liouville equation (6.6).

Exercise 163 Single-Particle Projection Operator
Show that the projection operator (6.25) associated with the single-particle distribution function and the generalized canonical ensemble can be written as

$$\mathcal{P}(f)A(z) = \sum_{j=1}^{N} \int \rho_f(z') \frac{f^{\Pi}(\boldsymbol{r}_j',\boldsymbol{p}_j';z)}{f(\boldsymbol{r}_j',\boldsymbol{p}_j')} A(z')\, d^{6N}z' - (N-1) \int \rho_f(z') A(z')\, d^{6N}z'. \qquad (7.51)$$

The jth term in the sum of (7.51) corresponds to replacing the factor $f(\boldsymbol{r}'_j, \boldsymbol{p}'_j)$ in the generalized canonical ensemble (7.50) by $f^{\Pi}(\boldsymbol{r}'_j, \boldsymbol{p}'_j; z)$.

Exercise 164 Natural Action of Space Transformations

Use the arguments of Section 6.1.6 to construct the natural action of space transformations on the single-particle distribution function $f(\boldsymbol{r}, \boldsymbol{p})$.

7.2.2 Energy and Entropy

To obtain the energy on the level of the single-particle distribution function from equation (6.48), we assume that the potential energy for particle interactions is negligible compared with the kinetic energy of the particles. Such an assumption is perfectly natural for the Grad limit in which collisions matter only for the time evolution but not for the instantaneous energy, as discussed in Section 7.1.1. We then have

$$E_0^{\text{approx}}(z) = \sum_{j=1}^{N} \left[\frac{\boldsymbol{p}_j^2}{2m} + \phi^{(\mathrm{e})}(\boldsymbol{r}_j) \right] = \int \left[\frac{\boldsymbol{p}^2}{2m} + \phi^{(\mathrm{e})}(\boldsymbol{r}) \right] f^{\Pi}(\boldsymbol{r}, \boldsymbol{p}; z) d^3r d^3p, \tag{7.52}$$

where the external potential $\phi^{(\mathrm{e})}(\boldsymbol{r})$ has been kept. This energy expression is of the form (6.50), that is, it is a functional of the slow variables. From this observation, or by performing the ensemble averaging of $E_0^{\text{approx}}(z)$ according to equation (6.48), we obtain as the first building block for the GENERIC formulation of Boltzmann's equation the energy functional

$$E[f] = \int \left[\frac{\boldsymbol{p}^2}{2m} + \phi^{(\mathrm{e})}(\boldsymbol{r}) \right] f(\boldsymbol{r}, \boldsymbol{p}) d^3r d^3p. \tag{7.53}$$

The functional derivative of $E[f]$ is given by

$$\frac{\delta E[f]}{\delta f(\boldsymbol{r}, \boldsymbol{p})} = \frac{\boldsymbol{p}^2}{2m} + \phi^{(\mathrm{e})}(\boldsymbol{r}). \tag{7.54}$$

Within the generalized canonical formalism, the entropy (6.55) evaluated for Boltzmann's equation with the help of equations (7.48) and (7.49) is given by

$$S[f] = -k_{\mathrm{B}} \int f(\boldsymbol{r}, \boldsymbol{p}) \ln \frac{f(\boldsymbol{r}, \boldsymbol{p})}{N} d^3r d^3p. \tag{7.55}$$

If N in the entropy expression (7.55) is considered as a functional of f, $N = \int f(\boldsymbol{r}, \boldsymbol{p}) d^3r d^3p$, then, according to Appendix C.3, the functional derivative of $S[f]$ is given by

$$\frac{\delta S[f]}{\delta f(\boldsymbol{r}, \boldsymbol{p})} = -k_{\mathrm{B}} \ln \frac{f(\boldsymbol{r}, \boldsymbol{p})}{N}. \tag{7.56}$$

If the functional dependence of the number of particles N on f is ignored, the entropy gradients are changed by an irrelevant constant only.

Exercise 165 Entropy and H Function

Express the entropy (7.55) in terms of Boltzmann's H function.

7.2.3 Poisson Matrix

To evaluate the Poisson matrix (6.68) for Boltzmann's kinetic equation, we first consider the Poisson bracket of the atomistic expression for the single-particle distribution function:

$$\begin{aligned}\{f^{\Pi}(\boldsymbol{r},\boldsymbol{p};z), f^{\Pi}(\boldsymbol{r}',\boldsymbol{p}';z)\} &= \sum_{j=1}^{N}\left[\frac{\partial f^{\Pi}(\boldsymbol{r},\boldsymbol{p};z)}{\partial \boldsymbol{r}_j}\cdot\frac{\partial f^{\Pi}(\boldsymbol{r}',\boldsymbol{p}';z)}{\partial \boldsymbol{p}_j}\right.\\ &\quad\left.-\frac{\partial f^{\Pi}(\boldsymbol{r},\boldsymbol{p};z)}{\partial \boldsymbol{p}_j}\cdot\frac{\partial f^{\Pi}(\boldsymbol{r}',\boldsymbol{p}';z)}{\partial \boldsymbol{r}_j}\right]\\ &= [-f^{\Pi}(\boldsymbol{r}',\boldsymbol{p};z)+f^{\Pi}(\boldsymbol{r},\boldsymbol{p}';z)]\frac{\partial\delta(\boldsymbol{r}-\boldsymbol{r}')}{\partial \boldsymbol{r}}\cdot\frac{\partial\delta(\boldsymbol{p}-\boldsymbol{p}')}{\partial \boldsymbol{p}}. \qquad (7.57)\end{aligned}$$

The linear occurrence of f^{Π} in (7.57) is very convenient. By performing the ensemble averaging required according to equation (6.68), we obtain

$$L(\boldsymbol{r},\boldsymbol{p};\boldsymbol{r}',\boldsymbol{p}') = [f(\boldsymbol{r},\boldsymbol{p}') - f(\boldsymbol{r}',\boldsymbol{p})]\frac{\partial\delta(\boldsymbol{r}-\boldsymbol{r}')}{\partial \boldsymbol{r}}\cdot\frac{\partial\delta(\boldsymbol{p}-\boldsymbol{p}')}{\partial \boldsymbol{p}}. \qquad (7.58)$$

To avoid the integrations over the primed variables when multiplying with the Poisson matrix, we can introduce the operator notation

$$L(\boldsymbol{r},\boldsymbol{p}) = \frac{\partial}{\partial \boldsymbol{p}}\cdot f(\boldsymbol{r},\boldsymbol{p})\frac{\partial}{\partial \boldsymbol{r}} - \frac{\partial}{\partial \boldsymbol{r}}\cdot f(\boldsymbol{r},\boldsymbol{p})\frac{\partial}{\partial \boldsymbol{p}}. \qquad (7.59)$$

This result is antisymmetric, and the corresponding Poisson bracket can be verified to satisfy the Jacobi identity. We have hence obtained a valid Poisson bracket by coarse-graining to the level of the single-particle distribution function. No approximation was required in deriving this Poisson bracket. Furthermore, it can be verified directly that, because the entropy gradient (7.56) depends on $\boldsymbol{r}$ and $\boldsymbol{p}$ through $f(\boldsymbol{r},\boldsymbol{p})$ only, the degeneracy condition (1.5) is fulfilled. The GENERIC building blocks E, S, L for the Boltzmann equation so far derived from first principles coincide with the ones given previously in a phenomenological attempt to formulate this kinetic equation within the GENERIC framework.[19]

The Poisson bracket associated with (7.59) can be understood as follows. In Section 6.1.1, we recognized the equivalence of the time evolution behind Hamilton's equations of motion and the Liouville equation, where the Liouville formulation admits a more general set of initial conditions (also smooth distributions in addition to δ-functions concentrated on given particle positions and momenta). The Poisson operator

$$\tilde{L}(f) = -\frac{\partial}{\partial z}\cdot f(z)L_0\cdot\frac{\partial}{\partial z} \qquad (7.60)$$

[19] Öttinger, J. Non-Equilib. Thermodyn. 22 (1997) 386.

associated with the cosymplectic matrix L_0 according to (6.155) may thus be regarded as a transformed version of the Poisson operator of classical mechanics. If we consider a single-particle system, the transformation provides the Poisson operator (7.59) on the level of single-particle distribution functions. With (7.60), we thus have a natural Poisson operator for any N-particle distribution function resulting from a transformation of the Poisson bracket of classical mechanics.

7.2.4 Friction Matrix

As a first step toward evaluating the friction matrix (6.122), we determine the action of the Liouville operator on the single-particle distribution function,

$$\begin{aligned} \{f^{\Pi}(\boldsymbol{r},\boldsymbol{p};z), E_0(z)\} &= \sum_{j=1}^{N}\left[\frac{\boldsymbol{p}_j}{m}\cdot\frac{\partial}{\partial \boldsymbol{r}_j} - \frac{\partial E_0(z)}{\partial \boldsymbol{r}_j}\cdot\frac{\partial}{\partial \boldsymbol{p}_j}\right] f^{\Pi}(\boldsymbol{r},\boldsymbol{p};z) \\ &= -\left(\frac{\boldsymbol{p}}{m}\cdot\frac{\partial}{\partial \boldsymbol{r}} - \frac{\partial \phi^{(\mathrm{e})}}{\partial \boldsymbol{r}}\cdot\frac{\partial}{\partial \boldsymbol{p}}\right) f^{\Pi}(\boldsymbol{r},\boldsymbol{p};z) \\ &\quad + \sum_{j,k=1}^{N}\frac{\partial \phi_{jk}}{\partial \boldsymbol{r}_j}\cdot\frac{\partial}{\partial \boldsymbol{p}}\delta(\boldsymbol{r}-\boldsymbol{r}_j)\delta(\boldsymbol{p}-\boldsymbol{p}_j). \end{aligned} \tag{7.61}$$

When we evaluate the rapidly fluctuating part of $i\mathcal{L}f^{\Pi}(\boldsymbol{r},\boldsymbol{p};z) = \{f^{\Pi}(\boldsymbol{r},\boldsymbol{p};z), E_0(z)\}$, the contribution linear in the relevant variables $f^{\Pi}(\boldsymbol{r},\boldsymbol{p};z)$ in (7.61) is eliminated. By assuming that the average single-particle distribution function does not vary appreciably over the range of interaction, we obtain by means of $\mathcal{Q} = 1 - \mathcal{P}$ and (7.51) (see Exercise 166)

$$\begin{aligned} \mathcal{Q}i\mathcal{L}f^{\Pi}(\boldsymbol{r}',\boldsymbol{p}';z) &= \frac{\partial}{\partial \boldsymbol{p}'}\cdot\sum_{j,k=1}^{N}\frac{\partial \phi_{jk}}{\partial \boldsymbol{r}_j}\delta(\boldsymbol{r}'-\boldsymbol{r}_j)\delta(\boldsymbol{p}'-\boldsymbol{p}_j) \\ &- \frac{\partial f(\boldsymbol{r}',\boldsymbol{p}')}{\partial \boldsymbol{p}'}\cdot\sum_{j=1}^{N}\frac{\partial \phi(|\boldsymbol{r}'-\boldsymbol{r}_j|)}{\partial \boldsymbol{r}'}. \end{aligned} \tag{7.62}$$

Because the second term in (7.62) does not contribute to the GENERIC time-evolution equation (due to the occurrence of $\partial f(\boldsymbol{r}',\boldsymbol{p}')/\partial \boldsymbol{p}'$) and because the force felt at the position $\boldsymbol{r}'$ is expected to vanish almost everywhere (except very close to a particular particle with known position), we use the simplified expression

$$\mathcal{Q}i\mathcal{L}f^{\Pi}(\boldsymbol{r}',\boldsymbol{p}';z) = \frac{\partial}{\partial \boldsymbol{p}'}\cdot\sum_{j,k=1}^{N}\frac{\partial \phi_{jk}}{\partial \boldsymbol{r}_j}\delta(\boldsymbol{r}'-\boldsymbol{r}_j)\delta(\boldsymbol{p}'-\boldsymbol{p}_j). \tag{7.63}$$

Note that we have assumed that the average single-particle distribution function f is essentially constant over a distance comparable to the size of a molecule, but no restrictions on the variation in f over distances comparable to the mean free path have been imposed. This is important for understanding the wide range of validity of Boltzmann's equation.

Exercise 166 Rapid Change of Single-Particle Distribution Function
Calculate the rapid contribution to the time evolution of the single-particle distribution function, $Qi\mathcal{L}f^{\Pi}(\boldsymbol{r}',\boldsymbol{p}';z)$, from (7.51) and (7.61) and show how (7.62) can be obtained.

With the help of (7.63), the friction matrix (6.122) then takes the specific form

$$M = \frac{1}{k_{\mathrm{B}}}\frac{\partial}{\partial \boldsymbol{p}'}\cdot\sum_{j,k,l=1}^{N}\int \rho_f(z)\frac{\partial \phi_{jk}}{\partial \boldsymbol{r}_j}\delta(\boldsymbol{r}'-\boldsymbol{r}_j)\delta(\boldsymbol{p}'-\boldsymbol{p}_j) \\ e^{Qi\mathcal{L}\tau}\delta(\boldsymbol{r}-\boldsymbol{r}_l)\delta(\boldsymbol{p}-\boldsymbol{p}_l)d^{6N}z, \tag{7.64}$$

where we simply write M for $M(\boldsymbol{r},\boldsymbol{p},\boldsymbol{r}',\boldsymbol{p}')$ in this subsection. We obtain a non-vanishing contribution to M only when an interaction between the particles j and k occurs, that is, while the particles j and k are undergoing a collision. For a sufficiently rarefied gas, there are no interactions between the pair j, k and further particles at the same time; therefore, the average force between the particles j and k vanishes unless $l=j$ or $l=k$. By exploiting the further fact that each pair of particles contributes in the same way, we obtain the simplified expression

$$M = \frac{N^2}{k_{\mathrm{B}}}\frac{\partial}{\partial \boldsymbol{p}'}\cdot\int \rho_f(z)\frac{\partial \phi_{12}}{\partial \boldsymbol{r}_1}\delta(\boldsymbol{r}'-\boldsymbol{r}_1)\delta(\boldsymbol{p}'-\boldsymbol{p}_1) \\ e^{Qi\mathcal{L}\tau}[\delta(\boldsymbol{r}-\boldsymbol{r}_1)\delta(\boldsymbol{p}-\boldsymbol{p}_1)+\delta(\boldsymbol{r}-\boldsymbol{r}_2)\delta(\boldsymbol{p}-\boldsymbol{p}_2)]d^{6N}z. \tag{7.65}$$

The rapid time evolution implied by $e^{i\mathcal{L}Q\tau}$ represents the effect of the rest of a collision between particles 1 and 2 that are already in the process of interacting, and the integrand is hence independent of the positions and momenta of all the other particles. When the canonical ensemble (7.50) is inserted, all integrations implied by $d^{6N}z$ can be performed, except those pertaining to particles 1 and 2. Instead of integrating over $\boldsymbol{r}_1$ and $\boldsymbol{r}_2$, we prefer to use the center-of-mass position and the difference vector $\boldsymbol{R}=\boldsymbol{r}_2-\boldsymbol{r}_1$ as integration variables. Assuming that ϕ represents a short-range potential, we keep the difference vector $\boldsymbol{R}$ only in the interaction force, whereas, in all other places, we neglect the differences between $\boldsymbol{r}_1$, $\boldsymbol{r}_2$ and the center-of-mass position. After symmetrizing in an alternative way with respect to the two colliding particles, we obtain

$$M = -\frac{\delta(\boldsymbol{r}'-\boldsymbol{r})}{k_{\mathrm{B}}}\frac{\partial}{\partial \boldsymbol{p}'}\cdot\int f(\boldsymbol{r},\boldsymbol{p}_1)f(\boldsymbol{r},\boldsymbol{p}_2)\frac{\partial \phi(R)}{\partial \boldsymbol{R}}[\delta(\boldsymbol{p}'-\boldsymbol{p}_1)-\delta(\boldsymbol{p}'-\boldsymbol{p}_2)] \\ e^{Qi\mathcal{L}\tau}\delta(\boldsymbol{p}-\boldsymbol{p}_1)d^3R\,d^3p_1\,d^3p_2. \tag{7.66}$$

In (7.66), we can see clearly the reason why we should not expect the friction matrix for Boltzmann's equation to be symmetric. It has been pointed out before that, to obtain symmetry, the ensemble used for averaging should not change appreciably over the time scale τ [see the corresponding remarks after (3.49) and (6.121)]. The

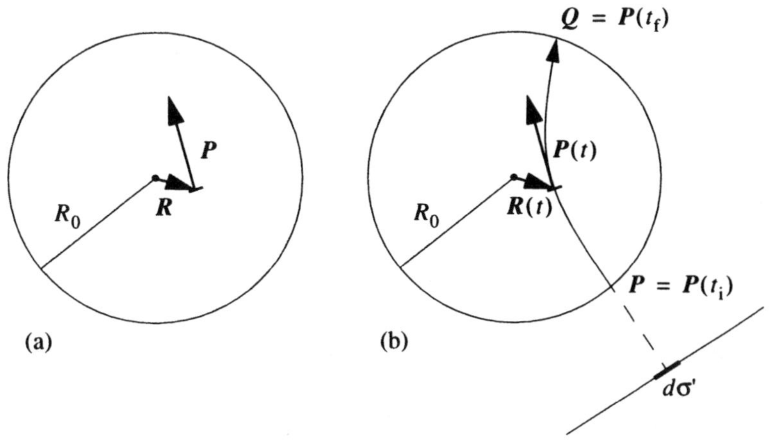

Fig. 7.7 Illustration of a change of variables from the relative positions and momenta of two colliding particles to trajectories and initial conditions.

product $f(\boldsymbol{r},\boldsymbol{p}_1)f(\boldsymbol{r},\boldsymbol{p}_2)$ arises from the ensemble (7.50) in (7.66) and, in general, it changes strongly during a collision. Only when $f(\boldsymbol{r},\boldsymbol{p}_1)f(\boldsymbol{r},\boldsymbol{p}_2)$ is an invariant of the collision, that is, for a Maxwellian velocity distribution, M must be symmetric (see Exercise 169). Otherwise, the ensemble associated with the single-particle distribution function of Boltzmann's kinetic theory is too strongly changing during a collision, and M cannot be expected to be symmetric.

We next introduce the total momentum, $\boldsymbol{p}_t = \boldsymbol{p}_1 + \boldsymbol{p}_2$, and the relative momentum, $\boldsymbol{P} = (\boldsymbol{p}_2 - \boldsymbol{p}_1)/2$, and we then obtain by a further transformation of variables

$$\begin{aligned} M &= -\frac{\delta(\boldsymbol{r}'-\boldsymbol{r})}{k_B}\frac{\partial}{\partial \boldsymbol{p}'}\cdot\int f(\boldsymbol{r},\boldsymbol{p}')f(\boldsymbol{r},\boldsymbol{p}_t-\boldsymbol{p}')\frac{\partial\phi(R)}{\partial \boldsymbol{R}}[\delta(\boldsymbol{p}'-\frac{\boldsymbol{p}_t}{2}+\boldsymbol{P}) \\ &- \delta(\boldsymbol{p}'-\frac{\boldsymbol{p}_t}{2}-\boldsymbol{P})]\delta(\boldsymbol{p}-\frac{\boldsymbol{p}_t}{2}+\boldsymbol{Q})d^3R\,d^3P\,d^3p_t, \end{aligned} \tag{7.67}$$

provided that the relative momentum at the end of the on-going collision is $\boldsymbol{Q}$. We hence must choose the intermediate time scale τ longer than the duration of a collision but short compared with the time between collisions, which leaves a wide range of equivalent choices.

At this point, the friction matrix (7.66) is expressed as an average in which no time-evolution operator occurs. Considering the integration over $\boldsymbol{R}$, due to the presence of the interaction force, we obtain nonzero contributions only during a collision process. The integration over $\boldsymbol{R}$ hence probes various phases of the collision process. For that reason, it is convenient to integrate along possible trajectories and to reintroduce time (see Figure 7.7). As discussed in Section 7.1.5, in terms of the variables $\boldsymbol{R}$ and $\boldsymbol{P}$, the relative motion during a collision is equivalent to the motion of an effective particle with reduced mass $m_r = m/2$ under the influence of the potential $\phi(R)$. Using the incompressibility of the corresponding effective-particle phase space for the reversible collision process, we can integrate over the initial values $\boldsymbol{P} = \boldsymbol{P}(t_i)$, the surface elements $d\sigma'$ of Figure 7.7, and the time t. By means of the differential cross section, $d\sigma'/d\Omega'$, the surface elements $d\sigma'$ can be related to the solid angles $d\Omega'$ into which the effective particles representing the relative motion are scattered in the collision process (see Section 7.1.5). Transforming to an angular integration at the expense of introducing the differential cross section is advantageous because the scattering angle is needed to calculate the final relative momentum $\boldsymbol{Q}$ from the initial relative momentum $\boldsymbol{P}$ in an elastic collision. The initial relative velocity, $v_{\rm rel} = |\boldsymbol{p}_2 - \boldsymbol{p}_1|/m = 2|\boldsymbol{P}|/m$ needs to be included as a factor to cover properly all the volume elements in the course of time. We then obtain

$$\begin{aligned} M &= \frac{\delta(\boldsymbol{r}'-\boldsymbol{r})}{k_{\rm B}} \frac{\partial}{\partial \boldsymbol{p}'} \cdot \int v_{\rm rel} f(\boldsymbol{r},\boldsymbol{p}') f(\boldsymbol{r},\boldsymbol{p}_{\rm t}-\boldsymbol{p}') \boldsymbol{I}_\phi(\boldsymbol{p}'-\frac{\boldsymbol{p}_{\rm t}}{2},\boldsymbol{P},\boldsymbol{Q}) \\ &\times \ \delta(\boldsymbol{p}-\frac{\boldsymbol{p}_{\rm t}}{2}+\boldsymbol{Q}) d\sigma' d^3P d^3p_{\rm t}. \end{aligned} \tag{7.68}$$

To obtain the final and explicit result for the M matrix, we still need to evaluate the time integral

$$\boldsymbol{I}_\phi(\boldsymbol{p},\boldsymbol{P},\boldsymbol{Q}) = \int_{t_{\rm i}}^{t_{\rm f}} \dot{\boldsymbol{P}}(t)[\delta(\boldsymbol{p}+\boldsymbol{P}(t)) - \delta(\boldsymbol{p}-\boldsymbol{P}(t))]dt, \tag{7.69}$$

where the direction of $\boldsymbol{Q}$ (with $|\boldsymbol{Q}| = |\boldsymbol{P}|$) indicates the dependence of the trajectory and hence of the integral on the relative position of the particles at the beginning of the collision.

Exercise 167 Symmetry Properties of Auxiliary Function
Establish the following useful symmetry properties of the auxiliary function $\boldsymbol{I}_\phi(\boldsymbol{p},\boldsymbol{P},\boldsymbol{Q})$:

$$\boldsymbol{I}_\phi(-\boldsymbol{p},\boldsymbol{P},\boldsymbol{Q}) = -\boldsymbol{I}_\phi(\boldsymbol{p},\boldsymbol{P},\boldsymbol{Q}), \tag{7.70}$$

$$\boldsymbol{I}_\phi(\boldsymbol{p},-\boldsymbol{P},-\boldsymbol{Q}) = \boldsymbol{I}_\phi(\boldsymbol{p},\boldsymbol{P},\boldsymbol{Q}), \tag{7.71}$$

and

$$\boldsymbol{I}_\phi(\boldsymbol{p},-\boldsymbol{Q},-\boldsymbol{P}) = -\boldsymbol{I}_\phi(\boldsymbol{p},\boldsymbol{P},\boldsymbol{Q}). \tag{7.72}$$

Note that, with the symmetries of Exercise 167, the $\boldsymbol{Q}$ in the δ function of equation (7.68) can be replaced by $-\boldsymbol{Q}$ or, with an overall minus sign, by $\boldsymbol{P}$ or $-\boldsymbol{P}$. An even more compact expression for the friction matrix can be obtained by performing the integration over $\boldsymbol{p}_{\rm t}$,

$$\begin{aligned} M &= -8\frac{\delta(\boldsymbol{r}'-\boldsymbol{r})}{k_{\rm B}} \frac{\partial}{\partial \boldsymbol{p}'} \cdot \int v_{\rm rel} f(\boldsymbol{r},\boldsymbol{p}') f(\boldsymbol{r},2\boldsymbol{p}-\boldsymbol{p}'+2\boldsymbol{P}) \\ &\times \ \boldsymbol{I}_\phi(\boldsymbol{p}'-\boldsymbol{p}-\boldsymbol{P},\boldsymbol{P},\boldsymbol{Q}) d\sigma' d^3P. \end{aligned} \tag{7.73}$$

As anticipated, this friction matrix is not symmetric in $\boldsymbol{p}$ and $\boldsymbol{p}'$ (see Exercise 168 for explicit proof).

When considering the divergence of $\boldsymbol{I}_\phi(\boldsymbol{p}, \boldsymbol{P}, \boldsymbol{Q})$ with respect to $\boldsymbol{p}$, we note the occurrence of a total time derivative, so that the integral can be carried out,

$$\frac{\partial}{\partial \boldsymbol{p}} \cdot \boldsymbol{I}_\phi(\boldsymbol{p}, \boldsymbol{P}, \boldsymbol{Q}) = \delta(\boldsymbol{p} + \boldsymbol{Q}) + \delta(\boldsymbol{p} - \boldsymbol{Q}) - \delta(\boldsymbol{p} + \boldsymbol{P}) - \delta(\boldsymbol{p} - \boldsymbol{P}). \quad (7.74)$$

Equation (7.74) is consistent with the symmetry properties (7.70)–(7.72). Note that $\boldsymbol{I}_\phi(\boldsymbol{p}, \boldsymbol{P}, \boldsymbol{Q})$ is not uniquely defined by equation (7.74) because any divergence-free function can be added to a particular solution. Although a particular solution of equation (7.74) can be found as easily as the electric field around four point charges, according to equation (7.69), the physical solution $\boldsymbol{I}_\phi(\boldsymbol{p}, \boldsymbol{P}, \boldsymbol{Q})$ should vanish unless $\boldsymbol{p}$ or $-\boldsymbol{p}$ lies on the one-dimensional curve of physical relative momentum vectors connecting the initial value $\boldsymbol{P}$ and the final value $\boldsymbol{Q}$ during a collision process. However, in the following, we do not need the explicit physical solution for a specific interaction potential. Equation (7.74) for the divergence and the symmetries of $\boldsymbol{I}_\phi(\boldsymbol{p}, \boldsymbol{P}, \boldsymbol{Q})$ are sufficient for deriving the desired results.

After inserting equations (7.54) and (7.73) and performing the substitution $\boldsymbol{p} \to \boldsymbol{p} - \boldsymbol{P}$, we can verify the degeneracy requirement in (1.22),

$$\frac{\delta E}{\delta x} \cdot M = \int \left[\frac{\boldsymbol{p}^2}{2m} + \phi^{(e)}(\boldsymbol{r}) \right] M(\boldsymbol{r}, \boldsymbol{p}, \boldsymbol{r}', \boldsymbol{p}') d^3r d^3p = 0, \quad (7.75)$$

by means of the symmetry properties (7.71), (7.72) and the conservation of energy in the collision process, $\boldsymbol{P}^2 = \boldsymbol{Q}^2$. A similar degeneracy requirement for the total momentum can be obtained from the conservation of momentum in the collision process.

In the special case of hard-sphere collisions, which we consider for illustration, the interaction force during the collision is in the direction of $\boldsymbol{Q} - \boldsymbol{P}$; thus, we obtain the more explicit expression

$$\boldsymbol{I}_\phi(\boldsymbol{p}, \boldsymbol{P}, \boldsymbol{Q}) = (\boldsymbol{Q} - \boldsymbol{P}) \int_0^1 [\delta(\boldsymbol{p} + (1-u)\boldsymbol{P} + u\boldsymbol{Q}) - \delta(\boldsymbol{p} - (1-u)\boldsymbol{P} - u\boldsymbol{Q})] du. \quad (7.76)$$

For hard spheres, the integral

$$\bar{\boldsymbol{I}}_\phi(\boldsymbol{p}, \boldsymbol{P}) = \int \boldsymbol{I}_\phi(\boldsymbol{p}, \boldsymbol{P}, \boldsymbol{Q}) d\sigma' \quad (7.77)$$

occurring in the expression (7.73) for the friction matrix can be evaluated because $(1-u)\boldsymbol{P} + u\boldsymbol{Q}$ provides a one-to-one parameterization of a sphere with radius $|\boldsymbol{P}|$ when the integration over du and $d\sigma'$ is carried out. The result is

$$\bar{\boldsymbol{I}}_\phi(\boldsymbol{p}, \boldsymbol{P}) = 4d^2 \left[\frac{(\boldsymbol{P} - \boldsymbol{p})(\boldsymbol{P} - \boldsymbol{p})}{|\boldsymbol{P} - \boldsymbol{p}|^4} - \frac{(\boldsymbol{P} + \boldsymbol{p})(\boldsymbol{P} + \boldsymbol{p})}{|\boldsymbol{P} + \boldsymbol{p}|^4} \right] \cdot \frac{\boldsymbol{P}}{|\boldsymbol{P}|} \quad (7.78)$$

for $|\boldsymbol{p}| \leq |\boldsymbol{P}|$ and $\bar{\boldsymbol{I}}_\phi(\boldsymbol{p}, \boldsymbol{P}) = 0$ otherwise. Note that this expression for $\bar{\boldsymbol{I}}_\phi(\boldsymbol{p}, \boldsymbol{P})$ has the proper symmetries

$$\bar{\boldsymbol{I}}_\phi(-\boldsymbol{p}, \boldsymbol{P}) = -\bar{\boldsymbol{I}}_\phi(\boldsymbol{p}, \boldsymbol{P}) \quad (7.79)$$

and

$$\bar{\boldsymbol{I}}_\phi(\boldsymbol{p},-\boldsymbol{P})=\bar{\boldsymbol{I}}_\phi(\boldsymbol{p},\boldsymbol{P}), \tag{7.80}$$

following from (7.70) and (7.71). The divergence of $\bar{\boldsymbol{I}}_\phi(\boldsymbol{p},\boldsymbol{P})$ with respect to $\boldsymbol{p}$ vanishes everywhere except on the sphere $|\boldsymbol{p}|=|\boldsymbol{P}|$, as expected from (7.74).

Exercise 168 Lacking Symmetry of Friction Matrix
Evaluate

$$\int M(\boldsymbol{r},\boldsymbol{p},\boldsymbol{r}',\boldsymbol{p}')\frac{\delta E[f]}{\delta f(\boldsymbol{r}',\boldsymbol{p}')}d^3r'd^3p'$$

for hard spheres at $\boldsymbol{p}=0$. Show that the friction matrix $M=M(\boldsymbol{r},\boldsymbol{p},\boldsymbol{r}',\boldsymbol{p}')$ obtained in (7.73) is not symmetric.

Exercise 169 Symmetric Friction Matrix at Local Equilibrium
Calculate the symmetric and positive semidefinite friction matrix for Boltzmann's kinetic theory under the assumption that the precollision ensemble does not change on the time scale τ. This situation arises for the local Maxwellian distribution (7.32), that is, for a local equilibration in momentum space.

7.2.5 Time-Evolution Equation

The irreversible contribution to the time evolution of the single-particle distribution function is given by

$$M\cdot\frac{\delta S}{\delta x}=\int M(\boldsymbol{r},\boldsymbol{p},\boldsymbol{r}',\boldsymbol{p}')\frac{\delta S[f]}{\delta f(\boldsymbol{r}',\boldsymbol{p}')}d^3r'd^3p'. \tag{7.81}$$

By inserting the entropy gradients (7.56) and the friction matrix (7.68), with $\boldsymbol{Q}$ in the δ function replaced by $\boldsymbol{P}$ according to the remark after Exercise 167, we obtain after an integration by parts

$$\begin{aligned}M\cdot\frac{\delta S}{\delta x} &= -\int v_{\rm rel}\frac{\partial f(\boldsymbol{r},\boldsymbol{p}')}{\partial \boldsymbol{p}'}\cdot f(\boldsymbol{r},\boldsymbol{p}_{\rm t}-\boldsymbol{p}')\boldsymbol{I}_\phi(\boldsymbol{p}'-\frac{\boldsymbol{p}_{\rm t}}{2},\boldsymbol{P},\boldsymbol{Q})\\ &\quad\times\ \delta(\boldsymbol{p}-\frac{\boldsymbol{p}_{\rm t}}{2}+\boldsymbol{P})d\sigma'd^3Pd^3p_{\rm t}d^3p'.\end{aligned} \tag{7.82}$$

The substitution $\boldsymbol{p}'\to\boldsymbol{p}_{\rm t}-\boldsymbol{p}'$ shows that, after the symmetry property (7.70) is used, this expression can be rewritten as

$$\begin{aligned}M\cdot\frac{\delta S}{\delta x} &= -\int \frac{v_{\rm rel}}{2}\frac{\partial f(\boldsymbol{r},\boldsymbol{p}')f(\boldsymbol{r},\boldsymbol{p}_{\rm t}-\boldsymbol{p}')}{\partial \boldsymbol{p}'}\cdot \boldsymbol{I}_\phi(\boldsymbol{p}'-\frac{\boldsymbol{p}_{\rm t}}{2},\boldsymbol{P},\boldsymbol{Q})\\ &\quad\times\ \delta(\boldsymbol{p}-\frac{\boldsymbol{p}_{\rm t}}{2}+\boldsymbol{P})d\sigma'd^3Pd^3p_{\rm t}d^3p',\end{aligned} \tag{7.83}$$

so that, after a further integration by parts and after insertion of equation (7.74), the $\boldsymbol{p}'$-integration can be carried out. By transforming back to the momenta in the laboratory system we obtain

$$M \cdot \frac{\delta S}{\delta x} = \int v_{\rm rel} \frac{d\sigma}{d\Omega} [f(\boldsymbol{r}, \boldsymbol{q}_1) f(\boldsymbol{r}, \boldsymbol{q}_2) - f(\boldsymbol{r}, \boldsymbol{p}_1) f(\boldsymbol{r}, \boldsymbol{p}_2)] \delta(\boldsymbol{p} - \boldsymbol{p}_1) d\Omega d^3 p_1 d^3 p_2, \tag{7.84}$$

where the differential cross section has been transformed from the center-of-mass system, $d\sigma'/d\Omega'$, to the laboratory system, $d\sigma/d\Omega$ (actually only the scattering angle and hence $d\Omega$ depends on the reference frame). After performing a further integration, we obtain the final result:

$$M \cdot \frac{\delta S}{\delta x} = \int v_{\rm rel} \frac{d\sigma}{d\Omega} [f(\boldsymbol{r}, \boldsymbol{q}) f(\boldsymbol{r}, \boldsymbol{q}') - f(\boldsymbol{r}, \boldsymbol{p}) f(\boldsymbol{r}, \boldsymbol{p}')] d\Omega d^3 p', \tag{7.85}$$

where the collision process maps the incoming momenta $\boldsymbol{p}$ and $\boldsymbol{p}'$ into $\boldsymbol{q}$ and $\boldsymbol{q}'$. Evaluating the reversible contribution $L \cdot \delta E/\delta x$ to GENERIC (1.1) by means of equations (7.54) and (7.59), we finally obtain the famous Boltzmann equation:

$$\begin{aligned} \frac{\partial f(\boldsymbol{r}, \boldsymbol{p})}{\partial t} &= -\left(\frac{\boldsymbol{p}}{m} \cdot \frac{\partial}{\partial \boldsymbol{r}} - \frac{\partial \phi^{(\rm e)}(\boldsymbol{r})}{\partial \boldsymbol{r}} \cdot \frac{\partial}{\partial \boldsymbol{p}} \right) f(\boldsymbol{r}, \boldsymbol{p}) \\ &+ \int v_{\rm rel} \frac{d\sigma}{d\Omega} [f(\boldsymbol{r}, \boldsymbol{q}) f(\boldsymbol{r}, \boldsymbol{q}') - f(\boldsymbol{r}, \boldsymbol{p}) f(\boldsymbol{r}, \boldsymbol{p}')] d\Omega d^3 p'. \end{aligned} \tag{7.86}$$

The building blocks obtained by a straightforward evaluation of the atomistic expressions in a generalized canonical ensemble on the level of single-particle distribution functions thus indeed reproduce the expected result of Boltzmann.

In summary, we have carried out an explicit coarse-graining from the phase space of an N-particle system to the level of single-particle distribution functions. The time-evolution equation on the coarser level of description is determined from the GENERIC building blocks, for which atomistic expressions in a generalized canonical ensemble are available. The evaluation of these building blocks is similar in spirit to the calculation of the Helmholtz free energy of an ideal gas at equilibrium and is almost as straightforward. Our final results are given by the energy (7.53), the entropy (7.55), the Poisson matrix (7.59), and the friction matrix (7.73). The energy, entropy, and Poisson matrix agree with those determined previously in a phenomenological approach. Although the friction matrix (7.73) is not symmetric, the famous Boltzmann equation is recovered from the atomistically derived building blocks of this section.

Exercise 170 Linearized Boltzmann Equation
Derive the linearized Boltzmann equation for $\hat{f}^{(1)}$ in the *ansatz*

$$f(\boldsymbol{r}, \boldsymbol{p}) = f^{\rm Maxw}_{\rho, \boldsymbol{v}, T}(\boldsymbol{r}, \boldsymbol{p}) \left[1 + \hat{f}^{(1)}(\boldsymbol{r}, \boldsymbol{p}) \right] \tag{7.87}$$

by using the symmetric local-equilibrium friction matrix of Exercise 169, by linearizing the entropy gradient (7.56) in $\hat{f}^{(1)}$, and by specializing to global equilibrium.
Note: To check for thermodynamic admissibility, approximations should always be made for the GENERIC building blocks rather than for the time-evolution equations.

Exercise 171 Phenomenological Friction Matrix
In a phenomenological approach,[20] the following friction matrix for Boltzmann's kinetic equation was proposed:

$$\begin{aligned} M &= \frac{\delta(\boldsymbol{r}'-\boldsymbol{r})}{k_\mathrm{B}} \int \delta(\boldsymbol{p}_1-\boldsymbol{p}) w(\boldsymbol{q}_1,\boldsymbol{q}_2|\boldsymbol{p}_1,\boldsymbol{p}_2) \frac{f(\boldsymbol{q}_1)f(\boldsymbol{q}_2)-f(\boldsymbol{p}_1)f(\boldsymbol{p}_2)}{\ln[f(\boldsymbol{q}_1)f(\boldsymbol{q}_2)]-\ln[f(\boldsymbol{p}_1)f(\boldsymbol{p}_2)]} \\ &\times [\delta(\boldsymbol{p}_1-\boldsymbol{p}')+\delta(\boldsymbol{p}_2-\boldsymbol{p}')-\delta(\boldsymbol{q}_1-\boldsymbol{p}')-\delta(\boldsymbol{q}_2-\boldsymbol{p}')] d^3p_1 d^3p_2 d^3q_1 d^3q_2, \end{aligned} \tag{7.88}$$

where the position argument $\boldsymbol{r}$ of f has been suppressed for convenience and the transition probabilities $w(\boldsymbol{q}_1,\boldsymbol{q}_2|\boldsymbol{p}_1,\boldsymbol{p}_2)$ are given by

$$w(\boldsymbol{q}_1,\boldsymbol{q}_2|\boldsymbol{p}_1,\boldsymbol{p}_2) = \delta(\boldsymbol{q}_1+\boldsymbol{q}_2-\boldsymbol{p}_1-\boldsymbol{p}_2)\delta(\boldsymbol{q}_1^2+\boldsymbol{q}_2^2-\boldsymbol{p}_1^2-\boldsymbol{p}_2^2)\frac{8}{m}\frac{d\sigma'}{d\Omega'}. \tag{7.89}$$

These transition probabilities have the symmetry properties (7.20), (7.21), corresponding to time reversal and particle exchange in the two-particle collisions. Show that the friction matrix (7.88) is positive semidefinite and symmetric. Verify the degeneracy (1.5) and check that Boltzmann's kinetic equation is recovered. With the expressions (7.73) and (7.88), we have an example of a nonunique friction matrix.

Exercise 172 Linearized Phenomenological Friction Matrix
Determine the friction matrix (7.88) for a single-particle distribution function of the form (7.87), where $\hat{f}^{(1)}$ describes a small deviation from a Maxwellian velocity distribution. Compare the result with the friction matrix found in Exercise 169.

7.3 CHAPMAN-ENSKOG EXPANSION

After formulating and deriving Boltzmann's kinetic equation, we now need to consider its time dependent solutions in some detail. First we provide several general remarks on expansion methods for solving this equation, and then we show how to obtain the equations of hydrodynamics from Boltzmann's equation, including molecular expressions for the transport coefficients.

7.3.1 Some Remarks on Expansions

The Boltzmann equation (7.86) is a nonlinear integro-differential equation for the single-particle distribution function, and it is generally impossible to find analytical solutions. It is hence natural to look for approximate solutions, in particular, to employ expansion techniques. A thorough discussion of such expansions, which yield inherently imprecise descriptions of a gas in terms of macroscopic fluid variables instead of the full single-particle distribution function, was given by Grad.[21] For the Chapman-

[20] Öttinger, J. Non-Equilib. Thermodyn. 22 (1997) 386.
[21] Grad, Phys. Fluids 6 (1963) 147.

Enskog expansion discussed in this section, these macroscopic fluid variables are the hydrodynamic fields. The solutions of the Boltzmann equation determined by the hydrodynamic fields are known as normal solutions. The Chapman-Enskog expansion picks out exactly those normal solutions. A more general class of solutions of the Boltzmann equation can be approximated only after the set of macroscopic state variables is extended, with each variable governed by partial differential equations similar to those of hydrodynamics; such a generalized expansion based on moments of the velocity distribution was proposed by Grad and is discussed in Section 7.4.

The small expansion parameter of the Chapman-Enskog approach (and related expansion techniques) typically represents the mean free path divided by the length scale of variations of the hydrodynamic fields in a given geometry, and this parameter, which is known as the Knudsen number, enters the Boltzmann equation in a singular manner. Such expansions are hence expected to be asymptotically valid except in a narrow boundary layer, a narrow initial layer, or any internal shock layer with a thickness of the order of the mean free path. Whenever there are significant changes in gas flow properties over a mean free path or over a collision time, then one should fall back on the Boltzmann equation.

To extract useful approximations from asymptotic expansions, the series must always be truncated. There is an optimum number of terms that decreases when the expansion parameter increases [see Figure 4.14, the expansion (4.173), and the corresponding discussion at the end of Section 4.3.2]. The Navier-Stokes equations obtained from the first-order Chapman-Enskog expansion has the largest range of validity. The Burnett equations obtained from the second-order Chapman-Enskog expansion and higher approximations, the so-called super-Burnett equations, "do *not* serve to extend the range of validity farther from the fluid regime. On the contrary, the higher approximations can only be used when the Navier-Stokes equations are very good (...) in order to give an ultra-refined fluid description."[22] For rough answers, extending to length and time scales as small as possible, one should use the Navier-Stokes equations.

As a final point, we emphasize that the Chapman-Enskog method is based on an expansion of equations rather than on solutions. The solutions of expanded equations may develop complicated functional dependencies on the expansion parameter so that we may expect the Chapman-Enskog theory to be more accurate than Hilbert's theory,[23] based on a similar expansion of solutions in powers of the mean free path. Although Hilbert's theory has a narrower scope (it is restricted to an arbitrary but fixed number of mean collision times), it is more universally valid due to the existence of the results for a larger class of solutions to Boltzmann's equation.[24] One needs to be careful about "small" terms in equations because they can change solutions drastically given enough time. For example, an arbitrarily small collision term cannot be neglected without further consideration; rather, some knowledge of the boundary

[22] See p. 149 of Grad, Phys. Fluids 6 (1963) 147.

[23] Hilbert, Math. Ann. 72 (1912) 562.

[24] See p. 177 of Grad, Phys. Fluids 6 (1963) 147.

conditions must be also considered. A small collision term can be neglected for stochastic boundary conditions; for specularly reflecting boundaries, however, the behavior for sufficiently large times will always be governed by collisions.

A deeper understanding of expansion techniques can be obtained by applying them to toy models of the Boltzmann equation. Illustrative examples have been discussed in articles by Grad[25] and by Karlin.[26]

From a GENERIC perspective, expansions should be done neither on time-evolution equations nor on their solutions but rather on the building blocks E, S, L, and M. At any stage of an expansion, one can then guarantee thermodynamic consistency by verifying the various properties of the building blocks. In the Chapman-Enskog approach, the basic variables would always be the hydrodynamic fields, but derivatives of increasingly higher order would occur in the building blocks. In Grad's approach, the list of variables increases (see Section 7.4.3).

7.3.2 Chapman-Enskog Solution Technique

As a starting point, we write Boltzmann's equation (7.17) or (7.86) in the absence of external forces in the form

$$\frac{\partial f(\boldsymbol{r},\boldsymbol{p})}{\partial t} = -\frac{\boldsymbol{p}}{m} \cdot \frac{\partial f(\boldsymbol{r},\boldsymbol{p})}{\partial \boldsymbol{r}} + \frac{1}{\epsilon_{\rm mfp}} \left[\frac{\partial f(\boldsymbol{r},\boldsymbol{p})}{\partial t}\right]_{\rm coll}^{ff}, \tag{7.90}$$

with the bilinear collision integral

$$\left[\frac{\partial f(\boldsymbol{r},\boldsymbol{p})}{\partial t}\right]_{\rm coll}^{f_1 f_2} = \int v_{\rm rel} \frac{d\sigma}{d\Omega} [f_1(\boldsymbol{r},\boldsymbol{q}) f_2(\boldsymbol{r},\boldsymbol{q}') - f_1(\boldsymbol{r},\boldsymbol{p}) f_2(\boldsymbol{r},\boldsymbol{p}')] d\Omega d^3p' \tag{7.91}$$

and where $\epsilon_{\rm mfp}$ is considered as a small expansion parameter. The parameter $\epsilon_{\rm mfp}$ is used to organize the calculation, and, in the end, it is set equal to unity. A small parameter $\epsilon_{\rm mfp}$ corresponds to a large differential cross section or a dominant effect of collisions, which, in turn, corresponds to a short mean free path. The dimensionless expansion parameter can thus be identified with the mean free path divided by the characteristic length scale for variations of the quantities of interest, which is known as the Knudsen number, so that the expansion for small $\epsilon_{\rm mfp}$ captures long-range features, and we expect an expansion in terms of $\epsilon_{\rm mfp}$ to be relevant to the hydrodynamic regime.

To develop the Chapman-Enskog expansion procedure described schematically in the diagram of Figure 7.8, we write

$$f(\boldsymbol{r},\boldsymbol{p}) = \sum_{j=0}^{\infty} \epsilon_{\rm mfp}^{j} f^{(j)}(\boldsymbol{p}; x^{\{j\}}(\boldsymbol{r})), \tag{7.92}$$

[25] See p. 177–178 of Grad, Phys. Fluids 6 (1963) 147.
[26] Karlin, J. Phys. A: Math. Gen. 33 (2000) 8037.

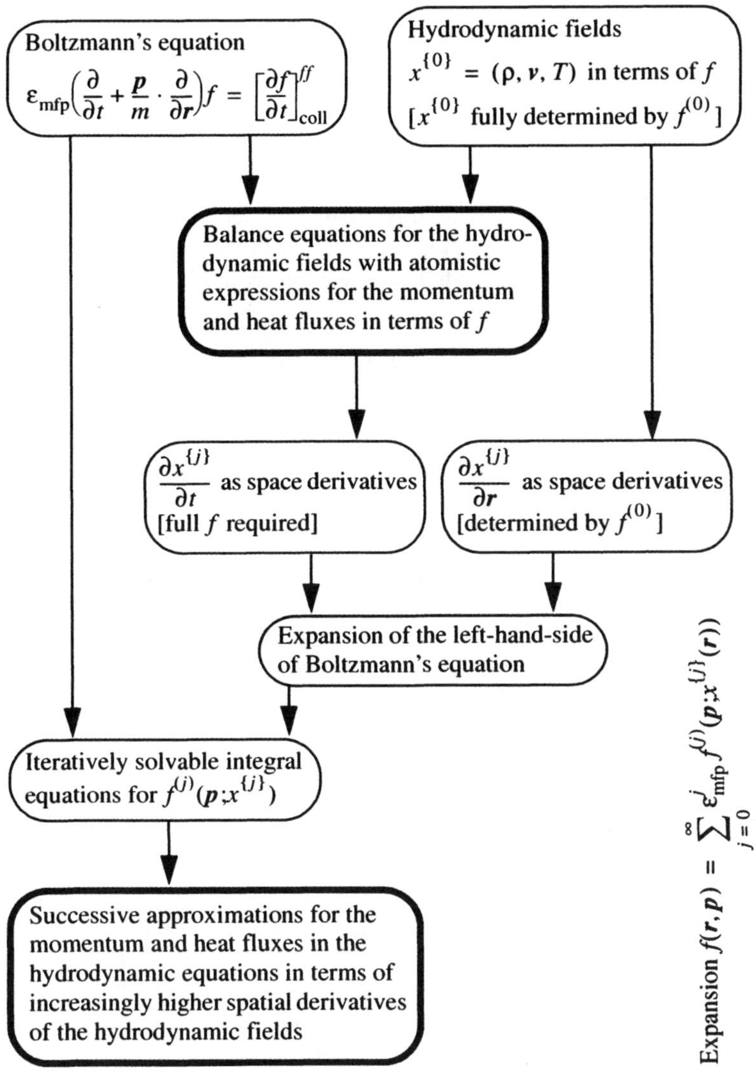

Fig. 7.8 Schematic explanation of the Chapman-Enskog procedure. The contribution to the single-particle distribution function of order $\epsilon_{\rm mfp}^j$ is given by $f^{(j)}$, which depends on the hydrodynamic fields $x^{\{0\}}(\boldsymbol{r})$ and their spatial derivatives of orders 1 to j, collectively denoted by $x^{\{j\}}(\boldsymbol{r})$. The gray background indicates where the expansion of the single-particle distribution function is relevant. The main results of the procedure are emphasized by thick lines around the corresponding boxes.

where the argument $x^{\{j\}}(\boldsymbol{r})$ in the expansion coefficient $f^{(j)}$ indicates that $f^{(j)}$ depends on the hydrodynamic fields and their spatial derivatives of orders 1 to j at the position $\boldsymbol{r}$, collectively denoted by $x^{\{j\}}(\boldsymbol{r})$; the list of hydrodynamic fields corresponds to $x^{\{0\}}(\boldsymbol{r})$. In the trial solutions (7.92), all position and time dependencies, are assumed to arise through the hydrodynamic fields and their derivatives; there is no explicit position or time dependence in $f^{(j)}$. Increasing powers of $\epsilon_{\rm mfp}$ come with increasingly higher-order space derivatives of the hydrodynamic fields and with products of lower-order derivatives; this reflects the fact that, in the dimensionless parameter $\epsilon_{\rm mfp}$, the mean free path is compared with the characteristic length scale for variations in the hydrodynamic fields. The hydrodynamic fields are given in terms of $f(\boldsymbol{r},\boldsymbol{p})$ as

$$\int m f(\boldsymbol{r},\boldsymbol{p}) d^3p = \rho(\boldsymbol{r}), \tag{7.93}$$

$$\int \boldsymbol{p} f(\boldsymbol{r},\boldsymbol{p}) d^3p = \boldsymbol{M}(\boldsymbol{r}) = \rho(\boldsymbol{r})\boldsymbol{v}(\boldsymbol{r}), \tag{7.94}$$

and

$$\int \frac{\boldsymbol{p}^2}{2m} f(\boldsymbol{r},\boldsymbol{p}) d^3p = \frac{1}{2}\rho(\boldsymbol{r})\boldsymbol{v}(\boldsymbol{r})^2 + \frac{3}{2}\frac{\rho(\boldsymbol{r})}{m} k_{\rm B} T(\boldsymbol{r}), \tag{7.95}$$

and the space derivatives of the hydrodynamic fields can then be expressed in terms of the space derivatives of $f(\boldsymbol{r},\boldsymbol{p})$; we can thus construct any $x^{\{j\}}(\boldsymbol{r})$ from $f(\boldsymbol{r},\boldsymbol{p})$. By means of Boltzmann's equation (7.90), we obtain the following expressions for the time derivatives of the hydrodynamic fields:

$$\frac{\partial \rho(\boldsymbol{r})}{\partial t} = -\frac{\partial}{\partial \boldsymbol{r}} \cdot [\boldsymbol{v}(\boldsymbol{r})\rho(\boldsymbol{r})], \tag{7.96}$$

$$\begin{aligned} \frac{\partial}{\partial t}[\boldsymbol{v}(\boldsymbol{r})\rho(\boldsymbol{r})] &= -\frac{\partial}{\partial \boldsymbol{r}} \cdot [\boldsymbol{v}(\boldsymbol{r})\boldsymbol{v}(\boldsymbol{r})\rho(\boldsymbol{r})] \\ &\quad -\frac{\partial}{\partial \boldsymbol{r}} \cdot \int \frac{[\boldsymbol{p} - m\boldsymbol{v}(\boldsymbol{r})][\boldsymbol{p} - m\boldsymbol{v}(\boldsymbol{r})]}{m} f(\boldsymbol{r},\boldsymbol{p}) d^3p, \end{aligned} \tag{7.97}$$

and

$$\begin{aligned} \frac{\partial T(\boldsymbol{r})}{\partial t} &= -\boldsymbol{v}(\boldsymbol{r}) \cdot \frac{\partial T(\boldsymbol{r})}{\partial \boldsymbol{r}} \\ &\quad -\frac{1}{\rho(\boldsymbol{r})\hat{c}_V} \int \frac{[\boldsymbol{p} - m\boldsymbol{v}(\boldsymbol{r})][\boldsymbol{p} - m\boldsymbol{v}(\boldsymbol{r})]}{m} f(\boldsymbol{r},\boldsymbol{p}) d^3p : \frac{\partial}{\partial \boldsymbol{r}}\boldsymbol{v}(\boldsymbol{r}) \\ &\quad -\frac{1}{\rho(\boldsymbol{r})\hat{c}_V}\frac{\partial}{\partial \boldsymbol{r}} \cdot \int \left[\frac{\boldsymbol{p}}{m} - \boldsymbol{v}(\boldsymbol{r})\right] \frac{[\boldsymbol{p} - m\boldsymbol{v}(\boldsymbol{r})]^2}{2m} f(\boldsymbol{r},\boldsymbol{p}) d^3p, \end{aligned} \tag{7.98}$$

where $\hat{c}_V = 3k_{\rm B}/(2m)$ and where the collision term does not contribute directly to these time derivatives in view of the local conservation laws for mass, momentum, and kinetic energy. The significance of the time-evolution equations (7.96)–(7.98) is twofold (see Figure 7.8). On the one hand, they constitute the final hydrodynamic equations once the expansion (7.92) for $f(\boldsymbol{r}, \boldsymbol{p})$ in terms of the hydrodynamic fields and their derivatives has been constructed. By comparing these equations for a rarefied gas to the general balance equations (2.10), (2.22), and (2.73), we actually recognize molecular expressions for the momentum and internal energy fluxes in terms of the single-particle distribution function:

$$\boldsymbol{\pi} = \int \frac{[\boldsymbol{p} - m\boldsymbol{v}(\boldsymbol{r})][\boldsymbol{p} - m\boldsymbol{v}(\boldsymbol{r})]}{m} f(\boldsymbol{r}, \boldsymbol{p}) d^3p, \tag{7.99}$$

and

$$\boldsymbol{j}^{\mathrm{q}} = \int \left[\frac{\boldsymbol{p}}{m} - \boldsymbol{v}(\boldsymbol{r})\right] \frac{[\boldsymbol{p} - m\boldsymbol{v}(\boldsymbol{r})]^2}{2m} f(\boldsymbol{r}, \boldsymbol{p}) d^3p. \tag{7.100}$$

On the other hand, (7.96)–(7.98) are needed to evaluate the time derivative of the single-particle distribution function when the expansion (7.92) in terms of the hydrodynamic fields and their derivatives is inserted into Boltzmann's equation (7.90) in order to find the equations for the expansion coefficients $f^{(j)}$ of the Chapman-Enskog procedure.

The subtle point to realize in the Chapman-Enskog expansion is that, when the series (7.92) is inserted into (7.90), the time derivatives of the hydrodynamic fields, and hence also of their space-derivatives, contain an expansion in ϵ_{mfp} due to the occurrence of $f(\boldsymbol{r}, \boldsymbol{p})$ in (7.97) and (7.98). To make the Chapman-Enskog procedure (involving the solution of an integral equation as seen in the detailed steps below) unique, we assume that the hydrodynamic fields (7.93)–(7.95) are fully determined by the terms $f^{(0)}$ and that the higher-order terms $f^{(j)}$ for $j = 1, 2, \ldots$ do not contribute to the corresponding integrals:

$$\int m f^{(j)}(\boldsymbol{p}; x^{\{j\}}(\boldsymbol{r})) d^3p = 0, \tag{7.101}$$

$$\int \boldsymbol{p} f^{(j)}(\boldsymbol{p}; x^{\{j\}}(\boldsymbol{r})) d^3p = 0, \tag{7.102}$$

and

$$\int \frac{\boldsymbol{p}^2}{2m} f^{(j)}(\boldsymbol{p}; x^{\{j\}}(\boldsymbol{r})) d^3p = 0. \tag{7.103}$$

In other words, the hydrodynamic fields ρ, $\boldsymbol{v}$, and T are not expanded in terms of ϵ_{mfp}.[27] It should be noted, however, that the higher-order terms $f^{(j)}$ do contribute to the time derivatives of the hydrodynamic fields in (7.97) and (7.98) and hence to the final hydrodynamic equations because uncontracted second and also certain third moments are involved.

We are now in the position to write successively the equations for $f^{(0)}$, $f^{(1)}$, ... by inserting the series (7.92) into Boltzmann's equation (7.90) and by using the time

[27] A Chapman-Enskog-like procedure can also be applied to a larger set of nonexpanded hydrodynamic fields, say by including the stress tensor and the heat flux into the list of coarse-grained variables; see Struchtrup & Torrilhon, Phys. Fluids 15 (2003) 2668 for details.

derivatives of the hydrodynamic fields and their space derivatives as obtained from (7.96)–(7.98). After multiplying by ϵ_{mfp}, one can compare terms order by order, and one obtains the equation for $f^{(j)}$ from the terms of order ϵ_{mfp}^j.

The equation for $f^{(0)}$ is given by

$$\left[\frac{\partial f(\boldsymbol{r},\boldsymbol{p})}{\partial t}\right]_{\text{coll}}^{f^{(0)}f^{(0)}} = 0, \tag{7.104}$$

the solution of which has been found in Section 7.1.4 to be the local Maxwellian,

$$f^{(0)}(\boldsymbol{p};x^{\{0\}}(\boldsymbol{r})) = f^{(0)}(\boldsymbol{p};\rho(\boldsymbol{r}),\boldsymbol{v}(\boldsymbol{r}),T(\boldsymbol{r})) = f_{\rho,\boldsymbol{v},T}^{\text{Maxw}}(\boldsymbol{r},\boldsymbol{p}). \tag{7.105}$$

According to (7.32), $f^{(0)}$ indeed depends only on the momenta and on the hydrodynamic fields. The moments of $f^{(0)}$ give the correct hydrodynamic fields (7.93)–(7.95), as can be verified with the results of Exercise 156. The hydrodynamic equations obtained by using $f^{(0)}$ in (7.96)–(7.98) are Euler's equations for an inviscid ideal gas, that is, the continuity equation (7.96), the momentum balance

$$\frac{\partial}{\partial t}[\boldsymbol{v}(\boldsymbol{r})\rho(\boldsymbol{r})] = -\frac{\partial}{\partial \boldsymbol{r}}\cdot[\boldsymbol{v}(\boldsymbol{r})\boldsymbol{v}(\boldsymbol{r})\rho(\boldsymbol{r})] - \frac{\partial}{\partial \boldsymbol{r}}\left[\frac{\rho(\boldsymbol{r})}{m}k_{\text{B}}T(\boldsymbol{r})\right], \tag{7.106}$$

and the temperature equation

$$\frac{\partial T(\boldsymbol{r})}{\partial t} = -\boldsymbol{v}(\boldsymbol{r})\cdot\frac{\partial T(\boldsymbol{r})}{\partial \boldsymbol{r}} - \frac{2T}{3}\frac{\partial}{\partial \boldsymbol{r}}\cdot\boldsymbol{v}(\boldsymbol{r}), \tag{7.107}$$

which should be compared with (2.22) and (2.73).

7.3.3 Transport Coefficients

As a next step, we determine the equation for $f^{(1)}$ by the procedure described in the preceding subsection. We obtain

$$\left(\frac{\partial}{\partial t} + \frac{\boldsymbol{p}}{m}\cdot\frac{\partial}{\partial \boldsymbol{r}}\right) f^{(0)}(\boldsymbol{p};x^{\{0\}}(\boldsymbol{r})) = \left[\frac{\partial f(\boldsymbol{r},\boldsymbol{p})}{\partial t}\right]_{\text{coll}}^{f^{(0)}f^{(1)}} + \left[\frac{\partial f(\boldsymbol{r},\boldsymbol{p})}{\partial t}\right]_{\text{coll}}^{f^{(1)}f^{(0)}}, \tag{7.108}$$

where the implied time derivatives of the hydrodynamic fields are to be evaluated with $f^{(0)}$, that is, Euler's equations are to be used. Equation (7.108) is an inhomogeneous linear integral equation for $f^{(1)}$. A similar integral equation, but with different inhomogeneous terms, needs to be solved for every $f^{(j)}$. The mathematical problem is furthermore related to the linearized Boltzmann equation (see Exercise 170) so that the steps involved in solving this linearized equation and in constructing the contributions to the Chapman-Enskog expansion are very similar. According to the theory of Fredholm equations, the conditions (7.101)–(7.103) make the solution unique because the invariants of the collision process give the only degenerate solutions.[28] By eliminating the time derivative with Euler's equations, we obtain the

[28] See p. 254–255 of Grad, Principles of the Kinetic Theory of Gases (1958).

explicit expression

$$\left(\frac{\partial}{\partial t}+\frac{\boldsymbol{p}}{m}\cdot\frac{\partial}{\partial \boldsymbol{r}}\right)f^{(0)} = \frac{f^{(0)}}{mk_{\rm B}T}\left[(\boldsymbol{p}-m\boldsymbol{v})(\boldsymbol{p}-m\boldsymbol{v})-\frac{(\boldsymbol{p}-m\boldsymbol{v})^2}{3}\mathbf{1}\right]:\frac{\partial}{\partial \boldsymbol{r}}\boldsymbol{v} + \frac{f^{(0)}}{mT}\left[\frac{(\boldsymbol{p}-m\boldsymbol{v})^2}{2mk_{\rm B}T}-\frac{5}{2}\right](\boldsymbol{p}-m\boldsymbol{v})\cdot\frac{\partial T}{\partial \boldsymbol{r}}, \qquad (7.109)$$

for the left-hand side of (7.108).

Exercise 173 Inhomogeneity in First-Order Equation
Derive (7.109).

By looking at (7.108) and (7.109), we realize that $f^{(1)}$ must be linear in the velocity and temperature gradients where the corresponding contributions can be determined separately. To form the scalar quantity $f^{(1)}$ out of these gradients, we need a tensor and a vector, to be formed out of $\boldsymbol{p}-m\boldsymbol{v}$. We hence arrive at the most general form

$$f^{(1)}(\boldsymbol{p};x^{\{1\}}) = f^{(0)}(\boldsymbol{p};x^{\{0\}})\left\{\frac{\bar{A}(|\boldsymbol{p}-m\boldsymbol{v}|)}{mT}(\boldsymbol{p}-m\boldsymbol{v})\cdot\frac{\partial T}{\partial \boldsymbol{r}} + \frac{\bar{B}(|\boldsymbol{p}-m\boldsymbol{v}|)}{mk_{\rm B}T}\left[(\boldsymbol{p}-m\boldsymbol{v})(\boldsymbol{p}-m\boldsymbol{v})-\frac{(\boldsymbol{p}-m\boldsymbol{v})^2}{3}\mathbf{1}\right]:\frac{\partial}{\partial \boldsymbol{r}}\boldsymbol{v}\right\}, \qquad (7.110)$$

where $\bar{A}$ and $\bar{B}$ are scalar functions of the scalar argument $|\boldsymbol{p}-m\boldsymbol{v}|$.[29] These scalar functions remain to be determined from the integral equations obtained by inserting (7.109) and (7.110) into (7.108), which is still a formidable exercise (see Exercise 175). From the consistency conditions (7.101)–(7.103), a constraint on $\bar{A}$ arises,

$$\int f^{\rm Maxw}_{\rho,0,T}(\boldsymbol{r},\boldsymbol{p})\bar{A}(|\boldsymbol{p}|)|\boldsymbol{p}|^2 d^3p = 0. \qquad (7.111)$$

The required multi-dimensional integrations in the collision integrals of the equations for $\bar{A}$ and $\bar{B}$ are usually done by expansions in terms of Sonine polynomials,[30] which also occur in solving the linearized Boltzmann equation. We can now see explicitly that $f^{(1)}$ depends not only on the hydrodynamic fields themselves but also on the gradients of the velocity and temperature fields, which are all summarized in $x^{\{1\}}$.

When the general form (7.110) is used in (7.99) and (7.100), even with undetermined functions $\bar{A}$ and $\bar{B}$, we obtain explicit expressions for the first-order contributions to the stress tensor and the heat flux in the Chapman-Enskog expansion. From

[29] See §7.3-d of Hirschfelder, Curtiss & Bird, *Molecular Theory of Gases and Liquids* (Wiley, 1954).

[30] Except for normalization, these Sonine polynomials are the same as the associated Laguerre polynomials.

(7.100) we obtain, after a shift in the integration variable and after exploiting the isotropy of the Maxwellian distribution,

$$\boldsymbol{j}^{\mathrm{q}(1)} = \frac{1}{6m^3T} \int f_{\rho,0,T}^{\mathrm{Maxw}}(\boldsymbol{r},\boldsymbol{p}) \bar{A}(|\boldsymbol{p}|)|\boldsymbol{p}|^4 d^3p \frac{\partial T}{\partial \boldsymbol{r}}. \tag{7.112}$$

By means of the Chapman-Enskog procedure, we have thus derived Fourier's law (2.37) for heat conduction with a molecular expression for the thermal conductivity,

$$\lambda^{\mathrm{q}} = -\frac{1}{6m^3T} \int f_{\rho,0,T}^{\mathrm{Maxw}}(\boldsymbol{r},\boldsymbol{p}) \bar{A}(|\boldsymbol{p}|)|\boldsymbol{p}|^4 d^3p. \tag{7.113}$$

The scalar function $\bar{A}$ needs to be determined from an integral equation in terms of the differential cross section for the interactions between gas molecules.

From (7.99) we obtain, after a shift in the integration variable,

$$\boldsymbol{\tau} = \frac{1}{m^2 k_\mathrm{B} T} \int f_{\rho,0,T}^{\mathrm{Maxw}}(\boldsymbol{r},\boldsymbol{p}) \bar{B}(|\boldsymbol{p}|)\boldsymbol{p}\boldsymbol{p} \left(\boldsymbol{p}\boldsymbol{p} - \frac{\boldsymbol{p}^2}{3}\boldsymbol{1}\right) d^3p : \frac{\partial}{\partial \boldsymbol{r}}\boldsymbol{v}. \tag{7.114}$$

The various fourth moments occurring in (7.114) can be related by again exploiting the isotropy of the Maxwellian distribution. We then arrive at Newton's stress tensor expression (2.38) with a vanishing bulk viscosity and a molecular expression for the shear viscosity,

$$\eta = -\frac{1}{15m^2 k_\mathrm{B} T} \int f_{\rho,0,T}^{\mathrm{Maxw}}(\boldsymbol{r},\boldsymbol{p}) \bar{B}(|\boldsymbol{p}|)|\boldsymbol{p}|^4 d^3p. \tag{7.115}$$

Like $\bar{A}$, the scalar function $\bar{B}$ needs to be determined from an integral equation in terms of the differential cross section for the interactions between gas molecules. Equations (7.113) and (7.115) are the more rigorous counterparts of the elementary kinetic theory results (7.14) and (7.12).

For the *Maxwell gas*, in which the interaction force is proportional to $1/R^5$ [see (7.45) for $n_\phi = 5$], the transport coefficients can be calculated more explicitly. A lengthy but straightforward calculation (see Exercise 175) shows that $\bar{A}$ and $\bar{B}$ are of the following forms:

$$\bar{A}(p) = -\frac{1}{\pi} \frac{1}{c^{\mathrm{Maxw}}} \sqrt{\frac{m}{2\hat{\phi}} \frac{m}{\rho}} \left(\frac{p^2}{2mk_\mathrm{B}T} - \frac{5}{2} \right) \tag{7.116}$$

and

$$\bar{B}(p) = \bar{B} = -\frac{2}{3\pi} \frac{1}{c^{\mathrm{Maxw}}} \sqrt{\frac{m}{2\hat{\phi}} \frac{m}{\rho}}, \tag{7.117}$$

where c^{Maxw} is a numerical constant. An approximate value can be found in Equation (209) of Boltzmann's lectures on gas theory,

$$c^{\mathrm{Maxw}} \approx \frac{2}{\pi} \cdot 1.3682 \approx 0.871, \tag{7.118}$$

where he states that the corresponding integral "was evaluated by Maxwell by mechanical quadrature."[31] With these expressions, one can evaluate the transport coefficients (7.113) and (7.115):

$$\lambda^{\mathrm{q}} = \frac{5}{2\pi} \frac{1}{c^{\mathrm{Maxw}}} \sqrt{\frac{m}{2\hat{\phi}}} k_{\mathrm{B}} T \frac{k_{\mathrm{B}}}{m} \tag{7.119}$$

and

$$\eta = \frac{2}{3\pi} \frac{1}{c^{\mathrm{Maxw}}} \sqrt{\frac{m}{2\hat{\phi}}} k_{\mathrm{B}} T. \tag{7.120}$$

For the Prandtl number of the Maxwell gas, we thus obtain the exact kinetic theory result

$$\frac{\eta \hat{c}_p}{\lambda^{\mathrm{q}}} = \frac{2}{3}, \tag{7.121}$$

which is in excellent agreement with the experimental findings listed in Table 7.1.3. The form of the expressions for the transport coefficients of the Maxwell gas can be compared more directly to the elementary kinetic theory results (7.12) and (7.14) if we introduce a characteristic size d of the Maxwell molecules. For dimensional reasons, we write

$$d^2 \approx \sqrt{\frac{\hat{\phi}}{k_{\mathrm{B}} T}}, \tag{7.122}$$

and we then observe the same functional form of the elementary and detailed kinetic theory results for the viscosity and the thermal conductivity.

The approximations obtained by continuing the Chapman-Enskog procedure and evaluating $f^{(2)}$ and $f^{(3)}$ are known as the Burnett and super-Burnett equations.[32] These equations, which involve higher derivatives of the hydrodynamic fields and products of lower-order derivatives, are not considered in this book. Usually, one considers only the structure of these equations in terms of the hydrodynamic fields and their derivatives because determinations of further scalar functions like $\bar{A}$ and $\bar{B}$ are beyond reach.

Exercise 174 Collision Integral for Maxwell Gas
Evaluate the collision integral for the Maxwell gas. In what respect is this interaction special?

Exercise 175 First-Order Distribution Function for Maxwell Gas
Derive the expressions (7.116) and (7.117) for the scalar functions $\bar{A}$ and $\bar{B}$.

Exercise 176 Transport Coefficients for Maxwell Gas
Derive the expressions (7.119) and (7.120) for the thermal conductivity and the viscosity.

[31] See p. 163 of Boltzmann, *Gastheorie, Teil I* (Barth, 1896); in the English translation [Boltzmann, *Lectures on Gas Theory* (Dover, 1995)], it was added by the translator that recalculations made in the early 20th century led to the slightly larger values 1.3704 and 1.3700 instead of 1.3682, implying $c^{\mathrm{Maxw}} \approx 0.872$. The author's own calculation gave $c^{\mathrm{Maxw}} \approx 0.87239$ (see solution to Exercise 175).

[32] See p. 206 of Kreuzer, *Nonequilibrium Thermodynamics & Statistical Foundations* (Oxford, 1981).

7.4 GRAD'S MOMENT METHOD

In the preceding section, by means of the Chapman-Enskog expansion, the detailed level of single-particle distribution functions was related to the much coarser level of conserved quantities or hydrodynamic fields. The purpose of this section is to show that one can find a hierarchy of intermediate levels, based on increasingly higher moments of the single-particle distribution function in momentum space. In doing so, one can approximate an increasingly larger set of solutions of the Boltzmann equation. As an application, we show how some features of extended irreversible thermodynamics can be motivated by means of the moment method from kinetic theory (see Section 5.1.6).

7.4.1 Basic Idea

In the Chapman-Enskog approach, increasingly higher-order derivatives and products of lower-order derivatives of the hydrodynamic fields occur in proceeding with the expansion. By including these higher-order derivatives, the range of validity is *not* extended to shorter length and time scales; one can only expect further improvements where the Navier-Stokes equations already work well. Additional features of the Boltzmann equation at short length and time scales can only be captured by considering further coarse-grained variables in addition to the hydrodynamic ones.

The idea underlying Grad's method is to use a class of trial solutions that contains a certain number of parameters in addition to the hydrodynamic fields ρ, $\boldsymbol{v}$, and T. These position-dependent parameters serve as the macroscopic fields of a coarse-grained level of description. Time-evolution equations for the macroscopic fields are obtained as follows. One chooses a number of low-order moments of the probability density in momentum space, and, for equal numbers of parameters and moments, one establishes a relationship between the parameters of the trial solutions and the moments in velocity space. The time-evolution equations for the moments obtained from Boltzmann's equation are closed because one can eliminate higher-order moments by means of the explicit form of the trial solutions. We thus arrive at time-evolution equations for the moments or, equivalently, for the parameters in the trial solutions, which are considered as the macroscopic fields. The form of the trial solutions and the set of moments to be reproduced are the basic inputs of the method. Unlike the Chapman-Enskog expansion, Grad's method is not based on a systematic expansion in a small parameter such as $\epsilon_{\rm mfp}$. As a consequence, the equations for the macroscopic fields in general depend on the chosen set of moments.[33] In the more abstract and more general *invariant manifold method* of Gorban and Karlin,[34] the deviation of the trial solutions from the exact solution of the time-evolution equation for a probability density is considered explicitly, and it is shown how it can be minimized iteratively.

[33] Velasco, Uribe & García-Colín, Phys. Rev. E 66 (2002) 032103.

[34] Gorban & Karlin, Physica A 190 (1992) 393; Transp. Theory Stat. Phys. 23 (1994) 559.

In Grad's work,[35] the trial functions are locally Maxwellian distributions multiplied by a factor whose velocity dependence is expanded in terms of Hermite polynomials.[36] This choice is motivated by the fact that the usual hydrodynamic fields are related to moments of degree zero (mass density), one (velocity), and two (temperature), as can be seen from (7.93)–(7.95).[37] Further terms of degree two and some terms of degree three are associated with the traceless pressure tensor and the heat-flux vector. The thirteen-moment method is hence related to "Extended Irreversible Thermodynamics,"[38] which is based on the fluxes of momentum and energy as independent variables in addition to the hydrodynamic ones. If some subtleties concerning the proper convected time derivatives are disregarded, Grad's method provides the kinetic theory behind extended irreversible thermodynamics in the same way as the first-order Chapman-Enskog expansion of Boltzmann's equation leads to hydrodynamics. In any case, the respective macroscopic equations are not necessarily restricted to the rarefied gases for which the kinetic theory was designed.

7.4.2 Thirteen-Moment Method

In Grad's original approach, the trial solutions for Boltzmann's equation are taken to be of the form

$$\begin{aligned} f(\boldsymbol{r},\boldsymbol{p}) &= f^{\rm Maxw}_{\rho,\boldsymbol{v},T}(\boldsymbol{r},\boldsymbol{p})\Bigg\{1+\frac{1}{mT}\left[\frac{(\boldsymbol{p}-m\boldsymbol{v})^2}{2mk_{\rm B}T}-\frac{5}{2}\right](\boldsymbol{p}-m\boldsymbol{v})\cdot\boldsymbol{w} \\ &- \frac{1}{2mk_{\rm B}T}\left[(\boldsymbol{p}-m\boldsymbol{v})(\boldsymbol{p}-m\boldsymbol{v})-\frac{(\boldsymbol{p}-m\boldsymbol{v})^2}{3}\boldsymbol{1}\right]:\mathring{\boldsymbol{c}}\Bigg\}, \end{aligned} \tag{7.123}$$

which should be compared with (7.110) with (7.116) and (7.117). The new variables introduced in (7.123), the vector $\boldsymbol{w}$ and the tensor $\mathring{\boldsymbol{c}}$ replacing the temperature and velocity gradients of (7.110), are, like the hydrodynamic fields, functions of position. Any antisymmetric or trace parts of $\mathring{\boldsymbol{c}}$ in the above *ansatz* are irrelevant, and we hence assume that $\mathring{\boldsymbol{c}}$ is traceless and symmetric to have an unambiguous representation. If a traceless tensor $\mathring{\boldsymbol{c}}$ is assumed, the isotropic term containing $\boldsymbol{1} : \mathring{\boldsymbol{c}}$ in (7.123) can be omitted.

The additional fields $\boldsymbol{w}$ and $\mathring{\boldsymbol{c}}$, which provide an intermediate level of description between kinetic theory and hydrodynamics, are associated with second-order and certain third-order terms in an expansion of the deviation from a local Maxwellian in

[35] Grad, Commun. Pure Appl. Math. 2 (1949) 331.

[36] Hermite polynomials are naturally associated with Gaussian probability distributions and provide convenient orthogonality and normalization conditions; see Section 8.95 of Gradshteyn & Ryzhik, *Tables* (Academic, 1980).

[37] Trivial expansion coefficients need to be chosen because the local Maxwellian contains the lowest-order terms in an exponentiated form.

[38] Jou et al., *Extended Irreversible Thermodynamics* (Springer, 1996).

powers of particle momentum. These variables are chosen such that they account for heat flow ($\boldsymbol{w}$) and for momentum flow ($\mathring{\boldsymbol{c}}$) with a vanishing bulk viscosity. In addition to the five hydrodynamic fields, the vector $\boldsymbol{w}$ has three components and the traceless symmetric tensor $\mathring{\boldsymbol{c}}$ has five independent components; we thus arrive at a total of 13 fields, and we hence refer to a thirteen-moment method.

Closer inspection of the term associated with $\boldsymbol{w}$ in (7.123) shows that a certain combination of third and first powers of the particle momentum occurs. From all possible third-order polynomials, those combinations allowing one a direct description of heat flow phenomena are chosen. First-order terms had to be included to keep the definition (7.94) for the velocity. The proper choice of polynomials is guided by the properties of Hermite polynomials, for which one has orthogonality and normalization relations when integrated with an isotropic Gaussian probability density. The observation of the relevance of Hermite polynomials becomes particularly important when the expansion (7.123) is continued to higher-order polynomials in the particle momentum, as proposed by Grad. In his original approach, the choice for the trial functions and for the moments to be reproduced are hence intimately related for reasons of mathematical elegance. However, we here restrict ourselves to the thirteen-moment method, and there is no need to go any further into the properties of Hermite polynomials. It should be noted that, a priori, the validity of such a moment-truncation procedure is unclear because a more rapid time evolution of the neglected higher moments compared with the first few moments cannot be established ("there is no proof of the efficacy of the method, nor of its accuracy at any given order of approximation"[39]). The invariant manifold method of Gorban and Karlin,[40] however, allows one to monitor the discrepancy between the exact and the approximate rates of change of the moments.

A serious drawback of Grad's expansion in terms of polynomials is the fact that the trial functions are no proper probability densities because they can become negative. Although one may argue that the consequences for the behavior of the moments are minor, major problems might arise for the thermodynamic admissibility. It is not even clear how one should evaluate the entropy (7.55) within the original Grad method.

Following the basic idea described in the preceding section, we now have to identify suitable second and third moments associated with $\mathring{\boldsymbol{c}}$ and $\boldsymbol{w}$ such that time-evolution equations can be obtained from Boltzmann's equation, exactly in the spirit of the moment expressions (7.93)–(7.95) and time-evolution equations (7.96)–(7.98), where closure is obtained by exploiting (7.123). By performing the respective Gaussian integrals, one obtains the following moment expressions for the parameters $\mathring{\boldsymbol{c}}$ and $\boldsymbol{w}$ in (7.123):[41]

$$\int \frac{(\boldsymbol{p} - m\boldsymbol{v})(\boldsymbol{p} - m\boldsymbol{v})}{m} f(\boldsymbol{r}, \boldsymbol{p}) d^3p = \frac{\rho}{m} k_{\mathrm{B}} T(\mathbf{1} - \mathring{\boldsymbol{c}}) \qquad (7.124)$$

[39] See p. 231 of de Groot et al., *Relativistic Kinetic Theory* (North-Holland, 1980).

[40] Gorban & Karlin, Transp. Theory Stat. Phys. 23 (1994) 559; Karlin et al., Phys. Rev. E 57 (1998) 1668.

[41] Although this choice is most natural, in principle, other moments could be chosen, and the resulting equations for $\mathring{\boldsymbol{c}}$ and $\boldsymbol{w}$ would be different; an explicit example has been discussed in Velasco, Uribe & García-Colín, Phys. Rev. E 66 (2002) 032103.

and

$$\int \left(\frac{\boldsymbol{p}}{m} - \boldsymbol{v}\right) \frac{(\boldsymbol{p} - m\boldsymbol{v})^2}{2m} f(\boldsymbol{r}, \boldsymbol{p}) d^3p = \frac{\rho}{m} k_B T \frac{5}{2} \frac{k_B}{m} \boldsymbol{w}. \tag{7.125}$$

A comparison with (7.99) and (7.100) shows that $\mathring{\boldsymbol{c}}$ and $\boldsymbol{w}$ are directly associated with the pressure tensor and the heat flux vector, respectively. These flux variables have thus been elevated to the status of independent variables in describing beyond-equilibrium states on a refined hydrodynamic level, which is precisely the basic idea of extended irreversible thermodynamics. We thus have the following reformulation of the equations (7.96)–(7.98) for the hydrodynamic fields,

$$\frac{\partial \rho}{\partial t} = -\frac{\partial}{\partial \boldsymbol{r}} \cdot (\boldsymbol{v}\rho), \tag{7.126}$$

$$\frac{\partial}{\partial t}(\boldsymbol{v}\rho) = -\frac{\partial}{\partial \boldsymbol{r}} \cdot (\boldsymbol{v}\boldsymbol{v}\rho) - \frac{\partial}{\partial \boldsymbol{r}} \cdot \left[\frac{\rho}{m} k_B T(\mathbf{1} - \mathring{\boldsymbol{c}})\right], \tag{7.127}$$

$$\frac{\partial T}{\partial t} = -\boldsymbol{v} \cdot \frac{\partial T}{\partial \boldsymbol{r}} - \frac{2}{3} T(\mathbf{1} - \mathring{\boldsymbol{c}}) : \boldsymbol{\kappa} - \frac{5}{3} \frac{1}{\rho m} \frac{\partial}{\partial \boldsymbol{r}} \cdot (\rho k_B T \boldsymbol{w}). \tag{7.128}$$

To obtain the time-evolution equations for $\mathring{\boldsymbol{c}}$ and $\boldsymbol{w}$ to close the system of equations, we derive the equation of change for the average of an arbitrary, possibly time dependent quantity $A = A(\boldsymbol{r}, \boldsymbol{p})$. From the Boltzmann equation (7.17), in the absence of external forces, we obtain the following expression, which may look unnecessarily complicated but is found to be most convenient:

$$\begin{aligned}
\frac{d}{dt} \int A f d^3p &= \int \left(\frac{\partial A}{\partial t} + \boldsymbol{v} \cdot \frac{\partial A}{\partial \boldsymbol{r}}\right) f d^3p - \frac{\partial}{\partial \boldsymbol{r}} \cdot \left(\boldsymbol{v} \int A f d^3p\right) \\
&+ \int \frac{\boldsymbol{p} - m\boldsymbol{v}}{m} \cdot \frac{\partial A}{\partial \boldsymbol{r}} f d^3p - \frac{\partial}{\partial \boldsymbol{r}} \cdot \left(\int \frac{\boldsymbol{p} - m\boldsymbol{v}}{m} A f d^3p\right) \\
&+ \int A \left[\frac{\partial f}{\partial t}\right]_{\text{coll}} d^3p.
\end{aligned} \tag{7.129}$$

In general, the contribution from collisions, that is, the last term in (7.129), cannot be evaluated rigorously. In evaluating this collision term according to (7.22) or (7.91), we obtain contributions linear and quadratic in the relative deviation $\hat{f}^{(1)}$ of (7.123) from the local Maxwellian distribution (cf. the notation introduced in Exercise 172). For Maxwell molecules, the results of Exercise 175 imply that the linear contributions correspond to

$$\begin{aligned}
\left[\frac{\partial f(\boldsymbol{p})}{\partial t}\right]_{\text{coll}} &= f^{\text{Maxw}}_{\rho,\boldsymbol{v},T}(\boldsymbol{p}) \int v_{\text{rel}} \frac{d\sigma}{d\Omega} f^{\text{Maxw}}_{\rho,\boldsymbol{v},T}(\boldsymbol{p}') [\hat{f}^{(1)}(\boldsymbol{q}) + \hat{f}^{(1)}(\boldsymbol{q}') \\
&\qquad - \hat{f}^{(1)}(\boldsymbol{p}) - \hat{f}^{(1)}(\boldsymbol{p}')] d\Omega d^3p' \\
&= \left\{ -\frac{1}{2mT} \left[\frac{(\boldsymbol{p} - m\boldsymbol{v})^2}{2mk_B T} - \frac{5}{2}\right] (\boldsymbol{p} - m\boldsymbol{v}) \cdot \boldsymbol{w} \right. \\
&+ \left. \frac{3}{8mk_B T} (\boldsymbol{p} - m\boldsymbol{v})(\boldsymbol{p} - m\boldsymbol{v}) : \mathring{\boldsymbol{c}} \right\} 2\pi c^{\text{Maxw}} \sqrt{\frac{2\hat{\phi}}{m} \frac{\rho}{m}} f^{\text{Maxw}}_{\rho,\boldsymbol{v},T}(\boldsymbol{p}).
\end{aligned}$$

(7.130)

With this expression for the collision integral, the general equation of change (7.129) can be applied to the moments (7.124) and (7.125), and, after a lengthy but straightforward calculation, we obtain

$$\begin{aligned}\frac{\partial \mathring{\boldsymbol{c}}}{\partial t} &= -\boldsymbol{v}\cdot\frac{\partial}{\partial \boldsymbol{r}}\mathring{\boldsymbol{c}} - \boldsymbol{\kappa}\cdot\mathring{\boldsymbol{c}} - \mathring{\boldsymbol{c}}\cdot\boldsymbol{\kappa}^T + \boldsymbol{\kappa} + \boldsymbol{\kappa}^T - \frac{2}{3}(\mathbf{1}-\mathring{\boldsymbol{c}})(\mathbf{1}-\mathring{\boldsymbol{c}}) : \boldsymbol{\kappa} \\ &+ \frac{1}{\rho m T}\left\{\frac{\partial}{\partial \boldsymbol{r}}(\rho k_B T \boldsymbol{w}) + \left[\frac{\partial}{\partial \boldsymbol{r}}(\rho k_B T \boldsymbol{w})\right]^T - \frac{2}{3}\frac{\partial}{\partial \boldsymbol{r}}\cdot(\rho k_B T\boldsymbol{w})\mathbf{1}\right\} \\ &+ \frac{5}{3}\frac{1}{\rho m T}\mathring{\boldsymbol{c}}\frac{\partial}{\partial \boldsymbol{r}}\cdot(\rho k_B T\boldsymbol{w}) - \frac{1}{\tau}\mathring{\boldsymbol{c}},\end{aligned} \tag{7.131}$$

for the tensor describing the momentum flux, and

$$\begin{aligned}\frac{\partial \boldsymbol{w}}{\partial t} &= -\boldsymbol{v}\cdot\frac{\partial}{\partial \boldsymbol{r}}\boldsymbol{w} - \frac{7}{5}\boldsymbol{\kappa}\cdot\boldsymbol{w} - \frac{2}{5}\boldsymbol{\kappa}^T\cdot\boldsymbol{w} - \frac{2}{5}\boldsymbol{w}\frac{\partial}{\partial \boldsymbol{r}}\cdot\boldsymbol{v} + \frac{2}{3}\boldsymbol{w}(\mathbf{1}-\mathring{\boldsymbol{c}}) : \boldsymbol{\kappa} \\ &+ \frac{5}{3}\boldsymbol{w}\frac{1}{\rho m T}\frac{\partial}{\partial \boldsymbol{r}}\cdot(\rho k_B T\boldsymbol{w}) - (\mathbf{1}-\mathring{\boldsymbol{c}})\cdot\frac{\partial T}{\partial \boldsymbol{r}} - \frac{2}{5}\frac{T}{\rho}\mathring{\boldsymbol{c}}\cdot\frac{\partial \rho}{\partial \boldsymbol{r}} \\ &+ \frac{2}{5}\frac{1}{\rho}(\mathbf{1}+\mathring{\boldsymbol{c}})\cdot\left[\frac{\partial}{\partial \boldsymbol{r}}\cdot(\rho T\mathring{\boldsymbol{c}})\right] - \frac{2}{3}\frac{1}{\tau}\boldsymbol{w},\end{aligned} \tag{7.132}$$

for the vector describing the heat flux, where

$$\tau = \frac{2}{3\pi}\frac{1}{c^{\mathrm{Maxw}}}\sqrt{\frac{m}{2\hat{\phi}}}\frac{m}{\rho} \tag{7.133}$$

is a fundamental relaxation time. Equations (7.126)–(7.128), (7.131), and (7.132) are the 13 time-evolution equations of Grad's thirteen-moment approximation. They can also be considered as the statistical foundation of extended irreversible thermodynamics.[42] It can be shown that the terms quadratic in the deviation from the Maxwellian velocity distribution, which had been neglected in (7.130), do not alter the result for Maxwell molecules.[43] For other interaction potentials, (7.131) and (7.132) provide approximate results. Grad found that the approximation based on the linearized collision integral is excellent, and he gave suitable expressions for the relaxation rates in the last contributions to (7.131) and (7.132) in terms of the differential cross section for the interaction between gas molecules.[44]

If the variables $\mathring{\boldsymbol{c}}$ and $\boldsymbol{w}$ relax rapidly compared with the time required for any appreciable macroscopic changes and if nonlinear terms in the gradients of the hydrodynamic fields and cross-couplings between the momentum and heat fluxes can be

[42] See Section 3.3 of Jou et al., *Extended Irreversible Thermodynamics* (Springer, 1996).
[43] See Section 28 of Grad, Principles of the Kinetic Theory of Gases (1958).
[44] See (28.21) of Grad, Principles of the Kinetic Theory of Gases (1958).

neglected, then (7.131) and (7.132) imply that $\mathring{c}$ and w become dependent variables characterized by

$$\boldsymbol{\kappa} + \boldsymbol{\kappa}^T - \frac{2}{3}\mathrm{tr}\boldsymbol{\kappa}\mathbf{1} = \frac{1}{\tau}\mathring{\boldsymbol{c}} \tag{7.134}$$

and

$$-\frac{\partial T}{\partial \boldsymbol{r}} = \frac{2}{3}\frac{1}{\tau}\boldsymbol{w}. \tag{7.135}$$

In view of (7.127) and (7.128), these equations reproduce Newton's expression for the momentum flux and Fourier's law for the heat flux, where the thermal conductivity and the viscosity,

$$\lambda^{\mathrm{q}} = \frac{15}{4}\frac{\rho k_{\mathrm{B}}^2 T}{m^2}\tau, \qquad \eta = \frac{\rho}{m}k_{\mathrm{B}}T\tau, \tag{7.136}$$

are given by (7.119) and (7.120). We thus again arrive at the equations of classical hydrodynamics. For Maxwell molecules, we found the same expressions for the transport coefficients as obtained by the Chapman-Enskog procedure. For other interaction potentials, the necessary approximations can naturally be made such that the Chapman-Enskog results are reproduced.

For steady homogeneous flows under isothermal conditions, (7.131) and (7.132) read $\boldsymbol{w} = 0$ and

$$-\boldsymbol{\kappa}\cdot\mathring{\boldsymbol{c}} - \mathring{\boldsymbol{c}}\cdot\boldsymbol{\kappa}^T + \boldsymbol{\kappa} + \boldsymbol{\kappa}^T - \frac{2}{3}(\mathbf{1} - \mathring{\boldsymbol{c}})(\mathbf{1} - \mathring{\boldsymbol{c}}) : \boldsymbol{\kappa} = \frac{1}{\tau}\mathring{\boldsymbol{c}}. \tag{7.137}$$

For the homogeneous shear flow characterized by the constant velocity gradient tensor (2.39), the explicit solution of (7.137) can be given (see Exercise 186). One obtains for the viscometric functions defined in (4.30)–(4.32)

$$\eta = \frac{\rho}{m}k_{\mathrm{B}}T\bar{x}, \quad \Psi_1 = -2\frac{\rho}{m}k_{\mathrm{B}}T\bar{x}\sqrt{\tau\bar{x}}, \quad \Psi_2 = 0, \tag{7.138}$$

where the auxiliary variable $\bar{x}$ is determined by the cubic equation

$$\bar{x}\left(1 + \frac{2}{3}\tau\bar{x}\dot{\gamma}^2\right)^2 = \tau. \tag{7.139}$$

For small shear rates, we recover the viscosity in (7.136), and we find a negative first normal-stress coefficient. The second normal-stress coefficient vanishes for all shear rates.

Grad's thirteen-moment method gives rise to a hyperbolic system of differential equations, that is, equations in which there is a finite speed of wave propagation. On the other hand, the equations of classical hydrodynamics are parabolic and exhibit an infinite speed of propagation of signals. For mathematical simplicity, a hyperbolic system is often preferable. However, the solution of hyperbolic systems can be interrupted by the appearance of shocks.

The time-evolution equations (7.131) and (7.132) should be supplemented by appropriate boundary conditions. The proper formulation of these boundary conditions is a subtle issue and, in view of the existence of zero speed characteristics, not even the required number of boundary conditions is obvious.[45]

Exercise 177 An Unusual Time Derivative
In (7.131), the time derivative

$$\frac{\partial \mathring{c}}{\partial t} + \boldsymbol{v} \cdot \frac{\partial \mathring{c}}{\partial \boldsymbol{r}} + \boldsymbol{\kappa} \cdot \mathring{c} + \mathring{c} \cdot \boldsymbol{\kappa}^T \tag{7.140}$$

occurs. Is this time derivative of the upper-convected or lower-convected type? Does it coincide with the Schowalter derivative (2.93) for a suitable choice of the slip parameter ξ?

7.4.3 Structured Moment Method

A fundamental problem with the standard expansion techniques used to derive hydrodynamic equations is that any truncation after a finite number of terms, in general, leads to coarse-grained equations that do not possess the proper structure of the time-evolution equations established for general nonequilibrium systems. In the course of time, higher-order terms do arise so that an arbitrary truncation after a finite number of terms is inconsistent with the time evolution. Although such expansions may be systematic to a certain order in an expansion parameter, thermodynamic consistency can make it necessary to keep certain higher-order terms. We here show how a moment expansion with GENERIC structure can be constructed.

In Chapter 6, we have shown how coarse-graining can be carried out without destroying the GENERIC structure of beyond-equilibrium thermodynamics. Here, however, we are not dealing with coarse-graining; we rather try to solve the thermodynamically admissible Boltzmann equation, and we want the approximate or truncated solution to be thermodynamically admissible, too. For that purpose we can consider the solution procedure as a transformation to the parameters of the trial solution. In the space of trial solutions, the transformation from the variables of the original problem to the parameters of the trial solutions can be considered as a one-to-one transformation, and the formulas of Section 1.2.4 lead us to a GENERIC solution.

Now, how should we choose the trial solutions of the Boltzmann equation in terms of moments? From a GENERIC perspective, Grad's choice (7.123) is not ideal for the time-structure invariance of reversible dynamics. This can be seen after recalling the arguments of Section 6.1.6, in particular, concerning the natural transformation behavior of the variables on different levels of description under space transformations. Although the action of space transformations on single-particle distribution functions is not needed to construct the Poisson bracket on the Boltzmann level, it is important for the extended set of hydrodynamic variables constructed from the single-particle distribution function because, as moments, these variables inherit their space-transformation behavior from the distribution functions, so that the construction

[45] See Section 29 of Grad, Principles of the Kinetic Theory of Gases (1958).

of extended Poisson brackets by Lie-Poisson reduction can be performed as described in Appendix B.3. Under such transformations, the isotropic Maxwellian distribution in (7.123) becomes anisotropic (see Exercise 164), and this observation is crucial for passing proper transformation behavior onto the moments of momentum. Although a small anisotropy could be expanded, we prefer to work with anisotropic Gaussians of the form

$$f(\boldsymbol{p}) = \frac{\rho}{m} \frac{[\det(\mathbf{1} - \mathring{\boldsymbol{c}})]^{-1/2}}{(2\pi m k_{\mathrm{B}} T)^{3/2}} \exp\left\{ -\frac{(\boldsymbol{p} - m\boldsymbol{v}) \cdot (\mathbf{1} - \mathring{\boldsymbol{c}})^{-1} \cdot (\boldsymbol{p} - m\boldsymbol{v})}{2 m k_{\mathrm{B}} T} \right\}, \tag{7.141}$$

where the normalization is such that the integral over momentum yields the particle number density. While ρ, $\boldsymbol{v}$, and T have their usual meanings as mass density, velocity, and temperature fields, the traceless tensor field $\mathring{\boldsymbol{c}}$ corresponds directly to the one in (7.123). More precisely, (7.141) reproduces the second moments (7.124), and (7.123) can be regarded as an expanded version of (7.141), where antisymmetric contributions related to heat transfer have generally been suppressed for the discussion of this subsection (we here assume $\boldsymbol{w} = 0$).

From a superficial point of view, one can choose a number of observables, and one can write the logarithm of the single-particle distribution function as a linear combination of the chosen observables to arrive at (7.141). This construction guarantees the positivity required for a probability density,[46] whereas the original function (7.123) typically becomes negative for large momenta. From a deeper point of view, such an expansion of $\ln f$ is obtained by maximizing the entropy (7.55) under the constraint that the averages of the chosen observables are prescribed. Entropy maximization then leads to a generalized canonical ensemble of the exponential form, as discussed in Section 6.1.2. From that perspective, it is most natural to base Grad's approach on trial functions of the exponential form and to use the averages of the observables occurring in the exponential as the macroscopic fields equivalent to the parameters, or Lagrange multipliers, in the trial functions. The anisotropic Gaussian (7.141) has previously been considered in this spirit by Kogan,[47] by Lewis,[48] and by Gorban and Karlin.[49] Maximizing the entropy for obtaining trial solutions in terms of a reduced number of variables is an important general concept, often associated with the keyword *quasi-equilibrium*.

We can now use the transformation formulas of Section 1.2.4 to obtain the GENERIC building blocks for a rarefied gas described on the level of the variables $y = (\rho, \boldsymbol{v}, T, \mathring{\boldsymbol{c}})$ from those for $x = f$. By evaluating the energy (7.53) with the single-particle distribution function (7.141), we obtain

$$E(y) = \int \left(\frac{1}{2} \rho v^2 + \frac{3}{2} \frac{\rho}{m} k_{\mathrm{B}} T \right) d^3 r, \tag{7.142}$$

[46] Note, however, that the integrability needs to be checked and, if necessary, restored by some regularization procedure.

[47] Kogan, J. Appl. Math. Mech. 29 (1965) 130.

[48] Lewis, J. Math. Phys. 8 (1967) 1448.

[49] Gorban & Karlin, Physica A 206 (1994) 401.

in which the kinetic energy and the internal energy of an ideal gas can be recognized. For the gradient of this energy function, we then find

$$\frac{\delta E(y)}{\delta y} = \begin{pmatrix} \frac{1}{2}\boldsymbol{v}^2 + \frac{3}{2}\frac{k_\mathrm{B}T}{m} \\ \rho\boldsymbol{v} \\ \frac{3}{2}\frac{\rho}{m}k_\mathrm{B} \\ 0 \end{pmatrix}. \tag{7.143}$$

Similarly, we can evaluate the entropy (7.55) with the single-particle distribution function (7.141),

$$S(y) = -k_\mathrm{B}\int \frac{\rho}{m}\left[\ln\frac{\rho}{mN} - \frac{1}{2}\ln\det(\mathbf{1} - \mathring{\boldsymbol{c}}) - \frac{3}{2}\ln(2\pi m k_\mathrm{B}T) - \frac{3}{2}\right] d^3r, \tag{7.144}$$

with the corresponding gradient

$$\frac{\delta S(y)}{\delta y} = \begin{pmatrix} \frac{k_\mathrm{B}}{2m}\ln[(T^3/\rho^2)\det(\mathbf{1} - \mathring{\boldsymbol{c}})] + \mathrm{const.} \\ 0 \\ \frac{3k_\mathrm{B}}{2m}\frac{\rho}{T} \\ -\frac{k_\mathrm{B}}{2m}\rho(\mathbf{1} - \mathring{\boldsymbol{c}})^{-1} \end{pmatrix}. \tag{7.145}$$

In calculating the functional derivatives in (7.143) and (7.145), we have neglected contributions resulting from constraints (see Appendix C.3), such as the conserved total mass or the traceless nature of $\mathring{\boldsymbol{c}}$, because they are irrelevant to the further calculations. Note that the calculation of the entropy (7.144) is simplified enormously by the fact that (7.141) is the exponential of a quadratic form. We will find below (see Exercise 181) that the entropy must have the functional form (7.144); this indicates that (7.141) is the proper choice for the distribution function.

To calculate the transformed Poisson and friction matrices (1.50) and (1.51), we need to calculate the functional derivatives $\delta y/\delta x$, which are the derivatives of the moments $(\rho, \boldsymbol{v}, T, \mathring{\boldsymbol{c}})$ in momentum space with respect to the single-particle distribution function. For any quantity $A(\boldsymbol{r},\boldsymbol{p})$, we define $\hat{A}(\boldsymbol{r},\boldsymbol{p})$ by

$$\hat{A}(\boldsymbol{r},\boldsymbol{p}) = \frac{\delta\int A(\boldsymbol{r}',\boldsymbol{p}')f(\boldsymbol{r}',\boldsymbol{p}')d^3r'd^3p'}{\delta f(\boldsymbol{r},\boldsymbol{p})}, \tag{7.146}$$

which can be different from A if A has a further dependence on averages such as $\boldsymbol{v}$. For evaluating the Poisson and friction matrices, we now have two options. Either we calculate the matrix entries directly for the independent variables of interest, or we calculate the Poisson and friction matrices for some simpler moments, say the averages of m, $\boldsymbol{p}$, and $\boldsymbol{pp}$, and afterward obtain these matrices for the variables of interest by means of the transformation formulas of Section 1.2.4. In the following two exercises, we develop some auxiliary results required for the more direct first option, which allows us to calculate each matrix element of the Poisson and friction matrices individually.

Exercise 178 Corrections for Observables Containing Averages

Evaluate $\hat{A}$ for quantities of the form $A(\rho(\boldsymbol{r}),\boldsymbol{p})$ and $A(\boldsymbol{v}(\boldsymbol{r}),\boldsymbol{p})$.

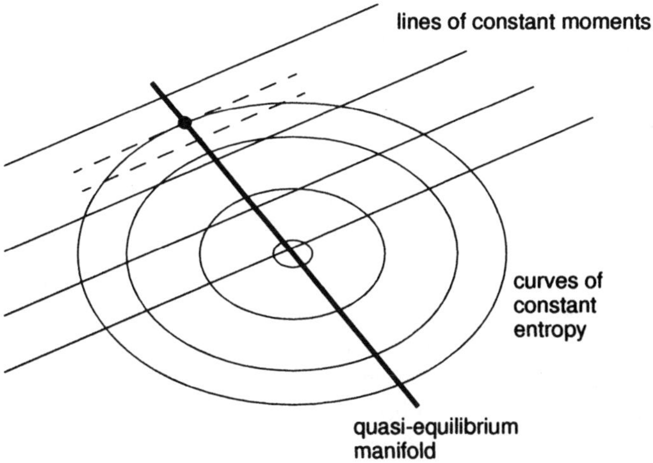

Fig. 7.9 Direction of entropy gradient for a quasi-equilibrium or maximum entropy manifold. The ellipses represent curves of constant entropy; on the continuous straight lines, the moments are constant. A gradient vector can be represented by a set of straight lines: its direction is perpendicular to the lines, and its magnitude is proportional to the density of level lines or inversely proportional to their distance. In any point of the quasi-equilibrium manifold of Gaussian distributions (thick line), the gradient of the entropy (represented by the dashed straight lines locally tangential to the ellipses) is a multiple of the gradient of the moments (represented by the continuous straight lines).

Exercise 179 Examples of Functional Derivatives

Evaluate $\hat{A}$ for the following quantities associated with ρ, $\boldsymbol{v}$, T, $\mathring{\boldsymbol{c}}$, respectively:

$$m, \quad \frac{\boldsymbol{p}}{\rho}, \quad \frac{(\boldsymbol{p}-m\boldsymbol{v})^2}{3\rho k_{\mathrm{B}}}, \quad \frac{(\boldsymbol{p}-m\boldsymbol{v})^2\mathbf{1}-3(\boldsymbol{p}-m\boldsymbol{v})(\boldsymbol{p}-m\boldsymbol{v})}{3\rho k_{\mathrm{B}}T}.$$

The functional derivatives in (7.146) are constructed without accounting for the constraint that the single-particle distribution function is a Gaussian. In view of the highly restrictive nature of the Gaussian constraint, the constrained derivatives would be significantly different from the unconstrained ones. For example, $\delta\mathring{\boldsymbol{c}}/\delta f$ as obtained in Exercise 179 is a second-order polynomial in the momenta; the corresponding constrained derivative, however, is a second-order polynomial multiplied by a Gaussian. The reason for choosing unconstrained derivatives can be explained with Figure 7.9. As emphasized before, the Gaussian distribution (7.141) can be obtained by maximizing the entropy (7.55). In any point of the quasi-equilibrium manifold, the gradient of the entropy can hence be expressed as a linear combination of the unconstrained derivatives of the moments, $\delta y/\delta f$, where the coefficients can actually be identified with the derivatives of the entropy with respect to the moments,

$\delta S/\delta y$,

$$\frac{\delta S}{\delta f} = \frac{\delta S}{\delta y} \cdot \frac{\delta y}{\delta f}. \tag{7.147}$$

Because the energy depends on f only through the moments y, one has a similar relation for the energy gradients:

$$\frac{\delta E}{\delta f} = \frac{\delta E}{\delta y} \cdot \frac{\delta y}{\delta f}. \tag{7.148}$$

From these transformation laws for energy and entropy and from the arguments of Section 1.2.4, it should be clear that also the Poisson and friction matrices must be transformed with the unconstrained derivatives $\delta y/\delta f$.

In the operator notation for Poisson matrices, which was discussed at the end of Section 2.2.2, the transformation formula (1.50) leads from (7.59) to the following element of the Poisson matrix associated with the observables A and B:

$$\begin{aligned} L_{AB}(\boldsymbol{r}) &= -\int \hat{A}(\boldsymbol{r},\boldsymbol{p})\frac{\partial}{\partial \boldsymbol{r}} \cdot f(\boldsymbol{r},\boldsymbol{p})\frac{\partial \hat{B}(\boldsymbol{r},\boldsymbol{p})}{\partial \boldsymbol{p}} d^3p \\ &\quad - \int f(\boldsymbol{r},\boldsymbol{p})\frac{\partial \hat{A}(\boldsymbol{r},\boldsymbol{p})}{\partial \boldsymbol{p}} \cdot \frac{\partial}{\partial \boldsymbol{r}}\hat{B}(\boldsymbol{r},\boldsymbol{p}) d^3p. \end{aligned} \tag{7.149}$$

Specializing this result to the pairs of components of $y = (\rho, \boldsymbol{v}, T, \mathring{\boldsymbol{c}})$, we obtain by means of the results of Exercise 179,

$$L(\boldsymbol{r}) = \begin{pmatrix} 0 & -\frac{\partial}{\partial \boldsymbol{r}} & 0 & 0 \\ -\frac{\partial}{\partial \boldsymbol{r}} & \frac{1}{\rho}\boldsymbol{\omega} & \frac{1}{\rho}\frac{\partial T}{\partial \boldsymbol{r}} - \frac{2}{3}\frac{1}{\rho}\frac{\partial}{\partial \boldsymbol{r}} \cdot (\mathbf{1} - \mathring{\boldsymbol{c}})T & L_{24} \\ 0 & -\frac{1}{\rho}\frac{\partial T}{\partial \boldsymbol{r}} - \frac{2}{3}T(\mathbf{1} - \mathring{\boldsymbol{c}}) \cdot \frac{\partial}{\partial \boldsymbol{r}}\frac{1}{\rho} & 0 & L_{34} \\ 0 & L_{42} & L_{43} & L_{44} \end{pmatrix}, \tag{7.150}$$

where $\boldsymbol{\omega}$ is the vorticity tensor

$$\boldsymbol{\omega} = \boldsymbol{\kappa}^T - \boldsymbol{\kappa} \quad \text{or} \quad \omega_{jk} = \frac{\partial v_k}{\partial r_j} - \frac{\partial v_j}{\partial r_k} \tag{7.151}$$

and the beyond-hydrodynamic entries in this Poisson matrix are

$$\begin{aligned} (L_{24})_{ikl} &= \frac{1}{\rho}\frac{\partial\, \mathring{c}_{kl}}{\partial r_i} + \frac{1}{\rho}\left\{\frac{\partial}{\partial r_k}(\mathbf{1} - \mathring{\boldsymbol{c}})_{il} + \frac{\partial}{\partial r_l}(\mathbf{1} - \mathring{\boldsymbol{c}})_{ki} \right. \\ &\quad \left. - \frac{2}{3}\left[\frac{\partial}{\partial \boldsymbol{r}} \cdot (\mathbf{1} - \mathring{\boldsymbol{c}})\right]_i (\mathbf{1} - \mathring{\boldsymbol{c}})_{kl}\right\}, \end{aligned} \tag{7.152}$$

$$\begin{aligned} (L_{42})_{ijk} &= -\frac{1}{\rho}\frac{\partial\, \mathring{c}_{ij}}{\partial r_k} + \left\{(\mathbf{1} - \mathring{\boldsymbol{c}})_{kj}\frac{\partial}{\partial r_i} + (\mathbf{1} - \mathring{\boldsymbol{c}})_{ik}\frac{\partial}{\partial r_j} \right. \\ &\quad \left. - \frac{2}{3}(\mathbf{1} - \mathring{\boldsymbol{c}})_{ij}\left[(\mathbf{1} - \mathring{\boldsymbol{c}}) \cdot \frac{\partial}{\partial \boldsymbol{r}}\right]_k \right\}\frac{1}{\rho}, \end{aligned} \tag{7.153}$$

$$L_{34} = -L_{43} = \frac{2m}{3\rho k_B}(\mathring{\boldsymbol{c}} \cdot \boldsymbol{\omega} - \boldsymbol{\omega} \cdot \mathring{\boldsymbol{c}}), \tag{7.154}$$

and

$$\begin{aligned}(L_{44})_{ijkl} &= \frac{m}{\rho k_B T}\left[\frac{2}{3}(\mathbf{1} - \mathring{\boldsymbol{c}})_{ij}(\mathring{\boldsymbol{c}} \cdot \boldsymbol{\omega} - \boldsymbol{\omega} \cdot \mathring{\boldsymbol{c}})_{kl} - \frac{2}{3}(\mathring{\boldsymbol{c}} \cdot \boldsymbol{\omega} - \boldsymbol{\omega} \cdot \mathring{\boldsymbol{c}})_{ij}(\mathbf{1} - \mathring{\boldsymbol{c}})_{kl}\right. \\ &+ \left.(\mathbf{1} - \mathring{\boldsymbol{c}})_{ik}\omega_{jl} + (\mathbf{1} - \mathring{\boldsymbol{c}})_{il}\omega_{jk} + (\mathbf{1} - \mathring{\boldsymbol{c}})_{jl}\omega_{ik} + (\mathbf{1} - \mathring{\boldsymbol{c}})_{jk}\omega_{il}\right].\end{aligned} \tag{7.155}$$

In determining the explicit Poisson matrix (7.150), in addition to up to second moments of the momentum distribution, it has only been assumed that all third moments of $\boldsymbol{p} - m\boldsymbol{v}$ vanish; the precise functional form of (7.141) has actually been irrelevant and, in particular, the same Poisson matrix would have been obtained with (7.123). As a consequence of the rigorous nature of the procedure, the Poisson bracket implied by (7.150) satisfies the Jacobi identity.

With (7.145) and (7.150), one can verify the degeneracy requirement (1.4). The precise functional form of (7.141) is important to get the entropy (7.144), and hence for the validity of the degeneracy requirement expressing the entropy conservation under reversible dynamics.

Exercise 180 Construction of Poisson Matrix

For understanding the construction of the Poisson matrix (7.150), calculate L_{34} from (7.149).

Exercise 181 Uniqueness of Entropy

Assume that the entropy expression is a generalized version of (7.144),

$$S(y) = -k_B \int \frac{\rho}{m}\left[\ln\frac{\rho}{mN} - s(\mathring{\boldsymbol{c}}) - \frac{3}{2}\ln(2\pi m k_B T) - \frac{3}{2}\right] d^3r. \tag{7.156}$$

Find the restriction on $s(\mathring{\boldsymbol{c}})$ resulting from the degeneracy requirement (1.4).

The reversible time evolution of $(\rho, \boldsymbol{v}, T, \mathring{\boldsymbol{c}})$ obtained by multiplying the energy gradient (7.143) by the Poisson matrix (7.150) coincides with the results of Section 7.4.2. More precisely, we fully recover (7.126)–(7.128) in the absence of $\boldsymbol{w}$ (one can formally use $\boldsymbol{w} = 0$ to eliminate all terms associated with $\boldsymbol{w}$), and we obtain all the terms of (7.131) except for the very last contribution, which is the irreversible relaxation term (a modified version of the relaxation term is formulated in Exercise 185). This agreement is not coincidental because the calculation of the reversible dynamics is rigorous in both approaches. Equation (7.123) would actually give the same energy and Poisson matrix as (7.141); however, (7.141) is preferable because it leads to the closed-form entropy expression (7.144) and hence to a strictly conserved entropy under reversible dynamics.

The fact that there are nonzero entries outside of the second row or column of the Poisson matrix (7.150) is remarkable because, as is clear from the underlying Boltzmann equation, the only reversible effect is of a convective nature. Convection can only have such an unexpected influence on the additional tensor variable $\mathring{\boldsymbol{c}}$, which can be regarded as an inertial effect, because $\mathring{\boldsymbol{c}}$ is a moment of the velocity distribution; if an additional tensor variable describes configurations in space, such nonzero entries outside of the second row or column cannot arise. This fundamental difference between momentum and configurational degrees of freedom is the reason why standard equations for viscoelastic liquids, such as Oldroyd's constitutive equation, are not easily mimicked by lattice Boltzmann simulations.[50]

We finally calculate the friction matrix by means of the transformation formula (1.51). From (7.68), we obtain for the matrix element associated with the observables A and B:

$$\begin{aligned} M_{AB}(\boldsymbol{r}) &= \frac{2\rho}{k_{\mathrm{B}} m^2} \int f_{\mathrm{t}}(\boldsymbol{r}, \boldsymbol{p}_{\mathrm{t}}) f_{\mathrm{r}}(\boldsymbol{r}, \boldsymbol{P}') \hat{A}(\boldsymbol{r}, \frac{\boldsymbol{p}_{\mathrm{t}}}{2} + \boldsymbol{P}) \\ & \quad |\boldsymbol{P}| \bar{\boldsymbol{I}}_{\phi}(\boldsymbol{P}', \boldsymbol{P}) \cdot \frac{\partial}{\partial \boldsymbol{P}'} \hat{B}(\boldsymbol{r}, \frac{\boldsymbol{p}_{\mathrm{t}}}{2} + \boldsymbol{P}') d^3P d^3P' d^3p_{\mathrm{t}}. \end{aligned} \tag{7.157}$$

As for the Poisson matrix, multiplication with this friction matrix does not include any spatial integrations. The probability densities for the total and relative momentum in (7.157) are given by

$$f_{\mathrm{t}}(\boldsymbol{p}_{\mathrm{t}}) = \frac{\rho}{m} \frac{[\det(\boldsymbol{1} - \mathring{\boldsymbol{c}})]^{-1/2}}{(4\pi m k_{\mathrm{B}} T)^{3/2}} \exp\left\{ - \frac{(\boldsymbol{p}_{\mathrm{t}} - 2m\boldsymbol{v}) \cdot (\boldsymbol{1} - \mathring{\boldsymbol{c}})^{-1} \cdot (\boldsymbol{p}_{\mathrm{t}} - 2m\boldsymbol{v})}{4 m k_{\mathrm{B}} T} \right\} \tag{7.158}$$

and

$$f_{\mathrm{r}}(\boldsymbol{P}') = \frac{[\det(\boldsymbol{1} - \mathring{\boldsymbol{c}})]^{-1/2}}{(\pi m k_{\mathrm{B}} T)^{3/2}} \exp\left\{ - \frac{\boldsymbol{P}' \cdot (\boldsymbol{1} - \mathring{\boldsymbol{c}})^{-1} \cdot \boldsymbol{P}'}{m k_{\mathrm{B}} T} \right\}. \tag{7.159}$$

The expression (7.157) implies that the first column of the friction matrix associated with the mass density, or $\hat{B} = m$, vanishes. For the first line, or $\hat{A} = m$, it is not so obvious that all entries vanish; this is a consequence of the lack of symmetry of the friction matrix (7.73). To construct the full rows of the friction matrix, we hence consider the integrals

$$\int |\boldsymbol{P}| \bar{\boldsymbol{I}}_{\phi}(\boldsymbol{P}', \boldsymbol{P}) d^3P, \quad \int \boldsymbol{P} |\boldsymbol{P}| \bar{\boldsymbol{I}}_{\phi}(\boldsymbol{P}', \boldsymbol{P}) d^3P, \quad \int \boldsymbol{P}\boldsymbol{P} |\boldsymbol{P}| \bar{\boldsymbol{I}}_{\phi}(\boldsymbol{P}', \boldsymbol{P}) d^3P. \tag{7.160}$$

We recall that, for repulsive potentials, the vector quantity $\bar{\boldsymbol{I}}_{\phi}(\boldsymbol{P}', \boldsymbol{P})$ vanishes for $|\boldsymbol{P}'| > |\boldsymbol{P}|$, has a vanishing divergence with respect to $\boldsymbol{P}'$ for $|\boldsymbol{P}'| < |\boldsymbol{P}|$, and has a jump along the sphere $|\boldsymbol{P}'| = |\boldsymbol{P}|$; the prototype of such a function is given in (7.78) for hard spheres. The integrations in (7.160) can thus be restricted to $|\boldsymbol{P}| > |\boldsymbol{P}'|$.

As a vector, the first integral in (7.160) must be a scalar multiple of $\boldsymbol{P}'$. Equations (7.74) and (7.77) actually imply that this integral must be a fixed multiple of $\boldsymbol{P}'/|\boldsymbol{P}'|^3$ and, because there should not be a singularity at $|\boldsymbol{P}'| = 0$, the integral must vanish

[50] Wagner, arXiv:cond-mat/0105067; see also p. 15 of Lallemand et al., Phys. Rev. E 67 (2003) 021203.

identically. For hard spheres, this result can be obtained by explicit integration over all orientations of $\boldsymbol{P}$ for constant length $|\boldsymbol{P}|$. The fact that the first integral in (7.160) vanishes implies that all entries in the first row of the friction matrix are zero.

In view of the antisymmetry property (7.80), the second integral in (7.160) vanishes; therefore, the second row of the friction matrix associated with the velocity vanishes for symmetry reasons, and the same is true also for the second column.

For the contraction of the first two indices of the third integral in (7.160), the same arguments as for the first integral can be applied, and we hence find that the third row of the friction matrix also vanishes. By now, we have established the following form of the friction matrix:

$$M(\boldsymbol{r}) = \begin{pmatrix} 0 & 0 & 0 & 0 \\ 0 & 0 & 0 & 0 \\ 0 & 0 & 0 & 0 \\ 0 & 0 & M_{43} & M_{44} \end{pmatrix}. \tag{7.161}$$

In view of its tensor structure, the third integral in (7.160) must be of the following form:

$$\left(\int \boldsymbol{P}\boldsymbol{P}|\boldsymbol{P}|\bar{\boldsymbol{I}}_\phi(\boldsymbol{P}', \boldsymbol{P})d^3P\right)_{ijk} = \frac{\hat{I}(P')}{P'^3}P'_k\left(3\frac{P'_iP'_j}{P'^2} - \delta_{ij}\right)$$
$$+ \left[\frac{P'^3}{2}\int(\sin\theta')^2 d\sigma' + \frac{1}{2}\frac{1}{P'^2}\frac{d\hat{I}(P')}{dP'}\right]\left(P'_i\delta_{jk} + P'_j\delta_{ik} - 2\frac{P'_iP'_jP'_k}{P'^2}\right). \tag{7.162}$$

Only a single function $\hat{I}(P') = \hat{I}(|\boldsymbol{P}'|)$ remains undetermined because the integral must vanish upon contraction of the indices i and j and because the divergence of $\bar{\boldsymbol{I}}_\phi(\boldsymbol{P}', \boldsymbol{P})$ with respect to $\boldsymbol{P}'$ is given by (7.74). The function $\hat{I}(P')$, which we refer to as "interaction impulse integral," depends on the particle interaction. A compact and explicit expression for the interaction impulse integral can be given in the form

$$\hat{I}(P') = \frac{1}{2}\frac{1}{P'}\int |\boldsymbol{P}|(\boldsymbol{P}\cdot\boldsymbol{P}')^2\boldsymbol{P}'\cdot\bar{\boldsymbol{I}}_\phi(\boldsymbol{P}', \boldsymbol{P})d^3P. \tag{7.163}$$

For hard spheres, we find by dimensional arguments that the interaction impulse integral $\hat{I}(P')$ must be of the form

$$\hat{I}(P') = c^{\mathrm{hs}}P'^6 d^2, \tag{7.164}$$

where c^{hs} is a dimensionless coefficient. However, if one tries to calculate c^{hs} from (7.78) and (7.163), one obtains a divergent integral expression for c^{hs}, indicating that the hard-sphere interaction is too singular for the present approach (see Exercise 182).

For the nonvanishing entries of the friction matrix (7.161), we now have

$$(M_{43})_{ij} = -\frac{4}{3m^3k_\mathrm{B}^3T}\int f_\mathrm{r}(\boldsymbol{P}')\frac{\hat{I}(P')}{P'}\left(3\frac{P'_iP'_j}{P'^2} - \delta_{ij}\right)d^3P' \tag{7.165}$$

and

$$\begin{aligned}(M_{44})_{ijkl} &= \frac{2}{m^3 k_B^3 T^2} \int f_r(\boldsymbol{P}') \left(\int \boldsymbol{P}\boldsymbol{P}|\boldsymbol{P}|\bar{\boldsymbol{I}}_\phi(\boldsymbol{P}', \boldsymbol{P}) d^3 P \right)_{ijn} \\ &\times \; (P'_k \delta_{ln} + P'_l \delta_{kn}) d^3 P' \\ &+ \; \frac{1}{T} (M_{43})_{ij} (\boldsymbol{1} - \mathring{\boldsymbol{c}})_{kl},\end{aligned} \tag{7.166}$$

where n is used as a summation index (from 1 to 3). The structure of M_{44} becomes more transparent if we introduce

$$\hat{\delta}_{ij} = \delta_{ij} - \frac{P'_i P'_j}{P'^2}. \tag{7.167}$$

By combining (7.162) and (7.166), we then obtain

$$\begin{aligned}(M_{44})_{ijkl} &= \frac{1}{m^3 k_B^3 T^2} \int f_r(\boldsymbol{P}') \Bigg\{ 4 \frac{\hat{I}(P')}{P'^3} \left(3 \frac{P'_i P'_j}{P'^2} - \delta_{ij} \right) P'_k P'_l \\ &+ \; \left[P'^3 \int (\sin\theta')^2 d\sigma' + \frac{1}{P'^2} \frac{d\hat{I}(P')}{dP'} \right] \\ &\qquad \times (P'_i \hat{\delta}_{jn} + P'_j \hat{\delta}_{in})(P'_k \delta_{ln} + P'_l \delta_{kn}) \Bigg\} d^3 P' \\ &+ \; \frac{1}{T} (M_{43})_{ij} (\boldsymbol{1} - \mathring{\boldsymbol{c}})_{kl},\end{aligned} \tag{7.168}$$

where we could equivalently write $\hat{\delta}_{ln}$ and $\hat{\delta}_{kn}$ instead of δ_{ln} and δ_{kn}, respectively. Equations (7.165) and (7.168) constitute our general result for the friction matrix (7.161). This completes our formulation of a Grad-type kinetic theory with GENERIC structure. However, we can still try to simplify the friction matrix without changing the time-evolution equation.

If we evaluate

$$\left(M \cdot \frac{\delta S}{\delta x} \right)_4 = M_{43} \frac{\delta S}{\delta T} + M_{44} : \frac{\delta S}{\delta \mathring{\boldsymbol{c}}}, \tag{7.169}$$

we obtain by means of (7.165) and (7.166)

$$\begin{aligned}\left(M \cdot \frac{\delta S}{\delta x} \right)_{4\,ij} &= -\frac{2\rho}{m^4 k_B^2 T^2} \int f_r(\boldsymbol{P}') \left(\int \boldsymbol{P}\boldsymbol{P}|\boldsymbol{P}|\bar{\boldsymbol{I}}_\phi(\boldsymbol{P}', \boldsymbol{P}) d^3 P \right)_{ijk} \\ &\times \; (\boldsymbol{1} - \mathring{\boldsymbol{c}})^{-1}_{kl} P'_l d^3 P'.\end{aligned} \tag{7.170}$$

For further simplification, it is important to note that, according to (7.159), the following building block occurs in (7.170):

$$(\boldsymbol{1} - \mathring{\boldsymbol{c}})^{-1}_{kl} P'_l = -\frac{1}{2} m k_B T \frac{\partial}{\partial P'_k} \ln f_r(\boldsymbol{P}'). \tag{7.171}$$

We thus obtain

$$\left(M \cdot \frac{\delta S}{\delta x}\right)_{4\,ij} = \frac{\rho}{m^3 k_B T} \int \left(\int \boldsymbol{P}\boldsymbol{P}|\boldsymbol{P}|\bar{\boldsymbol{I}}_\phi(\boldsymbol{P}', \boldsymbol{P}) d^3 P \right)_{ijk} \frac{\partial f_r(\boldsymbol{P}')}{\partial P'_k} d^3 P'. \tag{7.172}$$

After integrating by parts, we thus find that only the divergence of $\bar{\boldsymbol{I}}_\phi(\boldsymbol{P}', \boldsymbol{P})$ with respect to $\boldsymbol{P}'$ matters, which can be evaluated by means of (7.74) and (7.77). After using the time reversal symmetry of the collision process, we then find

$$\left(M \cdot \frac{\delta S}{\delta x}\right)_4 = \frac{2\rho}{m^3 k_B T} \int \int (\boldsymbol{P}\boldsymbol{P} - \boldsymbol{Q}\boldsymbol{Q}) d\sigma' f_r(\boldsymbol{P}) P d^3 P. \tag{7.173}$$

The result of integrating the dyadic $\boldsymbol{QQ}$ of the final relative momentum after the collision over all impact conditions $d\sigma'$ can depend only on the initial conditions $\boldsymbol{PP}$ and the unit tensor. We can hence further simplify (7.173) to obtain the final result

$$\left(M \cdot \frac{\delta S}{\delta x}\right)_4 = \frac{\rho}{m^3 k_B T} \int \int (\sin\theta')^2 d\sigma' (3\boldsymbol{P}\boldsymbol{P} - P^2 \boldsymbol{1}) f_r(\boldsymbol{P}) P d^3 P, \tag{7.174}$$

where θ' is the scattering angle in the center-of-mass frame. Equation (7.174) describes the full nonlinear relaxation behavior of the tensor $\mathring{\boldsymbol{c}}$.

Because only the divergence of $\bar{\boldsymbol{I}}_\phi(\boldsymbol{P}', \boldsymbol{P})$ is relevant in (7.172), all terms containing the interaction impulse integral in (7.162) cancel out in the time-evolution equation. It is hence unnecessary to evaluate the interaction impulse integral, and we can equivalently set $M_{43} = 0$ and

$$\begin{aligned} (M_{44})_{ijkl} &= \frac{1}{m^3 k_B^3 T^2} \int \int (\sin\theta')^2 d\sigma' P'^3 f_r(\boldsymbol{P}') \\ &\quad \times (P'_i \hat{\delta}_{jn} + P'_j \hat{\delta}_{in})(P'_k \hat{\delta}_{ln} + P'_l \hat{\delta}_{kn}) d^3 P'. \end{aligned} \tag{7.175}$$

The symmetry $(M_{44})_{ijkl} = (M_{44})_{klij}$ of this expression is obvious; the formula (7.175) leads to quadratic expressions, and thus the positive semidefinite nature of the friction matrix follows.

In contrast to (7.131), the friction matrix (7.175) multiplied by the entropy gradient (7.145) leads to a highly nonlinear relaxation behavior (7.174) of $\mathring{\boldsymbol{c}}$. The linearized relaxation term is obtained by linearizing the entropy gradient in $\mathring{\boldsymbol{c}}$ and by evaluating the fiction matrix for $\mathring{\boldsymbol{c}} = 0$, that is, with a local Maxwellian depending only on the length of $\boldsymbol{P}'$. From (7.168), we then obtain

$$(M_{44})_{ijkl} = \bar{M}_{44}(\delta_{ik}\delta_{jl} + \delta_{il}\delta_{jk} - \frac{2}{3}\delta_{ij}\delta_{kl}) \tag{7.176}$$

with

$$\bar{M}_{44} = \frac{8\pi}{5 m^3 k_B^3 T^2} \int_0^\infty \int (\sin\theta')^2 d\sigma' P'^7 f_r(P') dP'. \tag{7.177}$$

Note that this coefficient $\bar{M}_{44}$ can be obtained in a straightforward manner from the full matrix M_{44} by double contraction, $\bar{M}_{44} = (M_{44})_{lkkl}/10$.

Equations (7.176) and (7.177) constitute our final result for the friction matrix at local equilibrium. Multiplying this friction matrix with the entropy gradient (7.145) linearized in $\mathring{\boldsymbol{c}}$, we get the following relaxation time in the evolution equation for $\mathring{\boldsymbol{c}}$,

$$\tau = \frac{1}{k_B}\frac{m}{\rho}\frac{1}{\bar{M}_{44}}. \tag{7.178}$$

According to (7.136), this relaxation time corresponds to the following formula for the viscosity,

$$\eta = \frac{T}{\bar{M}_{44}} = \frac{5}{8\pi}(mk_BT)^3 \left[\int_0^\infty \int (\sin\theta')^2 d\sigma' P'^7 f_r(P')dP'\right]^{-1}. \tag{7.179}$$

This expression for the relaxation rate in (7.178), which is referred to as "exact for Maxwellian molecules and a good approximation in general" by Grad,[51] arises naturally in the GENERIC approach.

The equations from the structured kinetic theory may not look very different from Grad's original equations. Only the admissibility of a linearized relaxation term is questioned by thermodynamics (see Exercises 184 and 185). However, the unique form of the entropy (7.144) provides deeper insight into Grad's equations.

If we want to include the heat flux into the discussion of this section, we can accommodate the additional variable $\boldsymbol{w}$ of Section 7.4.2 in the single-particle distribution function by multiplying (7.141) with a factor of the form

$$1 + \sum_{j=0}^\infty f_j[\boldsymbol{w}\cdot(\boldsymbol{1}-\mathring{\boldsymbol{c}})^{-1}\cdot(\boldsymbol{p}-m\boldsymbol{v})]^{2j+1}, \tag{7.180}$$

where the coefficients f_j are functions of $(\boldsymbol{p}-m\boldsymbol{v})\cdot(\boldsymbol{1}-\mathring{\boldsymbol{c}})^{-1}\cdot(\boldsymbol{p}-m\boldsymbol{v})/(2mk_BT)$. For computational convenience, it is advantageous to write f_j in terms of Hermite polynomials. Whereas the factor (7.180) does not affect even moments, one needs to choose f_j carefully such that the definition of the velocity field is not affected.

Exercise 182 Interaction Impulse Integral for Hard Spheres

Derive an integral expression for the numerical factor c^{hs} in (7.164) and show that the integral diverges.

Exercise 183 Validity of Friction Matrix

Show that the simplified friction matrix (7.161) with $M_{43} = 0$ and M_{44} given by (7.175) reproduces the nonlinear relaxation term (7.174).

Exercise 184 Inadmissibility of Grad's Original Approach

We have learned that the local-equilibrium friction matrix (7.176) multiplying entropy gradients linearized in $\mathring{\boldsymbol{c}}$ reproduces Grad's original equation (7.131). What does the quadratic entropy term look like? Do we obtain the full GENERIC structure?

[51] See (28.21) on p. 271 of Grad, Principles of the Kinetic Theory of Gases (1958).

Exercise 185 Minimal Modification of Grad's Moment Equations

Formulate the "simplest possible" thermodynamically admissible moment equations. In other words, find the minimal modification of Grad's moment expansion that leads to GENERIC equations.

Exercise 186 Viscometric Functions From Grad's Moment Method

Elaborate the difference between the viscometric functions (7.138) for Grad's original approach and for the thermodynamically admissible moment equations formulated in Exercise 185.

Exercise 187 Grad's Method and Phenomenological Friction Matrix

Find the counterparts of (7.157), (7.161), and (7.168) by basing the GENERIC modification of Grad's approach on the phenomenological friction matrix (7.88) of Exercise 171 instead of the friction matrix (7.68) derived by the projection-operator technique. Compare the resulting relaxation term with (7.173). Evaluate the friction matrix at equilibrium.

7.5 TWO-PARTICLE DISTRIBUTION FUNCTIONS

The kinetic theory of rarefied gases presented in the previous sections is based entirely on the single-particle distribution function. Interaction effects are neglected in the energy expression (7.52) and in the pressure tensor (7.99). It is well-known from equilibrium statistical mechanics that, for dense fluids, pair correlations become important. If we move from the single-particle to the two-particle distribution function with the atomistic definition

$$f_2^{\Pi}(\boldsymbol{r},\boldsymbol{p},\boldsymbol{r}',\boldsymbol{p}';z) = \sum_{\substack{j,k=1\\ j\neq k}}^{N} \delta(\boldsymbol{r}-\boldsymbol{r}_j)\delta(\boldsymbol{p}-\boldsymbol{p}_j)\delta(\boldsymbol{r}'-\boldsymbol{r}_k)\delta(\boldsymbol{p}'-\boldsymbol{p}_k), \tag{7.181}$$

as the basic variable of kinetic theory, interaction effects can be incorporated in a rigorous way, as we briefly discuss in this section.

7.5.1 Internal Energy

Because the single-particle distribution function (7.47) can be obtained by integration from the two-particle distribution function,

$$f^{\Pi}(\boldsymbol{r},\boldsymbol{p};z) = \frac{1}{N-1}\int f_2^{\Pi}(\boldsymbol{r},\boldsymbol{p},\boldsymbol{r}',\boldsymbol{p}';z)d^3r'd^3p', \tag{7.182}$$

the kinetic energy and the potential energy in an external field can be evaluated by means of (7.52) or, after averaging, according to (7.53). The exact energy (7.46), including two-particle interactions, can be expressed as

$$\begin{aligned} E_0(z) &= \int \left[\frac{\boldsymbol{p}^2}{2m} + \phi^{(e)}(\boldsymbol{r})\right] f^{\Pi}(\boldsymbol{r},\boldsymbol{p};z)d^3rd^3p \\ &+ \frac{1}{2}\int \phi(|\boldsymbol{r}-\boldsymbol{r}'|) f_2^{\Pi}(\boldsymbol{r},\boldsymbol{p},\boldsymbol{r}',\boldsymbol{p}';z)d^3rd^3pd^3r'd^3p'. \end{aligned} \tag{7.183}$$

The two-particle distribution function gives us rigorous access to the total energy of any system in which only pair interactions are relevant. Equation (7.183) provides an important example of an energy expression that is linear in the relevant variables.

7.5.2 Rigorous Formula for the Pressure Tensor

From Hamilton's equations of motion, we obtain the following rigorous time-evolution equation for the atomistically defined single-particle distribution function (7.47):

$$\begin{aligned}
\frac{\partial}{\partial t} f^{\Pi}(\boldsymbol{r}, \boldsymbol{p}; z(t)) &= -\sum_{j=1}^{N} \left(\frac{d\boldsymbol{r}_j(t)}{dt} \cdot \frac{\partial}{\partial \boldsymbol{r}} + \frac{d\boldsymbol{p}_j(t)}{dt} \cdot \frac{\partial}{\partial \boldsymbol{p}} \right) \\
&\quad \times \delta(\boldsymbol{r} - \boldsymbol{r}_j(t)) \delta(\boldsymbol{p} - \boldsymbol{p}_j(t)) \\
&= -\sum_{j=1}^{N} \left(\frac{\boldsymbol{p}}{m} \cdot \frac{\partial}{\partial \boldsymbol{r}} - \frac{\partial \phi^{(e)}(\boldsymbol{r})}{\partial \boldsymbol{r}} \cdot \frac{\partial}{\partial \boldsymbol{p}} - \sum_{\substack{k=1 \\ k \neq j}}^{N} \frac{\partial \phi(|\boldsymbol{r} - \boldsymbol{r}_k(t)|)}{\partial \boldsymbol{r}} \cdot \frac{\partial}{\partial \boldsymbol{p}} \right) \\
&\quad \times \delta(\boldsymbol{r} - \boldsymbol{r}_j(t)) \delta(\boldsymbol{p} - \boldsymbol{p}_j(t)) \\
&= \left(-\frac{\boldsymbol{p}}{m} \cdot \frac{\partial}{\partial \boldsymbol{r}} + \frac{\partial \phi^{(e)}(\boldsymbol{r})}{\partial \boldsymbol{r}} \cdot \frac{\partial}{\partial \boldsymbol{p}} \right) f^{\Pi}(\boldsymbol{r}, \boldsymbol{p}; z(t)) \\
&\quad + \int \frac{\partial \phi(|\boldsymbol{r} - \boldsymbol{r}'|)}{\partial \boldsymbol{r}} \cdot \frac{\partial}{\partial \boldsymbol{p}} f_2^{\Pi}(\boldsymbol{r}, \boldsymbol{p}, \boldsymbol{r}', \boldsymbol{p}'; z(t)) d^3r' d^3p'. \qquad (7.184)
\end{aligned}$$

By averaging over initial conditions, we obtain an identical exact time-evolution equation for $f(\boldsymbol{r}, \boldsymbol{p})$ that involves $f_2(\boldsymbol{r}, \boldsymbol{p}, \boldsymbol{r}', \boldsymbol{p}')$. For any system with two-particle interactions, the time-evolution equation for the k-particle distribution function involves also the $(k+1)$-particle distribution function.[52] One thus arrives at the BBGKY hierarchy of equations for contracted distribution functions introduced independently by Bogolyubov (1946), Born and Green (1946), Kirkwood (1946), and Yvon (1937). If there were additional three-particle interactions, the $(k+2)$-particle distribution function would also arise in the time-evolution equation for the k-particle distribution function.

By considering the average momentum density as a function of position, we then obtain the following rigorous generalization of the momentum balance equation (7.97):

$$\frac{\partial}{\partial t}[\boldsymbol{v}(\boldsymbol{r})\rho(\boldsymbol{r})] = -\frac{\partial}{\partial \boldsymbol{r}} \cdot [\boldsymbol{v}(\boldsymbol{r})\boldsymbol{v}(\boldsymbol{r})\rho(\boldsymbol{r})]$$

[52] There may be rare exceptions, like Vlassov's mean-field kinetic equation for δ-function-type initial conditions, for which a closed equation for the single-particle distribution function exists.

$$-\frac{\partial}{\partial \boldsymbol{r}} \cdot \int \frac{[\boldsymbol{p} - m\boldsymbol{v}(\boldsymbol{r})][\boldsymbol{p} - m\boldsymbol{v}(\boldsymbol{r})]}{m} f(\boldsymbol{r}, \boldsymbol{p}) d^3p,$$
$$+\rho(\boldsymbol{r})\boldsymbol{g}(\boldsymbol{r}) + \rho(\boldsymbol{r})\tilde{\boldsymbol{g}}(\boldsymbol{r}), \tag{7.185}$$

with the external force field

$$m\boldsymbol{g}(\boldsymbol{r}) = -\frac{\partial \phi^{(e)}(\boldsymbol{r})}{\partial \boldsymbol{r}} \tag{7.186}$$

and the additional force density

$$\rho(\boldsymbol{r})\tilde{\boldsymbol{g}}(\boldsymbol{r}) = -\int \frac{\partial \phi(|\boldsymbol{r} - \boldsymbol{r}'|)}{\partial \boldsymbol{r}} f_2(\boldsymbol{r}, \boldsymbol{p}, \boldsymbol{r}', \boldsymbol{p}') d^3p d^3r' d^3p'. \tag{7.187}$$

Whereas $\boldsymbol{g}(\boldsymbol{r})$, as the external force per unit mass, represents a true body force, $\tilde{\boldsymbol{g}}(\boldsymbol{r})$ results from particle interactions and hence should be expressed as the divergence of a pressure tensor. Such a reformulation, which was originally achieved by Irving and Kirkwood via a formal Taylor expansion of δ-functions,[53] can be done by means of a formal trick, which is based on the chain rule and the translational invariance of the interaction forces. If we introduce the tensor field

$$\boldsymbol{\tau}(\boldsymbol{r}) = -\frac{1}{2}\int_0^1 \left[\int \boldsymbol{R}\frac{\partial \phi(R)}{\partial \boldsymbol{R}} f_2(\boldsymbol{r} + (1-\lambda)\boldsymbol{R}, \boldsymbol{p}, \boldsymbol{r} - \lambda\boldsymbol{R}, \boldsymbol{p}') d^3R d^3p d^3p'\right] d\lambda, \tag{7.188}$$

then we find the following expression for its divergence:

$$\begin{aligned}
\frac{\partial}{\partial \boldsymbol{r}} \cdot \boldsymbol{\tau}(\boldsymbol{r}) &= -\frac{1}{2}\int_0^1 \left[\int \frac{\partial \phi(R)}{\partial \boldsymbol{R}} \boldsymbol{R} \cdot \frac{\partial}{\partial \boldsymbol{r}} \right. \\
&\qquad \left. f_2(\boldsymbol{r} + (1-\lambda)\boldsymbol{R}, \boldsymbol{p}, \boldsymbol{r} - \lambda\boldsymbol{R}, \boldsymbol{p}') d^3R d^3p d^3p'\right] d\lambda \\
&= \frac{1}{2}\int_0^1 \left[\int \frac{\partial \phi(R)}{\partial \boldsymbol{R}} \frac{d}{d\lambda} \right. \\
&\qquad \left. f_2(\boldsymbol{r} + (1-\lambda)\boldsymbol{R}, \boldsymbol{p}, \boldsymbol{r} - \lambda\boldsymbol{R}, \boldsymbol{p}') d^3R d^3p d^3p'\right] d\lambda.
\end{aligned} \tag{7.189}$$

After performing the straightforward integration with respect to λ, we can use the substitution $\boldsymbol{R} \to -\boldsymbol{R}$ in one of the terms and the symmetry of f_2 in the two particle positions to obtain

$$\begin{aligned}
\frac{\partial}{\partial \boldsymbol{r}} \cdot \boldsymbol{\tau}(\boldsymbol{r}) &= \frac{1}{2}\int \frac{\partial \phi(R)}{\partial \boldsymbol{R}} [f_2(\boldsymbol{r}, \boldsymbol{p}, \boldsymbol{r} - \boldsymbol{R}, \boldsymbol{p}') - f_2(\boldsymbol{r} + \boldsymbol{R}, \boldsymbol{p}, \boldsymbol{r}, \boldsymbol{p}')] d^3R d^3p d^3p' \\
&= -\int \frac{\partial \phi(R)}{\partial \boldsymbol{R}} f_2(\boldsymbol{r}, \boldsymbol{p}, \boldsymbol{r} + \boldsymbol{R}, \boldsymbol{p}') d^3R d^3p d^3p' \\
&= -\rho(\boldsymbol{r})\tilde{\boldsymbol{g}}(\boldsymbol{r}),
\end{aligned} \tag{7.190}$$

[53] Irving & Kirkwood, J. Chem. Phys. 18 (1950) 817; the corresponding integral expression similar to (7.188) can be found in the appendix of that paper.

so that (7.188) provides an exact expression for the pressure tensor contribution due to interactions, to be added to the kinetic pressure tensor contribution appearing in the second line of (7.185). This is a famous result of Irving and Kirkwood.

Note that the pressure tensor expression (7.188) was derived without any assumption that the interaction forces must be short-ranged. This may be surprising in view of the fact that the pressure tensor represents contact forces between volume elements. The structure of the formula (7.188) shows that the pressure tensor, or the momentum flux, is determined by the forces and the distance over which momentum is exchanged by their action. The factor of $1/2$ is needed because the stress contribution of each interacting pair is counted twice. For the assumed central interaction forces depending on the distance between particles only, the pressure tensor expression is symmetric. Long-range forces have been incorporated by means of "smearing out" the stress contribution on the line between the interacting particles by means of the auxiliary variable λ. Although the Irving-Kirkwood trick for achieving the divergence form of $\rho\tilde{\boldsymbol{g}}$ in terms of a pressure tensor is exact in the bulk, it fails near the boundary (within a distance of the order of the range of the pair interaction), so that we are still forced to assume short-range contact forces to justify the concept of the pressure tensor near the boundary, which is what we need for calculating the stresses on the surface of a material.

Equations (7.183) and (7.188) show that, for many properties, only the position dependence of the two-particle distribution function matters. It is hence natural to introduce the pair correlation, which is frequently used in equilibrium statistical mechanics, by integrating over the momentum variables.

An expression analogous to (7.188) for the momentum flux can also be given for the heat flux. The exact generalization of (7.100) is given by[54]

$$\begin{aligned} \boldsymbol{j}^{\mathrm{q}} &= \int \left[\frac{\boldsymbol{p}}{m} - \boldsymbol{v}(\boldsymbol{r})\right] \frac{[\boldsymbol{p} - m\boldsymbol{v}(\boldsymbol{r})]^2}{2m} f(\boldsymbol{r},\boldsymbol{p}) d^3p \\ &+ \frac{1}{2}\int \left[\frac{\boldsymbol{p}}{m} - \boldsymbol{v}(\boldsymbol{r})\right] \phi(|\boldsymbol{r} - \boldsymbol{r}'|) f_2(\boldsymbol{r},\boldsymbol{p},\boldsymbol{r}',\boldsymbol{p}') d^3p d^3r' d^3p' \\ &- \frac{1}{4}\int_0^1 \left[\int \frac{\boldsymbol{R}\boldsymbol{R}}{R} \cdot \frac{\boldsymbol{p} + \boldsymbol{p}' - 2m\boldsymbol{v}(\boldsymbol{r})}{m} f_2(\boldsymbol{r} + (1-\lambda)\boldsymbol{R}, \boldsymbol{p}, \boldsymbol{r} - \lambda\boldsymbol{R}, \boldsymbol{p}') \right. \\ &\qquad \left. d^3R d^3p d^3p' \right] d\lambda. \end{aligned} \tag{7.191}$$

Whereas the energy can be expressed rigorously in terms of the two-particle distribution function, the formulation of the other GENERIC building blocks is less straightforward. The natural choice of the Poisson operator is given by (7.60) for $N = 2$. The formulation of the entropy requires an analysis of the generalized canonical ensemble for the relevant variables (7.181). The resulting ensemble is a product of two-particle contributions, one for each pair of particles, in the spirit of Kirkwood's

[54] See (7.52), (7.58), and (7.62) of Kreuzer, *Nonequilibrium Thermodynamics & Statistical Foundations* (Oxford, 1981).

superposition approximation.[55] The main challenge, however, is the introduction of relaxation processes in a dense fluid by a suitable friction matrix.

Once the generalized canonical ensemble associated with the two-particle distribution function is constructed, universal expressions—independent of a particular Hamiltonian—are available for all higher distribution functions. In particular, one can obtain an expression for the three-particle distribution function that can be used to close the BBGKY hierarchy. To carry out this program, Bugaenko et al.[56] constructed a formal expansion of the generalized canonical ensemble and, in particular, the three-particle distribution function in terms of a parameter that switches on the two-particle correlations, so that an expansion with respect to the number of pairs of correlated particles is obtained. In view of the combinatorial complexity of the recursive calculations, they developed a convenient diagram technique and then derived a number of closure approximations, including Kirkwood's superposition approximation, in a systematic manner. More advanced closure schemes can be developed to any order.

Even at equilibrium, understanding the pair correlation function in a dense fluid is a major challenge. One usually decomposes the pair correlation into short- and long-range parts, where the detailed form of this decomposition is expressed in the *Ornstein-Zernike equation*.[57] The underlying idea is that the short-range part, which is called the direct correlation function, can be modeled more conveniently. For example, the *Percus-Yevick approximation* for the direct correlation function provides an extremely successful description of the pair correlation function and the density dependence of the pressure obtained from molecular dynamics simulations.[58]

A common tool in the development of a molecular theory of *dynamics* in dense systems is known as *mode coupling*. Although "mode-coupling theory" refers to a family of techniques that were invented to explain anomalous properties related to critical dynamics and "long-time tails" of correlation functions, with a strong effect on transport coefficients, we here understand it as a general method to calculate time correlation functions in dense systems.[59] To evaluate the correlation function in the friction matrix (6.72) by means of the mode-coupling theory for the two-particle distribution function $f_2^{\Pi}(\boldsymbol{r}, \boldsymbol{p}, \boldsymbol{r}', \boldsymbol{p}'; z)$ as the set of slow variables $\Pi_j(z)$, for example, one first needs to identify the main decay channels, which are believed to be most important for the decay of the correlation function. As a characteristic feature of the mode-coupling approach, decay channels are taken as products of two variables,[60]

[55] Kirkwood, J. Chem. Phys. 3 (1935) 300.

[56] Bugaenko, Gorban & Karlin, Theor. Math. Phys. 88 (1991) 977.

[57] See Section 11.E of Reichl, *Modern Course in Statistical Physics* (University of Texas, 1980).

[58] See Sections 11.E and 11.G of Reichl, *Modern Course in Statistical Physics* (University of Texas, 1980); a more comprehensive discussion of correlation function theories can be found in Chapter 5 of Hansen & McDonald, *Theory of Simple Liquids* (Academic, 1986).

[59] See Sections 2.9, 3.6 and 6.7 of Boon & Yip, *Molecular Hydrodynamics* (Dover, 1991) or Section 9.5 of Hansen & McDonald, *Theory of Simple Liquids* (Academic, 1986); see also Section 9.1 of Zwanzig, *Nonequilibrium Statistical Mechanics* (Oxford, 2001).

[60] This is the simplest choice once one realizes that the most tempting single-mode channels do not contribute to the correlation function.

one of which typically is $Qi\mathcal{L}\Pi_j$ [which appears in the correlation function in (6.72)] and the other one may be Π_j itself or some hydrodynamic mode. As a second step, the correlation function of interest is expressed in terms of correlation functions of decay channels, including proper static coupling constants. In the third and final step, the new correlation functions involving four factors are factorized in the time domain while, at the same time, the projected time evolution in (6.72) is replaced by the free time evolution. According to Boon and Yip, the approximations of the mode-coupling theory are difficult to justify on purely theoretical grounds, and perhaps the most convincing argument is that such calculations have been found to give good numerical results with respect to all dynamical features of time correlation functions. In particular, the theory is inductive rather than deductive because it does not decide which decay channels are important.

8

Simulations

What is the goal of simulations of beyond-equilibrium systems, and how should they be performed? Which are the available tools, and how can they be employed most efficiently? One would expect that these questions have been answered; however, we should revisit them because some of the answers are affected in an essential way by the knowledge of a framework of beyond-equilibrium thermodynamics and statistical mechanics. Conversely, the enormous potential of the theory of beyond-equilibrium systems can only blossom to full beauty when fertilized by simulations. In this chapter, we hence address the above questions in great detail.

The goal of this chapter is to present the foundations and tools for "good simulations," that is, for useful simulations focused on deeper understanding and thus contributing seriously to progress in some field of application. The emphasis is on guidance for future work rather than on the discussion of existing simulations and their results. A key concept is that of beyond-equilibrium ensembles, which correspond to the generalized microcanonical, canonical, and mixed ensembles that we encountered in Chapter 6. Many of the suggestions for "good simulations" arising here in the context of beyond-equilibrium thermodynamics remain to be developed and made concrete, some of them will probably be modified or even abandoned. This is natural because simulations of beyond-equilibrium systems do not only constitute a field of great interest and potential but also a subject of intensive ongoing research.

In an attempt to clarify the situation and to classify the goals in simulating beyond-equilibrium systems, we make a distinction between simulations for establishing coarser models and those for directly solving the practical problems of ultimate interest. We further distinguish between two fundamentally different approaches, *brute-force simulations* and *thermodynamically guided simulations*, where the latter are based on GENERIC and its projection-operator justification. Brute-force simulations can be thought of as computer experiments mimicking the problem of interest directly on a computer; thermodynamically guided simulations rely on a statistical ensemble, which is defined in terms of the variables of some coarse-grained description of the nonequilibrium system of interest. The availability of an appropriate coarse-grained level of description is thus absolutely crucial for thermodynamically guided simulations. Brute-force simulations require fewer assumptions and less insight, which may be considered as good or bad. For reasons explained below (the wide range of time scales from atomistic motions to processing), I find a negative connotation more appropriately applies to brute-force simulations—otherwise, I had referred to them as *ab initio* simulations.

We first discuss the kind of results to be obtained from simulations, and we try to elaborate the general importance of simulations and the significance of the different approaches to simulation. A toolkit of thermodynamically guided simulation techniques for a variety of purposes of practical relevance is sketched. Because Monte Carlo simulations are the method of choice for obtaining all the static material properties of a system (that is, everything but the friction matrix), we present the underlying theory of Markov chains and an example of practical relevance in considerable detail. Concerning the dynamic material properties or the friction matrix, we discuss some fundamental issues associated with Green-Kubo-type formulas, and we explain and illustrate Brownian dynamics and molecular dynamics.

8.1 SIMULATION PHILOSOPHY

Before getting into any details of the various simulation techniques or of the simulations for specific systems, we shall pause and think about the general issues in beyond-equilibrium simulations. Simulations should be the workhorse necessary to understand realistic systems because actual coarse-graining is performed and because simulations provide the information relevant to characterizing beyond-equilibrium systems. We discuss how one can proceed in several steps from atomistic models to practical problems, and we explain which specific simulation techniques should be useful for each of the steps.

8.1.1 Understanding Through Simplicity

As pointed out briefly in the preface, there are different concepts of understanding. In beyond-equilibrium thermodynamics, we pursue the idea of "understanding through simplicity," which means that we look for the coarsest possible description for phenomena of interest in science and engineering. All detailed solutions to practical problems should be based on such coarsest possible models, but we need to avoid oversimplification. As illustrated in Figure 8.1, this concept is drastically different from the idea of "understanding through completeness," in which one is striving for a rigorous and complete "theory of everything." Although these extreme concepts

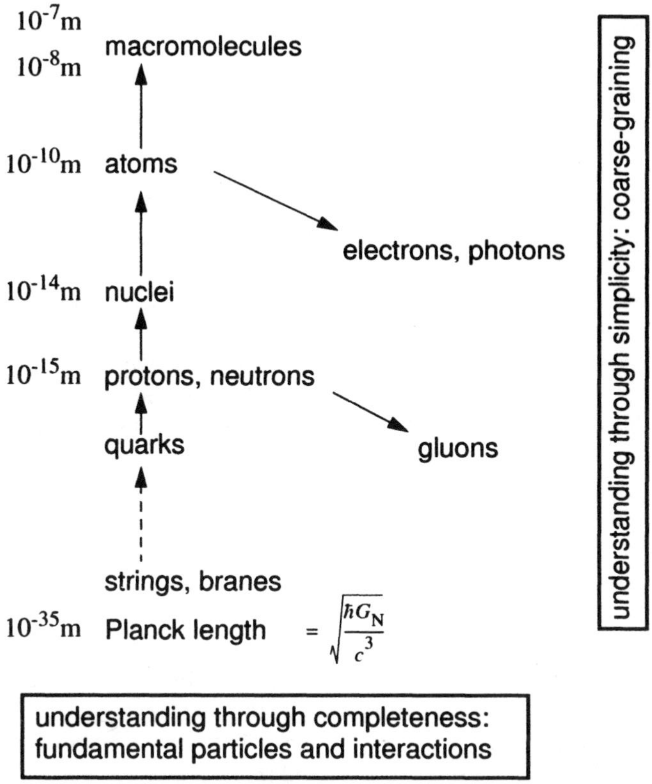

Fig. 8.1 Illustration of different concepts of understanding.

of understanding may be equally appealing, beyond-equilibrium thermodynamics clearly deals with coarse-graining.

Beyond-equilibrium thermodynamics is the road to understanding through simplicity, and simulations are the most versatile vehicle for moving along this road. The road may be steep and sometimes bumpy, and thus strong and convenient vehicles need to be designed carefully; perhaps even the roads need some further construction to become viable for some types of vehicles.

Our primary goal is to establish coarse-grained models and to determine the free parameters or functions occurring in the model equations. Within the GENERIC framework, this means that we need simulations for obtaining the generators $E(x)$, $S(x)$ and the matrices or linear operators $L(x)$, $M(x)$, with all the properties to be imposed on them, for a chosen list of independent variables x. Finding these building blocks is our big goal, while we also need to simultaneously validate that the chosen variables are sufficient for a closed-form description of the phenomena of our interest. We can then use the coarse-grained models for solving practical problems in science and engineering or, if necessary, as an input for further coarse-graining.

8.1.2 Relevance of Functional Forms

Assuming that the GENERIC building blocks are the target of our simulations, we actually need to find the functional form of the generators $E(x)$, $S(x)$ and the matrices $L(x)$, $M(x)$. The time-evolution equation (1.1) and the complementary degeneracy requirements (1.4) and (1.5) involve the gradients $\delta E(x)/\delta x$ and $\delta S(x)/\delta x$ of energy and entropy, and a check of the Jacobi identity requires information formally corresponding to $\delta L(x)/\delta x$ [see (1.14)], so that $E(x)$, $S(x)$ and $L(x)$ need to be sufficiently smooth functions of x. The functions $E(x)$ and $S(x)$ themselves do not actually occur in the GENERIC framework, but their existence implies integrability conditions for their gradients occurring in the equations. The need for functional forms is well known from equilibrium thermodynamics where, however, the functions of interest, that is, the thermodynamic potentials, depend only on a few variables. Beyond-equilibrium simulations involving many variables, even fields, thus pose a major challenge.

As a first possibility, we mention the most convenient situation that arises when the GENERIC building blocks on the target level are given functions with only a few free parameters, for example, obtained from some phenomenological model, experienced guessing, experimental results, or, more likely, from combinations of all of those. For this situation to occur, one needs a considerable a priori understanding of the problem of interest.

A second possibility is to determine the building blocks only for particular values of the independent variables x or, more frequently, for the corresponding Lagrange multipliers λ (at least, for some of the variables). Rather than calculating $\delta E(x)/\delta x$ and $\delta S(x)/\delta x$ through differences, one should try to obtain these gradients directly. For example, the relation (6.56) between entropy gradients and Lagrange multipliers shows how $\delta S(x)/\delta x$ can be obtained conveniently. For such pointwise simulations, it is crucial to select a proper set of values of x or λ. Often, one is interested in a sequence of parameter values, in particular, as they are occurring in the course of time for some problem of interest. If one is interested in steady states, for example, in steady homogeneous flows of complex fluids, one needs to estimate the Lagrange multipliers that lead to a generalized canonical ensemble of configurations characteristic of a given flow type and rate. The choice of proper Lagrange multipliers to define a beyond-equilibrium ensemble again requires a priori information. By an iterative procedure to obtain self-consistency, one can try to adjust the Lagrange multipliers so that they reproduce the steady-state situation of interest, thus reducing the dependence on a priori information to the choice of initial conditions for the iteration process.

In a third possible scenario, the functional form of the building blocks is expressed in terms of versatile black box functions rather than explicit mathematical formulas. For example, a promising option is the indirect formulation of functions in terms of neural networks. Actually, such neural-network formulations can be considered as a special case of the first scenario, where the parameters of the neural network are obtained by training the network. For successful training, the number of network parameters should be kept as small as possible. Alternatively, one can consider neural networks as a tool for interpolating and extrapolating the results for a finite set of values of x or λ as occurring in the second scenario. The potential of neural networks and other classes of black box fit functions can be particularly useful for obtaining self-consistency in an iterative approach because they offer the possibility to perform mathematical operations on the simulation results.

8.1.3 Overview of Simulation Techniques

At equilibrium, obtaining systematic thermodynamic information from atomistic modeling via statistical mechanics is a well-established option. For thermodynamic systems of practical relevance, analytical calculations are usually impossible, and numerical simulations are an indispensable tool. Integrations or summations in spaces with an extremely large number of degrees of freedom are required, and Monte Carlo simulations are a key technique for handling such problems.[1] Such Monte Carlo simulations are ideally suited for equilibrium problems because they aim directly at generating a set of microstates representing the proper statistical ensemble for the independent variables of choice; detailed balance in Monte Carlo moves is the core idea for generating representative ensembles of states, and the physical time evolution is entirely irrelevant.

Beyond equilibrium, one would intuitively expect that one is forced into simulation techniques accounting for the physical time evolution and that Monte Carlo might be obsolete. We would then immediately and unavoidably be faced with a major problem resulting from the many orders of magnitude separating the atomistic time scale from typical time scales of practical interest, say in the processing of materials. This wide separation of time scales, ranging from femtoseconds to minutes or even hours, is the crux of the direct approach previously referred to as brute-force simulations. For example, the restriction of atomistic flow simulations to extremely high flow rates is symptomatic for this problem. The obvious way out of this dilemma is to establish one or several intermediate levels of description, to establish successful models on these intermediate levels, and to find the phenomenological parameters of intermediate-scale or macroscopic models from more detailed levels of description. One is then faced with three general types of problems or three steps in the development or, in short, with the ABC of simulations for beyond-equilibrium systems illustrated in Figure 8.2:

Step A [A like "ab initio" or "atomistic simulations"] Establishing coarse-grained levels of description and finding the proper formulation or parameters of the corresponding models from atomistic principles.

Step B [B like "bridging gaps"] Establishing increasingly more coarse-grained levels of description and finding the proper formulation or parameters of the corresponding models from the more detailed but nonatomistic models arising from Step A, or formulated empirically.

[1] Landau & Binder, *A Guide to Monte Carlo Simulations* (Cambridge, 2000); Binder & Heermann, *Monte Carlo Simulation in Statistical Physics* (Springer, 1992).

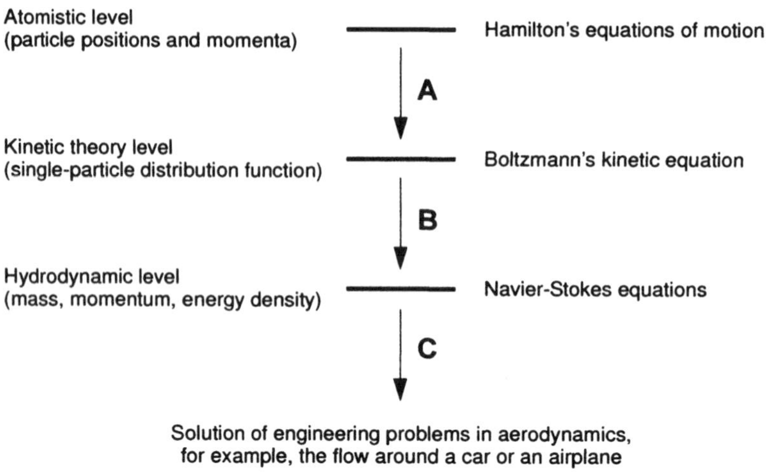

Fig. 8.2 Illustration of the ABC of simulations in the context of gases.

Step C [C like "concrete solutions"] Solving practical problems of ultimate interest with the coarse-grained models resulting from Steps A or B.

The distinction between steps B and C is not entirely unambiguous. For example, a viscosity calculation for some intermediate-level model of a complex fluid might be considered as the solution of a problem of direct practical relevance, but it might also be considered as the determination of an input parameter for some coarse-grained rheological model. Nevertheless, the distinction between several steps is found to be useful in clarifying and organizing the tasks and tools for beyond-equilibrium simulations.

For thermodynamically consistent modeling, Steps A and B are structured in the sense that the GENERIC building blocks E, S, L, and M are to be determined. As shown in Section 6.1.4, three of these building blocks do not contain any dynamic material information. The quantities E and L describing reversible dynamics are obtained as ensemble averages such as (6.48), (6.67), where Monte Carlo simulations are the ideal tool for calculating such averages. Although the entropy S in (6.53) or in (6.55) is not an ensemble average, it clearly does not contain any dynamic material information. Various Monte Carlo methods suitable for entropy calculations have been designed for equilibrium problems, as will be summarized below.

Motivated by an interest in the dynamic material properties contained in M, as given by the formula (6.72), it has sometimes been argued that the Monte Carlo time is proportional to the real time, that is, the sequence of microstates generated by Monte Carlo moves is a reasonable representation of the physical time evolution. To avoid the fundamental time-scale dilemma, however, *efficient Monte Carlo moves should be as unphysical as possible*, so that the configuration space can be sampled without following the slow physical dynamics. Ideally, one should have Monte Carlo moves where a single step leads to a conformational change for which the longest relaxation time of the system would be needed in a physical realization. The invention of "perverse" Monte Carlo moves is thus the most important task in determining the building blocks E, S, and L efficiently. Physical insight is needed to find proper coarse-grained variables, and the same insight should be turned into maximally "perverted" moves. Cutting and reassembling polymer molecules in a melt is an example of a useful unphysical move changing the entanglement structure. When ingenious Monte Carlo moves are used, there is no time-scale dilemma for the determination of E, S, and L. This dilemma arises only in the determination of M.

The calculation of friction matrices requires time dependent trajectories for evaluating two-time correlations. This is how dynamic material information enters the picture. This is why the range of relevant time scales needs to be taken into account. When starting from an atomistic level of description according to Step A (see Figure 8.2), the expression (6.72) for the friction matrix highlights the importance of solving Newton's or Hamilton's equations of motion, that is, of molecular dynamics. When starting from some coarser level of description according to Step B, where fast variables have been eliminated in favor of noise, the expression (6.181) for the friction matrix shows the relevance of solving stochastic differential equations of motion, that is, of Brownian dynamics. Even when GENERIC is employed without noise, Brownian dynamics can be important, namely, when one is working on a level of description involving configurational distribution functions and hence stochastic processes. Efficient methods to solve deterministic and stochastic differential equations of motion are thus needed. In particular, stiff problems involving a wide range of time scales are typical. Any knowledge about the abstract structure of time-evolution equations from a thermodynamic framework can be helpful for increasing the efficacy of numerical integration methods. For example, the symplectic structure of Hamilton's equations of classical mechanics suggests to benefit from symplectic integrators.

So far we have focused our discussion on the *determination of the GENERIC building blocks* for a given set of variables. *Model validation* is an equally important step. In the general setting, the important question is whether the phenomena of interest can be fully captured in terms of the chosen relevant variables. If the model equations resulting for the variables of choice after finding the GENERIC building blocks are insufficient to describe the phenomena of interest, the set of variables needs to be altered. In many cases of practical importance, one has specific ideas about the possible functional form of the building blocks coming with a certain choice of variables. Then, only the parameters in these expressions need to be determined by simulation techniques. Because these parameters can often be extracted in various ways from simulation results, the model validation should include a check of the internal consistency of model parameters obtained in different ways.

Thermodynamically guided simulations rely on the expressions for the GENERIC building blocks in terms of averages over beyond-equilibrium ensembles. The indispensibility of coarse-grained variables and the corresponding Lagrange multipliers for thermodynamically guided Monte Carlo or dynamic simulations might appear as a drawback, but it is the key to overcoming the time-scale problem, the crux of brute-force simulations. Actually, the importance of coarse-grained variables in statistical ensembles is well known at equilibrium, where one is so familiar with these variables that one is usually unaware of their significance for the procedure. In any "first-principles" equilibrium simulation, one must know the energy in terms of the atomistic configurations and, in the definition of canonical and grand canonical statistical ensembles, the Lagrange multiplier "temperature" occurs in combination with this energy function. Other macroscopic observables entering the definition of equilibrium ensembles can be the volume or the pressure, particle numbers, or chemical potentials. Whereas the choice of equilibrium variables is universal for all materials, the zoo of coarse-grained variables for nonequilibrium systems is much richer in number and variety. Their selection requires physical understanding of specific systems, and their merit for simulations is the bridging of the wide gap of times scales. This is the origin of both the challenging and the rewarding nature of beyond-equilibrium thermodynamics, which is admittedly less rigorous than equilibrium thermodynamics (see Section 1.1.1).

The *ultimate goal of beyond-equilibrium thermodynamics* is the description of processes observed in nature, including processes in living systems, the simulation of processing operations for complex fluids, the simulation of any kind of flow from air to rock in natural and technical systems, chemical reactions and population dynamics, etc. In solving practical problems with coarse-grained models according to Step C, the solution of partial differential equations often is the key issue. Finite elements are an important tool, but also spectral elements, finite volumes, or finite differences can be useful alternatives. For increasingly realistic modeling of dynamic material behavior, it can be very useful to combine stochastic simulation techniques with finite elements or other techniques to solve partial differential equations.[2]

Sophisticated methods to solve practical problems according to Step C are the subject of current research, with a particular emphasis on multiscale modeling. One can try to combine models on different levels of description, for example, by using a more detailed model in small regions of space with large gradients and better spatial resolution and a less detailed model otherwise. Such combinations of different levels of modeling for performing Step C have a potential going even far beyond that of multigrid approaches, in which different levels of resolution are considered for a single underlying model. One can also try to use an "unnecessarily" detailed model in small and widely separated "patches" in the region of interest for short periods of time, to interpolate the results in space, and to proceed by large time steps according to the time derivatives estimated in the short time simulations.[3] In doing so, one can actually avoid the potentially difficult step of formulating a closed-form coarser model, the mere existence of which in view of the associated separation in time scales, however, is crucial for the success of the procedure. In the terminology of Figure 8.2, one avoids the explicit performance of Step B, but one still tries to bridge time and length scales.

[2] Laso & Öttinger, J. Non-Newtonian Fluid Mech. 47 (1993) 1; Hulsen, van Heel & van den Brule, J. Non-Newtonian Fluid Mech. 70 (1997) 79; Öttinger, van den Brule & Hulsen, J. Non-Newtonian Fluid Mech. 70 (1997) 255.

[3] Kevrekidis et al., Comm. Math. Sci. 1 (2003) 715.

The theory of coarse-graining, as the guide to performing systematic and efficient simulations for beyond-equilibrium systems, is based on the projection-operator approach. This approach, in turn, is based on Hamilton's equations of motion for many-particle systems. Therefore, we are naturally led to particle-based simulations. As a fundamental alternative, one can reformulate the building blocks of interest in terms of field and develop *field-theoretic simulation techniques*.[4] Field-theoretic methods are particularly useful at equilibrium, where one has well-established procedures for reformulating partition functions in terms of fields. Compared with particle-based simulations, field-theoretic methods offer major advantages for dense systems for which most particle methods become inefficient. Field-theoretic simulations, which provide the deviations from mean-field-type theories, benefit from the success of mean-field approximations for dense systems. Although the usefulness of field-theoretic simulations is not obvious for beyond-equilibrium systems, they might benefit from the focus on the evaluation of well-defined building blocks, to be expressed in terms of fields, and from the guidance by coarse-grained models, for which the simulations should provide deviations from mean-field-type or Gaussian approximations, including related dynamical theories.

In summary, we have sketched the role of Monte Carlo simulations, molecular dynamics, Brownian dynamics, finite elements, and other methods for solving partial differential equations for the ultimate usefulness of beyond-equilibrium thermodynamics. Each of these techniques is associated with certain levels of coarse-graining and with either static or dynamic material information. A key feature of thermodynamically guided simulations is the need for an ensemble, that is, the proper identification of the coarse-grained variables for describing a system and explicit expressions for these relevant variables in terms of more detailed configurations. For a given problem of interest, the highest level of understanding is reached if one has established a hierarchy of different levels of description and connected the GENERIC building blocks on the various levels by statistical mechanics. When the hierarchy ranges from atoms to continuum mechanics, all sorts of questions can be addressed on the most appropriate level of description.

8.2 MONTE CARLO SIMULATIONS

The development of Monte Carlo simulations can be based on the theory of Markov chains,[5] which we therefore present here in some detail. A similar but more detailed presentation of the theory of Markov chains can be found in Feller's classical textbook on probability theory.[6] To generate states corresponding to the probability distribution of a canonical ensemble by means of Markov chains, we then introduce the concept of detailed balance. As an illustrative example, we discuss the simulation of polymer

[4] Fredrickson et al., Macromolecules 35 (2002) 16.

[5] See, for example, Section 2.2.2 of Mouritsen, *Computer Studies of Phase Transitions* (Springer, 1984).

[6] See Chapter XV of Feller, *Probability Theory 1* (Wiley, 1968).

melts. In particular, we discuss a number of simplifying assumptions of quite general nature that are needed to make complex fluids tractable.

8.2.1 Markov Chains

A *Markov chain* is a sequence of random variables $X_0, X_1, X_2, \ldots$ with the following properties. (1) The random variables X_0, X_1, X_2, ... assume values in a certain finite or denumerably infinite set, which, for convenience, we choose as the integers $\{1, 2, \ldots, N\}$ or $\{1, 2, \ldots\}$. (2) Each random variable X_n depends on all the previous random variables X_0, X_1, ... X_{n-1} only through X_{n-1}. With property (1), we require the state space to be *discrete*, and property (2) is known as the *Markov property* expressing a lack of memory.

A Markov chain is fully determined by the initial probabilities, $\pi_j^{(0)} = P(X_0 = j)$ and the transition probabilities $P(X_{n+1} = k|X_n = j)$. We here assume *stationary* transition probabilities, that is, the conditional probability measure

$$P(X_{n+1} = k|X_n = j) = P_{jk} \tag{8.1}$$

is independent of n. The matrix P is a *stochastic matrix*, generally defined by the positivity and normalization conditions

$$P_{jk} \geq 0, \qquad \sum_k P_{jk} = 1. \tag{8.2}$$

Two-step transition probabilities can be calculated by summing over all intermediate states,

$$P(X_{n+2} = k|X_n = j) = \sum_l P_{jl} P_{lk} = P_{jk}^2, \tag{8.3}$$

and transition probabilities for larger steps can similarly be obtained as the corresponding higher powers of the stochastic matrix P.

As an example of a Markov chain, we take X_n as the number of occurrences of "heads" after tossing a coin n times. The initial probability is concentrated at 0, and the stochastic matrix is given by

$$P = \begin{pmatrix} 1-\tilde{p} & \tilde{p} & 0 & 0 & \\ 0 & 1-\tilde{p} & \tilde{p} & 0 & \cdots \\ 0 & 0 & 1-\tilde{p} & \tilde{p} & \\ & & \vdots & & \ddots \end{pmatrix}, \tag{8.4}$$

where the probability $\tilde{p}$ (with $0 < \tilde{p} < 1$) for the occurrence of "head" may safely be assumed to be close to $1/2$. In each step, the number of "heads" increases by one (with probability $\tilde{p}$) or remains unchanged (with probability $1 - \tilde{p}$).

There is a number of important questions to be addressed in the theory of Markov chains: Starting from a state j, what is the probability $F_{jk}^{(n)}$ for reaching another given state k for the first time after n steps? How often is the state j visited? Is the number of visits of j, N_j, finite or infinite? Can some state k be reached from some other state j? What is the probability for finding the state j after many steps? The last question is of special importance for understanding Monte Carlo simulations because we want to use Markov chains to generate states with the probabilities given by statistical mechanics.

The probabilities $F_{jk}^{(n)}$ can be calculated recursively. We have

$$F_{jk}^{(1)} = P_{jk}, \tag{8.5}$$

and, if k is not reached in the first step,

$$F_{jk}^{(n)} = \sum_{l \neq k} P_{jl} F_{lk}^{(n-1)}. \tag{8.6}$$

Furthermore, we have a *first entrance theorem*,

$$P_{jk}^{n} = \sum_{l=1}^{n} F_{jk}^{(l)} P_{kk}^{n-l}. \tag{8.7}$$

The probability for reaching k from j in any number of steps is given by

$$F_{jk} = \sum_{n=1}^{\infty} F_{jk}^{(n)}. \tag{8.8}$$

By summing the recursion formula (8.6) from $n = 2$ to ∞ and adding (8.5), we obtain the identity

$$F_{jk} = P_{jk} + \sum_{l \neq k} P_{jl} F_{lk}. \tag{8.9}$$

For fixed k, the kth column vector of the matrix F thus is a solution of the following system of linear equations,

$$P_{jk} F_{kk} + \sum_{l} (\delta_{jl} - P_{jl}) F_{lk} = P_{jk}. \tag{8.10}$$

In words, the kth column vector of the matrix $1 - P$ multiplying F is replaced by the kth column vector of the unit matrix. The solution of (8.10) for the kth column vector of F is not necessarily unique (see Exercise 189).

The probability for a certain number of visits of the state k, given that one starts at $j \neq k$, can be expressed in terms of the probabilities F_{jk} and F_{kk},

$$P(N_k = n) = \begin{cases} 1 - F_{jk} & \text{for } n = 0 \\ F_{jk} F_{kk}^{n-1} (1 - F_{kk}) & \text{for } n \geq 1; \end{cases} \tag{8.11}$$

$P(N_k = 0)$ is the probability for not reaching k from j, and in $P(N_k = n)$ for $n \geq 1$, we recognize the probabilities for reaching k for the first time, for returning $n - 1$

times, and for never returning to k after n visits. We define a *potential matrix* R with entries R_{jk} as the expected number of visits of k, given that we start at the state j,

$$R_{jk} = \sum_{n=1}^{\infty} n F_{jk} F_{kk}^{n-1} (1 - F_{kk}) = \frac{F_{jk}}{1 - F_{kk}}. \tag{8.12}$$

This expression for R_{jk} holds only for $j \neq k$. For $j = k$, we need to add the initial visit to the expectation of N_k,

$$R_{kk} = 1 + \frac{F_{kk}}{1 - F_{kk}} = \frac{1}{1 - F_{kk}}, \tag{8.13}$$

which is equivalent to dropping the factor F_{jk} in (8.12) because an initial transition to k is not required. In (8.12) and (8.13), we use the convention $R_{jk} = R_{kk} = \infty$ for $F_{kk} = 1$, except for $F_{jk} = 0$, which leads to $R_{jk} = 0$ even for $F_{kk} = 1$. If the probability for returning to k is equal to unity, one expects an infinite number of visits of k, unless k cannot be reached at all.

The key to evaluating the probabilities for reaching certain states is the observation that the expected number of states can be expressed directly in terms of the transition probabilities,

$$R_{jk} = \sum_{n=0}^{\infty} P_{jk}^{n}. \tag{8.14}$$

This expression for the potential matrix in terms of the stochastic matrix of the Markov chain implies the useful relationship

$$R \cdot (1 - P) = 1, \tag{8.15}$$

where 1 represents the unit matrix. Given P_{jk}, we can thus evaluate first R_{jk} from (8.15) and then F_{jk} from (8.12), (8.13). In view of (8.2), the matrix $1 - P$ has a zero eigenvalue associated with the eigenvector $(1, 1, 1, \ldots)$ so that, in general, (8.15) cannot be solved in a straightforward way to obtain R. This degeneracy corresponds to the fact that we expect entries of ∞ in the potential matrix R associated with the recurrent states. In the solution to Exercise 189, we discuss a regularization that allows us to calculate R_{jk} and F_{jk} from P_{jk} for a finite Markov chain. For Markov chains in denumerably infinite sets, it is possible that all states are transient and that we can calculate the potential matrix R from (8.15) without any regularization (see Exercise 190).

Exercise 188 Calculation of Potential Matrix
Derive the formula (8.12) for the potential matrix by performing the summation.

To classify the states of a Markov chain, we introduce the following definitions. A state j is either *transient* or *recurrent*, depending on whether $R_{jj} < \infty$ or $R_{jj} = \infty$ or, according to (8.13) equivalently, depending on whether $F_{jj} < 1$ or $F_{jj} = 1$. A recurrent state j is called a *nullstate* if the mean waiting time for returning to it is

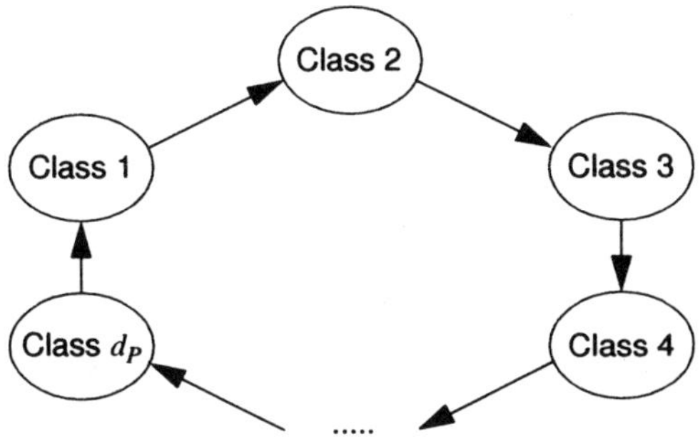

Fig. 8.3 Cyclic classes of states of a periodic Markov chain with period d_P.

infinite, that is, $\sum_{n=1}^{\infty} nF_{jj}^{(n)} = \infty$. A recurrent state j is *periodic* if there exists an integer $d_P > 1$ such that $F_{jj}^{(n)} = 0$ unless n is a multiple of d_P; the largest integer d_P with that property is called the period. We say that the state k is a *consequent* of the state j, and we introduce the symbol $j \to k$ for this transitive relation, if k can be reached from j, that is, if there exists an $n \geq 0$ such that $P_{jk}^n > 0$ or, equivalently, $R_{jk} > 0$. We further introduce an *equivalence relation* $j \leftrightarrow k$ if both $j \to k$ and $k \to j$, that is, if j and k are mutually reachable. We refer to states with $j \leftrightarrow k$ as *communicating*. A subset C of the discrete state space is *closed* if, for all states in C, all consequents are also contained in C. A Markov chain is called *irreducible* if the full state space is the only closed set.

Note that all consequents of a recurrent state are also recurrent (see Exercise 191). The set of recurrent states is thus a closed set. Moreover, when the set of recurrent states is partitioned into equivalence classes $C_1, C_2, \ldots$ of the equivalence relation $j \leftrightarrow k$, these classes of communicating states are closed sets, and the restrictions of the Markov chain to $C_1, C_2, \ldots$ are irreducible Markov chains. The chain can be studied separately on the different equivalence classes, and we hence focus on irreducible Markov chains from now on.

For an irreducible Markov chain, all states are either aperiodic or periodic with the same period d_P. In the latter case, there are d_P cyclic classes of states that are successively visited by the Markov chain, as illustrated in Figure 8.3. Furthermore, either all or none of the states of an irreducible Markov chain are nullstates; in particular, if the state space is finite, there cannot exist any nullstates.

Exercise 189 A Finite Markov Chain

Formulate the stochastic matrix for the Markov chain given in Figure 8.4. Calculate the matrices R_{jk} and F_{jk}. What is the closed set of recurrent states? Connect the results of the calculations with your intuitive expectations from looking at the figure. What are the relative frequencies for visiting the recurrent states over a long period of time?

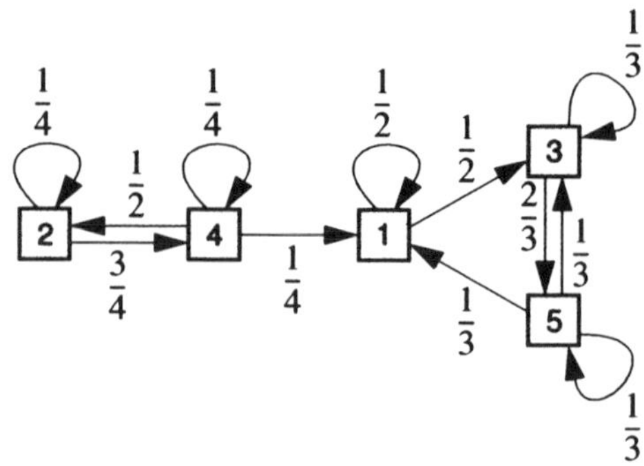

Fig. 8.4 Example of a Markov chain with the finite state space $\{1, 2, 3, 4, 5\}$ and transition probabilities as indicated on the arrows.

Exercise 190 A Chain with only Transient States
Consider the stochastic matrix (8.4) describing the number of occurrences of "heads" in tossing a coin. Calculate the potential matrix and show that all states are transient.

Exercise 191 Consequents of Recurrent States
Show that all consequents of a recurrent state are also recurrent.

In the remainder of this subsection, we study the limiting distribution

$$\pi_k^\infty = \lim_{n\to\infty} P_{jk}^n \tag{8.16}$$

for irreducible Markov chains. To extend the subsequent results to the periodic case, one can generalize the definition of the limiting distribution as follows,

$$\pi_k^\infty = \lim_{n\to\infty} \frac{1}{n} \sum_{l=1}^{n} P_{jk}^l, \tag{8.17}$$

so that periodic variations are averaged out. It is plausible to expect that the limiting probability π_k^∞ is given by the inverse of the mean waiting time $\sum_{n=1}^\infty nF_{kk}^{(n)}$ between two occurrences of k. We hence need to exclude nullstates from the following considerations. Two important theorems from the theory of Markov chains can now be formulated.

Theorem 1. Consider an aperiodic irreducible Markov chain with a stochastic matrix P_{jk}. All states are non-nullstates if, and only if, the linear system

$$\pi_k = \sum_j \pi_j P_{jk}, \quad \sum_k \pi_k = 1, \tag{8.18}$$

has a solution π_k.

The solution π_k of (8.18), which is an invariant distribution of the Markov chain, is given by the limiting distribution π_k^∞ introduced in (8.16) which, for non-nullstates, is equal to the inverse waiting time,

$$\pi_k^\infty = \frac{1}{\sum_{n=1}^\infty nF_{kk}^{(n)}}. \tag{8.19}$$

This expression is the basis for the next theorem.

Theorem 2 (Ergodic Theorem). Consider an aperiodic irreducible Markov chain X_0, X_1, X_2, ... of non-nullstates with invariant distribution π_k. For every bounded function g on the integers and for almost all realizations of the Markov chain (that is, with probability 1), one has

$$\sum_k g(k)\pi_k = \lim_{n\to\infty} \frac{1}{n}\sum_{l=1}^n g(X_l). \tag{8.20}$$

This ergodic theorem provides the mathematical foundation for Monte Carlo simulations. In a simulation, one evaluates the arithmetic average of a sequence of simulation results generated by the underlying Markov chain on the right-hand side of (8.20) for large n. If the Markov chain is constructed such that some given distribution of interest π_k, typically some canonical ensemble, occurs as its limiting distribution, then the simulation reproduces the average performed with the distribution π_k on the left-hand side of (8.20). A systematic procedure for constructing a Markov chain with a prescribed limiting distribution is developed in the next subsection.

8.2.2 Importance Sampling and Detailed Balance

We now have the mathematical background to explain Monte Carlo simulations. For the purpose of illustration, we assume that we are interested in a system with a finite number of states $\{1, 2, \ldots, N\}$. In an underlying statistical ensemble, each state j has a given probability π_j, and our goal is to calculate averages of the form

$$\langle g \rangle = \sum_{j=1}^N g(j)\pi_j \tag{8.21}$$

for certain observables g. We do that by generating a sequence of states X_0, X_1, X_2, ... by means of a Markov chain, where $\bar{\pi}_j$ is the relative frequency of occurrence or the limiting probability of the state j in the Markov chain. By applying the ergodic theorem (8.20) to $g(j)\pi_j/\bar{\pi}_j$ for a Markov chain with limiting distribution $\bar{\pi}_j$, we obtain

$$\langle g \rangle = \sum_{j=1}^N g(j)\pi_j = \lim_{n\to\infty} \frac{1}{n}\sum_{l=1}^n g(X_l)\frac{\pi_{X_l}}{\bar{\pi}_{X_l}}. \tag{8.22}$$

How shall we choose the relative frequency $\bar{\pi}_j$? To answer this question, we consider the scattering of the simulation results for the observable $g(j)\pi_j/\bar{\pi}_j$ to be averaged in the simulation according to the right-hand side of (8.22). Of course, we wish to minimize the scattering. For a nonnegative observable $g(j)$, we could choose

$$\bar{\pi}_j \propto g(j)\pi_j. \tag{8.23}$$

This choice would make all simulation results equal and thus reduce their variance to zero—a possibility that is too good to be true. The catch is that the constant of proportionality suppressed in (8.23) is given by the average that we are actually looking for, as we realize by rewriting (8.23) as an equality,

$$\bar{\pi}_j = \frac{g(j)\pi_j}{\langle g \rangle}, \tag{8.24}$$

so that optimum variance reduction is possible if, and only if, we know the complete solution to the problem of interest, $\langle g \rangle$. Then, (8.22) is trivial.

In general, and certainly if we are interested in the averages of several quantities, we clearly cannot compensate for the variations in the factor $g(j)$, but we can eliminate the usually much stronger variations in the factor $\pi_j/\bar{\pi}_j$ by choosing $\bar{\pi}_j = \pi_j$. This choice offers a major advantage compared with the naive choice $\bar{\pi}_j = 1/N$ which corresponds to treating all states equal in performing averages. For $\bar{\pi}_j = 1/N$, enormous variations in the factor $\pi_j/\bar{\pi}_j$ occur due to the typically exponential form of π_j in a canonical ensemble; most states carry an exponentially suppressed weight and thus make a negligible contribution to the average of interest. For the choice $\bar{\pi}_j = \pi_j$, which expresses the idea of *importance sampling*, the simulation efforts are rather focused on the states contributing to the averages of interest. Importance sampling is the key idea leading to the success of Monte Carlo simulations.

From now on, we focus on the choice $\bar{\pi}_j = \pi_j$. However, one should keep in mind that, if some information such as an approximate solution to the problem of interest is available, further variance reduction for particular observables may be possible by choosing an approximate version of (8.24) and thus eliminating some of the variations in the factor $g(j)$.[7]

Our next problem is the construction of a Markov chain with a given limiting distribution π_j, as characterized by the linear equations (8.18). This construction is based on the idea of *detailed balance*,

$$\pi_j P_{jk} = \pi_k P_{kj}, \tag{8.25}$$

for all states j, k. For $P_{kj} > 0$, this condition can be written as

$$\frac{P_{jk}}{P_{kj}} = \frac{\pi_k}{\pi_j}. \tag{8.26}$$

[7] See Section 4.1.5 of Öttinger, *Stochastic Processes in Polymeric Fluids* (Springer, 1996).

For detailed balance, the transition rates between two states are proportional to the probabilities of these states. Assuming detailed balance, the condition (8.18) for the invariant distribution follows immediately from (8.2),

$$\sum_j \pi_j P_{jk} = \sum_j \pi_k P_{kj} = \pi_k. \tag{8.27}$$

Assume that a set of rules for changing configurations are given such that, for every possible single-step transition, the direct reverse transition is also possible by these rules. There exist two widely used procedures for constructing transition probabilities with these rules such that detailed balance is achieved. In the *Metropolis method*,[8] the probability for accepting an offered transition from j to k is chosen as follows:

$$\begin{array}{ll} 1 & \text{for } \pi_k \geq \pi_j \\ \pi_k/\pi_j & \text{for } \pi_k < \pi_j. \end{array} \tag{8.28}$$

For the mutual transitions between j and k, one always accepts the transition from the less to the more probable state, and the reverse transition occurs with a probability given by the ratio of the invariant probabilities. If a move from j to k is rejected, the state j is kept, and it counts another time in evaluating arithmetic averages according to the ergodic theorem. The *heat bath method*,[9] which is best suited for lattice theories, is based on the idea that every lattice variable is successively touched by a heat bath. One thus considers the set of all states k that can be reached from a state j for a certain rule of changing states locally at or around a lattice site, including j itself, and one defines

$$P_{jk} \propto \pi_k. \tag{8.29}$$

The Metropolis method can be put into an abstract context to get a complete and elegant formulation. We assume that a symmetric, but otherwise arbitrary, stochastic matrix P^{mov} describing the possible Monte Carlo moves be given [see (8.2)]. The lengthy superscript is chosen to indicate that this stochastic matrix describes the possible configurational changes through Monte Carlo moves, including their relative importance. It was emphasized in Section 8.1.3 that the invention of efficient moves is a most important task in Monte Carlo simulations. The Metropolis method for associating probabilities with the possible moves is incorporated by defining a nonsymmetric stochastic matrix P satisfying detailed balance,

$$P_{jk} = \begin{cases} P_{jk}^{\text{mov}} & \text{for } j \neq k,\ \pi_k \geq \pi_j \\ P_{jk}^{\text{mov}} \pi_k/\pi_j & \text{for } j \neq k,\ \pi_k < \pi_j \\ 1 - \sum\limits_{l \neq j} P_{jl} & \text{for } j = k. \end{cases} \tag{8.30}$$

It should be mentioned once more that the Metropolis method involves only ratios of the invariant probabilities π_k.

[8] Metropolis et al., J. Chem. Phys. 21 (1953) 1087.

[9] Creutz, Phys. Rev. D 21 (1980) 2308.

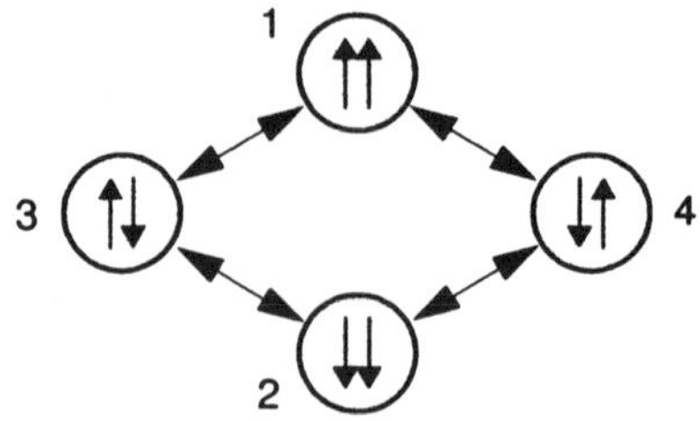

Fig. 8.5 Possible transitions for a two-spin system with four states.

As a concrete example, we consider the two-spin system with four states illustrated in Figure 8.5. We assume that the interaction between the spins favors parallel spins, and the probabilities for the four states are hence given by the Boltzmann factors,

$$\pi_1 = \pi_2 = \frac{1}{Z}, \quad \pi_3 = \pi_4 = \frac{e^{-\Delta}}{Z}, \quad Z = 2(1 + e^{-\Delta}), \tag{8.31}$$

with a dimensionless energy difference $\Delta \geq 0$. The rule for changing the spin configuration is to offer flips of the left or right spin randomly (with probability $1/2$ each). For this rule, the Metropolis method leads to the following stochastic matrix:

$$P = \begin{pmatrix} 1 - e^{-\Delta} & 0 & \frac{1}{2}e^{-\Delta} & \frac{1}{2}e^{-\Delta} \\ 0 & 1 - e^{-\Delta} & \frac{1}{2}e^{-\Delta} & \frac{1}{2}e^{-\Delta} \\ \frac{1}{2} & \frac{1}{2} & 0 & 0 \\ \frac{1}{2} & \frac{1}{2} & 0 & 0 \end{pmatrix}. \tag{8.32}$$

On the other hand, the heat bath method leads to

$$P = \frac{1}{Z} \begin{pmatrix} 2 & 0 & e^{-\Delta} & e^{-\Delta} \\ 0 & 2 & e^{-\Delta} & e^{-\Delta} \\ 1 & 1 & 2e^{-\Delta} & 0 \\ 1 & 1 & 0 & 2e^{-\Delta} \end{pmatrix}. \tag{8.33}$$

For both stochastic matrices, (8.32) and (8.33), the linear equation (8.10) has the unique solution $F_{jk} = 1$ for all states j and k. In words, all transitions occur with probability unity so that we have introduced two irreducible Markov chains of recurrent states. We have constructed two different Markov chains with the same limiting distribution (8.31); a third one is constructed in Exercise 193.

In the Monte Carlo rule of the two-spin example, we offered a flip for a randomly selected spin. As an alternative, we can systematically offer a flip to each spin of the system; this procedure is referred to as a "sweep" through the system. One then takes simulation data only after a sweep is completed. In our example, we would offer flips to the left and to the right spins in an alternating manner.

Exercise 192 Incorporation of the Metropolis Method

Formulate the matrix P^{mov} associated with Figure 8.5 and use it in the construction shown in (8.30).

Exercise 193 Sweep for a Two-Spin System
Formulate the stochastic matrices P_l and P_r for flips of the left and right spins of the two-spin system illustrated in Figure 8.5 according to the Metropolis method. Show that P_l and P_r describe reducible Markov chains. Calculate the stochastic matrix P for a sweep consisting of a flip of the left spin followed by a flip of the right spin. Show that P does not satisfy the detailed balance condition but, nevertheless, describes an irreducible Markov chain with invariant distribution (8.31).

The ergodic theorem guarantees the validity of Monte Carlo simulations based on detailed balance for aperiodic irreducible Markov chains. As indicated before, there is no need to worry about periodicity because the limiting distribution (8.16) can be replaced by (8.17). Moreover, because the construction of transition probabilities is such that states may remain unchanged in a single-step transition, our Markov chains are aperiodic. The crucial condition is that of irreducibility. It needs to be verified for any Monte Carlo simulation that all states are communicating, that is, that the rules for changing configurations are such that all possible states of the system can evolve from any initial state.

At this point, we have elaborated on how to estimate averages for given statistical ensembles. We can now evaluate the GENERIC building blocks E and L as the ensemble averages (6.48) and (6.67). The calculation of the entropy S from (6.53) or (6.55), which is not an ensemble average, is much harder. However, Monte Carlo methods for estimating entropies have been developed. For example, entropy calculations can be based on

- thermodynamic integration of (6.56);[10]
- the iterative flat-histogram technique, particularly suited for rough energy landscapes;[11]
- the insertion of test particles, including biased and gradual methods;[12]
- the simulation of a modified reference system by "umbrella sampling" to explore the relevant regions of configuration space for both the reference system and the original system of interest;[13]
- counting the number of possibilities in the "scanning method" for constructing single[14] and interacting[15] polymer molecules.

[10] See Section 5.6.1 of Landau & Binder, *A Guide to Monte Carlo Simulations* (Cambridge, 2000).
[11] Wang & Landau, Phys. Rev. Lett. 86 (2001) 2050; Phys. Rev. E 64 (2001) 056101; Faller & de Pablo, J. Chem. Phys. 119 (2003) 4405.
[12] See Section 6.1.6 of Landau & Binder, *A Guide to Monte Carlo Simulations* (Cambridge, 2000).
[13] See Section 7.1.2 of Landau & Binder, *A Guide to Monte Carlo Simulations* (Cambridge, 2000).
[14] Meirovitch, Macromolecules 16 (1983) 249; J. Phys. A: Math. Gen. 15 (1982) L735.
[15] Öttinger, Macromolecules 18 (1985) 93.

Five different methods for calculating entropies, based on thermodynamic integration or a perturbation formula for the deviation from a reference system have been compared in the context of molecular dynamics simulations of liquid water.[16]

As indicated in Section 8.1.2, one is often interested in varying a parameter, and one thus needs a number of Monte Carlo simulations. Instead of performing independent simulations of the ensembles corresponding to the different parameters, it can be advantageous to simulate all ensembles in parallel and to allow for swaps of configurations between different ensembles with slight variations of the parameters.[17] This idea, which is known as *parallel tempering*, can increase the efficiency of Monte Carlo simulations enormously by enhancing transitions that would have a very low acceptance rate in a fixed ensemble. While evolving, a configuration "diffuses" by "swap" moves through the parallel ensembles. For a further enhancement of parallel tempering, a so-called multicanonical implementation has been proposed.[18] The art of Monte Carlo simulation includes the selection of both the most powerful *technique* and the most efficient *moves* for a particular problem.

8.2.3 Example: Unentangled Polymer Melts

As an application of Monte Carlo simulations to a beyond-equilibrium problem of practical relevance, we consider a polymer melt subject to elongational flow. We start from an atomistic representation of a polyethylene melt by a multiple-chain system, and, considering an example of Step A ("ab initio") in the classification scheme of Section 8.1.3, we need to introduce a coarse-grained level of description to define a beyond-equilibrium ensemble. In addition to the hydrodynamic fields, we need configurational variables describing the structure of our complex fluid. We choose the molecular weight of the linear polyethylene molecules sufficiently small so that they remain unentangled. The individual chains move in an effective medium formed by the surrounding chains, and we hence assume that the Rouse model of Section 4.3.1 provides a promising level of description for unentangled linear polymers. If we focus on sufficiently low flow rates, only the slowest modes of the Rouse chain are kept from relaxing to their equilibrium configurations. As an idealization, we hence consider a dumbbell model corresponding to the slowest Rouse mode in the single configuration tensor representation (see Sections 4.2.6 and 4.3.1).

Our Monte Carlo simulations of unentangled polymer melts address the following questions, which are built on the exploration of the conformation-tensor level in Section 4.2.6: How does the entropy depend on the conformation tensor? More concretely, what is the functional form of the "effective spring constant" (4.97)? Is the FENE-P model appropriate for unentangled polymers in a melt? Can configurational energy effects be neglected? Does the single conformation tensor level provide a consistent description, for example, are the atomistic stresses reproduced on the

[16] Peter et al., J. Chem. Phys. 120 (2004) 2652.

[17] Marinari & Parisi, Europhys. Lett. 19 (1992) 451; Tesi et al., J. Stat. Phys. 82 (1996) 155.

[18] Faller, Yan & de Pablo, J. Chem. Phys. 116 (2002) 5419.

Table 8.1 Bond Potentials for Polyethylene

Interaction	Form of Potential	Parameters
Bending	$\frac{1}{2} H_{\rm bend} k_{\rm B} (\theta - \theta_0)^2$	$H_{\rm bend} = 57950\,{\rm K\,rad}^{-2}$, $\theta_0 = 112°{\rm C}$
Torsion	$k_{\rm B} \sum_{j=0}^{5} \alpha_j (\cos\varphi)^j$	$\alpha_0 = 1116\,{\rm K}$, $\alpha_1 = 1462\,{\rm K}$, $\alpha_2 = -1578\,{\rm K}$ $\alpha_3 = -368\,{\rm K}$, $\alpha_4 = 3156\,{\rm K}$, $\alpha_5 = -3788\,{\rm K}$

coarse-grained level? By answering these questions we learn something about the rheological material functions in elongational flow as predicted by the coarse-grained model, and we thus go all the way from atomistic details to the flow properties of ultimate interest. This might seem to be at variance with the general comments on the time-scale problem in Section 8.1. However, we here do not determine the friction matrix, and we hence do not need to perform any dynamic simulations. All the conclusions of this subsection are based on Monte Carlo simulations, and, by unphysical Monte Carlo moves corresponding to conformational changes on all physical time scales, we can indeed bridge wide gaps. An a priori understanding of the single-tensor level (see Section 4.2.6) and ideas about the general form of the friction matrix are the key to the success of the present Monte Carlo simulations.

The atomistic simulations of linear polyethylene are actually based on a united-atom description for the methylene and methyl groups. Carbon-carbon bond lengths are held constant at 1.54 Å by an infinitely stiff spring, bond angles are assumed flexible subject to a bending potential, and an additional torsional potential is taken into account (see Table 8.1). A regularized finite-range modification of the Lennard-Jones potential,

$$\phi(r) = 4\epsilon_{\rm LJ} \left[\left(\frac{\sigma_{\rm LJ}}{r} \right)^{12} - \left(\frac{\sigma_{\rm LJ}}{r} \right)^{6} \right], \tag{8.34}$$

with a cutoff at $r = 2.3\sigma_{\rm LJ}$ and a quintic spline approximation between $r = 1.45\sigma_{\rm LJ}$ and $r = 2.3\sigma_{\rm LJ}$, allowing discontinuities only in third- and higher-order derivatives of the potential, is employed for the interactions between the united atoms of a multiple-chain configuration, where the same Lennard-Jones parameters $\epsilon_{\rm LJ}/k_{\rm B} = 49.3\,{\rm K}$ and $\sigma_{\rm LJ} = 3.94\,$Å are used for the methylenes and the methyls. For the thermodynamic quantities obtained from the simulations, the effect of the cutoff in the potential is estimated and corrected for by a direct integration procedure for the attractive tail contributions.[19] All of the details on the various potentials defining the classical mechanics of the atomistic system have been taken from Table 1 of the work by Mavrantzas and Theodorou.[20] The volume of interest is filled with such united-atom chains, where the chain length need not be exactly the same for all polymer molecules or, in other words, we allow for some polydispersity.

[19] See Section 2.8 of Allen & Tildesley, *Computer Simulation of Liquids* (Clarendon, 1987).

[20] Mavrantzas & Theodorou, Macromolecules 31 (1998) 6310.

As explained above, appropriate coarse-grained variables are the mass density, the momentum density, the energy density, and the second moment of the slowest Rouse mode, all evaluated as local averages around a position $\boldsymbol{r}$. We subsequently discuss two simplifications that are important to make our system tractable by simulations. Because these simplifying steps are of a rather general nature and can thus be repeated in a similar way for a huge class of models of complex fluids, we discuss them separately and in considerable detail.

(i) Homogeneous Systems for Understanding Local Fluid Theories. A major simplification arises from the fact that we expect the building blocks for our complex fluid to be of a local nature. The essence of a *local theory* can be explained best in terms of an example, such as the model of a complex fluid with a tensor and a scalar as configurational variables. According to (4.34) and (4.35), the energy and entropy are space integrals of energy and entropy densities, where these densities at the position $\boldsymbol{r}$ depend only on the independent variables at the position $\boldsymbol{r}$. Being interested in these generators, we can thus limit ourselves to situations where only the variables at a given point are relevant or, in other words, we may consider a homogeneous system without any loss of generality.

In an atomistic simulation, we cannot work with intensive variables only; we also need to specify the size of the system. We hence consider $N_{\rm p}$ chains consisting of $N_1, N_2, \ldots N_{N_{\rm p}}$ united atoms; thus, $N = \sum_{k=1}^{N_{\rm p}} N_k$ is the total number of united atoms. Because we do not need any spatial resolution, we introduce position-independent Lagrange multipliers for controlling the fundamental variables of the system: a pressure p to control the volume and thus the mass density, a vectorial Lagrange multiplier $\boldsymbol{\lambda}_M$ closely related to the velocity for the momentum, a temperature T for the energy, and a tensorial multiplier $\boldsymbol{\lambda}_c$ for the conformation tensor. For working with a single configuration tensor, it is important that the chain lengths $N_1, N_2, \ldots N_{N_{\rm p}}$ are sharply peaked around the average chain length because longer chains are affected more severely by a flow field than shorter ones. For a wide distribution of chain lengths, the Lagrange multiplier $\boldsymbol{\lambda}_c$ would have to depend on chain length and further Rouse modes would need to be considered for the longer chains, so that the situation would become considerably more involved. Our single-tensor ensemble for nearly monodisperse chains is given by

$$\begin{aligned} \rho_x(\boldsymbol{r}_1, \boldsymbol{p}_1, \ldots \boldsymbol{r}_N, \boldsymbol{p}_N) &= \frac{1}{Z} \exp \Bigg\{ -\frac{p}{k_{\rm B}T} V(\boldsymbol{r}_1, \ldots \boldsymbol{r}_N) - \boldsymbol{\lambda}_M \cdot \sum_{j=1}^{N} \boldsymbol{p}_j \\ &\quad - \frac{1}{k_{\rm B}T} \left[\sum_{j=1}^{N} \frac{\boldsymbol{p}_j^2}{2m} + \phi_{\rm tot}(\boldsymbol{r}_1, \ldots \boldsymbol{r}_N) \right] - \boldsymbol{\lambda}_c : \sum_{k=1}^{N_{\rm p}} \boldsymbol{c}_k \Bigg\}, \end{aligned} \tag{8.35}$$

where $V(\boldsymbol{r}_1, \ldots \boldsymbol{r}_N)$ is the volume occupied by the N united atoms, m is the mass of a united atom, $\phi_{\rm tot}(\boldsymbol{r}_1, \ldots \boldsymbol{r}_N)$ is the total potential, and $\boldsymbol{c}_k$ is the atomistic expression for the conformation tensor of the kth chain. For example, if the united atoms of the first chain are located at the positions $\boldsymbol{r}_1, \dots \boldsymbol{r}_{N_1}$, we have from (4.131) and (4.137)

$$\boldsymbol{c}_1 = \alpha \boldsymbol{Q}'\boldsymbol{Q}', \qquad \boldsymbol{Q}' = \sqrt{\frac{2}{N_1}} \sum_{j=1}^{N_1-1} (\boldsymbol{r}_{j+1} - \boldsymbol{r}_j) \sin \frac{j\pi}{N_1}, \tag{8.36}$$

where the normalization factor α has been introduced to obtain a unit tensor at equilibrium, in accordance with the normalization of the conformation tensor used in Section 4.2. We expect α to be independent of the chain length and to depend only weakly on temperature and pressure; in general, α needs to be determined from equilibrium simulations.

The assumption of a local theory may be more subtle to implement when one is interested in the Poisson and friction matrices. In a Poisson bracket such as (4.44), we typically have first-order derivatives in addition to the independent fields, indicating the relevance of convective fluxes in the evolution equations. Further gradients related to heat conduction and diffusion occur through the friction matrix, but any higher nonlocal effects are absent.

(ii) Elimination of Momenta. As a next step, we note that the analytical form of the kinetic energy is well known and that the entropy does not depend on the momentum density or velocity. We hence eliminate the momentum variables. By integrating out the momentum variables, we obtain an ensemble in configuration space,

$$\begin{aligned} \rho_x &= \rho_x(\boldsymbol{r}_1, \dots \boldsymbol{r}_N) \\ &= \frac{1}{Z'} \exp\left\{ -\frac{p}{k_B T} V(\boldsymbol{r}_1, \dots \boldsymbol{r}_N) - \frac{1}{k_B T} \phi_{\text{tot}}(\boldsymbol{r}_1, \dots \boldsymbol{r}_N) - \boldsymbol{\lambda}_c : \sum_{k=1}^{N_p} \boldsymbol{c}_k \right\}, \end{aligned} \tag{8.37}$$

with

$$Z' = Z(2\pi m k_B T)^{-3N/2} \exp\left\{ -\frac{1}{2} N m k_B T \boldsymbol{\lambda}_M^2 \right\}. \tag{8.38}$$

The kinetic contribution to the internal energy in the absence of constraints[21] is given by $(3/2)Nk_BT$, and the entropy follows from the general formula (6.55) after performing the momentum integrations,

$$\begin{aligned} S &= S^{\text{ideal}}(T, \langle V \rangle_x, N) \\ &+ k_B \left[\ln\left(\frac{Z'}{\langle V \rangle_x^N} \right) + \frac{p}{k_B T} \langle V \rangle_x + \frac{1}{k_B T} \langle \phi_{\text{tot}} \rangle_x + \boldsymbol{\lambda}_c : \sum_{k=1}^{N_p} \langle \boldsymbol{c}_k \rangle_x \right], \end{aligned} \tag{8.39}$$

where S^{ideal} is the entropy of an ideal gas of noninteracting united atoms, and the pointed brackets indicate averages performed with the ensemble (8.37).

[21] Very stiff potentials instead of rigid constraints are allowed.

In the presence of constraints, one actually needs to be more careful. The internal kinetic energy becomes smaller [namely, $k_{\rm B}T/2$ times the number of degrees of freedom, which is $3N$ only for unconstrained systems], and an additional metric correction potential occurs.[22]

After these general considerations and simplifications, we can now proceed to the actual simulations and the data analyses. From (6.56) we obtain

$$\begin{aligned} dS &= k_{\rm B}\left(\frac{p}{k_{\rm B}T}d\langle V\rangle_x + \boldsymbol{\lambda}_M \cdot d\left\langle\sum_{j=1}^{N}\boldsymbol{p}_j\right\rangle_x + \frac{1}{k_{\rm B}T}d\left\langle\sum_{j=1}^{N}\frac{\boldsymbol{p}_j^2}{2m}\right\rangle_x\right. \\ &+ \left.\frac{1}{k_{\rm B}T}d\langle\phi_{\rm tot}\rangle_x + \boldsymbol{\lambda}_c : d\left\langle\sum_{k=1}^{N_{\rm p}}\boldsymbol{c}_k\right\rangle_x\right). \end{aligned} \tag{8.40}$$

By performing the momentum averages with the Gaussian distribution (8.35), we obtain the "fundamental form"

$$dS - \frac{1}{T}d\left(\frac{3}{2}Nk_{\rm B}T + \langle\phi_{\rm tot}\rangle_x\right) = \frac{p}{T}d\langle V\rangle_x + k_{\rm B}\boldsymbol{\lambda}_c : d\left\langle\sum_{k=1}^{N_{\rm p}}\boldsymbol{c}_k\right\rangle_x . \tag{8.41}$$

If we go back to the notation of Section 4.2 and neglect the additional scalar variable of the phenomenological approach, we obtain the following partial derivative of the Helmholtz free energy (4.58) with respect to the conformation tensor at constant ρ (equivalent to constant $\langle V\rangle_x$ for our fixed number of united atoms) and ϵ,

$$\frac{\partial\epsilon_{\rm c}}{\partial\boldsymbol{c}} - T\rho\frac{\partial\hat{s}_{\rm c}}{\partial\boldsymbol{c}} = \frac{\partial f_{\rm c}}{\partial\boldsymbol{c}} = -n_{\rm p}k_{\rm B}T\boldsymbol{\lambda}_c. \tag{8.42}$$

We have thus identified the Lagrange multiplier for our ensemble in terms of the configuration dependence of energy and entropy. This modification is necessary because, in Section 4.2, we have used ϵ as an independent variable, whereas here the total energy $\epsilon + \epsilon_{\rm c}$ occurs among the variables. It actually is the configurational dependence of the Helmholtz free energy in terms of the previously used variables that determines $\boldsymbol{\lambda}_c$ in (8.42), and it is this combination that drives the time evolution (4.79),

$$\frac{d\boldsymbol{c}}{dt} = \boldsymbol{\kappa}\cdot\boldsymbol{c} + \boldsymbol{c}\cdot\boldsymbol{\kappa}^T - \frac{\xi}{2}(\boldsymbol{c}\cdot\dot{\boldsymbol{\gamma}} + \dot{\boldsymbol{\gamma}}\cdot\boldsymbol{c}) + k_{\rm B}n_{\rm p}\boldsymbol{R}_2 : \boldsymbol{\lambda}_c, \tag{8.43}$$

and that determines the pressure tensor (4.81),

$$\boldsymbol{\pi}_c = 2n_{\rm p}k_{\rm B}T(1-\xi)\boldsymbol{c}\cdot\boldsymbol{\lambda}_c. \tag{8.44}$$

The discussion of dilute polymer solutions in Section 4.2.6 suggests the functional form

$$f_{\rm c} = -n_{\rm p}k_{\rm B}T\left[s^{(\phi)}({\rm tr}\boldsymbol{c}) + \frac{1}{2}\ln(\det\boldsymbol{c})\right], \tag{8.45}$$

[22] See § 12.3 of Bird, Curtiss, Armstrong & Hassager, *DPL 2, Kinetic Theory* (Wiley, 1987).

implying

$$\boldsymbol{\lambda}_c = \frac{1}{2}[\boldsymbol{c}^{-1} - h\boldsymbol{1}], \tag{8.46}$$

where the dimensionless "effective spring constant" h has been defined in (4.97). Although this result has originally been obtained under the assumption of purely entropic effects, the incorporation of energetic corrections into the effective spring constant h is unproblematic.

Cumulant expansion techniques suggest that the functional form (8.45) of the Helmholtz free energy might be too restrictive. If one assumes that the dominating correction to the free energy stems from the second-order cumulant[23] and that all higher cumulants vanish, there is a further dependence of the Helmholtz free energy on a fourth moment of the connector vector, $\langle (\boldsymbol{Q}^2)^2 \rangle$. We therefore allow a dependence on an invariant quadratic in the conformation tensor, where the smallness of the effect compared with the dominating trace contribution leads us to consider a quadratic invariant of the traceless conformation tensor,

$$\mathrm{tr}\left(\boldsymbol{c} - \frac{1}{3}\mathrm{tr}\boldsymbol{c}\boldsymbol{1}\right)^2 = \mathrm{tr}\boldsymbol{c}^2 - \frac{1}{3}(\mathrm{tr}\boldsymbol{c})^2, \tag{8.47}$$

as a further variable on which the free energy (8.45) could depend. We then obtain a more general expression for the Lagrange multiplier $\boldsymbol{\lambda}_c$ in (8.46):

$$\boldsymbol{\lambda}_c = \frac{1}{2}\left[\boldsymbol{c}^{-1} - h\boldsymbol{1} - \tilde{h}\left(\boldsymbol{c} - \frac{1}{3}\mathrm{tr}\boldsymbol{c}\boldsymbol{1}\right)\right], \tag{8.48}$$

in which h and $\tilde{h}$ are partial derivatives of the free energy with respect to the invariants. If $\tilde{h}$ is small compared with h, as we expect for a dominating dependence of the free energy on $\mathrm{tr}\boldsymbol{c}$, then h keeps its meaning as an effective spring constant.

Equations (8.42) and (8.44) show the *kinds of results that one can expect from thermodynamically guided beyond-equilibrium Monte Carlo simulations* for melts of unentangled polymers. We can obtain the pressure tensor in addition to the conformation tensor in dependence on the tensorial Lagrange multiplier $\boldsymbol{\lambda}_c$, and we can thus find the relationship between stress and conformation. Such a relation is, for example, useful for connecting stresses with optical properties. If the pressure tensor has several independent components, we can plot these components against each other, or we can determine ratios of stresses, in particular, in the limit of vanishing deformation rate. Equation (8.42) suggests a further study of the dependence of the internal energy on the configuration tensor in order to find the relative importance of energetic and entropic stress contributions on the coarse-grained level of description. Although the partial derivatives with respect to the conformation tensor are of most interest, the conformation dependence of the entropy can be obtained by

[23] See Ilg, Karlin & Öttinger, Phys. Rev. E 60 (1999) 5783 and Section 2.3 of Gorban et al., J. Non-Newtonian Fluid Mech. 96 (2001) 203.

thermodynamic integration, where (8.41) suggests to consider the Massieu function

$$S - \frac{1}{T}\left(\frac{3}{2}Nk_BT + \langle\phi_{tot}\rangle_x\right) - \frac{p}{T}\langle V\rangle_x - k_B\boldsymbol{\lambda}_c : \left\langle\sum_{k=1}^{N_p} \boldsymbol{c}_k\right\rangle_x \tag{8.49}$$

for the actual integration with respect to the Lagrange multipliers T, p and $\boldsymbol{\lambda}_c$. Finally, the validity of the coarse-grained level of description can be checked by comparing the pressure tensor (8.44) to the atomistic one, which can be obtained in the spirit of the Kramers tensor (4.124) or of the Irving-Kirkwood expression (7.188) from the forces and differences between all pairs of interacting particles in the system, in addition to a kinetic contribution of the type (7.99).[24]

For starting a simulation, we need to *choose the values of the tensorial Lagrange multiplier* $\boldsymbol{\lambda}_c$ corresponding to a given steady flow situation of interest, producing the proper conformation tensor $\boldsymbol{c} = \boldsymbol{c}(\boldsymbol{\lambda}_c)$ observed in that flow. Strictly speaking, the identification of the Lagrange multiplier $\boldsymbol{\lambda}_c$ requires the friction matrix or, in other words, dynamic material information that is beyond Monte Carlo simulations. An explicit expression for the tensorial Lagrange multiplier $\boldsymbol{\lambda}_c$ in steady flows involving the friction matrix can be obtained from (8.43),

$$\boldsymbol{\lambda}_c = -\frac{1}{k_B n_p}\boldsymbol{R}_2^{-1} : \left[\boldsymbol{\kappa}\cdot\boldsymbol{c} + \boldsymbol{c}\cdot\boldsymbol{\kappa}^T - \frac{\xi}{2}(\boldsymbol{c}\cdot\dot{\boldsymbol{\gamma}} + \dot{\boldsymbol{\gamma}}\cdot\boldsymbol{c})\right], \tag{8.50}$$

where $\boldsymbol{R}_2$ describes the relaxation of the conformation tensor in the friction matrix. If we neglect slip effects ($\xi = 0$), and if we assume that the fourth-rank tensor $\boldsymbol{R}_2$ can equivalently be replaced by the second-rank tensor $\hat{\boldsymbol{R}}_2$ (see Exercise 64), then we find

$$\boldsymbol{\lambda}_c = -\frac{1}{k_B n_p}\hat{\boldsymbol{R}}_2^{-1}\cdot(\boldsymbol{\kappa}\cdot\boldsymbol{c} + \boldsymbol{c}\cdot\boldsymbol{\kappa}^T). \tag{8.51}$$

We next introduce the expressions for $\boldsymbol{R}_2$ and $\hat{\boldsymbol{R}}_2$ given in (4.88) and Exercise 69, which are inspired by the Hookean dumbbell model but are expected to be more generally valid for describing the relaxation of tensor variables due to *isotropic* friction.[25] We then obtain

$$\boldsymbol{\lambda}_c = -\frac{1}{2}\tau(\boldsymbol{c}^{-1}\cdot\boldsymbol{\kappa}\cdot\boldsymbol{c} + \boldsymbol{\kappa}^T), \tag{8.52}$$

where the time scale τ of the dumbbell should be taken as the relaxation time of the slowest Rouse mode (8.36), that is, τ_1 of (4.133). For potential flows, that is, for symmetric velocity gradient tensors ($\boldsymbol{\kappa} = \boldsymbol{\kappa}^T$), the conformation tensor, being a function of $\boldsymbol{\kappa}$ only, must commute with $\boldsymbol{\kappa}$, and we obtain

$$\boldsymbol{\lambda}_c = -\tau\boldsymbol{\kappa}. \tag{8.53}$$

[24] For a detailed discussion of the atomistic expression for the pressure tensor, including subtle consequences of periodic boundary conditions used in simulations, see Theodorou et al., Makromol. Chem., Theory Simul. 2 (1993) 191, in particular, (1.2) and (4.6) of that paper.

[25] For example, when incorporating nonlinear spring behavior into the conformation tensor model, it is also assumed that $\hat{\boldsymbol{R}}_2$ is proportional to $\boldsymbol{c}$; see Hütter, Karlin & Öttinger, Phys. Rev. E 68 (2003) 016115.

The result (8.53) is of a more general nature than one might anticipate from its derivation, and it holds even for nonlinear connector forces; it follows from the quadratic "potential" that can be associated with the homogeneous velocity field for symmetric velocity gradient tensors. In the presence of hydrodynamic interactions, however, the definition of the slowest mode would need to be changed.[26] For *anisotropic* friction, (8.52) and (8.53) become invalid, and one needs to rely on (8.50) or (8.51).

The above expressions for $\boldsymbol{\lambda}_c$ show that *Monte Carlo and molecular dynamics simulations are closely interwoven*. On the one hand, Monte Carlo simulations depend on molecular dynamics simulations because the Lagrange multipliers needed for Monte Carlo can only be determined with the help of the friction matrix, the determination of which requires a dynamic simulation technique. On the other hand, the initial ensembles for molecular dynamics are most efficiently generated by Monte Carlo methods.

If we need the friction matrix to identify the Lagrange multiplier associated with the conformation tensor, *how can stand-alone Monte Carlo simulations without molecular dynamics be useful?* This question is important because, as discussed in Section 8.1.3, the time-scale dilemma is much less severe for Monte Carlo than for molecular dynamics simulations. If we use an approximate expression for the friction matrix, we can perform Monte Carlo simulations in the neighborhood of certain flows of interest. We can then establish explicit or black-box functional forms for the dependence of energy and entropy on the conformation tensor, we can estimate the relative importance of energetic and entropic effects, and we can investigate the relationship between stresses and conformations. Although these results are obtained only in the neighborhood of the flow of interest, functional forms and general conclusions can be extrapolated to the actual flow situation of interest.

We here illustrate these possibilities by a Monte Carlo simulation of the model of polyethylene introduced in the beginning of the subsection. All simulations described here are performed in a cubic box with periodic boundary conditions containing 80 chains of mean molecular length C_{78}, which is comparable to the entanglement chain length of some 60 backbone atoms (according to Table 4.2), so that our molecules are not significantly entangled. Efficient sampling of oriented melt conformations was made possible through a robust, connectivity-changing *end-bridging Monte Carlo* algorithm, in which a trimer is cut out of one chain near the end of another chain, to be connected with one of the two subchains obtained after cutting.[27] For the sake of efficient Monte Carlo sampling, a polydisperse system with a uniform probability distribution for the chain length between $78 - (78/2) = 39$ and $78 + (78/2) = 117$ is considered.[28] As stated before, the chain length distribution should be sharply peaked. Because this is not really the case in our simulations, polydispersity might

[26] See (13.2-14) and (16.2-10) of Bird, Curtiss, Armstrong & Hassager, *DPL 2, Kinetic Theory* (Wiley, 1987).

[27] Mavrantzas et al., Macromolecules 32 (1999) 5072.

[28] This corresponds to a ratio of mass-weighted to number-weighted average molecular weights, or *polydispersity index*, around 1.08.

Table 8.2 Results from a Monte Carlo Simulation of Unentangled Polyethylene in Elongational Flow

$-\lambda_{xx}$	c_{xx}	c_{yy}	h	$\tilde{h}$
0	0.98(6)	0.99(6)	1.01(5)	–
0.1	1.18(6)	0.95(6)	0.98(5)	0.4(5)
0.2	1.46(8)	0.89(6)	0.98(5)	0.3(2)
0.3	1.79(10)	0.80(5)	1.02(5)	0.21(11)

have a noticeable effect on the results. The temperature is fixed to $T = 450\,\mathrm{K}$, and the pressure to $p = 1$ atm, corresponding to the melt phase of polyethylene.

We here focus on simple elongational flow [see Exercise 52]. For symmetry reasons, we expect tensor variables of the form

$$\boldsymbol{c} = \begin{pmatrix} c_{xx} & 0 & 0 \\ 0 & c_{yy} & 0 \\ 0 & 0 & c_{yy} \end{pmatrix}, \qquad \boldsymbol{\lambda}_c = \begin{pmatrix} \lambda_{xx} & 0 & 0 \\ 0 & \lambda_{yy} & 0 \\ 0 & 0 & \lambda_{yy} \end{pmatrix}. \tag{8.54}$$

In accordance with (8.53), we assume that $\boldsymbol{\lambda}_c$ is traceless,

$$\lambda_{yy} = -\frac{1}{2}\lambda_{xx}, \tag{8.55}$$

where $-\lambda_{xx}$ corresponds to the parameter α of the publication on which the present subsection is based.[29] As discussed before, instead of assuming a traceless $\boldsymbol{\lambda}_c$, we should rather have a physically motivated subblock $\boldsymbol{R}_2$ or $\hat{\boldsymbol{R}}_2$ of the friction matrix or, even better, evaluate that subblock self-consistently by dynamic simulation techniques. Exercise 74 illustrates that, for anisotropic frictional properties, we cannot expect $\boldsymbol{\lambda}_c$ to be traceless even for incompressible fluids, in contrast to the expressions (8.52) and (8.53). We hence cannot assume that we simulate polymer conformations rigorously corresponding to simple elongational flow, but we are certainly close to elongational flow.

The actual simulations have been performed with four conformation tensors corresponding to the four slowest Rouse modes, but it was found that a single-conformation-tensor model considered here, accounting for the slowest mode only, is completely sufficient. According to the Rouse model, the xx-component of the conformation tensor for the second mode can be obtained from the component c_{xx} of the slowest mode as $4c_{xx}/(3c_{xx}+1)$, which suggests a 12% effect on the second mode for the strongest flow considered here; an even smaller effect was found in the actual simulations.

[29] Mavrantzas & Öttinger, Macromolecules 35 (2002) 960.

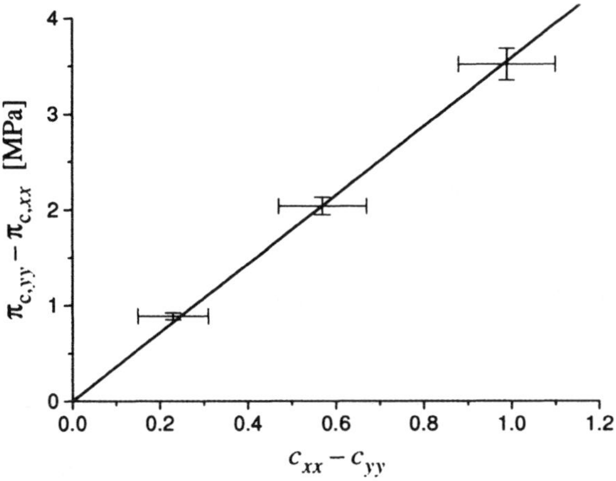

Fig. 8.6 Test of the stress-optical rule by means of atomistic Monte Carlo simulations for polyethylene (C_{78}) in elongational flow.

The simulation results for $\boldsymbol{c}$ as a function of λ_{xx} are summarized in Table 8.2.3. Based on the entropy-gradient expression (8.48), we obtain the parameters h and $\tilde{h}$:

$$h = \frac{1}{3}\left(\mathrm{tr}\boldsymbol{c}^{-1} - 2\mathrm{tr}\boldsymbol{\lambda}_c\right), \qquad \tilde{h} = 3\frac{c_{xx}^{-1} - h - 2\lambda_{xx}}{2c_{xx} - c_{yy} - c_{zz}}, \tag{8.56}$$

as listed in Table 8.2.3. Although the parameter $\tilde{h}$ is found to be close to zero (within the error bars), the results for $-\lambda_{xx} = 0.2$ and, more clearly, for $-\lambda_{xx} = 0.3$ are inconsistent with $\tilde{h} = 0$ and thus suggest that the entropy-gradient expression (8.46) is too restricted. In view of the smallness of $\tilde{h}$, however, we can still interpret h as an effective spring constant, and our simulation results imply that the effective spring constant is independent of the flow rate, even when c_{xx} is considerably larger than its equilibrium value, $c_{\mathrm{eq},xx} = 1$. The small values found for $\tilde{h}$ further imply that, except for an isotropic term, the pressure tensor $\boldsymbol{\pi}_c$ in (8.44) should be approximately proportional to the conformation tensor $\boldsymbol{c}$. This *stress-optical rule* is verified in Figure 8.6. It should be noted that the stresses in Figure 8.6 are given in direct physical units (MPa).

In a similar Monte Carlo simulation, Mavrantzas and Theodorou[30] separated the energetic and entropic contributions to the Helmholtz free energy, and hence to the Lagrange multiplier (8.42) and to the pressure tensor (8.44), by keeping track of the average potential energy in the simulations. They found that the melt response is purely entropic for long chains and small Lagrange multipliers. A significant energetic contribution develops for short chains and large Lagrange multipliers, where the shape of the chains becomes elongated and attractive lateral interchain interactions

[30] Mavrantzas & Theodorou, Macromolecules 31 (1998) 6310.

are intensified. As anticipated in Section 4.3.2, these authors furthermore found an excellent agreement between the stresses obtained on the coarser level from the mesoscopic formula (8.44) and on the atomistic level from the virial theorem.[31]

8.3 BROWNIAN DYNAMICS

The Monte Carlo technique developed and illustrated in Section 8.2 provides a powerful universal tool to calculate the GENERIC building blocks E, S and L. Such calculations are fully analogous to their equilibrium counterparts, with the only exception that the underlying generalized ensembles introduce more variables than one needs at equilibrium. Any calculation of friction matrices, which are needed to complete the list of GENERIC building blocks and to identify the Lagrange multipliers in Monte Carlo simulations, requires the dynamic simulation techniques, which are the topic of this and the following sections on Brownian dynamics and molecular dynamics.

8.3.1 Stochastic Differential Equations

Brownian dynamics simulations are based on the theory of stochastic differential equations. A cornerstone of that theory is the relationship between *diffusion equations* or *Fokker-Planck equations* of the form

$$\frac{\partial f(x,t)}{\partial t} = -\frac{\partial}{\partial x} \cdot [A(x,t)f(x,t)] + \frac{1}{2}\frac{\partial}{\partial x}\frac{\partial}{\partial x} : [D(x,t)f(x,t)], \tag{8.57}$$

on the one hand and *Langevin equations* or *stochastic differential equations* of the form

$$dX_t = A(X_t,t)dt + B(X_t,t) \cdot dW_t \tag{8.58}$$

with

$$B(x,t) \cdot B(x,t)^T = D(x,t) \tag{8.59}$$

on the one hand. In the second-order partial differential equation (8.57), the probability density f depends on a number of components, $x = (x_1, x_2, \ldots, x_n)$, in addition to time, t. By analogy with the continuity equation (2.10), the n-component function $A(x,t)$ represents a "drift velocity" contribution or a deterministic contribution to the rate of change of x, whereas the diffusion equation (2.46) for incompressible fluids further provides the interpretation of the $n \times n$-matrix $D(x,t)$ as a diffusion matrix.

Whereas x occurs as the argument of functions in the Fokker-Planck equation (8.57), in particular as an argument of the probability density f, X_t describes a stochastic process in the stochastic differential equation (8.58). In (8.58), the drift velocity occurs as the rate of change, and the diffusion is represented in terms of a

[31] See Figures 15 and 16 of Mavrantzas & Theodorou, Macromolecules 31 (1998) 6310.

matrix $B(x,t)$ satisfying (8.59), which need not be quadratic, and the multicomponent Wiener process, W_t. In Section 1.2.5, the Wiener process was introduced as a Gaussian process with first and second moments given by (1.58) or (1.59).

The Fokker-Planck equation (8.57) and the stochastic differential equation (8.58) are related in the sense that the probability density of the random variable X_t obtained by solving (8.58) is given by the function $f(x,t)$ obtained by solving (8.57) for all times t, provided that such a relationship holds for the respective initial conditions. This statement implies that averages can be evaluated equivalently either as integrals with the probability density $f(x,t)$ or as ensemble averages from an ensemble of trajectories constructed by solving a stochastic differential equation.[32] Brownian dynamics and related stochastic simulation techniques rely on the construction of discretized solutions of stochastic differential equations of motion for calculating averages.

A numerical solution of the stochastic differential equation (8.58) at discrete times $t_0 < t_1 < t_2 < \ldots$ is obtained by the explicit Euler scheme,

$$X_{t_{j+1}} = X_{t_j} + A(X_{t_j}, t_j)(t_{j+1} - t_j) + B(X_{t_j}, t_j) \cdot (W_{t_{j+1}} - W_{t_j}), \tag{8.60}$$

where, according to (1.58), the differences $W_{t_{j+1}} - W_{t_j}$ are independent Gaussian random variables with vanishing first moments and second moments given by

$$\langle (W_{t_{j+1}} - W_{t_j})(W_{t_{j+1}} - W_{t_j})^T \rangle = (t_{j+1} - t_j - t_j + t_j)1 = (t_{j+1} - t_j)1. \tag{8.61}$$

For this integration scheme, which, in the context of stochastic differential equations, is known as the Euler-Maruyama scheme, it can be shown that averages of sufficiently regular functions converge to first order in the time step. To obtain this order of convergence, the random variables $W_{t_{j+1}} - W_{t_j}$ need not even be Gaussian; it is only important that the first moments vanish and that the second moments are given by (8.61). For higher-order algorithms, accordingly higher moments become relevant.

At this point, one should recognize a most important subtlety associated with stochastic differential equations. Equation (8.61) can loosely be written as an order-of-magnitude estimate,

$$W_{t_{j+1}} - W_{t_j} \sim \sqrt{t_{j+1} - t_j}, \tag{8.62}$$

so that the integration scheme

$$\begin{aligned} X_{t_{j+1}} &= X_{t_j} + A(X_{t_j}, t_j)(t_{j+1} - t_j) \\ &\quad + \frac{1}{2}[B(X_{t_j}, t_j) + B(X_{t_{j+1}}, t_{j+1})] \cdot (W_{t_{j+1}} - W_{t_j}) \end{aligned} \tag{8.63}$$

[32] If averages involving the process at two or more times need to be evaluated, for example, in applying a Green-Kubo formula, one needs to consider the Fokker-Planck equation as an equation for transition probabilities; the Markov property of the solution to (8.58) allows us to construct all joint distributions from the initial probability density and the transition probabilities.

would lead to an entirely different stochastic process. Namely, by using a Taylor expansion of $B(X_{t_{j+1}}, t_{j+1})$ and keeping only the leading order terms, we obtain

$$\begin{aligned} X_{t_{j+1}} &\approx X_{t_j} + A(X_{t_j}, t_j)(t_{j+1} - t_j) + B(X_{t_j}, t_j) \cdot (W_{t_{j+1}} - W_{t_j}) \\ &\quad + \frac{1}{2}(X_{t_{j+1}} - X_{t_j}) \cdot \frac{\partial}{\partial x} B(X_{t_j}, t_j) \cdot (W_{t_{j+1}} - W_{t_j}) \\ &\approx X_{t_j} + A(X_{t_j}, t_j)(t_{j+1} - t_j) + B(X_{t_j}, t_j) \cdot (W_{t_{j+1}} - W_{t_j}) \\ &\quad + \frac{1}{2}\left[B(X_{t_j}, t_j)^T \cdot \frac{\partial}{\partial x}\right] \cdot B(X_{t_j}, t_j)^T (t_{j+1} - t_j), \end{aligned} \tag{8.64}$$

where we have replaced the second-order terms in $W_{t_{j+1}} - W_{t_j}$ by their averages (8.61), which are linear in the time step. The replacement

$$(W_{t_{j+1}} - W_{t_j})(W_{t_{j+1}} - W_{t_j})^T \to (t_{j+1} - t_j)1 \tag{8.65}$$

may be regarded as the most fundamental rule of stochastic calculus, namely, as a loose but equivalent formulation of the Itô formula (F.56) further discussed in the solution to Exercise 194. The integration scheme (8.63) thus leads to a modified drift term and hence to the modified Fokker-Planck equation, which can be rewritten in the compact form

$$\frac{\partial f(x,t)}{\partial t} = -\frac{\partial}{\partial x} \cdot [A(x,t) f(x,t)] + \frac{1}{2}\frac{\partial}{\partial x} \cdot \left\{ B(x,t) \frac{\partial}{\partial x} : [B(x,t) f(x,t)] \right\}. \tag{8.66}$$

The discretization scheme (8.60) and the Fokker-Planck equation (8.57) are associated with the name of *Itô*, whereas the discretization scheme (8.63) and the corresponding Fokker-Planck equation (8.66) go under the name of *Stratonovich*. The discrepancy between the Itô scheme (8.60) and the Stratonovich scheme (8.63) keeps us from treating stochastic differential equations in a naive manner like their deterministic counterparts. In particular, one needs to be careful in designing more sophisticated integration schemes such as derivative-free Runge-Kutta schemes or multistep methods, implicit or semi-implicit methods, or predictor-corrector schemes.[33] The following exercises illustrate how stochastic differential equations can be handled, and they establish the equivalence of Fokker-Planck equations and stochastic differential equations in an intuitive way by considering the convergence of averages and the special class of linear stochastic differential equations.

Exercise 194 Time Evolution of Averages

Derive the rate of change of the average $\langle g(X_t) \rangle$ for a sufficiently smooth but otherwise arbitrary function g both from the Fokker-Planck equation (8.57) and from the stochastic differential equation (8.58).

Exercise 195 Linear Stochastic Differential Equations

Consider the linear stochastic differential equation

$$dX_t = [a(t) + A(t) \cdot X_t]dt + B(t) \cdot dW_t. \tag{8.67}$$

[33] See Section 3.4.4 of Öttinger, *Stochastic Processes in Polymeric Fluids* (Springer, 1996).

Why is the solution X_t Gaussian, provided that the initial random variable X_{t_0} is Gaussian? Formulate time-evolution equations for the first moments and the covariances of the solution.

Exercise 196 Gaussian Solutions of Fokker-Planck Equations

Formulate the Gaussian probability density associated with the solution of the linear stochastic differential equation in Exercise 195 and verify that it satisfies the corresponding Fokker-Planck equation (8.57).

Exercise 197 Exceptional Linear Equations

Show that, for the linear differential equation (8.67), the unusual situation occurs that the stochastic equation can be solved in exactly the same way as its deterministic counterpart.

8.3.2 Example: Dilute Polymer Solutions

For the illustration of Brownian dynamics simulations, we choose an example of Step C ("concrete solutions") according to the classification scheme of Section 8.1.3. We start with a dumbbell model of finitely extensible polymers in dilute solution as a coarse-grained molecular model, the GENERIC structure of which on the level of configurational distribution functions has been established in Section 4.3.1. As the practical problem of ultimate interest, to be solved by Brownian dynamics, we consider the calculation of the polymer contribution to the viscosity as a function of shear rate.

More precisely, we consider the Fokker-Planck equation (4.127) with the FENE spring potential

$$\phi^{(S)}(\boldsymbol{Q}) = -\frac{1}{2} H Q_0^2 \ln\left(1 - \frac{\boldsymbol{Q}^2}{Q_0^2}\right), \tag{8.68}$$

where Q_0 is the maximum spring extension and H is the Hookean spring constant in the limit of small extensions, $\boldsymbol{Q}^2/Q_0^2 \ll 1$. The full spring force is given by the nonlinear function

$$-\frac{\partial \phi^{(S)}}{\partial \boldsymbol{Q}} = -\frac{H\boldsymbol{Q}}{1 - \boldsymbol{Q}^2/Q_0^2}. \tag{8.69}$$

In a homogeneous flow situation, the Fokker-Planck equation (4.127) corresponds to the stochastic differential equation

$$d\boldsymbol{Q}_t = \left(\boldsymbol{\kappa} \cdot \boldsymbol{Q}_t - \frac{2H}{\zeta} \frac{\boldsymbol{Q}_t}{1 - \boldsymbol{Q}_t^2/Q_0^2}\right) dt + 2\sqrt{\frac{k_B T}{\zeta}} d\boldsymbol{W}_t, \tag{8.70}$$

where $\boldsymbol{\kappa}$ is the transposed velocity gradient tensor and ζ is the bead friction coefficient. If we make all quantities dimensionless by choosing the fundamental units of length and time such that

$$\sqrt{\frac{k_B T}{H}} = 1 \quad \text{and} \quad \tau_0 = \frac{\zeta}{4H} = 1, \tag{8.71}$$

then we obtain a simpler version of (8.70):

$$d\boldsymbol{Q}_t = \left(\boldsymbol{\kappa} \cdot \boldsymbol{Q}_t - \frac{1}{2} \frac{\boldsymbol{Q}_t}{1 - \boldsymbol{Q}_t^2/b}\right) dt + d\boldsymbol{W}_t, \tag{8.72}$$

with the dimensionless finite extensibility parameter

$$b = \frac{HQ_0^2}{k_B T}. \tag{8.73}$$

Typically, b is a large number, namely, of the order of the number of monomers in the segment represented by a FENE spring.

Exercise 198 From FENE to FENE-P

Derive the time-evolution equation for the dimensionless second-moment tensor $\langle \boldsymbol{Q}_t \boldsymbol{Q}_t \rangle$ for the FENE model. Formulate the Peterlin moment decoupling approximation leading to the FENE-P equations (4.97) and (4.99).

The polymer contribution to the viscosity, η_p, for FENE dumbbells can be evaluated either by means of the Kramers expression (4.124) or from the Giesekus expression (4.130) for the pressure tensor, giving

$$\eta_p = \frac{n_p k_B T \tau_0}{\tau_0 \dot{\gamma}} \left\langle \frac{Q_x Q_y}{1 - \boldsymbol{Q}^2/b} \right\rangle \quad \text{or} \quad \eta_p = n_p k_B T \tau_0 \left\langle Q_y Q_y \right\rangle, \tag{8.74}$$

respectively, where $\boldsymbol{Q}$ is the dimensionless solution of (8.72) in steady shear flow. In the numerical integration of (8.72), the singularity in the FENE spring-force law requires some special precautions. If the Euler-Maruyama scheme is applied to FENE dumbbells, for any finite time step, there is a nonzero probability that the maximum allowed spring extension is exceeded, at least when Gaussian random numbers are used. The simplest possibility to avoid this problem is to reject moves in which the maximum extension is exceeded. A more efficient possibility to avoid configurations in the unphysical range, however, is to use an implicit algorithm. As a further benefit, larger time steps may be used for such an algorithm without losing stability or much accuracy. We here consider a semi-implicit algorithm in which only the spring-force law is treated implicitly, consisting of an Euler-Maruyama step to construct an intermediate configuration $\overline{\boldsymbol{Q}}_{t_{j+1}}$,

$$\overline{\boldsymbol{Q}}_{t_{j+1}} = \boldsymbol{Q}_{t_j} + \left[\boldsymbol{\kappa}(t_j) \cdot \boldsymbol{Q}_{t_j} - \frac{1}{2} \frac{\boldsymbol{Q}_{t_j}}{1 - \boldsymbol{Q}_{t_j}^2/b} \right] \Delta t_j + \Delta \boldsymbol{W}_j, \tag{8.75}$$

with $\Delta t_j = t_{j+1} - t_j$, $\Delta \boldsymbol{W}_j = \boldsymbol{W}_{j+1} - \boldsymbol{W}_j$, and of a nonlinear equation for constructing the final dumbbell configuration $\boldsymbol{Q}_{t_{j+1}}$,

$$\begin{aligned} \left[1 + \frac{1}{4} \frac{\Delta t_j}{1 - \boldsymbol{Q}_{t_{j+1}}^2/b} \right] \boldsymbol{Q}_{t_{j+1}} &= \boldsymbol{Q}_{t_j} \\ &+ \frac{1}{2} \left[\boldsymbol{\kappa}(t_{j+1}) \cdot \overline{\boldsymbol{Q}}_{t_{j+1}} + \boldsymbol{\kappa}(t_j) \cdot \boldsymbol{Q}_{t_j} - \frac{1}{2} \frac{\boldsymbol{Q}_{t_j}}{1 - \boldsymbol{Q}_{t_j}^2/b} \right] \Delta t_j + \Delta \boldsymbol{W}_j. \end{aligned} \tag{8.76}$$

The direction of $\boldsymbol{Q}_{t_{j+1}}$ is given by the direction of the right-hand side of (8.76). The length of $\boldsymbol{Q}_{t_{j+1}}$ can be determined from a cubic equation, which, for arbitrary length

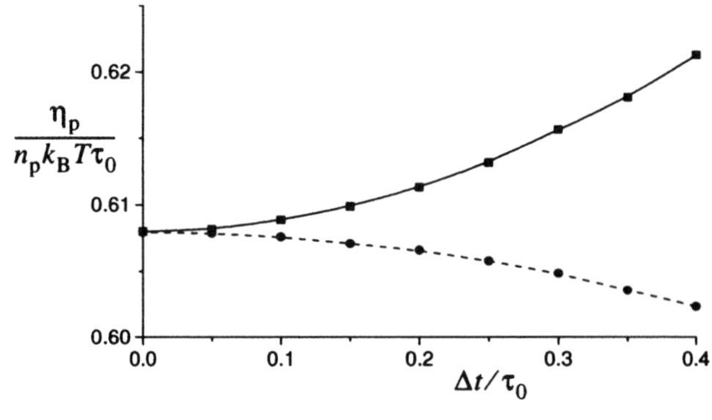

Fig. 8.7 Polymer contribution to the viscosity for $b = 50$ and $\tau_0\dot{\gamma} = 5$ as a function of time step; the squares connected by a continuous line represent the viscosity from the Kramers expression, whereas the circles connected by a dashed line correspond to the Giesekus expression.

of the vector on the right-hand side of (8.76), has a unique solution between 0 and $\sqrt{b}$.[34] Note that the same random numbers $\Delta \boldsymbol{W}_j$ occur in (8.75) and (8.76). This semi-implicit algorithm is designed to avoid unphysical spring extensions through the implicit equation for the length of $\boldsymbol{Q}_{t_{j+1}}$ and to produce averages with only second-order corrections in the time step through the additional use of $\overline{\boldsymbol{Q}}_{t_{j+1}}$. In the simulation, we use non-Gaussian random numbers $\Delta \boldsymbol{W}_j$ with correct second and fourth moments (and vanishing odd moments) to preserve the second-order nature of our algorithm.

Our simulations with the integrator (8.75), (8.76) are based on the previously developed program FENESI.[35] We choose $b = 50$, and we use eight different time steps for every shear rate. For small and intermediate shear rates ($\tau_0\dot{\gamma} \leq 10$), we use the time steps $\Delta t = k \cdot 0.05\tau_0$ with $k = 1, 2, \ldots 8$. The total simulation time for each step is $10^8\tau_0$, and the full trajectory is split into 10000 blocks for estimating the statistical errors from the fluctuations of the individual block results.[36] The initial configuration is obtained from some arbitrary connector vector, such as $\boldsymbol{Q} = 0$ or the final configuration of a simulation with another time step, by evolving it over a period of $10\tau_0$. Figure 8.7 confirms both the second order of convergence expected for our semi-implicit algorithm and the equivalence of the Kramers and Giesekus expressions for the viscosity in the limit of continuous time evolution. For the intermediate shear rate $\tau_0\dot{\gamma} = 5$, one would expect that the time step should be smaller than some $0.2\tau_0$ (so that we do not exceed a unit shear deformation per time step); however, even

[34] See Exercise 4.33 of Öttinger, *Stochastic Processes in Polymeric Fluids* (Springer, 1996).

[35] Öttinger, *Stochastic Processes in Polymeric Fluids* (Springer, 1996).

[36] See, for example, Section 8.3 of Honerkamp, *Stochastic Dynamical Systems* (VCH, 1994), Section 6.4.1 of Allen & Tildesley, *Computer Simulation of Liquids* (Clarendon, 1987) or p. 206 of Öttinger, *Stochastic Processes in Polymeric Fluids* (Springer, 1996).

Table 8.3 Viscosity Function for the FENE Model ($b = 50$)

$\tau_0\dot{\gamma}$	$\frac{\eta_p}{n_p k_B T \tau_0}$ Kramers	Giesekus
0.2	0.90787(40)	0.90719(6)
0.5	0.89715(20)	0.89713(11)
1	0.86770(15)	0.86760(8)
2	0.79030(11)	0.79036(7)
5	0.60800(6)	0.60794(4)
10	0.45220(5)	0.45219(7)
20	0.31734(6)	0.31736(4)
50	0.18839(3)	0.18837(2)
100	0.12370(2)	0.12374(1)

time steps considerably larger than $0.2\tau_0$ can be used in the semi-implicit algorithm. For high shear rates ($\tau_0\dot{\gamma} = 20, 50, 100$), we reduced the time steps and the total simulation time by a factor of 10.

For every shear rate, the results for the different time steps have been extrapolated to $\Delta t = 0$ by means of the fully automatized extrapolation program `TEXTRA`.[37] The results of the extrapolation are summarized in Table 8.3.2 and plotted in Figure 8.8, where the error bars for the simulation, as indicated by the numbers in parentheses in the table, are much smaller than the size of the symbols. According to Table 8.3.2, the error bars for the Gieskus expression for the viscosity are somewhat smaller than for the Kramers expression, in particular, for the smaller shear rates. From (8.74), it is clear that, due to the shear rate in the denominator of the Kramers expression for the viscosity, the advantage of the Giesekus expression would be really dramatic in approaching zero shear rate. Otherwise, the trends in the error bars are somewhat unsystematic. This is a consequence of the fact that we have not adapted the range of time steps to every single shear rate so that, in the extrapolation to zero time step, different degrees of the fit polynomials and different numbers of data points have been used. The results for the FENE-P approximation included in Figure 8.8 have been obtained from Exercise 75, where an explicit formula for the inverse function $\dot{\gamma}(\eta_p)$ has been elaborated. Figure 8.8 shows that, over the entire range of shear rates, the FENE-P model provides an excellent approximation to the viscosity of the FENE model. However, this remarkable agreement is limited to the steady-state viscometric functions while, for example, the time dependent viscometric functions in start-up of shear flow at high shear rates are overpredicted by more than a factor of two by the FENE-P approximation.[38]

[37] See Exercise 4.11 of Öttinger, *Stochastic Processes in Polymeric Fluids* (Springer, 1996).
[38] Herrchen & Öttinger, J. Non-Newtonian Fluid Mech. 68 (1997) 17.

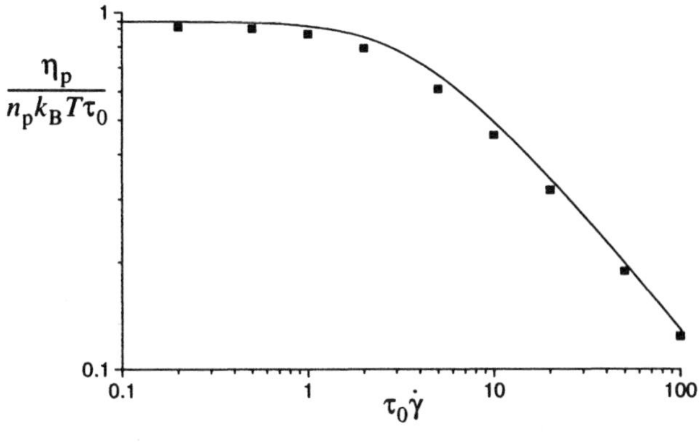

Fig. 8.8 Polymer contribution to the viscosity as a function of shear rate for the FENE model (squares) and the FENE-P approximation for $b = 50$.

The example of FENE dumbbells illustrates how one can obtain high-precision results for relevant material properties by means of Brownian dynamics. Particularly remarkable features of simulation techniques based on stochastic differential equations of motion are their enormous generality and the striking simplicity of the corresponding computer programs.

8.4 MOLECULAR DYNAMICS

A calculation of friction matrices from first principles requires molecular dynamics simulations. Before discussing and illustrating molecular dynamics in detail, we go into some basic considerations concerning the evaluation of friction matrices.

8.4.1 Expressions for the Friction Matrix

Friction matrices can be calculated from the Green-Kubo-type formula (6.72) or (6.73), where we go back to the more symbolic notation of (3.47),

$$M_{jk}(x) = \frac{1}{k_B} \int_0^{\tau} \left\langle \dot{x}_j^f(t) \dot{x}_k^f(0) \right\rangle_x dt. \tag{8.77}$$

We remind the reader that τ is an intermediate time scale separating the slow degrees of freedom from the fast ones, $\dot{x}^f$ is the rapidly fluctuating part of the time derivative of the atomistic expressions for the slow variables x, and the average indicated by the pointed brackets is over an ensemble of atomistic trajectories consistent with the coarse-grained state x at $t = 0$ and evolved according to the (projected or free) atomistic dynamics to the time t.

We assume that the correlation function in (8.77) decays sufficiently fast with increasing time difference t so that the integral is finite. Typically, we expect a power-law decay or, ideally, an exponential decay. To get a converged expression

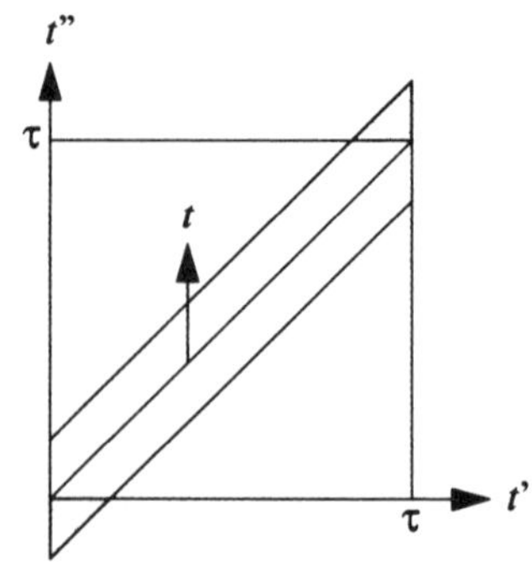

Fig. 8.9 Graphic justification of the change of integration variables in (8.78) for large τ.

for the friction matrix, we would like to take τ large, but then we are faced with two problems: (1) the separation of time scales may not be sufficiently pronounced, and (2) the evaluation of (8.77) in simulations leads to large error bars. As a consequence of the first problem, the slow variables begin to change in the time range considered in the integral (8.77), and one obtains corrections depending on the ratio of fast to slow time scales.[39] If one would like to consider large τ, then one should modify the slow time evolution by a control mechanism that artificially keeps x constant, where the effect of the control mechanism should be kept of the same order as the unavoidable corrections, depending on the ratio of fast to slow time scales. The best known control mechanisms of this type are the thermostats for maintaining a constant temperature (for generalizations, see Section 8.4.3).

Before we discuss the second problem, that is, the size of the statistical errors for large τ, we derive some alternative versions of the Green-Kubo expression (8.77). By assuming that the system is invariant under time translations and that we know its bare Onsager-Casimir symmetry, we obtain

$$\begin{aligned} [1+\varepsilon(x_j)\varepsilon(x_k)]M_{jk}(x) &= \frac{1}{k_B}\int_{-\tau}^{\tau}\left\langle \dot{x}_j^{\rm f}(t)\dot{x}_k^{\rm f}(0)\right\rangle_x dt \\ &= \frac{1}{k_B\tau}\int_0^{\tau}\int_{-\tau}^{\tau}\left\langle \dot{x}_j^{\rm f}(t'+t)\dot{x}_k^{\rm f}(t')\right\rangle_x dt dt' \\ &= \frac{1}{k_B\tau}\int_0^{\tau}\int_0^{\tau}\left\langle \dot{x}_j^{\rm f}(t'')\dot{x}_k^{\rm f}(t')\right\rangle_x dt'' dt'. \quad (8.78) \end{aligned}$$

For the last step, which is valid for large τ only, we have exploited the fact that only the small time differences indicated by the narrow stripe in Figure 8.9 contribute to

[39] Geigenmüller et al., Physica A 119 (1983) 53.

the integral. Equation (8.78) can be rewritten as

$$[1+\varepsilon(x_j)\varepsilon(x_k)]M_{jk}(x) = \frac{\tau}{k_B}\left\langle \left\{\frac{1}{\tau}\int_0^\tau \dot{x}_j^{\rm f}(t'')dt''\right\}\left\{\frac{1}{\tau}\int_0^\tau \dot{x}_k^{\rm f}(t')dt'\right\}\right\rangle_x . \tag{8.79}$$

The notation $\dot{x}_k^{\rm f}$ suggests that this quantity is closely related to a time derivative, and thus one is tempted to carry out the integrations over t' and t'' in (8.79). For a careful interpretation of $\dot{x}_k^{\rm f}$, we go back to (6.158) as a more precise version of (8.77). With the projected time-evolution operator in (6.158), we thus arrive at

$$\Delta_\tau \Pi_k(z) = \int_0^\tau \dot{x}_k^{\rm f}(t')dt' = (e^{\tilde{Q}i\mathcal{L}\tau}-1)\Pi_k(z) = \Pi_k(z(\tau)) - \Pi_k(z(0)). \tag{8.80}$$

Note that (8.80) implies the exact identity $\tilde{Q}\Delta_\tau\Pi_k(z) = \Delta_\tau\Pi_k(z)$. Therefore, if we would like to use the unprojected free or any other approximate time-evolution operator in (6.158), we should write

$$\Delta_\tau \Pi_k(z) = \tilde{Q}(e^{i\mathcal{L}\tau}-1)\Pi_k(z) = \tilde{Q}e^{i\mathcal{L}\tau}\Pi_k(z) = \tilde{Q}\Pi_k(z(\tau)). \tag{8.81}$$

We thus have the integrated version of (8.79),

$$[1+\varepsilon(x_j)\varepsilon(x_k)]M_{jk}(x) = \frac{1}{k_B\tau}\left\langle \Delta_\tau\Pi_j(z)\Delta_\tau\Pi_k(z)\right\rangle_x . \tag{8.82}$$

Equation (8.82) is a generalized version of a standard result.[40] In the context of diffusion, (8.77) corresponds to an expression for the diffusion coefficient in terms of the velocity autocorrelation function associated with the names of Green and Kubo, whereas (8.82) corresponds to a diffusion coefficient in terms of the mean square displacement often associated with the name of Einstein. Hess et al.[41] proposed to calculate transport coefficients from second moments of time averages according to the intermediate formula (8.79). Instead of working with the formula (8.82), it may be advantageous to use the alternative formula[42]

$$[1+\varepsilon(x_j)\varepsilon(x_k)]M_{jk}(x) = \frac{1}{k_B}\frac{d}{d\tau}\left\langle \Delta_\tau\Pi_j(z)\Delta_\tau\Pi_k(z)\right\rangle_x , \tag{8.83}$$

which is also associated with the name of Einstein. For calculating the viscosity of water by molecular dynamics,[43] it has been shown that it may be advantageous to use (8.83) rather than (8.82) because the asymptotic value is reached for smaller τ.

To get a feeling for the performance of the various expressions for the friction matrix, we discuss the simple example of one-dimensional diffusion, where we use

[40] See, for example, (2.108) and (2.118) of Allen & Tildesley, *Computer Simulation of Liquids* (Clarendon, 1987).
[41] Hess & Evans, Phys. Rev. E 64 (2001) 011207; Hess et al., Phys. Rev. E 67 (2003) 042201.
[42] See Section 6.5.5 of Allen & Tildesley, *Computer Simulation of Liquids* (Clarendon, 1987).
[43] Guo, Zhang & Zhao, Phys. Rev. E 67 (2003) 043101.

the slow variables of equilibrium thermodynamics so that we can discuss the problem of statistical errors without having the additional problem that the slow variables change in time. The evolution equations for our one-dimensional toy problem are taken from Exercise 12:

$$\frac{dQ}{dt} = \frac{P}{m} \tag{8.84}$$

and

$$\frac{dP}{dt} = -\zeta\frac{P}{m} + \sqrt{2k_{\rm B}T\zeta}\frac{dW_t}{dt}. \tag{8.85}$$

Although our toy model does not really fit into the setting of (8.77), we can use it for illustration of the evolution of variable $x = Q$ in the limit of a short momentum relaxation time m/ζ. In that limit, P and hence dx/dt become rapidly fluctuating, and we thus obtain the rapidly fluctuating part

$$\dot{x}^{\rm f}(t) = \frac{P(t)}{m} = \dot{x}^{\rm f}(0)e^{-\zeta t/m} + \frac{\sqrt{2k_{\rm B}T\zeta}}{m}\int_0^t e^{-\zeta(t-t')/m}dW_{t'}. \tag{8.86}$$

Because (8.85) is a linear stochastic differential equation, it has been solved by the methods for deterministic differential equations to obtain the explicit integral expression (8.86) [see Exercise 197]. We assume that $\dot{x}^{\rm f}(0)$ is a Gaussian random variable independent of the Gaussian increments dW_t of the Wiener process modeling the noise. Because the average $\langle\dot{x}^{\rm f}(t)\rangle$ for large times goes to zero for any initial condition, we assume

$$\langle\dot{x}^{\rm f}(0)\rangle = 0, \tag{8.87}$$

so that the average $\langle\dot{x}^{\rm f}(t)\rangle$ is identically zero for all times. We further obtain from (1.59)

$$\langle\dot{x}^{\rm f}(t)^2\rangle = \frac{k_{\rm B}T}{m} + \left[\langle\dot{x}^{\rm f}(0)^2\rangle - \frac{k_{\rm B}T}{m}\right]e^{-2\zeta t/m}, \tag{8.88}$$

so that we achieve a steady-state situation for

$$\langle\dot{x}^{\rm f}(0)^2\rangle = \frac{k_{\rm B}T}{m}. \tag{8.89}$$

Equations (8.87) and (8.89) fully characterize the Gaussian random variable $\dot{x}^{\rm f}(0)$. These moments are consistent with the Maxwellian velocity distribution (7.32) at equilibrium.

According to the Green-Kubo formula (8.77), the expression (8.79) proposed by Hess and Evans, and the Einstein expression (8.83), the diffusion coefficient can be obtained by averaging either of the following fluctuating quantities:

$$D_{\rm GK} = \dot{x}^{\rm f}(0)\int_0^{\tau}\dot{x}^{\rm f}(t)dt, \tag{8.90}$$

$$D_{\rm HE} = \frac{1}{2\tau}\left(\int_0^{\tau}\dot{x}^{\rm f}(t)dt\right)^2, \tag{8.91}$$

or

$$D_{\rm E} = \frac{1}{2}\frac{d}{d\tau}\left(\int_0^\tau \dot{x}^{\rm f}(t)dt\right)^2. \tag{8.92}$$

For our analytical discussion of one-dimensional diffusion, (8.79) and (8.82) are manifestly equivalent and, by comparing (8.90) with (8.92) after differentiation, we further expect $D_{\rm GK}$ and $D_{\rm E}$ to be equivalent; we hence focus on the comparison of $D_{\rm GK}$ and $D_{\rm HE}$. For the further arguments, the stochastic integral expression

$$\int_0^\tau \dot{x}^{\rm f}(t)dt = \dot{x}^{\rm f}(0)\frac{m}{\zeta}\left(1 - e^{-\zeta\tau/m}\right) + \sqrt{\frac{2k_{\rm B}T}{\zeta}}\int_0^\tau \left[1 - e^{-\zeta(\tau-t)/m}\right] dW_t, \tag{8.93}$$

for the increment previously considered in (8.80) thus plays an important role. From (8.90), (8.93) and the stochastic independence of $\dot{x}^{\rm f}(0)$ from the increments dW_t of the Wiener process, we obtain

$$\langle D_{\rm GK}\rangle = \frac{k_{\rm B}T}{\zeta}\left(1 - e^{-\zeta\tau/m}\right), \tag{8.94}$$

which, for increasing τ, converges exponentially fast to the correct diffusion coefficient $k_{\rm B}T/\zeta$. The same limit for the diffusion coefficient is obtained from the alternative expression (8.91) for large τ (see Exercise 200),

$$\langle D_{\rm HE}\rangle = \frac{k_{\rm B}T}{\zeta}\left[1 - \frac{m}{\zeta\tau}\left(1 - e^{-\zeta\tau/m}\right)\right]. \tag{8.95}$$

For finding the error bars in simulations, it is useful to look at the respective variances (see Exercise 201),

$$\langle D_{\rm GK}^2\rangle - \langle D_{\rm GK}\rangle^2 = \left(\frac{k_{\rm B}T}{\zeta}\right)^2\left(\frac{2\zeta\tau}{m} + e^{-2\zeta\tau/m} - 1\right) \tag{8.96}$$

and

$$\langle D_{\rm HE}^2\rangle - \langle D_{\rm HE}\rangle^2 = 2\langle D_{\rm HE}\rangle^2. \tag{8.97}$$

Useful insight can now be gained from a comparison of the averages and covariances of $D_{\rm GK}$ and $D_{\rm HE}$. For this toy example, $D_{\rm GK}$ seems to be much better suited to calculate the diffusion coefficient than $D_{\rm HE}$ because it converges exponentially fast with increasing τ. However, exponential convergence is rather atypical. For example, due to "back-flow effects" or long-lived hydrodynamic modes, the velocity autocorrelation function of a particle in a fluid of many interacting identical particles is proportional to $t^{-3/2}$, and this "long-time tail" implies that $\langle D_{\rm GK}\rangle$ converges as $\tau^{-1/2}$, where we expect a similarly slow convergence for $\langle D_{\rm HE}\rangle$. One is thus forced into considering large values of τ. Then, a crucial difference between the variances in (8.96) and (8.97) becomes clear. Whereas the variance of $D_{\rm HE}$ approaches a constant for large τ, the variance of $D_{\rm GK}$ increases linearly with τ. For keeping the statistical error on a given level, the ensemble size in a simulation has to be increased in proportion to τ. This observation reflects the fact that it is extremely time consuming to find the tail of the velocity autocorrelation function with sufficient precision when evaluating the integral in (8.90). A detailed comparison of the Green-Kubo, Hess-Evans, and Einstein expressions for the viscosity of water[44] confirms that the Green-Kubo and Einstein expressions are equally useful in molecular dynamics simulations. While the formula proposed by Hess and Evans leads to smaller statistical error bars than the other two expressions, it produces a slower approach to the viscosity and thus requires larger τ. In practice, one needs to find a proper compromise between systematic and statistical errors where, regarding the systematic error, it should be kept in mind that friction matrices are well defined only within the ratio of fast and slow time scales.

Our discussion of one-dimensional diffusion is based on the stochastic differential equation (8.85) or the stochastic integral expression (8.86) as a model for the rapid fluctuations of the relevant dynamical variable. To extract useful information about correlation functions and transport coefficients from molecular dynamics simulations, it has more generally been proposed to model the fluctuations of dynamical variables as autoregressive stochastic processes.[45] The autoregressive model is not a model for the correlation function itself but for the underlying stochastic process. By increasing the order of the model, the correlation function can be systematically extended to longer times without changing the short-time behavior. Algebraic long-time tails, however, cannot be captured by the autoregressive model. Nevertheless, reliable estimates for transport coefficients can be obtained. The ideas have been tested successfully for liquid argon as a model system.

Exercise 199 Variance of a Stochastic Integral

Derive the identity

$$\left\langle \left\{ \int_0^\tau \left[1 - e^{-\zeta(\tau - t)/m}\right] dW_t \right\}^2 \right\rangle = \tau - \frac{2m}{\zeta}\left(1 - e^{-\zeta\tau/m}\right) + \frac{m}{2\zeta}\left(1 - e^{-2\zeta\tau/m}\right). \tag{8.98}$$

Exercise 200 Diffusion Coefficient From Averages

Derive the expression (8.95).

Exercise 201 Variance of Diffusion Coefficient

Derive (8.96) and (8.97).

8.4.2 Friction Matrix for a Local Field Theory

We next consider how the friction matrix for a local field theory can be evaluated from the atomistic trajectories generated by molecular dynamics, where we try to identify simplifications similar to those for the ensembles employed in the Monte Carlo calculation of the static building blocks of GENERIC (see Section 8.2.3). Without loss

[44] Guo, Zhang & Zhao, Phys. Rev. E 67 (2003) 043101.
[45] Kneller & Hinsen, J. Chem. Phys. 115 (2001) 11097.

of generality we consider a conformation tensor as the only structural variable, which is invariant under time reversal.[46] Equation (8.82) then gives

$$M\delta(\boldsymbol{r}-\boldsymbol{r}') = \frac{1}{2k_{\mathrm{B}}\tau_{\mathrm{s}}} \langle \Delta_{\tau_{\mathrm{s}}} \boldsymbol{c}(\boldsymbol{r};z) \Delta_{\tau_{\mathrm{s}}} \boldsymbol{c}(\boldsymbol{r}';z) \rangle_x , \tag{8.99}$$

for the block of the fiction matrix related to the conformation tensor, where $\Delta_{\tau_{\mathrm{s}}} \boldsymbol{c}(\boldsymbol{r};z)$ is the fluctuating part of the increment of the atomistic expression for the conformation tensor at position $\boldsymbol{r}$ over a period τ_{s} and x describes the coarse-grained state in terms of the hydrodynamic fields and the averaged conformation tensor. We have used the symbol τ_{s} for the intermediate time scale separating fast and slow variables because τ is here reserved for the relaxation time for the average conformation tensor. On the left-hand side of (8.99), M is a fourth-rank tensor, and the occurrence of a factor $\delta(\boldsymbol{r}-\boldsymbol{r}')$ has been anticipated for a local field theory. The same assumption has been used in the phenomenological approach; for example in Section 4.2.3, the friction matrix acts by matrix multiplication without any integration over space. By averaging the $\boldsymbol{r}'$ dependence over a volume V and assuming that, for homogeneous conditions in the volume V, there is no explicit $\boldsymbol{r}$ dependence of the $\boldsymbol{r}'$ average so that we can further average over $\boldsymbol{r}$ without changing the result, we obtain

$$M = \frac{V}{2k_{\mathrm{B}}\tau_{\mathrm{s}}} \langle \Delta_{\tau_{\mathrm{s}}} \boldsymbol{c}(z) \Delta_{\tau_{\mathrm{s}}} \boldsymbol{c}(z) \rangle_x , \tag{8.100}$$

where $\boldsymbol{c}(z)$ is the atomistic expression for the conformation tensor averaged over the local volume V over which homogeneous conditions can be assumed. Equation (8.100) is our fundamental result for local fluid theories.

By expressing $\boldsymbol{c}(z)$ for melts of unentangled polymers in terms of the conformation tensors of the individual chains, we find

$$M = \frac{V}{2k_{\mathrm{B}}\tau_{\mathrm{s}}N_{\mathrm{p}}^2} \sum_{k,l=1}^{N_{\mathrm{p}}} \langle \Delta_{\tau_{\mathrm{s}}} \boldsymbol{c}_k \Delta_{\tau_{\mathrm{s}}} \boldsymbol{c}_l \rangle_x , \tag{8.101}$$

where $\boldsymbol{c}_k$ is the conformation tensor for the kth polymer chain defined according to (8.36) and N_{p} is the total number of chains in the volume V. If the different polymer molecules move independently, all terms with $k \neq l$ vanish and we get the simplified formula

$$M = \frac{\alpha^2}{2n_{\mathrm{p}}k_{\mathrm{B}}\tau_{\mathrm{s}}} \langle \Delta_{\tau_{\mathrm{s}}}(\boldsymbol{Q}'\boldsymbol{Q}') \Delta_{\tau_{\mathrm{s}}}(\boldsymbol{Q}'\boldsymbol{Q}') \rangle_x , \tag{8.102}$$

where n_{p} is the polymer number density, $\boldsymbol{Q}' = \boldsymbol{Q}'(z)$ is the slowest mode for any chain, and α is the normalization factor in (8.36); in practice, one averages over all chains. The straightforward GENERIC expression for the friction matrix, (8.101) or (8.102), involves fourth moments of the slowest mode $\boldsymbol{Q}'$, which is a consequence of using conformation tensors for the coarse-grained model and thus provides the full

[46] The straightforward generalization to configurational distribution functions is described in Section 8.4.6.

index structure of the fourth-rank tensor M. Here, we have chosen to elaborate the expression (8.82) for a local field theory with a conformation tensor variable; however, the Green-Kubo and Einstein formulas (8.77) and (8.83) can also be simplified in an analogous manner.

For a heuristic estimate of the friction matrix (8.102) for unentangled polymers, we assume that, on the intermediate time scale τ_s, there is no systematic change of $\boldsymbol{Q}'$ yet but only the noise effect resulting from the fast variables,

$$\boldsymbol{Q}'(\tau_s) = \boldsymbol{Q}'(0) + 2\sqrt{\frac{k_B T}{\zeta}}(\boldsymbol{W}_{\tau_s} - \boldsymbol{W}_0), \tag{8.103}$$

where we have borrowed this form from the dumbbell equation of motion (8.70). In the spirit of stochastic differential equations, the intermediate time τ_s is so small compared with the relaxation time that, for the short-time behavior, the stochastic term dominates according to (8.62) or, in other words, only the rapidly fluctuating part survives the projection in (8.81). The rapid dynamics is thus modeled by the noise in the stochastic differential equation of motion, which, according to our insights in Section 6.4.2, actually is a general feature. From (8.103) we have

$$\Delta_{\tau_s}(\boldsymbol{Q}'\boldsymbol{Q}') = 2\sqrt{\frac{k_B T}{\zeta}}[\boldsymbol{Q}'(0)(\boldsymbol{W}_{\tau_s} - \boldsymbol{W}_0) + (\boldsymbol{W}_{\tau_s} - \boldsymbol{W}_0)\boldsymbol{Q}'(0)], \tag{8.104}$$

where we have neglected an irrelevant second-order term in $\boldsymbol{W}_{\tau_s} - \boldsymbol{W}_0$.[47] With this expression, we obtain the fourth-rank tensor M from (8.102):

$$\begin{aligned} M &= \frac{2\alpha^2 k_B T}{n_p k_B \tau_s \zeta}\langle[\boldsymbol{Q}'(0)(\boldsymbol{W}_{\tau_s} - \boldsymbol{W}_0) + (\boldsymbol{W}_{\tau_s} - \boldsymbol{W}_0)\boldsymbol{Q}'(0)] \\ &\times [\boldsymbol{Q}'(0)(\boldsymbol{W}_{\tau_s} - \boldsymbol{W}_0) + (\boldsymbol{W}_{\tau_s} - \boldsymbol{W}_0)\boldsymbol{Q}'(0)]\rangle. \end{aligned} \tag{8.105}$$

Because the increment $\boldsymbol{W}_{\tau_s} - \boldsymbol{W}_0$ is independent of the initial condition $\boldsymbol{Q}'(0)$, we can factorize the fourth moments and we obtain by means of $\langle\alpha\boldsymbol{Q}'(0)\boldsymbol{Q}'(0)\rangle = \boldsymbol{c}$ and $\langle(\boldsymbol{W}_{\tau_s} - \boldsymbol{W}_0)(\boldsymbol{W}_{\tau_s} - \boldsymbol{W}_0)\rangle = \tau_s\boldsymbol{1}$

$$M_{ijkl} = \frac{2\alpha k_B T}{n_p k_B \zeta}(c_{ik}\delta_{jl} + c_{il}\delta_{jk} + \delta_{ik}c_{jl} + \delta_{il}c_{jk}). \tag{8.106}$$

For Hookean dumbbells, we have the normalization factor $\alpha = H/(k_B T)$ and the relaxation time $\tau = \zeta/(4H)$ (see Exercise 81) so that we recover

$$M_{ijkl} = \frac{1}{2n_p k_B \tau}(c_{ik}\delta_{jl} + c_{il}\delta_{jk} + \delta_{ik}c_{jl} + \delta_{il}c_{jk}), \tag{8.107}$$

which is a more symmetrized version of (4.88).

[47] As expressed in (8.65), the neglected term is dominated by its average and is eliminated by the projection operator in (8.81).

Although our heuristic discussion of the friction matrix for unentangled polymers demonstrates the significance and usefulness of (8.102) or the corresponding Einstein-type formula, in practice, the choice of the intermediate time scale τ_s is rather delicate. As we know from the discussion of (4.133), the time scale for the second Rouse mode is $\tau/4$ and hence not very different from that for the slowest mode, τ. If we choose $\tau_s \approx \tau/2$, then $\boldsymbol{Q}'$ is already changing on the time scale τ_s, and the second mode is not really fast enough to be considered as noise. One therefore could analyze the decay of the slowest mode and of a few further modes carefully in order to extract the fundamental Rouse time scale with high precision, and the contribution of all the higher modes to any material property is then calculated by means of the well-known Rouse model with the spectrum (4.133). However, in the general spirit of GENERIC, to evaluate (8.102) with $\tau_s \approx \tau/2$ is the best we can do on the level of a single conformation tensor, without making any assumptions about the dynamics on the coarse level. If better accuracy is required in a systematic approach, several modes need to be taken into account through several conformation tensors. As a general rule, detailed model assumptions should only be used to check the consistency of the coarser level. As long as one works with a single conformation tensor, it may be better to work with the end-to-end vector rather than the slowest mode because the finite extensibility effects become more natural and transparent and because, in contrast to the construction of the slowest mode, this choice does not depend on any properties of the Rouse model.

8.4.3 Variostats and Multiplostats

To calculate friction matrices by molecular dynamics, we need to generate initial configurations according to some ensemble, and we then need to evolve these initial configurations according to Newton's or Hamilton's equations of motion. During the time evolution of nonequilibrium initial configurations, the ensemble is expected to change slowly because the relevant variables defining the ensemble evolve slowly. We need to evaluate the configurations only over some intermediate time scale τ; thus, the change of the ensemble is only of minor importance. This change reflects the intrinsic inaccuracy of beyond-equilibrium thermodynamics, which is based on assuming a clear separation of time scales (see Section 1.1.1). For simulation purposes, it may be advantageous to modify the time-evolution equations such that the initial ensemble remains strictly invariant and such that arbitrarily long trajectories can be generated and analyzed. By doing so, one solves the problem of generating proper initial configurations because they are automatically available at all times. The intrinsic inaccuracy associated with time-scale separation is then shifted from the slow change of the ensemble to a minor modification of the time-evolution equations. Of course, "minor changes" of time-evolution equations can have subtle effects, in particular, in the long run. It should however be kept in mind that the friction matrix is estimated from short pieces of the trajectory of duration τ.

We have introduced generalized microcanonical and canonical ensembles; thus, there are two different tasks for keeping ensembles invariant under time evolution. For generalized microcanonical ensembles, the values of the slow relevant variables are prescribed, and we thus need to keep these slow variables at constant values. We refer to a modification of the equations of motion that keep the slow *vari*ables constant as a *variostat*. For a generalized canonical ensemble, we need to keep the Lagrange *multipl*iers constant; we hence refer to a corresponding modification of the equations of motion as a *multiplostat*. Note that keeping the Lagrange multipliers constant comes hand in hand with the preservation of the exponential form of the canonical ensemble (6.15). In this section, we discuss some basic ideas for the construction of variostats and multiplostats. The distinction between variostats and multiplostats is not always straightforward. For example, control of the temperature in molecular dynamics could naturally be classified as a multiplostat mechanism because temperature occurs as a Lagrange multiplier. In practice, however, one often keeps the kinetic energy constant so that one actually deals with a variostat. The same situation occurs when one keeps the pressure (as another Lagrange multiplier) constant by actually fixing the value of its virial expression. Of course, the subsequently described ideas for constructing variostats and multiplostats need to be combined if one uses mixed ensembles.

We first consider variostats. To keep an observable $\Pi(z)$ constant, we modify Hamilton's equations of motion (6.1) for $z = (\boldsymbol{r}_1, \boldsymbol{p}_1, \ldots, \boldsymbol{r}_N, \boldsymbol{p}_N)$ as follows:[48]

$$\frac{dz}{dt} = L_0 \cdot \frac{\partial E_0}{\partial z} - \frac{\partial \Pi}{\partial z} \cdot L_0 \cdot \frac{\partial E_0}{\partial z} \left(\frac{\partial \Pi}{\partial z} \cdot \frac{\partial \Pi}{\partial z} \right)^{-1} \frac{\partial \Pi}{\partial z}. \tag{8.108}$$

By means of the chain rule, we then obtain for rate of change of Π under the variostatted time evolution (8.108),

$$\frac{d\Pi}{dt} = \frac{\partial \Pi}{\partial z} \cdot \frac{dz}{dt} = 0. \tag{8.109}$$

Note that $(\partial\Pi/\partial z) \cdot L_0 \cdot (\partial E_0/\partial z)$ is the rate of change of Π under Hamilton's equations of motion, which is assumed to be small for a relevant slow variable. In that sense, the additional term in (8.108) is small. The construction of the variostat (8.108) may be considered as an example of the motion under constraints considered in Exercise 7. In particular, we then realize that the variostatted equation has an underlying symplectic structure in the constrained space.

With the variostat (8.108) we have solved the problem of keeping a variable strictly constant, which, according to the equations of motion, should change slowly. There is the additional numerical problem, well known from strictly conserved quantities like energy, in which, after discretizing (8.108), the conservation property might be lost. To avoid that numerical problem, one can choose the variostat proportional to $\partial\Pi/\partial z$ as in (8.108) but determine the prefactor such that Π is strictly conserved for the time discretized dynamics.

The construction of multiplostats is less obvious.[49] Maintaining the Lagrange multipliers in the course of the time evolution actually means that one needs to preserve

[48] Galea & Attard, Phys. Rev. E 66 (2002) 041207.

[49] Explicit reversible multiplostats have been reviewed by Martyna et al., Mol. Phys. 87 (1996) 1117; in particular, that review contains a description of the powerful operator factorization technique for deriving integrators and many listings of pseudocode.

a generalized canonical ensemble. More ambitiously, one actually wants to sample according to a generalized canonical ensemble along almost any trajectory. Our goal is to modify Hamilton's equations of motion such that an ensemble of the following form is sampled [see Exercise 127 and (8.35)]:

$$\rho_x \propto \exp\left\{ -\frac{1}{k_B T}\left[\sum_{j=1}^{N}\frac{(\boldsymbol{p}_j - m\boldsymbol{v}(\boldsymbol{r}_j))^2}{2m} + \phi_{\rm tot}(\boldsymbol{r}_1,\ldots \boldsymbol{r}_N)\right] - \frac{p}{k_B T}V(\boldsymbol{r}_1,\ldots \boldsymbol{r}_N) - \lambda\Pi(z)\right\}. \tag{8.110}$$

The Lagrange multipliers associated with the continuous extensive energy and volume variables of equilibrium thermodynamics, the temperature T, and the pressure p, are introduced in (8.110), whereas we assume the discrete number of particles to be constant. We further introduced a velocity field $\boldsymbol{v}(\boldsymbol{r})$ so that systems under flow could be discussed, and we let it depend on $\boldsymbol{r}$ to facilitate the comparison of our procedure to previous molecular dynamics simulations under flow. Finally, the Lagrange multiplier λ associated with some observable Π describes a typical slow structural variable of beyond-equilibrium thermodynamics. If Π depends on momenta, it should satisfy the conditions

$$\left\langle \lambda\frac{\partial\Pi(z)}{\partial\boldsymbol{p}_j}\right\rangle_{\boldsymbol{r}} = 0, \qquad \sum_{j=1}^{N}\left\langle \lambda\boldsymbol{p}_j\cdot\frac{\partial\Pi(z)}{\partial\boldsymbol{p}_j}\right\rangle_{\boldsymbol{r}} = 0, \tag{8.111}$$

where $\langle\ldots\rangle_{\boldsymbol{r}}$ indicates the conditional ensemble average of a property given that particle j is at position $\boldsymbol{r}$, so that the roles of $\boldsymbol{v}(\boldsymbol{r})$ as the velocity field and of T as the temperature remain valid.[50]

Exercise 202 Velocity and Temperature as Averages

Show that the conditions formulated in (8.111) imply

$$\left\langle\frac{\boldsymbol{p}_j}{m}\right\rangle_{\boldsymbol{r}} = \boldsymbol{v}(\boldsymbol{r}) \tag{8.112}$$

and

$$\frac{1}{N}\sum_{j=1}^{N}\left\langle\frac{\boldsymbol{p}_j^2}{2m}\right\rangle_{\boldsymbol{r}} = \frac{1}{2}m\boldsymbol{v}(\boldsymbol{r})^2 + \frac{3}{2}k_B T. \tag{8.113}$$

There are several strategies for sampling canonical ensembles by modifying molecular dynamics simulations. For a constant temperature ensemble, according to the original idea of Nosé,[51] one introduces an extra pair of degrees of freedom associated with a rescaling of time, and one postulates canonical Hamiltonian equations with

[50] When $\boldsymbol{v}$ is independent of $\boldsymbol{r}$, one can consider unconditional averages.

[51] Nosé, Mol. Phys. 52 (1984) 255.

an extended Hamiltonian involving the additional degrees of freedom. Physically speaking, the rescaling of time corresponds to an exchange of energy with the environment. The drawback of Nosé's equations is that a scaling factor is involved in the relationship between the rate of change of the positions and the physical momenta, which corresponds to a nontrivial transformation of time. By transforming time and combining variables, an appealing real-time formulation of Nosé's equations was found by Hoover,[52] however, at the expense of losing the symplectic structure for the resulting Nosé-Hoover equations as a consequence of transforming the time. A real-time formulation with symplectic structure can be based on Poincaré transformations of the time scale,[53] and a complete separation of variables in all kinetic and potential energy terms within this Nosé-Poincaré method has been recently achieved.[54]

To work at constant pressure in addition to constant temperature, one needs to let the volume fluctuate in time. This can be achieved by scaling the positions, as proposed by Andersen.[55] As a further generalization, Parrinello and Rahman[56] allowed for changes not only of the size but also of the shape of a simulation box so that structural changes in the solid state could be studied. The Nosé-Poincaré approach can be employed to obtain real-time equations with symplectic structure also for an ensemble at constant temperature and pressure.[57]

Preserving the symplectic structure of Hamilton's equations of motion in the modified equations for reproducing canonical ensembles has important consequences when the time-discretized equations are formulated as a sequence of canonical transformations, that is, when we use *symplectic integrators*. The key observation is that the discrete solution is rigorously equivalent to the exact continuous-time solution of a Hamiltonian system in which the original Hamiltonian is modified by a small time-step dependent term, which can be constructed by a formal Taylor expansion in the time step.[58] In a discrete-time simulation, one hence stays on the level surfaces of the modified Hamiltonian and, assuming structural stability (which seems to be essential for the success of any kind of numerical integration), those are close to the level surfaces of the original Hamiltonian so that the energy is nearly conserved over long periods of time. An accumulation of errors in time is avoided. A clear and thoughtful discussion of the advantages of symplectic integrators based on a quantitative exponential estimate for the norm of the difference between flows rather than a formal expansion has been offered by Benettin and Giorgilli.[59] Their arguments explain nicely why even the simplest symplectic algorithms often outperform more

[52] Hoover, Phys. Rev. A 31 (1995) 1695.

[53] Bond, Leimkuhler & Laird, J. Comput. Phys. 151 (1999) 114.

[54] Leimkuhler, Comput. Phys. Commun. 148 (2002) 206.

[55] Andersen, J. Chem. Phys. 72 (1980) 2384.

[56] Parrinello & Rahman, Phys. Rev. Lett. 45 (1980) 1196; J. Appl. Phys. 52 (1981) 7182; J. Chem. Phys. 76 (1982) 2662.

[57] Leimkuhler, Comput. Phys. Commun. 148 (2002) 206.

[58] See Section 3.(e) of Moser, Mem. Am. Math. Soc. 81 (1968) 1; a nice explicit discussion of the modified Hamiltonian based on a one-to-one correspondence between rooted trees and the expressions appearing in the Taylor expansion can be found in Hairer, Ann. Numer. Math. 1 (1994) 107.

[59] Benettin & Giorgilli, J. Stat. Phys. 74 (1994) 1117.

sophisticated or higher-order nonsymplectic schemes. Although the advantages of symplectic algorithms are appealing, one should, however, keep in mind that the projected dynamics (6.123), (6.124) appearing in the projection-operator expression for the friction matrix (6.121) don't possess the obvious symplectic structure of Hamilton's equations of motion.

We note in passing that an elegant alternative approach to symplectic integrators can be based on Hamilton's variational principle. Symplectic integrators can be obtained as the stationary points of a discrete action associated with a discrete Lagrangian without any need for an underlying geometric structure.[60] However, this approach seems to be restricted to nondegenerate Poisson brackets.

Following the spirit of the Nosé-Poincaré method for constant temperature and pressure ensembles, our strategy to obtain a dynamics leading to the generalized canonical ensemble (8.110) is based on introducing the canonical variables

$$\boldsymbol{r}'_1, \boldsymbol{p}'_1, \ldots, \boldsymbol{r}'_N, \boldsymbol{p}'_N, \quad q_{\rm L}, p_{\rm L}, \quad q_{\rm T}, p_{\rm T}, \tag{8.114}$$

where the canonical particle positions and momenta, $\boldsymbol{r}'_j, \boldsymbol{p}'_j$, are related to the physical ones, $\boldsymbol{r}_j, \boldsymbol{p}_j$, by

$$\boldsymbol{r}_j = q_{\rm L} \boldsymbol{r}'_j, \quad \boldsymbol{p}_j = \frac{1}{q_{\rm L} q_{\rm T}} \boldsymbol{p}'_j. \tag{8.115}$$

The scaling factors for length and time, $q_{\rm L}$ and $q_{\rm T}$, are introduced together with the corresponding canonical momenta, $p_{\rm L}$ and $p_{\rm T}$; whereas $q_{\rm L}$ and $q_{\rm T}$ are restricted to the positive real axis, $p_{\rm L}$ and $p_{\rm T}$ can take any real values. The reference to $\boldsymbol{r}_j$ and $\boldsymbol{p}_j$ as physical variables needs to be interpreted with care because we are going to modify Hamilton's equations of motion. These variables are quasi-physical in the sense that, at any time, they are distributed according to (8.110) and their time evolution is governed by a minor modification of the physical equations of motion [however, for example, $d\boldsymbol{r}_j/dt$ may in general deviate from $\boldsymbol{p}_j/m$, as we have previously noted for the projected time evolution in (6.123)].

Assuming that the variables (8.114) are canonical, we only need to specify a Hamiltonian to obtain all the time-evolution equations that constitute the multiplostatted version of (6.1). Continuing in the spirit of the Nosé-Poincaré method, we choose the Hamiltonian

$$\begin{aligned} H' &= q_{\rm T} \left\{ \sum_{j=1}^{N} \frac{[\boldsymbol{p}_j - m\boldsymbol{v}(\boldsymbol{r}_j)]^2}{2m} + \phi_{\rm tot}(\boldsymbol{r}_1, \ldots \boldsymbol{r}_N) + k_{\rm B} T \lambda \Pi(z) \right. \\ &+ \left. \frac{p_{\rm L}^2}{2M_{\rm L}} + \frac{p_{\rm T}^2}{2M_{\rm T}} + \phi_{\rm L}(q_{\rm L}) + \bar{f} k_{\rm B} T \ln q_{\rm T} - H_0 \right\}, \end{aligned} \tag{8.116}$$

where the physical positions and momenta should be expressed in terms of their canonical counterparts by means of (8.115) and the constant H_0 is chosen such that the initial value of H' is zero. The parameters $M_{\rm L}$ and $M_{\rm T}$ describe the inertia of

[60] Marsden & West, Acta Numerica 10 (2001) 357.

the scale factors, and $\phi_L(q_L)$ is some potential function to be related to pressure and volume. The number of degrees of freedom, $\bar{f}$, is close to $3N$, where the precise choice of $\bar{f}$ is discussed below. The logarithmic form of the potential for the scale factor q_T is crucial because it eventually leads to the exponential in the generalized canonical distribution (8.110); formally, it suggestively appears as an entropic potential for the heat bath. The Hamiltonian (8.116) implies all the time-evolution equations for the canonical variables (8.114), and, after transforming to the physical variables (8.115), we obtain the modified equations of motion (see Exercise 203),

$$\frac{d\boldsymbol{r}_j}{dt} = \frac{\boldsymbol{p}_j}{m} - \boldsymbol{v}(\boldsymbol{r}_j) + k_B T \lambda \frac{\partial \Pi(z)}{\partial \boldsymbol{p}_j} + \boldsymbol{r}_j \frac{d \ln q_L}{dt}, \tag{8.117}$$

$$\frac{d\boldsymbol{p}_j}{dt} = -\frac{\partial \phi_{\text{tot}}}{\partial \boldsymbol{r}_j} + [\boldsymbol{p}_j - m\boldsymbol{v}(\boldsymbol{r}_j)] \cdot \left[\frac{\partial}{\partial \boldsymbol{r}_j}\boldsymbol{v}(\boldsymbol{r}_j)\right]^T - k_B T \lambda \frac{\partial \Pi(z)}{\partial \boldsymbol{r}_j} - \boldsymbol{p}_j \frac{d \ln q_L q_T}{dt}, \tag{8.118}$$

along with the auxiliary equations of the scale factors q_L and q_T,

$$\frac{d \ln q_L}{dt} = \frac{q_T}{q_L} \frac{p_L}{M_L}, \tag{8.119}$$

$$\frac{d \ln q_T}{dt} = \frac{p_T}{M_T}, \tag{8.120}$$

which can be used to eliminate the time derivatives on the right-hand sides of (8.117) and (8.118), and the dynamic equations for the conjugate momenta p_L and p_T,

$$\begin{aligned} \frac{dp_L}{dt} &= -q_T \frac{\partial \phi_L(q_L)}{\partial q_L} + \frac{q_T}{q_L} \sum_{j=1}^{N} \left\{ \frac{\boldsymbol{p}_j^2}{m} - \boldsymbol{r}_j \cdot \frac{\partial \phi_{\text{tot}}}{\partial \boldsymbol{r}_j} - \boldsymbol{v}(\boldsymbol{r}_j) \cdot \boldsymbol{p}_j \right. \\ &+ \left. [\boldsymbol{p}_j - m\boldsymbol{v}(\boldsymbol{r}_j)]\boldsymbol{r}_j : \frac{\partial}{\partial \boldsymbol{r}_j}\boldsymbol{v}(\boldsymbol{r}_j) + k_B T \lambda \left[\boldsymbol{p}_j \cdot \frac{\partial \Pi(z)}{\partial \boldsymbol{p}_j} - \boldsymbol{r}_j \cdot \frac{\partial \Pi(z)}{\partial \boldsymbol{r}_j}\right] \right\}, \end{aligned} \tag{8.121}$$

$$\frac{dp_T}{dt} = \sum_{j=1}^{N} \left[\frac{\boldsymbol{p}_j^2}{m} - \boldsymbol{v}(\boldsymbol{r}_j) \cdot \boldsymbol{p}_j + k_B T \lambda \boldsymbol{p}_j \cdot \frac{\partial \Pi(z)}{\partial \boldsymbol{p}_j} \right] - \bar{f} k_B T. \tag{8.122}$$

To obtain the last equation, we have used the fact that the value of H' is given by zero at all times; in view of the conservation of H' as the generator for the Hamiltonian evolution of the canonical variables, this follows from the choice of H_0 in (8.116). Note that (8.121) can formally be simplified to

$$\frac{dp_L}{dt} = -q_T \frac{\partial \phi_L(q_L)}{\partial q_L} + \frac{q_T}{q_L} \left(\frac{p_T}{M_T} + \frac{d}{dt} \right) \sum_{j=1}^{N} \boldsymbol{r}_j \cdot \boldsymbol{p}_j. \tag{8.123}$$

After all the transformations used to write the time-evolution equations for the physical particle positions and momenta, the symplectic structure of these equations is no longer manifest. If one wishes to benefit from the advantages of symplectic integrators, one should fall back on the equations for the canonical particle positions and momenta given in the solution to Exercise 203. Note that, in the absence of Andersen's barostat, one can express the influence of the thermostat in terms of the single variable p_{T}, just as in the Nosé-Hoover approach.

At this point, we want to investigate under what conditions (8.117) and (8.118) are small modifications of the original equations of motion. The terms involving the scale factors can be kept small by choosing the parameters M_{L} and M_{T} sufficiently large; in practice, these parameters should be chosen such that the scale factors change on the time scale of the slowest relevant variables, thus allowing us to explore the physically achievable values of energy and volume on that time scale.[61] Equation (8.118) leads us to the conclusion that the velocity gradients need to be small compared with typical momentum relaxation rates, whereas (8.117) requires the average flow velocity to be small compared with the individual particle velocities. We hence use a coordinate system in which the initial total momentum vanishes [this choice is also suggested by (6.123)] and the origin coincides with the initial center of mass of the particles. We further restrict ourselves to linear velocity fields of the form

$$\boldsymbol{v}(\boldsymbol{r}) = \boldsymbol{\kappa} \cdot \boldsymbol{r}, \tag{8.124}$$

so that (8.117) and (8.118) can be rewritten as

$$\frac{d\boldsymbol{r}_j}{dt} = \frac{\boldsymbol{p}_j}{m} - \boldsymbol{\kappa} \cdot \boldsymbol{r}_j + k_{\mathrm{B}} T \lambda \frac{\partial \Pi(z)}{\partial \boldsymbol{p}_j} + \frac{q_{\mathrm{T}}}{q_{\mathrm{L}}} \frac{p_{\mathrm{L}}}{M_{\mathrm{L}}} \boldsymbol{r}_j, \tag{8.125}$$

$$\frac{d\boldsymbol{p}_j}{dt} = -\frac{\partial \phi_{\mathrm{tot}}}{\partial \boldsymbol{r}_j} + \boldsymbol{\kappa}^T \cdot (\boldsymbol{p}_j - m\boldsymbol{\kappa} \cdot \boldsymbol{r}_j) - k_{\mathrm{B}} T \lambda \frac{\partial \Pi(z)}{\partial \boldsymbol{r}_j} - \left(\frac{p_{\mathrm{T}}}{M_{\mathrm{T}}} + \frac{q_{\mathrm{T}}}{q_{\mathrm{L}}} \frac{p_{\mathrm{L}}}{M_{\mathrm{L}}} \right) \boldsymbol{p}_j. \tag{8.126}$$

For the flow terms in a multiplostat to be small, we thus obtain the further condition that the simulated system can be taken sufficiently small so that the variations of the flow velocity across the system are small compared with thermal velocities.[62] Finally, the smallness of the terms maintaining a constant Lagrange multiplier λ associated with some general variable $\Pi(z)$ needs to be checked for every particular choice. For example, for a polymeric liquid described by the second-moment tensor of the end-to-end vector of the polymer molecules (see Sections 4.3.1 and 8.2.3), $\partial\Pi/\partial\boldsymbol{p}_j$ vanishes and $\partial\Pi/\partial\boldsymbol{r}_j$ leads to a repulsive linear spring force on the chain ends, which prevents the polymers from configurational relaxation; the spring force is inversely proportional to molecular weight so that, for sufficiently large molecules, we indeed deal with a small modification of the equations of motion. Note that,

[61] For a discussion of the choice of M_{T} see § 6 of a review article on deterministic thermostats, Nosé, Prog. Theor. Phys. Suppl. 103 (1991) 1; see also Section 2.5 of Martyna et al., Mol. Phys. 87 (1996) 1117, and references therein.

[62] For a detailed discussion of the corresponding Lees-Edwards modification of periodic boundary conditions under flow see p. 133–137 of Evans & Morriss, *Statistical Mechanics of Nonequilibrium Liquids* (Academic, 1990).

if the relevant variable $\Pi(z)$ is invariant under spatial translations, the vanishing initial total momentum is conserved by (8.126); the momentum conservation for the original time-evolution equations is thus inherited by the multiplostatted equations. According to (8.125), the center of mass stays at the origin only if we further require $\sum_{j=1}^{N} \lambda \partial \Pi / \partial \boldsymbol{p}_j = 0$.

Exercise 203 Multiplostatted Time-Evolution Equations

Derive the time-evolution equations (8.117)—(8.122) from the Hamiltonian (8.116).

To verify that the modifications (8.125), (8.126) of Hamilton's equations of motion, together with the kinematic relations (8.119), (8.120) and the simplified versions of (8.121), (8.122) for linear flow fields,

$$\begin{aligned} \frac{dp_{\mathrm{L}}}{dt} &= -q_{\mathrm{T}} \frac{\partial \phi_{\mathrm{L}}(q_{\mathrm{L}})}{\partial q_{\mathrm{L}}} + \frac{q_{\mathrm{T}}}{q_{\mathrm{L}}} \sum_{j=1}^{N} \left\{ \frac{\boldsymbol{p}_j^2}{m} - m(\boldsymbol{\kappa} \cdot \boldsymbol{r}_j)^2 - \boldsymbol{r}_j \cdot \frac{\partial \phi_{\mathrm{tot}}}{\partial \boldsymbol{r}_j} \right. \\ &+ \left. k_{\mathrm{B}} T \lambda \left[\boldsymbol{p}_j \cdot \frac{\partial \Pi(z)}{\partial \boldsymbol{p}_j} - \boldsymbol{r}_j \cdot \frac{\partial \Pi(z)}{\partial \boldsymbol{r}_j} \right] \right\}, \end{aligned} \tag{8.127}$$

$$\frac{dp_{\mathrm{T}}}{dt} = \sum_{j=1}^{N} \left[\frac{\boldsymbol{p}_j^2}{m} - \boldsymbol{p}_j \cdot \boldsymbol{\kappa} \cdot \boldsymbol{r}_j + k_{\mathrm{B}} T \lambda \boldsymbol{p}_j \cdot \frac{\partial \Pi(z)}{\partial \boldsymbol{p}_j} \right] - \bar{f} k_{\mathrm{B}} T, \tag{8.128}$$

indeed provide the desired multiplostat, we need to study the resulting invariant ensemble. In the variables of (8.114), by construction, we have a canonical Hamiltonian system so that H' is fixed to its initial value, chosen to be zero. We thus expect a microcanonical distribution in the canonical variables of (8.114), given by the δ-function

$$\begin{aligned} \delta \Bigg(q_{\mathrm{T}} \Bigg[\sum_{j=1}^{N} \frac{[\boldsymbol{p}_j - m\boldsymbol{v}(\boldsymbol{r}_j)]^2}{2m} + \phi_{\mathrm{tot}}(\boldsymbol{r}_1, \ldots \boldsymbol{r}_N) + k_{\mathrm{B}} T \lambda \Pi(z) \\ + \frac{p_{\mathrm{L}}^2}{2M_{\mathrm{L}}} + \frac{p_{\mathrm{T}}^2}{2M_{\mathrm{T}}} + \phi_{\mathrm{L}}(q_{\mathrm{L}}) + \bar{f} k_{\mathrm{B}} T \ln q_{\mathrm{T}} - H_0 \Bigg] \Bigg), \end{aligned} \tag{8.129}$$

where we have already mentioned the possibility of further conservation laws, e.g., for total momentum (to be treated separately). The probability density for finding the physical variables (8.115) instead of the canonical particle positions and momenta in the list (8.114) requires the inclusion of the change of volume elements in $6N$-dimensional phase space, $(q_{\mathrm{L}} q_{\mathrm{T}})^{3N} / q_{\mathrm{L}}^{3N}$, so that the probability density in the physical positions and momenta, $q_{\mathrm{L}}, p_{\mathrm{L}}$, and $q_{\mathrm{T}}, p_{\mathrm{T}}$ is obtained as

$$\begin{aligned} q_{\mathrm{T}}^{3N} \delta \Bigg(q_{\mathrm{T}} \Bigg[\sum_{j=1}^{N} \frac{[\boldsymbol{p}_j - m\boldsymbol{v}(\boldsymbol{r}_j)]^2}{2m} + \phi_{\mathrm{tot}}(\boldsymbol{r}_1, \ldots \boldsymbol{r}_N) + k_{\mathrm{B}} T \lambda \Pi(z) \\ + \frac{p_{\mathrm{L}}^2}{2M_{\mathrm{L}}} + \frac{p_{\mathrm{T}}^2}{2M_{\mathrm{T}}} + \phi_{\mathrm{L}}(q_{\mathrm{L}}) + \bar{f} k_{\mathrm{B}} T \ln q_{\mathrm{T}} - H_0 \Bigg] \Bigg). \end{aligned} \tag{8.130}$$

As a next step, we integrate over q_T. The only contributing value of $\ln q_\mathrm{T}$ is obtained by setting the square bracket in (8.130) equal to zero. By using the following property of δ-functions,

$$\int \delta(g(q))dq = \left|\frac{dg(q_0)}{dq_0}\right|^{-1}, \tag{8.131}$$

for monotonic functions g on the real line with $g(q_0) = 0$, we obtain for the integral of (8.130) with respect to q_T, except for an unimportant normalization factor $\bar{f} k_\mathrm{B} T$,

$$\begin{aligned} q_{\mathrm{T},0}^{3N} &= \exp\left\{ -\frac{3N}{\bar{f}}\frac{1}{k_\mathrm{B}T}\left[\sum_{j=1}^{N}\frac{[\boldsymbol{p}_j - m\boldsymbol{v}(\boldsymbol{r}_j)]^2}{2m} + \phi_\mathrm{tot}(\boldsymbol{r}_1,\dots\boldsymbol{r}_N)\right]\right. \\ &\left. - \lambda\Pi(z) - \frac{1}{k_\mathrm{B}T}\left[\frac{p_\mathrm{L}^2}{2M_\mathrm{L}} + \frac{p_\mathrm{T}^2}{2M_\mathrm{T}} + \phi_\mathrm{L}(q_\mathrm{L}) - H_0\right]\right\}. \end{aligned} \tag{8.132}$$

It is now clear why the number of degrees of freedom, $\bar{f}$, should be chosen as $3N$. After integrating over p_T and p_L, we finally obtain for the probability to find $\boldsymbol{r}_1, \boldsymbol{p}_1, \dots, \boldsymbol{r}_N, \boldsymbol{p}_N$ and q_L

$$\rho_x \propto \exp\left\{\frac{-1}{k_\mathrm{B}T}\left[\sum_{j=1}^{N}\frac{[\boldsymbol{p}_j - m\boldsymbol{v}(\boldsymbol{r}_j)]^2}{2m} + \phi_\mathrm{tot}(\boldsymbol{r}_1,\dots\boldsymbol{r}_N)\right] - \frac{\phi_\mathrm{L}(q_\mathrm{L})}{k_\mathrm{B}T} - \lambda\Pi(z)\right\} \tag{8.133}$$

which should be compared with (8.110). If we identify the third power of the scale factor for length, q_L^3, with the volume V and set $\phi_\mathrm{L}(q_\mathrm{L})$ equal to $p q_\mathrm{L}^3$, where p is the pressure, we have indeed reproduced the desired general canonical distribution (8.110) with an extra volume variable.

In our derivation of the generalized canonical distribution (8.133) for the modified equations of motion, we have assumed ergodicity on the constant-energy surface, $H' = 0$. However, ergodicity cannot be taken for granted. For example, the conservation of momentum restricts us to a lower-dimensional surface so that in (8.130) a factor q_T^{3N-3} arises instead of q_T^{3N}. According to (8.132), this modification can be compensated by using the actual number of degrees of freedom, $\bar{f} = 3N - 3$, instead of $3N$. Possible further conserved quantities must be identified and handled in the same way by reducing the number of degrees of freedom. We then obtain a mixed ensemble in which some of the coarse-grained variables are treated microcanonically, others canonically. If there occurs an additional unphysical lack of ergodicity, or if one encounters difficulties with long periodic oscillations in the total kinetic energy, one can use chains of thermostats, in which the thermostats themselves are thermostatted. This idea has been introduced in the context of Nosé-Hoover thermostats,[63] but also Nosé chains and Nosé-Poincaré chains have been developed.[64] Nosé-Hoover

[63]Martyna et al., J. Chem. Phys. 101 (1994) 4177.

[64]Leimkuhler & Sweet, J. Chem. Phys. 121 (2004) 108.

thermostats have also been generalized in a different direction by the introduction of several temperature control variables for different degrees of freedom.

In the presence of *constraints*, a thermostat based on the additional variables q_T, p_T can be generalized by further reducing the number of degrees of freedom, $\bar{f}$, in the Hamiltonian (8.116) and by introducing constraint forces. Also Andersen's barostat based on the additional variables q_L, p_L needs to be modified in the presence of constraints. For example, it is possible to divide the system into small groups of particles that share common constraints and to apply the rescaling by q_L only to the center-of-mass positions and momenta.[65]

The problem of thermostatting is more severe for molecular dynamics than for Brownian dynamics where, in view of the fluctuation-dissipation theorem of the second kind, the interplay of friction and noise produces a canonical ensemble. Actually, this observation suggests that friction and noise could be added even to Hamilton's equations of motion to thermostat a molecular dynamics simulation.[66] This is an example of a stochastic thermostat. Instead of continuous noise, one can introduce stochastic collisions at discrete times and thus use Monte Carlo methods for thermostatting.[67]

Exercise 204 Multiplostatted Rouse Model

According to Section 4.3.1, the equations of motion for a Rouse chain in flow can be written as [see also (8.70)]

$$\frac{d\boldsymbol{Q}'_j}{dt} = \boldsymbol{\kappa} \cdot \boldsymbol{Q}'_j - \frac{1}{2\tau_j}\boldsymbol{Q}'_j + \sqrt{\frac{1}{\tau_j}}\frac{d\boldsymbol{W}'_j}{dt}, \tag{8.134}$$

for the eigenmodes $\boldsymbol{Q}'_j$, $j = 1, \ldots, N-1$, where we have used units of length such that $k_B T/H = 1$. We take (8.134) as a toy model of atomistic time-evolution equations, where we assume vanishing velocity gradients $\boldsymbol{\kappa}$ to obtain the analogue of the unmodified equations of motion. With a coarse-grained level of description based on a single conformation tensor in mind, we use the conformation tensor $\boldsymbol{Q}'_1\boldsymbol{Q}'_1$ formed of the slowest mode as the relevant variable Π. Show that a meaningful analogue of the multiplostatted equations (8.117), (8.118) is given by

$$\frac{d\boldsymbol{Q}'_1}{dt} = -\frac{1}{2\tau_1}\boldsymbol{Q}'_1 - \frac{1}{2\tau_1}(\boldsymbol{\lambda}_c + \boldsymbol{\lambda}_c^T) \cdot \boldsymbol{Q}'_1 + \sqrt{\frac{1}{\tau_1}}\frac{d\boldsymbol{W}'_1}{dt}. \tag{8.135}$$

Introduce the explicit form of the tensorial Lagrange multiplier $\boldsymbol{\lambda}_c$ for the Rouse model and consider the resulting moment equations for further interpretation of this equation.

8.4.4 Verlet-Type Integrators

If we are interested in first-principle or "*ab initio*" calculations of friction matrices by means of the Green-Kubo-type formula (8.77), or any of the alternative expressions

[65] Ryckaert & Ciccotti, J. Chem. Phys. 78 (1983) 7368.

[66] An important example of a molecular dynamics simulation based on such a thermostat can be found in the work of Kremer & Grest, J. Chem. Phys. 92 (1990) 5057.

[67] This idea was introduced by Andersen, J. Chem. Phys. 72 (1980) 2384; for a more recent reference, see Attard, J. Chem. Phys. 116 (2002) 9616.

discussed in Section 8.4.1, we need to solve Newton's or Hamilton's equations of motion for an atomistic system. We therefore need to define an appropriate Hamiltonian for the system of interest. Before presenting detailed results for Lennard-Jones gases and polyethylene melts, we discuss some general issues for molecular dynamics simulations.

In molecular dynamics simulations, one usually solves Hamilton's equations of motion (1.26) or, equivalently, Newton's equations of motion,

$$m_j \frac{d^2 \boldsymbol{r}_j}{dt^2} = -\frac{\partial E}{\partial \boldsymbol{r}_j} = \boldsymbol{F}_j, \tag{8.136}$$

for a system of N interacting particles, where $\boldsymbol{r}_j$ is the position of the jth particle, which has the mass m_j and feels the forces $\boldsymbol{F}_j$ resulting from interactions with all the other particles. These potential forces are known functions of the particle positions. In a simulation, the equations of motion are solved by a finite difference approach.[68] For example, the simplest discretization of (8.136) with time step Δt is given by

$$m_j \frac{\boldsymbol{r}_j(t+\Delta t) - 2\boldsymbol{r}_j(t) + \boldsymbol{r}_j(t-\Delta t)}{\Delta t^2} = \boldsymbol{F}_j(t), \tag{8.137}$$

which leads us to the integration scheme

$$\boldsymbol{r}_j(t+\Delta t) = 2\boldsymbol{r}_j(t) - \boldsymbol{r}_j(t-\Delta t) + \Delta t^2 \frac{\boldsymbol{F}_j(t)}{m_j}. \tag{8.138}$$

This integration scheme is known as the *Verlet algorithm*, or as *Störmer's rule*, and is one of the most widely used methods in molecular dynamics. Note the symmetric roles played by $\boldsymbol{r}_j(t+\Delta t)$ and $\boldsymbol{r}_j(t-\Delta t)$ in (8.137), which make the Verlet algorithm time reversible. Using Taylor expansions for $\boldsymbol{r}_j(t \pm \Delta t)$ around $\boldsymbol{r}_j(t)$, one finds that the Verlet method provides a third-order algorithm in Δt. The velocities or momenta do not occur in the Verlet algorithm. If they are needed, for example, for evaluating or even controlling the kinetic energy, the velocities are constructed from

$$\boldsymbol{v}_j(t) = \frac{\boldsymbol{r}_j(t+\Delta t) - \boldsymbol{r}_j(t-\Delta t)}{2\Delta t}, \tag{8.139}$$

which leads to a first-order algorithm for calculating $\boldsymbol{v}_j(t)$. Note that to initialize the algorithm (8.138) at time t_0, in addition to $\boldsymbol{r}_j(t_0)$, values for $\boldsymbol{r}_j(t_0) - \boldsymbol{r}_j(t_0 - \Delta t)$ or for the initial velocities need to be assumed. From (8.138) and (8.139) we obtain

$$\sum_{j=1}^{N} m_j[\boldsymbol{v}_j(t+\Delta t) - \boldsymbol{v}_j(t)] =$$

$$\sum_{j=1}^{N} m_j \frac{\boldsymbol{r}_j(t+2\Delta t) - \boldsymbol{r}_j(t+\Delta t) - \boldsymbol{r}_j(t) + \boldsymbol{r}_j(t-\Delta t)}{2\Delta t} =$$

[68] The following presentation of molecular dynamics is based on Section 3.2 of Allen & Tildesley, *Computer Simulation of Liquids* (Clarendon, 1987), where further details can be found.

$$\sum_{j=1}^{N} \frac{1}{2}[\boldsymbol{F}_j(t+\Delta t) - \boldsymbol{F}_j(t)]\Delta t = 0, \tag{8.140}$$

so that the total momentum is strictly conserved by the Verlet algorithm when the sum of all forces vanishes for all times according to Newton's third law. Furthermore, the algorithm has been shown to have excellent energy-conserving properties even for large Δt.

Taking the difference of two large numbers of order Δt^0 and adding a small number of order Δt^2 in (8.138) may lead to some numerical imprecision, which can be avoided in a modification of the Verlet algorithm known as the half-step *leap-frog scheme*,

$$\boldsymbol{v}_j(t + \tfrac{1}{2}\Delta t) = \boldsymbol{v}_j(t - \tfrac{1}{2}\Delta t) + \Delta t \frac{\boldsymbol{F}_j(t)}{m_j}, \tag{8.141}$$

$$\boldsymbol{r}_j(t + \Delta t) = \boldsymbol{r}_j(t) + \Delta t \boldsymbol{v}_j(t + \tfrac{1}{2}\Delta t), \tag{8.142}$$

which is initialized by specifying $\boldsymbol{v}_j(t_0 - \frac{1}{2}\Delta t)$ and $\boldsymbol{r}_j(t_0)$. The velocity at time t may be evaluated as

$$\boldsymbol{v}_j(t) = \frac{1}{2}[\boldsymbol{v}_j(t + \tfrac{1}{2}\Delta t) + \boldsymbol{v}_j(t - \tfrac{1}{2}\Delta t)]. \tag{8.143}$$

The leap-frog scheme is known to be symplectic;[69] that is, it respects the Hamiltonian structure of the equations of motion. Although symplectic integrators do not conserve the energy exactly, they are known to be very good at conserving it approximately.[70] An even more satisfactory handling of the velocities is obtained by the *velocity Verlet algorithm*, which is of the form

$$\boldsymbol{r}_j(t + \Delta t) = \boldsymbol{r}_j(t) + \Delta t \boldsymbol{v}_j(t) + \frac{1}{2}\Delta t^2 \frac{\boldsymbol{F}_j(t)}{m_j}, \tag{8.144}$$

$$\boldsymbol{v}_j(t + \Delta t) = \boldsymbol{v}_j(t) + \frac{\Delta t}{2m_j}[\boldsymbol{F}_j(t) + \boldsymbol{F}_j(t + \Delta t)]. \tag{8.145}$$

The most time-consuming step in molecular dynamics simulations is the calculation of forces, and thus the accuracy and stability for large time steps are important issues. For speeding up force calculations, one may use a cutoff for short-range interaction potentials at some sufficiently large distance and keep neighbor lists for each particle; for larger systems, one can introduce a cell structure and linked lists. Cell structures are particularly suitable for parallel computers. For complicated potentials, look-up tables with interpolations or spline fit potentials can be useful. In many cases, one can split the forces into rapidly changing primary and smaller, slowly varying

[69] For one degree of freedom, symplectic schemes preserve the area in two-dimensional phase space; for more than one degree of freedom, a sum of oriented areas in phase space does not change in an integration step.

[70] See remarks on p. 368 above, as well as p. 107 and Section 10.3.3 of Sanz-Serna & Calvo, *Numerical Hamiltonian Problems* (Chapman & Hall, 1994).

secondary forces, so that the secondary forces need to be evaluated less frequently than the primary ones; this approach is known as the *multiple time step method*. More details on these and many further tricks can be found in Chapter 5 of the book by Allen and Tildesley.[71]

Exercise 205 Algebraic Equivalence of Verlet Algorithms
Show that the leap-frog scheme and the velocity Verlet algorithm are algebraically equivalent to the original Verlet algorithm (8.138).

8.4.5 Example: Rarefied Lennard-Jones Gas

As a simple example of a molecular dynamics simulation under flow, we consider a gas of particles with Lennard-Jones interactions.[72] This provides an example of Step A ("*ab initio*" or "atomistic simulations") according to the classification scheme of Section 8.1.3, where we start from Hamilton's equations of motion, and we take the structured version of Grad's ten-moment method (see Section 7.4.3) as the coarser target level. Because the analytical forms of the energy E, the entropy S, and the Poisson matrix L for the structured ten-moment method are available, we here focus on the friction matrix M. Evaluating M essentially corresponds to determining the basic collision time scale for Lennard-Jones interactions, but also the relaxation due to collisions in anisotropic systems created by flow is of interest. Once the friction matrix is determined, all building blocks on the coarse-grained level are known. As illustrated in Exercise 186, we can then calculate the viscometric functions on the Grad level of description, which is an example of Step C ("concrete calculations"). In other words, the viscosity is not an immediate target of the simulations; it can be calculated once the friction matrix has been determined by simulations. Our procedure is thus fundamentally different from traditional molecular dynamics simulations.

Because the ten-moment level of description is coarser than the level of single-particle distribution functions behind the Boltzmann equation, we can only hope to get useful simulation results for the rarefied Lennard-Jones gas. Moreover, because there is no clear separation of time scales available to justify Grad's approach, simulations based on the ten-moment target level may not be ideal for practical purposes. For educational purposes, however, we obtain an illustrative toy example of a molecular dynamics simulation, so to say, "the ideal gas problem of molecular dynamics." The reader should hence concentrate on conceptual issues rather than on results. To obtain more interesting results, in particular for denser Lennard-Jones systems, one needs to choose a more sophisticated target level for the thermodynamically guided simulations. To account for interaction effects, one could use the well-established pair correlation function of equilibrium statistical mechanics (see Section 7.5.2) or,

[71] Allen & Tildesley, *Computer Simulation of Liquids* (Clarendon, 1987).

[72] The work summarized in this section was motivated by the wish to present a balanced picture of beyond-equilibrium thermodynamics in this book, including simulation results. Preliminary results have previously been published in Kröger & Öttinger, J. Non-Newtonian Fluid Mech. 120 (2004) 175.

maybe more elegantly, only the total potential energy. The potential energy could be combined either with the ten-moment level or with the level of single-particle distribution functions, where the latter possibility is suggested by the work of Garrido et al.[73]

We first introduce the tensor variable $\boldsymbol{\Pi}(z)$, such that $\langle \boldsymbol{\Pi} \rangle = \mathring{\boldsymbol{c}}$ is the normalized traceless second-moment tensor in velocity space. We assume isothermal conditions and hence ignore the heat flux. By matching the probability density (7.141) of the structured moment method with the generalized canonical ensemble (8.110), we obtain

$$2mk_{\mathrm{B}}T\boldsymbol{\lambda}_c : \boldsymbol{\Pi}(z) = [(\mathbf{1} - \mathring{\boldsymbol{c}})^{-1} - \mathbf{1}] : \sum_{j=1}^{N} [\boldsymbol{p}_j - m\boldsymbol{v}(\boldsymbol{r}_j)][\boldsymbol{p}_j - m\boldsymbol{v}(\boldsymbol{r}_j)]. \quad (8.146)$$

To write out the multiplostatted equations of motion, we evaluate the expressions

$$k_{\mathrm{B}}T \frac{\partial}{\partial \boldsymbol{p}_j} \boldsymbol{\lambda}_c : \boldsymbol{\Pi}(z) = [(\mathbf{1} - \mathring{\boldsymbol{c}})^{-1} - \mathbf{1}] \cdot \left(\frac{\boldsymbol{p}_j}{m} - \boldsymbol{\kappa} \cdot \boldsymbol{r}_j \right) \quad (8.147)$$

and

$$k_{\mathrm{B}}T \frac{\partial}{\partial \boldsymbol{r}_j} \boldsymbol{\lambda}_c : \boldsymbol{\Pi}(z) = -\boldsymbol{\kappa}^T \cdot [(\mathbf{1} - \mathring{\boldsymbol{c}})^{-1} - \mathbf{1}] \cdot (\boldsymbol{p}_j - m\boldsymbol{\kappa} \cdot \boldsymbol{r}_j), \quad (8.148)$$

where we have assumed a homogeneous flow field. With these expressions, the multiplostatted equations of motion (8.125), (8.126) can be written as

$$\frac{d\boldsymbol{r}_j}{dt} = (\mathbf{1} - \mathring{\boldsymbol{c}})^{-1} \cdot \left(\frac{\boldsymbol{p}_j}{m} - \boldsymbol{\kappa} \cdot \boldsymbol{r}_j \right), \quad (8.149)$$

$$\frac{d\boldsymbol{p}_j}{dt} = -\frac{\partial \phi_{\mathrm{tot}}}{\partial \boldsymbol{r}_j} + \boldsymbol{\kappa}^T \cdot (\mathbf{1} - \mathring{\boldsymbol{c}})^{-1} \cdot (\boldsymbol{p}_j - m\boldsymbol{\kappa} \cdot \boldsymbol{r}_j) - \frac{p_{\mathrm{T}}}{M_{\mathrm{T}}} \boldsymbol{p}_j, \quad (8.150)$$

where we have left out Andersen's barostat because we here consider a gas in a constant volume. The auxiliary variable for the Nosé-Hoover-type thermostat is obtained from (8.128):

$$\frac{dp_{\mathrm{T}}}{dt} = \sum_{j=1}^{N} \boldsymbol{p}_j \cdot (\mathbf{1} - \mathring{\boldsymbol{c}})^{-1} \cdot \left(\frac{\boldsymbol{p}_j}{m} - \boldsymbol{\kappa} \cdot \boldsymbol{r}_j \right) - f k_{\mathrm{B}} T. \quad (8.151)$$

The multiplostatted equations of motion (8.149)–(8.151) are consistent with the probability density (7.141) for all choices of $\boldsymbol{\kappa}$ and $\mathring{\boldsymbol{c}}$.

From (7.99) and (7.124), we further obtain the pressure tensor on the coarse-grained level as

$$\boldsymbol{\pi} = \frac{\rho}{m} k_{\mathrm{B}} T (\mathbf{1} - \mathring{\boldsymbol{c}}). \quad (8.152)$$

[73] Garrido, Goldstein & Lebowitz, Phys. Rev. Lett. 92 (2004) 050602.

For a consistency check, we can evaluate the stresses according to the Irving-Kirkwood expression (7.188) from the forces and distances between all pairs of interacting particles in the system in a molecular dynamics simulation. For a rarefied gas, these stresses due to interactions must be small compared with the kinetic stresses (8.152).

With the equations of motion (8.149)–(8.151), we want to evaluate the block of the friction matrix that describes the relaxation of $\mathring{\boldsymbol{c}}$,

$$M_{44} = \frac{V}{2k_{\mathrm{B}}\tau_{\mathrm{s}}} \left\langle \Delta_{\tau_{\mathrm{s}}}\boldsymbol{\Pi}\Delta_{\tau_{\mathrm{s}}}\boldsymbol{\Pi}\right\rangle_{\boldsymbol{x}}, \tag{8.153}$$

where V is the volume and τ_{s} is the separating time scale. The time scale τ_{s} must be small compared with the characteristic relaxation time τ occurring in the time-evolution equation for the slow variable $\mathring{\boldsymbol{c}}$, which can be interpreted as the time between collisions, and τ_{s} should be large compared with the duration of a collision. The expression for the fourth-rank tensor M_{44} is found by the same arguments that lead us to (8.100). If we are interested in the behavior of the Lennard-Jones gas in a steady-state flow, that is, for given velocity gradients $\boldsymbol{\kappa}$, we need to find the appropriate tensor $\mathring{\boldsymbol{c}}$ occurring in the equations of motion, and we need to specify the explicit form of $\boldsymbol{\Pi}(z)$ occurring in the friction matrix [so far, we have only considered the contraction $\boldsymbol{\lambda}_c : \boldsymbol{\Pi}(z)$ given in (8.146)]. The natural choice for $\boldsymbol{\Pi}(z)$ can be obtained from our analysis of the structured ten-moment method (see Exercise 179),

$$\boldsymbol{\Pi}(z) = \frac{1}{mk_{\mathrm{B}}T}\frac{1}{N}\sum_{j=1}^{N}\left[\frac{1}{3}(\boldsymbol{p}_j - m\boldsymbol{\kappa}\cdot\boldsymbol{r}_j)^2\mathbf{1} - (\boldsymbol{p}_j - m\boldsymbol{\kappa}\cdot\boldsymbol{r}_j)(\boldsymbol{p}_j - m\boldsymbol{\kappa}\cdot\boldsymbol{r}_j)\right], \tag{8.154}$$

and the Lagrange multiplier $\boldsymbol{\lambda}_c$ is obtained from the known gradient of the entropy (7.145),[74]

$$\boldsymbol{\lambda}_c = \frac{1}{k_{\mathrm{B}}}\frac{\partial S}{\partial \mathring{\boldsymbol{c}}} = \frac{1}{k_{\mathrm{B}}}V\frac{\delta S}{\delta \mathring{\boldsymbol{c}}} = -\frac{N}{2}(\mathbf{1} - \mathring{\boldsymbol{c}})^{-1}. \tag{8.155}$$

With the expressions (8.154) and (8.155), the identity (8.146) is not satisfied because, in general, the trace of the tensor $[(\mathbf{1} - \mathring{\boldsymbol{c}})^{-1} - \mathbf{1}]$ is nonzero. This inconsistency is related to the fact that, for the choice (8.146), the second identity in (8.111) is violated, indicating that the variable $\boldsymbol{\Pi}(z)$ in (8.154) cannot be introduced separately after defining the temperature; these closely related variables need to be introduced together. Strictly speaking, the assumption (8.146) leading to the multiplostatted equations of motion is inconsistent with the expressions (8.154) and (8.155); nevertheless, we use the resulting equations of motion because they reproduce the desired probability density (7.141) exactly. Even if we introduce the variable $\boldsymbol{\Pi}(z)$ separately from the temperature instead of jointly with it, the corresponding contributions

[74] Note the difference between the functional derivative of the entropy with respect to the local conformation tensor at position $\boldsymbol{r}$ and the partial derivative with respect to the mean conformation tensor obtained by averaging over V.

to the multiplostatted equations of motion combine properly. For a more consistent derivation of the multiplostat for a rarefied gas, see Exercise 206.

With the help of (8.155), we can finally establish the relationship between the velocity gradients $\boldsymbol{\kappa}$ and the second-moment tensor $\mathring{\boldsymbol{c}}$. For homogeneous steady flows, the time-evolution equation for $\mathring{\boldsymbol{c}}$ [cf. (7.131)] can be written as

$$\frac{k_B\rho}{2m} M_{44} : (\mathbf{1} - \mathring{\boldsymbol{c}})^{-1} = \boldsymbol{\kappa} \cdot (\mathbf{1} - \mathring{\boldsymbol{c}}) + (\mathbf{1} - \mathring{\boldsymbol{c}}) \cdot \boldsymbol{\kappa}^T - \frac{2}{3}(\mathbf{1} - \mathring{\boldsymbol{c}})(\mathbf{1} - \mathring{\boldsymbol{c}}) : \boldsymbol{\kappa}. \quad (8.156)$$

Equation (8.156) is similar to (8.50) in structure and significance; it provides the required relationship between $\boldsymbol{\kappa}$ and $\mathring{\boldsymbol{c}}$. We need to find a self-consistent solution for these two tensors, where M_{44} in (8.156) is to be determined according to (8.153) from a molecular dynamics simulation involving $\boldsymbol{\kappa}$ and $\mathring{\boldsymbol{c}}$ in the equations of motion (8.149)–(8.151). Once a consistent solution is found, we have also solved the flow problem of interest on the coarse-grained level, and (8.152) can be used to obtain the full pressure tensor describing the viscoelastic response of our rarefied gas (note that the use of Grad's variables for the coarse-grained level of description allows us to study also elastic effects due to momentum relaxation).

For small velocity gradients, that is, near equilibrium, $\mathring{\boldsymbol{c}}$ is small, and (8.156) with an estimate of the equilibrium friction matrix (see Exercise 207) implies

$$\mathring{\boldsymbol{c}} \approx \tau \left(\boldsymbol{\kappa} + \boldsymbol{\kappa}^T - \frac{2}{3} \mathrm{tr}\boldsymbol{\kappa} \mathbf{1} \right), \quad (8.157)$$

which, with a rough estimate of the collision time scale τ, can be used as our initial choice of $\mathring{\boldsymbol{c}}$. After performing a multiplostatted molecular dynamics simulation for a given $\boldsymbol{\kappa}$ and an approximate $\mathring{\boldsymbol{c}}$ and evaluating M_{44} from (8.153), we calculate an improved version of $\mathring{\boldsymbol{c}}$ from (8.156), where the previous $\mathring{\boldsymbol{c}}$ is used for evaluating the right-hand side of (8.156). The procedure of evaluating M_{44} by molecular dynamics and solving (8.156) to get the second-moment tensor in velocity space corresponding to a given flow is iterated until we achieve convergence. In an analytical calculation for given M_{44}, this iteration procedure could be used to calculate the perturbation expansion of $\mathring{\boldsymbol{c}}$ in terms of $\boldsymbol{\kappa}$, so that we expect rapid convergence in the near-equilibrium situations to which we are restricted by the multiplostat.

As discussed before, for the multiplostatted equations (8.149), (8.150) to be close to Hamilton's equations of motion, the velocity gradients need to be small compared with the inverse momentum relaxation time so that, according to (8.157), $\mathring{\boldsymbol{c}}$ is small. The further limitation on the flow rate, discussed after (8.126), can be expressed as follows. Whereas the mean free path divided by the size of the simulated box should certainly be smaller than unity, this ratio must be larger than the dimensionless flow rate, that is, the typical entries of $\mathring{\boldsymbol{c}}$, which hence need to be small. Therefore, the main modification of the equations of motion stems from the explicit flow terms in (8.125), (8.126), whereas the structural variable $\Pi(z)$ causes only a minor modification through $\mathring{\boldsymbol{c}}$ in (8.149), (8.150). In view of this discussion, the main restriction is

$$\text{dimensionless flow rate} \ll \frac{\text{mean free path}}{\text{size of simulation box}} \ll 1, \quad (8.158)$$

and the proposed beyond-equilibrium molecular dynamics for rarefied gases should only be used for sufficiently small flow rates so that, at best, only the onset of non-Newtonian, viscoelastic behavior can be investigated.

Our detailed simulation in simple shear flow is based on previous work,[75] where the only difference is that here we do not impose a vanishing second normal-stress difference, $\Psi_2 = 0$, as a constraint (see Exercise 186 for motivation). These simulations have been repeated and extended by D. Keffer,[76] whose comments provided invaluable help in formulating the remainder of this subsection. As for the Monte Carlo simulations of Section 8.2.3, we use a finite-range modification of the Lennard-Jones potential (8.34), now with a cutoff at $r = 2.5\sigma_{\rm LJ}$. For argon, one has the explicit Lennard-Jones parameters $\epsilon_{\rm LJ}/k_{\rm B} = 120\,{\rm K}$ and $\sigma_{\rm LJ} = 3.4\,\text{Å}$, but we here consider dimensionless combinations of quantities whenever possible.

To select a proper number density of gas particles, n, we consider the number of particles N in a cubic cell of the size of the mean free path, which is the minimum system size for a meaningful simulation. From the expression (7.6) for the mean free path, we have

$$N = l_{\rm mfp}^3 n = \frac{1}{(\sqrt{2}\pi)^3 d^6 n^2}, \tag{8.159}$$

where the particle radius, $d/2$, is of the order of the Lennard-Jones parameter $\sigma_{\rm LJ}$. We hence cannot afford a really small value of n because the number of particles to be simulated and the simulation box would become extremely large. For our simulations, we hence provide the fixed volume $10\sigma_{\rm LJ}^3$ per particle, which is not really sufficient for having a rarefied gas. By working at a sufficiently high temperature, $k_{\rm B}T/\epsilon_{\rm LJ} = 2.75$, however, we try to make sure that the Grad level of description is nevertheless reasonable. For the chosen state point, the contribution to the pressure tensor resulting from interactions turns out to be less than 10% of the kinetic contribution. At the chosen density, $N = 512$ particles are enough for obtaining a simulation cell considerably larger than the mean free path. We use Lees-Edwards boundary conditions, that is, if a particle transverses the simulation cell in the direction of the velocity gradient we change not only its velocity but also its position according to the time dependent shearing of the cell.

For describing the details of the molecular dynamics simulation, we introduce a characteristic collision time scale $\tau_{\rm LJ}$ for our Lennard-Jones gas as the slow time scale in our simulation. A reasonable estimate can be given in terms of the mean free path and the thermal velocity,

$$\tau_{\rm LJ} \approx l_{\rm mfp}\sqrt{\frac{m}{k_{\rm B}T}}. \tag{8.160}$$

[75] Kröger & Öttinger, J. Non-Newtonian Fluid Mech. 120 (2004) 175.

[76] Keffer, private communication (March 2004).

For the chosen state point, $l_{\rm mfp}$ is only slightly larger than $\sigma_{\rm LJ}$ and, as we work at an elevated temperature,

$$\tau_{\rm LJ} = \sigma_{\rm LJ}\sqrt{\frac{m}{\epsilon_{\rm LJ}}}, \tag{8.161}$$

which is just the conventional reference time scale for Lennard-Jones systems, should provide a fairly decent estimate of the characteristic time scale for our simulated system, as verified from our simulation results. For argon, $\tau_{\rm LJ}$ is about 2 ps. For the inertia of the thermostat we have chosen $M_{\rm T} = 20Nm\sigma_{\rm LJ}^2$, so that the characteristic time scale for the thermostat has the value $20\tau_{\rm LJ}$, large even compared with the slow time scale.

The equations of motion (8.149)–(8.151) are solved with the velocity Verlet algorithm (8.144), (8.145) with $\Delta t = 0.005\tau_{\rm LJ}$ where, formally, the right-hand sides of the equations of motion are taken as the velocities and forces so that we do not obtain a symplectic integrator. However, the resulting integration scheme is practical, stable, efficient, and easily implemented, and no significant differences were found compared with a simulation based on an extended symplectic scheme for this toy problem.

The initial conditions for the particle positions are not important; initial positions on a regular grid and random initial positions lead to the same results, as long as

$$\sum_{j=1}^{N} \boldsymbol{r}_j = 0. \tag{8.162}$$

Equation (8.132) suggests that we select the initial $\boldsymbol{p}_j - m\boldsymbol{\kappa}\cdot\boldsymbol{r}_j$ and $p_{\rm T}$ according to Gaussian distributions, where we impose the constraints

$$\sum_{j=1}^{N} \boldsymbol{p}_j = 0, \quad \sum_{j=1}^{N} \boldsymbol{p}_j \cdot (\boldsymbol{1} - \mathring{\boldsymbol{c}})^{-1} \cdot \left(\frac{\boldsymbol{p}_j}{m} - \boldsymbol{\kappa}\cdot\boldsymbol{r}_j\right) = f k_{\rm B} T. \tag{8.163}$$

These conditions are realized by pairwise Monte Carlo moves among the velocities. The reasons for choosing the origin at the center of mass and a vanishing total momentum have been discussed in the paragraph before (8.124). In view of the conservation of the total momentum, we use $f = 3N - 3$ as the number of degrees of freedom. Choosing $p_{\rm T} = 0$ instead of a Gaussian random number works equally well as an initial condition. As mentioned before, we can take (8.157) with $\tau = \tau_{\rm LJ}$ as an initial condition for $\mathring{\boldsymbol{c}}$. The fact that the Lees-Edwards boundary conditions do not exactly conserve the center-of-mass position has no significant influence on the simulation results as long as the conditions in (8.162) and (8.163) are satisfied by the initial conditions.

An important characteristic of the simulation is the total simulation time, $\tau_{\rm sim}$, used for each shear rate. We have always used $\tau_{\rm sim} = 2000\tau_{\rm LJ}$, except for an initial trial run with $\tau_{\rm sim} = 5000\tau_{\rm LJ}$ at the dimensionless shear rate $\tau_{\rm LJ}\dot{\gamma} = 0.01$. The separating time scale in the expression (8.153) for the friction matrix must be small compared with the slow time scale $\tau_{\rm LJ}$ but, for our not truly rarefied system, the

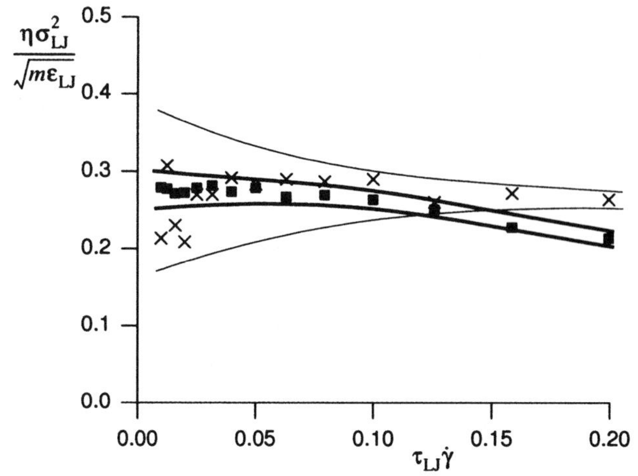

Fig. 8.10 Viscosity as a function of shear rate for a Lennard-Jones gas with particle number density $n = 0.1/\sigma_{LJ}^3$ and temperature $k_B T = 2.75\epsilon_{LJ}$. The squares represent the results of our thermodynamically guided simulations based on the Grad level, whereas the crosses correspond to a conventional nonequilibrium molecular dynamics simulation. The corresponding error bars are roughly indicated by the pairs of thick and thin lines.

duration of a collision is not by orders of magnitude smaller than τ_{LJ}. We hence choose $\tau_s = 0.2\tau_{LJ}$, and we have verified that $\tau_s = 0.1\tau_{LJ}$ gives identical results.

The total simulation time τ_{sim} is divided into intervals over which the estimates for the friction matrix are averaged and then used to update the estimate of $\mathring{c}$ at a fixed flow rate by means of (8.156). A trial run with updates after every $500\tau_{LJ}$ has shown that, even for deliberately poor initial conditions, a self-consistent solution is obtained after less than half a dozen iterations. In the further production runs, an update of $\mathring{c}$ was obtained after every interval of length τ_{LJ}, so that a converged (but fluctuating) result for $\mathring{c}$ is obtained almost immediately. The entire trajectory of length $\tau_{sim} = 2000\tau_{LJ}$ can then be used for averaging the quantities of interest, in particular, the pressure tensor (8.152).

For comparison, we have performed conventional nonequilibrium molecular dynamics simulations in shear flow driven by Lees-Edwards boundary conditions. This reference simulation has been carried out with a velocity Verlet algorithm and velocity rescaling to conserve the total kinetic energy. The same amount of CPU time is used for the conventional and for the thermodynamically guided simulations.

Our simulation results for the viscosity are shown in Figure 8.10. At low shear rates, the thermodynamically guided simulations produce significantly smaller error bars than the conventional simulations. At dimensionless shear rates around $\tau_{LJ}\dot{\gamma} = 0.2$, systematic discrepancies between the two simulation methodologies begin to arise. This should be expected in view of the restriction (8.158). According to (7.136), a relaxation time scale τ can be associated with the zero shear rate viscosity. Within

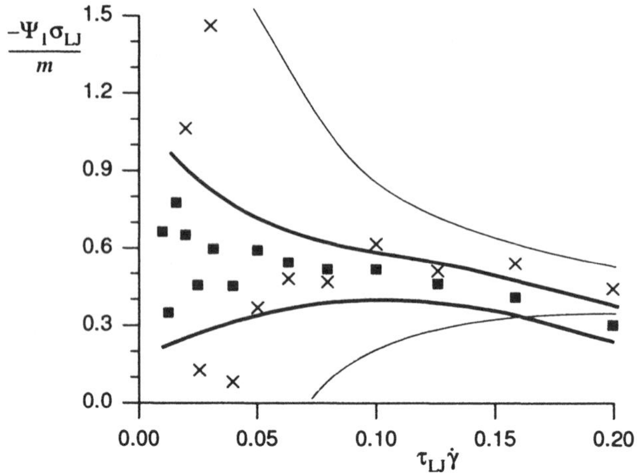

Fig. 8.11 First normal-stress coefficient as a function of shear rate for a Lennard-Jones gas with particle number density $n = 0.1/\sigma_{\rm LJ}^3$ and temperature $k_{\rm B}T = 2.75\epsilon_{\rm LJ}$. The squares represent the results of our thermodynamically guided simulations based on the Grad level, whereas the crosses correspond to a conventional nonequilibrium molecular dynamics simulation. The corresponding error bars are roughly indicated by the pairs of thick and thin lines.

the error bars, our simulation result for τ coincides with our a priori estimate (8.161), $\tau = \tau_{\rm LJ}$.

Figure 8.11 shows that the first normal-stress coefficient is negative. Meaningful simulations at dimensionless shear rates below $\tau_{\rm LJ}\dot{\gamma} = 0.1$ can be obtained only with the thermodynamically guided simulation method.

For the second normal-stress coefficient at low shear rates, conventional simulation results are not available, whereas the thermodynamically guided simulation produces the results shown in Figure 8.12. The second normal-stress coefficient is also negative, as expected according to Exercise 186. Within the error bars, our simulation results agree with the zero-shear-rate predictions of Exercise 186,

$$\left(\frac{\eta\sigma_{\rm LJ}^2}{\sqrt{m\epsilon_{\rm LJ}}}\right)^2 = \frac{1}{2}\,n\sigma_{\rm LJ}^3\,\frac{k_{\rm B}T}{\epsilon_{\rm LJ}}\left(\frac{-\Psi_1\sigma_{\rm LJ}}{m}\right), \qquad \Psi_2 = \frac{1}{2}\Psi_1. \tag{8.164}$$

In conclusion, the thermodynamically guided simulations provide useful information at low shear rates, where conventional simulations are plagued by large statistical error bars.

Exercise 206 Consistent Construction of Multiplostat
Use the variable

$$\boldsymbol{\Pi}'(z) = \frac{1}{mN}\sum_{j=1}^{N}(\boldsymbol{p}_j - m\boldsymbol{\kappa}\cdot\boldsymbol{r}_j)(\boldsymbol{p}_j - m\boldsymbol{\kappa}\cdot\boldsymbol{r}_j) \tag{8.165}$$

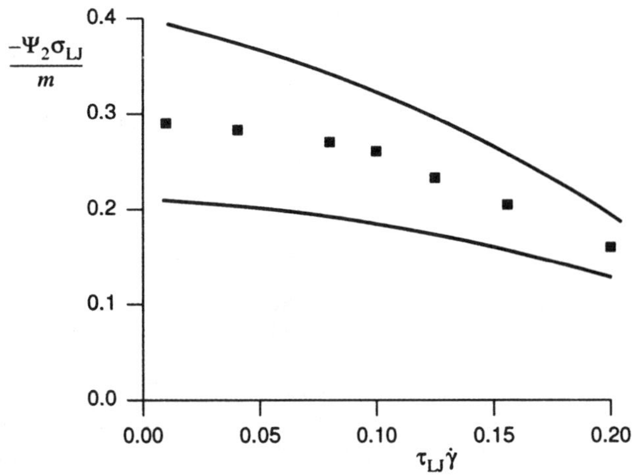

Fig. 8.12 Second normal-stress coefficient as a function of shear rate for a Lennard-Jones gas with particle number density $n = 0.1/\sigma_{LJ}^3$ and temperature $k_B T = 2.75\epsilon_{LJ}$. The squares represent the results of our thermodynamically guided simulations based on the Grad level, and the error bars are roughly indicated by the pair of thick lines.

to introduce the variables T and $\mathring{c}$ jointly and to construct the multiplostatted equations of motion consistently. How does the properly simplified starting point (8.110) for constructing multiplostats read? Identify the tensorial Lagrange multiplier $\boldsymbol{\lambda}_c'$ corresponding to $\boldsymbol{\Pi}'(z)$ and evaluate the ensemble average of $\boldsymbol{\Pi}'(z)$ to identify T and $\mathring{c}$. Write out the proper modification of the multiplostat (8.125), (8.126), and (8.128) and compare the result with (8.149)–(8.151).

Exercise 207 Estimate of Friction Matrix
Estimate the magnitude of the friction matrix (8.153) at equilibrium and discuss the functional form resulting for the viscosity.

8.4.6 Example: Entangled Polymer Melts

To elaborate a number of important issues, we consider a second, more relevant example of a molecular dynamics simulation. The example of a melt of entangled linear polymer molecules, described by the coarse-grained reptation model in Section 4.3.2, illustrates the following points.

- The coarse-grained level of description enters thermodynamically guided simulations in two different ways: (1) in the definition of the beyond-equilibrium ensemble and (2) by determining the basic information to be extracted from a simulation. Even at equilibrium, where the ensemble is given in terms of universal equilibrium variables, the methodology provides useful guidance in choosing the quantities to evaluate (the concept of the "primitive path" as a relevant object *implies* that one should study the diffusion along the "primitive path").

- When the structure of the GENERIC building blocks is analyzed, the fundamental mechanisms contained in a coarse-grained model can be revealed (the structure of the friction matrix for reptation models *suggests* that there is not only diffusion along the "primitive path" but also noise on the chain orientation, which can be interpreted as the "constraint release" associated with "double reptation").
- When a relevant variable reflecting physical insight is expressed in terms of the atomistic degrees of freedom, further relevant variables may arise naturally in the course of the construction (in defining a "primitive path," information about a "tube radius" and an anisotropic "tube cross section" arises *automatically*).
- If a coarse-grained model contains parameters of a fundamental physical nature, they enter through corresponding parameters in the atomistic expressions for the relevant variables, and a dynamic formulation through relevant variables rather than parameters should be considered (in the original reptation model, the number of "entanglements" occurs as a parameter, and there *must be* a corresponding smoothing parameter in the atomistic definition of "primitive paths;" a physical identification of "entanglements" would be more appealing).

To explain these points in detail, we now return to the example of polyethylene melts for which we have already defined a Hamiltonian for the Monte Carlo simulations of Section 8.2.3. If we consider a melt of increasingly longer polymer molecules, there occurs a crossover from unentangled to entangled behavior. The Rouse model becomes inappropriate and we should switch to the reptation model of Section 4.3.2, for which the configurational distribution function $f(\boldsymbol{u}, \sigma)$ for the orientation along a smoothed chain ("primitive path") is the proper structural variable. If $f^{\Pi}(\boldsymbol{u}, \sigma; z)$ is the atomistic expression for $f(\boldsymbol{u}, \sigma)$ evaluated in a homogeneous system of volume V, the arguments of Section 8.4.2 can be adapted from conformation tensors to the reptation model, and we obtain the following analogue of the expression (8.100):

$$M = M(\boldsymbol{u}, \sigma, \boldsymbol{u}', \sigma') = \frac{V}{2k_{\mathrm{B}}\tau_{\mathrm{s}}} \left\langle \Delta_{\tau_{\mathrm{s}}} f^{\Pi}(\boldsymbol{u}, \sigma; z) \Delta_{\tau_{\mathrm{s}}} f^{\Pi}(\boldsymbol{u}', \sigma'; z) \right\rangle_{x}. \quad (8.166)$$

We further consider the situation in which $f^{\Pi}(\boldsymbol{u}, \sigma; z)$ is given by an ensemble of orientation vectors and position labels, obtained by coarse-graining the N_{p} chains in the volume V to the length scale of entanglements (Z segments per chain). In view of the probability densities in (8.166), it is convenient to multiply M with arbitrary functions $g_1(\boldsymbol{u}, \sigma)$ and $g_2(\boldsymbol{u}', \sigma')$, and to integrate over all arguments. If we introduce averages over the ensemble of orientation vectors and position labels constructed from an atomistic configuration,

$$\bar{g}_j(z) = \int g_j(\boldsymbol{u}, \sigma) f^{\Pi}(\boldsymbol{u}, \sigma; z) d^2u d\sigma, \quad (8.167)$$

we can write

$$\int g_1(\boldsymbol{u}, \sigma) M g_2(\boldsymbol{u}', \sigma') d^2u d\sigma d^2u' d\sigma' = \frac{V}{2k_{\mathrm{B}}\tau_{\mathrm{s}}} \left\langle \Delta_{\tau_{\mathrm{s}}} \bar{g}_1(z) \Delta_{\tau_{\mathrm{s}}} \bar{g}_2(z) \right\rangle_{x}. \quad (8.168)$$

We hence need to understand what happens to the ensemble average of an arbitrary function in the course of time. As long as the configurations $(\boldsymbol{u}(z), \sigma(z))$ change continuously, our previous experience has shown that we need to consider only the noise effects on $\boldsymbol{u}$ and σ to estimate the leading-order behavior of the bilinear expression in the increments on the right-hand side of (8.166). If we neglect correlations between the noise effects on $\boldsymbol{u}$ and σ, a careful analysis of the changes of $\bar{g}_1$ and $\bar{g}_2$ due to noise, after Taylor expansion in terms of the noise on $\boldsymbol{u}$ and σ, leads us to the following expression for the friction matrix,

$$\begin{aligned} M &= M(\boldsymbol{u},\sigma,\boldsymbol{u}',\sigma') \\ &= \frac{1}{2n_{\mathrm{p}}k_{\mathrm{B}}\tau_{\mathrm{s}}}\left\{ \left\langle(\Delta_{\tau_{\mathrm{s}}}\sigma)^2\right\rangle_x \left[\frac{1}{Z}\left(-\frac{\partial}{\partial\sigma}f(\boldsymbol{u},\sigma)\frac{\partial}{\partial\sigma}\delta(\boldsymbol{u}-\boldsymbol{u}')\delta(\sigma-\sigma')\right.\right.\right. \\ &+ \left.\left.\frac{\partial}{\partial\sigma}f(\boldsymbol{u},\sigma)\delta(\sigma-\sigma')f(\boldsymbol{u}',\sigma')\frac{\partial}{\partial\sigma'}\right) + \frac{\partial f(\boldsymbol{u},\sigma)}{\partial\sigma}\frac{\partial f(\boldsymbol{u}',\sigma')}{\partial\sigma'}\right] \\ &- \left.\frac{1}{Z}\frac{\partial}{\partial\boldsymbol{u}}\cdot\langle\Delta_{\tau_{\mathrm{s}}}\boldsymbol{u}\Delta_{\tau_{\mathrm{s}}}\boldsymbol{u}\rangle_x\cdot f(\boldsymbol{u},\sigma)\frac{\partial}{\partial\boldsymbol{u}}\delta(\boldsymbol{u}-\boldsymbol{u}')\delta(\sigma-\sigma')\right\}, \end{aligned} \tag{8.169}$$

where $\Delta_{\tau_{\mathrm{s}}}\sigma$ is an estimate of the random increment of the position label and $\Delta_{\tau_{\mathrm{s}}}\boldsymbol{u}$ represents the fluctuations of the orientation vector, potentially depending on σ, over the time period τ_{s}. The structure of (8.169) is considerably richer than that of the phenomenological expression (4.148), which corresponds exactly to the second line of (8.169) and provides the formula

$$\frac{1}{\tau_{\mathrm{rept}}} = \frac{\pi^2}{2\tau_{\mathrm{s}}}\left\langle(\Delta_{\tau_{\mathrm{s}}}\sigma)^2\right\rangle_x, \tag{8.170}$$

as the straightforward GENERIC recipe for evaluating the relaxation time for reptative diffusion. The evaluation of (8.170) implies that we need to identify the reptative motion as one-dimensional diffusion along the chain in a molecular dynamics simulation of long polymer molecules, so that the formalism actually forces us to analyze the applicability of the coarse-grained reptation model. Once the choice of variables is based on the idea of "primitive paths," diffusion along those paths is the natural phenomenon to look at. In the third line of (8.169), our analysis suggests additional terms proportional to $\left\langle(\Delta_{\tau_{\mathrm{s}}}\sigma)^2\right\rangle_x$; although formally larger by a factor of Z, these additional terms have been missed in the phenomenological expression (4.148) because they contribute neither to the time-evolution equation nor to the degeneracy requirement. In the last line of (8.169), we naturally find a diffusion term associated with the orientation vector, which corresponds to a previously proposed mechanism of constraint release.[77]

Note that the intermediate time scale τ_{s} in (8.170) should be chosen small compared with the reptation time τ_{rept} but large compared with all other relaxation processes

[77] Öttinger, Phys. Rev. E 50 (1994) 4891.

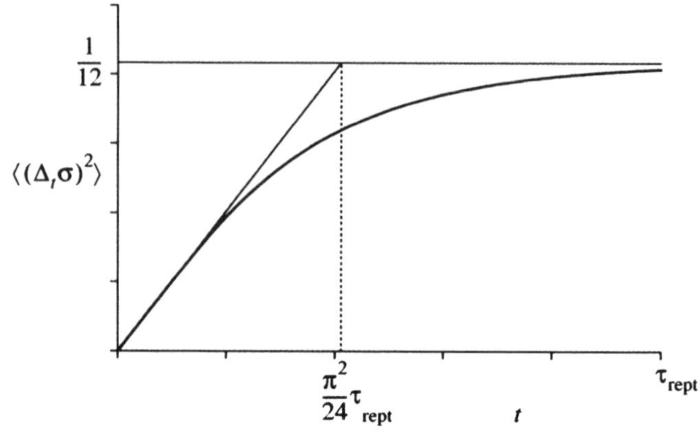

Fig. 8.13 Diffusion along the primitive path according to the reptation model.

occurring in entangled polymer melts. The result of diffusion along the primitive path according to (4.138), starting at $\sigma_0 = 1/2$ and imposing reflecting boundary conditions at $\sigma = 0$ and 1, is given by

$$\langle(\Delta_t\sigma)^2\rangle = \frac{1}{12} + \sum_{j=1}^{\infty} \frac{(-1)^j}{j^2\pi^2} e^{-4j^2 t/\tau_{\text{rept}}}, \tag{8.171}$$

as can be shown by by switching from sines to cosines in going from absorbing to reflecting boundaries.[78] The function (8.171) is shown in Figure 8.13, together with its initial slope given by (8.170) and the plateau value resulting from the reflecting boundaries at 0 and 1. It is for $t = \tau_s \lesssim 0.2\tau_{\text{rept}}$ that we expect to identify the initial slope, and thus the reptation time scale, as the relevant dynamic material information. In practice, one expects a different slope at short times (of the order of the Rouse time) rather than the ideal linear behavior in Figure 8.13 (cf. Figure 8.16 below). We then need to determine the slope in the range $\tau_{\text{Rouse}} \lesssim t \lesssim 0.2\tau_{\text{rept}}$ in order to extract the dynamical material information from molecular dynamics simulations for entangled polymer molecules. Such considerations show why an expression like that shown in (8.83), based on the slope, is preferable over that shown in (8.82).

We still need to define the function $f^{\Pi}(\boldsymbol{u}, \sigma; z)$ that maps an atomistic configuration z into the slow configurational variable of the reptation model.[79] As mentioned before, we can alternatively construct an ensemble of orientation vectors and position labels that are representative of $f^{\Pi}(\boldsymbol{u}, \sigma; z)$. Starting from an atomistic configuration of N_{p} chains, we construct a configuration of N_{p} smooth wormlike chains or primitive

[78] See Example 3.35 of Öttinger, *Stochastic Processes in Polymeric Fluids* (Springer, 1996).

[79] The construction of this mapping was motivated by the wish to present a balanced picture of beyond-equilibrium thermodynamics in this book, including simulation results. The ideas have previously been published and analyzed in Kröger, Ramírez & Öttinger, Polymer 43 (2002) 477 and Öttinger, J. Non-Newtonian Fluid Mech. 120 (2004) 207.

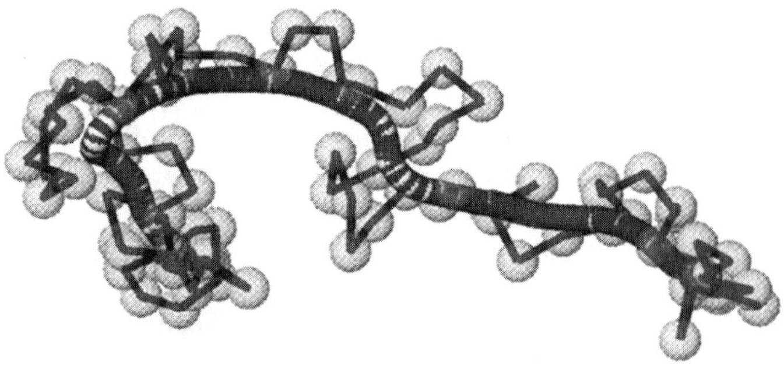

Fig. 8.14 Competing-spring construction of a smooth primitive path from an atomistic chain consisting of 85 beads ($\xi = 25$). [Figure courtesy of M. Kröger, ETH Zürich.]

paths (see Section 4.3.2), as illustrated in Figure 4.11. For every atomistic chain we construct the corresponding worm as follows. If the chain consists of N united atoms, the worm consists of N beads connected by Hookean springs with a spring constant H. Each bead of the worm is connected to the corresponding united atom with the same index $j = 1, 2, \ldots N$ by a Hookean spring with a spring constant $\bar{H}$. When $\bar{H}$ is large compared to H, the worm is forced to follow the atomistic chain; when, however, H is large compared to $\bar{H}$, the worm is forced to be smooth. By choosing the ratio $\xi = H/\bar{H}$ of the spring constants for the competing springs, we can dial in the smoothness of the worm following the original atomistic chain. The parameter ξ is thus directly related to the number of entanglements per chain, Z. Note that, like Z in the reptation model, ξ is treated as a parameter rather than a dynamical variable. As an illustration of the competing-spring construction, Figure 8.14 shows the result of coarse-graining an atomistic chain of 85 beads with the parameter $\xi = 25$. A rough estimate of ξ for long chains, based on the coarse-graining of a simple model of an atomistic chain, is given by[80]

$$\xi = \left[\frac{1}{4} \frac{(N-1)L_{\rm e}^2}{\langle R^2 \rangle} \right]^2, \tag{8.172}$$

where $L_{\rm e}$ is the entanglement length scale (see Table 4.2) and $\langle R^2 \rangle$ is the mean-square end-to-end distance of the polymer molecules, which is proportional to $N - 1$. A typical value for polyethylene is $\xi = 220$.[80] Also the estimation of ξ for short chains is described in the cited paper. It turns out that our final results depend only weakly on the choice of ξ, so that there is no need for a more sophisticated procedure of choosing ξ. For a smooth primitive path, orientation vectors are naturally defined all along the path.

Looking again at Figure 8.14, it should be clear that the coarse-graining procedure does not only offer information about the orientation of the primitive path

[80] Öttinger, J. Non-Newtonian Fluid Mech. 120 (2004) 207.

but also describes the distribution of atoms around the primitive path. The ensemble of displacement vectors of the atoms thus offers the natural possibility to introduce a coarse-grained variable characterizing the tube diameter and a potential tube anisotropy under deformation. Such atomistic ideas could fertilize previous attempts to study anisotropic tube cross sections phenomeologically.[81]

The subsequent example of molecular dynamics simulations for polyethylene relies on the insight developed in a series of papers by Harmandaris et al.[82] and further work in progress. The parameters for the various potentials of the united atom model are taken exactly as for the Monte Carlo simulations in Section 8.2.3, but the bonds are now assumed to be rigid rather than being held constant at 1.54 Å by an infinitely stiff spring.[83] The equations of motion are integrated with the velocity Verlet algorithm, and the constraints of fixed bond lengths are enforced with the SHAKE method, in which one adjusts the coordinates to satisfy the constraints one by one, iterating the procedure until all the constraints are satisfied to within a given tolerance.[84] A reversible multiple time step method is employed, where bond bending and torsional vibrations are considered to be fast and nonbonded Lennard-Jones interactions are treated as slow. The smaller time step is taken as 2 fs; the larger one as 10 fs. Typical overall simulation times are of the order of some hundred nanoseconds, depending on the chain length of the simulated polymer molecules.

To evaluate the reptation time scale by means of (8.170), we need to identify the mean-square change of the position label σ during the time interval τ_s as a result of diffusion along the primitive path. The chains studied in the literature turned out to be too short for a proper identification of motions along a primitive path. The author is hence grateful to N. Ch. Karayiannis and V. G. Mavrantzas for providing results of excessive equilibrium molecular dynamics simulations for a system of $N_p = 8$ polyethylene chains with $N = 1000$ carbon atoms per chain prior to publication for a preliminary analysis. Conformations were available every 100 fs, for a total simulation time of 0.7 μs. This total simulation time is larger than the Rouse time of 0.41 μs, as estimated from molecular dynamics simulations of shorter chains, but much shorter than the expected reptation time scale (see also the estimates for τ_{rept} in Table 8.4.6 below). It is hence not possible to estimate τ_{rept} by conventional techniques, which require a simulation over a time period of the order of τ_{rept}. For thermodynamically guided simulations, however, the time step τ_s in (8.170) should be small compared with τ_{rept} (smaller than $0.2\tau_{\text{rept}}$ according to Figure 8.13), so that we still have a chance to estimate τ_{rept} reliably.

[81] Ianniruberto & Marrucci, J. Non-Newtonian Fluid Mech. 79 (1998) 225; Marrucci & Ianniruberto, J. Non-Newtonian Fluid Mech. 82 (1999) 275; Öttinger, J. Non-Newtonian Fluid Mech. 89 (2000) 165.

[82] Harmandaris et al., Macromolecules 31 (1998) 7934; 33 (2000) 8062; 36 (2003) 1376.

[83] Because rigid bonds and infinitely stiff springs are not equivalent, a so-called Fixman potential has been introduced when using rigid bonds to get results identical to those for infinitely stiff springs; for further references and a detailed discussion in the present context, see Section 2 of Harmandaris et al., Macromolecules 31 (1998) 7934.

[84] See Section 3.4 of Allen & Tildesley, *Computer Simulation of Liquids* (Clarendon, 1987).

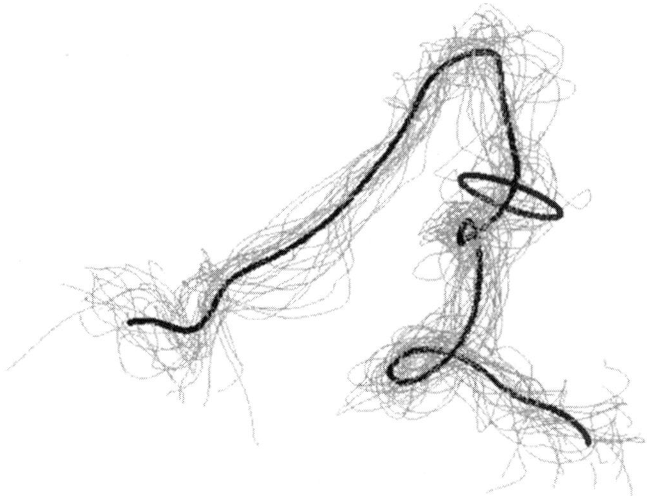

Fig. 8.15 Primitive path for a polyethylene chain with $N = 1000$ carbon atoms at 25 different times (in equal time steps of 20 ns). The solid line represents the averaged primitive path, used as a reference for defining the position label σ at any time, and the circle around its center indicates the size of the tube cross section. [Figure courtesy of M. Kröger, ETH Zürich.]

In motivating (8.169), we have assumed that the configurations $(\boldsymbol{u}(z), \sigma(z))$ change continuously and slowly in time, in accordance with the projection-operator derivation of GENERIC, which is based on the idea that the relevant variables are the slowly changing ones. In a reptation model, we observe large jumps in the orientation vector $\boldsymbol{u}$ when, due to reptation or chain retraction, a chain moves out of the corresponding tube segment and explores other orientations. Such a process, which is modeled through the boundary condition (4.139) of the reptation model, is not covered by the projection-operator formalism. Although GENERIC in its current formulation can provide a proper Fokker-Planck equation, boundary conditions have to be supplied as external information. It would clearly be desirable to include the description of large jumps into the framework. The transition state theory, as for example used for the atomistically based description of chemical reactions, might provide useful guidelines.

The inaccessibility of boundary conditions in the present coarse-graining approach has important consequences for the analysis of simulations because the restriction of the position label σ to the interval $[0, 1]$ is related to the boundary conditions. We need to construct σ for any atomistic configuration without referring to the chain ends, and we hence need to define an intrinsic scale for σ. This scale is provided by a reference configuration, which, for example, could be taken as the primitive path at a particular time. For each chain, we here obtain a reference configuration by averaging the primitive path over 0.5 μs, which is of the order of the Rouse time, and small compared with the reptation time scale. Figure 8.15 shows that we obtain a well-defined scale for σ in the inner part of the chain. This scale is used to construct

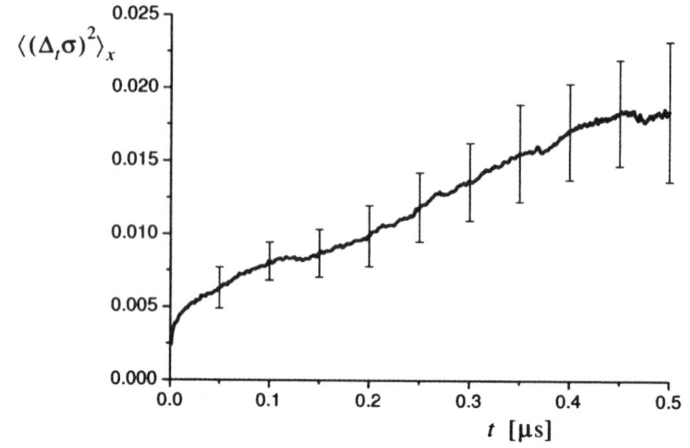

Fig. 8.16 Diffusion along the primitive path constructed for a polyethylene chain with $N = 1000$ carbon atoms with smoothing parameter $\xi = 220$. [Figure courtesy of M. Kröger, ETH Zürich.]

a representative ensemble for $f^{\Pi}(\boldsymbol{u}, \sigma; z)$ at any time, and it allows us to identify the differences $\Delta_t\sigma$ for any two times separated by t without any reference to boundary conditions. The actual identification of $\Delta_t\sigma$ has been achieved by means of the ring shown in Figure 8.15, placed at the center of the reference chain ($\sigma = 0.5$). The radius of the ring is estimated from the fluctuations of the primitive path at different times around the reference path, where the central third of the reference chain has been used for this analysis. As a consequence of smoothing (with the smoothing parameter $\xi = 220$ for polyethylene), almost every primitive path has a unique intersection point with the ring. If we move the ring to $\sigma = 0.6$, our results remain unchanged.

Figure 8.16 shows the results for $\langle(\Delta_t\sigma)^2\rangle_x$ as a function of t, where all possible pairs of chain conformations separated by t have been considered for all the eight chains in the system to improve the statistics. The error bars in the figure have been estimated from the variations in the results for the eight different chains. By comparing the scale for $\langle(\Delta_t\sigma)^2\rangle_x$ with the one in Figure 8.13, one expects to be in the linear regime. Figure 8.16 shows the expected nonlinear behavior at short times. However, well below the Rouse time scale of $0.41\ \mu$s, we observe a nicely linear behavior. This is a consequence of the high tension and the rather uniform distribution of beads in the primitive path that filter out almost purely reptative behavior for surprisingly short times.

Table 8.4.6 shows the results for τ_{rept} obtained from the slopes in Figure 8.16, where only the data for $t \leq \tau_{\text{s}}$ have been used to explore the minimum duration of a simulation to extract meaningful results. The numbers in parentheses indicate the statistical uncertainty in the last figure(s). Our best estimate is $\tau_{\text{rept}} = 6.8(2)\ \mu$s. It is remarkable that we can extract a meaningful estimate for τ_{rept} although we have simulation results only for eight polymer chains and some 10% of a reptation time.

Table 8.4 Reptation Time from Molecular Dynamics

τ_s [μs]	τ_{rept} [μs]	τ_s [μs]	τ_{rept} [μs]
0.05	5.6(12)	0.30	7.3(5)
0.10	5.5(13)	0.35	6.9(4)
0.15	7.5(13)	0.40	6.7(3)
0.20	8.5(12)	0.45	6.6(2)
0.25	8.1(7)	0.50	6.8(2)

This example hence illustrates the enormous potential of thermodynamically guided simulation techniques.

A number of consistency checks for the Rouse and reptation models in the unentangled and entangled regimes can be performed in molecular dynamics simulations, as suggested in the work of Harmandaris et al.[85] In the unentangled regime, one can check the dynamic structure factor for a range of scattering vectors in order to verify the validity of the Rouse modes. The Rouse modes can alternatively be verified in the relaxation of configurations corresponding to slightly deformed coarse-grained states, such as those generated by nonequilibrium Monte Carlo simulations in Section 8.2.3. In the entangled regime, one can consider the diffusion of primitive path points that, on time scales smaller than the reptation time, probes higher modes, which, according to the reptation model, should correspond to the relaxation times τ_{rept}/j^2 with integer j [we have first encountered this spectrum for the reptation model in (4.153) where, due to uniform initial conditions, only odd values of j contribute]. A further check of the crossover picture from unentangled to entangled behavior can be obtained from the molecular-weight dependence of the center-of-mass diffusion coefficient. The time dependent segment diffusion allows us to test reptation ideas, including short-time features that are considered to be typical for reptation but are not described by the equations for the reptation model given in Section 4.3.2.

In this section, we have learned how the analysis of equilibrium molecular dynamics simulations is guided by the ideas of beyond-equilibrium statistical mechanics and thermodynamics. The simulation results need to be analyzed in the light of a coarse-grained model. Also for traditional equilibrium simulations, the search for reptation has led researchers to study coarse-grained polymer conformations and to look for diffusion within tubes.[86] However, the thermodynamically guided simulations go much further by using nonequilibrium ensembles defined in terms of coarse-grained variables, as explained for the rarefied Lennard-Jones gas in Section 8.4.5. Far beyond equilibrium, the use of a fixed parameter ξ, such as the ratio of spring constants used to adjust the number of entanglements at equilibrium, is no longer satisfactory

[85] Harmandaris et al., Macromolecules 31 (1998) 7934; 33 (2000) 8062; 36 (2003) 1376.

[86] See, for example, Figures 30–33 in the pioneering work of Kremer & Grest, J. Chem. Phys. 92 (1990) 5057.

because the number of entanglements should be expected to become an important dynamic variable. The characteristic length scale for the curvature of primitive paths should not be set by an external parameter ξ, but it should be found by identifying entanglements in a multichain system. Simulations should breathe life into a generalized class of reptation models with the number of entanglements and/or the tube diameter as dynamic variables by providing concrete GENERIC building blocks. Following these and similar lines for polymer melts and other complex fluids, significant new results should be expected from thermodynamically guided beyond-equilibrium molecular dynamics simulations.

Appendices

Appendix A Crash-Course on Equilibrium Thermodynamics

To provide the background for beyond-equilibrium thermodynamics, we here outline the fundamentals of equilibrium thermodynamics. Equilibrium thermodynamics must not only be obtained as a special case of any acceptable nonequilibrium generalization but, through its shining example, it also elucidates the wide scope, the emphasis on universal relationships, and the advantages of a systematic postulational approach to beyond-equilibrium thermodynamics. This appendix is based on an article[1] that was written with exactly the same motivation, providing the background of equilibrium thermodynamics for a special issue of the *Journal of Non-Newtonian Fluid Mechanics* on nonequilibrium thermodynamics.[2] The article, in turn, was based on the famous textbook on equilibrium thermodynamics by Callen.[3] Callen's book, finally, is based on authority. The present crash-course on thermodynamics hence

[1] Jongschaap & Öttinger, J. Non-Newtonian Fluid Mech. 96 (2001) 5.
[2] See issue no. 1–2 in J. Non-Newtonian Fluid Mech. 96 (2001).
[3] Callen, *Thermodynamics* (Wiley, 1985).

follows Callen's clearly structured and widely accepted postulational approach as closely as possible.

A.1 APPROACH AND SCOPE

In this appendix, we summarize the key elements of equilibrium thermodynamics. Although this subject is covered by many textbooks and is a standard part of the education in most branches of science and engineering and despite the fact that the treatment of common applications such as the calculation of the efficiency of engines, phase behavior, chemical equilibria, and so forth are well understood and frequently used, we nevertheless undertake this effort because there still seems to be no consensus on the treatment of the fundamentals of thermodynamics. It is not only for pedagogical reasons that in the various textbooks many starting points and many routes of the further developments can be found. For example, one could take the traditional considerations of Carnot, Kelvin, and Clausius, which led to the famous formulations of the second law as a starting point, or Boltzmann's famous statistical expression of the entropy, $S = k_B \ln \Omega$, or the asymmetrical time evolution of irreversible processes, or the topology of the space of equilibrium states, or the information theoretical method, or the treatment based on the Clausius-Duhem inequality of rational thermodynamics.

We do not attempt to discuss any of the above-mentioned approaches here. We just present a concise and consistent outline of the fundamentals of equilibrium thermodynamics. For that purpose, we follow the well-known textbook of Callen, which is widely accepted as a brilliant source on the foundations of equilibrium thermodynamics. The present appendix is a selection of annotated quotations from this book.

We follow the treatment of Callen as closely as possible, with a special emphasis on fundamental concepts and principles. We focus on elucidating the structural elements of thermodynamics and the principles of statistical mechanics. How can we make sure that our selection of the key elements of thermodynamics and statistical mechanics indeed is in the spirit of Callen? For thermodynamics, we have Chapter 12 of his book as a valuable guide, in which the basic principles are summarized and generalized to systems including magnetic, electric, elastic, and other effects. The increasing importance that he ascribed to statistical mechanics is evident from the inclusion of six chapters on this topic in going from the first edition (1960) to the second edition (1985) of his book. For further details on specific subjects, the reader is of course referred to Callen's book or to alternative sources, such as Reichl's comprehensive textbook.[4]

Although we assume that most of that readers of this book are familiar with equilibrium thermodynamics to an extent that he or she would feel that a crash-course on thermodynamics is hardly necessary, we would still like to offer some basic remarks

[4] Reichl, *Modern Course in Statistical Physics* (University of Texas, 1980).

in the spirit of Callen's book by citing from his introduction. He describes the scope of thermodynamics as follows:

> Thermodynamics (...) neither claims a unique domain of systems over which it asserts primacy, nor does it introduce a new fundamental law analogue to Newton's or Maxwell's equations. In contrast to the specialty of mechanics and electromagnetism, the hallmark of thermodynamics is generality. Generality first in the sense that thermodynamics applies to all types of systems in macroscopic aggregation, and second in the sense that thermodynamics does not predict specific numerical values for observable quantities. Instead thermodynamics sets limits (*in*equalities) on permissible physical processes, and it establishes relationships among apparently unrelated properties.

Callen characterizes thermodynamics as a universal framework for the macroscopic description of matter:

> whereas thermodynamics is not based on a new and particular law of nature, it instead reflects a commonality of universal feature of *all* laws. In brief, *thermodynamics is the study of restriction on the possible properties of matter that follows from the symmetry properties of the fundamental laws of physics.*

Extrapolating Callen's point of view, beyond-equilibrium thermodynamics should be a general theory of the properties of matter, including the dynamical ones. By virtue of its nature of commonality, beyond-equilibrium thermodynamics should provide relationships between various static and dynamic material properties on the macroscopic or any other coarse-grained level of description. When statistical mechanics comes into the game, beyond-equilibrium thermodynamics may be regarded as the *theory of coarse-graining*: it should provide the structure of the equations for describing coarse-grained systems, and statistical mechanics should provide microscopic interpretations and expressions for the inputs of the phenomenological thermodynamic approach, so that the "macroscopic postulates are precisely and clearly the theorems of statistical mechanics." The reader should verify that the GENERIC approach used in this book for presenting a unified view of beyond-equilibrium thermodynamics fulfills all these requirements.

A.2 EQUILIBRIUM STATES

A very fundamental notion in thermodynamics is the *state* concept. The state of an isolated thermodynamic system is described by a finite set of *state variables* (X_i). Citing Callen,

> Systems tend to subside to very simple states, independent of their specific history. ... *in all systems there is a tendency to evolve toward states in which the properties are determined by intrinsic factors and not by previously applied external influences. Such simple terminal states are, by definition, time independent. They are called equilibrium states.* [p. 13]

Next, the *internal energy* U needs to be established as a state variable. Callen introduces it as a macroscopic manifestation of definite conservation principles of energy and summarizes the fundamental properties of equilibrium states in his first postulate:

> **Postulate I.** *There exist particular states (called equilibrium states) that, macroscopically, are characterized completely by the specification of the internal energy U and a set of parameters X_1, X_2, ... X_t later to be specifically enumerated.* [p. 283]

In many applications of thermodynamics, the parameters or state variables X_i are the volume V and the mole numbers N_1, N_2, ...N_r of the chemically pure components of which a system is a mixture. These parameters, with the property that their value in a composite system is the sum of their values in each of the subsystems, are called *extensive*. In general, "the choice of variables in terms of which a given problem is formulated, while a seemingly innocuous step, is often the most crucial step in the solution" [p. 465]. If the choice of variables is so crucial, how can we understand which variables are relevant to describing the time independent states of a system? Callen demonstrates in the final part of his book on fundamentals that, for equilibrium systems, the answer to this question is intimately related to the *symmetries of the laws of nature*.

The conserved quantities are clearly the most natural variables to describe time independent states. According to Noether's theorem, such conserved quantities result from the continuous symmetries of the dynamical behavior of a system. For example, symmetry under time translation implies the conservation of energy, and the conservation of electric charge is a consequence of the gauge symmetry of Maxwell's equations. Another possible reason for variables to become relevant to the description of time independent states lies in the observation that certain dynamic modes in the limit of long wavelengths acquire a vanishing frequency.

> An important general consequence of broken symmetry is formulated in the Goldstone theorem. It asserts that *any system with broken symmetry (and with certain weak restrictions on the atomic interactions) has a spectrum of excitations for which the frequency approaches zero as the wavelength becomes infinitely large.* [p. 464]

For a crystalline solid, for example, there exist infinitely many equally possible positions, slightly displaced by an arbitrary fraction of a lattice constant, which are related through the broken symmetry of translations. The fact that these positions are equally possible leads to a zero-frequency mode and to the occurrence of the volume as a state variable.

> To return to the general formalism, we thus recognize that *all symmetries* [leading to conservation laws and Goldstone excitations] *must be taken into account in specifying the relevant state space of a system.* [p. 471]

These symmetry considerations are limited to equilibrium thermodynamics. In complex fluids, the choice of the relevant state variables will be more complicated. In fact, a fundamental challenge in applying beyond-equilibrium thermodynamics is to identify in some way the structural state variables X_i for a particular (class of) system(s). The relevant variables correspond to the slowest modes of a system. In beyond-equilibrium thermodynamics, one thus needs to separate the slow and the fast variables and, after keeping only the slow as the relevant variables, one expects corrections to be governed by the ratio of the associated time scales, whereas the theory of equilibrium states deals with the truly time independent variables associated with symmetries. In that sense, beyond-equilibrium thermodynamics has an intrinsically approximate character absent in its equilibrium counterpart.

Considerations of the state concept introduced in Postulate I in connection with various forms of energy exchange lead to the definition of heat and work, as well as a balance equation for the internal energy U:

$$dU = đQ + đW. \tag{A.1}$$

The symbol $đ$ is used instead of d since $đQ$ and $đW$ are not differentials of state variables (that is, not differentials of functions of $U, X_1, X_2, \ldots X_t$) but are small amounts of heat and work. In Callen's words [p. 20], "Heat, like work, is only a form of energy *transfer* ... the energy U of a state *cannot* be considered as the sum of 'work' and 'heat' components." In general, $đQ$ and $đW$ are process-dependent quantities, and there are no corresponding state variables in the sense defined above. Equation (A.1) can be considered as a formal representation of the balance equation of internal energy derived from phenomenological considerations.

The use of the notions of *thermodynamic systems* and *equilibrium states* in this section deserves a few words of explanation. A thermodynamic system should actually be *defined* through a set of state variables and the accessible domain in the corresponding state space. Although one can formally define equilibrium states as time independent states of isolated systems, this definition is not very helpful for practical purposes because very long time scales might occur. For practical purposes, Callen hence proposes a rather indirect, somewhat provocative operational explanation [p. 15]: "In practice the criterion for equilibrium is circular. *Operationally, a system is in an equilibrium state if its properties are consistently described by thermodynamic theory!*"

A.3 BASIC PROBLEM OF THERMODYNAMICS

To keep a system in an equilibrium state, *constraints* are needed. If some of these constraints are released, the system will evolve to other states. What occurs then is, in fact, what Callen calls *the basic problem of thermodynamics*:

> *The single, all-encompassing problem of thermodynamics is the determination of the equilibrium state that eventually results after the removal of internal constraints in a closed, composite system.* [p. 26]

It should be noted that Callen uses the term *closed* for systems that we nowadays usually refer to as *isolated*. The solution of the basic problem of thermodynamics is achieved through the entropy postulates, which "are, in fact, the most natural guess that we might make, providing the *simplest conceivable formal solution* to the basic problem . . . the most economical form for the equilibrium criterion would be in terms of an extremum principle":

> **Postulate II.** *There exists a function (called the entropy) of the extensive parameters, defined for all equilibrium states, and having the following property. The values assumed by the extensive parameters in the absence of a constraint are those that maximize the entropy over the manifold of constrained equilibrium states.* [p. 284]

> **Postulate III.** *The entropy of a composite system is additive over the constituent subsystems (whence the entropy of each constituent system is a homogeneous first-order function of the extensive parameters). The entropy is continuous and differentiable and is a monotonically increasing function of the energy.* [p. 284]

> **Postulate IV.** *The entropy of any system vanishes in the state for which* $T \equiv (\partial U/\partial S)_{X_1, X_2, \ldots} = 0$. [p. 284]

The function introduced in Postulate II is called a *fundamental equation* of the system:

$$S = S(U, X_1, X_2, \ldots X_t). \tag{A.2}$$

This is the entropy representation. By solving with respect to U—this is possible in view of the last part of Postulate III—one obtains the energy representation

$$U = U(S, X_1, X_2, \ldots X_t). \tag{A.3}$$

Fundamental equations in thermodynamics deserve their name because they contain the information about all thermodynamic properties (equations of state, heat capacities, compressibilities, expansion coefficients, phase behavior, etc.) of the system. This, however, is primarily of theoretical interest, because for most systems no analytical expressions of the fundamental equations are available. The mere existence of fundamental equations, however, is already very important to establish relationships between various thermodynamic properties; this emphasis on relationships is the essence of thermodynamics.

The total differential of equation (A.3) is

$$dU = TdS + P_1 dX_1 + P_2 dX_2 + \ldots P_t dX_t, \tag{A.4}$$

with the temperature

$$T = \left(\frac{\partial U}{\partial S}\right)_{X_1, X_2, \ldots, X_t} = T(S, X_1, X_2, \ldots X_t), \tag{A.5}$$

and the thermodynamic forces

$$P_i = \left(\frac{\partial U}{\partial X_i}\right)_{S, X_1, X_2, \ldots, X_{i-1}, X_{i+1}, \ldots X_t} = P_i(S, X_1, X_2, \ldots X_t). \tag{A.6}$$

The functional dependencies implied by (A.5) and (A.6) are constitutive equations, known as *thermodynamic equations of state*. Contrary to fundamental equations like

(A.3), an equation of state does not contain the full information about the thermodynamic properties of the system but only part of it.

It should be emphasized that, despite the formal similarity of (A.1) and (A.4), these expressions are fundamentally different. Equation (A.1) is a balance equation for U, expressed in *process-dependent* quantities, describing an energy transfer, whereas (A.4) is the total differential of the dependent state variable U, expressed as a function of a complete set of state variables.

It should also be noted that, in the treatment of Callen, at this point it is not evident that T is in fact the usual temperature and how the P_i are related to the usual mechanical variables, such as pressure. This, in fact, turns out to be a consequence of the entropy maximum principle, expressed in Callen's second postulate. By considering the equilibrium of a system with external reservoirs, it follows that in sufficiently slow, *quasi-static* processes

$$đQ = TdS, \tag{A.7}$$

and

$$P_i = P_i^e, \tag{A.8}$$

where $-P_i$ is a thermodynamic force (such as the thermodynamic pressure) and $-P_i^e$ the corresponding external, mechanical force (such as the mechanical pressure). *Quasi-static* processes play an important role in the determination of thermodynamic quantities. These processes may be considered as trajectories in the manifold of equilibrium states. It should be noted, however, that quasi-static processes are an idealization because real processes always involve nonequilibrium intermediate states.

As mentioned above, the variables U, S and $X_1 \dots X_t$ are extensive, that is, proportional to the size of the system. As a consequence, (A.3) is homogeneous of the first order:

$$U(\lambda S, \lambda X_1, \lambda X_2, \dots \lambda X_t) = \lambda U(S, X_1, X_2, \dots X_t). \tag{A.9}$$

By differentiating (A.9) with respect to λ and setting λ equal to unity, one can derive the *Euler equation*,

$$U = TS + \sum_{i=1}^{t} P_i X_i, \tag{A.10}$$

and the so-called *Gibbs-Duhem equation*,

$$SdT + \sum_{i=1}^{t} X_i dP_i = 0. \tag{A.11}$$

Note, however, that in the modeling of complex fluids in general not all of the structural variables will be extensive. Structural parameters, like configuration tensors etc., which describe the internal state of a system, are an example of this. In such cases, one cannot directly generalize the expressions (A.10) and (A.11).

A classical example in thermodynamics is a system of N moles of a gas or liquid in a volume V. In this case, the total differential (A.4) of $U = U(S, V, N)$ becomes

$$dU = TdS - PdV + \mu dN, \tag{A.12}$$

where P is the thermodynamic pressure and μ is the chemical potential. The Gibbs-Duhem relation (A.11) becomes

$$SdT - VdP + Nd\mu = 0, \tag{A.13}$$

probably better known in the molar form

$$d\mu = -\check{s}dT + \check{v}dP, \tag{A.14}$$

with $\check{s} = S/N$ and $\check{v} = V/N$.

A.4 THERMODYNAMIC POTENTIALS

Equations (A.2) and (A.3) are probably the most fundamental but certainly not the most practical equations of thermodynamics. The reason is that, in many situations, the variables used in these expressions are not the ones controlled or measured in practice. Processes with temperature and pressure control are for instance more common than processes with controlled entropy and volume. Fortunately, there exist transformations by which fundamental equations can be transformed with conservation of their fundamental nature, that is, the property that they contain all information about the thermodynamic properties of the system. These transformations are the so-called *Legendre transformations*, and the new functions, which are the result of the transformation, are called thermodynamic potentials. By means of Legendre transformations, all extensive and intensive state variables $(U, S, X_1, \ldots, X_t, T, P_1, \ldots, P_t)$ are elevated to an equal footing.

The relevance of these transformations in thermodynamics should not be underestimated. To cite Callen:

> Multiple equivalent formulations also appear in Mechanics—Newtonian, Lagrangian, and Hamiltonian formulations are tautologically equivalent. Again certain problems are more tractable in a Lagrangian formalism than in a Newtonian formalism, or vice versa. But the difference in convenience of the different formalisms is enormously greater in thermodynamics. It is for this reason that *the general theory of transformations among equivalent representations is here incorporated as a fundamental aspect of thermostatistical theory.* [p. 131]

A partial Legendre transformation can be made by replacing the variables $X_0, X_1, \ldots X_s$ by $P_0, P_1, \ldots P_s$ (where $s < t$, $X_0 = S$, $P_0 = T$, and the first s variables, instead of any s variables, are selected without loss of generality). The Legendre transformed function is

$$U[P_0, P_1, \ldots, P_s] = U - \sum_{i=0}^{s} P_i X_i. \tag{A.15}$$

Following Callen, we here employ the notation $U[P_0, P_1, \ldots, P_s]$ to denote the function obtained by making a Legendre transformation with respect to $X_0, X_1, \ldots X_s$

on the function $U(X_1, X_2, \ldots X_t)$. Thus $U[P_0, P_1, \ldots, P_s]$ is a function of the independent variables $P_0, P_1, \ldots, P_s, X_{s+1}, X_{s+2} \ldots X_t$. The natural derivatives of this function are

$$\frac{\partial U[P_0, P_1, \ldots, P_s]}{\partial P_k} = -X_k \qquad (k = 0, 1, \ldots, s) \tag{A.16}$$

$$\frac{\partial U[P_0, P_1, \ldots, P_s]}{\partial X_k} = P_k \qquad (k = s+1, s+2, \ldots, t) \tag{A.17}$$

and consequently

$$dU[P_0, P_1, \ldots, P_s] = \sum_{i=0}^{s}(-X_i)dP_k + \sum_{i=s+1}^{t} P_i dX_i. \tag{A.18}$$

For the example of a gas or a fluid, the following well-known thermodynamic potentials are obtained by Legendre transformation: *the Helmholtz potential* or *Helmholtz free energy*,

$$F = U[T] = U - TS, \tag{A.19}$$

the enthalpy,

$$H = U[P] = U + PV, \tag{A.20}$$

and *the Gibbs potential* or *Gibbs free energy*,

$$G = U[T, P] = U - TS + PV. \tag{A.21}$$

The fundamental equations of these functions in differential form read:

$$\begin{cases} dF &= -SdT - PdV + \mu dN, \\ dH &= TdS + VdP + \mu dN, \\ dG &= -SdT + VdP + \mu dN. \end{cases} \tag{A.22}$$

The coefficients in these expressions are functions of the corresponding variables. In this way, a set of equations of state is associated with each of the potentials, for example

$$\begin{cases} S(T, V, N) &= -\left(\frac{\partial F}{\partial T}\right)_{V,N}, \\ P(T, V, N) &= -\left(\frac{\partial F}{\partial V}\right)_{T,N}, \\ \mu(T, V, N) &= \left(\frac{\partial F}{\partial N}\right)_{T,V}, \end{cases} \tag{A.23}$$

and so on.

For the Helmholtz potential function $F(T, V, N)$ an extremum principle similar to Callen's Postulate II applies. This result, which can be derived from Postulate II by considering a small system in thermal contact with a large reservoir of temperature T^r, reads:

> **Helmholtz Potential Minimum Principle.** *The equilibrium value of any unconstrained internal parameter in a system in diathermal contact with a heat reservoir minimizes the Helmholtz potential over the manifold of states for which* $T = T^r$. [p. 155]

Similar results may be derived for the other thermodynamic potentials.

There exists an alternative to thermodynamic potentials to condense all the information about a thermodynamic system into a single function that is useful in the context of developing beyond-equilibrium thermodynamics:

> Whereas the most common functions definable in terms of Legendre transformations are those mentioned ... [above], another set can be defined by performing the Legendre transformation on the entropy rather than on the energy. That is, the fundamental relation in the form $S = S(U, V, N_1, N_2, \ldots)$ can be taken as the relation on which the transformation is performed. Such Legendre transforms of the entropy were invented by Massieu in 1869 and actually predated the transforms of the energy introduced by Gibbs in 1875. We refer to the transforms of the entropy as *Massieu functions,* as distinguished from the *thermodynamic potentials* transformed from the energy. The Massieu funtions will turn out to be particularly useful in the theory of irreversible thermodynamics, and they also arise naturally in statistical mechanics and in the theory of thermal fluctuations. [p. 151]

A.5 MAXWELL RELATIONS AND STABILITY CRITERIA

The material quantities that are of primary interest in applications to thermodynamics are the equations of state, like (A.23), and in particular derivatives such as $\left(\frac{\partial S}{\partial T}\right)_{V,N}$. It should be noted that, because the equations of state are obtained by differentiation of fundamental equations, these quantities are second derivatives of a thermodynamic potential. By using the rule of interchanging the order of partial differentiation, one readily obtains the so-called *Maxwell relations.* For example, one has

$$\frac{\partial^2 F}{\partial T \partial V} = \frac{\partial^2 F}{\partial V \partial T} \tag{A.24}$$

and hence

$$\left(\frac{\partial P}{\partial T}\right)_{V,N} = \left(\frac{\partial S}{\partial V}\right)_{T,N}. \tag{A.25}$$

These interrelations are heavily used in thermodynamic calculations. Maxwell, whose turn of mind was geometrical, obtained (A.25), and similar relations, from the properties of a parallelogram; he concluded that such a relation "is a merely geometrical truth, and does not depend upon thermodynamic principles," whereas he offered the formulation in the language of differential calculus only as a footnote.[5]

[5] See p. 169 of Maxwell, *Theory of Heat* (Dover, 2001).

Besides derivatives of the equations of state, one also encounters derivatives like $\left(\frac{\partial S}{\partial V}\right)_{P,T}$, $\left(\frac{\partial U}{\partial P}\right)_{G,N}$, and so on. In simple cases, a reduction of quantities of this type to more familiar expressions is possible by the use of some identities of calculus. Callen describes a general procedure in five steps for the reduction of derivatives by using such identities and the Maxwell relations [p. 186–189]. In addition, a more general method based on Jacobian representations of partial derivatives is available.

The second derivatives of the thermodynamic potentials are not only the basis for the Maxwell relations; they are also fundamental in considerations about thermodynamic stability. The starting point is, again, the maximum entropy postulate. A consequence of this postulate is that the hypersurface $S = S(U, X_1 \ldots, X_t)$ in the thermodynamic configuration space should have the property that it lies everywhere below its tangent planes. This implies conditions for the derivatives of the fundamental relation. For example, in the case $S = S(U, V, N)$, we have

$$\left(\frac{\partial^2 S}{\partial U^2}\right)_{V,N} \leq 0, \tag{A.26}$$

$$\left(\frac{\partial^2 S}{\partial V^2}\right)_{U,N} \leq 0, \tag{A.27}$$

and

$$\frac{\partial^2 S}{\partial U^2}\frac{\partial^2 S}{\partial V^2} - \left(\frac{\partial^2 S}{\partial U \partial V}\right)^2 \geq 0. \tag{A.28}$$

The positiveness of physical quantities like specific heats, compressibilities, and expansion coefficients is a direct consequence of these thermodynamic stability requirements. Chemical equilibria for systems consisting of several components can be treated accordingly.

If criteria of stability are not satisfied, a system breaks up into two or more phases. The molar Gibbs potential of each component is then equal in each phase. The dimension $\bar{f}$ of the thermodynamic space in which a given number M of phases can exist for a system of r components is given by the *Gibbs phase rule*

$$\bar{f} = r - M + 2. \tag{A.29}$$

A.6 STATISTICAL MECHANICS

All the thermodynamic properties of a system are determined by a single fundamental equation, and equilibrium thermodynamics implies very general relationships between the various thermodynamic properties. To understand a particular system, however, we are still faced with the problem of determining an appropriate fundamental equation. One option would be to collect experimental information about various equations of state, which could then be used to construct a fundamental equation. Another option would be to calculate a fundamental equation from an atomistic understanding of the system by means of statistical mechanics. The key to this route to fundamental equations is a microscopic understanding of the concept of entropy. In Callen's words,

> ... the entropy is one of a set of extensive parameters, together with the energy, volume, mole numbers and magnetic moment. As these latter quantities each have clear and fundamental physical interpretations it would be strange if the entropy alone were to be exempt from physical interpretation.
>
> The subject of statistical mechanics provides the physical interpretation of the entropy, and it accordingly provides a heuristic justification for the extremum principle of thermodynamics. [p. 329]

The basis of statistical mechanics can actually be established in a few lines, provided that one formulates one more postulate, the *fundamental postulate of statistical mechanics*:

> The assumption of equal probability of all permissible microstates is the fundamental postulate of statistical mechanics [a permissible microstate being one consistent with the external constraints] ....
>
> Suppose now that some external constraint is removed—such as the opening of a valve permitting the system to expand into a larger volume. From the microphysical point of view the removal of the constraint activates the possibility of many microstates that previously had been precluded. Transitions occur into these newly available states. After some time the system will have lost all distinction between the original and the newly available states, and the system will thenceforth make random transitions that sample the *augmented* set of states with equal probability. *The number of microstates among which the system undergoes transitions, and which thereby share uniform probability of occupation, increases to the maximum permitted by the imposed constraints.*
>
> This statement is strikingly reminiscent of the entropy postulate of thermodynamics, according to which the entropy increases to the maximum permitted by the imposed constraints. It suggests that the entropy can be identified with the number of microstates consistent with the imposed macroscopic constraints.
>
> One difficulty arises: The entropy is additive (extensive), whereas the number of microstates is multiplicative. The number of microstates available to two systems is the *product* of the numbers available to each (...). To interpret the entropy, then, we require an *additive* quantity that measures the number of microstates available to a system. The (unique!) answer is to *identify the entropy with the logarithm of the number of available microstates* (the logarithm of a product being the sum of the logarithms). Thus
>
> $$S = k_B \ln \Omega$$
>
> where Ω is the number of microstates consistent with the macroscopic constraints. The constant prefactor merely determines the scale of S; it is chosen to obtain agreement with the Kelvin scale of temperature ...
>
> ... this single postulate is dramatic in its brevity, simplicity, and completeness. The statistical mechanical formalism that derives directly from it is one in which we 'simply' calculate the logarithm of the number of states available to the system, thereby obtaining S as a function of the constraints U, V, and N. That is, it is statistical mechanics in the entropy formulation, or, in the parlence of the field, it is statistical mechanics in the *microcanonical formalism*. [p. 331–332]

The fundamental postulate of statistical mechanics can be supported by symmetry arguments. Following Callen, for any two states i and j, "the interchangeability of future and past, or the time reversibility of physical laws, associates the transitions $i \to j$ and $j \to i$ and leads directly to the equality ... of the transition probabilities" [p. 468]. This principle of detailed balance leads to the principle of equal probabilities of states in the microcanonical ensemble, so that "*the equal probabilities of permissible states for a closed system in equilibrium is a consequence of time reversal symmetry of the relevant quantum mechanical laws*" [p. 468]. Again, symmetry plays an important role for a fundamental understanding of principles. Ergodic theory offers an alternative approach to justify the fundamental postulate of statistical mechanics.

According to the well-established microcanonical formalism, the calculation of the fundamental equation for the entropy amounts to counting the number of microstates for given extensive variables (more precisely, one needs to consider intervals around a given energy to find any states with the prescribed energy, where the entropy is essentially independent of the size of the interval). Whereas the microcanonical formalism solves the fundamental problem of statistical mechanics, it may be very useful for practical purposes to work in other formalisms, in particular, in the so-called *canonical formalism*:

> As in thermodynamics, the entropy representation is not always the most convenient representation. For statistical mechanical calculations it is frequently so inconvenient that it is analytically intractable. [p. 332]

The canonical formalism can be derived by considering the system of interest in contact with a thermal reservoir of a given temperature rather than an isolated system. The "system plus the reservoir constitute a *closed system*, to which the principle of equal probability of microstates again applies" [p. 349]. For the system alone, however, microstates then do not have the same probability. Rather, the probabilities are proportional to the Boltzmann factors, and the normalization factor for the probabilities is given by the *canonical partition function*, Z, as a function of the temperature and the extensive mechanical state variables. The Helmholtz free energy is given by the famous relation

$$F = -k_B T \ln Z = -k_B T \ln \sum_j e^{-E_j/(k_B T)}, \tag{A.30}$$

"which should be committed to memory" [p. 351]. Equation (A.30) is the basis for most first-principles calculations of thermodynamic properties, in particular, by Monte Carlo methods.

In the canonical approach, the probabilities of all the microstates are known, so that not only averages but also fluctuations are accessible. One thus obtains the *theory of fluctuations* which establishes a close relationship between thermodynamic material properties and the magnitude of fluctuations (both are given by the second-order derivatives of thermodynamic potentials).

Callen's approach to statistical mechanics is actually based on quantum systems, for which we have a discrete set of microstates. The quantum approach has a number of advantages. The natural concept of indistinguishable particles avoids the classical overestimate of the number of states leading to a nonextensive entropy (this problem is associated with the famous key words "Gibbs' paradox" and "correct Boltzmann counting"). The existence of (weakly degenerate) ground states implies that the entropy goes to zero when the absolute temperature approaches zero. Finally, only by means of Planck's constant can one naturally form the dimensionless quantities required for taking logarithms.

A.7 PERSPECTIVES

Equilibrium thermodynamics is in perfect shape. The all-encompassing problem of equilibrium thermodynamics has been identified, a clearly structured elegant formalism has been developed within a postulational approach, the postulates can be supported by statistical mechanics, the universal features have been elaborated, successful sets of state variables have been identified and corroborated for many problems, the rules for atomistic calculations of fundamental equations have been discovered, transformations for simplifying practical calculations have been found, and an enormous number of problems in various fields of science and engineering have been solved. Beyond-equilibrium thermodynamics should now be shaped to a similar level of perfection.

Appendix B
Mechanics and Geometry

For the distinction between reversible and irreversible contributions to dynamics, one should have the intuitive notion that reversible means "under mechanistic control," whereas irreversible is associated with "the uncontrollable rest." As discussed in Section 1.2.1, such an intuitive picture is the counterpart of the fundamental distinction between "work" and "heat" in equilibrium thermodynamics. It is therefore important to have clear ideas, condensed into a well-defined structure, associated with the notion of "mechanistic," a structure that can be generally recognized in mechanical and other reversible systems, and that is so restrictive and specific that irreversible dynamics could never share it. We then obtain both a clear-cut distinction between reversible and irreversible and powerful guidelines for modeling in specific nonequilibrium situations. The fundamental structure associated with the notion of "under mechanistic control" is well-established in the mathematical literature on mechanics, and it relies on elegant and powerful geometrical concepts: The geometric theory of Hamiltonian dynamical systems.

We here sketch and relate the different arenas that are relevant to mechanics: Symplectic, Poisson, and Dirac structures. Each of these structures allows us to introduce dynamical systems in terms of Hamiltonians. By realizing that these structures are deeply rooted in geometry and symmetry one obtains powerful tools to construct reversible dynamics and to introduce geometric structures on coarse-grained levels of description. Symplectic structures will be recognized as too restrictive, not leaving any room for an "uncontrollable rest." Poisson and Dirac structures are the proper generalizations. Instead of "Mechanics and Geometry," this appendix could hence be entitled "How to construct the most general Hamiltonian systems."

The emphasis in this appendix is on the basic concepts and their relationships, as well as on some most useful results for beyond-equilibrium thermodynamics. Only the reduction procedure, which allows us to construct geometric structures for particular applications and is hence crucial for our purposes, is discussed in considerable detail. The real power of the geometric formulation stems from the fact that it can handle infinite-dimensional systems rigorously, so that we can extend the "mechanistic" analysis from particle systems to field theories. To facilitate understanding, however, we give many of the equations in terms of a finite number of coordinates.

After reading this appendix, the reader should be aware of three situations in which Poisson brackets arise naturally, so that there is no need to check the Jacobi identity.

- The canonical bracket of classical mechanics naturally exists on any cotangent bundle, that is, on any manifold considered together with the vector space of one-forms, or differentials, on the manifold.

- For every Lie group, the dual of the associated Lie algebra is equipped with a natural Poisson bracket.

- If a Lie group acts on a vector space, the product of the duals of the Lie algebra and that vector space is equipped with a natural Poisson bracket.

In this appendix, we elaborate on these brief descriptions and show that these three situations are intimately related. The Poisson brackets on duals of Lie algebras are obtained by reduction of the canonical Poisson bracket on the cotangent bundle of the Lie group. If a Lie group acts on a vector space, the semidirect product is itself a Lie group, so that the third situation is a special case of the second one. In using these scenarios for complex fluids, the fundamental Lie group is the group of space transformations, the associated Lie algebra can be interpreted as the space of velocity fields, and the dual of the Lie algebra consists of momentum density fields.

The crash-course on mechanics given in this appendix is written such that it may be unsatisfactory to hard-core mathematicians because many concepts are explained in a rather informal way and almost all proofs are omitted; however, it should not be overly annoying to mathematicians in the sense of being misleading or even incorrect. This appendix is mainly based on two books, which we refer to as Marsden & Ratiu[1] and Dorfman,[2] and to which we refer the reader for all the details. The notation in this appendix is slightly richer than in the main parts of the book: For clarity, we use script symbols for manifolds and sans serif letters for vectors and vector spaces.

[1] Marsden & Ratiu, *Mechanics and Symmetry* (Springer, 1999).

[2] Dorfman, *Dirac Structures and Integrability* (Wiley, 1993).

B.1 BASIC NOTIONS FROM GEOMETRY

Before we can explain the geometric concepts for describing mechanical systems, we need to introduce the basic notions of manifolds, tensors, Lie algebras, and differential forms. A further important concept, in particular for introducing the most general Dirac structures, is that of differential forms, which are alternating tensor fields.

Manifolds

An n-dimensional *manifold* is a set $\mathcal{P}$ of points p covered by a collection of compatible overlapping charts by which, simply speaking, smooth local coordinates $(\xi^1, \ldots, \xi^n)$ can be introduced. *Differentiable functions* from manifolds to manifolds can be defined in terms of coordinates, where the assumed smoothness of transformations between different possible coordinates in overlapping regions of a manifold guarantees that the definition of differentiable functions is independent of the particular choice of coordinates.

In each point p of a manifold, one can introduce tangent vectors (namely, as equivalence classes of curves with equal derivatives through p), which form a vector space called the *tangent space*, $T_p\mathcal{P}$. For each tangent vector $\mathsf{v} \in T_p\mathcal{P}$, one can introduce the components $(\mathsf{v}^1, \ldots, \mathsf{v}^n)$ depending on the chosen coordinates. The union of all tangent spaces at all the points $p \in \mathcal{P}$ is known as the *tangent bundle*, $T\mathcal{P}$. The set $T\mathcal{P}$ is a $2n$-dimensional manifold, where the coordinates $(\xi^1, \ldots, \xi^n, \mathsf{v}^1, \ldots, \mathsf{v}^n)$ can be used to characterize a point in $T\mathcal{P}$, consisting of a point in $\mathcal{P}$ and a tangent vector at that point. A *vector field* on a manifold $\mathcal{P}$ is a map that assigns to each point of $\mathcal{P}$ a vector from the tangent space at that point, and the infinite-dimensional space of vector fields on $\mathcal{P}$ is denoted by $\mathsf{V}(\mathcal{P})$.

Note that vectors at different points of a manifold live in different vector spaces. For example, the concept of a constant vector field hence, in general, does not make any sense. To introduce such a concept, one needs to define a *connection*.[3] This can be done by specifying the differentials $d\mathsf{v}^k$ in terms of $d\xi^j$ for constant vector fields in a particular coordinate system or consistently in any coordinate system. An alternative approach to connections relies on the definition of parallel transport along curves. Once one knows which vector fields are constant along curves, by comparison, one can also try to define derivatives of vector fields along curves or along vector fields. Natural connections between the vector spaces at different points can be introduced if a manifold is equipped with a metric or if a group acts on a manifold.

The *derivative* of a differentiable function between two manifolds in a point can be introduced as a linear map between the corresponding tangent spaces. For a real-valued function $f : \mathcal{P} \to \mathbb{R}$, the derivative in p is a linear map $df(p) : T_p\mathcal{P} \to \mathbb{R}$, that is, this linear map or one-form $df(p)$ is the *differential* of f in p and, if it is evaluated for some vector $\mathsf{v} \in T_p\mathcal{P}$, we obtain the *directional derivative* of f in the direction v. The differential $df(p)$ is an element of $T_p^*\mathcal{P}$, which is the vector space of linear maps

[3] See Sections 54 and 55 of Burke, *Applied Differential Geometry* (Cambridge, 1985).

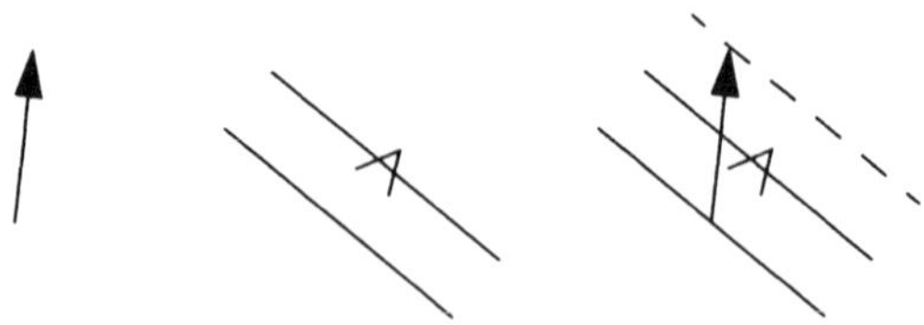

Fig. B.1 Coordinate-free visualization of vectors, covectors, and the meaning of the natural pairing (with a value of two in the example).

on $T_p\mathcal{P}$ and thus an example of the general construction of a *dual* vector space. The elements of the dual vector space are sometimes referred to as *covectors*. In analogy with the tangent bundle, the *cotangent bundle*, $T^*\mathcal{P}$, is the $2n$-dimensional manifold obtained as the union of all, with natural coordinates $(\xi^1, \ldots, \xi^n, \xi_1^*, \ldots, \xi_n^*)$, where the components ξ_j^* characterize a differential in terms of the base vectors $d\xi^1, \ldots, d\xi^n$ in the space of one-forms. The vector space of one-forms on a manifold $\mathcal{P}$ is denoted by $\mathsf{V}^*(\mathcal{P})$, which is dual to the space of vector fields, $\mathsf{V}(\mathcal{P})$. For the dual basis in $\mathsf{V}(\mathcal{P})$, we use the suggestive notation $\partial/\partial\xi^1, \ldots, \partial/\partial\xi^n$. To gain intuition, a vector can be visualized as a line with an arrowhead on one end of it, and a covector can be represented by two parallel hyperplanes corresponding to level hypersurfaces of a locally linearized real-valued map, with an arrow indicating the direction of growth (see Figure B.1).[4] The natural pairing between vectors and covectors can be evaluated by counting the number of hyperplanes crossed by a vector, just as a mountain climber would determine the rate of ascent from his speed, the direction of the trail, and the level lines on his map.

Because tangent vectors are associated with curves, there is a natural concept of *flow* along these curves for all the points $p \in \mathcal{P}$ associated with a vector field. A smooth vector field thus defines a *dynamical system*. The goal of introducing geometric structures in Section B.2 is to associate vector fields with real-valued Hamiltonians, so that we obtain Hamiltonian dynamical systems.

Tensors

Vectors and differentials of real-valued functions provide examples of tensors. In general, we introduce a *tensor* t *of covariant rank* l *and contravariant rank* m at the point p of a manifold $\mathcal{P}$ as a multilinear map

$$\mathsf{t} : \underbrace{T_p^*\mathcal{P} \times \ldots \times T_p^*\mathcal{P}}_{m \text{ factors}} \times \underbrace{T_p\mathcal{P} \times \ldots \times T_p\mathcal{P}}_{l \text{ factors}} \to \mathbb{R}, \tag{B.1}$$

that is, t is linear in each of its $l + m$ arguments. A corresponding tensor field is defined by assigning a tensor at each $p \in \mathcal{P}$. The vector space of all tensor fields of covariant rank l and contravariant rank m is denoted by $\mathsf{T}_l^m(\mathcal{P})$. The vector space of

[4] See Section 1 of Burke, *Applied Differential Geometry* (Cambridge, 1985).

one-forms on a manifold $\mathcal{P}$ corresponds to $m = 0$ and $l = 1$, so that $\mathsf{V}^*(\mathcal{P}) = \mathsf{T}^0_1(\mathcal{P})$ is the space of covariant tensor fields of rank 1. Contravariant tensor fields of rank 1 define a linear map on $T^*_p\mathcal{P}$ at each point p, that is, an element of the dual of $T^*_p\mathcal{P}$, which can be identified with $T_p\mathcal{P}$ (the linear maps of differential forms can be obtained by evaluating the forms for a particular vector); we thus have $\mathsf{V}(\mathcal{P}) = \mathsf{T}^1_0(\mathcal{P})$. Furthermore, $\mathsf{T}^0_0(\mathcal{P})$ is the space of smooth real-valued functions $\mathsf{F}(\mathcal{P})$ on $\mathcal{P}$. The *tensor product* $\mathsf{t} \otimes \bar{\mathsf{t}}$ of any two tensors $\mathsf{t}, \bar{\mathsf{t}}$ can be defined by the product of the two multilinear maps, so that the covariant and contravariant ranks of the factors add up separately.[5] Tensors can be contracted with vectors or covectors by inserting them into the proper slots of the map (B.1). More general *contractions* or *inner products* between tensors can be defined by expanding them in tensor products of vectors and covectors and employing linearity.

Like vectors, tensors at different points of a manifold live in different spaces. Tensor fields can be defined by assigning to each point a tensor from the corresponding tensor space.

For given coordinates ξ^j, a tensor t of covariant rank l and contravariant rank m can be represented by the components

$$\mathsf{t}^{j_1\ldots j_m}_{k_1\ldots k_l} = \mathsf{t}\left(d\xi^{j_1},\ldots,d\xi^{j_m},\frac{\partial}{\partial\xi^{k_1}},\ldots,\frac{\partial}{\partial\xi^{k_l}}\right). \tag{B.2}$$

If we use different coordinates $\bar{\xi}^{\bar{j}}$, the relationships

$$d\bar{\xi}^{\bar{j}} = \frac{\partial\bar{\xi}^{\bar{j}}}{\partial\xi^j}d\xi^j, \qquad \frac{\partial}{\partial\bar{\xi}^{\bar{k}}} = \frac{\partial\xi^k}{\partial\bar{\xi}^{\bar{k}}}\frac{\partial}{\partial\xi^k}, \tag{B.3}$$

where we have used Einstein's summation convention for repeated indices, imply the following transformation rules for the components of a tensor,

$$\bar{\mathsf{t}}^{\bar{j}_1\ldots\bar{j}_m}_{\bar{k}_1\ldots\bar{k}_l} = \frac{\partial\bar{\xi}^{\bar{j}_1}}{\partial\xi^{j_1}}\cdots\frac{\partial\bar{\xi}^{\bar{j}_m}}{\partial\xi^{j_m}}\frac{\partial\xi^{k_1}}{\partial\bar{\xi}^{\bar{k}_1}}\cdots\frac{\partial\xi^{k_l}}{\partial\bar{\xi}^{\bar{k}_l}}\mathsf{t}^{j_1\ldots j_m}_{k_1\ldots k_l}. \tag{B.4}$$

Let us consider the simplest case of a connection, in which a distinguished coordinate system ξ is specified for which a locally constant vector field has constant components. There is a natural concept of derivative associated with such a connection, which turns a tensor of any rank into a tensor with the same contravariant rank and a covariant rank increased by one. We consider the following quantities in the distinguished coordinate system:

$$\mathsf{t}^{j_1\ldots j_m}_{k_1\ldots k_l;n} = \frac{\partial}{\partial\xi^n}\mathsf{t}^{j_1\ldots j_m}_{k_1\ldots k_l}. \tag{B.5}$$

To obtain a tensor with these components in the distinguished coordinate system ξ, we need to transform these components into any other coordinate system by means

[5] An alternative approach to general tensors can be based on such tensor products of vectors and covectors.

of (B.4). The result can be rewritten as [compare (5.131)]

$$\begin{aligned} \bar{t}^{\bar{j}_1\ldots\bar{j}_m}_{\bar{k}_1\ldots\bar{k}_l;n} &= \frac{\partial}{\partial\bar{\xi}^n}\bar{t}^{\bar{j}_1\ldots\bar{j}_m}_{\bar{k}_1\ldots\bar{k}_l} + \sum_{\bar{m}=1}^{m}\bar{\Gamma}^{\bar{j}_{\bar{m}}}_{nj}\,\bar{t}^{\bar{j}_1\ldots\bar{j}_{\bar{m}-1}j\,\bar{j}_{\bar{m}+1}\ldots\bar{j}_m}_{\bar{k}_1\ldots\bar{k}_l} \\ &\quad - \sum_{\bar{l}=1}^{l}\bar{\Gamma}^{k}_{n\bar{k}_{\bar{l}}}\,\bar{t}^{\bar{j}_1\ldots\bar{j}_m}_{\bar{k}_1\ldots\bar{k}_{\bar{l}-1}k\,\bar{k}_{\bar{l}+1}\ldots\bar{k}_l}, \end{aligned} \tag{B.6}$$

with

$$\bar{\Gamma}^j_{kl} = \frac{\partial\bar{\xi}^j}{\partial\xi^i}\frac{\partial^2\xi^i}{\partial\bar{\xi}^k\partial\bar{\xi}^l}. \tag{B.7}$$

The quantities $\bar{\Gamma}^j_{kl}$ are not the components of a tensor. An important step in specifying the derivative (B.6) associated with a connection is the intrinsic formulation of $\bar{\Gamma}^j_{kl}$ in terms of the properties characterizing the connection rather than by explicit use of the distinguished coordinate system ξ in (B.7).

In general, there is not only one distinguished coordinate system but a class of distinguished coordinate systems; the definition of the derivative (B.6) must then be independent of the chosen member of the class. For example, if the connection results from a metric, the distinguished coordinate systems are those for which the components of the metric tensor are given by the unit matrix.

Lie Algebras

A *Lie algebra* A is a vector space equipped with a bilinear, antisymmetric bracket $[\mathsf{v}, \bar{\mathsf{v}}]$ satisfying the *Jacobi identity*,

$$[\mathsf{v}, [\bar{\mathsf{v}}, \bar{\bar{\mathsf{v}}}]] + [\bar{\mathsf{v}}, [\bar{\bar{\mathsf{v}}}, \mathsf{v}]] + [\bar{\bar{\mathsf{v}}}, [\mathsf{v}, \bar{\mathsf{v}}]] = 0, \tag{B.8}$$

for all vectors $\mathsf{v}, \bar{\mathsf{v}}, \bar{\bar{\mathsf{v}}} \in \mathsf{A}$. For example, $\mathbb{R}^3$ with the ordinary vector cross product is a Lie algebra.

Two important examples of Lie algebras arise naturally for manifolds and, in particular, for continuous groups. For the vector fields on a manifold, a natural bracket can be introduced as follows. Consider two vector fields $\mathsf{v}, \bar{\mathsf{v}} \in \mathsf{V}(\mathcal{P})$ and an arbitrary smooth real-valued function f. If we denote the operation of differentials of real-valued functions on vectors by a dot, then the definition of the *Jacobi-Lie bracket* $[\mathsf{v}, \bar{\mathsf{v}}]$ between vector fields on the manifold relies on the following identity,

$$d(df \cdot \bar{\mathsf{v}}) \cdot \mathsf{v} - d(df \cdot \mathsf{v}) \cdot \bar{\mathsf{v}} = df \cdot [\mathsf{v}, \bar{\mathsf{v}}], \tag{B.9}$$

which, loosely speaking, expresses the fact that the second-order derivatives of f on the left-hand side cancel so that the right-hand side is linear in the differential of f. In coordinates, we obtain for the finite-dimensional case,

$$[\mathsf{v}, \bar{\mathsf{v}}]^j = \mathsf{v}^k\frac{\partial\bar{\mathsf{v}}^j}{\partial\xi^k} - \bar{\mathsf{v}}^k\frac{\partial\mathsf{v}^j}{\partial\xi^k}. \tag{B.10}$$

Manifolds with a group multiplication that is a smooth function when expressed in terms of coordinates are known as *Lie groups*. Because Lie groups are manifolds, the space of vector fields is equipped with the Jacobi-Lie bracket. In view of the possibility to perform left or right translations by means of the group transformation for establishing a connection between tangent spaces at different points, it is possible and natural to restrict the Jacobi-Lie bracket to the tangent space of the group identity, which then becomes the *Lie algebra of the Lie group*. A full vector field associated with a vector from the tangent space of the group identity can be constructed by left translations, so that the Lie algebra arises from the group multiplication and actually characterizes it. From these ideas, the bracket for the Lie algebra of the Lie group can be computed by means of the following useful formula [Marsden & Ratiu, p. 275–276],

$$[\mathsf{g}, \bar{\mathsf{g}}] = \frac{d}{dt}\frac{d}{d\bar{t}}\, g(t)\bar{g}(\bar{t})g(t)^{-1}\bigg|_{t=\bar{t}=0}, \tag{B.11}$$

where $g(t)$ is a representative curve of the vector g in the tangent space of the group identity, that is, at $t = 0$, $g(t)$ is the group identity, and the derivative of $g(t)$ with respect to t is given by g; for $\bar{\mathsf{g}}$, one chooses an analogous representative curve $\bar{g}(\bar{t})$. For each t, the right-hand side of (B.11) defines a curve through the group identity. Note that the role of $g(t)$ can be interpreted as a parallel transport according to a connection implied by the group multiplication.

In a naive way, one can consider a matrix representation of the group $\mathcal{G}$ for better understanding of (B.11). For small $t, \bar{t}$, one could introduce the representations $g(t) = \exp\{\mathsf{g}t\} \approx 1 + \mathsf{g}t$ and $\bar{g}(\bar{t}) = \exp\{\bar{\mathsf{g}}\bar{t}\} \approx 1 + \bar{\mathsf{g}}\bar{t}$, where also the elements of the Lie algebra are represented by matrices. By extracting the terms proportion to $t\bar{t}$, in the matrix representation, (B.11) becomes $[\mathsf{g}, \bar{\mathsf{g}}] = \mathsf{g} \cdot \bar{\mathsf{g}} - \bar{\mathsf{g}} \cdot \mathsf{g}$, so that the Lie bracket (B.11) can be regarded as an abstraction of the commutator of matrices. Such arguments, which are useful in practical calculations, may be underpinned by rigorous mathematics [Marsden & Ratiu, p. 273–276].

Differential Forms

The main idea behind differential forms is to provide a generalization of the basic operations of vector calculus (divergence, gradient and curl) and the integral theorems of Green, Gauss and Stokes to manifolds of arbitrary dimension [Marsden & Ratiu, Section 4.2]. We have already met a basic example of one-forms, namely, the differential $df(p)$ of a real-valued function f. More generally, a *k-form*, sometimes called a *differential form of degree k*, on a manifold $\mathcal{P}$ is an alternating or skew tensor field of covariant rank k (that is, it changes sign whenever two of its arguments are interchanged). For the vector space of k-forms, $\mathsf{D}_k(\mathcal{P})$, we thus have $\mathsf{D}_k(\mathcal{P}) \subseteq \mathsf{T}^0_k(\mathcal{P})$.

For example, introducing coordinates $(\xi^1, \ldots, \xi^n)$ in the manifold $\mathcal{P}$ and the corresponding components $(\mathsf{v}^1, \ldots, \mathsf{v}^n)$ in $T_p\mathcal{P}$ for any $p \in \mathcal{P}$, a two-form can be characterized by

$$\Omega : (\mathsf{v}, \bar{\mathsf{v}}) \mapsto \Omega_{jk} \mathsf{v}^j \mathsf{v}^k \quad \text{with} \quad \Omega_{jk} = \Omega \left(\frac{\partial}{\partial \xi^j}, \frac{\partial}{\partial \xi^k} \right), \tag{B.12}$$

where the antisymmetric matrix Ω_{jk}, like the alternating bilinear map Ω, depends on the point p. The determinant of the $n \times n$-matrix formed by the components of n vectors provides an important example of an n-form on an n-dimensional manifold.

A convenient tool to introduce differentials of k-forms is given by the concept of the *wedge product* of two differential forms $\omega \in \mathsf{D}_k(\mathcal{P})$ and $\bar{\omega} \in \mathsf{D}_{\bar{k}}(\mathcal{P})$,

$$\omega \wedge \bar{\omega} = \frac{(k + \bar{k})!}{k! \bar{k}!} \mathcal{A}(\omega \otimes \bar{\omega}), \tag{B.13}$$

where the alternation operator $\mathcal{A}$ extracts the alternating part of $\omega \otimes \bar{\omega}$ (by summing over all permutations of the arguments and introducing the sign of the permutations; the normalization is such that $\mathcal{A}$ leaves an alternating tensor unchanged). The wedge product is bilinear, anticommutative, and associative. In terms of the dual basis $d\xi^j$, any k-form ω can be written as

$$\omega = \frac{1}{k!} \omega_{j_1 \ldots j_k} d\xi^{j_1} \wedge \ldots \wedge d\xi^{j_k}, \tag{B.14}$$

where a unique representation is, for example, obtained by choosing $\omega_{j_1 \ldots j_k}$ of the alternating type, or $\omega_{j_1 \ldots j_k} = 0$ unless $j_1 < j_2 < \ldots j_k$. Equation (B.13) implies

$$d\xi^{j_1} \wedge \ldots \wedge d\xi^{j_k} = \sum_{\text{permutations } \pi} \operatorname{sgn}(\pi) \, d\xi^{\pi(j_1)} \otimes \ldots \otimes d\xi^{\pi(j_k)}. \tag{B.15}$$

The *exterior derivative* of the differential form ω with representation (B.14) is locally given by

$$d\omega = \frac{1}{k!} \frac{\partial \omega_{j_1 \ldots j_k}}{\partial \xi^j} d\xi^j \wedge d\xi^{j_1} \wedge \ldots \wedge d\xi^{j_k}. \tag{B.16}$$

If $\omega_{j_1 \ldots j_k}$ is chosen to be of the alternating type, we have

$$\omega = \omega_{j_1 \ldots j_k} d\xi^{j_1} \otimes \ldots \otimes d\xi^{j_k}, \tag{B.17}$$

and, by proper antisymmetrization of the component matrix,

$$d\omega = \left[\frac{\partial \omega_{j_1 \ldots j_k}}{\partial \xi^j} + \sum_{l=1}^{k} (-1)^l \frac{\partial \omega_{j j_1 \ldots \not{j_l} \ldots j_k}}{\partial \xi^{j_l}} \right] d\xi^j \otimes d\xi^{j_1} \otimes \ldots \otimes d\xi^{j_k}, \tag{B.18}$$

where $\not{j_l}$ indicates that j_l is missing in the list of indices.

In view of the commutativity of partial derivatives and the skewness of the wedge product, (B.16) implies the fundamental result for iterated exterior derivatives,

$$dd\omega = 0. \tag{B.19}$$

Equation (B.19) means that, if a differential form Ω is of the type $\Omega = d\omega$, then one has $d\Omega = 0$. The following lemma shows that the reverse statement is also true [Marsden & Ratiu, p. 133].

Poincaré Lemma: If Ω is a *closed* form, that is, $d\Omega = 0$, then Ω is locally *exact*, that is, there is a differential form ω in a neighborhood around each point for which $\Omega = d\omega$.

The condition $d\Omega = 0$ can thus be regarded as an integrability condition, ensuring that there is a "potential" ω for Ω. For example, for a one-form Ω, the Poincaré Lemma becomes

$$\frac{\partial \Omega_k}{\partial \xi^j} d\xi^j \wedge d\xi^k = 0, \text{ i.e. } \frac{\partial \Omega_j}{\partial \xi^k} = \frac{\partial \Omega_k}{\partial \xi^j} \quad \Rightarrow \quad \Omega_j = \frac{\partial \omega}{\partial \xi^j}, \tag{B.20}$$

which is the classical result for the existence of a scalar potential for a (co)vector field guaranteed by integrability conditions.

Although the wedge product is a convenient tool to introduce the exterior derivative of differential forms, such an additional multiplicative structure is not really necessary for building up a consistent theory. One can introduce a more abstract theory of differential forms [Dorfman, Section 2.2]. In this book, this abstraction is only needed for introducing Dirac structures, so that a reader interested mostly in a basic understanding of the Poisson structures used in GENERIC may skip the rest of this subsection on differential forms. To understand the abstraction, we develop algebraic rules for calculating the exterior derivative of a differential form, instead of the suggestive coordinate representation (B.16). For a form ω of degree zero, we have

$$d\omega = \frac{\partial \omega}{\partial \xi^j} d\xi^j, \tag{B.21}$$

and hence

$$d\omega(\mathsf{v}) = \frac{\partial \omega}{\partial \xi^j} \mathsf{v}^j = \mathsf{v}^j \frac{\partial}{\partial \xi^j} \omega = \mathsf{v}\omega, \tag{B.22}$$

which emphasizes the role of v as a directional derivative. For a one-form ω, we have

$$d\omega = \frac{\partial \omega_k}{\partial \xi^j} d\xi^j \wedge d\xi^k = \frac{\partial \omega_k}{\partial \xi^j} (d\xi^j \otimes d\xi^k - d\xi^k \otimes d\xi^j), \tag{B.23}$$

and thus

$$\begin{aligned} d\omega(\bar{\mathsf{v}}, \mathsf{v}) &= \frac{\partial \omega_k}{\partial \xi^j} (\bar{\mathsf{v}}^j \mathsf{v}^k - \bar{\mathsf{v}}^k \mathsf{v}^j) \\ &= \bar{\mathsf{v}}^j \frac{\partial (\omega_k \mathsf{v}^k)}{\partial \xi^j} - \mathsf{v}^j \frac{\partial (\omega_k \bar{\mathsf{v}}^k)}{\partial \xi^j} + \omega_k \left(\mathsf{v}^j \frac{\partial \bar{\mathsf{v}}^k}{\partial \xi^j} - \bar{\mathsf{v}}^j \frac{\partial \mathsf{v}^k}{\partial \xi^j} \right) \\ &= \bar{\mathsf{v}}\omega(\mathsf{v}) - \mathsf{v}\omega(\bar{\mathsf{v}}) + \omega([\mathsf{v}, \bar{\mathsf{v}}]), \end{aligned} \tag{B.24}$$

where (B.10) has been used. The occurrence of the Jacobi-Lie bracket is related to the product or Leibniz rule or, in other words, it expresses the derivation character of the operator d. By means of (B.22), we obtain $\mathsf{v}\omega(\bar{\mathsf{v}}) = d[\omega(\bar{\mathsf{v}})](\mathsf{v})$, where $\bar{\mathsf{v}}$ is inserted into the one-form ω *before* we operate with d on the resulting form of degree zero, whereas the arguments of $d\omega(\bar{\mathsf{v}}, \mathsf{v})$ on the left-hand side of (B.24) are inserted *after* the two-form $d\omega$ has been calculated. It is hence convenient to define an "insertion operator" i_{v}, or an "inner product," by which (B.24) can more transparently be rewritten as

$$(i_{\mathsf{v}} i_{\bar{\mathsf{v}}} d + i_{\mathsf{v}} d i_{\bar{\mathsf{v}}} - i_{\bar{\mathsf{v}}} d i_{\mathsf{v}})\omega = i_{[\mathsf{v},\bar{\mathsf{v}}]}\omega, \tag{B.25}$$

where ω is a one-form. This equation is a generalization of (B.9), which is recovered by choosing $\omega = df$. Although (B.25) holds only for one-forms ω, it can be generalized to arbitrary differential forms,

$$(i_{\mathsf{v}} i_{\bar{\mathsf{v}}} d + i_{\mathsf{v}} d i_{\bar{\mathsf{v}}} - i_{\bar{\mathsf{v}}} d i_{\mathsf{v}} - d i_{\bar{\mathsf{v}}} i_{\mathsf{v}})\omega = i_{[\mathsf{v},\bar{\mathsf{v}}]}\omega. \tag{B.26}$$

We thus have an algebraic identity that allows us to reduce the exterior derivative of a k-form to the derivatives of $(k-1)$-forms and $(k-2)$-forms. This identity thus allows us to compute exterior derivatives recursively, without making any use of coordinates or wedge products. Only the differential of forms of degree zero needs to be defined.

Exercise 208 Algebraic Construction of Exterior Derivatives

Derive (B.26).

We now have collected all ingredients required to define abstract differential forms. Let us consider a Lie algebra G, the elements of which play the role of vector fields in the abstract setting, together with an infinite series of vector spaces $(\mathsf{D}_0, \mathsf{D}_1, \mathsf{D}_2, \ldots)$, where D_k is going to be interpreted as the vector space of abstract k-forms. Two kinds of linear operators need to be introduced: an *exterior derivative* d,

$$d : \mathsf{D}_k \to \mathsf{D}_{k+1}, \tag{B.27}$$

and an *inner product* operator i_{g} for any $\mathsf{g} \in \mathsf{G}$,

$$i_{\mathsf{g}} : \mathsf{D}_k \to \mathsf{D}_{k-1}, \quad \text{for } k \geq 1, \text{ and} \quad \mathsf{D}_0 \to \{0\}, \tag{B.28}$$

depending linearly on g. The following characteristic properties are then imposed on these operators:

$$d^2 = 0, \tag{B.29}$$

$$i_{\mathsf{g}} i_{\bar{\mathsf{g}}} = -i_{\bar{\mathsf{g}}} i_{\mathsf{g}}, \tag{B.30}$$

$$i_{\mathsf{g}} i_{\bar{\mathsf{g}}} d + i_{\mathsf{g}} d i_{\bar{\mathsf{g}}} - i_{\bar{\mathsf{g}}} d i_{\mathsf{g}} - d i_{\bar{\mathsf{g}}} i_{\mathsf{g}} = i_{[\mathsf{g},\bar{\mathsf{g}}]}, \tag{B.31}$$

for all $\mathsf{g}, \bar{\mathsf{g}} \in \mathsf{G}$.

Equation (B.29) is the characteristic property (B.19) of exterior derivatives on which the idea of integrability criteria is based (see the Poincaré Lemma). Equation (B.30) expresses the alternating character of differential forms, as becomes clear when we interpret the inner product operator i_{g} as the "insertion of the vector field g as the first argument of a differential form." For $\omega \in \mathsf{D}_k$, iterated use of this interpretation leads us to an alternating multilinear form

$$\mathsf{G} \times \ldots \times \mathsf{G} \to \mathsf{D}_0, \ (\mathsf{g}_1, \ldots, \mathsf{g}_k) \mapsto \omega(\mathsf{g}_1 \ldots \mathsf{g}_k) = i_{\mathsf{g}_k} i_{\mathsf{g}_{k-1}} \cdots i_{\mathsf{g}_1} \omega. \tag{B.32}$$

In particular, (B.32) shows that for every $\omega \in \mathsf{D}_1$ there exists a linear map $\mathsf{g} \mapsto i_{\mathsf{g}}\omega = \omega(\mathsf{g}) = \omega \cdot \mathsf{g}$. The map $\omega(\mathsf{g})$ uniquely characterizes ω if we assume that $\omega(\mathsf{g}) = 0$ for all $\mathsf{g} \in \mathsf{G}$ implies $\omega = 0$. If we conversely assume that $\omega(\mathsf{g}) = 0$ for all $\omega \in \mathsf{D}_1$ implies $\mathsf{g} = 0$, then we call the pairing between G and D_1 nondegenerate. Equation (B.31) is motivated by the algebraic rule (B.26) and thus expresses the derivation character of d. In view of the recursion relation (B.31), it is sufficient to define d on D_0.

The standard setting of differential forms on manifolds is recovered for the vector fields $\mathsf{G} = \mathsf{V}(\mathcal{P})$, the observables $\mathsf{D}_0 = \mathsf{F}(\mathcal{P})$, and the differential forms $\mathsf{D}_1 = \mathsf{V}^*(\mathcal{P})$; vector fields act as directional derivatives on observables. The concept of tensor fields of arbitrary rank can thus be generalized naturally to the abstract setting by realizing that, in the definition (B.1) of a tensor, G plays the role of $T_p\mathcal{P}$, and D_1 that of $T_p^*\mathcal{P}$.

The prototypical example of abstract differential forms is based on a Lie algebra G with a representation on some vector space D_0 [Dorfman, p. 12–14]. If R_{g} describes the action of g on D_0, we have the requirement

$$R_{\mathsf{g}} R_{\bar{\mathsf{g}}} \omega - R_{\bar{\mathsf{g}}} R_{\mathsf{g}} \omega = R_{[\mathsf{g},\bar{\mathsf{g}}]} \omega \quad \text{for all } \mathsf{g}, \bar{\mathsf{g}} \in \mathsf{G}, \ \omega \in \mathsf{D}_0. \tag{B.33}$$

We can then take D_k as the space of all alternating k-linear maps with the interpretation of i_{g} as the "insertion operator" according to (B.32), and we define the exterior derivative

$$d\omega(\mathsf{g}) = R_{\mathsf{g}} \omega \quad \text{for all } \omega \in \mathsf{D}_0. \tag{B.34}$$

The definition of the exterior derivative on all D_k with $k \geq 1$ is then obtained from the recursion relation (B.31). A special case of this construction is given by the adjoint action, that is, for $\mathsf{D}_0 = \mathsf{G}$ with

$$R_{\mathsf{g}} \bar{\mathsf{g}} = [\mathsf{g}, \bar{\mathsf{g}}] = d\bar{\mathsf{g}}(\mathsf{g}). \tag{B.35}$$

For this special case, (B.33) corresponds to the Jacobi identity, and D_1 is the space of linear maps $\mathsf{G} \to \mathsf{G}$, so that D_1 is much larger than G. In particular, $\mathsf{D}_0 = \mathsf{G}$ can be a vector space of matrices or linear operators equipped with the commutator as the Lie algebra bracket (for the relationship of this construction to quantum mechanics, see Exercise 8).

In the infinite-dimensional case, it is natural to equip all vector spaces with an inner product. An inner product implies a norm, and we assume the space to be complete in this norm (every Cauchy sequence has a limit in the space). Vector spaces with inner products or norms are known as Hilbert spaces or Banach spaces, respectively.

Because the abstract setting of vector fields and differential forms does not rely on an underlying manifold, how are abstract vector fields associated with flows of dynamical systems? As mentioned before, D_0, as the generalization of the vector space $\mathsf{D}_0(\mathcal{P}) = \mathsf{F}(\mathcal{P})$ of real-valued functions on $\mathcal{P}$, is interpreted as the space of physical observables. The flow or rate of change of an observable $\omega \in \mathsf{D}_0$ generated by a vector field $\mathsf{g} \in \mathsf{G}$ is naturally given by the directional derivative $d\omega(\mathsf{g}) = i_{\mathsf{g}} d\omega$. In general, we define a Lie derivative

$$d_{\mathsf{g}} = i_{\mathsf{g}} d + d i_{\mathsf{g}}, \tag{B.36}$$

which generalizes the concept of the directional derivative from D_0 to arbitrary D_k and thus defines the flow of any k-form generated by a vector field $\mathsf{g} \in \mathsf{G}$. This generalization is chosen such that the Lie derivative inherits the Lie algebra structure from G by considering the commutator,

$$d_{\mathsf{g}} d_{\bar{\mathsf{g}}} - d_{\bar{\mathsf{g}}} d_{\mathsf{g}} = d_{[\mathsf{g},\bar{\mathsf{g}}]}, \tag{B.37}$$

as can be verified explicitly by means of (B.29)–(B.31). For vector fields, one can define

$$d_{\mathsf{g}} \bar{\mathsf{g}} = [\mathsf{g}, \bar{\mathsf{g}}], \tag{B.38}$$

and for arbitrary tensor fields we obtain a definition of d_{g} by considering suitable contractions and imposing the product rule. Moreover, (B.38) is consistent with (B.35) for the special case $\mathsf{D}_0 = \mathsf{G}$. The concept of Lie derivatives is associated with the action of vector fields on the underlying manifold $\mathcal{P}$, which defines a connection on $\mathcal{P}$.

Exercise 209 Lie Derivative of Vector Fields
Motivate the definition (B.38) by means of a product rule for Lie derivatives.

Exercise 210 Exterior Derivative of a Two-Form
Show that the exterior derivative of a two-form Ω is given by

$$\begin{aligned} d\Omega(\mathsf{g}, \bar{\mathsf{g}}, \bar{\bar{\mathsf{g}}}) &= d_{\mathsf{g}}[\Omega(\bar{\mathsf{g}}, \bar{\bar{\mathsf{g}}})] + d_{\bar{\mathsf{g}}}[\Omega(\bar{\bar{\mathsf{g}}}, \mathsf{g})] + d_{\bar{\bar{\mathsf{g}}}}[\Omega(\mathsf{g}, \bar{\mathsf{g}})] \\ &\quad + \Omega(\mathsf{g}, [\bar{\mathsf{g}}, \bar{\bar{\mathsf{g}}}]) + \Omega(\bar{\mathsf{g}}, [\bar{\bar{\mathsf{g}}}, \mathsf{g}]) + \Omega(\bar{\bar{\mathsf{g}}}, [\mathsf{g}, \bar{\mathsf{g}}]). \end{aligned} \tag{B.39}$$

Exercise 211 Lie Derivative of a Two-Form
Show that the Lie derivative of a two-form Ω is given by

$$d_{\mathsf{g}} \Omega(\bar{\mathsf{g}}, \bar{\bar{\mathsf{g}}}) = d_{\mathsf{g}}[\Omega(\bar{\mathsf{g}}, \bar{\bar{\mathsf{g}}})] + \Omega(\bar{\mathsf{g}}, [\bar{\bar{\mathsf{g}}}, \mathsf{g}]) + \Omega(\bar{\bar{\mathsf{g}}}, [\mathsf{g}, \bar{\mathsf{g}}]). \tag{B.40}$$

Express this formula for the standard setting of differential forms in the component form

$$(d_{\mathsf{v}} \Omega)_{jk} = \mathsf{v}^l \frac{\partial \Omega_{jk}}{\partial \xi^l} + \Omega_{jl} \frac{\partial \mathsf{v}^l}{\partial \xi^k} + \frac{\partial \mathsf{v}^l}{\partial \xi^j} \Omega_{lk}. \tag{B.41}$$

B.2 SYMPLECTIC, POISSON, AND DIRAC STRUCTURES

We now have the basic tools to discuss the geometric structures behind mechanics. All these structures allow us to introduce dynamical systems in terms of Hamiltonians.

We present three different arenas in the order of increasing generality and elaborate the relationships between them. In studying the various geometric structures, one is interested in *canonical transformations*, which leave the respective structures invariant. The latter concept is the key to the implementation of the structure of reversible dynamics in *time discretized dynamical systems*, clearly important for computer simulations: every discrete time step must be described by a canonical transformation.

Symplectic Structures

Definition: A *symplectic manifold* is a manifold $\mathcal{P}$ endowed with a closed, weakly nondegenerate two-form Ω. The two-form Ω is referred to as a *symplectic form* or *symplectic structure*.

For further explanation, we formulate these defining properties for a symplectic form in terms of coordinates [Marsden & Ratiu, p. 65, 66, 147–148]. If we choose a particular coordinate system ξ^j on a finite-dimensional manifold $\mathcal{P}$, according to (B.14), the two-form Ω can be expressed in terms of the basic differentials as $\Omega = (1/2)\Omega_{jk} d\xi^j \wedge d\xi^k$, with an antisymmetric matrix,

$$\Omega_{jk} = -\Omega_{kj}, \tag{B.42}$$

and, according to (B.39), the following differential equation expresses that Ω is closed,

$$\frac{\partial \Omega_{jk}}{\partial \xi^l} + \frac{\partial \Omega_{lj}}{\partial \xi^k} + \frac{\partial \Omega_{kl}}{\partial \xi^j} = 0. \tag{B.43}$$

The weak nondegeneracy can be formulated as follows:

$$\mathsf{v}^j \Omega_{jk} \bar{\mathsf{v}}^k = 0 \text{ for all } \bar{\mathsf{v}} \in \mathbb{R}^n \quad \Rightarrow \quad \mathsf{v}^j = 0. \tag{B.44}$$

For an alternative characterization of symplectic structures, consider a linear operator

$$\bar{\Omega} : \mathsf{V}(\mathcal{P}) \to \mathsf{V}^*(\mathcal{P}), \tag{B.45}$$

which defines a bilinear form by

$$\Omega(\mathsf{v}, \bar{\mathsf{v}}) = i_{\bar{\mathsf{v}}} \bar{\Omega}(\mathsf{v}) = \bar{\Omega}(\mathsf{v}) \cdot \bar{\mathsf{v}} \text{ for all } \mathsf{v}, \bar{\mathsf{v}} \in \mathsf{V}(\mathcal{P}), \tag{B.46}$$

where, as before, the dot indicates the pairing between vector fields and one-forms as dual objects. The operator $\bar{\Omega}$ is called a *symplectic operator* if, and only if, (B.46) defines a closed two-form. In other words, a symplectic operator must be skew-symmetric and, in the coordinate representation with $(\bar{\Omega}(\mathsf{v}))_k = \mathsf{v}^j \Omega_{jk}$, it must satisfy (B.43) [Dorfman, p. 2]. If a symplectic operator (B.45) is injective, it defines a weakly nondegenerate form, so that a symplectic structure can be characterized either by a symplectic form or by an injective symplectic operator. If, moreover, the symplectic operator is surjective, the form Ω is strongly nondegenerate, and we speak of a *strong symplectic manifold*. In finite dimensions, every symplectic manifold must be of the strong type.

Exercise 212 Integrability Condition for Symplectic Operator
Show that the coordinate-free condition for a skew-symmetric operator to be symplectic is

$$i_{\bar{\mathsf{v}}} d_{\bar{\bar{\mathsf{v}}}} \bar{\Omega}(\mathsf{v}) + i_{\bar{\bar{\mathsf{v}}}} d_{\mathsf{v}} \bar{\Omega}(\bar{\mathsf{v}}) + i_{\mathsf{v}} d_{\bar{\mathsf{v}}} \bar{\Omega}(\bar{\bar{\mathsf{v}}}) = 0. \tag{B.47}$$

A vector field v_H on a symplectic manifold $\mathcal{P}$ is called a *Hamiltonian vector field* if there is a differentiable real-valued function $H : \mathcal{P} \to \mathbb{R}$ such that [Marsden & Ratiu, p. 157]

$$\Omega(\mathsf{v}_H, \mathsf{v}) = i_{\mathsf{v}} dH = dH \cdot \mathsf{v} \quad \text{for all } \mathsf{v} \in \mathsf{V}(\mathcal{P}), \tag{B.48}$$

or, equivalently,

$$\bar{\Omega}(\mathsf{v}_H) = dH. \tag{B.49}$$

The function H then plays the role of a *Hamiltonian*, and *Hamilton's equations* are the evolution equations

$$\frac{dp}{dt} = \mathsf{v}_H(p). \tag{B.50}$$

Equation (B.49) makes clear that dH lies in the image space of $\bar{\Omega}$, so that there is no a priori guarantee that, for a given H, a Hamiltonian vector field v_H exists in the infinite-dimensional case. Following Marsden & Ratiu [p. 159], we assume the existence of v_H for any smooth H and verify it in examples without going any further into technicalities; v_H is unique because $\bar{\Omega}$ is injective.

The prototypical example of a symplectic structure is given by the phase space of classical mechanics or, more generally, by cotangent bundles $T^*\mathcal{P}$. In classical mechanics, the natural coordinates $(\xi^1, \ldots, \xi^n, \xi_1^*, \ldots, \xi_n^*)$ correspond to the positions and momenta of the particles. The *symplectic matrix*

$$(\Omega_{jk}) = \begin{pmatrix} 0 & 1 \\ -1 & 0 \end{pmatrix}, \tag{B.51}$$

where 1 represents a $n \times n$ unit matrix and 0 is the $n \times n$ matrix in which all entries are zero, defines a symplectic structure on $T^*\mathcal{P}$. The symplectic form is given by

$$\Omega = d\xi^j \wedge d\xi_j^*, \tag{B.52}$$

and Hamilton's evolution equations are of the canonical form

$$\begin{aligned} \frac{d\xi^j}{dt} &= \frac{\partial H}{d\xi_j^*}, \\ \frac{d\xi_j^*}{dt} &= -\frac{\partial H}{d\xi^j}. \end{aligned} \tag{B.53}$$

The importance of the symplectic matrix (B.51) stems from the following result, which is known as *Darboux' Theorem* [Marsden & Ratiu, p. 148–149]: if $\mathcal{P}$ is a strong symplectic manifold then, for each point p in $\mathcal{P}$, there exist local coordinates around

p for which Ω is constant. In particular, a finite-dimensional symplectic manifold must be even-dimensional and, around any point p, there exist local coordinates $(\xi^1, \ldots, \xi^n, \xi_1^*, \ldots, \xi_n^*)$ such that the symplectic form Ω is given by the symplectic matrix (B.51).

The proof of Darboux' Theorem relies heavily on the closed nature of Ω, so that this property may be considered to be essentially equivalent to the possibility of introducing *canonical coordinates* $(\xi^1, \ldots, \xi^n, \xi_1^*, \ldots, \xi_n^*)$. This observation moreover explains why (B.43) is often referred to as an *integrability condition*: because (B.43) expresses the condition that Ω is closed, that is, $d\Omega = 0$, Ω can be written as $\Omega = d\omega$ for some one-form ω.

Cotangent bundles consisting of particle positions and momenta are the natural arena for the Hamiltonian formulation of classical mechanics. The alternative formulation of classical mechanics in terms of Lagrangians involves particle positions and velocities, so that tangent bundles provide the natural arena. Note that the Lagrangian formulation on tangent bundles is intimately related to a variational principle.

Poisson Structures

Definition: A *Poisson manifold* is a manifold $\mathcal{P}$ endowed with a bilinear, antisymmetric bracket operation $\{\,,\}$ on the set of smooth real-valued functions $\mathsf{F}(\mathcal{P})$ on $\mathcal{P}$, which satisfies the *Jacobi identity*,

$$\{A, \{B, C\}\} + \{B, \{C, A\}\} + \{C, \{A, B\}\} = 0, \tag{B.54}$$

and *Leibniz' rule*,

$$\{AB, C\} = A\{B, C\} + B\{A, C\}, \tag{B.55}$$

for all $A, B, C \in \mathsf{F}(\mathcal{P})$. The bracket operation $\{\,,\}$ is referred to as a *Poisson bracket*, or *Poisson structure* [Marsden & Ratiu, p. 328].

Note that, in particular, a Poisson bracket $\{\,,\}$ turns the vector space of smooth functions $\mathsf{F}(\mathcal{P})$ on $\mathcal{P}$ into a Lie algebra. The additional Leibniz rule makes it possibile to represent the bracket in terms of differentials,

$$\{A, B\} = i_{L(dB)} dA = dA \cdot L(dB), \tag{B.56}$$

where the skew-symmetric linear operator

$$L : \mathsf{V}^*(\mathcal{P}) \to \mathsf{V}(\mathcal{P}), \tag{B.57}$$

is called a *Poisson operator*, provided that (B.56) satisfies the Jacobi identity. A Poisson structure can thus be characterized either by a Poisson bracket or by a Poisson operator, where a Poisson operator is the most convenient tool to introduce dynamic systems because it directly maps differentials (of Hamiltonians) into vector fields. In the finite-dimensional case, the Jacobi identity can be rewritten as

$$L^{il} \frac{\partial L^{jk}}{\partial \xi^l} + L^{jl} \frac{\partial L^{ki}}{\partial \xi^l} + L^{kl} \frac{\partial L^{ij}}{\partial \xi^l} = 0, \tag{B.58}$$

for $L^{jk} = d\xi^j \cdot L(d\xi^k)$. Because our systems of interest are usually described by a number of fields, we clearly need the mathematical tools for the infinite-dimensional case, in which $\mathsf{F}(\mathcal{P})$ is a space of functionals.

Every symplectic manifold is a Poisson manifold, because a Poisson bracket can be introduced in terms of the symplectic form and the Hamiltonian vector fields (B.48) assumed to be associated with any $A, B \in \mathsf{F}(\mathcal{P})$,

$$\{A, B\} = \Omega(\mathsf{v}_A, \mathsf{v}_B). \tag{B.59}$$

In finite dimensions, the bracket definition (B.59) can be expressed as

$$L^{jk} = (\Omega^{-1})^{kj}. \tag{B.60}$$

In the nondegenerate case, L^{jk} is the negative inverse of the sympletic matrix Ω_{jk}. By means of (B.60), the Jacobi identity (B.58) follows from the integrability condition (B.43) which expresses the fact that the symplectic form is closed; therefore, the Jacobi identity is sometimes referred to as an integrability condition. In canonical coordinates, the Poisson bracket on a cotangent bundle is given by

$$\{A, B\} = \frac{\partial A}{\partial \xi^j}\frac{\partial B}{\partial \xi^*_j} - \frac{\partial A}{\partial \xi^*_j}\frac{\partial B}{\partial \xi^j}, \tag{B.61}$$

which is a generalization of (1.28).

From the above, we realize that the concept of Poisson structures is more general than that of symplectic structures because one allows for the degeneracy excluded in the definition of symplectic forms. This degeneracy is crucial for introducing entropy and hence irreversible terms into dynamic systems. Note that, for a finite-dimensional Poisson manifold, the rank of the Poisson operator can change from point to point on the manifold.

On Poisson manifolds, time evolution can naturally be associated with any Hamiltonian $H \in \mathsf{F}(\mathcal{P})$,

$$\frac{dp}{dt} = \mathsf{v}_H(p) = L(dH), \tag{B.62}$$

which, by comparison with (B.49), makes the role of L as the negative inverse of $\bar{\Omega}$ on the image of $\bar{\Omega}$ obvious. This formulation of *Hamilton's equations* further explains why Dorfman [p. 2] refers to Poisson operators as Hamiltonian operators.

It is natural to ask how the Hamiltonian vector field associated with $H = \{A, B\}$ is related to the Hamiltonian vector fields associated with A and B, individually? The answer can be obtained by observing that (B.56) and (B.62) can be combined into $\{A, B\} = dA \cdot \mathsf{v}_B$. With this identity, we can rewrite the Jacobi identity (B.54) for $C = f$ as

$$\{df \cdot \mathsf{v}_B, A\} - \{df \cdot \mathsf{v}_A, B\} + df \cdot \mathsf{v}_{\{A,B\}} = 0, \tag{B.63}$$

and further

$$d(df \cdot \mathsf{v}_B) \cdot \mathsf{v}_A - d(df \cdot \mathsf{v}_A) \cdot \mathsf{v}_B + df \cdot \mathsf{v}_{\{A,B\}} = 0. \tag{B.64}$$

Equation (B.9) thus shows that the Hamiltonian vector field associated with the Poisson bracket is given by the Jacobi-Lie bracket of the individual Hamiltonian vector fields,

$$\mathsf{v}_{\{A,B\}} = -[\mathsf{v}_A, \mathsf{v}_B]. \tag{B.65}$$

The minus sign in (B.65) is related to the common sign convention or the Jacobi-Lie bracket. The Lie algebras on the space of smooth functions $\mathsf{F}(\mathcal{P})$ and on the space of Hamiltonian vector fields, which is a subspace of the space of all vector fields $\mathsf{V}(\mathcal{P})$, are thus intimately related.

To understand the general structure of Poisson manifolds, we introduce the concept of Casimir functions. A smooth function $C \in \mathsf{F}(\mathcal{P})$ is called a *Casimir function* if $\{A, C\} = 0$ for all $A \in \mathsf{F}(\mathcal{P})$, that is, if the observable C is constant under any Hamiltonian evolution or C generates trivial dynamics. The existence of Casimir functions, which form a subspace of the vector space $\mathsf{F}(\mathcal{P})$, expresses the degeneracy of the Poisson bracket. Within the GENERIC framework, the existence of Casimir functions, and their determination, is of crucial importance because the entropy for generating irreversible dynamics must be chosen from the space of Casimir functions. By prescribing the values on a basis for the space of Casimir functions, we can decompose a Poisson manifold into symplectic submanifolds on which all Casimir functions are constant. For finite-dimensional Poisson manifolds, this idea leads to a rigorous *Symplectic Stratification Theorem* [Marsden & Ratiu, p. 345], which implies a generalization of *Darboux' Theorem for Poisson manifolds* [Marsden & Ratiu, p. 348]. When the rank of the Poisson operator is constant near some point p in $\mathcal{P}$, there exist local coordinates $(\xi^1, \ldots, \xi^n, \xi_1^*, \ldots, \xi_n^*, c^1, \ldots, c^l)$ in a neighborhood of p such that the Poisson operator is given by the matrix

$$(L^{jk}) = \left(\begin{array}{c|c|c} \overbrace{\begin{matrix} & & & \\ & & & \\ & 0 & & \\ & & & \end{matrix}}^{n} & \overbrace{\begin{matrix} 1 & 0 & \cdots & 0 \\ 0 & 1 & \ddots & \vdots \\ \vdots & \ddots & \ddots & 0 \\ 0 & \cdots & 0 & 1 \end{matrix}}^{n} & \overbrace{\begin{matrix} & & \\ & 0 & \\ & & \end{matrix}}^{l} \\ \hline \begin{matrix} -1 & 0 & \cdots & 0 \\ 0 & -1 & \ddots & \vdots \\ \vdots & \ddots & \ddots & 0 \\ 0 & \cdots & 0 & -1 \end{matrix} & 0 & 0 \\ \hline 0 & 0 & \begin{matrix} 0 & \cdots & 0 \\ \vdots & \ddots & \vdots \\ 0 & \cdots & 0 \end{matrix} \end{array}\right). \tag{B.66}$$

In the absence of degeneracy ($l = 0$), we observe that the matrix (L^{jk}) in (B.66) coincides with the symplectic matrix (B.51) [see also Exercise 5]. Because Poisson structures are sometimes alternatively called *cosymplectic structures*, one should better refer to (L^{jk}) for $l = 0$ as a cosymplectic matrix, although it coincides with the symplectic matrix, which happens to be the transpose of its inverse [see also (B.60)].

In the abstract setting of differential forms, $(\mathsf{D}_0, \mathsf{D}_1, \ldots, \mathsf{G}, d, i_\mathsf{g})$, Hamiltonians, like other observables, can be taken as elements of D_0. A Poisson operator L is a linear operator $\mathsf{D}_1 \to \mathsf{G}$ such that the Poisson bracket (B.56) is antisymmetric and satisfies the Jacobi identity (B.54) for all observables $A, B, C \in \mathsf{D}_0$. The time evolution of an observable A generated by a Hamiltonian H is given by

$$\frac{dA}{dt} = \{A, H\} = d_{L(H)} A, \tag{B.67}$$

that is, by the Lie derivative along the Hamiltonian vector field $L(H)$.

Dirac Structures

The key difference between symplectic and Poisson structures is related to degeneracy. The Dirac structures to be introduced next are particularly convenient for models with constraints, and they are thus convenient for incorporating an interconnection structure between variables, which is important in control theory.[6] Moreover, they are useful to describe driven systems.[7] Readers interested only in a basic understanding of the Poisson structures used in GENERIC may skip this subsection on Dirac structures. Following Dorfman,[8] the formulation of Dirac structures in the abstract setting of differential forms $(\mathsf{D}_0, \mathsf{D}_1, \ldots, \mathsf{G}, d, i_\mathsf{g})$ is based on the idea of formulating properties of the graphs of linear operators such as symplectic operators (B.45) or Poisson operators (B.57), which consist of pairs of vectors from G and D_1, or elements of the *direct sum* $\mathsf{G} \oplus \mathsf{D}_1$ (with componentwise addition and scalar multiplication of all components to obtain another vector space). The theory of Dirac structures for the special case of vector fields and differential forms on finite-dimensional manifolds was developed by Courant.[9] In the context of beyond-equilibrium thermodynamics, Dirac structures have a promising potential for the formulation of a unified theory of closed and driven systems.[10]

Definition: A *Dirac structure* is a linear subspace D of $\mathsf{G} \oplus \mathsf{D}_1$ such that

$$i_\mathsf{g}\bar{\omega} + i_{\bar{\mathsf{g}}}\omega = 0, \tag{B.68}$$

for all $(\mathsf{g} \oplus \omega), (\bar{\mathsf{g}} \oplus \bar{\omega}) \in \mathsf{D}$, where

$$\text{the dimension of } \mathsf{D} \text{ is maximum}, \tag{B.69}$$

[6] See Introduction of van der Schaft & Maschke, J. Geom. Phys. 42 (2002) 166.
[7] Jongschaap & Öttinger, J. Non-Newtonian Fluid Mech. 120 (2004) 3.
[8] See Section 2.4 of Dorfman, *Dirac Structures and Integrability* (Wiley, 1993).
[9] Courant, Trans. Am. Math. Soc. 319 (1990) 631.
[10] Jongschaap & Öttinger, J. Non-Newtonian Fluid Mech. 120 (2004) 3.

that is, if (B.68) is satisfied for all $(\mathsf{g} \oplus \omega), (\bar{\mathsf{g}} \oplus \bar{\omega}) \in \bar{\mathsf{D}} \supseteq \mathsf{D}$, then $\bar{\mathsf{D}} = \mathsf{D}$, and, furthermore,

$$i_{\bar{\mathsf{g}}} d_{\bar{\bar{\mathsf{g}}}} \omega + i_{\bar{\bar{\mathsf{g}}}} d_{\mathsf{g}} \bar{\omega} + i_{\mathsf{g}} d_{\bar{\mathsf{g}}} \bar{\bar{\omega}} = 0, \tag{B.70}$$

for all $(\mathsf{g} \oplus \omega), (\bar{\mathsf{g}} \oplus \bar{\omega}), (\bar{\bar{\mathsf{g}}} \oplus \bar{\bar{\omega}}) \in \mathsf{D}$.

A linear subspace satisfying (B.68) is called isotropic, so that a Dirac structure is required to be maximally isotropic.

Exercise 213 Reformulation of Defining Properties for Dirac Structures
Show that (B.68) and (B.70) can be rewritten as

$$\bar{\omega}(\mathsf{g}) + \omega(\bar{\mathsf{g}}) = 0, \tag{B.71}$$

and

$$d_{\mathsf{g}} \bar{\omega}(\bar{\bar{\mathsf{g}}}) + d_{\bar{\mathsf{g}}} \bar{\bar{\omega}}(\mathsf{g}) + d_{\bar{\bar{\mathsf{g}}}} \omega(\bar{\mathsf{g}}) + \omega([\bar{\mathsf{g}}, \bar{\bar{\mathsf{g}}}]) + \bar{\omega}([\bar{\bar{\mathsf{g}}}, \mathsf{g}]) + \bar{\bar{\omega}}([\mathsf{g}, \bar{\mathsf{g}}]) = 0, \tag{B.72}$$

respectively.

To show that symplectic structures and Poisson structures lead to Dirac structures, we assume that the pairing between G and D_1 is nondegenerate, which is the case in the standard situation of symplectic or Poisson manifolds $\mathcal{P}$ with dual spaces $\mathsf{G} = \mathsf{V}(\mathcal{P})$ of vector fields and $\mathsf{D}_1 = \mathsf{V}^*(\mathcal{P})$ of one-forms. Let us consider the graph of a linear operator $\bar{\Omega} : \mathsf{G} \to \mathsf{D}_1$,

$$\mathsf{D} = \{\mathsf{g} \oplus \bar{\Omega}(\mathsf{g}) | \mathsf{g} \in \mathsf{G}\} \subseteq \mathsf{G} \oplus \mathsf{D}_1, \tag{B.73}$$

as a candidate for a Dirac structure. Equation (B.68) expresses the skew-symmetry of Ω, that is, $\bar{\Omega}(\bar{\mathsf{g}}) \cdot \mathsf{g} = -\bar{\Omega}(\mathsf{g}) \cdot \bar{\mathsf{g}}$ for all $\mathsf{g}, \bar{\mathsf{g}} \in \mathsf{G}$. For a nondegenerate pairing, the maximality shown in (B.69) is satisfied automatically. According to (B.47), the property (B.70) expresses the closedness of the two-form $\bar{\Omega}(\mathsf{g}) \cdot \bar{\mathsf{g}}$. Alternatively, one can compare (B.39) and (B.72). The characteristic property of symplectic operators is thus that their graphs are Dirac structures.

We can also consider the graph of a linear operator $L : \mathsf{D}_1 \to \mathsf{G}$,

$$\mathsf{D} = \{L(\omega) \oplus \omega | \omega \in \mathsf{D}_1\} \subseteq \mathsf{G} \oplus \mathsf{D}_1, \tag{B.74}$$

as a candidate for a Dirac structure. Equation (B.68) expresses the skew-symmetry of L and, again, for a nondegenerate pairing, the maximality (B.69) is satisfied automatically. Finally, when ω, $\bar{\omega}$ and $\bar{\bar{\omega}}$ are chosen as differentials of coordinates, $\omega = d\xi^i$, $\bar{\omega} = d\xi^j$, and $\bar{\bar{\omega}} = d\xi^k$, the property (B.70) is found to coincide with the Jacobi identity (B.58). In summary, we have thus shown that the graphs of symplectic and Poisson operators provide examples of Dirac structures.

Equations (B.73) and (B.74) are examples of representations of Dirac structures. There is a variety of further representations, which actually form an important part of the theory of Dirac structures. For example, a Dirac structure can be represented by a set of linear equations or, in other words, as the kernel of a linear operator (kernel representation).

Exercise 214 Dirac Structure for Two Systems Exchanging Heat and Volume
Consider the two systems exchanging heat and volume introduced in Exercise 3 with properly chosen mass and size of the wall (see Exercise 4). Specify a set of base vectors for the corresponding Dirac structure D. Construct a linear mapping such that the Dirac structure D is obtained as the kernel of this mapping.

Because the relationship between the differentials of Hamiltonians and Hamiltonian vector fields has been expressed by symplectic or Poisson operators, and Dirac structures are intended to characterize their graphs, we arrive at the following definition.

Definition: Given a Dirac structure D, a Hamiltonian vector field $\mathsf{g}_H \in \mathsf{G}$ with Hamiltonian $H \in \mathsf{D}_0$ is defined by $(\mathsf{g}_H \oplus dH) \in \mathsf{D}$.

We can thus use Dirac structures to introduce Hamiltonian dynamical systems. If we restrict the space of observables to be the space of Hamiltonians, $\mathsf{O}_\mathsf{D} \subset \mathsf{D}_0$, then, in agreement with (B.56), we can introduce the Poisson bracket

$$\{A, B\}_\mathsf{D} = dA(\mathsf{g}_B) = -dB(\mathsf{g}_A), \tag{B.75}$$

for all $A, B \in \mathsf{O}_\mathsf{D}$.

For Poisson structures, the Poisson operator relates $\mathsf{V}(\mathcal{P})$ to $\mathsf{V}^*(\mathcal{P})$. In this standard setting, a Dirac structure must be defined for every point of the underlying manifold $\mathcal{P}$, as already noted in the solution to Exercise 214. Dirac structures generalize such a relationship to more general pairs of Lie algebras G, to be interpreted as vector fields, and vector spaces D_1, which consist of the linear maps from G to some vector space of observables D_0. According to the abstract theory of differential forms, the only relationship assumed between G and D_1 is that the Lie algebra G acts on D_0. For Dirac structures, no underlying manifold $\mathcal{P}$ is required, and there is no need for the vector space of observables D_0 to be the space of smooth real-valued functions on a manifold, $\mathsf{F}(\mathcal{P})$. In particular, quantum systems can thus be handled.

The strength of the concept of Dirac structures becomes clear if we wish to introduce constraints on a $2n$-dimensional symplectic manifold $\mathcal{P}$ with symplectic operator $\bar{\Omega} : \mathsf{V}(\mathcal{P}) \to \mathsf{V}^*(\mathcal{P})$. We assume that k independent constraints of the form $\Pi_j(p) = c_j$, $j = 1 \dots k$, define a $(2n - k)$-dimensional manifold for any choice of the constants c_j.[11] A Dirac structure appropriate to introduce Hamiltonian dynamics on the constrained manifold can then be defined by restricting the vector field v in the Dirac representation of the symplectic structure,

$$\mathsf{D} = \{\mathsf{v} \oplus \omega | \mathsf{v} \in \mathsf{V}(\mathcal{P}), \omega = \bar{\Omega}(\mathsf{v})\}, \tag{B.76}$$

[11] For reasons given in Exercise 7, we restrict ourselves to nonholonomic constraints.

to the tangent space of the constrained manifold and, in exchange, to enrich the corresponding differential form ω by transverse differentials to maintain maximality,

$$\begin{aligned}\mathsf{D}_\mathrm{c} &= \{\mathsf{v}\oplus\omega|\mathsf{v}\in\mathsf{V}(\mathcal{P}) \text{ with } d\Pi_1(\mathsf{v}) = \ldots = d\Pi_k(\mathsf{v}) = 0,\\ &\qquad \omega = \bar{\Omega}(\mathsf{v}) + \sum_{j=1}^{k}\lambda_j d\Pi_j \text{ with } \lambda_1\ldots\lambda_k\in\mathsf{F}(\mathcal{P})\}. \end{aligned} \tag{B.77}$$

We thus have a coordinate-free representation of a structure containing only tangential vector fields. The antisymmetry (B.68) follows from $\omega(\bar{\mathsf{v}}) = \Omega(\mathsf{v},\bar{\mathsf{v}})$, and the maximum property of D_c, intuitively clear as we merely shift k dimensions from the space of vector fields to the space of differential forms, follows from the fact that

$$[\bar{\omega} - \bar{\Omega}(\bar{\mathsf{v}})]\cdot\mathsf{v} + \sum_{j=1}^{k}\lambda_j d\Pi_j(\bar{\mathsf{v}}) = 0 \tag{B.78}$$

for all tangential $\mathsf{v}\in\mathsf{V}(\mathcal{P})$ and all $\lambda_j\in\mathsf{F}(\mathcal{P})$ implies that $\bar{\mathsf{v}}\oplus\bar{\omega}$ belongs to D_c. Equation (B.72) follows from the fact that the extra contribution to ω in (B.77) vanishes for tangential vector fields and that the Jacobi-Lie bracket of two tangential vector fields is again tangential to the constrained manifold, as can be seen in (B.9) for $f = \Pi_j$.

If one wishes to incorporate constraints in terms of symplectic rather than Dirac structures, one can introduce a *constrained bracket*, also known as *Dirac bracket* [Dorfman, p. 4–5]. To construct the Dirac bracket from D_c, we consider the Hamiltonian vector field v^c_A associated with the Hamiltonian A, as implied by the Dirac structure D_c according to our general definition,

$$dA = \bar{\Omega}(\mathsf{v}^\mathrm{c}_A) + \sum_{j=1}^{k}\lambda^A_j d\Pi_j, \tag{B.79}$$

where the superscript c indicates that v^c_A is constrained to be tangential

$$d\Pi_j(\mathsf{v}^\mathrm{c}_A) = 0 \quad \text{for } j = 1\ldots k, \tag{B.80}$$

by adjusting the Lagrange multipliers λ^A_j. The vector field v^c_A is different from the unconstrained v_A given by $dA = \bar{\Omega}(\mathsf{v}_A)$. By applying $\bar{\Omega}^{-1}$ to (B.79) and contracting with any $d\Pi_l$, we obtain with (B.56), (B.60), and the condition (B.80),

$$\{A,\Pi_l\} = \sum_{j=1}^{k}\lambda^A_j\{\Pi_j,\Pi_l\}, \tag{B.81}$$

which determines λ^A_j if the matrix of Poisson brackets $\{\Pi_j,\Pi_l\}$ can be inverted, as we assume in the following. We similarly have

$$dB = \bar{\Omega}(\mathsf{v}^\mathrm{c}_B) + \sum_{j=1}^{k}\lambda^B_j d\Pi_j, \tag{B.82}$$

and

$$\{B, \Pi_l\} = \sum_{j=1}^{k} \lambda_j^B \{\Pi_j, \Pi_l\}. \tag{B.83}$$

If we now evaluate $\{A, B\}$ by means of (B.56) and (B.60), we obtain

$$\{A, B\} = \Omega(\mathsf{v}_A^{\mathrm{c}}, \mathsf{v}_B^{\mathrm{c}}) + \sum_{j,l=1}^{k} \lambda_j^A \{\Pi_j, \Pi_l\} \lambda_l^B. \tag{B.84}$$

One can now introduce $\Omega(\mathsf{v}_A^{\mathrm{c}}, \mathsf{v}_B^{\mathrm{c}})$ as the constrained or Dirac bracket of A and B. Equation (B.84), together with (B.81) and (B.83), defines the Dirac bracket in terms of unconstrained Poisson brackets [Marsden & Ratiu, p. 243]. In (B.84), it is assumed that A and B are functions on $\mathcal{P}$. Dirac showed that, if A and B are given on the constrained manifold only and extended to $\mathcal{P}$, the definition of the Dirac bracket $\Omega(\mathsf{v}_A^{\mathrm{c}}, \mathsf{v}_B^{\mathrm{c}})$ is independent of the choice of the extension. This result reflects the fact that the structure D_{c} in (B.77) is fully determined by the constraint equations.

Dirac's approach is restricted to holonomic constraints. For a wide class of constraints, it has been shown that the Jacobi identity for the bracket in a Hamiltonian formulation of the constrained dynamics is valid if, and only if, the constraints are holonomic.[12]

B.3 REDUCTION OF GEOMETRIC STRUCTURES

For obtaining a variety of realizations of geometric structures, it is important to construct new structures from given ones. The goal of reduction is to provide a general tool for obtaining new structures on smaller spaces, typically with the help of symmetry groups. We here focus on *Poisson reduction*, but we also mention *symplectic reduction*.[13]

In most general terms, reduction of a geometric structure on a manifold $\mathcal{P}$ is based on partitioning the manifold into disjoint subsets, which can be regarded as equivalence classes. The partition must be such that a manifold of equivalence classes arises, that is, that the classes can be parametrized smoothly. The manifold of equivalence classes is the reduced manifold $\mathcal{P}_{\mathrm{red}}$ on which we want to introduce a geometric structure. The physical observables on $\mathcal{P}_{\mathrm{red}}$ correspond to those real-valued functions on $\mathcal{P}$ that are constant on equivalence classes, so that they can be considered as functions on $\mathcal{P}_{\mathrm{red}}$. If, for example, the Poisson bracket of any two observables on $\mathcal{P}$, which are constant on equivalence classes, is also constant on equivalence classes, then we obtain a corresponding Poisson bracket for the observables on $\mathcal{P}_{\mathrm{red}}$. If, moreover, the Hamiltonian is also constant on equivalence classes, then we can define a Hamiltonian for the reduced setting. Under these conditions, the time evolution of observables that

[12] van der Schaft & Maschke, Rep. Math. Phys. 34 (1994) 225.

[13] See Section 2.8 of Marsden, *Lectures on Mechanics* (Cambridge, 1992).

are constant on equivalence classes is rigorously reproduced on the reduced manifold with the reduced geometric structure and Hamiltonian. The possibility of fulfilling these conditions is usually related to symmetries which leave a geometric structure and a Hamiltonian invariant.

For a more concrete construction, we start with a Poisson manifold $\mathcal{P}$ on which a Lie group $\mathcal{G}$ acts by canonical transformations; that is, the action leaves the Poisson bracket on $\mathcal{P}$ invariant. We then decompose $\mathcal{P}$ into equivalence classes, where two elements are equivalent if they are mapped into each other by the action of some group element. We assume that we can select a representative of each equivalence class in such a way that $\mathcal{P}_{\text{red}} = \mathcal{P}/\mathcal{G}$ is a manifold, on which we can define a reduced Poisson structure. For defining a Poisson bracket on $\mathcal{P}_{\text{red}}$, consider any two $A_{\text{red}}, B_{\text{red}} \in \mathsf{F}(\mathcal{P}_{\text{red}})$. Construct $A, B \in \mathsf{F}(\mathcal{P})$ from $A_{\text{red}}, B_{\text{red}}$ by mapping $p \in \mathcal{P}$ to the value of $A_{\text{red}}, B_{\text{red}}$ on the equivalence class of p. The function $C = \{A, B\}$ is constant on equivalence because the bracket is invariant under group actions, and A, B are constant on equivalence classes. We thus obtain a well-defined Poisson bracket $\{A_{\text{red}}, B_{\text{red}}\}_{\text{red}}$ on $\mathcal{P}_{\text{red}}$, which inherits all the defining properties from the Poisson bracket on $\mathcal{P}$. This construction of a Poisson structure is known as the *Poisson Reduction Theorem* [Marsden & Ratiu, p. 349].

If a Hamiltonian H on $\mathcal{P}$ is invariant under $\mathcal{G}$, it must be constant on equivalence classes so that it defines a function H_{red} on $\mathcal{P}_{\text{red}}$. The Hamiltonian system given by H on $\mathcal{P}$ then reduces to that given by H_{red} on $\mathcal{P}_{\text{red}}$. The invariance of the Hamiltonian is indispensable for establishing the smaller space $\mathcal{P}_{\text{red}}$ as an appropriate setting for reversible dynamics.

If we start from a symplectic manifold $\mathcal{P}$ on which a Lie group $\mathcal{G}$ acts canonically, that is, the action leaves the symplectic form on $\mathcal{P}$ invariant, we could try the above construction for Poisson manifolds because there exists a Poisson structure on any symplectic manifold. However, a degeneracy problem arises so that one needs to select a symplectic leaf. To that end, one can identify the Casimir functions of the reduced Poisson bracket, which are related to the Hamiltonians corresponding to infinitesimal actions of the group. The symplectic leaves can be labeled by the elements of the dual of the Lie algebra, $\mathsf{g}^* \in \mathsf{G}^*$ (the explicit construction can be done in terms of momentum maps to be introduced below). We thus obtain a symplectic leaf $\mathcal{P}_{\text{red}}(\mathsf{g}^*)$ for each $\mathsf{g}^* \in \mathsf{G}^*$ as the arena for symplectic reduction.

The most natural starting point for reduction is the canonical symplectic structure available on any cotangent bundle. If a group acts on a manifold $\mathcal{P}$, the action can be lifted to the cotangent bundle $T^*\mathcal{P}$ such that it leaves the symplectic form on $T^*\mathcal{P}$ invariant. In canonical coordinates, the idea is to extend the action of some group element, $\xi^j \mapsto \bar{\xi}^j$, by an *extended point transformation* to

$$\xi^j \mapsto \bar{\xi}^j, \quad \xi_j^* \mapsto \bar{\xi}_j^* = \frac{\partial \xi^k}{\partial \bar{\xi}^j} \xi_k^*, \tag{B.85}$$

because we then obtain

$$d\bar{\xi}^j \wedge d\bar{\xi}_j^* = \frac{\partial \bar{\xi}^j}{\partial \xi^k} \frac{\partial \xi^l}{\partial \bar{\xi}^j} d\xi^k \wedge d\xi_l^* = \delta_k^l d\xi^k \wedge d\xi_l^* = d\xi^j \wedge d\xi_j^*, \tag{B.86}$$

so that such an extended point transformation leaves the symplectic form invariant.

Exercise 215 Canonical Extended Point Transformation
The dependence of $\bar{\xi}^*_j$ on ξ^j in addition to ξ^*_k has been neglected in the transformation (B.86). Show that the resulting term indeed vanishes.

In particular, the action of a Lie group $\mathcal{G}$ on itself by left or right translation can be lifted to $T^*\mathcal{G}$. If we apply the Poisson Reduction Theorem to $\mathcal{P} = T^*\mathcal{G}$ with the Poisson bracket implied by the canonical symplectic form, then $\mathcal{P}_{\rm red} = T^*\mathcal{G}/\mathcal{G}$ can naturally be identified with the cotangent space at the group identity, which is the dual of the Lie algebra, $\mathcal{P}_{\rm red} = \mathsf{G}^*$. More precisely, the equivalence class of $(\bar{g}, \bar{\mathsf{v}}^*) \in T^*\mathcal{G}$ can be characterized by $\mathsf{g}^* \in \mathsf{G}^*$ according to

$$\mathsf{g}^*(\mathsf{g}) = \bar{\mathsf{v}}^* \left(\frac{d}{dt} g(t)\bar{g} \bigg|_{t=0} \right), \tag{B.87}$$

where $\mathsf{g} \in \mathsf{G}$ is the dummy argument of the form g^*, and $g(t)$ is a curve through the group identity representing the vector g. Equation (B.87) corresponds to a parallel transport associated with the group action. We here choose right translations by $\bar{g}$ in (B.87) rather than left translations because this situation arises more naturally in describing fluid motion.

By means of (B.87), functions $A_{\rm red}, B_{\rm red}$ of g^* can be used to define functions A, B of $(\bar{g}, \bar{\mathsf{v}}^*)$, that is, on $T^*\mathcal{G}$, where we have a canonical Poisson bracket. To construct the reduced Poisson bracket, we start from a coordinate-free version of (B.61),

$$\{A, B\} = \frac{\partial A}{\partial \bar{g}} \cdot \frac{\partial B}{\partial \bar{\mathsf{v}}^*} - \frac{\partial A}{\partial \bar{\mathsf{v}}^*} \cdot \frac{\partial B}{\partial \bar{g}}, \tag{B.88}$$

so that we need the derivatives of A and B with respect to $\bar{g}$ and $\bar{\mathsf{v}}^*$. We assume that the curves $g_A(t)$ and $g_B(t')$ represent the vectors $\partial A_{\rm red}/\partial \mathsf{g}^*$ and $\partial B_{\rm red}/\partial \mathsf{g}^*$. Then the chain rule and (B.87) lead to

$$\frac{\partial B}{\partial \bar{\mathsf{v}}^*} = \frac{\partial \mathsf{g}^*}{\partial \bar{\mathsf{v}}^*} \left(\frac{\partial B_{\rm red}}{\partial \mathsf{g}^*} \right) = \frac{d}{dt'} g_B(t')\bar{g} \bigg|_{t'=0}, \tag{B.89}$$

which is an element of the tangent vector space in the point $\bar{g}$, as it should be. The change of g^* resulting from a small change of $\bar{g}$ along a curve from $\bar{g}(0)$ to $\bar{g}(\delta t)$ in (B.87) leads to

$$\delta A = \delta \mathsf{g}^* \left(\frac{\partial A_{\rm red}}{\partial \mathsf{g}^*} \right) = \bar{\mathsf{v}}^* \left(\frac{d}{dt} g_A(t)\bar{g}(\delta t) \bigg|_{t=0} \right) - \bar{\mathsf{v}}^* \left(\frac{d}{dt} g_A(t)\bar{g}(0) \bigg|_{t=0} \right), \tag{B.90}$$

and hence, together with (B.89),

$$\frac{\partial A}{\partial \bar{g}} \cdot \frac{\partial B}{\partial \bar{\mathsf{v}}^*} = \frac{1}{\delta t} \left[\bar{\mathsf{v}}^* \left(\frac{d}{dt} g_A(t) g_B(\delta t)\bar{g} \bigg|_{t=0} \right) - \bar{\mathsf{v}}^* \left(\frac{d}{dt} g_A(t) g_B(0)\bar{g} \bigg|_{t=0} \right) \right], \tag{B.91}$$

where the limit $\delta t \to 0$ is to be taken. Formally, we then obtain by inserting into (B.88)

$$\{A,B\} = \bar{\mathsf{v}}^* \left(\frac{d}{dt}\frac{d}{dt'} g_A(t) g_B(t') \bar{g} \bigg|_{t=t'=0} - \frac{d}{dt}\frac{d}{dt'} g_B(t') g_A(t) \bar{g} \bigg|_{t=t'=0} \right), \tag{B.92}$$

which can be reformulated more rigorously in terms of the Lie algebra bracket of $\partial A_{\rm red}/\partial \mathsf{g}^*$ and $\partial B_{\rm red}/\partial \mathsf{g}^*$, represented by $g_{AB}(t)$,

$$\{A_{\rm red}, B_{\rm red}\}_{\rm red} = \{A,B\} = \bar{\mathsf{v}}^* \left(\frac{d}{dt} g_{AB}(t) \bar{g} \bigg|_{t=0} \right) = \mathsf{g}^* \left(\left[\frac{\partial A_{\rm red}}{\partial \mathsf{g}^*}, \frac{\partial B_{\rm red}}{\partial \mathsf{g}^*} \right] \right), \tag{B.93}$$

where, once more, (B.87) has been used. The most direct way to verify the steps from (B.91) to (B.93) is by means of a matrix representation of the group, as discussed after (B.11), and to note that the initial and final results do not make any use of that representation. If we had used left instead of right translations for our construction, we would have obtained a negative sign in (B.93). For simplicity, we rewrite our abstract general result (B.93) in the following theorem in terms of coordinates.

Lie-Poisson Reduction Theorem: If the Lie algebra G of a Lie group $\mathcal{G}$ has the bracket

$$[\mathsf{g}, \bar{\mathsf{g}}]^j = \mathsf{C}^j_{kl} \mathsf{g}^k \bar{\mathsf{g}}^l, \tag{B.94}$$

for all $\mathsf{g}, \bar{\mathsf{g}} \in \mathsf{G}$, then

$$\{A,B\} = \mathsf{C}^j_{kl} \mathsf{g}^*_j \frac{\partial A}{\partial \mathsf{g}^*_k} \frac{\partial B}{\partial \mathsf{g}^*_l} \tag{B.95}$$

defines a Poisson bracket on the dual of the Lie algebra G^*.

Note that the Lie-Poisson bracket (B.95) is linear in g^* in the sense that the corresponding Poisson operator, given by $L_{kl} = \mathsf{C}^j_{kl} \mathsf{g}^*_j$, is linear in g^*. Conversely, all such linear brackets can be obtained as Lie-Poisson brackets [Marsden & Ratiu, p. 364].

Corollary for Extending Poisson Brackets by Lie Poisson Reduction: Assume that the Poisson bracket $\{A,B\}$ on some configuration space with elements x is the Lie-Poisson bracket associated with a Lie group $\mathcal{G}$ [that is, the configuration space is a vector space or, more precisely, the dual G^* of the Lie algebra of $\mathcal{G}$]. A Lie-Poisson bracket on an extended configuration space of elements $(x, \bar{x})$, where $\bar{x}$ is from a vector space V^* which we consider as the dual of a vector space V, can be constructed as follows. Given a right action or representation[14] of the Lie group $\mathcal{G}$ on the vector space V, we have a linear operator R_{g} as a representation of any vector g from the Lie algebra G, and

$$\{A,B\}_{\rm extended} = \{A,B\} + \bar{x} \cdot \left\{ R_{\frac{\partial B}{\partial x}} \left(\frac{\partial A}{\partial \bar{x}} \right) - R_{\frac{\partial A}{\partial x}} \left(\frac{\partial B}{\partial \bar{x}} \right) \right\} \tag{B.96}$$

[14] Actions by linear transformations on Banach spaces, that is, vector spaces with norms, are called representations.

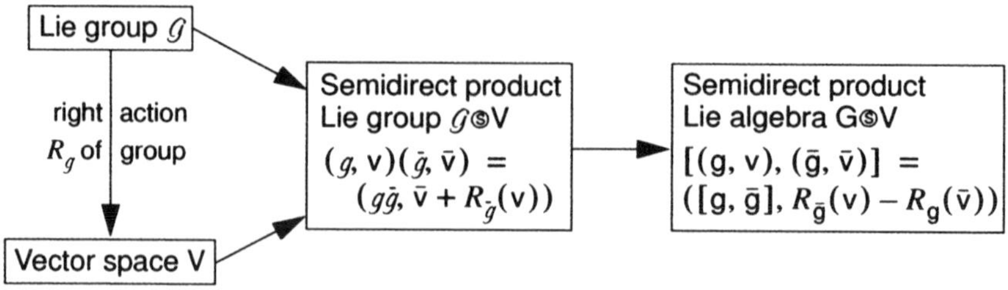

Fig. B.2 The concept of semidirect products associated with right actions.

is a Lie-Poisson bracket on the extended state space. The dot in (B.96) indicates the canonical pairing between the vector space V and its dual V^*.

This corollary follows by applying the Lie-Poisson Reduction Theorem to the semidirect product of $\mathcal{G}$ and V, which is a Lie group [Marsden & Ratiu, p. 325; the concept of semidirect products is illustrated in Figure B.2]. It is of fundamental importance for constructing Poisson brackets in complex fluid dynamics (see Section B.4). In that context, the underlying Lie group is the group of space transformations, with the Lie algebra of velocity fields and the configuration space of dual velocity fields. This configuration space is extended by two hydrodynamic scalar density fields, say mass and internal energy density and, for complex fluids, a number of further configurational fields. One only needs to construct a natural action of the group of space transformations on the additional variables in order to arrive at a Poisson bracket. In many cases, the elegant concepts of material entrainment and body tensors are most useful to find natural actions (see Section B.4); in other cases, natural actions can be found by coarse-graining of atomistic systems (see Section 6.1.6).

An alternative approach to reduction can be based on the concept of momentum maps.[15] This approach puts Lie-Poisson brackets into a wider context, and it provides further insight into the nature of Casimir functions. If a Lie group $\mathcal{G}$ acts on a Poisson manifold $\mathcal{P}$ by canonical transformations, then the elements of the Lie algebra G, as infinitesimal generators of the action of the group $\mathcal{G}$ on $\mathcal{P}$, define vector fields on $\mathcal{P}$. The problem that arises naturally in this situation is, for a given infinitesimal generator $\mathsf{g} \in \mathsf{G}$, to find a "Hamiltonian" H_g that generates the dynamics associated with that vector field, that is, we are looking for a map $\tilde{J} : \mathsf{G} \to \mathsf{F}(\mathcal{P}), \mathsf{g} \mapsto H_\mathsf{g}$. As the real-valued functional H_g on $\mathcal{P}$ is expected to be linear in g, it must be of the form

$$H_\mathsf{g} = J \cdot \mathsf{g}. \tag{B.97}$$

This observation leads us to the concept of a *momentum map* associated with the Poisson manifold $\mathcal{P}$ on which the Lie group $\mathcal{G}$ acts. To have a natural pairing indicated by the "·" in (B.97), J must take values in the dual of the Lie algebra, $J : \mathcal{P} \to$

[15] The remainder of this section is not required for continuing with Section B.4.

G^*. As J corresponds to momentum for infinitesimal translations, and to angular momentum for infinitesimal rotations, the abstract construction of J according to (B.97) is generally referred to as the momentum map [Marsden & Ratiu, p. 367]. We are interested in infinitesimally equivariant momentum maps for which the associated map $\tilde{J}$ satisfies the condition

$$\tilde{J}([\mathsf{g}, \bar{\mathsf{g}}]) = \{\tilde{J}(\mathsf{g}), \tilde{J}(\bar{\mathsf{g}})\} \qquad \text{for all } \mathsf{g}, \bar{\mathsf{g}} \in \mathsf{G}. \tag{B.98}$$

This condition should be compared with (B.65). Whereas (B.98) is formulated as a condition for Hamiltonians associated with vector fields, (B.65) is the corresponding condition for vector fields associated with Hamiltonians; the different sign conventions have already been pointed out after (B.65).

Noether's Theorem: If a Lie group $\mathcal{G}$ acts by canonical transformations on a Poisson manifold and admits a momentum map J, and if the Hamiltonian H is invariant under $\mathcal{G}$ (that is, $\mathcal{G}$ is a symmetry group of H), then all components of the covector-valued momentum map J are constants of motion for the dynamics generated by H [Marsden & Ratiu, p. 372].

This important result can be established in a few lines. The existence of a momentum map implies that $L(dH_{\mathsf{g}})$ with the Hamiltonian (B.97) describes the action of any $\mathsf{g} \in \mathsf{G}$ on $\mathcal{P}$. This action leaves H invariant, that is, $dH \cdot L(dH_{\mathsf{g}}) = \{H, H_{\mathsf{g}}\} = 0$. By expressing H_{g} in terms of the momentum map, we obtain $\{H, J\} \cdot \mathsf{g} = 0$ for arbitrary $\mathsf{g} \in \mathsf{G}$, and hence $\{J, H\} = 0$, which is Noether's theorem.

Our typical construction of a space suitable to describe complex fluids starts from the Lie group $\mathcal{G}$ of space transformations, with the Lie algebra G of velocity fields, and with the dual Lie algebra G^* of dual velocity fields, naturally interpreted as the vector space of momentum density fields. We then consider the infinite-dimensional vector space G^* supplemented by a vector space V^* of further fields, where $\mathcal{G}$ acts by canonical transformations on the dual V of V^*, equipped with the Lie-Poisson bracket of the semidirect product. For hydrodynamics, the fields in V^* are the mass density and the internal energy or entropy density; for complex fluids, we have further fields in V^*, such as a conformation tensor field. Note that, in general, the Lie group $\tilde{\mathcal{G}}$ of Noether's theorem does not coincide with the group $\mathcal{G}$ of space transformations, and that $\tilde{\mathcal{G}}$ acts on the Poisson manifold $\mathcal{P} = \mathsf{G}^* \times \mathsf{V}^*$ equipped with the Lie-Poisson bracket, which we use as the configuration space for modeling Newtonian or complex fluids. For example, $\tilde{\mathcal{G}}$ could be the group of Galilean transformations. If we would like to choose $\tilde{\mathcal{G}} = \mathcal{G}$, we need to find a canonical action of space transformations on the Poisson manifold $\mathcal{P} = \mathsf{G}^* \times \mathsf{V}^*$. For the construction of the Lie-Poisson bracket on our configuration space $\mathcal{P} = \mathsf{G}^* \times \mathsf{V}^*$, we have already used a right action of $\mathcal{G}$ on the vector space V, from which we naturally obtain a right action of $\mathcal{G}$ on V^* by postulating that the natural pairing is invariant under the simultaneous action of a group element on V and V^* (this means, the two representations are related by going to the adjoint operators for the inverse group elements). The natural right action of $\mathcal{G}$ on itself is given by conjugation, that is,

$$\bar{g} \mapsto g^{-1}\bar{g}g \quad \text{for all } \bar{g} \in \mathcal{G}. \tag{B.99}$$

By differentiating conjugation at the group identity, we obtain the corresponding adjoint action on the Lie algebra G (according to (B.11), the infinitesimal action of g on G is then given by $\bar{\mathsf{g}} \mapsto -[\mathsf{g}, \bar{\mathsf{g}}]$ for all $\bar{\mathsf{g}} \in \mathsf{G}$), and, by keeping the natural pairing invariant, we find the *coadjoint action* of $\mathcal{G}$ on G^* [Marsden & Ratiu, p. 311 and p. 375]. After identifying the natural action of the group of space transformations $\mathcal{G}$ on all components of our configuration space $\mathcal{P} = \mathsf{G}^* \times \mathsf{V}^*$, which has previously been studied in detail (but on a more intuitive level) to construct Poisson brackets,[16] one can verify that the momentum map is given by a simple projection,

$$J : \mathsf{G}^* \times \mathsf{V}^* \to \mathsf{G}^*, \; (\mathsf{g}^*, \mathsf{v}^*) \mapsto \mathsf{g}^*. \tag{B.100}$$

The corresponding map $\tilde{J}(\mathsf{g})$ is given by

$$\tilde{J}(\mathsf{g}) : \mathcal{P} \to \mathbb{R}, (\mathsf{g}^*, \mathsf{v}^*) \mapsto \mathsf{g}^* \cdot \mathsf{g} = \mathsf{g}^*(\mathsf{g}), \tag{B.101}$$

so that, in view of (B.93), this momentum map has the equivariance property (B.98). The simplicity of this momentum map indicates that the essence of convection has been built into the construction of the configuration space $\mathcal{P} = \mathsf{G}^* \times \mathsf{V}^*$. Or, in other words, this simplicity expresses the relevance of momentum maps in the reduction process.

We obtained the Lie-Poisson Reduction Theorem by reducing the canonical Poisson bracket on the symplectic manifold $T^*\mathcal{G}$ to $T^*\mathcal{G}/\mathcal{G} = \mathsf{G}^*$; momentum maps can reduce Poisson brackets from general Poisson manifolds to the Lie-Poisson bracket on G^*. Under quite general conditions, such a relationship is provided by a *Theorem on Canonical Momentum Maps* [Marsden & Ratiu, p. 403].

Momentum Maps and Casimir Functions: We consider a Lie group $\mathcal{G}$ acting on a Poisson manifold $\mathcal{P}$ and admitting an equivariant momentum map J. Let $C : \mathsf{G}^* \to \mathbb{R}$ be a function that is invariant under the coadjoint action. Then, C is a Casimir function of the Lie-Poisson bracket. Furthermore, the function $C \circ J : \mathcal{P} \to \mathbb{R}$ is constant on the equivalence classes of $\mathcal{P}/\mathcal{G}$, and the corresponding function C_{red} : $\mathcal{P}/\mathcal{G} \to \mathbb{R}$ is a Casimir function of the reduced bracket on $\mathcal{P}/\mathcal{G}$.

B.4 BODY TENSORS

In the description of fluid motion in space, the velocity field plays a central role. The velocity fields can be considered as the Lie algebra of the group of space transformations, which can also be regarded as the possible embeddings of a material into space. In constructing Poisson brackets for modeling complex fluids, we can hence follow two different philosophies:

- *Focus on space and Lie algebra* (Eulerian approach): We select the space fields required in addition to the velocity field such that a consistent description of

[16] See Sections II.D and III.B of Öttinger & Grmela, Phys. Rev. E 56 (1997) 6633.

the complex fluid is expected to be possible. We then look for an action of the group of space transformations on the additional fields, where the action of a space transformation can only depend on the total deformation, not on how it is realized in time or broken up into small steps. The proper formulation of the action is left to intuition, and it directly leads to a valid Poisson bracket. The energy is formulated independently in terms of the velocity field and the additional fields.

- *Focus on material and Lie group* (Lagrangian approach): We model a complex fluid on the level of the group of space transformations describing all possible embeddings of a material into space. In such a Lagrangian approach, we introduce structural variables defined throughout the material body. These variables determine the material properties and must hence be invariant under particle relabeling. This particle relabeling symmetry is then used to reduce the material description to the Lie algebra level of the space description. In particular, the Hamiltonian on the Lagrangian level should depend only on the structural variables and hence possess the particle relabeling symmetry, so that the reversible dynamics of the Lagrangian description can be reduced to the Eulerian description.

Whereas the first approach has been used in our original work on the GENERIC framework,[17] we here follow the second approach, which is based on the concept of body tensors.

The body tensor formalism has been developed by Arthur Lodge[18] as a powerful tool to formulate the rheological behavior of simple or complex fluids in terms of manifestly intrinsic properties of a material. In that formalism, a fluid is treated as a continuous body that fills a region of space at each instant of time. The *body manifold* $\mathcal{P}$ is a manifold of so-called *particles* p corresponding to the fluid elements. The formalism allows one to formulate constitutive equations that are admissible, that is, independent of any coordinate system, of position in space, and of the translational or rotational motions of a fluid element. Such admissibility criteria are known as the principle of *frame indifference* or *material objectivity*. In this section, we show that the body tensor formalism for introducing structural variables is deeply related to symmetry and the geometry behind Hamiltonian systems.

We first consider the fluid motion from the material or Lagrangian perspective. We introduce coordinates (ξ^1, ξ^2, ξ^3) in the body manifold and coordinates (r^1, r^2, r^3) in space. The motion of a deforming body in space can then be specified through smooth functions,

$$r^j = g^j(\xi^1, \xi^2, \xi^3, t), \qquad j = 1, 2, 3, \tag{B.102}$$

assigning positions in space to all the particles of the body, where we exclude discontinuities, cavitation, and fluid interpenetration.

[17] Öttinger & Grmela, Phys. Rev. E 56 (1997) 6633.

[18] Lodge, *Body Tensor Fields* (Academic, 1974).

To describe complex fluids, we wish to introduce further variables to describe the internal structure of the body (see Section 1.1.3). These variables are most naturally chosen as body tensor fields of a certain co- and contravariant rank. To be specific, we consider a second-rank covariant body tensor field and a scalar density field. When the general definition (B.1) of tensors is expressed in terms of the coordinates ξ^j, a second-rank body tensor is given by its components b_{jk}. According to (B.4), the characteristic of a covariant tensor is the transformation rule

$$\bar{b}_{jk} = \frac{\partial \xi^l}{\partial \bar{\xi}^j} \frac{\partial \xi^n}{\partial \bar{\xi}^k} b_{ln}, \tag{B.103}$$

for the tensor components $\bar{b}_{jk}$ in an alternative body coordinate system $\bar{\xi}^j$. The transformation rule for a scalar density field s is given by

$$\bar{s} = \left| \det \frac{\partial \xi}{\partial \bar{\xi}} \right| s, \tag{B.104}$$

where the change in the density through a change in the volume has been taken into account. In the new coordinate system, we obtain $\boldsymbol{r}$ as

$$\boldsymbol{r} = g(\xi, t) = \bar{g}(\bar{\xi}, t) = \bar{g}(f(\xi), t), \tag{B.105}$$

or

$$g_t = \bar{g}_t \circ f, \quad \dot{g}_t = \dot{\bar{g}}_t \circ f, \quad g_t^{-1} = f^{-1} \circ \bar{g}_t^{-1}, \quad f \circ g_t^{-1} = \bar{g}_t^{-1}, \tag{B.106}$$

where the relationship $\bar{\xi} = f(\xi)$ between the body coordinates ξ and $\bar{\xi}$ has been used. We thus have all the transformation rules for particle relabeling.

To construct Poisson brackets by means of the Lie-Poisson Reduction Theorem, we need to identify the Lie group behind our Lagrangian description of fluid motion. As both the deformation of the body and the particle relabeling are described by smooth transformations in $\mathbb{R}^3$, it is natural to consider the group of such transformations. The fact that (B.102) leads from body to space coordinates, whereas f relates different body coordinates, does not cause any problems because these different types of coordinate systems can be identified as, at any time t, there is a smooth one-to-one correspondence between the particles and the points which they occupy. According to (B.106), particle relabeling corresponds to a right translation for the composition of transformations. In short, we have identified transformations in $\mathbb{R}^3$ as the relevant Lie group to describe fluid motion. Because the labeling of the particles must be irrelevant to all physical properties, in particular it should not affect the energy of the body, we expect the *particle relabeling symmetry* to be the key to successful reduction.

The embedding of the body into space leads to the existence of a metric tensor in the body manifold. In particular, we can use Cartesian space coordinates to label the particles, and such a distinguished coordinate system introduces a connection in the body manifold. The natural derivative associated with this connection according to (B.6) is known as the *covariant derivative*, which is characterized by the condition

$$\gamma_{kl;i} = 0 \tag{B.107}$$

for the components of the metric tensor. Equation (B.107) implies an explicit expression for the *Christoffel symbols* Γ^j_{kl} defined in (B.7) in terms of the components of the metric tensor in any coordinate system [see (5.159)],

$$\Gamma^j_{kl} = \frac{1}{2}\gamma^{ij}\left(\frac{\partial \gamma_{il}}{\partial \xi^k} + \frac{\partial \gamma_{ki}}{\partial \xi^l} - \frac{\partial \gamma_{kl}}{\partial \xi^i}\right), \tag{B.108}$$

where γ^{ij} denotes the components of the inverse of the metric tensor.

Exercise 216 Derivation of Christoffel symbols
Derive (B.108).

We can now develop a spatial or Eulerian description of fluid motion. The velocity of a particle labeled by ξ is obtained as the partial derivative of $g(\xi, t)$ with respect to time,

$$\boldsymbol{v}_\xi(t) = \frac{\partial g(\xi, t)}{\partial t} = \dot{g}_t(\xi), \tag{B.109}$$

and a velocity field can then be defined at each point $\boldsymbol{r}$ in space and at any time t by identifying the particle $g_t^{-1}(\boldsymbol{r})$ occupying that point at time t,

$$\boldsymbol{v}(\boldsymbol{r}, t) = \dot{g}_t(g_t^{-1}(\boldsymbol{r})). \tag{B.110}$$

Equation (B.106) implies

$$\boldsymbol{v}(\boldsymbol{r}, t) = \dot{g}_t(g_t^{-1}(\boldsymbol{r})) = \dot{\bar{g}}_t(f(g_t^{-1}(\boldsymbol{r}))) = \dot{\bar{g}}_t(\bar{g}_t^{-1}(\boldsymbol{r})), \tag{B.111}$$

so that the velocity field is independent of the chosen body coordinate system. We have thus verified the expected particle relabeling symmetry of the velocity field. If we label the particles of the body by the positions they occupy at $t = 0$, such that g_0 is the identity, the velocity fields (B.110) can be identified with the Lie algebra of the transformation group in $\mathbb{R}^3$. The momentum density fields, $\boldsymbol{M}(\boldsymbol{r}) = \delta E/\delta \boldsymbol{v}(\boldsymbol{r})$, then form the dual of the Lie algebra, so that the situation of the Lie-Poisson Reduction Theorem arises most naturally.

According to (B.106), equivalent transformations differ by right translations associated with particle relabeling. The Lie-Poisson Reduction Theorem in the form (B.93) thus gives the following reduced bracket on the dual of the Lie algebra,

$$\{A, B\} = \int \boldsymbol{M}(\boldsymbol{r}) \cdot \left[\frac{\delta A}{\delta \boldsymbol{M}(\boldsymbol{r})}, \frac{\delta B}{\delta \boldsymbol{M}(\boldsymbol{r})}\right] d^3 r, \tag{B.112}$$

where the bracket corresponds to the Lie algebra associated with the group of transformations of $\mathbb{R}^3$. This Lie algebra consists of vector fields, and, for the most common sign conventions chosen here, the corresponding Lie algebra bracket is the negative of the Jacobi-Lie bracket (B.10) [Marsden & Ratiu, p. 20; see also Exercise 217]. We thus obtain

$$\{A, B\} = -\int M_k \left(\frac{\delta A}{\delta M_j}\frac{\partial}{\partial r^j}\frac{\delta B}{\delta M_k} - \frac{\delta B}{\delta M_j}\frac{\partial}{\partial r^j}\frac{\delta A}{\delta M_k}\right) d^3 r, \tag{B.113}$$

which coincides with (2.86).

Exercise 217 Lie Algebra of Space Transformations
Show that the Lie bracket associated with the group of transformations of $\mathbb{R}^3$ is given by the negative of the Jacobi-Lie bracket (B.10).

As a next step, we introduce the space fields $\boldsymbol{c}(\boldsymbol{r},t)$ and $s(\boldsymbol{r},t)$ associated with the covariant body tensor field (B.103) and the scalar density field (B.104) as additional structural variables. To that end, we again identify the particle ξ that is at the position $\boldsymbol{r}$ at time t in an arbitrary body coordinate system, and we introduce special body coordinates $\bar{\xi}^j$ by labeling the particles with the coordinates r^j of the points that they occupy at time t (rather than 0). We then obtain from (B.103) and (B.104)

$$c_{jk}(\boldsymbol{r},t) = \frac{\partial \xi^l}{\partial r^j}\frac{\partial \xi^n}{\partial r^k} b_{ln}(\xi), \qquad \text{with } \xi = g_t^{-1}(\boldsymbol{r}) \tag{B.114}$$

and

$$s(\boldsymbol{r},t) = \left|\det \frac{\partial \xi}{\partial \boldsymbol{r}}\right| s(\xi), \qquad \text{with } \xi = g_t^{-1}(\boldsymbol{r}). \tag{B.115}$$

In view of (B.103) and (B.104), these fields are invariant under particle relabeling. According to (B.106), $g_t = \bar{g}_t \circ f$ describes particle relabeling, but we can also consider the reverse composition $g_t^f = f \circ g_t$ as the action associated with the transformation f. With $\xi = g_t^{-1}(\boldsymbol{r}) = (g_t^f)^{-1}(f(\boldsymbol{r}))$, the effect on the fields (B.114) and (B.115) is

$$\frac{\partial f^l(\boldsymbol{r})}{\partial r^j}\frac{\partial f^n(\boldsymbol{r})}{\partial r^k} c_{ln}^f(f(\boldsymbol{r}),t) = c_{jk}(\boldsymbol{r},t) \tag{B.116}$$

and

$$\left|\det \frac{\partial f(\boldsymbol{r})}{\partial \boldsymbol{r}}\right| s^f(f(\boldsymbol{r}),t) = s(\boldsymbol{r},t). \tag{B.117}$$

By looking at infinitesimal transformations of the type $f(\boldsymbol{r}) = \boldsymbol{r} + \boldsymbol{v}(\boldsymbol{r})$, we obtain the infinitesimal actions

$$(R_{\boldsymbol{v}}(\boldsymbol{c}))_{jk} = -v^l \frac{\partial c_{jk}}{\partial r^l} - \frac{\partial v^l}{\partial r^j} c_{lk} - c_{jl}\frac{\partial v^l}{\partial r^k} \tag{B.118}$$

and

$$R_{\boldsymbol{v}}(s) = -\frac{\partial}{\partial \boldsymbol{r}} \cdot (\boldsymbol{v}s). \tag{B.119}$$

According to (B.116), the space transformation f implies how tensors at different points in space are related. We thus encounter the concept of a connection implied by the action of a group, and it is naturally associated with a concept of derivative. Because the corresponding Lie algebra is that of vector fields, the representation that we have constructed is nothing but the Lie derivative.

An explicit calculation for the actions in (B.118) and (B.119) shows

$$R_{\boldsymbol{v}} R_{\bar{\boldsymbol{v}}} - R_{\bar{\boldsymbol{v}}} R_{\boldsymbol{v}} = R_{-[\boldsymbol{v},\bar{\boldsymbol{v}}]}, \tag{B.120}$$

where $[\boldsymbol{v}, \bar{\boldsymbol{v}}]$ is the Jacobi-Lie bracket (B.9) for vector fields. As pointed out before, the Jacobi-Lie bracket is the negative of the Lie algebra bracket associated with the group of transformations of $\mathbb{R}^3$. Therefore, (B.120) implies that we have identified the natural left action on the structural variables; this should be clear because it is based on left multiplication, $g_t^f = f \circ g_t$. In view of the above-mentioned connection with Lie derivatives, (B.120) is a special case of (B.37). To apply the corollary for extending Poisson brackets by Lie Poisson reduction in the form given above, we need the corresponding right action of the Lie algebra on the duals c^{*jk}, s^* of our structural variables c_{jk}, s, which is given by

$$(R_{\boldsymbol{v}}^*(\boldsymbol{c}^*))^{jk} = \frac{\partial}{\partial r^l}(v^l c^{*jk}) - \frac{\partial v^j}{\partial r^l} c^{*lk} - c^{*jl} \frac{\partial v^k}{\partial r^l} \tag{B.121}$$

and

$$R_{\boldsymbol{v}}^*(s^*) = \boldsymbol{v} \cdot \frac{\partial s^*}{\partial \boldsymbol{r}}. \tag{B.122}$$

Equation (B.96) then leads to the following extensions of the Poisson bracket, associated with covariant body tensor and scalar density fields:

$$\begin{aligned}
&\int c_{kj} \frac{\partial}{\partial r^l} \left(\frac{\delta A}{\delta c_{jk}} \frac{\delta B}{\delta M_l} - \frac{\delta B}{\delta c_{jk}} \frac{\delta A}{\delta M_l} \right) d^3 r \\
&- \int c_{kl} \left(\frac{\delta A}{\delta c_{jk}} \frac{\partial}{\partial r^j} \frac{\delta B}{\delta M_l} - \frac{\delta B}{\delta c_{jk}} \frac{\partial}{\partial r^j} \frac{\delta A}{\delta M_l} \right) d^3 r \\
&- \int c_{lj} \left(\frac{\delta A}{\delta c_{jk}} \frac{\partial}{\partial r^k} \frac{\delta B}{\delta M_l} - \frac{\delta B}{\delta c_{jk}} \frac{\partial}{\partial r^k} \frac{\delta A}{\delta M_l} \right) d^3 r,
\end{aligned} \tag{B.123}$$

and

$$- \int s \left(\frac{\delta A}{\delta M_j} \frac{\partial}{\partial r^j} \frac{\delta B}{\delta s} - \frac{\delta B}{\delta M_j} \frac{\partial}{\partial r^j} \frac{\delta A}{\delta s} \right) d^3 r, \tag{B.124}$$

respectively. These results are in agreement with (2.91) and (2.87).

The body tensor formalism offers a natural action of the group of space transformations on structural variables used to describe complex fluids. This group action is the key to introducing Poisson brackets and hence to Hamiltonian dynamics for systems with additional structural variables. The construction of geometric structures by reduction based on the particle relabeling symmetry in the body tensor formalism may be regarded as the deep expression of the ideas behind frame indifference or material objectivity. The resulting Poisson brackets lead to the upper and lower convected derivatives which are usually considered as a hallmark of material objectivity. If the Hamiltonian (B.97) with the momentum map (B.100) expresses the kinetic energy, then the corresponding time evolution naturally leads to contributions of the form

$$\frac{\partial \boldsymbol{c}}{\partial t} - R_{\boldsymbol{v}}(\boldsymbol{c}) = \frac{\partial \boldsymbol{c}}{\partial t} + d_{\boldsymbol{v}} \boldsymbol{c} = \frac{\partial \boldsymbol{c}}{\partial t} + \boldsymbol{v} \cdot \frac{\partial \boldsymbol{c}}{\partial \boldsymbol{r}} + \boldsymbol{\kappa}^T \cdot \boldsymbol{c} + \boldsymbol{c} \cdot \boldsymbol{\kappa}, \tag{B.125}$$

where $R_{\boldsymbol{v}}(\boldsymbol{c})$ is the left action of (B.118) naturally associated with the Lie derivative and $\boldsymbol{\kappa}$ is the transposed velocity gradient tensor. The proper sign for the Lie derivative has been chosen by comparison with (B.41). Modified time derivatives of the type (B.125) are known as convected time derivatives.

Our construction does not yield the more general Schowalter derivatives often discussed in connection with material objectivity [see (2.93) and (4.77)]. For a deeper discussion of this point, we consider the concept of an *entrainment law*, which was introduced by Bolder[19] and nicely summarized and illustrated by Jongschaap.[20] Equation (B.114) can be used to establish a relationship between the spatial components of a tensor when it is transferred or "entrained" with a particle ξ that moves from the point $\boldsymbol{r}'$ at time t' to $\boldsymbol{r}$ at t,

$$c_{jk}(\boldsymbol{r},t) = \frac{\partial r'^l}{\partial r^j}\frac{\partial r'^n}{\partial r^k} c_{ln}(\boldsymbol{r}',t'), \tag{B.126}$$

where, to evaluate the derivatives $\partial r'^l/\partial r^j$, we need to consider also particles in the neighborhood of ξ. Equation (B.126) can also be obtained as a special case of (B.116), where f corresponds to the transformation of the body from $\boldsymbol{r}$ back to the previous configuration $\boldsymbol{r}'$.

In the nomenclature of Bolder, (B.126) would be a special entrainment type for material tensors expressed in a chart fixed in space. Following Bolder, one can consider more general entrainment laws than (B.126), and Jongschaap has elaborated how a general class of convected derivatives can be constructed in terms of entrainment laws. For the geometric construction of Poisson brackets by reduction, however, it is necessary that the entrainment depends on the total deformation only, not on the details of its realization between t' and t, because we need a unique representation of a group element. In other words, we are forced into material entrainment laws, which lead to upper and lower convected derivatives depending on the contravariant or covariant nature of the underlying body tensors used to formulate the entrainment.

More general Poisson brackets can be constructed by considering body fields that depend on additional variables $q = (q^l)$, for example, a scalar density field $s(\xi, q)$. Most important is the case in which $s(\xi, q)$ is a configurational distribution function, that is, a probability density in the space of internal configurations q (see Section 4.3). We then need to formulate a transformation rule for the variables q under a change of body coordinates from ξ to $\bar{\xi} = f(\xi)$,

$$\bar{q} = F(q, \xi, f), \tag{B.127}$$

where F is a function of q and ξ and a functional of the one-to-one map f. To establish the equivalence of different body coordinate systems, it is important that the transformation rule F satisfies the consistency conditions

$$F(F(q, \xi, f_1), f_1(\xi), f_2) = F(q, \xi, f_2 \circ f_1) \tag{B.128}$$

and

$$F(q, \xi, \mathrm{id}) = q. \tag{B.129}$$

[19] Bolder, Arch. Rational Mech. Anal. 35 (1969) 321.

[20] See Appendix 2 of Jongschaap, Rep. Prog. Phys. 53 (1990) 1.

By choosing $f_1 = f$, $f_2 = f^{-1}$ in (B.128) and using (B.129), we find the inverse transformation of (B.127) to be given by

$$q = F(\bar{q}, \bar{\xi}, f^{-1}). \tag{B.130}$$

The consistency condition (B.128) implies that there is an enormous amount of redundancy contained in $F(q, \xi, f)$. It is sufficient to know $F(q, \xi, f)$ for transformations f close to the identity because any transformation can be decomposed into small steps. All the relevant information is hence contained in the infinitesimal generator

$$\Phi^l_j(q, \xi, \xi') = \left. \frac{\delta F^l(q, \xi, f)}{\delta f^j(\xi')} \right|_{f=\mathrm{id}}. \tag{B.131}$$

We expect that the transformation of q depends only on the deformation of the neighborhood of ξ; thus, the generator $\Phi^l_j(q, \xi, \xi')$ must be of the local form

$$\Phi^l_j(q, \xi, \xi') = \mathcal{L}^l_j(q, \xi)\delta(\xi - \xi'), \tag{B.132}$$

where $\mathcal{L}^l_j(q, \xi)$ is a differential operator with respect to its second argument, ξ. By differentiating the fundamental consistency condition (B.128) with respect to f_1 and then setting $f_1 = \mathrm{id}$, $f_2 = f$, we obtain the following differential equation for reconstructing $F(q, \xi, f)$ from the infinitesimal generator $\Phi^l_j(q, \xi, \xi')$:

$$\frac{\partial f^k(\xi')}{\partial \xi'^j} \frac{\delta F^l(q, \xi, f)}{\delta f^k(\xi')} = \frac{\partial F^l(q, \xi, f)}{\partial q^n} \Phi^n_j(q, \xi, \xi') + \frac{\partial F^l(q, \xi, f)}{\partial \xi^j} \delta(\xi - \xi'). \tag{B.133}$$

A unique reconstruction of $F(q, \xi, f)$, independent of the particular decomposition of f into small steps, imposes certain integrability conditions on $\Phi^l_j(q, \xi, \xi')$. We here do not elaborate these integrability conditions involving second functional derivatives in detail because, for our applications, the validity of (B.128) can typically be checked directly in a one-line calculation.[21]

[21] For an infinitesimal generator of the local form (B.132), the operator $\mathcal{L}^l_j(q, \xi)$ is narrowed down tremendously to the form $\tilde{\mathcal{L}}^{lk}_j(q)\partial/\partial\xi^k$ of a first-order differential operator with coefficients depending on q only and further satisfying the condition

$$\tilde{\mathcal{L}}^{l'j'}_k \frac{\partial \tilde{\mathcal{L}}^{lk'}_j}{\partial q^{l'}} - \tilde{\mathcal{L}}^{lj'}_j \delta^{k'}_k = \tilde{\mathcal{L}}^{l'k'}_j \frac{\partial \tilde{\mathcal{L}}^{lj'}_k}{\partial q^{l'}} - \tilde{\mathcal{L}}^{lk'}_k \delta^{j'}_j .$$

A general class of solutions is given by

$$\tilde{\mathcal{L}}^{lk}_j = \delta^l_j q^k + g(q)\frac{q^j q^k q^l}{q^2} + c[1 + g(q)]\delta^k_j q^l;$$

alternative solutions are

$$\tilde{\mathcal{L}}^{lk}_j = g(q)\delta^k_j q^l \qquad \text{and} \qquad \tilde{\mathcal{L}}^{lk}_j = \delta^{lk} q^j + c\delta^k_j q^l,$$

where $g(q)$ is an arbitrary function and c an arbitrary constant.

Exercise 218 Codeformations and Corotations

For $q = (q^1, q^2, q^3)$, define the codeformational transformation rule

$$\bar{F}^l(q, \xi, f) = \frac{\partial f^l(\xi)}{\partial \xi^j} q^j \tag{B.134}$$

and the corotational transformation rule

$$F^l(q, \xi, f) = \frac{\bar{F}^l(q, \xi, f)}{|\bar{F}(q, \xi, f)|}, \tag{B.135}$$

with

$$|\bar{F}(q, \xi, f)| = \sqrt{\bar{F}^l(q, \xi, f)\bar{F}^l(q, \xi, f)}. \tag{B.136}$$

Show that the transformation rules (B.134) and (B.135) satisfy the consistency condition (B.128) and construct the corresponding generators $\bar{\mathcal{L}}^l_j$ and $\mathcal{L}^l_j$ implied by (B.131) and (B.132).

We can now introduce body fields with an additional dependence on q and well-defined transformation behavior. For example, (B.104) can be generalized to

$$\bar{s}(\bar{\xi}, \bar{q}) = \left| \det \frac{\partial(\xi, q)}{\partial(\bar{\xi}, \bar{q})} \right| s(\xi, q) \tag{B.137}$$

or

$$\bar{s}(\bar{\xi}, \bar{q}) = \left| \det \frac{\partial q}{\partial \bar{q}} \right| s(\xi, q), \tag{B.138}$$

depending on whether s is a density in both position and configuration space or in configuration space only. The two cases are realized for the configurational distribution functions in (4.114) and (4.140), respectively. Proceeding as before with scalar densities, we introduce the space field generalizing (B.115):

$$s(\boldsymbol{r}, Q, t) = \left| \det \frac{\partial(\xi, q)}{\partial(\boldsymbol{r}, Q)} \right| s(\xi, q), \quad \text{with } \xi = g_t^{-1}(\boldsymbol{r}) \text{ and } q = F(Q, \boldsymbol{r}, g_t^{-1}). \tag{B.139}$$

Under an infinitesimal transformation of the type $f(\boldsymbol{r}) = \boldsymbol{r} + \boldsymbol{v}(\boldsymbol{r})$, the change of Q is

$$\delta Q^l = \int \left. \frac{\delta F^l(Q, \boldsymbol{r}, f)}{\delta f^j(\boldsymbol{r}')} \right|_{f=\text{id}} v^j(\boldsymbol{r}') d^3 r' = \mathcal{L}^l_j(Q, \boldsymbol{r}) v^j(\boldsymbol{r}), \tag{B.140}$$

and hence we obtain the following generalization of the left action (B.119) on s,

$$R_{\boldsymbol{v}}(s) = -\frac{\partial}{\partial \boldsymbol{r}} \cdot (\boldsymbol{v} s) - \frac{\partial}{\partial Q^l}(s \mathcal{L}^l_j v^j). \tag{B.141}$$

If we use the definition (B.138) instead of (B.137), we obtain the action

$$R_{\boldsymbol{v}}(s) = -\boldsymbol{v} \cdot \frac{\partial}{\partial \boldsymbol{r}} s - \frac{\partial}{\partial Q^l}(s \mathcal{L}^l_j v^j). \tag{B.142}$$

With this construction, in the time-evolution equation for s, terms of the following form arise:

$$\frac{\partial s}{\partial t} - R_{\boldsymbol{v}}(s) = \frac{\partial s}{\partial t} + \boldsymbol{v} \cdot \frac{\partial s}{\partial \boldsymbol{r}} + \frac{\partial}{\partial Q^l}(s \mathcal{L}^l_j v^j). \tag{B.143}$$

Note that, with the operators $\bar{\mathcal{L}}^l_j$ and $\mathcal{L}^l_j$ of Exercise 218, we can reproduce the convection mechanisms in the time-evolution equations (4.127) and (4.138) for the configurational distribution functions characterizing dilute polymer solutions and reptating polymer chains in a melt. The idea of using the action of space transformations on configurational variables in a probability density to obtain a Poisson bracket has been the basis for constructing a thermodynamically consistent reptation model with anisotropic tube cross sections.[22]

Finally, we note that the body tensor formalism is a continuum approach based on a smooth body manifold. Clearly, the particles of a body manifold cannot be atoms or molecules but should rather be considered as fluid elements. The deformations of a body in space describe average motions of fluid elements and neglect the motion of atoms or molecules within the body manifold. Nevertheless, the influence of atomistic effects can be taken into account within the body tensor formalism by "smearing," as has been shown for the effect of molecular diffusion on the stress tensor.[23]

[22] Öttinger, J. Non-Newtonian Fluid Mech. 89 (2000) 165.
[23] Öttinger, J. Rheol. 35 (1991) 1275.

Appendix C
Functional Derivatives

In the formulation of the GENERIC framework of beyond-equilibrium thermodynamics, functional derivatives of energy and entropy play a central role. The occurrence of functional derivatives is related to the minimization or maximization of certain functionals in the approach of isolated systems to equilibrium. In this appendix, we first develop the basic formulas for calculating functional derivatives, then we present some conditions for the existence of extrema, and finally we describe how constraints on the independent variables can be incorporated into such calculations. After the presentation of the basic definitions, all the developments are performed in terms of specific examples.

C.1 DEFINITIONS AND EXAMPLES

In many beyond-equilibrium situations, the independent variables are fields defined on a region of space. For example, in describing Newtonian fluids, we can use the hydrodynamic fields $x = (\rho(\boldsymbol{r}), \boldsymbol{M}(\boldsymbol{r}), s(\boldsymbol{r}))$ as independent variables, where $\boldsymbol{r}$ varies in some region of space, say $V \subset \mathbb{R}^3$. In such a situation, the real-valued generators of GENERIC, the energy $E(x)$ and the entropy $S(x)$, provide examples of functionals, that is, maps taking functions into real numbers, of which derivatives with respect to the hydrodynamic fields need to be taken to obtain time-evolution equations. The concept of a *functional derivative*, say $\delta E/\delta\rho(\boldsymbol{r})$, is based on an infinite-dimensional analogue of the chain rule in which an integral occurs instead of the sum for functions of a finite number of variables,

$$\left.\frac{d}{d\lambda}E(\rho+\lambda\delta\rho,\boldsymbol{M},s)\right|_{\lambda=0} = \int_V \delta\rho(\boldsymbol{r})\frac{\delta E(\rho,\boldsymbol{M},s)}{\delta\rho(\boldsymbol{r})}d^3r, \tag{C.1}$$

where λ is a dummy real variable and where the function $\delta\rho(\boldsymbol{r})$, when multiplied by the small amplitude λ, defines a variation of the field $\rho(\boldsymbol{r})$. Such functional derivatives are also known as *Volterra derivatives*. From the definition (C.1), it is clear that $\delta E/\delta\rho(\boldsymbol{r})$ is a real-valued function on the underlying region of space. Similarly, the functional derivative $\delta E/\delta\boldsymbol{M}(\boldsymbol{r})$, defined by

$$\left.\frac{d}{d\lambda}E(\rho,\boldsymbol{M}+\lambda\delta\boldsymbol{M},s)\right|_{\lambda=0} = \int_V \delta\boldsymbol{M}(\boldsymbol{r})\cdot\frac{\delta E(\rho,\boldsymbol{M},s)}{\delta\boldsymbol{M}(\boldsymbol{r})}d^3r, \tag{C.2}$$

is a vector-valued function. Functional derivatives describe how the value of a functional changes when the functions in its arguments change a little. In the definition of functional derivatives, the inner product between the derivative and the variation, for example of the functions $\delta E/\delta\rho(\boldsymbol{r})$ and $\delta\rho(\boldsymbol{r})$, plays an important role. We here use the canonical inner product for functions, but, in some applications, it may be convenient to use a different inner product. For the uniqueness of the definition of the functional derivative, however, the inner product must be nondegenerate. For example, in the covariant formulation of relativistic hydrodynamics, it is useful to include the determinant of the metric tensor in the definition of the inner product [see (5.132)].

Exercise 219 Function Evaluation at a Point

Show that the derivative of the functional obtained by evaluating the function ρ at the fixed position $\boldsymbol{r}'$, $\rho\mapsto\rho(\boldsymbol{r}')$, is given by

$$\frac{\delta\rho(\boldsymbol{r}')}{\delta\rho(\boldsymbol{r})} = \delta(\boldsymbol{r}-\boldsymbol{r}'). \tag{C.3}$$

If we focus on the dependence on the function $\rho(\boldsymbol{r})$ for fixed $\boldsymbol{M}(\boldsymbol{r})$ and $s(\boldsymbol{r})$, the energy function (2.50) for hydrodynamics is of the form

$$E(\rho) = \int e(\rho(\boldsymbol{r}))d^3r. \tag{C.4}$$

In such a situation, we have

$$\begin{aligned}\left.\frac{d}{d\lambda}E(\rho+\lambda\delta\rho)\right|_{\lambda=0} &= \int \left.\frac{de(\rho(\boldsymbol{r})+\lambda\delta\rho(\boldsymbol{r}))}{d\lambda}\right|_{\lambda=0} d^3r \\ &= \int \delta\rho(\boldsymbol{r})\left.\frac{de(\rho)}{d\rho}\right|_{\rho=\rho(\boldsymbol{r})} d^3r,\end{aligned} \tag{C.5}$$

so that, according to the basic definition (C.1), the Volterra derivative for a functional of the form shown in (C.4) is obtained by ordinary differentiation:

$$\frac{\delta E(\rho)}{\delta\rho(\boldsymbol{r})} = \left.\frac{de(\rho)}{d\rho}\right|_{\rho=\rho(\boldsymbol{r})}. \tag{C.6}$$

The evaluation of the gradient (2.51), as well as all the other gradients of energy and entropy for hydrodynamics and most functional derivatives for complex fluids throughout this book, is based on the formula (C.6). The agreement between functional and ordinary derivatives in (C.6) should, however, not veil the conceptual difference.

More generally, a functional can also depend on the spatial derivatives of its arguments, which are fields, for example,

$$E(\rho) = \int e\left(\rho(\boldsymbol{r}), \frac{\partial\rho(\boldsymbol{r})}{\partial\boldsymbol{r}}\right) d^3r. \tag{C.7}$$

Possible origins of derivatives in a functional can lie in nonlocal energy contributions but also in the derivatives occurring in Poisson brackets. For example, for

$$A = \frac{1}{2}\int [\rho(\boldsymbol{r})]^2 d^3r, \qquad B = \frac{1}{2}\int [\boldsymbol{M}(\boldsymbol{r})]^2 d^3r, \tag{C.8}$$

the Poisson bracket (2.58) of hydrodynamics is given by the functional

$$\{A, B\} = \int \rho(\boldsymbol{r})\boldsymbol{M}(\boldsymbol{r}) \cdot \frac{\partial\rho(\boldsymbol{r})}{\partial\boldsymbol{r}} d^3r, \tag{C.9}$$

which is of the form (C.7). By applying the definition (C.1), we obtain

$$\begin{aligned}
\left.\frac{d\{A, B\}}{d\lambda}\right|_{\lambda=0} &= \left.\frac{d}{d\lambda}\int [\rho(\boldsymbol{r}) + \lambda\delta\rho(\boldsymbol{r})]\boldsymbol{M}(\boldsymbol{r}) \cdot \frac{\partial[\rho(\boldsymbol{r}) + \lambda\delta\rho(\boldsymbol{r})]}{\partial\boldsymbol{r}} d^3r\right|_{\lambda=0} \\
&= \int \left[\delta\rho(\boldsymbol{r})\boldsymbol{M}(\boldsymbol{r}) \cdot \frac{\partial\rho(\boldsymbol{r})}{\partial\boldsymbol{r}} + \rho(\boldsymbol{r})\boldsymbol{M}(\boldsymbol{r}) \cdot \frac{\partial}{\partial\boldsymbol{r}}\delta\rho(\boldsymbol{r})\right] d^3r.
\end{aligned} \tag{C.10}$$

By assuming $\delta\rho(\boldsymbol{r}) = 0$ at the boundaries and performing an integration by parts in the second term in the square brackets, we further obtain

$$\left.\frac{d\{A, B\}}{d\lambda}\right|_{\lambda=0} = \int \delta\rho(\boldsymbol{r}) \left\{\boldsymbol{M}(\boldsymbol{r}) \cdot \frac{\partial\rho(\boldsymbol{r})}{\partial\boldsymbol{r}} - \frac{\partial}{\partial\boldsymbol{r}} \cdot [\rho(\boldsymbol{r})\boldsymbol{M}(\boldsymbol{r})]\right\} d^3r. \tag{C.11}$$

According to the definition (C.1), the term in curly brackets represents the functional derivative, which can be simplified by means of the product rule,

$$\frac{\delta\{A, B\}}{\delta\rho(\boldsymbol{r})} = -\rho(\boldsymbol{r})\frac{\partial}{\partial\boldsymbol{r}} \cdot \boldsymbol{M}(\boldsymbol{r}). \tag{C.12}$$

The same result can be obtained more directly by rewriting (C.9) as

$$\{A, B\} = \frac{1}{2} \int \boldsymbol{M}(\boldsymbol{r}) \cdot \frac{\partial}{\partial \boldsymbol{r}} [\rho(\boldsymbol{r})]^2 d^3 r = -\frac{1}{2} \int [\rho(\boldsymbol{r})]^2 \frac{\partial}{\partial \boldsymbol{r}} \cdot \boldsymbol{M}(\boldsymbol{r}) d^3 r \tag{C.13}$$

and applying the basic formula (C.6). A further example of a functional of the type (C.7) is given by the total entropy production rate $\Sigma(T)$ in (3.31) [see Exercise 221].

The ideas employed to evaluate the functional derivative of $\{A, B\}$ in (C.10) and (C.11) can be used to derive the Volterra derivative of any functional of the form (C.7) by taking into account the additional contribution from the variation of $\partial\rho(\boldsymbol{r})/\partial\boldsymbol{r}$ and integrating by parts,

$$\frac{\delta E(\rho)}{\delta \rho(\boldsymbol{r})} = \left.\frac{\partial e(\rho, \boldsymbol{d})}{\partial \rho}\right|_{\rho=\rho(\boldsymbol{r}), \boldsymbol{d}=\partial\rho(\boldsymbol{r})/\partial\boldsymbol{r}} - \frac{\partial}{\partial \boldsymbol{r}} \cdot \left\{ \left.\frac{\partial e(\rho, \boldsymbol{d})}{\partial \boldsymbol{d}}\right|_{\rho=\rho(\boldsymbol{r}), \boldsymbol{d}=\partial\rho(\boldsymbol{r})/\partial\boldsymbol{r}} \right\}. \tag{C.14}$$

The formula (C.14) generalizes (C.6) in the presence of gradients. A dependence of the functional E on higher-order derivatives of ρ can be treated in an analogous way by repeated integrations by part, provided that proper boundary conditions on the derivatives of $\delta\rho(\boldsymbol{r})$ are imposed. More details, including second functional derivatives and the functional chain rule known as the DuBois-Reymond lemma of the calculus of variations, can be found in a review article by Morrison.[1] On the more general calculus of variations, there exists a classical monograph by Gelfand and Fomin.[2] The following exercise shows that a formula analogous to (C.14) exists also for partial derivatives and that the fundamental difference between functional and partial derivatives is the occurrence of integrals and derivatives in continuous labels instead of sums and differences in discrete labels.

Exercise 220 Comparison of Functional and Partial Derivatives

The entropic spring potential in the Rouse model of polymer dynamics (see Section 4.3.1) is given by

$$\phi^{(S)}(\boldsymbol{r}_1, \ldots, \boldsymbol{r}_N) = \frac{1}{2} \sum_{j=1}^{N-1} H(\boldsymbol{r}_{j+1} - \boldsymbol{r}_j)^2 = \sum_{j=1}^{N-1} f(\boldsymbol{r}_j, \boldsymbol{r}_{j+1} - \boldsymbol{r}_j), \tag{C.15}$$

where H is the entropic spring constant, possibly depending on position through temperature. Calculate the partial derivatives $\partial\phi^{(S)}/\partial\boldsymbol{r}_k$ for arbitrary functions f of the bead position vectors $\boldsymbol{r}_j$ and their differences.

Exercise 221 Functional Derivative of Entropy Production Rate

Calculate the functional derivative of the entropy production rate (3.31), which is of the form

$$\Sigma(T) = \int f(T(\boldsymbol{r})) \frac{\partial T(\boldsymbol{r})}{\partial \boldsymbol{r}} \cdot \frac{\partial T(\boldsymbol{r})}{\partial \boldsymbol{r}} d^3 r, \tag{C.16}$$

[1] See Section III.A of Morrison, Rev. Mod. Phys. 70 (1998) 467.

[2] Gelfand & Fomin, *Calculus of Variations* (Prentice-Hall, 1963).

with respect to the temperature field $T(\boldsymbol{r})$.

C.2 EXTREMIZATION OF FUNCTIONALS

The main objective of the general calculus of variations is the extremization of functionals. The principle of minimum entropy production considered in Section 3.1.5 provides an example of a situation in which the minimization of functionals is important. The candidates for the maxima and minima of functionals are the points, that is functions, with vanishing functional derivatives. From the experience with finite-dimensional calculus, one expects to find sufficient conditions for maxima and minima by studying the second derivatives at points with vanishing first functional derivatives. We therefore introduce second functional derivatives as illustrated by the following examples:

$$\frac{d^2}{d\lambda d\lambda'} E(\rho + \lambda\delta\rho + \lambda'\delta\rho', \boldsymbol{M}, s)\bigg|_{\lambda=0,\lambda'=0} = \int\limits_V \delta\rho \left[\frac{\delta^2 E(\rho, \boldsymbol{M}, s)}{\delta\rho\delta\rho}\delta\rho'\right] d^3r \tag{C.17}$$

and

$$\frac{d^2}{d\lambda d\lambda'} E(\rho + \lambda\delta\rho, \boldsymbol{M} + \lambda'\delta\boldsymbol{M}, s)\bigg|_{\lambda=0,\lambda'=0} = \int\limits_V \delta\rho \left[\frac{\delta^2 E(\rho, \boldsymbol{M}, s)}{\delta\rho\delta\boldsymbol{M}} \cdot \delta\boldsymbol{M}\right] d^3r, \tag{C.18}$$

where the second functional derivatives are linear operators acting on $\delta\rho'$ and $\delta\boldsymbol{M}$, respectively. These definitions show that interchanging the order of the derivatives corresponds to using the adjoint linear operator. We can now formulate the following sufficient condition for establishing that a function $\rho(\boldsymbol{r})$ for which the functional derivative of $E(\rho)$ vanishes is a minimum:

$$\int \delta\rho \left[\frac{\delta^2 E(\rho)}{\delta\rho\delta\rho}\delta\rho\right] d^3r = \frac{d^2}{d\lambda^2} E(\rho + \lambda\delta\rho)\bigg|_{\lambda=0} \geq c\|\delta\rho\|^2, \tag{C.19}$$

where c is a positive constant and $\|\cdot\|$ is a norm defined on the domain V. The weaker condition

$$\int \delta\rho \left[\frac{\delta^2 E(\rho)}{\delta\rho\delta\rho}\delta\rho\right] d^3r = \frac{d^2}{d\lambda^2} E(\rho + \lambda\delta\rho)\bigg|_{\lambda=0} \geq 0, \tag{C.20}$$

is necessary but not sufficient. These and further necessary and sufficient conditions for extrema have been discussed by Gelfand and Fomin.[3]

Exercise 222 Second Functional Derivative of Entropy Production Rate
Show that the second functional derivative of the entropy production rate shown in (3.31) at a

[3] See Chapters 5 and 6 of Gelfand & Fomin, *Calculus of Variations* (Prentice-Hall, 1963).

point with vanishing first functional derivative can be written as

$$\begin{aligned}\frac{\delta^2\Sigma(T)}{\delta T \delta T} &= -2f^{\alpha}\frac{\partial}{\partial \boldsymbol{r}} \cdot f^{1-2\alpha}\frac{\partial}{\partial \boldsymbol{r}} f^{\alpha} \\ &+ \left[(1-\alpha-2\alpha^2)\frac{1}{f}\left(\frac{df}{dT}\right)^2 - (1-2\alpha)\frac{d^2 f}{dT^2}\right]\left(\frac{\partial T}{\partial \boldsymbol{r}}\right)^2, \end{aligned} \tag{C.21}$$

where α is an arbitrary constant.

C.3 INCORPORATION OF CONSTRAINTS

To obtain unique expressions for functional derivatives from identities such as (C.1) or (C.2), in addition to the nondegeneracy of the involved inner product, it is important that the variations $\delta\rho(\boldsymbol{r})$ or $\delta\boldsymbol{M}(\boldsymbol{r})$ are arbitrary functions. If the variations are constrained, the functional derivative is not defined unambiguously by (C.1) or (C.2). For example, if we perform a variation under the constraint that the total mass in a given volume V is constant,

$$\int_V \delta\rho(\boldsymbol{r}) d^3r = 0, \tag{C.22}$$

then an arbitrary constant can be added to a given expression for $\delta E/\delta\rho(\boldsymbol{r})$ without changing the identity (C.1). This arbitrary constant is most naturally fixed by imposing orthogonality of $\delta E/\delta\rho(\boldsymbol{r})$ to the functional derivative of the total mass,

$$\int_V \frac{\delta E(\rho, \boldsymbol{M}, s)}{\delta\rho(\boldsymbol{r})} d^3r = 0. \tag{C.23}$$

For example, the formula (C.6) for the Volterra derivative of a functional of the form (C.4), in the presence of the constraint (C.22), is modified into

$$\frac{\delta E(\rho)}{\delta\rho(\boldsymbol{r})} = \left.\frac{de(\rho)}{d\rho}\right|_{\rho=\rho(\boldsymbol{r})} - \frac{1}{V}\int_V \left.\frac{de(\rho)}{d\rho}\right|_{\rho=\rho(\boldsymbol{r}')} d^3r'. \tag{C.24}$$

In a fully analogous way, the normalization of the configurational distribution function has been taken into account as a constraint in evaluating the functional derivative (4.144) of the entropy of the reptation model. However, it should be noted that in (C.22) the normalization is upon integrating over position whereas, for the reptation model, we have a normalization for the configurational variables at any point in space. Therefore, "constant" after (C.22) means independent of position, whereas it means a position-dependent function independent of the configurational variables for reptation models or other models with configurational distribution functions as structural variables. In the functional derivative (7.56) of the Boltzmann entropy, the normalization constraint for the single-particle distribution function is taken into account explicitly through the dependence of the normalization factor N on the single-particle distribution function.

As a further example, we consider an incompressibility constraint in the functional derivative with respect to the momentum density field,

$$\frac{\partial}{\partial \boldsymbol{r}} \cdot \delta \boldsymbol{M}(\boldsymbol{r}) = 0, \tag{C.25}$$

for the variation of $\boldsymbol{M}(\boldsymbol{r})$. Whereas (C.22) is a global constraint for a single functional, namely the total mass, (C.25) may be considered as a local constraint, or as a set of constraints for an infinite number of local functionals. Under the constraints (C.25), when assuming suitable boundary conditions, the divergence of an arbitrary function can be added to a given $\delta E/\delta \boldsymbol{M}(\boldsymbol{r})$. A unique functional derivative is naturally obtained by imposing orthogonality conditions for all constraining functionals or the local condition

$$\frac{\partial}{\partial \boldsymbol{r}} \cdot \frac{\delta E(\rho, \boldsymbol{M}, s)}{\delta \boldsymbol{M}(\boldsymbol{r})} = 0. \tag{C.26}$$

We thus obtain the following functional derivative under the incompressibility constraint (C.25):

$$\frac{\delta}{\delta \boldsymbol{M}(\boldsymbol{r})} \int e(\boldsymbol{M}(\boldsymbol{r}'))d^3r' = \left. \frac{\partial e(\boldsymbol{M})}{\partial \boldsymbol{M}} \right|_{\boldsymbol{M}=\boldsymbol{M}(\boldsymbol{r})} - \frac{\partial}{\partial \boldsymbol{r}} f(\boldsymbol{r}), \tag{C.27}$$

where (C.26) leads to a Poisson equation for the function $f(\boldsymbol{r})$. This result was used in evaluating the functional derivative (3.43). More details on evaluating functional derivatives in the presence of the incompressibility constraint can be found in the book of Beris and Edwards.[4]

[4] See Section 5.4 of Beris & Edwards, *Thermodynamics of Flowing Systems* (Oxford, 1994).

Appendix D
Quantum Systems

Starting from Hamilton's equations of classical mechanics, we have derived the GENERIC form of coarse-grained time-evolution equations for the slow dynamic variables by means of the standard projection-operator technique (see Section 6.2). It is the goal of this appendix to derive GENERIC from a quantum mechanical description of an isolated microscopic system.[1] Whereas the probability density approach of Section 6.3 for classical mechanics includes a description of fluctuations in terms of stochastic differential equations with multiplicative noise, the present derivation for quantum systems is limited to situations with negligible fluctuation effects. In the final remarks, we comment on the possibility of including fluctuations and on the expected form of the corresponding fluctuation-dissipation theorem.

We here consider the following situation. The relevant variables of an isolated nonequilibrium system with Hamilton operator H be given by the self-adjoint Hilbert space operators A_k, where k is a discrete or continuous label. Our goal is to determine time-evolution equations for the expectation values x_k of these generally noncommuting observables A_k by means of the projection-operator technique.

[1] The work summarized in this appendix was motivated by the wish to present a balanced picture of beyond-equilibrium thermodynamics in this book. The results have previously been published in Öttinger, Phys. Rev. E 62 (2000) 4720.

There are several fundamental differences between our approach and previous work on the projection-operator approach, as elegantly presented by Grabert.[2] First, we express the projected time-evolution equations in terms of two separate generators for reversible and irreversible contributions, whereas previously only a single generator was used, either the free energy or the entropy. Second, we neither treat the Hamiltonian separately in constructing projectors nor insist on including it explicitly in the list of relevant variables, thus avoiding the concept of a global temperature constant typically occurring in one-generator theories and possible redundancies among the relevant variables. A further difference concerns the time evolution assumed in correlation functions occurring in Green-Kubo-type formulas.

D.1 RELEVANT DENSITY MATRIX

A key step in the projection-operator formalism is the identification of the relevant density matrix $\bar{\rho}$ when the features of interest in a given system can be fully described in terms of the time dependent variables $x = (x_k)$. We here assume that the Hilbert space operator $\bar{\rho} = \bar{\rho}(x)$ has the following properties:

$$\mathrm{tr}\{\bar{\rho}(x)\} = 1 \tag{D.1}$$

$$\mathrm{tr}\{\bar{\rho}(x) A_k\} = x_k, \tag{D.2}$$

and that $\bar{\rho}(x)$ maximizes the entropy

$$S[\bar{\rho}] = -k_B \mathrm{tr}\{\bar{\rho} \ln \bar{\rho}\}, \tag{D.3}$$

where k_B is Boltzmann's constant. We thus obtain the quantum generalization of the generalized canonical ensemble (6.15) and (6.16),

$$\bar{\rho}(x) = Z(x)^{-1} \exp[-\sum_k \lambda_k(x) A_k], \tag{D.4}$$

$$Z(x) = \mathrm{tr}\{\exp[-\sum_k \lambda_k(x) A_k]\}, \tag{D.5}$$

where the Lagrange multipliers $\lambda_k = \lambda_k(x)$ are determined by (D.2). As before, we define the entropy $S(x)$ by inserting the density matrix (D.4) into (D.3) for $S[\bar{\rho}]$, thus recovering the entropy (6.55) of the generalized canonical ensemble:

$$S(x) = k_B[\ln Z(x) + \sum_k \lambda_k(x) x_k]. \tag{D.6}$$

Equations (D.2), (D.5), and (D.6) imply that the interpretation of the Lagrange multipliers λ_k as the conjugates of the state variables x_k or as the entropy gradient (6.56) remains valid. Of course, the total energy of the quantum system is given by

$$E(x) = \mathrm{tr}\{\bar{\rho}(x) H\}, \tag{D.7}$$

[2] Grabert, *Projection Operators* (Springer, 1982).

instead of the classical expression (6.49).

D.2 PROJECTION OPERATOR

The fundamental projection operator for quantum systems in terms of the relevant density matrix can be obtained as a straightforward modification of (6.25),

$$\mathcal{P}(x)F = \operatorname{tr}\{\bar{\rho}(x)F\} + \sum_k (A_k - x_k) \operatorname{tr}\left\{\frac{\partial\bar{\rho}(x)}{\partial x_k}F\right\}, \tag{D.8}$$

where F is an arbitrary observable. As before, we introduce the complementary projector $\mathcal{Q}(x) = 1 - \mathcal{P}(x)$, and we use the notation $\mathcal{P}(t) = \mathcal{P}(x(t))$, $\mathcal{Q}(t) = \mathcal{Q}(x(t))$. The rigorous projection-operator identity (6.40) remains valid if we replace the Liouville operator by the generator of time translations for quantum systems. In the Heisenberg picture, this generator operates on observables F as the commutator with the Hamilton operator divided by Planck's constant,

$$\mathcal{L}F = \frac{1}{\hbar}[H, F]. \tag{D.9}$$

D.3 EXACT TIME-EVOLUTION EQUATION

By acting with both sides of (6.40) on $i\mathcal{L}A_j$ and averaging with respect to the initial density matrix $\bar{\rho}(x(0))$, we obtain an exact time-evolution equation for $x(t)$:

$$\frac{dx_j(t)}{dt} = \operatorname{tr}\{\bar{\rho}(x(t))i\mathcal{L}A_j\} + \int_0^t \operatorname{tr}\{\bar{\rho}(x(u))i\mathcal{L}\mathcal{Q}(u)G(u,t)i\mathcal{L}A_j\}du. \tag{D.10}$$

In deriving (D.10), we assumed that the exact initial density matrix is of the relevant form (D.4). The first term on the right-hand side of (D.10) is referred to as reversible, and the second term as irreversible. The reversible term can be written as

$$\left[\frac{dx_j(t)}{dt}\right]_{\text{rev}} = \frac{i}{\hbar}\operatorname{tr}\{\bar{\rho}(x(t))[H, A_j]\}. \tag{D.11}$$

The energy, as a conserved quantity, should certainly be among the slow variables. However, contrary to previous work, here we do not necessarily include H explicitly into the list of relevant observables because this list typically contains internal energy and momentum densities, in terms of which the total energy can be expressed. The accessibility of the energy through the relevant variables is hence expressed through the formal assumption

$$\mathcal{P}(t)H = H. \tag{D.12}$$

With this assumption one obtains the final expression for the reversible contribution to the time evolution,

$$\left[\frac{dx_j(t)}{dt}\right]_{\text{rev}} = \sum_k L_{jk}(x(t))\frac{\partial E(x(t))}{\partial x_k}, \tag{D.13}$$

with the Poisson matrix

$$L_{jk}(x) = \frac{1}{i\hbar}\text{tr}\{\bar{\rho}(x)[A_j, A_k]\}. \tag{D.14}$$

By means of the operator identity

$$[e^{-A}, B] = e^{-A}\int_0^1 e^{\xi A}[B, A]e^{-\xi A}d\xi \tag{D.15}$$

we obtain the following still exact result for the irreversible contribution in (D.10):

$$\begin{aligned}\left[\frac{dx_j(t)}{dt}\right]_{\text{irr}} &= \left(\frac{i}{\hbar}\right)^2\int_0^t \text{tr}\{[\bar{\rho}(x(u)), H]\mathcal{Q}(u)G(u,t)[H, A_j]\}du \\ &= \left(\frac{i}{\hbar}\right)^2\int_0^t \lambda_k(x(u))\,\text{tr}\{\bar{\rho}(x(u))[H, A_k]^{(s)}_{x(u)}\mathcal{Q}(u) \\ &\qquad\qquad \times G(u,t)[H, A_j]\}du,\end{aligned} \tag{D.16}$$

with

$$[H, A_k]^{(s)}_x = \int_0^1 e^{\xi\sum_l \lambda_l(x)A_l}[H, A_k]e^{-\xi\sum_l \lambda_l(x)A_l}d\xi. \tag{D.17}$$

D.4 MARKOVIAN APPROXIMATION

The crucial assumption of the projection-operator formalism is that there should exist a clear separation of time scales. The relevant variables x_k evolve on a large time scale compared with some intermediate scale τ, and all other variables evolve rapidly compared with τ. In view of the occurrence of the projectors $\mathcal{Q}$ (projecting on the fast variables) in the integral of (D.16), the integrand should decay rapidly, and the integral is expected to be dominated by values of u between $t-\tau$ and t. Then, all slow variables in the integral can be evaluated at time t and, in particular, $\lambda_k(x(u))$ can be pulled out of the integral as $\lambda_k(x(t))$. With (6.56), we then obtain the following approximate version of (D.16):

$$\left[\frac{dx_j(t)}{dt}\right]_{\text{irr}} = \sum_k M_{jk}(x(t))\frac{\partial S(x(t))}{\partial x_k} \tag{D.18}$$

with the friction matrix

$$M_{jk}(x) = \frac{1}{k_B}\left(\frac{i}{\hbar}\right)^2\int_0^\tau \text{tr}\{\bar{\rho}(x)[H, A_k]^{(s)}_x\mathcal{Q}(x)e^{i\mathcal{L}\mathcal{Q}(x)u}\mathcal{Q}(x)[H, A_j]\}du. \tag{D.19}$$

It is customary to neglect the projector $Q(x)$ occurring in the exponential of this expression, which then leads to

$$M_{jk}(x) = \frac{1}{k_B}\left(\frac{i}{\hbar}\right)^2\int_0^\tau \text{tr}\{\bar{\rho}(x)[H, A_k]^{(s)}_x\mathcal{Q}(x)e^{iHu/\hbar}\mathcal{Q}(x)[H, A_j]e^{-iHu/\hbar}\}du. \tag{D.20}$$

However, this latter approximation is not necessary in order to obtain the decomposition (D.18).

Equations (D.19) and (D.20) can be expressed conveniently in terms of the *canonical nonequilibrium correlation*

$$\langle A; B\rangle_x = \int_0^1 \mathrm{tr}\left\{\bar{\rho}(x) e^{u\sum_k \lambda_k(x) A_k} A e^{-u\sum_k \lambda_k(x) A_k} B\right\} du, \tag{D.21}$$

which is the natural generalization of the analogous canonical correlation of Kubo et al.[3] (the original canonical correlation is based on a density matrix proportional to $\exp\{-H/(k_B T)\}$) and closely related to Grabert's generalized canonical correlation (as mentioned before, we here avoid the distinguished role of H in the list of relevant variables). This generalization is necessary for going from near-equilibrium to far-from-equilibrium situations. The two-fold role of the Hamiltonian in the time evolution and in the density matrix, which leads to considerable mathematical simplifications near equilibrium (see, for example, temperature Green's functions[4]), is lost in moving far away from equilibrium.

In summary, we have derived the fundamental time-evolution equation (1.1) of GENERIC for quantum systems, where the building blocks S, E, L, M are given by the microscopic expressions (D.6), (D.7), (D.14), and (D.19), respectively. In this equation, memory effects are eliminated in favor of nonlinearities in x.

D.5 GENERIC PROPERTIES

As in Section 6.1.5, various properties of the GENERIC building blocks can immediately be derived from the microscopic expressions, such as the antisymmetry of the matrix L_{jk} from (D.14). The fact that $\bar{\rho}(x)$ commutes with $\sum_k \lambda_k(x) A_k$ implies the degeneracy requirement

$$\sum_k L_{jk}(x) \frac{\partial S(x)}{\partial x_k} = 0, \tag{D.22}$$

which is an important part of the GENERIC framework. The other basic degeneracy requirement,

$$\sum_k M_{jk}(x) \frac{\partial E(x)}{\partial x_k} = 0, \tag{D.23}$$

can only be shown when the possibly approximate assumption (D.12) expressing the accessibility of energy may be used.

With the probability density approach of Section 6.3, we have been able to include thermal fluctuations in the fundamental time-evolution equation for classical systems, which turned out to be the stochastic differential equation (1.56) with the fluctuation-dissipation theorem (1.57). Also for quantum systems, it would be convenient to

[3] See (2.9.45) of Kubo et al., *Statistical Physics II* (Springer, 1991).

[4] See Section 5.6 of Kubo et al., *Statistical Physics II* (Springer, 1991).

describe thermal fluctuations by stochastic differential equations for the expectations of the operators A_k, rather than by time-evolution equations for some density matrix[5] because one would like to have a self-contained description on the coarse-grained level and because efficient integration schemes are known for stochastic differential equations. It is natural to assume that even in the quantum case we recover the equations (1.56) and (1.57) for the thermal noise. As in equilibrium thermodynamics, the structure of the coarse-grained equations should not depend on the classical or quantum nature of the microscopic system, whereas the explicit expressions for the building blocks (a thermodynamic potential for an equilibrium system; E, S, L, M for a beyond-equilibrium system) should occur as natural generalizations in going from classical to quantum systems. Moreover, canonical correlations naturally occur in the fluctuation-dissipation theorem.[6] Although the generalized canonical ensemble clearly becomes inappropriate for evaluating the Poisson and friction matrices (D.14) and (D.19) in the presence of large fluctuations, it is not obvious what density matrix should then be used in these expressions. A generalized microcanonical ensemble with sharp values for the observables A_k does not exist in a quantum mechanical system because, in general, the observables A_k do not commute. The proper coupling of the quantum fluctuations resulting from Heisenberg's uncertainty principle and of the thermal fluctuations resulting from coarse-graining should be achieved by constructing a quantum-microcanonical nonequilibrium ensemble with uniformly as well-defined values of the observables A_k as allowed by their commutation relations. Wigner distribution functions[7] as relevant variables would provide a convenient mathematical tool for developing the corresponding projection-operator formalism.

[5] See Chapter 5 of Grabert, *Projection Operators* (Springer, 1982).
[6] See Section 4.2.2 of Kubo et al., *Statistical Physics II* (Springer, 1991).
[7] See, for example, Section 5.9.4 of Toda et al., *Statistical Physics I* (Springer, 1992), Mori, Oppenheim & Ross, Some Topics in Quantum Statistics (1962), or Hillery et al., Phys. Rep. 106 (1984) 121.

Appendix E
List of Applications of Beyond-Equilibrium Thermodynamics

The usefulness of thermodynamics must be judged by the number and relevance of its successful applications. For linear irreversible thermodynamics, we have compiled a list of applications in Section 3.1.1. Although the linear theory is part of the GENERIC framework of beyond-equilibrium thermodynamics, we do not repeat those applications here. It is the goal of this appendix to extend the previous list by applications from the nonlinear domain of beyond-equilibrium thermodynamics. The subsequent list is far from complete and focuses on those applications that employ the full GENERIC structure. It should be stressed that there will be a large number of promising new applications in the near future, in particular, based on thermodynamically guided simulation techniques.

E.1 COMPLEX FLUIDS

- Reptation models for the *rheology of melts of entangled linear polymer molecules* have been analyzed and further developed through the GENERIC formalism. The established way of avoiding the independent alignment approximation has been found too be inadmissible, and thermodynamically consistent formulations have been proposed (see Section 4.3.2). Moreover, the thermodynamic framework has been used to formulate a model incorporating constraint release, including convective constraint release, chain stretching, and anisotropic tube cross sections.[1] A comparison with experimental data shows that the predictions of the thermodynamically consistent single-segment theory with chain stretching are as good as those of a full-chain model.

- The thermodynamic admissibility of the differential and integral *pompon models for melts of branched polymer molecules* has been established (see Section 4.2.7).[2] The formulation of the model in terms of GENERIC building blocks naturally suggests promising modifications of the original model.

- Existing models for the *rheology of polymer blends* have been formulated in a thermodynamically consistent way, and more general new models have been proposed. After the Doi-Ohta model for polymer blends has been cast into the GENERIC form, natural modifications have been suggested.[3] A two-tensor model accounting for the influence of matrix viscoelasticity has been developed and thoroughly analyzed in shear, elongational, and mixed flows.[4]

- The Smoluchowski equation for *colloidal suspensions* has been developed and analyzed through the GENERIC formalism, together with the corresponding pressure tensor for colloidal particles on the pair distribution level, including the effects of nontrivial convection due to hydrodynamic interactions.[5] Suspensions of fibers in viscoelastic fluids have also been analyzed through the same formalism.[6]

- *Solid-liquid two-phase systems* including viscous stresses, heat conduction in the two phases as well as heat exchange through the interface, *and phase change* have been modeled within the GENERIC framework.[7] The model incorporates the Schneider rate equations for nucleation and growth into two-phase models. An analysis of Minkowski functionals as morphological variables for two-phase systems has been carried out.[8]

[1] Öttinger, J. Rheol. 43 (1999) 1461; Fang, Kröger & Öttinger, J. Rheol. 44 (2000) 1293.
[2] See also van Meerveld, J. Non-Newtonian Fluid Mech. 108 (2002) 291.
[3] Wagner, Öttinger & Edwards, AIChE J. 45 (1999) 1169; Grmela et al., Rheol. Acta 40 (2001) 560.
[4] Dressler & Edwards, Rheol. Acta 43 (2004) 257.
[5] Wagner, J. Non-Newtonian Fluid Mech. 96 (2001) 177.
[6] Grmela et al., J. Chem. Phys. 109 (1998) 6973.
[7] Hütter, Phys. Rev. E 64 (2001) 11209.
[8] Hütter, J. Non-Equilib. Thermodyn. 27 (2002) 349.

- Thermodynamically consistent *hydrodynamic equations for a fluid able to display liquid-vapor coexistence* have been developed within the GENERIC approach.[9] Hydrodynamic equations of a phase-separating fluid mixture have also been derived from the microscopic dynamics of the system by separating the short- and long-range parts of the potential in the spirit of the original work of van der Waals.[10]

E.2 RELATIVISTIC HYDRODYNAMICS AND COSMOLOGY

- A thermodynamically consistent set of *hydrodynamic equations for relativistic fluids* has been developed (see Section 5.2). The GENERIC formalism leads us to introduce equations similar in structure to the well-established Israel-Stewart theory. For self-gravitating fluids, minimal coupling or a covariant formulation of GENERIC leads to the proper equations for matter and field in a general relativistic setting (see Section 5.4).

- The thermodynamically consistent hydrodynamic equations have been applied in the study of *bulk viscous cosmology*. A detailed comparison to previous theories reveals the small influence of bulk viscous effects and the absence of a rapid expansion or high entropy production phase in bulk viscous cosmology (see Section 5.5).

E.3 DISCRETE FORMULATIONS OF HYDRODYNAMICS FOR SIMULATIONS

Spatially discrete formulations of hydrodynamics are important for solving flow problems for Newtonian fluids numerically. Frequently used discrete models have limitations that have been alleviated with the help of the GENERIC framework of beyond-equilibrium thermodynamics:[11]

- thermal fluctuations have been incorporated into *smoothed particle hydrodynamics*;

- any desired local-equilibrium thermodynamic behavior has been incorporated into *dissipative particle dynamics*.

A modified fluid particle model that is both a thermodynamically consistent version of smoothed particle hydrodynamics and a version of dissipative particle dynamics,

[9] Español, J. Chem. Phys. 115 (2001) 5392.

[10] Español & Thieulot, J. Chem. Phys. 118 (2003) 9109.

[11] Español, Serrano & Öttinger, Phys. Rev. Lett. 83 (1999) 4542; Serrano & Español, Phys. Rev. E 64 (2001) 046115.

capturing the best features of both methods (inclusion of fluctuations, clear-cut connection to Navier-Stokes equations), has been developed.[12] An additional elastic variable for capturing complex fluid behavior has been included into the fluid particle model.[13]

E.4 THERMODYNAMICALLY GUIDED SIMULATIONS

- *Monte Carlo simulations* have been performed for an *unentangled polyethylene melt* in a generalized canonical beyond-equilibrium ensemble for elongational flow based on single and multiple conformation tensor viscoelastic models. The stress-optical rule and the independence of the stress tensor of the level of description have been confirmed (see Section 8.2.3).

- The statistical expressions for the GENERIC building blocks on the level of the single-segment orientational distribution function for describing reptation have been analyzed, thus revealing the natural occurrence of a noise term describing constraint release and suggesting additional effects due to changes in the tube cross section. *Equilibrium molecular dynamics simulations of entangled polyethylene molecules* have been focused on further evaluating the expression for the friction matrix, with a need for considering diffusion along the primitive path. As a result, a reliable estimate of the reptation time has been obtained from simulations covering only 10% of a relaxation time (see Section 8.4.6).

- A *rarefied Lennard-Jones gas* has been investigated by *beyond-equilibrium molecular dynamics* in shear flow, based on Grad's ten-moment method for defining the coarse-grained target level. From a self-consistent calculation of moments, Lagrange multipliers, velocity gradients, and the corresponding friction matrix, reliable results for the viscosity and normal-stress coefficients at low shear rates have been obtained (see Section 8.4.5).

[12] Español & Revenga, Phys. Rev. E 67 (2003) 026705.

[13] Ellero, Español & Flekkøy, Phys. Rev. E 68 (2003) 041504.

Appendix F
Solutions to Exercises

F.1 INTRODUCTION

Exercise 1

From (1.2), we obtain the following matrix of second-order derivatives:

$$\begin{pmatrix} \frac{\partial^2 s(\rho,\epsilon)}{\partial\rho^2} & \frac{\partial^2 s(\rho,\epsilon)}{\partial\rho\partial\epsilon} \\ \frac{\partial^2 s(\rho,\epsilon)}{\partial\epsilon\partial\rho} & \frac{\partial^2 s(\rho,\epsilon)}{\partial\epsilon^2} \end{pmatrix} = -\frac{3}{2}\frac{k_{\mathrm{B}}}{m}\rho \begin{pmatrix} \frac{5}{3}\frac{1}{\rho^2} & -\frac{1}{\rho\epsilon} \\ -\frac{1}{\rho\epsilon} & \frac{1}{\epsilon^2} \end{pmatrix},$$

which is negative definite. On the other hand, according to (1.3),

$$\begin{pmatrix} \frac{\partial^2 s(\rho,T)}{\partial\rho^2} & \frac{\partial^2 s(\rho,T)}{\partial\rho\partial T} \\ \frac{\partial^2 s(\rho,T)}{\partial T\partial\rho} & \frac{\partial^2 s(\rho,T)}{\partial T^2} \end{pmatrix} = -\frac{3}{2}\frac{k_{\mathrm{B}}}{m}\rho \begin{pmatrix} \frac{2}{3}\frac{1}{\rho^2} & -\frac{1}{\rho T} \\ -\frac{1}{\rho T} & \frac{1}{T^2} \end{pmatrix},$$

has two eigenvalues of opposite sign (as one can check by looking at the determinant).

Exercise 2

By using the definition (1.6) twice, we obtain for constant L

$$\{A,\{B,C\}\} = \sum_{ijkl} L_{ij}L_{kl}\frac{\delta A}{\delta x_i}\left(\frac{\delta^2 B}{\delta x_j \delta x_k}\frac{\delta C}{\delta x_l} + \frac{\delta B}{\delta x_k}\frac{\delta^2 C}{\delta x_j \delta x_l}\right), \tag{F.1}$$

and similarly

$$\{B, \{C, A\}\} = \sum_{ijkl} L_{ij} L_{kl} \frac{\delta B}{\delta x_i} \left(\frac{\delta^2 C}{\delta x_j \delta x_k} \frac{\delta A}{\delta x_l} + \frac{\delta C}{\delta x_k} \frac{\delta^2 A}{\delta x_j \delta x_l} \right), \tag{F.2}$$

$$\{C, \{A, B\}\} = \sum_{ijkl} L_{ij} L_{kl} \frac{\delta C}{\delta x_i} \left(\frac{\delta^2 A}{\delta x_j \delta x_k} \frac{\delta B}{\delta x_l} + \frac{\delta A}{\delta x_k} \frac{\delta^2 B}{\delta x_j \delta x_l} \right). \tag{F.3}$$

The sum of the second-order derivatives of C in (F.1) and (F.2) is, after renaming the indices in the second contribution,

$$\sum_{ijkl} L_{ij} L_{kl} \frac{\delta A}{\delta x_i} \frac{\delta B}{\delta x_k} \frac{\delta^2 C}{\delta x_j \delta x_l} + \sum_{ijkl} L_{kl} L_{ji} \frac{\delta A}{\delta x_i} \frac{\delta B}{\delta x_k} \frac{\delta^2 C}{\delta x_l \delta x_j} = 0,$$

where the antisymmetry $L_{ij} = -L_{ji}$ has been used. In a similar way, there is a pairwise cancellation of the second-order derivatives of A and B in the sum of (F.1)–(F.3), so that the Jacobi identity is verified.

It should go without saying that, according to the above calculation, also for x-*dependent* L the second-order derivative terms cancel (because the terms involving second-order derivatives of A, B, and C are exactly the same as in the present calculation).

Exercise 3

The situation described in the exercise suggests

$$E(x) = \frac{1}{2} m v^2 + E_1 + E_2,$$

$$S(x) = S(E_1, q A_c, N) + S(E_2, (2L_g - q) A_c, N),$$

where A_c is the cross section and $2L_g$ is the length of the container. We then obtain

$$\frac{\partial E(x)}{\partial x} = \begin{pmatrix} 0 \\ mv \\ 1 \\ 1 \end{pmatrix},$$

$$\frac{\partial S(x)}{\partial x} = \begin{pmatrix} \left(\frac{p_1}{T_1} - \frac{p_2}{T_2} \right) A_c \\ 0 \\ \frac{1}{T_1} \\ \frac{1}{T_2} \end{pmatrix}.$$

The Poisson matrix is obtained by the observation that the variables q and v correspond to a one-dimensional motion of a single particle [see also (1.27)]; pressure terms need to be included to satisfy the degeneracy requirement (1.4), while observing the antisymmetry

$$L = \frac{1}{m}\begin{pmatrix} 0 & 1 & 0 & 0 \\ -1 & 0 & p_1 A_c & -p_2 A_c \\ 0 & -p_1 A_c & 0 & 0 \\ 0 & p_2 A_c & 0 & 0 \end{pmatrix}.$$

One can show that this Poisson matrix indeed satisfies the Jacobi identity because the pressure functions p_1 and p_2 are independent of the velocity v of the wall (see Exercise 4).

The symmetric friction matrix is chosen as

$$M = \begin{pmatrix} 0 & 0 & 0 & 0 \\ 0 & 0 & 0 & 0 \\ 0 & 0 & \alpha & -\alpha \\ 0 & 0 & -\alpha & \alpha \end{pmatrix},$$

describing merely the heat exchange through the parameter α and satisfying the degeneracy requirement (1.5). The only nonzero eigenvalue of M is 2α, corresponding to $(0, 0, 1, -1)$, so that M is positive semidefinite for $\alpha \geq 0$. The resulting time-evolution equations are

$$\frac{dq}{dt} = v,$$

$$\frac{dv}{dt} = \frac{1}{m}(p_1 - p_2)A_c,$$

$$\frac{dE_1}{dt} = -p_1 A_c v + \alpha\left(\frac{1}{T_1} - \frac{1}{T_2}\right),$$

$$\frac{dE_2}{dt} = p_2 A_c v - \alpha\left(\frac{1}{T_1} - \frac{1}{T_2}\right).$$

An analysis of the equations linearized around equilibrium shows that the system always relaxes to equilibrium. By another argument, all oscillations have to be damped because they induce (time dependent) temperature differences and hence a heat flux with entropy production.

Exercise 4

If, using the result of Exercise 2, we neglect all terms containing second-order derivatives of the functions A and B, then we obtain

$$\begin{aligned} \{A, \{B, C\}\} &= -\frac{\partial A}{\partial v}\left(\frac{\partial}{\partial q} - p_1\frac{\partial}{\partial E_1} + p_2\frac{\partial}{\partial E_2}\right)\{B, C\} \\ &= \frac{\partial A}{\partial v}\left(\frac{\partial p_1}{\partial q} - p_1\frac{\partial p_1}{\partial E_1} + p_2\frac{\partial p_1}{\partial E_2}\right)\left(\frac{\partial B}{\partial E_1}\frac{\partial C}{\partial v} - \frac{\partial B}{\partial v}\frac{\partial C}{\partial E_1}\right) \\ &- \frac{\partial A}{\partial v}\left(\frac{\partial p_2}{\partial q} - p_1\frac{\partial p_2}{\partial E_1} + p_2\frac{\partial p_2}{\partial E_2}\right)\left(\frac{\partial B}{\partial E_2}\frac{\partial C}{\partial v} - \frac{\partial B}{\partial v}\frac{\partial C}{\partial E_2}\right). \end{aligned}$$

Note that, in each term, two first-order derivatives with respect to v occur. Then, when the combination $\{A, \{B, C\}\} + \{B, \{C, A\}\} + \{C, \{A, B\}\}$ with cyclic permutations is formed, all terms resulting from each of the two lines cancel separately. The Jacobi identity thus is a result of the fact that the pressure functions p_1 and p_2 are independent of the velocity v of the wall.

Exercise 5

The 6×6 cosymplectic matrix for a single particle is given by

$$\begin{pmatrix} 0 & 0 & 0 & 1 & 0 & 0 \\ 0 & 0 & 0 & 0 & 1 & 0 \\ 0 & 0 & 0 & 0 & 0 & 1 \\ -1 & 0 & 0 & 0 & 0 & 0 \\ 0 & -1 & 0 & 0 & 0 & 0 \\ 0 & 0 & -1 & 0 & 0 & 0 \end{pmatrix}.$$

The full cosymplectic matrix for the N-particle system with $z = (\boldsymbol{r}_1, \boldsymbol{p}_1, \ldots, \boldsymbol{r}_N, \boldsymbol{p}_N)$ has N such blocks along the diagonal, one for each particle. If we had listed the coordinates in the alternative order $z = (\boldsymbol{r}_1, \boldsymbol{r}_2, \ldots, \boldsymbol{r}_N, \boldsymbol{p}_1, \boldsymbol{p}_2, \ldots, \boldsymbol{p}_N)$, we would get a bigger version of the above 6×6 matrix with plus/minus the $3N \times 3N$ unit matrix in the upper-right/lower-left corner.

Exercise 6

The Hamiltonian is

$$E(z) = \frac{\boldsymbol{p}^2}{2m} + \phi^{(\mathrm{e})}(\boldsymbol{r}).$$

Equation (1.26) leads to

$$\begin{aligned} \dot{\boldsymbol{r}} &= \frac{\boldsymbol{p}}{m} \\ \dot{\boldsymbol{p}} &= -\frac{\partial \phi^{(\mathrm{e})}(\boldsymbol{r})}{\partial \boldsymbol{r}} \end{aligned} \tag{F.4}$$

and thus

$$m\ddot{\boldsymbol{r}} = \dot{\boldsymbol{p}} = -\frac{\partial \phi^{(\mathrm{e})}(\boldsymbol{r})}{\partial \boldsymbol{r}},$$

which is Newton's equation of motion.

The occurrence of the second-order time derivative in Newton's equation of motion can be regarded as an explicit expression of a memory effect in differential form (the further evolution is not determined by $\boldsymbol{r}$ alone; information about the rate of change of $\boldsymbol{r}$, and hence about $\boldsymbol{r}$ at previous times, is required). This explicit memory effect is eliminated in Hamilton's formalism by introducing the additional variable $\boldsymbol{p}$ (the further evolution is fully determined by $\boldsymbol{r}$ and $\boldsymbol{p}$).

Exercise 7

The key step is the introduction of generalized or conjugate momenta,

$$P_j = \frac{\partial L^{\mathrm{cm}}(q_1 \dots q_d, \dot{q}_1 \dots \dot{q}_d)}{\partial \dot{q}_j},$$

to obtain the canonical coordinates $(q_1 \dots q_d, P_1 \dots P_d)$. The Hamiltonian is obtained by Legendre transformation,

$$H(q_1 \dots q_d, P_1 \dots P_d) = \sum_{j=1}^{d} \dot{q}_j P_j - L^{\mathrm{cm}}(q_1 \dots q_d, \dot{q}_1 \dots \dot{q}_d),$$

where $\dot{q}_j = \dot{q}_j(q_1 \dots q_d, P_1 \dots P_d)$ is to be inserted as a function of the canonical coordinates. We then obtain

$$\frac{\partial H}{\partial P_k} = \dot{q}_k + \sum_{j=1}^{d} \frac{\partial \dot{q}_j}{\partial P_k} P_j - \sum_{j=1}^{d} \frac{\partial L^{\mathrm{cm}}}{\partial \dot{q}_j} \frac{\partial \dot{q}_j}{\partial P_k} = \dot{q}_k,$$

where the definition of the conjugate momenta has been used. We further find

$$\frac{\partial H}{\partial q_k} = \sum_{j=1}^{d} \frac{\partial \dot{q}_j}{\partial q_k} P_j - \frac{\partial L^{\mathrm{cm}}}{\partial q_k} - \sum_{j=1}^{d} \frac{\partial L^{\mathrm{cm}}}{\partial \dot{q}_j} \frac{\partial \dot{q}_j}{\partial q_k} = -\frac{\partial L^{\mathrm{cm}}}{\partial q_k} = -\dot{P}_k,$$

where the Euler-Lagrange equations (1.29) have been used in addition to the definition of the conjugate momenta. The last two equations are of the canonical Hamiltonian form.

Exercise 8

Leibniz' rule (1.8) is obtained as follows:

$$\begin{aligned} [AB, C]_{\mathrm{c}} &= ABC - CAB \\ &= ABC - ACB + ACB - CAB = A[B, C]_{\mathrm{c}} + [A, C]_{\mathrm{c}} B. \end{aligned}$$

By applying the definition (1.30) twice, we obtain

$$[A, [B, C]_{\mathrm{c}}]_{\mathrm{c}} = ABC - ACB - BCA + CBA,$$

and similarly

$$[B, [C, A]_{\mathrm{c}}]_{\mathrm{c}} = BCA - BAC - CAB + ACB,$$

$$[C, [A, B]_{\mathrm{c}}]_{\mathrm{c}} = CAB - CBA - ABC + BAC.$$

By adding up these three terms, we obtain the Jacobi identity (1.9).

Exercise 9

With the transformation matrix

$$Q^{\mathrm{G}} = \begin{pmatrix} 1 & 0 & 0 & 0 \\ 0 & 1 & 0 & 0 \\ \frac{p_1}{T_1} A_{\mathrm{c}} & 0 & \frac{1}{T_1} & 0 \\ -\frac{p_2}{T_2} A_{\mathrm{c}} & 0 & 0 & \frac{1}{T_2} \end{pmatrix}$$

we can construct L' from

$$L = \frac{1}{m} \begin{pmatrix} 0 & 1 & 0 & 0 \\ -1 & 0 & p_1 A_{\mathrm{c}} & -p_2 A_{\mathrm{c}} \\ 0 & -p_1 A_{\mathrm{c}} & 0 & 0 \\ 0 & p_2 A_{\mathrm{c}} & 0 & 0 \end{pmatrix}$$

according to (1.50) as

$$L' = Q^{\mathrm{G}} \cdot L \cdot Q^{\mathrm{G}\,T}.$$

The explicit result is

$$L' = \frac{1}{m} \begin{pmatrix} 0 & 1 & 0 & 0 \\ -1 & 0 & 0 & 0 \\ 0 & 0 & 0 & 0 \\ 0 & 0 & 0 & 0 \end{pmatrix}.$$

The reversible dynamics does not affect the entropies of the two subsystems, and hence the Poisson matrix L' has a particularly simple form. Checking the Jacobi identity associated with L' is trivial (see Exercise 2), and the Jacobi identity for the bracket of Exercise 4 can hence be obtained very conveniently by a transformation. Except for the factor $1/m$, L' is of the canonical form (B.66). A natural set of canonical coordinates is hence given by (q, mv, S_1, S_2), consisting of a position-momentum pair and two degenerate variables.

Exercise 10

From the chain rule and (1.1), we obtain the time-evolution equation

$$\frac{dx'}{dt} = \frac{\delta x'}{\delta x} \cdot \frac{dx}{dt} = \frac{\delta x'}{\delta x} \cdot \left[L(x) \cdot \frac{\delta E(x)}{\delta x} + M(x) \cdot \frac{\delta S(x)}{\delta x} \right]. \tag{F.5}$$

Once more using the chain rule, we obtain from (1.48) and (1.49)

$$\frac{\delta E(x)}{\delta x} = \frac{\delta x'}{\delta x} \cdot \frac{\delta E'(x')}{\delta x'}, \tag{F.6}$$

$$\frac{\delta S(x)}{\delta x} = \frac{\delta x'}{\delta x} \cdot \frac{\delta S'(x')}{\delta x'}. \tag{F.7}$$

After introducing these expressions into (F.5), we obtain with (1.50) and (1.51) the desired result,

$$\frac{dx'}{dt} = L'(x') \cdot \frac{\delta E'(x')}{\delta x'} + M'(x') \cdot \frac{\delta S'(x')}{\delta x'}.$$

Exercise 11

The transformation rules (1.48)–(1.51) and the degeneracy requirements (1.4) and (1.5) imply

$$L'(x')\cdot\frac{\delta S'(x')}{\delta x'}=\frac{\delta x'}{\delta x}\cdot L(x)\cdot\frac{\delta x'}{\delta x}\cdot\frac{\delta S'(x')}{\delta x'}=\frac{\delta x'}{\delta x}\cdot L(x)\cdot\frac{\delta S(x)}{\delta x}=0,$$

$$M'(x')\cdot\frac{\delta E'(x')}{\delta x'}=\frac{\delta x'}{\delta x}\cdot M(x)\cdot\frac{\delta x'}{\delta x}\cdot\frac{\delta E'(x')}{\delta x'}=\frac{\delta x'}{\delta x}\cdot M(x)\cdot\frac{\delta E(x)}{\delta x}=0,$$

where (F.6) and (F.7) have been used.

Exercise 12

The total energy and entropy as generators are given by

$$E=\frac{P^2}{2m}+E_{\mathrm{e}}(S_{\mathrm{e}}),\qquad S=S_{\mathrm{e}},$$

with the gradients

$$\frac{\partial E}{\partial x}=\begin{pmatrix}0\\ \frac{P}{m}\\ T\end{pmatrix},\qquad \frac{\partial S}{\partial x}=\begin{pmatrix}0\\0\\1\end{pmatrix},$$

where the temperature T is defined as $\partial E_{\mathrm{e}}/\partial S_{\mathrm{e}}$. For the Poisson matrix, we generalize (1.27),

$$L=\begin{pmatrix}0&1&0\\-1&0&0\\0&0&0\end{pmatrix}.$$

The friction matrix is constructed from the frictional force in M_{23}, where the other nonvanishing entries are found from the symmetry of M and the degeneracy requirement (1.5),

$$M=\begin{pmatrix}0&0&0\\0&\zeta T&-\zeta\frac{P}{m}\\0&-\zeta\frac{P}{m}&\frac{\zeta}{T}\left(\frac{P}{m}\right)^2\end{pmatrix},$$

which can be written in the alternative form

$$M=\frac{\zeta}{T}\begin{pmatrix}0\\T\\-\frac{P}{m}\end{pmatrix}\begin{pmatrix}0&T&-\frac{P}{m}\end{pmatrix},$$

suggesting

$$B=\sqrt{\frac{2k_{\mathrm{B}}\zeta}{T}}\begin{pmatrix}0\\T\\-\frac{P}{m}\end{pmatrix}.$$

Equation (1.56) thus implies the following stochastic equations of motion for a diffusing particle:

$$dQ=\frac{P}{m}dt,$$

$$dP = -\zeta \frac{P}{m} dt + \sqrt{2k_{\rm B}T\zeta}\, dW_t,$$
$$dS_{\rm e} = \frac{\zeta}{T}\left(\frac{P}{m}\right)^2 dt - \sqrt{\frac{2k_{\rm B}\zeta}{T}} \frac{P}{m} dW_t.$$

Note that these equations would be useless without the noise terms because P would decay to 0.

Exercise 13

For obtaining a first auxiliary result, we write

$$\begin{aligned}
\frac{\delta}{\delta x'} \cdot M' &= \left(\frac{\delta x}{\delta x'} \cdot \frac{\delta}{\delta x}\right) \cdot \left(\frac{\delta x'}{\delta x} \cdot M \cdot \frac{\delta x'}{\delta x}\right) \\
&= \frac{\delta x}{\delta x'} : \frac{\delta^2 x'}{\delta x \delta x} \cdot M \cdot \frac{\delta x'}{\delta x} + \frac{\delta}{\delta x} \cdot \left(M \cdot \frac{\delta x'}{\delta x}\right) \\
&= \frac{\delta x}{\delta x'} : \frac{\delta^2 x'}{\delta x \delta x} \cdot M \cdot \frac{\delta x'}{\delta x} + \left[\frac{\delta}{\delta x} \cdot M\right] \cdot \frac{\delta x'}{\delta x} + M : \frac{\delta^2 x'}{\delta x \delta x}.
\end{aligned}$$

With this auxiliary result we can rewrite (1.63) as

$$\begin{aligned}
dx' &= L' \cdot \frac{\delta E'}{\delta x'} dt + M' \cdot \frac{\delta S'}{\delta x'} dt + k_{\rm B} \frac{\delta}{\delta x'} \cdot M' dt + B' \cdot dW_t \\
&\quad - k_{\rm B} \frac{\delta x}{\delta x'} : \frac{\delta^2 x'}{\delta x \delta x} \cdot M \cdot \frac{\delta x'}{\delta x} dt. \qquad \text{(F.8)}
\end{aligned}$$

The term in the second line of (F.8) destroys the form invariance of GENERIC with fluctuations. With a second auxiliary result,

$$\frac{\delta x}{\delta x'} : \frac{\delta^2 x'}{\delta x \delta x} = \frac{\delta}{\delta x} \ln \left| \det \frac{\delta x'}{\delta x} \right|,$$

we obtain

$$\frac{\delta x}{\delta x'} : \frac{\delta^2 x'}{\delta x \delta x} \cdot M \cdot \frac{\delta x'}{\delta x} = \frac{\delta x'}{\delta x} \cdot M \cdot \frac{\delta}{\delta x} \ln \left| \det \frac{\delta x'}{\delta x} \right| = -M' \cdot \frac{\delta}{\delta x'} \ln \left| \det \frac{\delta x}{\delta x'} \right|.$$

When this result is used in (F.8), we can restore the form invariance of GENERIC with fluctuations by choosing the modified transformation law (1.64) for the entropy.

F.2 HYDRODYNAMICS

Exercise 14

If we start from the fundamental relationship (A.12), $dE = TdS - pdV + \mu' dN$, where we use μ' for the chemical potential per particle and $\mu = \mu'/m$ for the chemical potential per unit mass, then the expression

$$E(S, V, N) = V\epsilon\left(m\frac{N}{V}, \frac{S}{V}\right)$$

leads to the following results:

$$T = \frac{\partial E}{\partial S} = V \frac{\partial \epsilon}{\partial s} \frac{1}{V},$$

$$\mu' = \frac{\partial E}{\partial N} = V \frac{\partial \epsilon}{\partial \rho} \frac{m}{V},$$

$$-p = \frac{\partial E}{\partial V} = \epsilon - V \frac{\partial \epsilon}{\partial \rho} \frac{mN}{V^2} - V \frac{\partial \epsilon}{\partial s} \frac{S}{V^2} = \epsilon - \mu\rho - Ts.$$

After rewriting the fundamental relationship as $dS = (1/T)dE + (p/T)dV - (\mu'/T)dN$, we similarly obtain from

$$S(E, V, N) = Vs\left(m\frac{N}{V}, \frac{E}{V}\right)$$

the following further results:

$$\frac{1}{T} = \frac{\partial S}{\partial E} = V \frac{\partial s}{\partial \epsilon} \frac{1}{V},$$

$$-\frac{\mu'}{T} = \frac{\partial S}{\partial N} = V \frac{\partial s}{\partial \rho} \frac{m}{V}.$$

For the entropy density (2.5), (2.4) leads to

$$\epsilon = \frac{3}{2} \frac{\rho}{m} k_B T$$

and

$$\mu\rho + Ts = \frac{5}{2} \frac{\rho}{m} k_B T,$$

respectively. With (2.3) we further obtain

$$p = \frac{\rho}{m} k_B T,$$

so that we have reproduced the equations of state for a monatomic ideal gas.

Exercise 15

By successively using (2.18) and (2.16), the continuity equations (2.10) and (2.12), and the definition (2.17), we obtain the following chain of equalities:

$$\begin{aligned}
\rho \frac{Dc_1}{Dt} &= \rho\left(\frac{\partial c_1}{\partial t} + \boldsymbol{v} \cdot \frac{\partial c_1}{\partial \boldsymbol{r}}\right) \\
&= \frac{\partial \rho_1}{\partial t} + \boldsymbol{v} \cdot \frac{\partial \rho_1}{\partial \boldsymbol{r}} - c_1\left(\frac{\partial \rho}{\partial t} + \boldsymbol{v} \cdot \frac{\partial \rho}{\partial \boldsymbol{r}}\right) \\
&= \frac{\partial \rho_1}{\partial t} + \boldsymbol{v} \cdot \frac{\partial \rho_1}{\partial \boldsymbol{r}} + \rho_1 \frac{\partial}{\partial \boldsymbol{r}} \cdot \boldsymbol{v} \\
&= -\frac{\partial}{\partial \boldsymbol{r}} \cdot (\boldsymbol{v}_1 \rho_1) + \frac{\partial}{\partial \boldsymbol{r}} \cdot (\boldsymbol{v} \rho_1) = -\frac{\partial}{\partial \boldsymbol{r}} \cdot \boldsymbol{j}_1^{\text{diff}}.
\end{aligned}$$

Exercise 16
The generalized equations are

$$\frac{d}{dt}\int_V \boldsymbol{M} d^3r = -\int_{\partial V} \boldsymbol{n}\cdot(\boldsymbol{v}\boldsymbol{M}+\boldsymbol{\pi})\,dA + \int_V \rho \boldsymbol{g} d^3r,$$

$$\frac{\partial \boldsymbol{M}}{\partial t} = -\frac{\partial}{\partial \boldsymbol{r}}\cdot(\boldsymbol{v}\boldsymbol{M}) - \frac{\partial}{\partial \boldsymbol{r}}\cdot\boldsymbol{\pi} + \rho\boldsymbol{g}.$$

The latter equation contains two flux terms (in the form of a divergence) and a source term. The flux terms are due to convection and due to internal motions and interactions, respectively. The external gravitational field acts as a source of momentum for the fluid.

Exercise 17
Two effects need to be considered: an additional contribution to $\partial s/\partial t$ from the chain rule due to the additional dependence of s on the intensive variable c_1 in (2.32),

$$-\frac{1}{\rho}\frac{\partial s}{\partial c_1}\left(\frac{\partial}{\partial \boldsymbol{r}}\cdot\boldsymbol{j}_1^{\mathrm{diff}} + \rho\boldsymbol{v}\cdot\frac{\partial c_1}{\partial \boldsymbol{r}}\right),$$

and a further contribution from the modified internal energy equation (2.31),

$$\frac{1}{T}(\boldsymbol{j}_1^{\mathrm{diff}}\cdot\boldsymbol{g}_1 + \boldsymbol{j}_2^{\mathrm{diff}}\cdot\boldsymbol{g}_2).$$

The contribution

$$-\frac{\partial s}{\partial c_1}\boldsymbol{v}\cdot\frac{\partial c_1}{\partial \boldsymbol{r}}$$

is actually needed to reproduce the full term $-\boldsymbol{v}\cdot\frac{\partial s}{\partial \boldsymbol{r}}$ in the second line of (2.34).
The quantity $\partial s/\partial c_1$ can be expressed as

$$\frac{\partial s}{\partial c_1} = (\mu_2-\mu_1)\frac{\rho}{T}.$$

We hence get the following generalizations of the flux and source terms in (2.34):

$$\boldsymbol{v}s + \boldsymbol{j}^{\mathrm{s}} = \boldsymbol{v}s + \frac{\boldsymbol{j}^{\mathrm{q}}}{T} - \frac{\mu_1}{T}\boldsymbol{j}_1^{\mathrm{diff}} - \frac{\mu_2}{T}\boldsymbol{j}_2^{\mathrm{diff}},$$

and

$$\sigma = \boldsymbol{j}^{\mathrm{q}}\cdot\frac{\partial}{\partial \boldsymbol{r}}\frac{1}{T} - \boldsymbol{j}_1^{\mathrm{diff}}\cdot\frac{\partial}{\partial \boldsymbol{r}}\frac{\mu_1}{T} - \boldsymbol{j}_2^{\mathrm{diff}}\cdot\frac{\partial}{\partial \boldsymbol{r}}\frac{\mu_2}{T} + \frac{1}{T}\boldsymbol{j}_1^{\mathrm{diff}}\cdot\boldsymbol{g}_1 + \frac{1}{T}\boldsymbol{j}_2^{\mathrm{diff}}\cdot\boldsymbol{g}_2 - \frac{1}{T}\boldsymbol{\tau} : \left(\frac{\partial}{\partial \boldsymbol{r}}\boldsymbol{v}\right)^T.$$

Exercise 18
The key observation is

$$\left[\frac{\partial}{\partial \boldsymbol{r}}\boldsymbol{v} + \left(\frac{\partial}{\partial \boldsymbol{r}}\boldsymbol{v}\right)^T - \frac{2}{3}\frac{\partial}{\partial \boldsymbol{r}}\cdot\boldsymbol{v}\boldsymbol{1}\right] : \frac{\partial}{\partial \boldsymbol{r}}\cdot\boldsymbol{v}\boldsymbol{1} = 0.$$

The viscous entropy production rate in (2.36) hence is

$$\sigma = \frac{\eta}{2T}\left[\frac{\partial}{\partial \boldsymbol{r}}\boldsymbol{v} + \left(\frac{\partial}{\partial \boldsymbol{r}}\boldsymbol{v}\right)^T - \frac{2}{3}\frac{\partial}{\partial \boldsymbol{r}}\cdot\boldsymbol{v}\boldsymbol{1}\right] : \left[\frac{\partial}{\partial \boldsymbol{r}}\boldsymbol{v} + \left(\frac{\partial}{\partial \boldsymbol{r}}\boldsymbol{v}\right)^T - \frac{2}{3}\frac{\partial}{\partial \boldsymbol{r}}\cdot\boldsymbol{v}\boldsymbol{1}\right] + \frac{\kappa}{T}\left(\frac{\partial}{\partial \boldsymbol{r}}\cdot\boldsymbol{v}\right)^2,$$

which is nonnegative for nonnegative viscosity and bulk viscosity.

Exercise 19

According to Exercise 17, the entropy production rate due to diffusion can be written as

$$\boldsymbol{j}_1^{\mathrm{diff}} \cdot \frac{\partial}{\partial \boldsymbol{r}}\frac{1}{\rho}\frac{\partial s}{\partial c_1}.$$

We hence assume

$$\boldsymbol{j}_1^{\mathrm{diff}} = D'\frac{\partial}{\partial \boldsymbol{r}}\frac{1}{\rho}\frac{\partial s}{\partial c_1},$$

where D' must be nonnegative. For an ideal mixture, the concentration dependence of the entropy density results exclusively from the entropy of mixing. The expression for the entropy of mixing per particle, usually expressed in the mole fractions x_k^{m} of the components rather than in the mass concentrations, can be found in any textbook on thermodynamics [see, for example, (3.40) in the book by Callen[1] or (3.32) in the book by Reichl[2]]. For the entropy of mixing per unit volume, this expression can be rewritten as

$$s^{\mathrm{mix}} = -k_{\mathrm{B}}\rho\left(\frac{c_1}{m_1}\ln x_1^{\mathrm{m}} + \frac{c_2}{m_2}\ln x_2^{\mathrm{m}}\right). \tag{F.9}$$

The mole fractions x_k^{m} in (F.9) can be expressed in terms of the concentrations c_j by noting the proportionality $c_j \propto m_j x_j^{\mathrm{m}}$, where the constant of proportionality is determined by the normalization conditions $c_1 + c_2 = x_1^{\mathrm{m}} + x_2^{\mathrm{m}} = 1$. Because of these conditions, c_2, x_1^{m}, and x_2^{m} are uniquely determined by c_1; for the derivatives, we obtain

$$\frac{\partial c_2}{\partial c_1} = -1, \quad \frac{\partial x_1^{\mathrm{m}}}{\partial c_1} = -\frac{\partial x_2^{\mathrm{m}}}{\partial c_1} = \frac{1}{m_1 m_2}\frac{m_1 x_1^{\mathrm{m}} + m_2 x_2^{\mathrm{m}}}{\frac{c_1}{m_1} + \frac{c_2}{m_2}}.$$

If s^{mix} is considered as a function of c_1 only, we obtain

$$\frac{\partial s^{\mathrm{mix}}}{\partial c_1} = -k_{\mathrm{B}}\rho\left(\frac{1}{m_1}\ln x_1^{\mathrm{m}} - \frac{1}{m_2}\ln x_2^{\mathrm{m}}\right).$$

By introducing this result into the above expression for the diffusion flow we obtain

$$\boldsymbol{j}_1^{\mathrm{diff}} = -k_{\mathrm{B}}D'\left(\frac{1}{m_1 x_1^{\mathrm{m}}} + \frac{1}{m_2 x_2^{\mathrm{m}}}\right)\frac{\partial x_1^{\mathrm{m}}}{\partial c_1}\frac{\partial c_1}{\partial \boldsymbol{r}},$$

[1] Callen, *Thermodynamics* (Wiley, 1985).

[2] Reichl, *Modern Course in Statistical Physics* (University of Texas, 1980).

and finally

$$j_1^{\text{diff}} = -\frac{k_B D'}{c_1 c_2 (c_1 m_2 + c_2 m_1)} \frac{\partial c_1}{\partial \boldsymbol{r}}.$$

By properly defining the diffusion coefficient D as the product and ratio of nonnegative quantities we obtain the desired result.

Note that we have

$$j_1^{\text{diff}} = D \frac{c_1 c_2 m_1 m_2}{k_B} \left(\frac{c_1}{m_1} + \frac{c_2}{m_2} \right) \rho \frac{\partial}{\partial \boldsymbol{r}} \frac{1}{\rho} \frac{\partial s}{\partial c_1}$$

as an alternative expression for the diffusion flow directly in terms of the diffusion coefficient D valid even for nonideal mixtures.

Exercise 20

The energy expression (2.48) remains unchanged. In particular, there is no kinetic energy associated with the relative motion between the two components because j_1^{diff} is not among the independent variables; this energy contribution is contained in the internal energy ϵ.

Exercise 21

The local entropy density $s = s(\rho, \epsilon, c_1)$ can depend on the additional concentration variable, and the entropy gradient becomes,

$$\frac{\delta S}{\delta x} = \begin{pmatrix} -\frac{\mu_1 c_1 + \mu_2 c_2}{T} \\ 0 \\ \frac{1}{T} \\ \frac{\rho(\mu_2 - \mu_1)}{T} \end{pmatrix}.$$

Assuming an ideal mixture, the concentration dependence of the entropy density can be obtained by adding the entropy of mixing (see the solution to Exercise 19 for the explicit functional form) to the usual entropy density function of hydrodynamics.

Exercise 22

A comparison of (2.49) and (2.54) shows that a meaningful definition of temperature arises only for the variable $x_3 = \epsilon(\boldsymbol{r})$, which implies that $\hat{T}_3 = T(\boldsymbol{r})$ coincides with the local equilibrium temperature (the occurrence of functional derivatives makes $\hat{T}$ a local concept).

In a fully analogous way, a comparison of (2.51) and (2.55) shows that a meaningful definition of temperature arises only for the variable $x_3' = s(\boldsymbol{r})$, which implies that $\hat{T}_3' = T(\boldsymbol{r})$ also coincides with the local equilibrium temperature.

Exercise 23

After an integration by parts, we obtain

$$\int \rho(\boldsymbol{r}') \frac{\partial \delta}{\partial \boldsymbol{r}'} \boldsymbol{v}(\boldsymbol{r}') d^3 r' = -\frac{\partial}{\partial \boldsymbol{r}} \cdot [\boldsymbol{v}(\boldsymbol{r}) \rho(\boldsymbol{r})],$$

and an analogous expression for the entropy density. We thus reproduce the reversible parts of (2.10) and (2.34).

For the momentum components, we similarly obtain

$$-\rho\frac{\partial}{\partial \boldsymbol{r}}\left(\mu-\frac{1}{2}\boldsymbol{v}^2\right)-\frac{\partial}{\partial \boldsymbol{r}}\cdot(\boldsymbol{v}\boldsymbol{M})-\left(\frac{\partial}{\partial \boldsymbol{r}}\boldsymbol{v}\right)\cdot\boldsymbol{M}-s\frac{\partial T}{\partial \boldsymbol{r}},$$

and, with (2.2),

$$-\frac{\partial}{\partial \boldsymbol{r}}\cdot(\boldsymbol{v}\boldsymbol{M})-\frac{\partial}{\partial \boldsymbol{r}}(\mu\rho+Ts-\epsilon).$$

With (2.3) we thus obtain the reversible part of (2.22).

Exercise 24

The most direct version is

$$\begin{aligned}
\{A,B\} &= \int\left[\frac{\delta B}{\delta\rho(\boldsymbol{r})}\frac{\partial}{\partial \boldsymbol{r}}\cdot\left(\frac{\delta A}{\delta\boldsymbol{M}(\boldsymbol{r})}\rho(\boldsymbol{r})\right)-\frac{\delta A}{\delta\rho(\boldsymbol{r})}\frac{\partial}{\partial \boldsymbol{r}}\cdot\left(\frac{\delta B}{\delta\boldsymbol{M}(\boldsymbol{r})}\rho(\boldsymbol{r})\right)\right]d^3r\\
&+ \int\left[\frac{\delta B}{\delta\boldsymbol{M}(\boldsymbol{r})}\frac{\partial}{\partial \boldsymbol{r}}:\left(\frac{\delta A}{\delta\boldsymbol{M}(\boldsymbol{r})}\boldsymbol{M}(\boldsymbol{r})\right)\right.\\
&\qquad\left.-\frac{\delta A}{\delta\boldsymbol{M}(\boldsymbol{r})}\frac{\partial}{\partial \boldsymbol{r}}:\left(\frac{\delta B}{\delta\boldsymbol{M}(\boldsymbol{r})}\boldsymbol{M}(\boldsymbol{r})\right)\right]d^3r\\
&+ \int\left[\frac{\delta B}{\delta\epsilon(\boldsymbol{r})}\frac{\partial}{\partial \boldsymbol{r}}\cdot\left(\frac{\delta A}{\delta\boldsymbol{M}(\boldsymbol{r})}\epsilon(\boldsymbol{r})\right)-\frac{\delta A}{\delta\epsilon(\boldsymbol{r})}\frac{\partial}{\partial \boldsymbol{r}}\cdot\left(\frac{\delta B}{\delta\boldsymbol{M}(\boldsymbol{r})}\epsilon(\boldsymbol{r})\right)\right]d^3r\\
&+ \int p(\boldsymbol{r})\left(\frac{\delta B}{\delta\epsilon(\boldsymbol{r})}\frac{\partial}{\partial \boldsymbol{r}}\cdot\frac{\delta A}{\delta\boldsymbol{M}(\boldsymbol{r})}-\frac{\delta A}{\delta\epsilon(\boldsymbol{r})}\frac{\partial}{\partial \boldsymbol{r}}\cdot\frac{\delta B}{\delta\boldsymbol{M}(\boldsymbol{r})}\right)d^3r,
\end{aligned}$$

but boundary terms could be added.

Exercise 25

From (2.56), we obtain

$$L'(\boldsymbol{r})=-\begin{pmatrix} 0 & \frac{\partial}{\partial \boldsymbol{r}}\rho(\boldsymbol{r}) & 0\\ \rho(\boldsymbol{r})\frac{\partial}{\partial \boldsymbol{r}} & \left[\frac{\partial}{\partial \boldsymbol{r}}\boldsymbol{M}(\boldsymbol{r})+\boldsymbol{M}(\boldsymbol{r})\frac{\partial}{\partial \boldsymbol{r}}\right]^T & s(\boldsymbol{r})\frac{\partial}{\partial \boldsymbol{r}}\\ 0 & \frac{\partial}{\partial \boldsymbol{r}}s(\boldsymbol{r}) & 0\end{pmatrix}.$$

Exercise 26

According to the transformation formula (1.50), the desired expression is obtained from the result of Exercise 25 as

$$\begin{pmatrix} 1 & 0 & 0\\ 0 & 1 & 0\\ -\frac{s}{\rho^2} & 0 & \frac{1}{\rho}\end{pmatrix}\cdot L'(\boldsymbol{r})\cdot\begin{pmatrix} 1 & 0 & -\frac{s}{\rho^2}\\ 0 & 1 & 0\\ 0 & 0 & \frac{1}{\rho}\end{pmatrix},$$

where the occurrence of matrix multiplications without integrations over position variables for the transformation law is a consequence of the local nature of the transformation. By performing the matrix multiplications, we obtain

$$-\begin{pmatrix} 0 & \frac{\partial}{\partial \boldsymbol{r}}\rho & 0 \\ \rho\frac{\partial}{\partial \boldsymbol{r}} & \left(\frac{\partial}{\partial \boldsymbol{r}}\boldsymbol{M} + \boldsymbol{M}\frac{\partial}{\partial \boldsymbol{r}}\right)^T & -\rho\frac{\partial}{\partial \boldsymbol{r}}\frac{s}{\rho^2} + s\frac{\partial}{\partial \boldsymbol{r}}\frac{1}{\rho} \\ 0 & -\frac{s}{\rho^2}\frac{\partial}{\partial \boldsymbol{r}}\rho + \frac{1}{\rho}\frac{\partial}{\partial \boldsymbol{r}}s & 0 \end{pmatrix}.$$

This expression can be rewritten more elegantly in terms of $\hat{s} = s/\rho$,

$$-\begin{pmatrix} 0 & \frac{\partial}{\partial \boldsymbol{r}}\rho(\boldsymbol{r}) & 0 \\ \rho(\boldsymbol{r})\frac{\partial}{\partial \boldsymbol{r}} & \left[\frac{\partial}{\partial \boldsymbol{r}}\boldsymbol{M}(\boldsymbol{r}) + \boldsymbol{M}(\boldsymbol{r})\frac{\partial}{\partial \boldsymbol{r}}\right]^T & -\frac{\partial \hat{s}(\boldsymbol{r})}{\partial \boldsymbol{r}} \\ 0 & \frac{\partial \hat{s}(\boldsymbol{r})}{\partial \boldsymbol{r}} & 0 \end{pmatrix}.$$

Note that $\partial\hat{s}(\boldsymbol{r})/\partial\boldsymbol{r}$ is merely the derivative of $\hat{s}$; the derivative no longer acts as an operator to the right (see footnote on notation on p. 56). Therefore, the antisymmetry in the corresponding entries of the Poisson matrix is more obvious because no integration by parts is involved. A comparison between this last expression and the result of Exercise 25 gives a feeling for the difference between a scalar density and a scalar; for a scalar density, the divergence of the velocity is relevant, so that the derivative operator must act also on the terms to be multiplied.

Exercise 27

We need to add a further row and column to (2.60) that correspond to the additional variable $c_1(\boldsymbol{r})$. The only nonzero new entries are L_{42} and L_{24}. Because $c_1(\boldsymbol{r})$ is a scalar rather than a scalar density, Exercise 26 teaches us to choose

$$L_{42} = -\frac{\partial c_1(\boldsymbol{r})}{\partial \boldsymbol{r}}, \quad L_{24} = \frac{\partial c_1(\boldsymbol{r})}{\partial \boldsymbol{r}}.$$

The expression (2.3) for the pressure remains unchanged as the degeneracy (1.4) still holds.

Exercise 28

Equation (2.47) can be written as

$$\underline{\underline{L}} \bullet \frac{\delta E}{\delta x} = \int L(\boldsymbol{r}, \boldsymbol{r}') \cdot \begin{pmatrix} \frac{\delta E}{\delta \rho(\boldsymbol{r}')} \\ \frac{\delta E}{\delta \boldsymbol{M}(\boldsymbol{r}')} \\ \frac{\delta E}{\delta \epsilon(\boldsymbol{r}')} \end{pmatrix} d^3r'.$$

With the Poisson operator of (2.60), we obtain the same result from

$$\underline{L} \cdot \frac{\delta E}{\delta x} = L(\boldsymbol{r}) \cdot \begin{pmatrix} \frac{\delta E}{\delta \rho(\boldsymbol{r})} \\ \frac{\delta E}{\delta \boldsymbol{M}(\boldsymbol{r})} \\ \frac{\delta E}{\delta \epsilon(\boldsymbol{r})} \end{pmatrix}.$$

Exercise 29

According to (2.19) and Exercise 19, we need to produce a dissipative contribution

$$-\frac{1}{\rho}\frac{\partial}{\partial \boldsymbol{r}} \cdot \boldsymbol{j}_1^{\text{diff}} = -\frac{1}{\rho}\frac{\partial}{\partial \boldsymbol{r}} \cdot D' \frac{\partial}{\partial \boldsymbol{r}}\frac{1}{\rho}\frac{\partial s}{\partial c_1}$$

to the time evolution of c_1. We hence introduce

$$M_{44} = -\frac{1}{\rho}\frac{\partial}{\partial \boldsymbol{r}} \cdot D' \frac{\partial}{\partial \boldsymbol{r}}\frac{1}{\rho},$$

which is symmetric and positive (the minus sign is related to an integration by parts). If we choose all other elements in the fourth row and column of the friction matrix to be zero, the degeneracy condition (1.5) remains valid because the energy does not depend on the composition (see Exercise 20).

If we put the building blocks of this exercise and Exercises 20, 21, and 27 together, we obtain the equation of continuity (2.10), diffusion equation (2.19), momentum equation (2.22), and energy equation (2.26).

Exercise 30

The generators can be written as

$$E = \int \left[\frac{1}{2}\frac{\boldsymbol{M}(\boldsymbol{r})^2}{\rho(\boldsymbol{r})} + \epsilon(\rho(\boldsymbol{r}), T(\boldsymbol{r}))\right] d^3r,$$

and

$$S = \int s(\rho(\boldsymbol{r}), T(\boldsymbol{r})) d^3r,$$

where $\epsilon(\rho, T)$ and $s(\rho, T)$ can be obtained by inverting $T(\rho, \epsilon)$ from (2.4) and $T(\rho, s)$ from (2.2). The Poisson and friction matrices are obtained by multiplying the result of Exercise 25 and (2.65) from the left by the transformation matrix

$$\begin{pmatrix} 1 & 0 & 0 \\ 0 & 1 & 0 \\ \frac{\partial T(\rho,s)}{\partial \rho} & 0 & \frac{\partial T(\rho,s)}{\partial s} \end{pmatrix},$$

and from the right with the transpose of this transformation matrix. The occurrence of matrix multiplications only, without integrations over position variables, for the transformation law is a consequence of the local nature of the transformation.

The result for the Poisson operator is

$$-\begin{pmatrix} 0 & \frac{\partial}{\partial \boldsymbol{r}}\rho(\boldsymbol{r}) & 0 \\ \rho(\boldsymbol{r})\frac{\partial}{\partial \boldsymbol{r}} & \left[\frac{\partial}{\partial \boldsymbol{r}}\boldsymbol{M}(\boldsymbol{r}) + \boldsymbol{M}(\boldsymbol{r})\frac{\partial}{\partial \boldsymbol{r}}\right]^T & -\frac{\partial T(\boldsymbol{r})}{\partial \boldsymbol{r}} + \frac{\partial}{\partial \boldsymbol{r}}\frac{1}{\alpha_s} \\ 0 & \frac{\partial T(\boldsymbol{r})}{\partial \boldsymbol{r}} + \frac{1}{\alpha_s}\frac{\partial}{\partial \boldsymbol{r}} & 0 \end{pmatrix},$$

with

$$\frac{1}{\alpha_s} = \rho\frac{\partial T(\rho, s)}{\partial \rho} + s\frac{\partial T(\rho, s)}{\partial s}.$$

The temperature is an intensive or scalar variable rather than a density, and α_s is the adiabatic thermal expansivity,

$$\alpha_s = -\frac{1}{V}\frac{\partial V(T, N, S)}{\partial T}, \quad \frac{1}{\alpha_s} = -V\frac{\partial T(V, N, S)}{\partial V} = -V\frac{\partial T(N/V, S/V)}{\partial V},$$

the inverse of which describes the change of temperature associated with relative volume changes due to reversible effects (a positive sign actually indicates a decrease of volume with temperature at constant entropy). For example, the famous relationship $VT^{3/2}$ = constant for isentropic volume changes of an ideal gas implies $\alpha_s = 3/(2T)$.

The result for the friction matrix is

$$\begin{pmatrix} 0 & 0 & 0 \\ 0 & M_{22} & M_{23} \frac{1}{T(\rho,s)} \frac{\partial T(\rho,s)}{\partial s} \\ 0 & \frac{1}{T(\rho,s)} \frac{\partial T(\rho,s)}{\partial s} M_{32} & \frac{1}{T(\rho,s)} \frac{\partial T(\rho,s)}{\partial s} M_{33} \frac{1}{T(\rho,s)} \frac{\partial T(\rho,s)}{\partial s} \end{pmatrix},$$

where M_{22}, M_{23}, M_{32}, and M_{33} are given by (2.67)–(2.70), and

$$\frac{1}{T(\rho,s)} \frac{\partial T(\rho,s)}{\partial s} = \frac{1}{\rho \hat{c}_V}$$

can be expressed in terms of the heat capacity per unit mass at constant volume,

$$\hat{c}_V = \frac{T}{mN} \frac{\partial S(T,V,N)}{\partial T} = \frac{T}{\rho} \frac{\partial s(\rho,T)}{\partial T}.$$

Exercise 31

From the results of Exercise 30, we obtain

$$\hat{T}_3'' = \frac{\frac{\partial \epsilon(\rho,T)}{\partial T}}{\frac{\partial s(\rho,T)}{\partial T}}$$

as the only meaningful candidate for a temperature. By considering $\epsilon(\rho,T) = \epsilon(\rho, s(\rho,T))$ we recognize that also $\hat{T}_3''$ coincides with the local equilibrium temperature $T = T(\boldsymbol{r})$ (cf. Exercise 22).

Exercise 32

The desired equation is obtained by combining the building blocks of Exercise 30.

Exercise 33

In evaluating the nested brackets in (1.9), we do not need to consider second-order derivatives of the functionals A, B, C (see Exercise 2). For $\{A, \{B,C\}_M\}_M$, we hence obtain

$$\int M_k \left[\frac{\delta A}{\delta M_j} \frac{\partial}{\partial r_j} \frac{\delta B}{\delta M_l} \frac{\partial}{\partial r_l} \frac{\delta C}{\delta M_k} - \frac{\delta B}{\delta M_l} \left(\frac{\partial}{\partial r_l} \frac{\delta C}{\delta M_j} \right) \left(\frac{\partial}{\partial r_j} \frac{\delta A}{\delta M_k} \right) \right] d^3r, \quad \text{(F.10)}$$

where terms with $B \leftrightarrow C$ and opposite signs have been suppressed. When the terms with cyclic permutations of A, B, and C are added according to (1.9), then (F.10) actually represents a total of six terms which result from all the $3! = 6$ permutations of A, B, and C, where a positive or negative sign occurs as the permutation is even or odd. Any permutation leads to an equally good representative of all six terms when its sign is taken into account properly.

From the first term in (F.10), two types of terms arise: products of first-order spatial derivatives and second-order spatial derivatives. The second-order derivatives are cancelled when the permutations are considered because it is symmetric under the odd permutation $A \leftrightarrow B$; there occurs a pairwise cancellation of the corresponding six terms represented by (F.10). The product of first-order spatial derivatives is cancelled by the second term in (F.10); this can be seen by performing the cyclic permutation $A \to C$, $B \to A$, $C \to B$ in the second term in (F.10) and by exchanging the summation indices j and l. Therefore, all terms cancel, and the Jacobi identity is established.

Exercise 34

The "new in old" terms vanish. The "old in new" terms are generated through permutations by

$$\int w_k \left(-\frac{\partial}{\partial r_j} \frac{\delta A}{\delta w_k} \frac{\delta B}{\delta M_l} \frac{\partial}{\partial r_l} \frac{\delta C}{\delta M_j} + \frac{\delta A}{\delta w_j} \frac{\partial}{\partial r_j} \frac{\delta B}{\delta M_l} \frac{\partial}{\partial r_l} \frac{\delta C}{\delta M_k} \right) d^3r, \tag{F.11}$$

whereas the "new in new" terms are generated by

$$\begin{aligned} &- \int w_k \frac{\partial}{\partial r_j} \frac{\delta A}{\delta M_j} \left(\frac{\partial}{\partial r_l} \frac{\delta B}{\delta w_k} \frac{\delta C}{\delta M_l} - \frac{\delta B}{\delta w_l} \frac{\partial}{\partial r_l} \frac{\delta C}{\delta M_k} \right) d^3r \\ &+ \int w_k \frac{\partial(\delta A/\delta M_k)}{\partial r_j} \left(\frac{\partial}{\partial r_l} \frac{\delta B}{\delta w_j} \frac{\delta C}{\delta M_l} - \frac{\delta B}{\delta w_l} \frac{\partial}{\partial r_l} \frac{\delta C}{\delta M_j} \right) d^3r. \end{aligned} \tag{F.12}$$

After a cyclic permutation, the first term in (F.12) becomes

$$-\int w_k \frac{\partial}{\partial r_j} \frac{\delta C}{\delta M_j} \frac{\partial}{\partial r_l} \frac{\delta A}{\delta w_k} \frac{\delta B}{\delta M_l} d^3r,$$

which can be rewritten as

$$-\int w_k \frac{\partial}{\partial r_j} \frac{\partial}{\partial r_l} \frac{\delta C}{\delta M_j} \frac{\delta A}{\delta w_k} \frac{\delta B}{\delta M_l} d^3r + \int w_k \frac{\partial}{\partial r_j} \frac{\delta A}{\delta w_k} \frac{\delta B}{\delta M_l} \frac{\partial}{\partial r_l} \frac{\delta C}{\delta M_j} d^3r,$$

where the first term vanishes for symmetry reasons ($B \leftrightarrow C$, $j \leftrightarrow l$), and the second term cancels the first one in (F.11). The leftover second term in (F.11) can then be rewritten as

$$\int w_k \frac{\delta A}{\delta w_j} \frac{\delta B}{\delta M_l} \frac{\partial}{\partial r_j} \frac{\partial}{\partial r_l} \frac{\delta C}{\delta M_k} d^3r, \tag{F.13}$$

plus a term that after a cyclic permutation of A, B, and C and after exchanging the summation indices j and l cancels the last term in (F.12).

Using similar steps, the leftover terms in (F.12) can be combined into

$$\int w_k \frac{\delta A}{\delta M_j} \frac{\delta B}{\delta w_l} \frac{\partial}{\partial r_j} \frac{\partial}{\partial r_l} \frac{\delta C}{\delta M_k} d^3r. \tag{F.14}$$

After A and B as well as j and l are exchanged, (F.13) and (F.14) cancel each other, so that the Jacobi identity for the bracket $\{A, B\}_{\mathrm{lcv}}$ in (2.90) is verified explicitly.

Exercise 35
Only matrices with one position label occur:

$$\underline{M} = \underline{C_M} \cdot \underline{D_M} \cdot \underline{C_M}^T,$$

$$\underline{C_M}^T \cdot \frac{\delta E}{\delta x} = 0,$$

$$\underline{B} = \underline{C_M} \cdot \sqrt{2k_{\mathrm{B}}\underline{D_M}}.$$

No integrations are involved in any of these equations.

Exercise 36
The concentration variable is decoupled from the hydrodynamic variables. We hence obtain the minimalistic modifications

$$C_M = \begin{pmatrix} 0 & 0 & 0 \\ \frac{\partial}{\partial \boldsymbol{r}} & 0 & 0 \\ \boldsymbol{\kappa} & \frac{\partial}{\partial \boldsymbol{r}} & 0 \\ 0 & 0 & \frac{1}{\rho}\frac{\partial}{\partial \boldsymbol{r}} \end{pmatrix},$$

$$C_M^T = \begin{pmatrix} 0 & -\frac{\partial}{\partial \boldsymbol{r}} & \boldsymbol{\kappa}^T & 0 \\ 0 & 0 & -\frac{\partial}{\partial \boldsymbol{r}} & 0 \\ 0 & 0 & 0 & -\frac{\partial}{\partial \boldsymbol{r}}\frac{1}{\rho} \end{pmatrix},$$

and

$$D_M = \begin{pmatrix} \eta T(\delta_{ik}\delta_{jl} + \delta_{il}\delta_{jk}) + \hat{\kappa} T \delta_{ij}\delta_{kl} & 0 & 0 \\ 0 & \lambda^{\mathrm{q}} T^2 \delta_{ik} & 0 \\ 0 & 0 & D'\delta_{ik} \end{pmatrix}.$$

Exercise 37
The additional contribution (2.77) can be written as

$$M^{\mathrm{diff}} = \begin{pmatrix} -\frac{\partial}{\partial \boldsymbol{r}} \\ -\frac{\partial}{\partial \boldsymbol{r}}\boldsymbol{v} \\ -\frac{\partial}{\partial \boldsymbol{r}}\alpha \end{pmatrix} D' \begin{pmatrix} \frac{\partial}{\partial \boldsymbol{r}} & \boldsymbol{v}\frac{\partial}{\partial \boldsymbol{r}} & \alpha\frac{\partial}{\partial \boldsymbol{r}} \end{pmatrix}.$$

F.3 LINEAR IRREVERSIBLE THERMODYNAMICS

Exercise 38
For two-component systems, the matrix C_M required for evaluating (3.3) has been

formulated in Exercise 36. We hence obtain for the concentration component of the generalized force

$$X_3^{\rm LIT} = \left(-C_M^T \cdot \frac{\delta S}{\delta x}\right)_{c_1} = \frac{\partial}{\partial \boldsymbol{r}} \frac{1}{\rho} \frac{\partial s^{\rm mix}}{\partial c_1} = \frac{\partial}{\partial \boldsymbol{r}} \frac{\mu_2 - \mu_1}{T},$$

where, for the last step, a result previously given in Exercise 17 has been used. According to (3.4) and Exercise 36, the Onsager coefficient is D', so that we obtain

$$J_3^{\rm LIT} = D' \frac{\partial}{\partial \boldsymbol{r}} \frac{\mu_2 - \mu_1}{T} = \boldsymbol{j}_1^{\rm diff}.$$

For the entropy production rate σ, we obtain from (3.5)

$$X_3^{\rm LIT} J_3^{\rm LIT} = \boldsymbol{j}_1^{\rm diff} \cdot \frac{\partial}{\partial \boldsymbol{r}} \frac{\mu_2 - \mu_1}{T} = -\boldsymbol{j}_1^{\rm diff} \cdot \frac{\partial}{\partial \boldsymbol{r}} \frac{\mu_1}{T} - \boldsymbol{j}_2^{\rm diff} \cdot \frac{\partial}{\partial \boldsymbol{r}} \frac{\mu_2}{T},$$

which is consistent with the result previously derived in Exercise 17.

For single-component fluids, the solution to Exercise 37 suggests

$$X^{\rm LIT} = -\frac{\partial}{\partial \boldsymbol{r}} \frac{\mu}{T} + \alpha \frac{\partial}{\partial \boldsymbol{r}} \frac{1}{T}, \quad J^{\rm LIT} = D' \left(-\frac{\partial}{\partial \boldsymbol{r}} \frac{\mu}{T} + \alpha \frac{\partial}{\partial \boldsymbol{r}} \frac{1}{T}\right) = \rho(\boldsymbol{v}_{\rm m} - \boldsymbol{v}), \tag{F.15}$$

where (2.81) has been used.

Exercise 39

The comparison of (3.14) with (F.15) suggests

$$\rho \boldsymbol{v} = D' \frac{1}{T} \frac{\boldsymbol{F}}{m} = \frac{D' \zeta}{T m} \boldsymbol{v},$$

and hence

$$\zeta = \frac{m \rho T}{D'}.$$

By inserting (2.85), we obtain the Nernst-Einstein equation (3.15).

Exercise 40

From the decomposition of the friction matrix given in Exercise 12, we have

$$X^{\rm LIT} = \frac{P}{m} = v, \qquad J^{\rm LIT} = \frac{1}{T} \zeta v.$$

The flux turns out to be the frictional force (divided by temperature), and the generalized force is the particle velocity. This counterintuitive situation of reversed roles of forces and fluxes is enlightened by Exercises 97 and 101.

Exercise 41

Equation (3.16) is obtained from the condition of positive semidefiniteness by multiplying $L^{\rm LIT}$ with a particular vector from the left and the right: all components of this vector should be zero except for the jth component, which is taken as unity. We next modify the test vector by changing its kth component from zero to a parameter λ, and we then obtain the condition

$$L_{jj}^{\rm LIT} + (L_{jk}^{\rm LIT} + L_{kj}^{\rm LIT})\lambda + L_{kk}^{\rm LIT}\lambda^2 \geq 0$$

for all λ. This quadratic form in λ has a unique minimum at

$$\lambda = -\frac{L_{jk}^{\rm LIT} + L_{kj}^{\rm LIT}}{2L_{kk}^{\rm LIT}},$$

where the minimum value of the form is

$$L_{jj}^{\rm LIT} - \frac{(L_{jk}^{\rm LIT} + L_{kj}^{\rm LIT})^2}{4L_{kk}^{\rm LIT}},$$

thus implying (3.17).

Exercise 42

According to (3.8) we need to show that the energy dissipation rate is $\boldsymbol{E} \cdot \boldsymbol{i}$. If we consider a piece of material of cross section $A_{\rm c}$ and length $L_{\rm g}$, the total energy dissipation per unit time in the material is $\boldsymbol{E} \cdot \boldsymbol{i} A_{\rm c} L_{\rm g}$; now, $|\boldsymbol{E}| L_{\rm g}$ is the voltage and $|\boldsymbol{i}| A_{\rm c}$ is the current (perpendicular to the cross section), and we recover the famous expression for the energy dissipation as the product of voltage and current.

Exercise 43

According to (1.60) and (2.96), the transformation behavior is expected to be

$$C'_M = \frac{\partial x'}{\partial x} \cdot C_M.$$

Together with (1.49), we obtain

$$C_M'^T \cdot \frac{\delta S'}{\delta x'} = C_M^T \cdot \frac{\partial x'}{\partial x} \cdot \frac{\delta S'}{\delta x'} = C_M^T \cdot \frac{\delta S}{\delta x},$$

which shows that the generalized force for each dissipative effect is invariant under changes of the independent variables.

Exercise 44

The discussion of the "principle" of minimum entropy production is usually based on the equation and inequality

$$\frac{\partial \Sigma}{\partial t} = 2 \int_V \sum_k J_k^{\rm LIT} \frac{\partial X_k^{\rm LIT}}{\partial t} d^3 r \leq 0, \tag{F.16}$$

where the equation part follows for a constant phenomenological matrix $L^{\rm LIT}$. de Groot and Mazur[3] stated explicitly that they proved the inequality in (F.16) only for

[3] See p. 54 of de Groot & Mazur, *Non-Equilibrium Thermodynamics* (Dover, 1984)

the example of heat conduction in a nonexpanding solid, which corresponds to the force-flux pair of (3.11). Prigogine[4] stated that an entire chapter is based on (F.16) but that "we shall not present a general proof of this inequality but shall only prove it for chemical reactions." In Chapter V, § 3, de Groot and Mazur actually considered the combination of these two effects, including diffusion, which corresponds to the force-flux pairs of (3.11) and (3.13). Relying on previous books, Kreuzer[5] simply stated that it can be shown that (F.16) holds. Although it was established that there is no problem with the force-flux pairs in (3.11) and (3.13), the extra factors of $1/T$ in the forces of (3.12) and (3.21) change the situation. From (3.12), we obtain for incompressible fluids

$$\begin{aligned}\int_V \sum_k J_2^{\rm LIT} \frac{\partial X_2^{\rm LIT}}{\partial t} d^3r &= \int_V \eta(\boldsymbol{\kappa} + \boldsymbol{\kappa}^T) : \frac{\partial}{\partial t}\left(\frac{1}{T}\frac{\partial}{\partial \boldsymbol{r}} \boldsymbol{v}\right) d^3r \\ &= -\int_V \frac{\partial T}{\partial t}\frac{\eta}{2T^2}\dot{\gamma} : \dot{\gamma} d^3r + \int_V \frac{\partial \boldsymbol{v}}{\partial t} \cdot \left(\frac{\partial}{\partial \boldsymbol{r}} \cdot \frac{\boldsymbol{\tau}}{T}\right) d^3r,\end{aligned}$$

whereas (3.11) similarly gives

$$\int_V \sum_k J_1^{\rm LIT} \frac{\partial X_1^{\rm LIT}}{\partial t} d^3r = -\int_V \frac{\partial T}{\partial t}\frac{1}{T^2}\frac{\partial}{\partial \boldsymbol{r}} \cdot \lambda^{\rm q} \frac{\partial T}{\partial \boldsymbol{r}}.$$

If we use the nonconvective parts of (2.22) and (2.73) we note that all terms produce manifestly nonpositive contributions except

$$\int_V \frac{\partial \boldsymbol{v}}{\partial t} \cdot \left(\frac{\partial}{\partial \boldsymbol{r}} \cdot \frac{\boldsymbol{\tau}}{T}\right) d^3r.$$

Exercise 45

With $\varepsilon(\varrho) = \varepsilon(\epsilon) = \varepsilon(c_1) = +1$ and $\varepsilon(\boldsymbol{M}) = -1$, we obtain for the bare symmetry of the friction matrix

$$\mathcal{OC}_{\rm bare}(M) = \begin{pmatrix} + & - & + & + \\ - & + & - & - \\ + & - & + & + \\ + & - & + & + \end{pmatrix}.$$

The particular form of the friction matrix found in Exercise 29 implies

$$\mathcal{OC}_{\rm dressed}(M) = \begin{pmatrix} \bigcirc & \bigcirc & \bigcirc & \bigcirc \\ \bigcirc & \oplus & \oplus & \bigcirc \\ \bigcirc & \oplus & \oplus & \bigcirc \\ \bigcirc & \bigcirc & \bigcirc & \oplus \end{pmatrix},$$

[4] See p. 96 of Prigogine, *Thermodynamics of Irreversible Processes* (Wiley, 1967)

[5] See p. 70 of Kreuzer, *Nonequilibrium Thermodynamics & Statistical Foundations* (Oxford, 1981)

which is consistent with a fully symmetric M matrix.

Exercise 46

The key to solving the problem is the generalization of the friction matrix of Exercise 29 due to the presence of $L^{\rm LIT}_{\rm qd}$ and $L^{\rm LIT}_{\rm dq}$. With the entropy gradient of Exercise 21 and equations (2.19) and (2.26), we obtain for the contribution to the friction matrix associated with the fluxes considered in this exercise

$$M = \begin{pmatrix} 0 & 0 & 0 & 0 \\ 0 & 0 & 0 & 0 \\ 0 & 0 & -\frac{\partial}{\partial \boldsymbol{r}} \cdot L^{\rm LIT}_{\rm qq} \frac{\partial}{\partial \boldsymbol{r}} & -\frac{\partial}{\partial \boldsymbol{r}} \cdot L^{\rm LIT}_{\rm qd} \frac{\partial}{\partial \boldsymbol{r}} \frac{1}{\rho} \\ 0 & 0 & -\frac{1}{\rho}\frac{\partial}{\partial \boldsymbol{r}} \cdot L^{\rm LIT}_{\rm dq} \frac{\partial}{\partial \boldsymbol{r}} & -\frac{1}{\rho}\frac{\partial}{\partial \boldsymbol{r}} \cdot L^{\rm LIT}_{\rm dd} \frac{\partial}{\partial \boldsymbol{r}} \frac{1}{\rho} \end{pmatrix}.$$

Because the $(3, 4)$ and $(4, 3)$ entries of this matrix are nonzero and independent of the velocity, the dressed symmetry of the full friction matrix is now more specific than in Exercise 45,

$$\mathcal{OC}_{\rm dressed}(M) = \begin{pmatrix} \bigcirc & \bigcirc & \bigcirc & \bigcirc \\ \bigcirc & \oplus & \oplus & \bigcirc \\ \bigcirc & \oplus & \oplus & \oplus \\ \bigcirc & \bigcirc & \oplus & \oplus \end{pmatrix}.$$

This dressed symmetry implies $L^{\rm LIT}_{\rm qd} = L^{\rm LIT}_{\rm dq}$ for these constant coefficients.

Exercise 47

We first need to extend the matrix C_M of Exercise 36 to account for the additional variable and the new transport process associated with it,

$$C_M = \begin{pmatrix} 0 & 0 & 0 & 0 \\ \frac{\partial}{\partial \boldsymbol{r}} & 0 & 0 & 0 \\ \boldsymbol{\kappa} & \frac{\partial}{\partial \boldsymbol{r}} & 0 & \frac{\partial \phi^{\rm el}}{\partial \boldsymbol{r}} \\ 0 & 0 & \frac{1}{\rho}\frac{\partial}{\partial \boldsymbol{r}} & 0 \\ 0 & 0 & 0 & \frac{\partial}{\partial \boldsymbol{r}} \end{pmatrix},$$

and

$$C_M^T = \begin{pmatrix} 0 & -\frac{\partial}{\partial \boldsymbol{r}} & \boldsymbol{\kappa}^T & 0 & 0 \\ 0 & 0 & -\frac{\partial}{\partial \boldsymbol{r}} & 0 & 0 \\ 0 & 0 & 0 & -\frac{\partial}{\partial \boldsymbol{r}}\frac{1}{\rho} & 0 \\ 0 & 0 & \frac{\partial \phi^{\rm el}}{\partial \boldsymbol{r}} & 0 & -\frac{\partial}{\partial \boldsymbol{r}} \end{pmatrix},$$

where (3.19) has been used. The friction matrix can now be evaluated by straightforward matrix multiplications according to (2.94), with $D_M = L^{\rm LIT}$ as given in (3.70). The result is of the form

$$M(\boldsymbol{r}) = \begin{pmatrix} 0 & 0 & 0 & 0 & 0 \\ 0 & M_{22} & M_{23} & 0 & 0 \\ 0 & M_{32} & M_{33} + M'_{33} & 0 & M'_{35} \\ 0 & 0 & 0 & M_{44} & 0 \\ 0 & 0 & M'_{53} & 0 & -\frac{\partial}{\partial \boldsymbol{r}} \cdot L^{\rm LIT}_{\rm ee} \frac{\partial}{\partial \boldsymbol{r}} \end{pmatrix},$$

where the contributions associated with momentum transport and diffusion,

$$M_{22} = -\left(\frac{\partial}{\partial \boldsymbol{r}}\eta T \frac{\partial}{\partial \boldsymbol{r}} + \mathbf{1}\frac{\partial}{\partial \boldsymbol{r}} \cdot \eta T \frac{\partial}{\partial \boldsymbol{r}}\right)^T - \frac{\partial}{\partial \boldsymbol{r}}\hat{\kappa} T \frac{\partial}{\partial \boldsymbol{r}},$$

$$M_{23} = \frac{\partial}{\partial \boldsymbol{r}} \cdot \eta T \dot{\boldsymbol{\gamma}} + \frac{\partial}{\partial \boldsymbol{r}} \frac{\hat{\kappa} T}{2} \mathrm{tr}\dot{\boldsymbol{\gamma}},$$

$$M_{32} = -\eta T \dot{\boldsymbol{\gamma}} \cdot \frac{\partial}{\partial \boldsymbol{r}} - \frac{\hat{\kappa} T}{2} \mathrm{tr}\dot{\boldsymbol{\gamma}} \frac{\partial}{\partial \boldsymbol{r}},$$

$$M_{33} = \frac{\eta T}{2} \dot{\boldsymbol{\gamma}} : \dot{\boldsymbol{\gamma}} + \frac{\hat{\kappa} T}{4} (\mathrm{tr}\dot{\boldsymbol{\gamma}})^2,$$

and

$$M_{44} = -\frac{1}{\rho}\frac{\partial}{\partial \boldsymbol{r}} \cdot D' \frac{\partial}{\partial \boldsymbol{r}} \frac{1}{\rho},$$

are taken from the previous results (2.67)–(2.70) and Exercise 29. Finally, all the remaining elements associated with thermoelectric effects contain contributions from both pure transport mechanisms and cross-effects:

$$M'_{33} = -\frac{\partial}{\partial \boldsymbol{r}} \cdot L^{\mathrm{LIT}}_{\mathrm{qq}} \frac{\partial}{\partial \boldsymbol{r}} + \frac{\partial}{\partial \boldsymbol{r}} \cdot L^{\mathrm{LIT}}_{\mathrm{qe}} \frac{\partial \phi^{\mathrm{el}}}{\partial \boldsymbol{r}} - \frac{\partial \phi^{\mathrm{el}}}{\partial \boldsymbol{r}} \cdot L^{\mathrm{LIT}}_{\mathrm{eq}} \frac{\partial}{\partial \boldsymbol{r}} + \frac{\partial \phi^{\mathrm{el}}}{\partial \boldsymbol{r}} \cdot L^{\mathrm{LIT}}_{\mathrm{ee}} \frac{\partial \phi^{\mathrm{el}}}{\partial \boldsymbol{r}},$$

$$M'_{35} = -\frac{\partial}{\partial \boldsymbol{r}} \cdot L^{\mathrm{LIT}}_{\mathrm{qe}} \frac{\partial}{\partial \boldsymbol{r}} - \frac{\partial \phi^{\mathrm{el}}}{\partial \boldsymbol{r}} \cdot L^{\mathrm{LIT}}_{\mathrm{ee}} \frac{\partial}{\partial \boldsymbol{r}},$$

and

$$M'_{53} = -\frac{\partial}{\partial \boldsymbol{r}} \cdot L^{\mathrm{LIT}}_{\mathrm{eq}} \frac{\partial}{\partial \boldsymbol{r}} + \frac{\partial}{\partial \boldsymbol{r}} \cdot L^{\mathrm{LIT}}_{\mathrm{ee}} \frac{\partial \phi^{\mathrm{el}}}{\partial \boldsymbol{r}}.$$

Note that, for the discussion of thermoelectric effects in Section 3.2.2, c_1 and ρ^{el} have been taken as redundant variables. The friction matrix formulated in the present exercise is inappropriate for this situation because a redundance of c_1 and ρ^{el} is clearly not echoed in the fourth and fifth rows and columns of M.

Exercise 48
Such people are in great danger because their minds must be affected whenever spoons are bent or bodies are moved. Please eat your soup slowly in the presence of people with pronounced psychokinetic abilities.

F.4 COMPLEX FLUIDS

Exercise 49
For $\dot{\gamma}(t) = \dot{\gamma}_0 \cos\omega t$, (4.1) implies

$$\tau_{xy}(t) = -\dot{\gamma}_0 \int_{-\infty}^{t} G(t-t')\cos(\omega t')dt',$$

which, by substitution, can be rewritten as

$$\tau_{xy}(t) = -\dot{\gamma}_0 \int_0^\infty G(t') \cos(\omega(t-t'))dt'.$$

From the basic relations for trigonometric functions, we obtain

$$\tau_{xy}(t) = -\dot{\gamma}_0 \int_0^\infty G(t')(\cos\omega t \cos\omega t' + \sin\omega t \sin\omega t')dt',$$

or, after inserting (4.7) and (4.8),

$$\tau_{xy}(t) = -\frac{\dot{\gamma}_0}{\omega}[G'(\omega)\sin\omega t + G''(\omega)\cos\omega t].$$

In words, if we apply an oscillatory shear rate, the stress response will also be oscillatory with the same angular frequency, but with a phase shift. The storage modulus $G'(\omega)$ characterizes the contribution in phase with the shear deformation (cf. Hooke's law), the loss modulus $G''(\omega)$ is found to be in phase with the shear rate (cf. Newton's law).

Exercise 50

After using $e^{-i\omega t} = \cos\omega t - i\sin\omega t$ in (4.10), we obtain

$$\eta'(\omega) = \int_0^\infty G(t)\cos\omega t dt,$$

$$\eta''(\omega) = \int_0^\infty G(t)\sin\omega t dt.$$

Exercise 51

According to (3.8) and (3.12), the local entropy production rate in oscillatory shear flow is

$$\begin{aligned}\sigma(t) &= -\frac{1}{T}\dot{\gamma}(t)\tau_{xy}(t) = \frac{\dot{\gamma}_0^2}{\omega T}\cos\omega t[G'(\omega)\sin\omega t + G''(\omega)\cos\omega t] \\ &= \frac{\dot{\gamma}_0^2}{2\omega T}[G'(\omega)\sin 2\omega t + G''(\omega)(1+\cos 2\omega t)],\end{aligned}$$

where the result of Exercise 49 for $\tau_{xy}(t)$ has been used. Note that the factors $\sin 2\omega t$ and $\cos 2\omega t$ oscillate around zero so that they vanish on average, and we thus obtain the desired result.

If one normally simplifies the calculations for harmonic oscillations by using a complex notation, then one must be careful in evaluating the entropy production rate because it is a bilinear quantity, so that real and imaginary parts get mixed.

Exercise 52

From (4.14), we have

$$\tau_{xx}(t) - \tau_{yy}(t) = -3\dot{\epsilon}\int_0^t G(t-t')dt' = -3\dot{\epsilon}\int_0^t G(t')dt'.$$

Equation (4.4) thus implies

$$\eta_{\mathrm{E}}^{+}(t) = 3\eta^{+}(t).$$

The factor of 3 between $\eta_{\mathrm{E}}^{+}(t)$ and $\eta^{+}(t)$ is known as the *Trouton ratio.*

Exercise 53

It is sufficient to consider a single mode $(\eta_j/\tau_j)\exp\{-t/\tau_j\}$; the η_∞ contribution corresponds to a vanishing relaxation time. With the indefinite integral

$$\int \frac{\eta_j}{\tau_j} e^{-t/\tau_j}\, dt = -\eta_j e^{-t/\tau_j},$$

we obtain $\eta^{+}(t)$ and $\eta^{-}(t)$ from (4.4) and (4.5), respectively. From the more general formula

$$\int \frac{\eta_j}{\tau_j} e^{-i\omega t - t/\tau_j}\, dt = -\frac{\eta_j}{1 + i\omega\tau_j} e^{-i\omega t - t/\tau_j},$$

$\eta^*(\omega)$ is obtained from (4.10). The real and imaginary parts $\eta'(\omega)$ and $\eta''(\omega)$ defined in (4.15) can then be read off from

$$\frac{\eta_j}{1 + i\omega\tau_j} = \frac{\eta_j}{1 + i\omega\tau_j}\,\frac{1 - i\omega\tau_j}{1 - i\omega\tau_j} = \eta_j \frac{1 - i\omega\tau_j}{1 + \omega^2\tau_j^2}.$$

According to (4.9) and (4.10), $G^*(\omega)$ is obtained as $i\omega\eta^*(\omega)$, and hence $G'(\omega) = \omega\eta''(\omega)$ and $G''(\omega) = \omega\eta'(\omega)$.

Exercise 54

After inserting the expression for $\eta'(\omega)$ from Table 4.1 into the Kramers-Kronig relation (4.23), we obtain

$$\eta''(\omega) = \frac{2\omega}{\pi} \sum_j \eta_j \frac{\tau_j^2}{1 + \omega^2\tau_j^2} \int_0^\infty \frac{1}{1 + \omega'^2\tau_j^2}\, d\omega'.$$

By substitution, we find the final result

$$\eta''(\omega) = \frac{2\omega}{\pi} \sum_j \eta_j \frac{\tau_j}{1 + \omega^2\tau_j^2} \int_0^\infty \frac{1}{1 + z^2}\, dz = \sum_j \eta_j \frac{\omega\tau_j}{1 + \omega^2\tau_j^2}.$$

Exercise 55

For single-mode relaxation, we have

$$\eta'(\omega) - \eta_\infty = \frac{\eta_1}{1 + \omega^2\tau_1^2}, \qquad \eta''(\omega) = \frac{\eta_1\omega\tau_1}{1 + \omega^2\tau_1^2},$$

and we hence obtain

$$\begin{aligned}
\left[\eta'(\omega) - \eta_\infty - \frac{\eta_1}{2}\right]^2 + \eta''(\omega)^2 &= \eta_1^2 \left(\frac{1}{1 + \omega^2\tau_1^2} - \frac{1}{2}\right)^2 + \eta_1^2 \frac{\omega^2\tau_1^2}{(1 + \omega^2\tau_1^2)^2} \\
&= \frac{\eta_1^2}{4} \left(\frac{1 - \omega^2\tau_1^2}{1 + \omega^2\tau_1^2}\right)^2 + \eta_1^2 \frac{\omega^2\tau_1^2}{(1 + \omega^2\tau_1^2)^2} = \frac{\eta_1^2}{4}.
\end{aligned}$$

The data points thus lie on a semicircle with radius $\eta_1/2$. For $\omega = 0$, we obtain $\eta'(\omega) - \eta_\infty = \eta_1$, $\eta''(\omega) = 0$, and in the limit $\omega \to \infty$ we find $\eta'(\omega) - \eta_\infty = 0$, $\eta''(\omega) = 0$, as indicated in Figure 4.3. The position of the maximum is found from

$$0 = \frac{d\eta''(\omega)}{d\omega} = \frac{\eta_1 \tau_1}{(1+\omega^2\tau_1^2)^2}(1 - \omega^2\tau_1^2).$$

With the expressions

$$G'(\omega) = \frac{\eta_1 \omega^2 \tau_1}{1+\omega^2\tau_1^2}, \qquad G''(\omega) = \eta_\infty \omega + \frac{\eta_1 \omega}{1+\omega^2\tau_1^2},$$

for $\eta_\infty = 0$, we similarly have,

$$G''(\omega)^2 + \left(G'(\omega) - \frac{\eta_1}{2\tau_1}\right)^2 = \frac{\eta_1^2\omega^2}{(1+\omega^2\tau_1^2)^2} + \left(\frac{\eta_1\omega^2\tau_1}{1+\omega^2\tau_1^2} - \frac{\eta_1}{2\tau_1}\right)^2 = \frac{\eta_1^2}{4\tau_1^2}.$$

In view of $\eta''(\omega) = \tau_1 G''(\omega)$, the maxima of $\eta''(\omega)$ and $G''(\omega)$ appear at the same frequency.

Exercise 56
By means of (4.19), we obtain

$$\begin{aligned} G(t;T) &= \sum_j \frac{\eta_j(T)}{\tau_j(T)} e^{-t/\tau_j(T)} + G_0 \\ &= \sum_j \frac{\eta_j(T_0)}{\tau_j(T_0)} e^{-t/[a_T \tau_j(T_0)]} + G_0 = G(t/a_T; T_0). \end{aligned}$$

Equation (4.7) then implies

$$\begin{aligned} G'(\omega;T) &= \omega \int_0^\infty G(t;T) \sin \omega t dt = \omega \int_0^\infty G(t/a_T; T_0) \sin \omega t dt \\ &= a_T \omega \int_0^\infty G(t'; T_0) \sin(a_T \omega t') dt' = G'(a_T\omega; T_0), \end{aligned}$$

where the substitution $t = a_T t'$ has been used. Equation (4.27) follows in the same way. Finally,

$$\eta(T) = \int_0^\infty G(t;T)dt = a_T \int_0^\infty G(t';T_0)dt' = a_T \eta(T_0).$$

Exercise 57
The value of the exponent that I can read off from Figure 4.5 is 0.64. However, the asymptotic regime may not yet be reached in Figure 4.5.

Exercise 58

From the expression

$$\tau_\Psi = \frac{\int_0^\infty tG(t)dt}{\int_0^\infty G(t)dt}$$

it is clear that τ_Ψ can be considered as an average weighted by the memory function of the stress–strain rate relationship. In terms of the spectrum (4.19), we have

$$\Psi_1 = 2\sum_j \frac{\eta_j}{\tau_j}\int_0^\infty te^{-t/\tau_j}dt = 2\sum_j \eta_j\tau_j,$$

and thus

$$\tau_\Psi = \frac{\sum_j \eta_j\tau_j}{\sum_j \eta_j}.$$

So, in the average τ_Ψ, each relaxation time is weighted by its viscosity contribution. For single-mode relaxation, we have $\tau_\Psi = \tau_1$.

Exercise 59

By straightforward differentiation we obtain

$$\frac{\delta E}{\delta x} = \begin{pmatrix} -\frac{1}{2}\boldsymbol{v}^2 + \frac{\partial \epsilon_c}{\partial \rho} \\ \boldsymbol{v} \\ 1 \\ \frac{\partial \epsilon_c}{\partial \boldsymbol{c}} \\ \frac{\partial \epsilon_c}{\partial \Lambda} \end{pmatrix}, \qquad \frac{\delta S}{\delta x} = \begin{pmatrix} \frac{\partial s}{\partial \rho} + \hat{s}_c \\ 0 \\ \frac{\partial s}{\partial \epsilon} \\ \rho\frac{\partial \hat{s}_c}{\partial \boldsymbol{c}} \\ \rho\frac{\partial \hat{s}_c}{\partial \Lambda} \end{pmatrix}.$$

Exercise 60

If one calculates the difference between the invariants evaluated for $\boldsymbol{c} + \delta\boldsymbol{c}$ and for $\boldsymbol{c}$ to lowest order in $\delta\boldsymbol{c}$, one obtains

$$\delta\mathrm{tr}\boldsymbol{c} = \mathrm{tr}\delta\boldsymbol{c} = \boldsymbol{1} : \delta\boldsymbol{c} = \boldsymbol{c}^0 : \delta\boldsymbol{c},$$

$$\begin{aligned}\delta \ln\det \boldsymbol{c} &= \ln\det(\boldsymbol{c}+\delta\boldsymbol{c}) - \ln\det\boldsymbol{c} = \ln\det(\boldsymbol{1} + \boldsymbol{c}^{-1}\cdot\delta\boldsymbol{c}) \\ &= \mathrm{tr}\ln(\boldsymbol{1} + \boldsymbol{c}^{-1}\cdot\delta\boldsymbol{c}) = \mathrm{tr}(\boldsymbol{c}^{-1}\cdot\delta\boldsymbol{c}) = \boldsymbol{c}^{-1} : \delta\boldsymbol{c},\end{aligned}$$

$$-\delta\mathrm{tr}\boldsymbol{c}^{-1} = -\mathrm{tr}\delta\boldsymbol{c}^{-1} = \mathrm{tr}(\boldsymbol{c}^{-1}\cdot\delta\boldsymbol{c}\cdot\boldsymbol{c}^{-1}) = \boldsymbol{c}^{-2} : \delta\boldsymbol{c},$$

where $\delta\boldsymbol{c}^{-1} = -\boldsymbol{c}^{-1}\cdot\delta\boldsymbol{c}\cdot\boldsymbol{c}^{-1}$ follows from $\delta(\boldsymbol{c}\cdot\boldsymbol{c}^{-1}) = 0$.

Exercise 61

By inserting the entropy gradient found in Exercise 59 into (4.44), we obtain

$$\begin{aligned}\{A, S\} &= \int \frac{\delta A}{\delta M_j}\left[\rho\frac{\partial \hat{s}_c}{\partial c_{kl}}\frac{\partial c_{kl}}{\partial r_j} + \rho\frac{\partial \hat{s}_c}{\partial \Lambda}\frac{\partial \Lambda}{\partial r_j}\right. \\ &\qquad \left. -\rho\frac{\partial \hat{s}_c}{\partial r_j} - \rho\frac{\partial}{\partial r_j}\frac{\partial s}{\partial \rho} - \epsilon\frac{\partial}{\partial r_j}\frac{\partial s}{\partial \epsilon}\right]d^3r \\ &+ \int\left[\pi_{jk}^{(S)}\frac{\partial s}{\partial \epsilon} - 2\rho c_{jl}\frac{\partial \hat{s}_c}{\partial c_{lk}} - \rho g_{jk}\frac{\partial \hat{s}_c}{\partial \Lambda}\right]\frac{\partial}{\partial r_j}\frac{\delta A}{\delta M_k}d^3r.\end{aligned}$$

In the first line, the terms containing $\hat{s}_c$ cancel due to the chain rule, and we thus obtain after an integration by parts

$$\begin{aligned}\{A, S\} &= \int \left[\left(\rho \frac{\partial s}{\partial \rho} + \epsilon \frac{\partial s}{\partial \epsilon} \right) \frac{\partial}{\partial r_j} \frac{\delta A}{\delta M_j} + \frac{\delta A}{\delta M_j} \frac{\partial s}{\partial r_j} \right] d^3r \\ &+ \int \left[\pi_{jk}^{(S)} \frac{\partial s}{\partial \epsilon} - 2\rho c_{jl} \frac{\partial \hat{s}_c}{\partial c_{lk}} - \rho g_{jk} \frac{\partial \hat{s}_c}{\partial \Lambda} \right] \frac{\partial}{\partial r_j} \frac{\delta A}{\delta M_k} d^3r,\end{aligned}$$

or

$$\begin{aligned}\{A, S\} &= \int \left[\pi_{jk}^{(S)} \frac{\partial s}{\partial \epsilon} - \left(s - \rho \frac{\partial s}{\partial \rho} - \epsilon \frac{\partial s}{\partial \epsilon} \right) \delta_{jk} - 2\rho c_{jl} \frac{\partial \hat{s}_c}{\partial c_{lk}} - \rho g_{jk} \frac{\partial \hat{s}_c}{\partial \Lambda} \right] \\ &\quad \times \frac{\partial}{\partial r_j} \frac{\delta A}{\delta M_k} d^3r.\end{aligned}$$

The degeneracy requirement (1.4) is equivalent to the condition $\{A, S\} = 0$ for all A, so that the term in square brackets in the last equation must be zero, and (4.45) is obtained.

Exercise 62

We can evaluate $\dot{\Lambda}_{\text{convect}}$ by means of the chain rule in terms of (4.36) and (4.50) to obtain

$$\dot{\Lambda}_{\text{convect}} = -\boldsymbol{v} \cdot \frac{\partial \Lambda}{\partial \boldsymbol{r}} + 2\boldsymbol{\kappa} : \left(\frac{\partial \Lambda}{\partial I_1} \boldsymbol{c} + \frac{\partial \Lambda}{\partial I_2} \boldsymbol{1} + \frac{\partial \Lambda}{\partial I_3} \boldsymbol{c}^{-1} \right).$$

By comparing with (4.37) and expressing the partial derivatives of $\Lambda = \Lambda(\hat{s}, I_1, I_2, I_3)$ in terms of partial derivatives of the inverse function $\hat{s}(\Lambda, I_1, I_2, I_3)$, we find the explicit representation

$$g_j = -2 \frac{\partial \hat{s}}{\partial I_j} \left(\frac{\partial \hat{s}}{\partial \Lambda} \right)^{-1},$$

for arbitrary $\hat{s}(\Lambda, I_1, I_2, I_3)$. Equations (4.47)–(4.49) can be verified by an explicit calculation of the required derivatives. For example, both sides of (4.47) turn out to be

$$4 \left(\frac{\partial \hat{s}}{\partial I_1} \frac{\partial^2 \hat{s}}{\partial \Lambda \partial I_2} - \frac{\partial \hat{s}}{\partial I_2} \frac{\partial^2 \hat{s}}{\partial \Lambda \partial I_1} \right) \left(\frac{\partial \hat{s}}{\partial \Lambda} \right)^{-2}.$$

Exercise 63

By identifying the mechanical contribution C_M in (3.3) from (4.65), with a convenient choice of sign, and using the entropy gradient of Exercise 59, we obtain

$$X_\Lambda^{\text{diff}} = -\frac{\partial}{\partial \boldsymbol{r}} \frac{1}{\rho T} \frac{\partial f_c}{\partial \Lambda}.$$

According to (3.6), the corresponding flux is given by

$$J_\Lambda^{\text{diff}} = -\Lambda \rho \boldsymbol{D}' \cdot \frac{\partial}{\partial \boldsymbol{r}} \frac{1}{\rho T} \frac{\partial f_c}{\partial \Lambda}.$$

Exercise 64

The key observation is that, as we work with a single symmetric second-rank conformation tensor $\boldsymbol{c}$ among the independent variables, all second-rank tensors constructed from $\boldsymbol{c}$ commute, and all fourth-rank tensors must be built from second-rank tensors. If $\boldsymbol{R}_2$ is of the form

$$(\boldsymbol{R}_2)_{ijkl} = (\boldsymbol{g}_1(\boldsymbol{c}))_{ik}(\boldsymbol{g}_2(\boldsymbol{c}))_{jl}, \tag{F.17}$$

then we have

$$\boldsymbol{R}_2 : \boldsymbol{g}(\boldsymbol{c}) = \boldsymbol{g}_1(\boldsymbol{c}) \cdot \boldsymbol{g}(\boldsymbol{c}) \cdot \boldsymbol{g}_2(\boldsymbol{c}) = \boldsymbol{g}_1(\boldsymbol{c}) \cdot \boldsymbol{g}_2(\boldsymbol{c}) \cdot \boldsymbol{g}(\boldsymbol{c}) = \hat{\boldsymbol{R}}_2 \cdot \boldsymbol{g}(\boldsymbol{c}),$$

where $\boldsymbol{g}_1(\boldsymbol{c})$, $\boldsymbol{g}_2(\boldsymbol{c})$, and $\boldsymbol{g}(\boldsymbol{c})$ are arbitrary second-rank tensor functions of $\boldsymbol{c}$. The argument goes through for all linear combinations of terms of the type (F.17), and it can be extended to include linear combinations of terms of the form

$$(\boldsymbol{R}_2)_{ijkl} = (\boldsymbol{g}_1(\boldsymbol{c}))_{il}(\boldsymbol{g}_2(\boldsymbol{c}))_{jk}, \tag{F.18}$$

which are actually equivalent to (F.17) in forming double contractions with symmetric tensors, such as $\boldsymbol{R}_2 : \boldsymbol{g}(\boldsymbol{c})$. However, the construction clearly fails for contributions of the form

$$(\boldsymbol{R}_2)_{ijkl} = (\boldsymbol{g}_1(\boldsymbol{c}))_{ij}(\boldsymbol{g}_2(\boldsymbol{c}))_{kl}. \tag{F.19}$$

Exercise 65

With the condition that $\tilde{\phi}$ can be chosen such that $\tilde{\phi}\partial\hat{s}_{\rm c}/\partial\boldsymbol{c}$ is independent of Λ, we obtain

$$\frac{d\tilde{\phi}}{d\Lambda}\frac{\partial\hat{s}_{\rm c}}{\partial\boldsymbol{c}} + \tilde{\phi}\frac{\partial^2\hat{s}_{\rm c}}{\partial\Lambda\partial\boldsymbol{c}} = 0,$$

which can be combined with the chain rule,

$$\frac{\partial}{\partial\boldsymbol{r}}\frac{\partial\hat{s}_{\rm c}}{\partial\Lambda} = \frac{\partial^2\hat{s}_{\rm c}}{\partial\Lambda^2}\frac{\partial\Lambda}{\partial\boldsymbol{r}} + \frac{\partial^2\hat{s}_{\rm c}}{\partial\Lambda\partial\boldsymbol{c}} : \frac{\partial\boldsymbol{c}}{\partial\boldsymbol{r}},$$

to obtain

$$-\frac{\partial}{\partial\boldsymbol{r}}\frac{\partial\hat{s}_{\rm c}}{\partial\Lambda} - \frac{1}{\tilde{\phi}}\frac{d\tilde{\phi}}{d\Lambda}\frac{\partial\boldsymbol{c}}{\partial\boldsymbol{r}} : \frac{\partial\hat{s}_{\rm c}}{\partial\boldsymbol{c}} = -\frac{\partial^2\hat{s}_{\rm c}}{\partial\Lambda^2}\frac{\partial\Lambda}{\partial\boldsymbol{r}}.$$

For $\epsilon_{\rm c} = 0$, we have $f_{\rm c} = -\rho T\hat{s}_{\rm c}$, and we thus obtain (4.85). Equations (4.76)–(4.78) suggest the thermodynamic force

$$X_\Lambda^{\rm diff} = -\left(\frac{\partial}{\partial\boldsymbol{r}}\frac{1}{\rho T}\frac{\partial f_{\rm c}}{\partial\Lambda} + \frac{1}{\tilde{\phi}\rho T}\frac{d\tilde{\phi}}{d\Lambda}\frac{\partial\boldsymbol{c}}{\partial\boldsymbol{r}} : \frac{\partial f_{\rm c}}{\partial\boldsymbol{c}}\right),$$

which, by means of (4.85), can be simplified to

$$X_\Lambda^{\rm diff} = \frac{\partial^2\hat{s}_{\rm c}}{\partial\Lambda^2}\frac{\partial\Lambda}{\partial\boldsymbol{r}}.$$

We still have

$$J_\Lambda^{\rm diff} = \Lambda\rho\boldsymbol{D}' \cdot X_\Lambda^{\rm diff}.$$

Exercise 66

For the probability density

$$f(\boldsymbol{x}) = \frac{1}{(2\pi)^{3/2}(\det \boldsymbol{c})^{1/2}} \exp\left\{-\frac{1}{2}\boldsymbol{x}\cdot\boldsymbol{c}^{-1}\cdot\boldsymbol{x}\right\},$$

we obtain the desired result (note that $\boldsymbol{c}_{\text{eq}} = \mathbf{1}$),

$$\begin{aligned} -k_{\text{B}}\int f(\boldsymbol{x})\ln\frac{f(\boldsymbol{x})}{f_{\text{eq}}(\boldsymbol{x})}d^3x &= \frac{k_{\text{B}}}{2}\int[\ln(\det \boldsymbol{c}) + \boldsymbol{x}\boldsymbol{x} : (\boldsymbol{c}^{-1} - \boldsymbol{c}_{\text{eq}}^{-1})]f(\boldsymbol{x})d^3x \\ &= \frac{k_{\text{B}}}{2}[\text{tr}(\mathbf{1} - \boldsymbol{c}) + \ln(\det \boldsymbol{c})]. \end{aligned}$$

Exercise 67

From (4.35), we have

$$\frac{\delta S}{\delta \boldsymbol{c}} = \rho\frac{\partial \hat{s}_{\text{c}}}{\partial \boldsymbol{c}},$$

and thus

$$\frac{\delta S}{\delta \boldsymbol{c}} = \frac{1}{2}\Lambda\frac{\rho}{m}k_{\text{B}}(\boldsymbol{c}^{-1} - \mathbf{1}).$$

Exercise 68

We perform the expansion in the coordinate system in which $\boldsymbol{c}$ is diagonal, where the eigenvalues c_1, c_2, c_3 are close to unity. We then have

$$\text{tr}(\mathbf{1} - \boldsymbol{c}) = 1 - c_1 + 1 - c_2 + 1 - c_3$$

and

$$\ln(\det \boldsymbol{c}) = \ln(c_1c_2c_3) = \ln c_1 + \ln c_2 + \ln c_3,$$

with

$$\ln c_j \approx (c_j - 1) - \frac{1}{2}(c_j - 1)^2.$$

By combining these results, we obtain

$$\frac{k_{\text{B}}}{2}[\text{tr}(\mathbf{1} - \boldsymbol{c}) + \ln(\det \boldsymbol{c})] \approx -\frac{k_{\text{B}}}{4}[(c_1 - 1)^2 + (c_2 - 1)^2 + (c_3 - 1)^2]$$

or the final result

$$\frac{k_{\text{B}}}{2}[\text{tr}(\mathbf{1} - \boldsymbol{c}) + \ln(\det \boldsymbol{c})] \approx -\frac{k_{\text{B}}}{4}(\mathbf{1} - \boldsymbol{c}) : (\mathbf{1} - \boldsymbol{c}). \tag{F.20}$$

Exercise 69

From (4.84) and (4.88), we obtain

$$\hat{\boldsymbol{R}}_2 = \frac{2m}{\rho\Lambda k_{\text{B}}}\frac{1}{\tau}\boldsymbol{c}$$

and thus

$$c \cdot \hat{R}_2^{-1} = \frac{\rho \Lambda k_B}{2m} \tau \mathbf{1}.$$

By inserting this result into (4.82), we obtain the desired simplification.

Exercise 70

In components, (4.93) for shear flow takes the form

$$\begin{aligned}\frac{\partial}{\partial t}\begin{pmatrix} c_{xx} & c_{xy} & 0 \\ c_{xy} & c_{yy} & 0 \\ 0 & 0 & c_{zz} \end{pmatrix} &= \begin{pmatrix} 2\dot{\gamma}c_{xy} & \dot{\gamma}c_{yy} & 0 \\ \dot{\gamma}c_{yy} & 0 & 0 \\ 0 & 0 & 0 \end{pmatrix} \\ &- \frac{1}{\tau}\begin{pmatrix} c_{xx}-1 & c_{xy} & 0 \\ c_{xy} & c_{yy}-1 & 0 \\ 0 & 0 & c_{zz}-1 \end{pmatrix}.\end{aligned}$$

We see that c_{yy} and c_{zz} cannot be affected by a shear flow, and they will always relax to their equilibrium values $c_{yy} = c_{zz} = 1$. We then obtain the time-evolution equations

$$\frac{\partial c_{xy}}{\partial t} = \dot{\gamma} - \frac{1}{\tau} c_{xy}, \tag{F.21}$$

$$\frac{\partial c_{xx}}{\partial t} = 2\dot{\gamma} c_{xy} - \frac{1}{\tau}(c_{xx} - 1). \tag{F.22}$$

Exercise 71

By solving the linear differential equation (F.21), we obtain

$$c_{xy}(t) = \int_{-\infty}^{t} e^{-(t-t')/\tau} \dot{\gamma}(t') dt',$$

and, from (4.87), the shear stress

$$\tau_{xy}^{(S)}(t) = -n_p k_B T \int_{-\infty}^{t} e^{-(t-t')/\tau} \dot{\gamma}(t') dt'.$$

According to (4.1), this implies the shear relaxation modulus

$$G(t) = n_p k_B T e^{-t/\tau}.$$

In view of (4.19), this relaxation modulus corresponds to a single-mode relaxation spectrum with the relaxation time $\tau_1 = \tau$ and the viscosity $\eta_1 = n_p k_B T \tau$. In addition, there is an instantaneous viscous solvent contribution to the shear relaxation modulus.

Exercise 72

Under steady shear flow conditions, (F.21) and (F.22) become

$$c_{xy} = \dot{\gamma}\tau$$

and

$$c_{xx} = 1 + 2(\dot{\gamma}\tau)^2,$$

in addition to $c_{yy} = c_{zz} = 1$. Equation (4.87) implies

$$\boldsymbol{\tau}^{(S)} = -n_{\text{p}} k_{\text{B}} T \begin{pmatrix} 2(\dot{\gamma}\tau)^2 & \dot{\gamma}\tau & 0 \\ \dot{\gamma}\tau & 0 & 0 \\ 0 & 0 & 0 \end{pmatrix},$$

and this leads to

$$\eta = \eta_{\text{s}} + n_{\text{p}} k_{\text{B}} T \tau,$$

where the additional solvent viscosity η_{s} has been taken into account,

$$\Psi_1 = 2 n_{\text{p}} k_{\text{B}} T \tau^2,$$

$$\Psi_2 = 0.$$

Exercise 73

Because no configurational contributions to energy exist, (4.86) implies

$$X_c^{\text{relax}} = -\frac{1}{T}\frac{\partial f_{\text{c}}}{\partial \boldsymbol{c}} = \rho \frac{\partial \hat{s}_{\text{c}}}{\partial \boldsymbol{c}} = -\frac{1}{2}\rho\Lambda\frac{k_{\text{B}}}{m}(\mathbf{1} - \boldsymbol{c}^{-1}) = -\frac{1}{2}\rho\Lambda\frac{k_{\text{B}}}{m}\boldsymbol{c}^{-1}\cdot(\boldsymbol{c} - \mathbf{1}).$$

With (4.60) and (4.88), we further obtain

$$J_c^{\text{relax}} = -\frac{1}{\tau}(\boldsymbol{c} - \mathbf{1}).$$

As noted in the general context, the flux J_c^{relax} is given by the relaxational contribution to $d\boldsymbol{c}/dt$, which is the last term in (4.93).

Exercise 74

For homogeneous steady-state flows, (4.98) and (4.101) imply

$$-\frac{k_{\text{B}}}{m}\Lambda\tau \text{tr}\boldsymbol{\kappa} = \text{tr}\left\{[(1-\alpha)\mathbf{1} + \alpha\boldsymbol{c}]\cdot\frac{\partial \hat{s}_{\text{c}}}{\partial \boldsymbol{c}}\right\}.$$

Therefore, for $\alpha = 0$, incompressibility implies traceless $\partial \hat{s}_{\text{c}}/\partial \boldsymbol{c}$. In general, we have

$$\alpha = -\text{tr}\left(\frac{\partial \hat{s}_{\text{c}}}{\partial \boldsymbol{c}}\right)\left\{\text{tr}\left[(\boldsymbol{c} - \mathbf{1})\cdot\frac{\partial \hat{s}_{\text{c}}}{\partial \boldsymbol{c}}\right]\right\}^{-1}. \tag{F.23}$$

Exercise 75

The form (4.103) is obtained as in the solution to Exercise 70, where h' is given by

$$h' = h(\text{tr}\boldsymbol{c}) = h\left(\frac{1}{h'}\left(3 + 2\frac{\dot{\gamma}^2\tau^2}{h'^2}\right)\right).$$

For the FENE-P model, the explicit form of this equation is

$$h'^3 - h'^2 = \frac{2\dot{\gamma}^2\tau^2}{b+3}.$$

The pressure tensor (4.102) generally leads to the viscometric functions

$$\eta = \eta_s + \frac{n_p k_B T \tau}{h'},$$

$$\Psi_1 = \frac{2 n_p k_B T \tau^2}{h'^2},$$

and

$$\Psi_2 = 0.$$

Exercise 76
From (4.45), we obtain

$$\boldsymbol{\tau}^{(S)} = \boldsymbol{\pi}^{(S)} - T\left(s - \rho\frac{\partial s}{\partial \rho} - \epsilon\frac{\partial s}{\partial \epsilon}\right)\mathbf{1} = T\rho\left(2\boldsymbol{c}\cdot\frac{\partial \hat{s}_c}{\partial \boldsymbol{c}} + \boldsymbol{g}\frac{\partial \hat{s}_c}{\partial \Lambda}\right),$$

and thus with (4.105)

$$\boldsymbol{\tau}^{(S)} = Z\frac{N_A \rho k_B T}{M_p}\left\{\mathbf{1} - 3\boldsymbol{S} - 3\boldsymbol{S}[\Lambda^2 - 1 - \Lambda(\Lambda - q)H(\Lambda - q)]\right\}.$$

After minor simplifications, we obtain the desired result.

Exercise 77
From the definition (4.104) of the orientation tensor, we obtain

$$\frac{\partial \boldsymbol{S}}{\partial t} = \frac{1}{\mathrm{tr}\boldsymbol{c}}\left(\frac{\partial \boldsymbol{c}}{\partial t} - \boldsymbol{S}\mathrm{tr}\frac{\partial \boldsymbol{c}}{\partial t}\right),$$

and a similar equation for $\partial\boldsymbol{S}/\partial\boldsymbol{r}$ (or, for the substantial time derivative). With (4.108), we then get

$$\frac{\partial \boldsymbol{S}}{\partial t} = -\boldsymbol{v}\cdot\frac{\partial \boldsymbol{S}}{\partial \boldsymbol{r}} + \boldsymbol{\kappa}\cdot\boldsymbol{S} + \boldsymbol{S}\cdot\boldsymbol{\kappa}^T - 2\boldsymbol{S}\mathrm{tr}(\boldsymbol{\kappa}\cdot\boldsymbol{S}) - \frac{1}{\tau_b}[\ldots] + \frac{1}{\tau_b}\boldsymbol{S}\mathrm{tr}[\ldots],$$

where [...] represents the tensor in square brackets in (4.108). After combining the terms for each of the coefficients α_j, we find (4.109).

Exercise 78
We are interested only in the lowest-order deviation of the orientation tensor from its equilibrium form, $\boldsymbol{S}' = \boldsymbol{S} - (1/3)\mathbf{1}$, where $\boldsymbol{S}'$ is linear in the velocity gradients. For homogeneous flows, (4.109) leads to the following linearized equation,

$$\frac{\partial \boldsymbol{S}'}{\partial t} = -\frac{1}{3}(\boldsymbol{\kappa} + \boldsymbol{\kappa}^T) - \frac{1}{\tau_b}(1 + \alpha_1 + 3\alpha_2)\boldsymbol{S}'.$$

From the solution

$$\boldsymbol{S}'(t) = -\frac{1}{3}\int_{-\infty}^{t} e^{-(1+\alpha_1+3\alpha_2)(t-t')/\tau_\text{b}}[\boldsymbol{\kappa}(t') + \boldsymbol{\kappa}(t')^T]dt'$$

and (4.106), we obtain

$$\boldsymbol{\tau}^{(S)}(t) = Z\frac{N_\text{A}\rho}{M_\text{p}}k_\text{B}T\int_{-\infty}^{t} e^{-(1+\alpha_1+3\alpha_2)(t-t')/\tau_\text{b}}[\boldsymbol{\kappa}(t') + \boldsymbol{\kappa}(t')^T]dt'.$$

According to (4.1), the shear relaxation modulus is

$$G(t) = Z\frac{N_\text{A}\rho}{M_\text{p}}k_\text{B}Te^{-(1+\alpha_1+3\alpha_2)t/\tau_\text{b}},$$

which represents single-mode relaxation with

$$\tau_1 = \frac{\tau_\text{b}}{1+\alpha_1+3\alpha_2}, \qquad \eta_1 = Z\frac{N_\text{A}\rho}{M_\text{p}}k_\text{B}T\frac{\tau_\text{b}}{1+\alpha_1+3\alpha_2}.$$

Exercise 79

In shear flow, the orientation tensor $\boldsymbol{S}$ must be of the form

$$3\boldsymbol{S} = \left[1 - \frac{1}{3}(\bar{y}+\bar{z})\dot{\gamma}^2\right]\boldsymbol{1} + \bar{x}(\boldsymbol{\kappa}+\boldsymbol{\kappa}^T) + \bar{y}\boldsymbol{\kappa}\cdot\boldsymbol{\kappa}^T + \bar{z}\boldsymbol{\kappa}^T\cdot\boldsymbol{\kappa}. \tag{F.24}$$

To second order in the shear rate, we then get

$$(3\boldsymbol{S})^2 = \left[1 - \frac{2}{3}(\bar{y}+\bar{z})\dot{\gamma}^2\right]\boldsymbol{1} + 2\bar{x}(\boldsymbol{\kappa}+\boldsymbol{\kappa}^T) + (\bar{x}^2+2\bar{y})\boldsymbol{\kappa}\cdot\boldsymbol{\kappa}^T + (\bar{x}^2+2\bar{z})\boldsymbol{\kappa}^T\cdot\boldsymbol{\kappa},$$

$$(3\boldsymbol{S})^{-1} = \left[1 + \frac{1}{3}(\bar{y}+\bar{z})\dot{\gamma}^2\right]\boldsymbol{1} - \bar{x}(\boldsymbol{\kappa}+\boldsymbol{\kappa}^T) + (\bar{x}^2-\bar{y})\boldsymbol{\kappa}\cdot\boldsymbol{\kappa}^T + (\bar{x}^2-\bar{z})\boldsymbol{\kappa}^T\cdot\boldsymbol{\kappa},$$

and

$$3(\boldsymbol{\kappa}\cdot\boldsymbol{S} + \boldsymbol{S}\cdot\boldsymbol{\kappa}^T) = (\boldsymbol{\kappa}+\boldsymbol{\kappa}^T) + 2\bar{x}\boldsymbol{\kappa}\cdot\boldsymbol{\kappa}^T.$$

By inserting these expressions into the steady-state version of (4.109),

$$\begin{aligned} 9(\boldsymbol{S}^2 - \boldsymbol{S}\text{tr}\boldsymbol{S}^2) + \alpha_0(3\boldsymbol{S}-\boldsymbol{1}) + \alpha_2\left(\boldsymbol{S}\text{tr}\boldsymbol{S}^{-1} - \boldsymbol{S}^{-1}\right) = \\ 3\tau_\text{b}[\boldsymbol{\kappa}\cdot\boldsymbol{S} + \boldsymbol{S}\cdot\boldsymbol{\kappa}^T - 2\boldsymbol{S}\text{tr}(\boldsymbol{\kappa}\cdot\boldsymbol{S})], \end{aligned} \tag{F.25}$$

we obtain by comparing the coefficients of $\boldsymbol{\kappa}+\boldsymbol{\kappa}^T$, $\boldsymbol{\kappa}\cdot\boldsymbol{\kappa}^T$, and $\boldsymbol{\kappa}^T\cdot\boldsymbol{\kappa}$,

$$\bar{x} = \frac{\tau_\text{b}}{1+\alpha_0+6\alpha_2} = \frac{1}{1+\alpha_1+3\alpha_2}\tau_\text{b},$$

$$\bar{y} = \frac{2\tau_\text{b}\bar{x} - (1-3\alpha_2)\bar{x}^2}{1+\alpha_0+6\alpha_2} = \frac{1+2\alpha_1+9\alpha_2}{(1+\alpha_1+3\alpha_2)^3}\tau_\text{b}^2,$$

and

$$\bar{z} = -\frac{(1-3\alpha_2)\bar{x}^2}{1+\alpha_0+6\alpha_2} = -\frac{1-3\alpha_2}{(1+\alpha_1+3\alpha_2)^3}\tau_{\mathrm{b}}^2,$$

respectively. For the zero-shear-rate limits of the viscosity and of the normal-stress coefficients, we obtain from (4.106) for $\Lambda = 1$

$$\eta = Z\frac{N_{\mathrm{A}}\rho}{M_{\mathrm{p}}}k_{\mathrm{B}}T\bar{x},$$

$$\Psi_1 = Z\frac{N_{\mathrm{A}}\rho}{M_{\mathrm{p}}}k_{\mathrm{B}}T(\bar{y}-\bar{z}) = Z\frac{N_{\mathrm{A}}\rho}{M_{\mathrm{p}}}k_{\mathrm{B}}T2\bar{x}^2,$$

and

$$\Psi_2 = Z\frac{N_{\mathrm{A}}\rho}{M_{\mathrm{p}}}k_{\mathrm{B}}T\bar{z}.$$

We have $\eta = \eta_1$ and $\Psi_1 = 2\eta\tau_1$, where η_1 and τ_1 are given in the solution to Exercise 78.

Exercise 80

We again start form the general form (F.24) of the orientation tensor in steady shear flow, from which we obtain the following exact generalizations of the corresponding expressions given in the solution to Exercise 79,

$$\begin{aligned}(3\boldsymbol{S})^2 &= \left[1-\frac{1}{3}(\bar{y}+\bar{z})\dot{\gamma}^2\right]^2 \mathbf{1} + 2\bar{x}\left[1+\frac{1}{6}(\bar{y}+\bar{z})\dot{\gamma}^2\right](\boldsymbol{\kappa}+\boldsymbol{\kappa}^T) \\ &+ \left[\bar{x}^2 + 2\bar{y}\left(1+\frac{1}{6}\bar{y}\dot{\gamma}^2 - \frac{1}{3}\bar{z}\dot{\gamma}^2\right)\right]\boldsymbol{\kappa}\cdot\boldsymbol{\kappa}^T \\ &+ \left[\bar{x}^2 + 2\bar{z}\left(1-\frac{1}{3}\bar{y}\dot{\gamma}^2 + \frac{1}{6}\bar{z}\dot{\gamma}^2\right)\right]\boldsymbol{\kappa}^T\cdot\boldsymbol{\kappa},\end{aligned}$$

$$\begin{aligned}(3\boldsymbol{S})^{-1} &= \frac{1}{1-\frac{1}{3}(\bar{y}+\bar{z})\dot{\gamma}^2}\left(1-\frac{\boldsymbol{\kappa}\cdot\boldsymbol{\kappa}^T}{\dot{\gamma}^2}-\frac{\boldsymbol{\kappa}^T\cdot\boldsymbol{\kappa}}{\dot{\gamma}^2}\right) \\ &- \frac{1}{\left(1-\frac{1}{3}\bar{y}\dot{\gamma}^2+\frac{2}{3}\bar{z}\dot{\gamma}^2\right)\left(1+\frac{2}{3}\bar{y}\dot{\gamma}^2-\frac{1}{3}\bar{z}\dot{\gamma}^2\right)-\bar{x}^2\dot{\gamma}^2}\Bigg[\bar{x}(\boldsymbol{\kappa}+\boldsymbol{\kappa}^T) \\ &- \left(1-\frac{1}{3}\bar{y}\dot{\gamma}^2+\frac{2}{3}\bar{z}\dot{\gamma}^2\right)\frac{\boldsymbol{\kappa}\cdot\boldsymbol{\kappa}^T}{\dot{\gamma}^2} - \left(1+\frac{2}{3}\bar{y}\dot{\gamma}^2-\frac{1}{3}\bar{z}\dot{\gamma}^2\right)\frac{\boldsymbol{\kappa}^T\cdot\boldsymbol{\kappa}}{\dot{\gamma}^2}\Bigg],\end{aligned}$$

and

$$3(\boldsymbol{\kappa}\cdot\boldsymbol{S}+\boldsymbol{S}\cdot\boldsymbol{\kappa}^T) = \left(1-\frac{1}{3}\bar{y}\dot{\gamma}^2+\frac{2}{3}\bar{z}\dot{\gamma}^2\right)(\boldsymbol{\kappa}+\boldsymbol{\kappa}^T) + 2\bar{x}\boldsymbol{\kappa}\cdot\boldsymbol{\kappa}^T.$$

From (F.25), we then obtain the following equations for $\bar{x}$, $\bar{y}$, $\bar{z}$:

$$\bar{x}\lambda_1 = \tau_{\mathrm{b}}\left[1-\frac{1}{3}(2\bar{x}^2+\bar{y}-2\bar{z})\dot{\gamma}^2\right], \tag{F.26}$$

$$\bar{y}(\lambda_1 - \bar{z}\dot{\gamma}^2) = 2\tau_b\bar{x}\left(1 - \frac{1}{3}\bar{y}\dot{\gamma}^2\right) - \bar{x}^2 + \lambda_2, \tag{F.27}$$

$$\bar{z}(\lambda_1 - \bar{y}\dot{\gamma}^2) = -\frac{2}{3}\tau_b\bar{x}\bar{z}\dot{\gamma}^2 - \bar{x}^2 + \lambda_2, \tag{F.28}$$

where the auxiliary quantities λ_1 and λ_2 are given by

$$\begin{aligned} \lambda_1 &= 1 - \frac{1}{3}(2\bar{x}^2 - \bar{y} - \bar{z})\dot{\gamma}^2 - \frac{2}{9}(\bar{y}^2 + \bar{z}^2 - \bar{y}\bar{z})\dot{\gamma}^4 \\ &+ \alpha_1 - (\alpha_2 + 3\alpha_3)\left[3 - \frac{1}{1 - \frac{1}{3}(\bar{y} + \bar{z})\dot{\gamma}^2}\right] \\ &+ \frac{5\alpha_2 + 6\alpha_3 + \left(\frac{1}{3}\alpha_2 + \alpha_3\right)(\bar{y} + \bar{z})\dot{\gamma}^2}{(1 - \frac{1}{3}\bar{y}\dot{\gamma}^2 + \frac{2}{3}\bar{z}\dot{\gamma}^2)(1 + \frac{2}{3}\bar{y}\dot{\gamma}^2 - \frac{1}{3}\bar{z}\dot{\gamma}^2) - \bar{x}^2\dot{\gamma}^2} \end{aligned} \tag{F.29}$$

and

$$\lambda_2 = \frac{3\alpha_2}{\dot{\gamma}^2}\left[\frac{1 + \frac{2}{3}(\bar{y} + \bar{z})\dot{\gamma}^2}{(1 - \frac{1}{3}\bar{y}\dot{\gamma}^2 + \frac{2}{3}\bar{z}\dot{\gamma}^2)(1 + \frac{2}{3}\bar{y}\dot{\gamma}^2 - \frac{1}{3}\bar{z}\dot{\gamma}^2) - \bar{x}^2\dot{\gamma}^2} - \frac{1}{1 - \frac{1}{3}(\bar{y} + \bar{z})\dot{\gamma}^2}\right]. \tag{F.30}$$

If we eliminate λ_2 by subtracting (F.27) and (F.28), and if we use the resulting equation to eliminate λ_1 from (F.26), we then obtain the compact general relationship

$$\frac{2\bar{x}^2}{\bar{y} - \bar{z}} = 1 - \frac{1}{3}(\bar{y} - 2\bar{z})\dot{\gamma}^2 \tag{F.31}$$

between $\bar{x}$, $\bar{y}$, and $\bar{z}$ for the pompon model.

We now turn to the special case $\alpha_2 = \alpha_3 = 0$. We then have $\lambda_2 = 0$, and we obtain

$$\bar{z}\bar{y}(\bar{y} - \bar{z})\dot{\gamma}^2 = 2\tau_b\bar{x}\bar{z} + (\bar{y} - \bar{z})\bar{x}^2$$

by multiplying (F.27) with $\bar{z}$ and (F.28) with $\bar{y}$ and taking the difference of the resulting equations. If we eliminate $\bar{x}^2$ by means of (F.31), and $\bar{x}$ after a further squaring, then we obtain a direct relationship between $\bar{y}$ and $\bar{z}$:

$$\frac{1}{2}(\bar{y} - \bar{z})\frac{\left[\frac{1}{2}(\bar{y} - \bar{z}) - \frac{1}{6}(\bar{y}^2 + 3\bar{y}\bar{z} + 2\bar{z}^2)\dot{\gamma}^2\right]^2}{1 - \frac{1}{3}(\bar{y} - 2\bar{z})\dot{\gamma}^2} = \bar{z}^2\tau_b^2. \tag{F.32}$$

With

$$\begin{aligned} \lambda_1 &= 1 + \alpha_1 - \frac{1}{3}(2\bar{x}^2 - \bar{y} - \bar{z})\dot{\gamma}^2 - \frac{2}{9}(\bar{y}^2 + \bar{z}^2 - \bar{y}\bar{z})\dot{\gamma}^4 \\ &= 1 + \alpha_1 + \frac{2}{3}\bar{z}\dot{\gamma}^2 - \frac{1}{9}\bar{y}(\bar{y} + \bar{z})\dot{\gamma}^4, \end{aligned}$$

and (F.26) in the modified form

$$\bar{x}\lambda_1 = \tau_b\left[1 - \frac{1}{3}(2\bar{y} - 3\bar{z})\dot{\gamma}^2 + \frac{1}{9}(\bar{y}^2 - 3\bar{y}\bar{z} + 2\bar{z}^2)\dot{\gamma}^4\right],$$

we obtain after forming the ratio and eliminating the remaining $\bar{x}$ by means of (F.31):

$$\frac{1}{2}(\bar{y}-\bar{z})\left[1-\frac{1}{3}(\bar{y}-2\bar{z})\dot{\gamma}^2\right]\left[\frac{1+\alpha_1+\frac{2}{3}\bar{z}\dot{\gamma}^2-\frac{1}{9}\bar{y}(\bar{y}+\bar{z})\dot{\gamma}^4}{1-\frac{1}{3}(2\bar{y}-3\bar{z})\dot{\gamma}^2+\frac{1}{9}(\bar{y}^2-3\bar{y}\bar{z}+2\bar{z}^2)\dot{\gamma}^4}\right]^2=\tau_b^2. \tag{F.33}$$

For $\alpha_2 = \alpha_3 = 0$, we thus have (F.31), (F.32), and (F.33) for determining $\bar{x}$, $\bar{y}$, and $\bar{z}$, and hence the orientation tensor, in steady shear flow. A closer consideration of the behavior for high shear rates suggests the following asymptotic results:

$$\bar{x}=\frac{1}{\dot{\gamma}}\left(\frac{9}{4}\frac{\alpha_1}{\tau_b\dot{\gamma}}\right)^{1/3},$$

$$\bar{y}=\frac{3}{\dot{\gamma}^2}-\frac{2}{\dot{\gamma}^2}\left(\frac{9}{4}\frac{\alpha_1}{\tau_b\dot{\gamma}}\right)^{2/3},$$

$$\bar{z}=-\frac{3}{2}\frac{1}{\dot{\gamma}^2}\frac{1}{\tau_b\dot{\gamma}}\left(\frac{9}{4}\frac{\alpha_1}{\tau_b\dot{\gamma}}\right)^{1/3}.$$

To understand the viscometric functions at high shear rates, we further need the chain stretch in steady shear flow. From (4.112) we obtain

$$\tau_s\Lambda\boldsymbol{\kappa}:\boldsymbol{S}=\frac{1}{3}\tau_s\Lambda\dot{\gamma}^2\bar{x}=\begin{cases}\Lambda-1 & \text{for } \Lambda\le q\\ q & \text{for } \Lambda>q\end{cases}. \tag{F.34}$$

Taking the condition $\tau_b \gg \tau_s$ into account, we conclude that for small and intermediate $\tau_b\dot{\gamma}$ the stretch remains close to unity, and we obtain

$$\eta=Z\frac{N_A\rho}{M_p}k_BT\bar{x}\propto\dot{\gamma}^{-4/3},$$

$$\Psi_1=Z\frac{N_A\rho}{M_p}k_BT(\bar{y}-\bar{z})\propto\dot{\gamma}^{-2},$$

and

$$\Psi_2=Z\frac{N_A\rho}{M_p}k_BT\bar{z}\propto\dot{\gamma}^{-10/3}.$$

At the high dimensionless shear rate

$$\tau_b\dot{\gamma}=\left(3\frac{q-1}{q}\right)^{3/2}\left(\frac{4}{9\alpha_1}\right)^{1/2}\left(\frac{\tau_b}{\tau_s}\right)^{3/2},$$

($\tau_b \gg \tau_s$), $\Lambda = q$ is reached and branch-point withdrawal sets in. To find a finite steady-state stretch at even higher shear rates, one then needs a relaxation time τ_s that decreases for increasing Λ.

Exercise 81

To obtain the previously used normalization, we define

$$\boldsymbol{c}'=\frac{H}{k_BT}\int\boldsymbol{Q}\boldsymbol{Q}f(\boldsymbol{r},\boldsymbol{Q})d^3Q,$$

to be used along with (4.114). By integrating (4.127) over $\boldsymbol{Q}$, we obtain

$$\frac{\partial n_{\text{p}}}{\partial t} = -\frac{\partial}{\partial \boldsymbol{r}} \cdot (\boldsymbol{v} n_{\text{p}}) + \frac{\partial}{\partial \boldsymbol{r}} \cdot \frac{k_{\text{B}}T}{2\zeta} \frac{\partial}{\partial \boldsymbol{r}} n_{\text{p}}.$$

After multiplying (4.127) by $\boldsymbol{QQ}$ and integrating over $\boldsymbol{Q}$, we obtain after several integrations by part

$$\frac{\partial \boldsymbol{c}'}{\partial t} = -\frac{\partial}{\partial \boldsymbol{r}} \cdot (\boldsymbol{v}\boldsymbol{c}') + \boldsymbol{\kappa} \cdot \boldsymbol{c}' + \boldsymbol{c}' \cdot \boldsymbol{\kappa}^T - \frac{4H}{\zeta}(\boldsymbol{c}' - n_{\text{p}}\boldsymbol{1}) + \frac{\partial}{\partial \boldsymbol{r}} \cdot \frac{k_{\text{B}}T}{2\zeta} \frac{\partial}{\partial \boldsymbol{r}} \boldsymbol{c}'.$$

By comparison with (4.91) and (4.92), we find

$$\boldsymbol{D} = \frac{k_{\text{B}}T}{2\zeta} \boldsymbol{1},$$

and

$$\tau = \frac{\zeta}{4H}.$$

Note that, for the comparison, the solvent density ρ was assumed to be independent of position.

Exercise 82

For all $j, k = 1, 2, \ldots N-1$, we have

$$2Q_{jk}^{\text{Rouse}} - Q_{j-1\,k}^{\text{Rouse}} - Q_{j+1\,k}^{\text{Rouse}} = 2\left(1 - \cos\frac{k\pi}{N}\right) Q_{jk}^{\text{Rouse}} = 4\left(\sin\frac{k\pi}{2N}\right)^2 Q_{jk}^{\text{Rouse}},$$

where standard formulas for trigonometric functions have been used;[6] for $j = 1$ and $j = N-1$, the terms $\sin(0k\pi/N) = 0$ and $\sin(Nk\pi/N) = 0$, respectively, have been subtracted in order to keep the same structure as for intermediate j. The column vectors of Q^{Rouse} are hence eigenvectors of the symmetric Rouse matrix. Because all eigenvalues are different, these eigenvectors must be orthogonal. The proper normalization follows from

$$\sum_{k=1}^{N-1} Q_{kj}^{\text{Rouse}} Q_{kj}^{\text{Rouse}} = \frac{2}{N} \sum_{k=1}^{N-1} \left(\sin\frac{jk\pi}{N}\right)^2 = 1$$

for all j.[7]

Exercise 83

From (4.3) and (4.33) we obtain with the relaxation modulus (4.135)

$$\eta = \eta_{\text{s}} + n_{\text{p}} k_{\text{B}} T \sum_{k=1}^{N-1} \tau_k,$$

[6] See 1.314.1 and 1.321.1 of Gradshteyn & Ryzhik, *Tables* (Academic, 1980).

[7] See 1.351.1 of Gradshteyn & Ryzhik, *Tables* (Academic, 1980).

and

$$\Psi_1 = 2n_\mathrm{p}k_\mathrm{B}T\sum_{k=1}^{N-1}\tau_k^2.$$

In spite of the complicated form of the τ_k given in (4.133), the sums in these expressions can be carried out in closed form. For small λ, we have the auxiliary equation[8]

$$\sum_{k=1}^{N-1}\left(\sin^2\frac{k\pi}{2N}+\frac{\lambda^2}{4}\right)^{-1} = \frac{2}{3}(N^2-1) - \frac{1}{45}(N^2-1)(2N^2+7)\lambda^2.$$

With the auxiliary equation for $\lambda = 0$ we obtain

$$\eta = \eta_\mathrm{s} + \frac{1}{3}(N^2-1)n_\mathrm{p}k_\mathrm{B}T\frac{\zeta}{4H}.$$

By comparing the terms proportional to λ^2 in the auxiliary equation we find

$$\Psi_1 = \frac{2}{45}(N^2-1)(2N^2+7)n_\mathrm{p}k_\mathrm{B}T\left(\frac{\zeta}{4H}\right)^2.$$

Exercise 84

The degeneracy requirement (1.4) implies the following condition for the configurational contribution to the pressure tensor:

$$\frac{\partial}{\partial \boldsymbol{r}}\cdot\frac{\boldsymbol{\tau}^{(S)}}{T} = \int_0^1\int\left(-L_{24}\frac{\delta S}{\delta f} + \rho\frac{N_\mathrm{A}}{M_\mathrm{e}}k_\mathrm{B}\frac{\partial}{\partial \boldsymbol{r}}f\ln f\right)d^2u d\sigma.$$

After inserting (4.146) for L_{24}, we obtain

$$\frac{\partial}{\partial \boldsymbol{r}}\cdot\frac{\boldsymbol{\tau}^{(S)}}{T} = \int_0^1\int\left(\frac{\delta S}{\delta f}\frac{\partial f}{\partial \boldsymbol{r}} + \frac{\partial}{\partial \boldsymbol{r}}\cdot f\boldsymbol{u}\boldsymbol{P}(\boldsymbol{u})\cdot\frac{\partial}{\partial \boldsymbol{u}}\frac{\delta S}{\delta f} + \rho\frac{N_\mathrm{A}}{M_\mathrm{e}}k_\mathrm{B}\ln f\frac{\partial f}{\partial \boldsymbol{r}}\right)d^2u d\sigma.$$

With the entropy gradient (4.144), the first and third terms in parentheses cancel, and we obtain

$$\frac{\partial}{\partial \boldsymbol{r}}\cdot\frac{\boldsymbol{\tau}^{(S)}}{T} = -\int_0^1\int\frac{\partial}{\partial \boldsymbol{r}}\cdot f\boldsymbol{u}\boldsymbol{P}(\boldsymbol{u})\cdot\frac{\partial}{\partial \boldsymbol{u}}\rho\frac{N_\mathrm{A}}{M_\mathrm{e}}k_\mathrm{B}\ln f d^2u d\sigma.$$

This leads to

$$\begin{aligned}\boldsymbol{\tau}^{(S)} &= -\rho\frac{N_\mathrm{A}}{M_\mathrm{e}}k_\mathrm{B}T\int_0^1\int\boldsymbol{u}\boldsymbol{P}(\boldsymbol{u})\cdot\frac{\partial}{\partial \boldsymbol{u}}f d^2u d\sigma \\ &= \rho\frac{N_\mathrm{A}}{M_\mathrm{e}}k_\mathrm{B}T\int_0^1\int f\frac{\partial}{\partial \boldsymbol{u}}\cdot\boldsymbol{P}(\boldsymbol{u})\boldsymbol{u}d^2u d\sigma \\ &= \rho\frac{N_\mathrm{A}}{M_\mathrm{e}}k_\mathrm{B}T\int_0^1\int f(\boldsymbol{1}-3\boldsymbol{u}\boldsymbol{u})d^2u d\sigma,\end{aligned}$$

[8] This useful auxiliary equation can, for example, be obtained by considering 1.382.2 of Gradshteyn & Ryzhik, *Tables* (Academic, 1980) for small x.

where we have used $\boldsymbol{u}^2 = 1$ after performing the differentiation of $\boldsymbol{P}(\boldsymbol{u})$ with respect to $\boldsymbol{u}$.

Exercise 85

For an arbitrary function $A(\boldsymbol{u})$, we have on the one hand

$$\frac{\partial}{\partial t} A(\widehat{\boldsymbol{u}}(\boldsymbol{u}', t, t')) = \int d^2u A(\boldsymbol{u}) \frac{\partial}{\partial t} \delta(\boldsymbol{u} - \widehat{\boldsymbol{u}}(\boldsymbol{u}', t, t')).$$

On the other hand, we have according to the chain rule

$$\begin{aligned} \frac{\partial}{\partial t} A(\widehat{\boldsymbol{u}}(\boldsymbol{u}', t, t')) &= \int d^2u \delta(\boldsymbol{u} - \widehat{\boldsymbol{u}}(\boldsymbol{u}', t, t')) \frac{\partial A(\boldsymbol{u})}{\partial \boldsymbol{u}} \cdot (\mathbf{1} - \boldsymbol{u}\boldsymbol{u}) \cdot \boldsymbol{\kappa} \cdot \boldsymbol{u} \\ &= -\int d^2u A(\boldsymbol{u}) \frac{\partial}{\partial \boldsymbol{u}} \cdot (\mathbf{1} - \boldsymbol{u}\boldsymbol{u}) \cdot \boldsymbol{\kappa} \cdot \boldsymbol{u} \delta(\boldsymbol{u} - \widehat{\boldsymbol{u}}(\boldsymbol{u}', t, t')). \end{aligned}$$

By comparing these two expressions, we obtain the the auxiliary equation (4.154).

Exercise 86

The desired expression is

$$\widehat{\boldsymbol{u}}(\boldsymbol{u}', t, t') = \frac{\left[\mathbf{1} + \int_{t'}^{t} \boldsymbol{\kappa}(t'')dt''\right] \cdot \boldsymbol{u}'}{\left|\left[\mathbf{1} + \int_{t'}^{t} \boldsymbol{\kappa}(t'')dt''\right] \cdot \boldsymbol{u}'\right|},$$

as can be verified by a one-line calculation. Namely, for any type of flow, we obtain

$$\frac{\partial \widehat{\boldsymbol{u}}(\boldsymbol{u}', t, t')}{\partial t} = [\mathbf{1} - \widehat{\boldsymbol{u}}(\boldsymbol{u}', t, t')\widehat{\boldsymbol{u}}(\boldsymbol{u}', t, t')] \cdot \boldsymbol{\kappa}(t) \cdot \frac{\boldsymbol{u}'}{\left|\left[\mathbf{1} + \int_{t'}^{t} \boldsymbol{\kappa}(t'')dt''\right] \cdot \boldsymbol{u}'\right|}. \tag{F.35}$$

For shear flow, the product $\boldsymbol{\kappa}(t) \cdot \boldsymbol{\kappa}(t'')$ vanishes so that the vector

$$\frac{\boldsymbol{u}'}{\left|\left[\mathbf{1} + \int_{t'}^{t} \boldsymbol{\kappa}(t'')dt''\right] \cdot \boldsymbol{u}'\right|}$$

in (F.35) can be replaced by $\widehat{\boldsymbol{u}}(\boldsymbol{u}', t, t')$. Thus, we have verified that the proposed expression for $\widehat{\boldsymbol{u}}(\boldsymbol{u}', t, t')$ satisfies the defining time-evolution equation.

Exercise 87

If we use the parameterization

$$\boldsymbol{u}' = \begin{pmatrix} \sin\theta \sin\varphi \\ \cos\theta \\ \sin\theta\cos\varphi \end{pmatrix},$$

the occurring integrations over φ can be performed by means of the auxiliary equation[9]

$$\int_0^{2\pi} \frac{\lambda_0 \sin^2\varphi + \lambda_1 \sin\varphi + \lambda_2}{\lambda_3 \sin\varphi + \lambda_4} \frac{d\varphi}{2\pi} = \frac{\lambda_1}{\lambda_3} - \frac{\lambda_0\lambda_4}{\lambda_3^2} + \left[\lambda_2 - \left(\frac{\lambda_1}{\lambda_3} - \frac{\lambda_0\lambda_4}{\lambda_3^2}\right)\lambda_4\right] \frac{1}{\sqrt{\lambda_3^2 - \lambda_4^2}}.$$

With the substitution $z = \cos\theta$, we thus obtain

$$\begin{aligned} Q_{xy}^{\mathrm{DE}}(\gamma) &= \frac{1}{2}\int_{-1}^{1} d\cos\theta \int_0^{2\pi} \frac{d\varphi}{2\pi} \frac{\sin\theta\cos\theta\sin\varphi + \gamma\cos^2\theta}{1 + \gamma^2\cos^2\theta + 2\gamma\sin\theta\cos\theta\sin\varphi} \\ &= \frac{1}{2\gamma}\int_0^1 \left[1 + \frac{\gamma^2 z^2 - 1}{\sqrt{(\gamma^4 + 4\gamma^2)z^4 - 2\gamma^2 z^2 + 1}}\right] dz, \end{aligned}$$

$$\begin{aligned} Q_{xx}^{\mathrm{DE}}(\gamma) &- Q_{yy}^{\mathrm{DE}}(\gamma) - \gamma Q_{xy}^{\mathrm{DE}}(\gamma) \\ &= \int_{-1}^{1} d\cos\theta \int_0^{2\pi} \frac{d\varphi}{2\pi} \frac{\sin^2\theta\sin^2\varphi + \gamma\sin\theta\cos\theta\sin\varphi - \cos^2\theta}{1 + \gamma^2\cos^2\theta + 2\gamma\sin\theta\cos\theta\sin\varphi} \\ &= \int_0^1 \frac{1}{4\gamma^2 z^2}\left[\gamma^2 z^2 - 1 - \frac{(\gamma^4 + 4\gamma^2)z^4 - 1}{\sqrt{(\gamma^4 + 4\gamma^2)z^4 - 2\gamma^2 z^2 + 1}}\right] dz \\ &= \frac{1}{4}\int_0^1 \frac{\partial}{\partial z}\left[z + \frac{1 - \sqrt{(\gamma^4 + 4\gamma^2)z^4 - 2\gamma^2 z^2 + 1}}{\gamma^2 z}\right] dz = 0, \end{aligned}$$

and

$$\begin{aligned} Q_{xx}^{\mathrm{DE}}(\gamma) &- 2Q_{yy}^{\mathrm{DE}}(\gamma) + Q_{zz}^{\mathrm{DE}}(\gamma) \\ &= \frac{1}{2}\int_{-1}^{1} d\cos\theta \int_0^{2\pi} \frac{d\varphi}{2\pi} \frac{1 + (\gamma^2 - 3)\cos^2\theta + 2\gamma\sin\theta\cos\theta\sin\varphi}{1 + \gamma^2\cos^2\theta + 2\gamma\sin\theta\cos\theta\sin\varphi} \\ &= \int_0^1 \left[1 - \frac{3z^2}{\sqrt{(\gamma^4 + 4\gamma^2)z^4 - 2\gamma^2 z^2 + 1}}\right] dz. \end{aligned}$$

The leading-order expressions of these integrals in γ are

$$Q_{xy}^{\mathrm{DE}}(\gamma) = \gamma \int_0^1 z^4 dz = \frac{1}{5}\gamma$$

[9] Obtained with 2.551.3 of Gradshteyn & Ryzhik, *Tables* (Academic, 1980).

and

$$Q_{xx}^{\rm DE}(\gamma) - 2Q_{yy}^{\rm DE}(\gamma) + Q_{zz}^{\rm DE}(\gamma) = \int_0^1 [1 - 3z^2 + 3\gamma^2(2z^6 - z^4)]dz = \frac{9}{35}\gamma^2.$$

Actually, (4.179) can be generalized to all orders in γ,[10]

$$Q_{xy}^{\rm DE}(\gamma) = \sum_{j=0}^{\infty}(-1)^j \frac{(2j+1)!!(2j+3)!!}{(4j+5)!!}\gamma^{2j+1}, \tag{F.36}$$

where $(2j+1)!!$ is the product of all odd integers from 1 to $2j+1$ (for any nonnegative integer j). This series converges for $|\gamma| < 2$.

Exercise 88
From (4.156) and (4.179), we obtain in the linear viscoelastic limit

$$\tau_{xy}^{(S)}(t) = -G(0) \int_{-\infty}^{t} dt' \left[\frac{8}{\pi^2 \tau_{\rm rept}} \sum_{\substack{j=1 \\ j \text{ odd}}}^{\infty} e^{-j^2(t-t')/\tau_{\rm rept}} \right] \int_{t'}^{t} \dot{\gamma}(t'')dt''.$$

With

$$e^{-j^2(t-t')/\tau_{\rm rept}} = \frac{\tau_{\rm rept}}{j^2} \frac{\partial}{\partial t'} e^{-j^2(t-t')/\tau_{\rm rept}},$$

we find after an integration by parts

$$\tau_{xy}^{(S)}(t) = -G(0) \int_{-\infty}^{t} dt' \left[\frac{8}{\pi^2} \sum_{\substack{j=1 \\ j \text{ odd}}}^{\infty} \frac{1}{j^2} e^{-j^2(t-t')/\tau_{\rm rept}} \right] \dot{\gamma}(t'),$$

which, according to (4.1), implies the shear relaxation modulus (4.157).

Exercise 89
From (4.157) we obtain[11]

$$\int_0^\infty G(t)dt = G(0)\frac{8}{\pi^2} \sum_{\substack{j=1 \\ j \text{ odd}}}^{\infty} \frac{\tau_{\rm rept}}{j^4} = \frac{\pi^2}{12} G(0)\tau_{\rm rept}$$

and[12]

$$\int_0^\infty tG(t)dt = G(0)\frac{8}{\pi^2} \sum_{\substack{j=1 \\ j \text{ odd}}}^{\infty} \frac{\tau_{\rm rept}^2}{j^6} = \frac{\pi^4}{120} G(0)\tau_{\rm rept}^2.$$

[10] See (3) of Öttinger, J. Non-Newtonian Fluid Mech. 37 (1990) 265.
[11] See 0.234.5 of Gradshteyn & Ryzhik, *Tables* (Academic, 1980).
[12] By expansion of 1.421.2 of Gradshteyn & Ryzhik, *Tables* (Academic, 1980) with respect to x^2, or from 0.233.5 and 9.71.

With (4.3), (4.33), and (4.158), we find (4.160) and (4.161). The results of Exercise 87 imply

$$\frac{Q_{yy}^{\rm DE}(\gamma) - Q_{zz}^{\rm DE}(\gamma)}{Q_{xx}^{\rm DE}(\gamma) - Q_{yy}^{\rm DE}(\gamma)} = -\frac{2}{7},$$

which leads to (4.164).

Exercise 90

From (4.167), we conclude that (4.145) and (4.146) need to be modified as follows:

$$L_{42} = \frac{\partial}{\partial \boldsymbol{r}} f + \frac{\partial}{\partial \boldsymbol{u}} \cdot f \boldsymbol{P} \boldsymbol{u} \cdot \frac{\partial}{\partial \boldsymbol{r}} + \frac{\partial}{\partial \sigma} f \boldsymbol{S}^{\rm drift} \cdot \frac{\partial}{\partial \boldsymbol{r}} - f \boldsymbol{S}^{\rm creation} \cdot \frac{\partial}{\partial \boldsymbol{r}}$$

and

$$L_{24} = f \frac{\partial}{\partial \boldsymbol{r}} - \frac{\partial}{\partial \boldsymbol{r}} \cdot f \boldsymbol{u} \boldsymbol{P} \cdot \frac{\partial}{\partial \boldsymbol{u}} - \frac{\partial}{\partial \boldsymbol{r}} \cdot f \boldsymbol{S}^{\rm drift} \frac{\partial}{\partial \sigma} - \frac{\partial}{\partial \boldsymbol{r}} \cdot f \boldsymbol{S}^{\rm creation}.$$

Exercise 91

From (4.156) and (4.177), we obtain

$$\eta(\dot\gamma) = 3\rho \frac{N_{\rm A}}{M_{\rm e}} k_{\rm B} T \frac{1}{\dot\gamma} \int_{-\infty}^{t} dt' \left[\frac{8}{\pi^2 \tau_{\rm rept}} \sum_{\substack{j=1 \\ j \text{ odd}}}^{\infty} e^{-j^2 (t-t')/\tau_{\rm rept}} \right] Q_{xy}^{\rm DE}(\dot\gamma(t-t')),$$

which can be further evaluated by means of (F.36) given in the solution to Exercise 87:

$$\begin{aligned} \eta(\dot\gamma) &= 24\rho \frac{N_{\rm A}}{M_{\rm e}} k_{\rm B} T \tau_{\rm rept} \sum_{j=0}^{\infty} (-1)^j \frac{(2j+1)!!(2j+3)!!}{(4j+5)!!} \\ &\times (\tau_{\rm rept}\dot\gamma)^{2j} \frac{(2j+1)!}{\pi^2} \sum_{\substack{k=1 \\ k \text{ odd}}}^{\infty} \frac{1}{k^{4j+4}}. \end{aligned}$$

The expansion (4.173) follows by means of a series expression for the Bernoulli numbers.[13]

F.5 RELATIVISTIC HYDRODYNAMICS

Exercise 92

The first three components are the usual ones for hydrodynamics given in (2.51). The last two components are obtained by the chain rule as

$$\frac{\partial \epsilon}{\partial \boldsymbol{w}} = \sigma^{*(1)} \boldsymbol{w} + \sigma^{*(2)} \boldsymbol{c} \cdot \boldsymbol{w} + \sigma^{*(3)} \boldsymbol{c}^2 \cdot \boldsymbol{w},$$

[13] See 0.233.5 of Gradshteyn & Ryzhik, *Tables* (Academic, 1980).

and

$$\frac{\partial \epsilon}{\partial \boldsymbol{c}} = \frac{1}{2}\left[\phi^{*(1)}\mathbf{1} + \phi^{*(2)}\boldsymbol{c} + \phi^{*(3)}\boldsymbol{c}^2 + \sigma^{*(2)}\boldsymbol{w}\boldsymbol{w} + \sigma^{*(3)}(\boldsymbol{w}\boldsymbol{w}\cdot\boldsymbol{c} + \boldsymbol{c}\cdot\boldsymbol{w}\boldsymbol{w})\right].$$

Exercise 93

By inverting (5.2), one obtains

$$s = s(\rho, \epsilon, \mathrm{tr}\boldsymbol{c}, \mathrm{tr}\boldsymbol{c}^2, \mathrm{tr}\boldsymbol{c}^3, \boldsymbol{w}^2, \boldsymbol{w}\cdot\boldsymbol{c}\cdot\boldsymbol{w}, \boldsymbol{w}\cdot\boldsymbol{c}^2\cdot\boldsymbol{w}).$$

From the total increment due to changes of all the possible arguments,

$$\begin{aligned} d\epsilon &= \mu^* d\rho + T^* ds + \frac{1}{2}\phi^{*(1)} d(\mathrm{tr}\boldsymbol{c}) + \frac{1}{4}\phi^{*(2)} d(\mathrm{tr}\boldsymbol{c}^2) + \frac{1}{6}\phi^{*(3)} d(\mathrm{tr}\boldsymbol{c}^3) \\ &+ \frac{1}{2}\sigma^{*(1)} d\boldsymbol{w}^2 + \frac{1}{2}\sigma^{*(2)} d(\boldsymbol{w}\cdot\boldsymbol{c}\cdot\boldsymbol{w}) + \frac{1}{2}\sigma^{*(3)} d(\boldsymbol{w}\cdot\boldsymbol{c}^2\cdot\boldsymbol{w}), \end{aligned}$$

we obtain after solving for ds,

$$\begin{aligned} -\frac{\mu^*}{T^*} &= \frac{\partial s}{\partial \rho}, \\ \frac{1}{T^*} &= \frac{\partial s}{\partial \epsilon}, \\ -\frac{\phi^{*(1)}}{T^*} &= 2\frac{\partial s}{\partial \mathrm{tr}\boldsymbol{c}}, \\ -\frac{\phi^{*(2)}}{T^*} &= 4\frac{\partial s}{\partial \mathrm{tr}\boldsymbol{c}^2}, \\ -\frac{\phi^{*(3)}}{T^*} &= 6\frac{\partial s}{\partial \mathrm{tr}\boldsymbol{c}^3}, \\ -\frac{\sigma^{*(1)}}{T^*} &= 2\frac{\partial s}{\partial \boldsymbol{w}^2} \\ -\frac{\sigma^{*(2)}}{T^*} &= 2\frac{\partial s}{\partial \boldsymbol{w}\cdot\boldsymbol{c}\cdot\boldsymbol{w}}, \\ -\frac{\sigma^{*(3)}}{T^*} &= 2\frac{\partial s}{\partial \boldsymbol{w}\cdot\boldsymbol{c}^2\cdot\boldsymbol{w}}. \end{aligned}$$

To express the entropy density (5.16) in terms of the proper variables, we use the auxiliary identities

$$\mathrm{tr}(\mathbf{1}-\boldsymbol{c})^{-1} = \left[3 - 2\mathrm{tr}\boldsymbol{c} + \frac{1}{2}(\mathrm{tr}\boldsymbol{c})^2 - \frac{1}{2}\mathrm{tr}\boldsymbol{c}^2\right][\det(\mathbf{1}-\boldsymbol{c})]^{-1},$$

and

$$\det(\mathbf{1}-\boldsymbol{c}) = 1 - \mathrm{tr}\boldsymbol{c} + \frac{1}{2}(\mathrm{tr}\boldsymbol{c})^2 - \frac{1}{2}\mathrm{tr}\boldsymbol{c}^2 - \frac{1}{6}(\mathrm{tr}\boldsymbol{c})^3 + \frac{1}{2}(\mathrm{tr}\boldsymbol{c})\mathrm{tr}\boldsymbol{c}^2 - \frac{1}{3}\mathrm{tr}\boldsymbol{c}^3,$$

which can be checked most conveniently by diagonalizing $\boldsymbol{c}$. For the partial derivative of s with respect to any variable x, we then obtain

$$\begin{aligned} -2T^* \frac{\partial s}{\partial x} &= \frac{n_p k_B T^*}{\det(\mathbf{1}-\boldsymbol{c})} \left\{ \frac{\partial}{\partial x} \left[-2\mathrm{tr}\boldsymbol{c} + \frac{1}{2}(\mathrm{tr}\boldsymbol{c})^2 - \frac{1}{2}\mathrm{tr}\boldsymbol{c}^2 \right] \right. \\ &+ \left. [1-\mathrm{tr}(\mathbf{1}-\boldsymbol{c})^{-1}] \frac{\partial \det(\mathbf{1}-\boldsymbol{c})}{\partial x} \right\}. \end{aligned}$$

By choosing $x = \mathrm{tr}\boldsymbol{c}^2$, for example, we obtain

$$\phi^{*(2)} = \frac{n_p k_B T^*}{\det(\mathbf{1}-\boldsymbol{c})} \{-1 + [1-\mathrm{tr}(\mathbf{1}-\boldsymbol{c})^{-1}](-1+\mathrm{tr}\boldsymbol{c})\}.$$

The final expression for $\boldsymbol{\phi}^*$ is

$$\boldsymbol{\phi}^* = n_p k_B T^* [(\mathbf{1}-\boldsymbol{c})^{-2} - (\mathbf{1}-\boldsymbol{c})^{-1}] = -2T^* \frac{\partial s}{\partial \boldsymbol{c}}.$$

Exercise 94

According to Section 2.3.1, the bracket associated with (5.17) without the $(4,3)$ and $(3,4)$ elements is indeed a valid bracket,

$$\begin{aligned} \{A,B\}_0 &= -\int \rho \left(\frac{\delta A}{\delta M_j} \frac{\partial}{\partial r_j} \frac{\delta B}{\delta \rho} - \frac{\delta B}{\delta M_j} \frac{\partial}{\partial r_j} \frac{\delta A}{\delta \rho} \right) d^3r \\ &- \int M_k \left(\frac{\delta A}{\delta M_j} \frac{\partial}{\partial r_j} \frac{\delta B}{\delta M_k} - \frac{\delta B}{\delta M_j} \frac{\partial}{\partial r_j} \frac{\delta A}{\delta M_k} \right) d^3r \\ &- \int s \left(\frac{\delta A}{\delta M_j} \frac{\partial}{\partial r_j} \frac{\delta B}{\delta s} - \frac{\delta B}{\delta M_j} \frac{\partial}{\partial r_j} \frac{\delta A}{\delta s} \right) d^3r \\ &+ \int w_j \frac{\partial}{\partial r_l} \left(\frac{\delta A}{\delta w_j} \frac{\delta B}{\delta M_l} - \frac{\delta B}{\delta w_j} \frac{\delta A}{\delta M_l} \right) d^3r \\ &- \int w_l \left(\frac{\delta A}{\delta w_j} \frac{\partial}{\partial r_j} \frac{\delta B}{\delta M_l} - \frac{\delta B}{\delta w_j} \frac{\partial}{\partial r_j} \frac{\delta A}{\delta M_l} \right) d^3r \\ &+ \int (c_{jk} - \delta_{jk}) \frac{\partial}{\partial r_l} \left(\frac{\delta A}{\delta c_{jk}} \frac{\delta B}{\delta M_l} - \frac{\delta B}{\delta c_{jk}} \frac{\delta A}{\delta M_l} \right) d^3r \\ &- \int (c_{lk} - \delta_{lk}) \left(\frac{\delta A}{\delta c_{jk}} \frac{\partial}{\partial r_j} \frac{\delta B}{\delta M_l} - \frac{\delta B}{\delta c_{jk}} \frac{\partial}{\partial r_j} \frac{\delta A}{\delta M_l} \right) d^3r \\ &- \int (c_{jl} - \delta_{jl}) \left(\frac{\delta A}{\delta c_{jk}} \frac{\partial}{\partial r_k} \frac{\delta B}{\delta M_l} - \frac{\delta B}{\delta c_{jk}} \frac{\partial}{\partial r_k} \frac{\delta A}{\delta M_l} \right) d^3r, \end{aligned}$$

where we have transformed from $\boldsymbol{c}$ to $\boldsymbol{c} - \mathbf{1}$. The additional term to be considered is

$$-\int \left(\frac{\delta A}{\delta w_j} \frac{\partial}{\partial r_j} \frac{\delta B}{\delta s} - \frac{\delta B}{\delta w_j} \frac{\partial}{\partial r_j} \frac{\delta A}{\delta s} \right) d^3r. \tag{F.37}$$

Following the ideas of Section 2.3.1, the terms of the types "new in old" and "new in new" vanish because there is no explicit dependence on the independent fields in (F.37). The terms of the type "old in new" are

$$-\int \left(\frac{\delta A}{\delta w_j} \frac{\partial}{\partial r_j} \frac{\delta\{B,C\}_0}{\delta s} - \frac{\delta\{B,C\}_0}{\delta w_j} \frac{\partial}{\partial r_j} \frac{\delta A}{\delta s} \right) d^3r,$$

plus two cyclic permutations or, after inserting the functional derivatives of $\{B,C\}_0$,

$$\int \frac{\delta A}{\delta w_k} \frac{\partial}{\partial r_k} \left(\frac{\delta B}{\delta M_j} \frac{\partial}{\partial r_j} \frac{\delta C}{\delta s} \right) d^3r$$

$$+ \int \left[\frac{\partial}{\partial r_j} \left(\frac{\delta A}{\delta s} \right) \right] \left[\frac{\partial}{\partial r_l} \left(\frac{\delta B}{\delta w_j} \frac{\delta C}{\delta M_l} \right) - \frac{\delta B}{\delta w_k} \frac{\partial}{\partial r_k} \frac{\delta C}{\delta M_j} \right] d^3r,$$

plus five permutations. After using the product rule for the differentiation with respect to r_k in the first integral, the two resulting terms cancel those of the second integral as seen by cyclic permutations (for one of the terms, after an integration by parts). We have thus verified the Jacobi identity.

Exercise 95

The idea behind the extra variables $\boldsymbol{w}$ and $\boldsymbol{c}$ is exactly that they should represent the heat flux and the viscous momentum flux, so that these effects should not arise directly from the friction matrix.

Exercise 96

After (5.23) is multiplied by the vector formed from the last three components of (5.11), the following conditions remain to be verified:

$$\frac{r_0}{\tau_0} + \frac{r_2}{\tau_2} - \frac{1}{\tau_0} \frac{1}{2} \boldsymbol{\phi}^* : \overline{\boldsymbol{c}} - \frac{1}{\tau_2} \frac{1}{2} \boldsymbol{\phi}^* : \mathring{\boldsymbol{c}} = 0$$

and

$$-\frac{T^*}{\tau_0} \overline{\boldsymbol{c}} - \frac{T^*}{\tau_2} \mathring{\boldsymbol{c}} + \frac{T^*}{\tau_0 r_0} \overline{\boldsymbol{c}} \frac{1}{2} \boldsymbol{\phi}^* : \overline{\boldsymbol{c}} + \frac{T^*}{\tau_2 r_2} \mathring{\boldsymbol{c}} \frac{1}{2} \boldsymbol{\phi}^* : \mathring{\boldsymbol{c}} = 0.$$

These conditions are trivially satisfied in view of the definitions (5.22).

Exercise 97

To identify the force-flux pairs, the basic concepts discussed in Section 3.1.1 need to be employed for the different relaxation mechanisms. From the contribution

$$\begin{pmatrix} \frac{1}{T^*} \boldsymbol{j}^{\mathrm{s}} \\ -\mathbf{1} \\ 0 \end{pmatrix} \cdot \frac{T^*}{\tau_1} \boldsymbol{\sigma}^{*-1} \cdot \left(\frac{1}{T^*} \boldsymbol{j}^{\mathrm{s}} \quad -\mathbf{1} \quad 0 \right),$$

to the friction matrix (5.23), we obtain

$$X_1^{\text{relax}} = -\frac{1}{T^*} \boldsymbol{j}^{\mathrm{s}}, \qquad J_1^{\text{relax}} = -\frac{1}{\tau_1} \boldsymbol{w}.$$

From the contribution

$$\begin{pmatrix} \frac{1}{2T^*}\boldsymbol{\phi}^* \\ 0 \\ -\mathbf{1} \end{pmatrix} \cdot \frac{T^*}{\tau_0 r_0} \overline{\boldsymbol{c}\boldsymbol{c}} \cdot \left(\frac{1}{2T^*}\boldsymbol{\phi}^* \quad 0 \quad -\mathbf{1} \right),$$

we obtain

$$X_0^{\mathrm{relax}} = -\frac{1}{2T^*}\boldsymbol{\phi}^*, \qquad J_0^{\mathrm{relax}} = -\frac{1}{\tau_0}\overline{\boldsymbol{c}},$$

and, similarly,

$$X_2^{\mathrm{relax}} = -\frac{1}{2T^*}\boldsymbol{\phi}^*, \qquad J_2^{\mathrm{relax}} = -\frac{1}{\tau_2}\mathring{\boldsymbol{c}}.$$

Exercise 98
The positive semidefiniteness follows immediately from the decompositions of the three additive contributions to the friction matrix, which were given in the solution to Exercise 97.

Exercise 99
By multiplying the second line of (5.17) with (5.11) to obtain the momentum balance equation, one finds

$$\boldsymbol{\pi} = (\rho\mu^* + sT^* - \epsilon)\mathbf{1} + \boldsymbol{j}^{\mathrm{s}}\boldsymbol{w} + \boldsymbol{\phi}^* \cdot (\boldsymbol{c} - \mathbf{1}).$$

After inserting (5.12)–(5.14), one obtains the manifestly symmetric pressure tensor (5.29).

Exercise 100
According to Exercise 93, we need to invert $\epsilon(\rho, s, \mathrm{tr}\boldsymbol{c}^2, \boldsymbol{w}^2)$ by solving for s. We do this inversion as an expansion in terms of $\boldsymbol{w}^2$ and $\mathrm{tr}\boldsymbol{c}^2$. To zeroth order, we obtain $s = s_0(\rho, \epsilon)$ by inverting $\epsilon_0(\rho, s)$ with respect to s. To first order, we have

$$\begin{aligned} s &= s_0\left(\rho, \epsilon - \frac{1}{2}(\alpha_1(\rho, s_0(\rho,\epsilon))\boldsymbol{w}^2 + \alpha_2(\rho, s_0(\rho,\epsilon))\mathrm{tr}\boldsymbol{c}^2)\right) \\ &= s_0(\rho,\epsilon) - \frac{1}{2T^*}(\alpha_1(\rho, s_0(\rho,\epsilon))\boldsymbol{w}^2 + \alpha_2(\rho, s_0(\rho,\epsilon))\mathrm{tr}\boldsymbol{c}^2), \end{aligned}$$

which implies $\alpha'_j = \alpha_j/T^*$. The assumption of quadratic terms in (5.34) and (5.42) are thus related in a nontrivial way. These assumptions are equivalent only for small $\boldsymbol{w}^2$ and $\mathrm{tr}\boldsymbol{c}^2$. Whether one should prefer (5.34) or (5.42) depends on the energetic or entropic origin of the underlying effects. For example, we have assumed (non-quadratic) additive entropic and no energetic effects of a conformation tensor in (4.86) for dilute polymer solutions and in (4.105) for the pompon model.

Exercise 101
In the limit of short relaxation times, the forces and fluxes exchange their roles. By

comparing with (3.11), we find

$$X_1^{\text{relax}} = -\frac{1}{T^{*2}} J_1^{\text{LIT}}, \qquad J_1^{\text{relax}} = -T^{*2} X_1^{\text{LIT}}.$$

We further have

$$X_0^{\text{relax}} = X_2^{\text{relax}} = \frac{1}{2T^*} J_2^{\text{LIT}},$$

where (3.12) has been used. With

$$J_0^{\text{relax}} = -\frac{2}{3}\frac{\partial}{\partial \boldsymbol{r}} \cdot \boldsymbol{v}\boldsymbol{1}$$

and

$$J_2^{\text{relax}} = -\left(\boldsymbol{\kappa} + \boldsymbol{\kappa}^T - \frac{2}{3}\frac{\partial}{\partial \boldsymbol{r}} \cdot \boldsymbol{v}\boldsymbol{1}\right),$$

we finally obtain

$$J_0^{\text{relax}} + J_2^{\text{relax}} = 2T^* X_2^{\text{LIT}},$$

where X_2^{LIT} has actually been symmetrized.

Exercise 102

When ϵ is used as an independent variable, the functions in the GENERIC pressure-tensor expression (5.29) must be evaluated with the formulas of Exercise 93. For the quadratic entropy expression (5.42), we thus obtain

$$\boldsymbol{\pi} = (\rho\mu^* + sT^* - \epsilon)\boldsymbol{1} + \alpha_1' T^* \boldsymbol{w}\boldsymbol{w} + 2\alpha_2' T^* \boldsymbol{c} \cdot (\boldsymbol{c} - \boldsymbol{1}).$$

For the quadratic energy expression (5.34) and s as an independent variable, on the other hand, we obtain from (5.29) and (5.3)–(5.10)

$$\boldsymbol{\pi} = (\rho\mu^* + sT^* - \epsilon)\boldsymbol{1} + \alpha_1 \boldsymbol{w}\boldsymbol{w} + 2\alpha_2 \boldsymbol{c} \cdot (\boldsymbol{c} - \boldsymbol{1}).$$

The relationship between the pressure tensors (concerning their tensorial structure) is thus more direct than the relation between the assumptions (5.34) and (5.42) or the functional forms of the scalar coefficients (see Exercise 100).

Exercise 103

From (5.43) and (5.57), we obtain

$$c_{j0} + \sum_{k=1}^{3} c_{jk} \frac{v_k}{c} = \frac{v_j}{c},$$

which leads to the desired result for $c_{j0} = c_{0j}$, and

$$c_{00} + \sum_{j=1}^{3} c_{0j} \frac{v_j}{c} = -1,$$

which then gives c_{00}. From (5.43) and (5.59), we obtain

$$\gamma w_0 + \gamma \frac{\boldsymbol{v}}{c} \cdot \boldsymbol{w} = -\frac{T_{\mathrm{f}}^*}{c},$$

which implies (5.60).

Exercise 104

By inserting (5.77) and (5.78) into the expression (5.86) for the pressure tensor, we obtain

$$\begin{aligned}
\pi_{\mu\nu} &= (c_{\mu\mu'} - \eta_{\mu\mu'})(\phi_{\mathrm{f}}^{*(1)} c^0 + \phi_{\mathrm{f}}^{*(2)} c^1 + \phi_{\mathrm{f}}^{*(3)} c^2)^{\mu'}{}_{\nu} + p_{\mathrm{f}} \eta_{\mu\nu} \\
&+ (\sigma_{\mathrm{f}}^{*(1)} - \sigma_{\mathrm{f}}^{*(2)}) w_\mu w_\nu + (\sigma_{\mathrm{f}}^{*(2)} - \sigma_{\mathrm{f}}^{*(3)})(w_\mu w^{\mu'} c_{\mu'\nu} + c_{\mu\mu'} w^{\mu'} w_\nu) \\
&+ \sigma_{\mathrm{f}}^{*(3)} (w_\mu w^{\mu'} c^2_{\mu'\nu} + c^2_{\mu\mu'} w^{\mu'} w_\nu + c_{\mu\mu'} w^{\mu'} w^{\nu'} c_{\nu'\nu}).
\end{aligned}$$

This expression is manifestly symmetric, and it can be recognized as a natural generalization of (5.29).

Exercise 105

We first combine (5.83), (5.85), and (5.87) into

$$\begin{aligned}
\frac{\partial T^{00}}{\partial x} &= \frac{c^2 + \mu_{\mathrm{f}}^*}{\gamma} \frac{\partial \rho}{\partial x} + \boldsymbol{v} \cdot \frac{\partial \boldsymbol{M}}{\partial x} + \frac{T_{\mathrm{f}}^*}{\gamma} \frac{\partial s}{\partial x} + \sum_{j=1}^{3} w_\mu \left(\sigma^{*\mu j} - \sigma^{*\mu 0} \frac{v^j}{c} \right) \frac{\partial w_j}{\partial x} \\
&+ \frac{1}{2} \phi^{*\mu\nu} \frac{\partial c_{\mu\nu}}{\partial x} - \frac{1}{c} \sum_{j,k=1}^{3} \left(\phi^{*0k} - \phi^{*00} \frac{v_k}{c} \right) (\delta_{jk} - c_{jk}) \frac{\partial v_j}{\partial x}. \qquad \text{(F.38)}
\end{aligned}$$

When differentiating with respect to the independent variables $x = (\rho, \boldsymbol{M}, s, \boldsymbol{w}, \boldsymbol{c})$, the terms in the first line are in a convenient form because they contain derivatives of the independent ρ, $\boldsymbol{M}$, s, and $\boldsymbol{w}$ only. The term $\partial c_{\mu\nu}/\partial x$ leads to contributions from $\partial c_{00}/\partial x$, $\partial c_{j0}/\partial x$, and $\partial c_{0k}/\partial x$ in addition to the desirable derivatives of independent variables, $\partial c_{jk}/\partial x$; in particular, additional derivatives of velocity occur. More precisely, (5.58) leads to

$$\frac{1}{2} \phi^{*\mu\nu} \frac{\partial c_{\mu\nu}}{\partial x} = \frac{1}{2} \hat{\boldsymbol{\phi}}^* : \frac{\partial \boldsymbol{c}}{\partial x} + \frac{1}{c} \sum_{j,k=1}^{3} \left(\phi^{*0k} - \phi^{*00} \frac{v_k}{c} \right) (\delta_{jk} - c_{jk}) \frac{\partial v_j}{\partial x},$$

so that the second line of (F.38) is simply given by $(1/2)\hat{\boldsymbol{\phi}}^* : \partial \boldsymbol{c}/\partial x$. The auxiliary equation (5.87), and hence the constitutive assumption (5.86), is crucial for this simplification to occur. The gradient (5.88) can now be read off directly.

Exercise 106

Starting from

$$\sum_{k=1}^{3} \hat{\sigma}^{*jk} \left(w_k - \gamma T_{\mathrm{f}}^* \frac{v_k}{c^2} \right) = \sum_{k=1}^{3} \left(\sigma^{*jk} - \frac{v_j}{c} \sigma^{*0k} \right) \left(w_k - \gamma T_{\mathrm{f}}^* \frac{v_k}{c^2} \right)$$

$$- \quad \frac{1}{c}\left(\sigma^{*j0} - \frac{v_j}{c}\sigma^{*00}\right)\left(\boldsymbol{v}\cdot\boldsymbol{w} - \gamma\frac{v^2}{c^2}T_{\rm f}^*\right)$$

and inserting

$$\boldsymbol{v}\cdot\boldsymbol{w} - \gamma\frac{v^2}{c^2}T_{\rm f}^* = -(cw_0 + \gamma T_{\rm f}^*), \tag{F.39}$$

we obtain

$$\begin{aligned}\sum_{k=1}^{3}\hat{\sigma}^{*jk}\left(w_k - \gamma T_{\rm f}^*\frac{v_k}{c^2}\right) &= \sum_{k=1}^{3}\left(\sigma^{*jk} - \frac{v_j}{c}\sigma^{*0k}\right)w_k + \left(\sigma^{*j0} - \frac{v_j}{c}\sigma^{*00}\right)w_0 \\ &+ \frac{T_{\rm f}^*}{c}\left(\frac{v_j}{c}\sigma^{*0\nu}u_\nu - \sigma^{*j\nu}u_\nu\right).\end{aligned}$$

The term in the second line vanishes in view of (5.57) and (5.77), and we thus have the desired relationship

$$\sum_{k=1}^{3}\hat{\sigma}^{*jk}\left(w_k - \gamma T_{\rm f}^*\frac{v_k}{c^2}\right) = \left(\sigma^{*j\mu} - \frac{v_j}{c}\sigma^{*0\mu}\right)w_\mu.$$

We further obtain

$$\boldsymbol{j}^{\rm s}\cdot\hat{\boldsymbol{\sigma}}^{*-1}\cdot\boldsymbol{j}^{\rm s} = \sum_{j=1}^{3}\left(w_j - u_j\frac{T_{\rm f}^*}{c}\right)\left(\sigma^{*j\nu'} - \frac{v_j}{c}\sigma^{*0\nu'}\right)w_{\nu'},$$

or, with (F.39),

$$\boldsymbol{j}^{\rm s}\cdot\hat{\boldsymbol{\sigma}}^{*-1}\cdot\boldsymbol{j}^{\rm s} = \sum_{j=1}^{3}\left(w_j - u_j\frac{T_{\rm f}^*}{c}\right)\sigma^{*j\nu'}w_{\nu'} + \left(w_0 - u_0\frac{T_{\rm f}^*}{c}\right)\sigma^{*0\nu'}w_{\nu'},$$

which, in view of (5.59) and (5.45), is (5.125).

Exercise 107

By repeated use of (5.57) we obtain on one hand

$$\partial_\nu(u^\mu u^{\mu'}c_{\mu\mu'}) = \partial_\nu(u^\mu u_\mu) = 0,$$

and, on the other hand, the product rule gives

$$\partial_\nu(u^\mu u^{\mu'}c_{\mu\mu'}) = u^\mu u^{\mu'}\partial_\nu c_{\mu\mu'} + 2u^{\mu'}c_{\mu\mu'}\partial_\nu u^\mu,$$

with

$$2u^{\mu'}c_{\mu\mu'}\partial_\nu u^\mu = 2u_\mu\partial_\nu u^\mu = \partial_\nu(u^\mu u_\mu) = 0.$$

Exercise 108

We evaluate the differential quotient explicitly by means of,

$$\begin{aligned}\Delta\ln|\det(g_{\mu\nu})| &= \ln|\det(g_{\mu\nu} + \Delta g_{\mu\nu})| - \ln|\det(g_{\mu\nu})| \\ &= \ln\det(\delta^\mu{}_\nu + g^{\mu\mu'}\Delta g_{\mu'\nu}) \\ &\approx \ln(1 + g^{\mu\mu'}\Delta g_{\mu'\mu}) \approx g^{\mu\mu'}\Delta g_{\mu'\mu}.\end{aligned}$$

Note that g is the absolute value of the determinant of $g_{\mu\nu}$ and that $g^{\mu\nu}$ is the inverse of $g_{\mu\nu}$. Further note that the argument is valid for general matrices.

Exercise 109

From $C_{\mu\nu} = c_{\mu\nu} - \eta_{\mu\nu}$, (5.56), and (5.60), we obtain

$$g^{\mu\nu} C_{\nu\mu} = \mathrm{tr}\boldsymbol{c}_{\mathrm{f}} - 3,$$

$$g^{\mu\nu} C_{\nu\mu'} g^{\mu'\nu'} C_{\nu'\mu} = \mathrm{tr}\boldsymbol{c}_{\mathrm{f}}^2 - 2\mathrm{tr}\boldsymbol{c}_{\mathrm{f}} + 3,$$

$$g^{\mu\nu} C_{\nu\mu'} g^{\mu'\nu'} C_{\nu'\mu''} g^{\mu''\nu''} = \mathrm{tr}\boldsymbol{c}_{\mathrm{f}}^3 - 3\mathrm{tr}\boldsymbol{c}_{\mathrm{f}}^2 + 3\mathrm{tr}\boldsymbol{c}_{\mathrm{f}} - 3,$$

$$w_\mu g^{\mu\nu} w_\nu = \boldsymbol{w}_{\mathrm{f}}^2 - \frac{T_{\mathrm{f}}^{*2}}{c^2},$$

$$w_\mu g^{\mu\nu} C_{\nu\mu'} g^{\mu'\nu'} w_{\nu'} = \boldsymbol{w}_{\mathrm{f}} \cdot \boldsymbol{c}_{\mathrm{f}} \cdot \boldsymbol{w}_{\mathrm{f}} - \boldsymbol{w}_{\mathrm{f}}^2,$$

$$w_\mu g^{\mu\nu} C_{\nu\mu'} g^{\mu'\nu'} C_{\nu'\mu''} g^{\mu''\nu''} w_{\nu''} = \boldsymbol{w}_{\mathrm{f}} \cdot \boldsymbol{c}_{\mathrm{f}}^2 \cdot \boldsymbol{w}_{\mathrm{f}} - 2\boldsymbol{w}_{\mathrm{f}} \cdot \boldsymbol{c}_{\mathrm{f}} \cdot \boldsymbol{w}_{\mathrm{f}} + \boldsymbol{w}_{\mathrm{f}}^2.$$

Exercise 110

Every factor of the type $g^{\mu''\nu''} C_{\nu''\mu'''}$ in the cyclic expressions (5.137) is changed into $g^{\mu''\nu} C_{\mu\mu'''}$ by the differential operator $C_{\mu\mu'}\partial/\partial C_{\nu\mu'}$. The operator $g^{\nu'\nu}\partial/\partial g^{\nu'\mu}$ has exactly the same effect on $g^{\mu''\nu''} C_{\nu''\mu'''}$, so that (5.141) is established for all the scalars in (5.137) by means of the product rule.

For the scalars in (5.138), we have an extra factor of the type $g^{\mu''\nu''} w_{\nu''}$ at the right end, which is turned into $g^{\mu''\nu} w_\mu$ by $g^{\nu'\nu}\partial/\partial g^{\nu'\mu}$. The operator $w_\mu \partial/\partial w_\nu$ has exactly the same effect on $g^{\mu''\nu''} w_{\nu''}$. The factor $1/2$ in (5.141) is required because there is another factor of w at the left end, which is on an equal footing with the one at the right end.

Exercise 111

For example, if we choose $A = A(n)$ in the fundamental equation (5.126), then we find

$$\{A, I\} = \int \sqrt{g} n \frac{\delta I}{\delta M_\mu} \partial_\mu \frac{\delta A}{\delta n} d^4x = 0,$$

or, after integrating by parts and using the fact that $A(n)$ is arbitrary, we obtain

$$\partial_\mu \left(\sqrt{g} n \frac{\delta I}{\delta M_\mu} \right) = 0.$$

If we apply (5.126) to an arbitrary functional $A = A(g)$ and if we remember that the vector field V^ν can be chosen arbitrarily, we obtain

$$\frac{\delta I}{\delta \Gamma^\nu_{\mu\mu'}} = 0.$$

We have thus derived (5.147) and (5.150).

Exercise 112
With the chain rule for the differentiation of q and the operator identity (5.141), we obtain after some rearrangements

$$
\begin{aligned}
(qu^\nu)_{;\nu} &= -\left(pg^\nu{}_\mu + 2g^{\nu\nu'}\frac{\partial q}{\partial g^{\nu'}{}_\mu}\right) u^\mu{}_{;\nu} + \frac{\partial q}{\partial n}(nu^\nu)_{;\nu} + \frac{\partial q}{\partial s}(su^\nu)_{;\nu} \\
&+ \frac{\partial q}{\partial C_{\mu\nu}}\left[u^{\mu'}(C_{\mu\nu;\mu'} - C_{\mu'\nu;\mu} - C_{\mu\mu';\nu}) \right.\\
&\qquad \left. + \left(u^{\nu'}C_{\nu'\nu}\right)_{;\mu} + \left(C_{\mu\mu'}u^{\mu'}\right)_{;\nu}\right] \\
&+ \frac{\partial q}{\partial w_\mu}\left[u^\nu(w_{\mu;\nu} - w_{\nu;\mu}) + (w_\nu u^\nu)_{;\mu}\right].
\end{aligned}
$$

With the field equations (5.161), (5.170), (5.171), and (5.172), we obtain the internal energy equation

$$
\left(qu^\nu + \frac{1}{c}\frac{\partial q}{\partial s}\frac{\partial q}{\partial w_\nu}\right)_{;\nu} + \left(pg^\nu{}_\mu + 2g^{\nu\nu'}\frac{\partial q}{\partial g^{\nu'}{}_\mu}\right) u^\mu{}_{;\nu} = 0. \tag{F.40}
$$

From (5.166) contracted with u^μ, that is, from

$$
u^\mu\left[(p+q)u^\nu u_\mu + pg^\nu{}_\mu + 2g^{\nu\nu'}\frac{\partial q}{\partial g^{\nu'}{}_\mu}\right]_{;\nu} = 0,
$$

we find

$$
\begin{aligned}
u^\mu u_\mu[(p+q)u^\nu]_{;\nu} + \frac{1}{2}(p+q)u^\nu(u^\mu u_\mu)_{;\nu} + \left[pu^\nu + 2g^{\nu\nu'}\frac{\partial q}{\partial g^{\nu'}{}_\mu}u^\mu\right]_{;\nu} \\
- \left(pg^\nu{}_\mu + 2g^{\nu\nu'}\frac{\partial q}{\partial g^{\nu'}{}_\mu}\right) u^\mu{}_{;\nu} = 0,
\end{aligned}
$$

from which we obtain the following result by once more using the operator identity (5.141) and adding (F.40),

$$
(1 + u^\mu u_\mu)[(p+q)u^\nu]_{;\nu} + \frac{1}{2}(p+q)u^\nu(u^\mu u_\mu)_{;\nu} = 0. \tag{F.41}
$$

Equation (F.41) is a consequence of the field equations. It shows that, if we have $u^\mu u_\mu = -1$ at some point, then we have also the evolution equation $u^\nu(u^\mu u_\mu)_{;\nu} = 0$.

Exercise 113
As the additional term depends only on g, (5.151) is the only equation to be changed in the list (5.147)–(5.153). As a consequence, the left-hand side of (5.164) is changed into

$$
R_{\mu\nu} - \frac{1}{2}g_{\mu\nu}g^{\mu'\nu'}R_{\mu'\nu'} + \Lambda g_{\mu\nu}.
$$

We thus obtain Einstein's field equation with a cosmological constant Λ.

Exercise 114

We find

$$\frac{\delta I_{\rm M}}{\delta g^{\mu\nu}} = \frac{1}{2} p g_{\mu\nu} + \frac{1}{2}(p+q) u_\mu u_\nu + \frac{\partial q}{\partial g^{\mu\nu}},$$

where these three terms result from the prefactor $\sqrt{g}$, the explicit occurrence of $g^{\mu\nu}$, and the further dependence of U_I, V_I on $g^{\mu\nu}$ in the action (5.136), respectively. Equations (5.140), (5.160), and (5.162) have been used to obtain the above expression, which, according to (5.167), confirms the result (5.169).

Exercise 115

We assume $0 \leq F < 1$, implying $F = \sqrt{F^2}$. According to (5.183), for Π to be proportional to F, the functional form of q must be

$$q(n, s, F^2) = q(n, s, 0) - 2 \left. \frac{\partial q(n, s, F^2)}{\partial F^2} \right|_{F^2=0} [F + \ln(1 - F)],$$

so that

$$\frac{\partial q(n, s, F^2)}{\partial F^2} = \frac{1}{2F} \frac{\partial q(n, s, F^2)}{\partial F} \propto \frac{1}{1-F}.$$

Only for small F we obtain a quadratic dependence of q on F. A logarithmic term becomes visible for larger F, which is natural in view of the fact that F is associated with the tensor $C_{\mu\nu}$ and entropy expressions such as (4.86) and (4.105). The entropy production is proportional to $F^2/(1-F)$, which implies that F and hence Π is bounded. This result actually implies that linear irreversible thermodynamics cannot be fully consistent with the GENERIC treatment of bulk viscous cosmology. In general, (5.183) implies that one cannot require Π to be proportional to F and an entropy production rate proportional to Π^2 or F^2 at the same time.

Exercise 116

From (5.159), we obtain

$$\Gamma^1_{01} = \Gamma^1_{10} = \Gamma^2_{02} = \Gamma^2_{20} = \Gamma^3_{03} = \Gamma^3_{30} = \frac{\dot{a}}{ca}$$

and

$$\Gamma^0_{11} = \Gamma^0_{22} = \Gamma^0_{33} = \frac{a\dot{a}}{c}$$

as the only nonvanishing Christoffel symbols. The Ricci tensor can now be evaluated by means of (5.139). We obtain

$$R_{00} = -\frac{3}{c^2}\frac{\ddot{a}}{a}, \quad R_{11} = R_{22} = R_{33} = \frac{1}{c^2}(a\ddot{a} + 2\dot{a}\dot{a}),$$

which can be rewritten as in (5.192) by noting that $h_{11} = h_{22} = h_{33} = a^2$ are the only nonzero components of $h_{\mu\nu}$.

With (5.192) we obtain for the left-hand side of (5.164)

$$R_{\mu\nu} - \frac{1}{2} g_{\mu\nu} g^{\mu'\nu'} R_{\mu'\nu'} = \frac{3}{c^2} u_\mu u_\nu \frac{\dot{a}^2}{a^2} - \frac{1}{c^2} h_{\mu\nu} \left(2\frac{\ddot{a}}{a} + \frac{\dot{a}^2}{a^2} \right),$$

which leads to (5.186) and (5.187) when the energy-momentum tensor (5.182) is used for the right-hand side of (5.164).

Exercise 117

From the general formula (5.131) and (5.178), we obtain

$$u_{\mu;\nu} = \Gamma^0_{\mu\nu},$$

from which (5.193) follows from the results for $\Gamma^0_{\mu\nu}$ given in Exercise 116.

By inserting (5.181) into (5.172), with a relaxation term as in (5.104), we obtain

$$u^{\mu'}[h_{\mu\nu}\partial_{\mu'}F - (F-1)u_{\mu'}(u_{\nu;\mu} + u_{\mu;\nu})] = -\frac{1}{c\tau_0}Fh_{\mu\nu},$$

where $\bar{c}_{\mu\nu} = Fh_{\mu\nu}$ and $\mathring{c} = 0$ have been found from (5.143) and (5.144). By means of (5.193) and $u^{\mu'}u_{\mu'} = -1$ we obtain (5.189).

Exercise 118

The Poisson bracket is

$$\begin{aligned}\{A, B\} &= n\left(\frac{\partial A}{\partial n}\frac{\partial B}{\partial H} - \frac{\partial A}{\partial H}\frac{\partial B}{\partial n}\right) + s\left(\frac{\partial A}{\partial s}\frac{\partial B}{\partial H} - \frac{\partial A}{\partial H}\frac{\partial B}{\partial s}\right) \\ &- \frac{2}{3}(1-F)\left(\frac{\partial A}{\partial F}\frac{\partial B}{\partial H} - \frac{\partial A}{\partial H}\frac{\partial B}{\partial F}\right).\end{aligned}$$

If, using the result of Exercise 2, we neglect all terms containing second-order derivatives of the functions A and B, then we obtain

$$\begin{aligned}\{A, \{B, C\}\} &= -n\frac{\partial A}{\partial H}\left(\frac{\partial B}{\partial n}\frac{\partial C}{\partial H} - \frac{\partial B}{\partial H}\frac{\partial C}{\partial n}\right) - s\frac{\partial A}{\partial H}\left(\frac{\partial B}{\partial s}\frac{\partial C}{\partial H} - \frac{\partial B}{\partial H}\frac{\partial C}{\partial s}\right) \\ &+ \frac{4}{9}(1-F)\frac{\partial A}{\partial H}\left(\frac{\partial B}{\partial F}\frac{\partial C}{\partial H} - \frac{\partial B}{\partial H}\frac{\partial C}{\partial F}\right).\end{aligned}$$

Due to the double occurrence of derivatives with respect to H in each of the terms, we find that the combination $\{A, \{B, C\}\} + \{B, \{C, A\}\} + \{C, \{A, B\}\}$ with cyclic permutations vanishes.

Exercise 119

By combining (5.205) and (5.209), we obtain

$$n\dot{\check{s}} = n\dot{\check{s}}_0 - n\frac{\Pi^2}{2}\frac{d}{dt}\left(\frac{\tau_0}{nT_0\kappa}\right) - \frac{\tau_0}{T_0\kappa}\Pi\dot{\Pi} = \frac{\Pi^2}{T_0(\kappa + \tau_*\Pi)}.$$

By inserting

$$n\dot{\check{s}}_0 = n\frac{\partial \check{s}_0}{\partial n}\dot{n} + n\frac{\partial \check{s}_0}{\partial \rho}\dot{\rho} = -\frac{1}{T_0}(\rho + p_0)\frac{\dot{n}}{n} + \frac{1}{T_0}\dot{\rho} = -\frac{3H\Pi}{T_0},$$

we arrive at (5.211). Equation (5.207) is obtained as the special case for $\tau_* = 0$.

Exercise 120
From (5.168), we find that p is a function of n and $\hat{q}(s/n, F^2)$. By inverting (5.212) for a fixed n, $\hat{q}(s/n, F^2)$ can be expressed in terms of q, so that p can be considered as a function of n and q.

Exercise 121
In the chain rule

$$\frac{\partial q(n, s(n,z,F^2), F^2)}{\partial z} = \frac{\partial q(n,s,F^2)}{\partial s}\frac{\partial s(n,z,F^2)}{\partial z},$$

the factor $\partial q/\partial s$ is the temperature, so that (5.217) implies

$$\frac{\partial s(n,z,F^2)}{\partial z} = \frac{k_B}{mc^2} z \frac{\partial q(n,z,F^2)}{\partial z},$$

and the integrated version (5.228). From

$$\frac{\partial q(n,z,F^2)}{\partial F^2} = \frac{\partial q(n,s,F^2)}{\partial F^2} + \frac{\partial q(n,s,F^2)}{\partial s}\frac{\partial s(n,z,F^2)}{\partial F^2},$$

we further obtain

$$\frac{\partial q(n,s,F^2)}{\partial F^2} = \rho_1(n,z) + \frac{1}{z}\int_z^\infty z' \frac{\partial \rho_1(n,z')}{\partial z'} dz' = -\frac{1}{z}\int_z^\infty \rho_1(n,z')dz'.$$

Exercise 122
From (5.223), we have

$$\dot{z}z\left(\frac{1}{z^2} - \frac{1}{3}\frac{d}{dz}\frac{K_1(z)}{K_2(z)}\right) = \frac{\dot{a}}{a}.$$

With the auxiliary equations[14]

$$\frac{K_1(z)}{K_2(z)} = \frac{z}{2} \qquad \text{for } z \ll 1 \tag{F.42}$$

and[15]

$$\frac{K_1(z)}{K_2(z)} = 1 - \frac{3}{2z} \qquad \text{for } z \gg 1, \tag{F.43}$$

we obtain

$$\frac{\dot{z}}{z} = \frac{\dot{a}}{a}, \qquad \text{or } z \propto a \qquad \text{for } z \ll 1,$$

[14] See, for example, Section 6.6 of Press et al., *Numerical Recipes* (Cambridge, 1992).
[15] See, for example, 8.451.6 of Gradshteyn & Ryzhik, *Tables* (Academic, 1980).

and

$$\frac{1}{2}\frac{\dot{z}}{z} = \frac{\dot{a}}{a}, \qquad \text{or } z \propto a^2 \qquad \text{for } z \gg 1.$$

By combing (5.186) and (5.219), we obtain

$$\dot{a}^2 = \frac{8\pi G_\text{N}}{3} mna^3 \left(\frac{3}{z} + \frac{K_1(z)}{K_2(z)} \right) \frac{1}{a},$$

where na^3 is a constant. We thus conclude

$$\dot{a} \propto \frac{1}{a}, \qquad \text{or } a \propto t^{1/2} \qquad \text{for } z \ll 1,$$

and

$$\dot{a}^2 \propto \frac{1}{a}, \qquad \text{or } a \propto t^{2/3} \qquad \text{for } z \gg 1.$$

If we rewrite (5.219) as

$$\rho = na^3 mc^2 \left(\frac{3}{z} + \frac{K_1(z)}{K_2(z)} \right) \frac{1}{a^3},$$

where na^3 is a constant, we obtain by once more using (F.42) and (F.43)

$$\rho \propto a^{-4} \qquad \text{for } z \ll 1,$$

and

$$\rho \propto a^{-3} \qquad \text{for } z \gg 1.$$

F.6 PROJECTION-OPERATOR METHOD

Exercise 123

The explicit expression follows directly from the last equality in the definition (6.2) when the Poisson bracket (1.28) is inserted.

Exercise 124

Using the explicit expression (6.5) for the Liouville operator, an integration by parts gives

$$\int A^*(z) \mathcal{L} B(z) d^{6N} z = - \int B(z) \mathcal{L} A^*(z) d^{6N} z,$$

where the second-order derivatives of energy cancel because of

$$\frac{\partial^2 E_0}{\partial \boldsymbol{r}_j \partial \boldsymbol{p}_j} = \frac{\partial^2 E_0}{\partial \boldsymbol{p}_j \partial \boldsymbol{r}_j}.$$

The sign is used for the complex conjugation of $\mathcal{L}$ to arrive at (6.4) with $\mathcal{L}^\dagger = \mathcal{L}$.

Exercise 125
The time dependent probability density is given by

$$\rho(z,t) = \int \delta(z - z(z_0,t))\rho(z_0,0)d^{6N}z_0,$$

where $z(z_0,t)$ is the solution of Hamilton's equations of motion with the initial condition z_0. By taking the time derivative, we obtain

$$\begin{aligned}\frac{\partial \rho(z,t)}{\partial t} &= -\int \frac{\partial z(z_0,t)}{\partial t} \cdot \frac{\partial}{\partial z}\delta(z - z(z_0,t))\rho(z_0,0)d^{6N}z_0 \\ &= -\int \sum_{j=1}^{N} \left[\frac{\partial \boldsymbol{r}_j(z_0,t)}{\partial t} \cdot \frac{\partial}{\partial \boldsymbol{r}_j} + \frac{\partial \boldsymbol{p}_j(z_0,t)}{\partial t} \cdot \frac{\partial}{\partial \boldsymbol{p}_j}\right] \\ &\quad \times \delta(z - z(z_0,t))\rho(z_0,0)d^{6N}z_0.\end{aligned}$$

In view of the δ function, the velocities and forces can be evaluated at z rather than with the configuration $z(z_0,t)$, and we obtain the desired result.

Exercise 126
From the explicit formula (6.15), we obtain

$$\begin{aligned}\frac{\delta \rho_x(z)}{\delta x_k} &= -\rho_x(z)\frac{\delta \ln Z}{\delta x_k} - \rho_x(z)\sum_l \frac{\delta \lambda_l(x)}{\delta x_k}\Pi_l(z) \\ &= -\rho_x(z)\sum_l \frac{\delta \lambda_l(x)}{\delta x_k}\left[\frac{\delta \ln \hat{Z}}{\delta \lambda_l} + \Pi_l(z)\right].\end{aligned}$$

With (6.17), we obtain the desired result (6.24). The symmetry of the matrix $\delta\lambda_l/\delta x_k$ follows from the symmetry of its inverse, which is obtained by differentiating (6.17) with respect to λ_l.

Exercise 127
The partition function is given by the Gaussian integral

$$\begin{aligned}\hat{Z} &= V^N \int \exp\left[-\boldsymbol{\lambda}_M \cdot \sum_{j=1}^{N} \boldsymbol{p}_j - \lambda_E \sum_{j=1}^{N} \frac{\boldsymbol{p}_j^2}{2m}\right] d^3p_1 \ldots d^3p_N \\ &= \left\{V \int \exp\left[-\boldsymbol{\lambda}_M \cdot \boldsymbol{p}_1 - \lambda_E \frac{\boldsymbol{p}_1^2}{2m}\right] d^3p_1\right\}^N,\end{aligned}$$

which can be evaluated to be

$$\hat{Z} = \left\{V\left(2\pi\frac{m}{\lambda_E}\right)^{3/2} \exp\left[\frac{m\boldsymbol{\lambda}_M^2}{2\lambda_E}\right]\right\}^N.$$

From (6.17), we obtain

$$\boldsymbol{P} = -\frac{\partial \ln \hat{Z}}{\partial \boldsymbol{\lambda}_M} = -\frac{Nm\boldsymbol{\lambda}_M}{\lambda_E}$$

and

$$E = -\frac{\partial \ln \hat{Z}}{\partial \lambda_E} = \frac{3}{2}\frac{N}{\lambda_E} + \frac{Nm\boldsymbol{\lambda}_M^2}{2\lambda_E^2},$$

so that the probability density is proportional to

$$\exp\left[-\lambda_E \sum_{j=1}^{N} \frac{(\boldsymbol{p}_j - \boldsymbol{P}/N)^2}{2m}\right]$$

where the relationship

$$E_{\text{internal}} = E - \frac{\boldsymbol{P}^2}{2Nm} = \frac{3}{2}N\frac{1}{\lambda_E}$$

suggests that λ_E is the inverse temperature of a moving ideal gas, $1/\lambda_E = k_B T$. From

$$\frac{\partial^2 \ln \hat{Z}}{\partial \boldsymbol{\lambda}_M \partial \boldsymbol{\lambda}_M} = \frac{Nm}{\lambda_E}\boldsymbol{1},$$

we conclude that the different spatial components of the momentum are uncorrelated, and, for the trace,[16] we obtain the following explicit form of the general condition (6.23):

$$E_{\text{internal}} \ll \frac{\boldsymbol{P}^2}{2m} = N\frac{\boldsymbol{P}^2}{2Nm}.$$

As further conditions, we obtain

$$\frac{\partial^2 \ln \hat{Z}}{\partial \lambda_E \partial \lambda_E} = \frac{2}{3}\frac{1}{N}\left(E - \frac{\boldsymbol{P}^2}{2Nm}\right)\left(E + \frac{\boldsymbol{P}^2}{2Nm}\right) \ll E^2$$

and

$$\frac{\partial^2 \ln \hat{Z}}{\partial \lambda_E \partial \boldsymbol{\lambda}_M} = \frac{2}{3}\frac{1}{N}\boldsymbol{P}\left(E - \frac{\boldsymbol{P}^2}{2Nm}\right) \ll \boldsymbol{P}E,$$

where, in the latter inequality, one should actually compare the absolute values of vector components. In all cases, in view of the extensive nature of E and $\boldsymbol{P}$, the inequalities are naturally fulfilled when N becomes large.

If we use the internal energy instead of the total energy as a macroscopic variable, we can transform from the integration over all particle momenta to the total momentum and relative momenta as integration variables. In contrast to the total energy, the

[16] Actually, there are similar conditions for each of the components of momentum.

internal energy depends only on the relative momentum variables so that the partition function contains a factor of the type

$$\int \exp\left[-\boldsymbol{\lambda}_M \cdot \boldsymbol{p}\right] d^3p.$$

As the exponential decreases in one half-plane and increases in the other one, the integral is infinite. When using momentum and internal energy as variables, the distribution is so broad that one cannot even normalize it. For further information on the existence of canonical ensembles, see Exercise 138.

Exercise 128

From (6.2) or (6.5), we have

$$i\mathcal{L}\boldsymbol{r}_j = \frac{\boldsymbol{p}_j}{m}$$

and hence from (6.29)

$$\int \rho_x(z) i\mathcal{L}\boldsymbol{r}_j d^{6N}z = \frac{1}{mN}\int \boldsymbol{M}(\boldsymbol{r})d^3r.$$

We have thus evaluated the first term in (6.25) for $A(z) = i\mathcal{L}\boldsymbol{r}_j$, and in the second term only derivatives with respect to $x_k = \boldsymbol{M}(\boldsymbol{r})$ occur (note that k is a continuous position label). With the atomistic momentum density (6.8), we then obtain

$$\begin{aligned}\mathcal{P}(x)i\mathcal{L}\boldsymbol{r}_j &= \frac{1}{mN}\int \boldsymbol{M}(\boldsymbol{r})d^3r + \frac{1}{mN}\int\left[\sum_{l=1}^{N}\boldsymbol{p}_l\delta(\boldsymbol{r}_l-\boldsymbol{r}) - \boldsymbol{M}(\boldsymbol{r})\right]d^3r \\ &= \frac{1}{N}\sum_{l=1}^{N}\frac{\boldsymbol{p}_l}{m},\end{aligned}$$

so that

$$\mathcal{Q}(x)i\mathcal{L}\boldsymbol{r}_j = i\mathcal{L}\boldsymbol{r}_j - \mathcal{P}(x)i\mathcal{L}\boldsymbol{r}_j$$

can be written in the form (6.30). Note that the slow part of the particle velocity is given by the particle's share of the conserved total momentum.

Exercise 129

By averaging (6.25), we obtain

$$\int \rho_x(z)\mathcal{P}(x)A(z)d^{6N}z = \int \rho_x(z')A(z')d^{6N}z', \tag{F.44}$$

where (6.9) and (6.10) have been used. From the definition (6.25) of the projector $\mathcal{P}(x)$, we further obtain

$$\begin{aligned}\int [\mathcal{P}(x)A(z)]\frac{\delta\rho_x(z)}{\delta x_k}d^{6N}z &= \int\left[\int \rho_x(z')A(z')d^{6N}z'\right]\frac{\delta\rho_x(z)}{\delta x_k}d^{6N}z \\ &+ \int\sum_l(\Pi_l(z)-x_l)\left[\int A(z')\frac{\delta\rho_x(z')}{\delta x_l}d^{6N}z'\right]\frac{\delta\rho_x(z)}{\delta x_k}d^{6N}z.\end{aligned}$$

With the auxiliary equation

$$0 = \frac{\delta}{\delta x_k} \int (\Pi_l(z) - x_l)\rho_x(z) d^{6N}z = -\frac{\delta x_l}{\delta x_k} + \int (\Pi_l(z) - x_l) \frac{\delta \rho_x(z)}{\delta x_k} d^{6N}z,$$

where again (6.9) and (6.10) have been used, we further obtain

$$\begin{aligned}
\int [\mathcal{P}(x)A(z)] \frac{\delta \rho_x(z)}{\delta x_k} d^{6N}z &= \sum_l \frac{\delta x_l}{\delta x_k} \int A(z') \frac{\delta \rho_x(z')}{\delta x_l} d^{6N}z' \\
&= \int A(z') \frac{\delta \rho_x(z')}{\delta x_k} d^{6N}z'. \qquad \text{(F.45)}
\end{aligned}$$

By combining the results (F.44) and (F.45), we obtain $\mathcal{P}(x)[\mathcal{P}(x)A] = \mathcal{P}(x)A$.

Exercise 130

As all involved quantities are real, we can evaluate $\int B(z)\mathcal{P}(x)A(z)d^{6N}z$ and $\int A(z)\mathcal{P}^\dagger(x)B(z)d^{6N}z$ by means of (6.25) and (6.28), respectively; the results indeed coincide for arbitrary A and B after exchanging the names of the integration variables, z and z'.

Exercise 131

According to the definition (6.4) of the adjoint operator, it is sufficient to show

$$\mathcal{P}^\dagger(x)(\rho_x B) = \rho_x \mathcal{P}(x) B.$$

With the expression (6.28), we obtain

$$\begin{aligned}
\mathcal{P}^\dagger(x)(\rho_x B) &= \rho_x \int \rho_x(z') B(z') d^{6N}z' \\
&+ \sum_k \frac{\delta \rho_x}{\delta x_k} \int (\Pi_k(z') - x_k)\rho_x(z') B(z') d^{6N}z',
\end{aligned}$$

whereas (6.25) implies

$$\rho_x \mathcal{P}(x) B = \rho_x \int \rho_x(z') B(z') d^{6N}z' + \sum_k (\Pi_k - x_k)\rho_x \int B(z') \frac{\delta \rho_x(z')}{\delta x_k} d^{6N}z'.$$

The results of Exercise 126 for the generalized canonical ensemble show that these two expressions indeed coincide.

Exercise 132

By differentiating (6.10) with respect to x_j, one obtains

$$\int \Pi_k(z) \frac{\delta \rho_x(z)}{\delta x_j} d^{6N}z = \delta_{jk},$$

and the desired relation follows immediately from (6.25) or (6.33).

Exercise 133
For the full projector (6.28), we obtain

$$\mathcal{P}^\dagger(x)\rho = \rho_x,$$

whereas, for (6.34), we find the less natural relationship

$$\mathcal{P}^\dagger(x)\rho = \sum_k x_k \frac{\delta \rho_x}{\delta x_k}.$$

From (6.28), we obtain

$$\begin{aligned}
\mathcal{P}^\dagger(x)\frac{d\rho}{dt} &= \rho_x \int \frac{d\rho(z')}{dt} d^{6N}z' + \sum_k \frac{\delta \rho_x}{\delta x_k} \int (\Pi_k(z') - x_k)\frac{d\rho(z')}{dt} d^{6N}z' \\
&= \sum_k \frac{\delta \rho_x}{\delta x_k} \int \Pi_k(z')\frac{d\rho(z')}{dt} d^{6N}z'.
\end{aligned}$$

The second part is obtained from the normalization condition (6.9), which implies $\int \frac{d\rho(z)}{dt} d^{6N}z = 0$, or directly by using the simplified projector (6.34). According to the chain rule, we obtain

$$\mathcal{P}^\dagger(x)\frac{d\rho}{dt} = \sum_k \frac{\delta \rho_x}{\delta x_k}\frac{dx_k}{dt} = \frac{d\rho_x}{dt}.$$

Exercise 134
The observable $\mathcal{P}(x)A$ is a linear combination of the slow variables Π_j and a constant. Any projection operator $\mathcal{P}(x')$ leaves the slow variables and constants invariant, so that the desired result is obtained. This result is based on the fact that the same functions Π_j are used to construct the projection operators $\mathcal{P}(x)$ for any choice of the macroscopic variables x.

Exercise 135
By taking the derivatives of $\mathcal{P}(t')\mathcal{P}(t) = \mathcal{P}(t)$ with respect to t and t' and subsequently setting $t = t'$, we obtain

$$\mathcal{P}(t)\dot{\mathcal{P}}(t) = \dot{\mathcal{P}}(t)$$

and

$$\dot{\mathcal{P}}(t)\mathcal{P}(t) = 0 = \dot{\mathcal{P}}(t) - \dot{\mathcal{P}}(t)\mathcal{Q}(t).$$

Factors of $\mathcal{P}(t)$ to the left and $\mathcal{Q}(t)$ to the right of $\dot{\mathcal{P}}(t)$ can be introduced or skipped without changing the result.

Exercise 136
The property $\tilde{\mathcal{P}}\tilde{\mathcal{P}} = \tilde{\mathcal{P}}$ follows from the fact that $\tilde{\mathcal{P}}$ leaves all phase-space functions

that depend only on $\Pi(z)$ invariant (applied to $\tilde{\mathcal{P}}A$); alternatively, this property may be shown by an explicit calculation.

$\tilde{\mathcal{P}}$ is self-adjoint because

$$\int A(z)\tilde{\mathcal{P}}B(z)d^{6N}z = \frac{1}{\Omega(x)}\int\int A(z)B(z')\delta(\Pi(z)-\Pi(z'))d^{6N}z d^{6N}z'$$

is symmetric in A and B.

Also

$$\int \rho_x(z)A(z)\tilde{\mathcal{P}}B(z)d^{6N}z = \langle A\rangle_x \langle B\rangle_x$$

is symmetric in A and B.

$\tilde{\mathcal{P}}\rho_x = \rho_x$ follows for any ensemble that depends on z only through $\Pi(z)$ and, because $\tilde{\mathcal{P}}$ is independent of time, $\tilde{\mathcal{P}}\frac{d\rho_x}{dt} = \frac{d\rho_x}{dt}$ then follows by differentiation.

Exercise 137

From the solution to Exercise 128 for a generalized microcanonical ensemble, we have

$$\langle i\mathcal{L}\boldsymbol{r}_j\rangle_x = \frac{1}{mN}\int \boldsymbol{M}(\boldsymbol{r})d^3r.$$

According to the definition (6.46), we need to insert the atomistic momentum density (6.8) to obtain

$$\tilde{\mathcal{P}}i\mathcal{L}\boldsymbol{r}_j = \frac{1}{mN}\int\sum_{l=1}^{N}\boldsymbol{p}_l\delta(\boldsymbol{r}_l-\boldsymbol{r})d^3r = \frac{1}{N}\sum_{l=1}^{N}\frac{\boldsymbol{p}_l}{m}.$$

Due to the direct linear coupling of the particle velocity to the conserved total momentum, the richer projection operator $\tilde{\mathcal{P}}$ thus produces the same separation of the slow and fast contributions to the particle velocity as $\mathcal{P}(x)$.

Exercise 138

The entropy expressions (6.54) and (6.55) can be inserted into (6.22) to obtain the desired result (6.61). For a sharply peaked distribution, the macroscopic state corresponds to the maximum of $f_x(x')$, and we obtain

$$\frac{\delta S^{\mathrm{mc}}(x')}{\delta x'_k} = k_{\mathrm{B}}\lambda_k(x) = \frac{\delta S^{\mathrm{c}}(x)}{\delta x_k},$$

where (6.56) for the canonical entropy has been used. Therefore, the entropies S^{mc} and S^{c} must coincide up to a constant, which is actually important to obtain the proper normalization of $f_x(x')$. Finally, (6.62) follows by Taylor expansion of $S^{\mathrm{mc}}(x')$.

Exercise 139

The desired result (6.63) follows by inserting (6.24) into (6.25) and then using (6.56). The additional statement follows from

$$\frac{\delta x_j}{\delta\lambda_k} = \frac{\delta}{\delta\lambda_k}\int \Pi_j(z)\frac{1}{\hat{Z}(\lambda)}\exp\left[-\sum_l \lambda_l\Pi_l(z)\right]d^{6N}z = x_j x_k - \langle\Pi_j\Pi_k\rangle_x .$$

Exercise 140
A straightforward calculation gives

$$\begin{aligned}
\sum_k \frac{\delta}{\delta x_k}\left\langle\frac{\partial \Pi_k}{\partial z_j}A\right\rangle_x &= \sum_k \int \frac{\partial \Pi_k(z)}{\partial z_j}A(z)\frac{\delta \rho_x(z)}{\delta x_k}d^{6N}z \\
&= \sum_k \int \frac{\partial \Pi_k(z)}{\partial z_j}A(z) \\
&\quad \times \left[\frac{1}{\Omega(x)}\frac{\delta}{\delta x_k}\delta(\Pi(z)-x) - \rho_x(z)\frac{\delta \ln \Omega(x)}{\delta x_k}\right] d^{6N}z \\
&= -\int A(z)\frac{1}{\Omega(x)}\frac{\partial}{\partial z_j}\delta(\Pi(z)-x)d^{6N}z \\
&\quad -\frac{1}{k_B}\sum_k \left\langle\frac{\partial \Pi_k}{\partial z_j}A\right\rangle_x \frac{\delta S}{\delta x_k} \\
&= \left\langle\frac{\partial A}{\partial z_j}\right\rangle_x - \frac{1}{k_B}\sum_k \left\langle\frac{\partial \Pi_k}{\partial z_j}A\right\rangle_x \frac{\delta S}{\delta x_k}.
\end{aligned}$$

Exercise 141
For the generalized canonical ensemble, one obtains the following equation by acting with Liouville's differential operator on (6.15):

$$k_B i\mathcal{L}\rho_x(z) = -\sum_k \rho_x(z) k_B \lambda_k(x) i\mathcal{L}\Pi_k(z),$$

and the desired result follows from (6.56).

For the generalized microcanonical ensemble, one obtains the following equation by acting with Liouville's differential operator on (6.11):

$$\begin{aligned}
k_B i\mathcal{L}\rho_x(z) &= -\frac{k_B}{\Omega(x)}\sum_k \left[\frac{\delta}{\delta x_k}\delta(\Pi(z)-x)\right] i\mathcal{L}\Pi_k(z) \\
&= -\frac{k_B}{\Omega(x)}\sum_k \left[\frac{\delta}{\delta x_k}\Omega(x)\rho_x(z)\right] i\mathcal{L}\Pi_k(z) \\
&= -k_B\sum_k \left[\frac{\delta \ln \Omega(x)}{\delta x_k}\rho_x(z) + \frac{\delta \rho_x(z)}{\delta x_k}\right] i\mathcal{L}\Pi_k(z),
\end{aligned}$$

and the desired result follows from (6.54).

Exercise 142
Because only constant momenta are added in Galilean transformations, derivatives are unchanged and (6.85) is fulfilled. Whereas the potential energy is invariant under Galilean transformations, the kinetic energy changes so that (6.84) is not satisfied.

However, because the energy difference

$$\frac{1}{2}\left(\sum_{j=1}^{N} m_j\right)(\Delta \boldsymbol{v})^2 - \Delta \boldsymbol{v} \cdot \left(\sum_{j=1}^{N} \boldsymbol{p}_j\right)$$

is an additive constant, it is irrelevant.

Exercise 143
By writing out the conditions

$$\begin{pmatrix} \boldsymbol{Q}_{rr} & \boldsymbol{Q}_{rp} \\ \boldsymbol{Q}_{pr} & \boldsymbol{Q}_{pp} \end{pmatrix} \cdot \begin{pmatrix} 0 & 1 \\ -1 & 0 \end{pmatrix} \cdot \begin{pmatrix} \boldsymbol{Q}_{rr}^T & \boldsymbol{Q}_{pr}^T \\ \boldsymbol{Q}_{rp}^T & \boldsymbol{Q}_{pp}^T \end{pmatrix} = \begin{pmatrix} 0 & 1 \\ -1 & 0 \end{pmatrix},$$

we obtain the following three equations:

$$\boldsymbol{Q}_{rr} \cdot \boldsymbol{Q}_{rp}^T = (\boldsymbol{Q}_{rr} \cdot \boldsymbol{Q}_{rp}^T)^T,$$

$$\boldsymbol{Q}_{pp} \cdot \boldsymbol{Q}_{pr}^T = (\boldsymbol{Q}_{pp} \cdot \boldsymbol{Q}_{pr}^T)^T,$$

$$\boldsymbol{Q}_{rr} \cdot \boldsymbol{Q}_{pp}^T - \boldsymbol{Q}_{rp} \cdot \boldsymbol{Q}_{pr}^T = \boldsymbol{1}.$$

For linear extended point transformations we have

$$\boldsymbol{Q}_{pp}^T = \boldsymbol{Q}_{rr}^{-1}, \quad \boldsymbol{Q}_{rp} = \boldsymbol{Q}_{pr} = \boldsymbol{0},$$

so that all three conditions are fulfilled.

Exercise 144
We first consider $i\mathcal{L}\mathcal{S}^g A$ for an arbitrary phase space function $A(z)$. With (6.2) and (6.98), we obtain

$$i\mathcal{L}\mathcal{S}^g A(z) = \frac{\partial A(z^g(z))}{\partial z} \cdot L_0 \cdot \frac{\partial E_0(z)}{\partial z},$$

and the invariance property (6.84) gives

$$i\mathcal{L}\mathcal{S}^g A(z) = \frac{\partial A(z^g(z))}{\partial z} \cdot L_0 \cdot \frac{\partial E_0(z^g(z))}{\partial z}.$$

With the chain rule and the invariance property (6.85), we finally obtain

$$i\mathcal{L}\mathcal{S}^g A(z) = \frac{\partial A(z^g)}{\partial z^g} \cdot L_0 \cdot \frac{\partial E_0(z^g)}{\partial z^g} = \mathcal{S}^g i\mathcal{L}A(z).$$

Instead of (6.99), we show the equivalent identity $\mathcal{P}(x)\mathcal{S}^g = \mathcal{S}^g\mathcal{P}(x^g)$. From (6.25), we have

$$\begin{aligned} \mathcal{P}(x)\mathcal{S}^g A(z) &= \int \rho_x(z')A(z^g(z'))d^{6N}z' \\ &+ \sum_k [\Pi_k(z) - x_k] \int A(z^g(z'))\frac{\delta \rho_x(z')}{\delta x_k} d^{6N}z'. \end{aligned}$$

By using the invariance of the ensemble (6.92), we obtain after changing the integration variable

$$\begin{aligned}\mathcal{P}(x)\mathcal{S}^g A(z) &= \int \rho_{x^g}(z')A(z')d^{6N}z' \\ &+ \sum_{kl}[\Pi_k(z) - x_k]\frac{\delta x_l^g}{\delta x_k}\int A(z')\frac{\delta \rho_{x^g}(z')}{\delta x_l^g}d^{6N}z',\end{aligned}$$

and, as a consequence of the linear nature of the mapping (6.90) for $x \to x^g(x)$ and of (6.89), we obtain

$$\mathcal{P}(x)\mathcal{S}^g A(z) = [\mathcal{P}(x^g)A](z^g(z)) = \mathcal{S}^g\mathcal{P}(x^g)A(z).$$

Finally, we have

$$\begin{aligned}\tilde{\mathcal{P}}\mathcal{S}^g A(z) &= \int \rho_{\Pi(z)}(z')A(z^g(z'))d^{6N}z' \\ &= \int \rho_{x^g(\Pi(z))}(z^g(z'))A(z^g(z'))d^{6N}z' \\ &= \int \rho_{\Pi(z^g(z))}(z')A(z')d^{6N}z' = \mathcal{S}^g\tilde{\mathcal{P}}A(z),\end{aligned}$$

where (6.89), (6.90), and (6.92) have been used, and the integration variables have been changed.

Exercise 145

By acting with $e^{i\mathcal{L}t}$ on the definition (6.25), we obtain

$$\begin{aligned}e^{i\mathcal{L}t}\mathcal{P}(t)A(z) &= e^{i\mathcal{L}t}\Big\{\int \rho_{x(t)}(z')A(z')d^{6N}z' \\ &+\sum_k[\Pi_k(z) - x_k(t)]\int A(z')\frac{\delta\rho_{x(t)}(z')}{\delta x_k}d^{6N}z'\Big\}.\end{aligned}$$

When the linear operator $e^{i\mathcal{L}t}$ acts on a constant independent of z it reproduces the constant, as can be seen by Taylor expansion of the exponential and the fact that $\mathcal{L}$ is a differential operator. We furthermore note that

$$\int \rho_{x(0)}(z)e^{i\mathcal{L}t}\Pi_k(z)d^{6N}z = x_k(t),$$

because $\rho_x(0)$ is the exact initial condition, $e^{i\mathcal{L}t}$ generates the exact time evolution of the observables $\Pi_k(z)$, and $x_k(t)$ is made to coincide with the exact time dependent averages of $\Pi_k(z)$. By putting together all these results, we obtain the desired identity.

We furthermore have

$$\begin{aligned}\int \rho_{x(0)}(z)e^{i\mathcal{L}t}\tilde{\mathcal{P}}A(z)d^{6N}z &= \int \frac{1}{\Omega(x(0))}\delta(\Pi(z) - x(0))\frac{1}{\Omega(\Pi(z(t)))} \\ &\times\ \delta(\Pi(z') - \Pi(z(t)))A(z')d^{6N}z'd^{6N}z.\end{aligned}$$

If $\Pi(z) = x(0)$ implies $\Pi(z(t)) = x(t)$, we can perform the integration over z to obtain

$$\int \rho_{x(0)}(z) e^{i\mathcal{L}t}\tilde{\mathcal{P}} A(z) d^{6N}z = \int \frac{1}{\Omega(x(t))}\delta(\Pi(z') - x(t)) A(z') d^{6N}z',$$

which is the result shown in (6.112).

Exercise 146
By averaging the definition (6.25), we obtain

$$\int \rho_x(z)\mathcal{P}(x)A(z)d^{6N}z = \int \rho_x(z')A(z')d^{6N}z',$$

where (6.9) and (6.10) have been used.

Exercise 147
If we consider the time dependence of $\mathcal{P}(x)A$ in (6.25) resulting from the time dependence of x, there are three terms associated with the occurrence of x in $\rho_x(z')$, x_k, and $\delta\rho_x(z')/\delta x_k$, respectively. By using the chain rule, it is found that the contributions associated with $\rho_x(z')$ and x_k cancel, and the remaining third contribution associated with $\delta\rho_x(z')/\delta x_k$ vanishes upon averaging.

Exercise 148
From the general definition (6.25), we obtain

$$\begin{aligned}[\tilde{\mathcal{P}}A](z) &= \int \tilde{\rho}_f(z')A(z')d^{6N}z' \\ &+ \int [\tilde{\Pi}_y(z) - f(y)]\frac{\delta}{\delta f(y)}\int A(z')\tilde{\rho}_f(z')d^{6N}z'd^Ky.\end{aligned}$$

With the auxiliary equation (6.149), we can simplify this expression to

$$\begin{aligned}[\tilde{\mathcal{P}}A](z) &= \int f(y)\langle A\rangle_y d^Ky + \int [\tilde{\Pi}_y(z) - f(y)]\langle A\rangle_y d^Ky \\ &= \int \tilde{\Pi}_y(z)\langle A\rangle_y d^Ky = \langle A\rangle_{\Pi(z)}\,.\end{aligned}$$

The key to the independence of $\tilde{\mathcal{P}}$ of f lies in the linear dependence of $\tilde{\rho}_f(z)$ on f.

Exercise 149
Starting from the identity

$$k_B \int \frac{\delta}{\delta x}\cdot\left[\exp\left\{\frac{S_1(x) - S_2(y)}{k_B}\right\} M_1(x)\cdot\frac{\delta\Pi(x)}{\delta x}\delta(\Pi(x) - y)\right] d^Kx = 0,$$

after performing all the derivatives, we obtain with (6.180)

$$\left\langle \frac{\delta\Pi}{\delta x}\cdot\left(M_1\cdot\frac{\delta S_1}{\delta x} + k_B\frac{\delta}{\delta x}\cdot M_1\right) + k_B M_1 : \frac{\delta^2\Pi}{\delta x\delta x}\right\rangle_y \quad =$$

$$k_{\rm B} \exp\left\{-\frac{S_2(y)}{k_{\rm B}}\right\} \frac{\delta}{\delta y} \cdot \left[\exp\left\{\frac{S_2(y)}{k_{\rm B}}\right\} M_2'(y)\right],$$

where the right-hand side can be rewritten as

$$M_2'(y) \cdot \frac{\delta S_2(y)}{\delta y} + k_{\rm B} \frac{\delta}{\delta y} \cdot M_2'(y).$$

With (6.168) and (6.171), one then obtains the desired result.

Exercise 150

With the explicit expression

$$\rho_{2,y}(z) = \frac{1}{Z_2(y)} \exp\left[-\sum_k \lambda_k(y)\Pi_{2,k}(z)\right],$$

we obtain from (6.186) and (6.187):

$$\rho_y(x) = \frac{\Omega_1(x)}{Z_2(y)} \exp\left[-\sum_k \lambda_k(y)\Pi_k(x)\right].$$

Exercise 151

We use the notation $\boldsymbol{c} = \langle \boldsymbol{QQ} \rangle = y$ for the configuration tensor and $\boldsymbol{\lambda}$ for the corresponding tensorial Lagrange multiplier. With $\Omega_1(\boldsymbol{Q}, N) = \exp\{S_1(\boldsymbol{Q}, N)/k_{\rm B}\}$, we obtain from (6.194) and (6.197) the Gaussian distribution

$$\rho_{(\boldsymbol{c},N)}(\boldsymbol{Q}) \propto \exp\left\{-\frac{1}{2}\boldsymbol{Q} \cdot \left(\frac{3}{Na^2}\boldsymbol{1} + 2\boldsymbol{\lambda}\right) \cdot \boldsymbol{Q}\right\},$$

from which we have

$$\boldsymbol{c}^{-1} = \frac{3}{Na^2}\boldsymbol{1} + 2\boldsymbol{\lambda},$$

and thus

$$\rho_{(\boldsymbol{c},N)}(\boldsymbol{Q}) = \frac{1}{\sqrt{(2\pi)^3 \det \boldsymbol{c}}} \exp\left\{-\frac{1}{2}\boldsymbol{Q} \cdot \boldsymbol{c}^{-1} \cdot \boldsymbol{Q}\right\}$$

turns out to be independent of N. From (6.196), we find

$$\begin{aligned} S_2(\boldsymbol{c}, N) &= k_{\rm B} \int \rho_{(\boldsymbol{c},N)}(\boldsymbol{Q}) \left[N \ln c - \frac{3}{2}\frac{\boldsymbol{Q}^2}{Na^2}\right. \\ &+ \left. \frac{1}{2} \ln \det \left(\frac{\boldsymbol{c}}{Na^2/3}\right) + \frac{1}{2}\boldsymbol{Q} \cdot \boldsymbol{c}^{-1} \cdot \boldsymbol{Q}\right] d^3Q, \end{aligned}$$

which can be integrated to get the final result

$$S_2(\boldsymbol{c}, N) = k_{\rm B} \left[N \ln c + \frac{1}{2}{\rm tr}\left(\boldsymbol{1} - \frac{\boldsymbol{c}}{Na^2/3}\right) + \frac{1}{2} \ln \det \left(\frac{\boldsymbol{c}}{Na^2/3}\right)\right].$$

Keeping the proper normalization of c in mind, the configurational part of the entropy coincides with the previosly given expression (4.86).

F.7 KINETIC THEORY OF GASES

Exercise 152

With the volume of 22.4 liters for a mole of gas under normal conditions, we obtain

$$\begin{aligned} \langle \boldsymbol{v}^2 \rangle &= 3\frac{1\,\mathrm{atm}}{(2\mathrm{g}/22.4\mathrm{l})} = 3\frac{1.01 \cdot 10^5\, 22.4 \cdot 10^{-3}}{2 \cdot 10^{-3}} \left(\frac{\mathrm{m}}{\mathrm{s}}\right)^2 \\ &= 1.5 \cdot 1.01 \cdot 2.24 \left(10^3 \frac{\mathrm{m}}{\mathrm{s}}\right)^2 = \left(1.84 \cdot 10^3 \frac{\mathrm{m}}{\mathrm{s}}\right)^2 . \end{aligned}$$

Exercise 153

The Gaussian distribution can be given in terms of the second moments,

$$\left(\frac{3}{2\pi \langle \boldsymbol{v}^2 \rangle}\right)^{3/2} \exp\left\{-\frac{3}{2}\frac{\boldsymbol{v}^2}{\langle \boldsymbol{v}^2 \rangle}\right\} .$$

We hence obtain

$$\begin{aligned} \langle |\boldsymbol{v}| \rangle &= \int_0^\infty v 4\pi v^2 \left(\frac{3}{2\pi \langle \boldsymbol{v}^2 \rangle}\right)^{3/2} \exp\left\{-\frac{3}{2}\frac{v^2}{\langle \boldsymbol{v}^2 \rangle}\right\} dv \\ &= -\int_0^\infty 2v^2 \left(\frac{3}{2\pi \langle \boldsymbol{v}^2 \rangle}\right)^{1/2} \frac{d}{dv} \exp\left\{-\frac{3}{2}\frac{v^2}{\langle \boldsymbol{v}^2 \rangle}\right\} dv \\ &= \int_0^\infty 4v \left(\frac{3}{2\pi \langle \boldsymbol{v}^2 \rangle}\right)^{1/2} \exp\left\{-\frac{3}{2}\frac{v^2}{\langle \boldsymbol{v}^2 \rangle}\right\} dv \\ &= -\int_0^\infty \left(\frac{8 \langle \boldsymbol{v}^2 \rangle}{3\pi}\right)^{1/2} \frac{d}{dv} \exp\left\{-\frac{3}{2}\frac{v^2}{\langle \boldsymbol{v}^2 \rangle}\right\} dv = \left(\frac{8 \langle \boldsymbol{v}^2 \rangle}{3\pi}\right)^{1/2} . \end{aligned}$$

The second part of (7.5) follows from $(1/2)m \langle \boldsymbol{v}^2 \rangle = (3/2)k_\mathrm{B}T$.

Exercise 154

The frequency of collisions between gas particles for hydrogen under normal conditions is given by

$$\frac{\langle |\boldsymbol{v}| \rangle}{l_\mathrm{mfp}} = \sqrt{\frac{8}{3\pi}} \frac{\sqrt{\langle \boldsymbol{v}^2 \rangle}}{l_\mathrm{mfp}} \approx \sqrt{\frac{8}{3\pi}} \frac{1840}{0.15 \cdot 10^{-6}} \frac{1}{\mathrm{s}} \approx 10^{10} \frac{1}{\mathrm{s}} .$$

Exercise 155

With $m = 14\mathrm{g}/N_\mathrm{A}$, $d = 2.5$ Å, and $T = 293$ K, we obtain $\eta = 1.9 \cdot 10^{-5}$ Pas from (7.12). The measured values are $\eta = 1.75 \cdot 10^{-5}$ Pas for nitrogen gas and $\eta = 1.81 \cdot 10^{-5}$ Pas for air.

Exercise 156

From the normalization and the first and second moments of Gaussian probability measures, we obtain:

$$\int f^{\text{Maxw}}_{\rho,\boldsymbol{v},T}(\boldsymbol{r},\boldsymbol{p})d^3p = \frac{\rho(\boldsymbol{r})}{m} = n(\boldsymbol{r}),$$

$$\int \boldsymbol{p} f^{\text{Maxw}}_{\rho,\boldsymbol{v},T}(\boldsymbol{r},\boldsymbol{p})d^3p = \frac{\rho(\boldsymbol{r})}{m} m\boldsymbol{v}(\boldsymbol{r}) = \rho(\boldsymbol{r})\boldsymbol{v}(\boldsymbol{r}) = \boldsymbol{M}(\boldsymbol{r}),$$

$$\int \frac{[\boldsymbol{p} - m\boldsymbol{v}(\boldsymbol{r})]^2}{2m} f^{\text{Maxw}}_{\rho,\boldsymbol{v},T}(\boldsymbol{r},\boldsymbol{p})d^3p = \frac{1}{2m}\frac{\rho(\boldsymbol{r})}{m} 3mk_{\text{B}}T(\boldsymbol{r}) = \frac{3}{2}n(\boldsymbol{r})k_{\text{B}}T(\boldsymbol{r}).$$

Exercise 157

We obtain $d\Delta H_{\text{B}}/dt = 0$ because, in the absence of external forces, the total number of particles, momentum, and energy are conserved for Boltzmann's equation.

Exercise 158

For $\boldsymbol{p}_1 = 0$, we have the total momentum $\boldsymbol{p}_{\text{t}} = \boldsymbol{p}_2$ and $\boldsymbol{P} = \boldsymbol{p}_2/2 = \boldsymbol{p}_{\text{t}}/2$. The momentum of Particle 2 changes during the collision from $\boldsymbol{p}_{\text{t}}/2 + \boldsymbol{P} = 2\boldsymbol{P} = \boldsymbol{p}_2$ to $\boldsymbol{p}_{\text{t}}/2 + \boldsymbol{Q} = \boldsymbol{P} + \boldsymbol{Q}$. We thus obtain from Figure 7.5

$$\frac{\tan\theta}{\tan\theta'} = \frac{\cos\theta'}{1+\cos\theta'}, \qquad \tan\theta = \frac{\sin\theta'}{1+\cos\theta'}$$

or

$$\cos\theta = \frac{1}{\sqrt{1+(\tan\theta)^2}} = \frac{1+\cos\theta'}{\sqrt{2(1+\cos\theta')}} = \sqrt{\frac{1+\cos\theta'}{2}} = \cos\frac{\theta'}{2}.$$

We thus find the simple relationship $\theta = \theta'/2$.

Exercise 159

From (7.44) and (7.42), we obtain

$$\theta = \pi - 2\int_{R_{\min}/b}^{\infty} \left(\frac{b}{R}\right)^2 \left[1 - \left(\frac{b}{R}\right)^2 - \frac{2m_{\text{r}}\hat{\phi}}{b^{n_\phi - 1}P^2}\left(\frac{b}{R}\right)^{n_\phi - 1}\right]^{-1/2} d\left(\frac{R}{b}\right),$$

and

$$\left(\frac{b}{R_{\min}}\right)^2 + \frac{2m_{\text{r}}\hat{\phi}}{b^{n_\phi - 1}P^2}\left(\frac{b}{R_{\min}}\right)^{n_\phi - 1} = 1.$$

For given n_ϕ, these equations imply the functional form

$$\theta = \theta\left(\frac{m_{\text{r}}\hat{\phi}}{b^{n_\phi - 1}P^2}\right)$$

and hence by inversion

$$b = \left(\frac{m_{\text{r}}\hat{\phi}}{P^2}\right)^{\frac{1}{n_\phi - 1}} \hat{b}(\theta),$$

where the dimensionless function $\hat{b}(\theta)$ can be determined from the first two equations after setting $m_r\hat{\phi}/P^2 = 1$, that is, after the substitution $u = b/R$,

$$\theta = \pi - 2\int_0^{u_{\max}} \frac{du}{\sqrt{1 - u^2 - 2(u/\hat{b})^{n_\phi - 1}}},$$

and

$$u_{\max}^2 + 2\left(\frac{u_{\max}}{\hat{b}}\right)^{n_\phi - 1} = 1.$$

For the differential cross section, we then obtain from (7.34)

$$\frac{d\sigma}{d\Omega} = \left(\frac{m_r\hat{\phi}}{P^2}\right)^{\frac{2}{n_\phi - 1}} \frac{\hat{b}(\theta)}{\sin\theta}\left|\frac{d\hat{b}(\theta)}{d\theta}\right|,$$

which gives the explicit power-law dependence on all the variables except θ.

For $n_\phi = 5$, the function $\theta(\hat{b})$ can be expressed as a complete elliptic integral of the first kind. We have the explicit expression

$$u_{\max}^2 = \frac{\hat{b}^2}{4}\left[(\hat{b}^4 + 8)^{1/2} - \hat{b}^2\right],$$

and, after the substitution $u/u_{\max} \to u$, we find,[17]

$$\begin{aligned}\theta(\hat{b}) &= \pi - 2\int_0^1 \frac{du}{\sqrt{(1 - u^2)[1 + 2u_{\max}^2(1 + u^2)/\hat{b}^4]}} \\ &= \pi - \frac{2\hat{b}}{(\hat{b}^4 + 8)^{1/4}}\boldsymbol{K}(k),\end{aligned} \tag{F.46}$$

where $\boldsymbol{K}$ is the complete elliptic integral of the first kind, and the square of its argument is given by

$$k^2 = \frac{1}{2}\left[1 - \frac{\hat{b}^2}{(\hat{b}^4 + 8)^{1/2}}\right].$$

For large $\hat{b}$, we obtain the asymptotic formula

$$\theta(\hat{b}) \approx \frac{3\pi}{2}\frac{1}{\hat{b}^4}.$$

Exercise 160

From Figure 7.6, we obtain the relationship

$$\sin\frac{\pi - \theta'}{2} = \cos\frac{\theta'}{2} = \frac{b}{d},$$

[17] See 3.152.4 and 8.112.1 of Gradshteyn & Ryzhik, *Tables* (Academic, 1980).

and the differential cross section can then be obtained as the ratio of the area to be hit in order to be scattered into a given solid angle and this solid angle [see (7.34)],

$$\frac{d\sigma'}{d\Omega'} = \frac{b}{\sin\theta'}\left|\frac{db}{d\theta'}\right|,$$

where b is the impact parameter (see Figure 7.6). From these equations, we obtain

$$\frac{d\sigma'}{d\Omega'} = \frac{b}{\sin\theta'}\frac{1}{2}d\sin\frac{\theta'}{2} = \frac{d^2}{4}.$$

The differential cross section for a hard-sphere collision in the center-of-mass frame is independent of the scattering angle. Integrating $d\sigma'/d\Omega'$ over all solid angles thus means multiplying by 4π. The resulting total cross section πd^2 is the area of the circle to be hit in order to have a scattering event. Note that, for two hard spheres, a collision occurs whenever the impact parameter is smaller than the sum of the radii.

Exercise 161

From (7.44), we obtain

$$\begin{aligned} \theta &= \pi - 2b\int_d^\infty \frac{1}{R^2\sqrt{1-(b/R)^2}}dR \\ &= \pi - 2\int_0^{b/d}\frac{du}{\sqrt{1-u^2}} = \pi - 2\arcsin\frac{b}{d}, \end{aligned}$$

and hence

$$\frac{b}{d} = \sin\frac{\pi-\theta}{2} = \cos\frac{\theta}{2},$$

which agrees with the result of Exercise 160.

Exercise 162

From (7.50), we obtain by the product rule

$$\frac{\partial\rho_f}{\partial t} = \rho_f\sum_{k=1}^{N}\frac{1}{f(\boldsymbol{r}_k,\boldsymbol{p}_k)}\frac{\partial f(\boldsymbol{r}_k,\boldsymbol{p}_k)}{\partial t},$$

and, after using (7.17) in the absence of collisions,

$$\begin{aligned} \frac{\partial\rho_f}{\partial t} &= -\rho_f\sum_{k=1}^{N}\frac{1}{f(\boldsymbol{r}_k,\boldsymbol{p}_k)}\left(\frac{\boldsymbol{p}_k}{m}\cdot\frac{\partial}{\partial\boldsymbol{r}_k} - \frac{\partial\phi^{(e)}(\boldsymbol{r}_k)}{\partial\boldsymbol{r}_k}\cdot\frac{\partial}{\partial\boldsymbol{p}_k}\right)f(\boldsymbol{r}_k,\boldsymbol{p}_k) \\ &= -\sum_{k=1}^{N}\left(\frac{\boldsymbol{p}_k}{m}\cdot\frac{\partial}{\partial\boldsymbol{r}_k} - \frac{\partial\phi^{(e)}(\boldsymbol{r}_k)}{\partial\boldsymbol{r}_k}\cdot\frac{\partial}{\partial\boldsymbol{p}_k}\right)\rho_f, \end{aligned}$$

which is the Liouville equation (6.6).

Exercise 163

From the definition (6.25), we obtain

$$\begin{aligned} \mathcal{P}(f)A(z) &= \int \rho_f(z')A(z')d^{6N}z' + \int d^3r d^3p[f^{\Pi}(\boldsymbol{r},\boldsymbol{p};z) - f(\boldsymbol{r},\boldsymbol{p})] \\ &\times \int A(z')\frac{\delta}{\delta f(\boldsymbol{r},\boldsymbol{p})}\left(\frac{1}{N^N}\prod_{j=1}^{N} f(\boldsymbol{r}'_j,\boldsymbol{p}'_j)\right) d^{6N}z'. \end{aligned}$$

The product rule yields a contribution from every factor,

$$\begin{aligned} \mathcal{P}(f)A(z) &= \int \rho_f(z')A(z')d^{6N}z' + \int d^3r d^3p[f^{\Pi}(\boldsymbol{r},\boldsymbol{p};z) - f(\boldsymbol{r},\boldsymbol{p})] \\ &\times \sum_{j=1}^{N}\int \rho_f(z')A(z')\frac{\delta(\boldsymbol{r}-\boldsymbol{r}'_j)\delta(\boldsymbol{p}-\boldsymbol{p}'_j)}{f(\boldsymbol{r}'_j,\boldsymbol{p}'_j)} d^{6N}z'. \end{aligned}$$

After carrying out the integrations over d^3r and d^3p, we obtain the desired result (7.51).

Exercise 164

By introducing the extended space transformations (6.86), (6.87) into the atomistically defined distribution function (7.47), we obtain the natural action of a space transformation g (instead of f, to avoid confusion with the single-particle distribution function) in exactly the same way as we constructed (6.105) and (6.107),

$$(f^{\Pi})^g(g(\boldsymbol{r}),[\partial g(\boldsymbol{r})/\partial \boldsymbol{r}]^{-1}\cdot \boldsymbol{p};z) = f^{\Pi}(\boldsymbol{r},\boldsymbol{p};z).$$

By averaging, we obtain the corresponding action on the coarse-grained level,

$$f^g(g(\boldsymbol{r}),[\partial g(\boldsymbol{r})/\partial \boldsymbol{r}]^{-1}\cdot \boldsymbol{p}) = f(\boldsymbol{r},\boldsymbol{p}).$$

In particular, this action transforms isotropic distributions in momentum space into anisotropic ones.

Exercise 165

By comparing (7.24) and (7.55), we obtain

$$S = -k_{\rm B}(H_{\rm B} - N\ln N) = -k_{\rm B}H_{\rm B} + \text{constant}.$$

Exercise 166

The contribution linear in the relevant variables $f^{\Pi}(\boldsymbol{r},\boldsymbol{p};z)$ in (7.61) without any further z-dependence is reproduced by the projection operator $\mathcal{P}(f)$ and hence eliminated by the complementary operator $\mathcal{Q}(f)$ (see Exercise 132). By acting with $\mathcal{P}(f)$ on the term in the last line of (7.61), we obtain

$$\sum_{j,k,l=1}^{N}\frac{\partial}{\partial \boldsymbol{p}}\cdot\int \rho_f(z')\frac{f^{\Pi}(\boldsymbol{r}'_l,\boldsymbol{p}'_l;z)}{f(\boldsymbol{r}'_l,\boldsymbol{p}'_l)}\frac{\partial\phi(|\boldsymbol{r}-\boldsymbol{r}'_k|)}{\partial \boldsymbol{r}}\delta(\boldsymbol{r}-\boldsymbol{r}'_j)\delta(\boldsymbol{p}-\boldsymbol{p}'_j)d^{6N}z'$$

$$-\frac{(N-1)^2}{N}\frac{\partial f(\boldsymbol{r},\boldsymbol{p})}{\partial \boldsymbol{p}}\cdot\int f(\boldsymbol{r}',\boldsymbol{p}')\frac{\partial\phi(|\boldsymbol{r}-\boldsymbol{r}'|)}{\partial \boldsymbol{r}}d^3r'd^3p'.$$

In the sum of the first line, where $j \neq k$, we consider the terms for $l = j$, for $l = k$, and for three different indices. We thus obtain the equivalent expression

$$\frac{N-1}{N}\frac{\partial f^{\Pi}(\boldsymbol{r},\boldsymbol{p};z)}{\partial \boldsymbol{p}}\cdot\int f(\boldsymbol{r}',\boldsymbol{p}')\frac{\partial\phi(|\boldsymbol{r}-\boldsymbol{r}'|)}{\partial \boldsymbol{r}}d^3r'd^3p'$$
$$+\frac{N-1}{N}\frac{\partial f(\boldsymbol{r},\boldsymbol{p})}{\partial \boldsymbol{p}}\cdot\int f^{\Pi}(\boldsymbol{r}',\boldsymbol{p}';z)\frac{\partial\phi(|\boldsymbol{r}-\boldsymbol{r}'|)}{\partial \boldsymbol{r}}d^3r'd^3p'$$
$$+\frac{(N-1)(N-2)-(N-1)^2}{N}\frac{\partial f(\boldsymbol{r},\boldsymbol{p})}{\partial \boldsymbol{p}}\cdot\int f(\boldsymbol{r}',\boldsymbol{p}')\frac{\partial\phi(|\boldsymbol{r}-\boldsymbol{r}'|)}{\partial \boldsymbol{r}}d^3r'd^3p',$$

which, for large N, becomes

$$\frac{\partial f^{\Pi}(\boldsymbol{r},\boldsymbol{p};z)}{\partial \boldsymbol{p}}\cdot\int f(\boldsymbol{r}',\boldsymbol{p}')\frac{\partial\phi(|\boldsymbol{r}-\boldsymbol{r}'|)}{\partial \boldsymbol{r}}d^3r'd^3p'$$
$$+\frac{\partial f(\boldsymbol{r},\boldsymbol{p})}{\partial \boldsymbol{p}}\cdot\int f^{\Pi}(\boldsymbol{r}',\boldsymbol{p}';z)\frac{\partial\phi(|\boldsymbol{r}-\boldsymbol{r}'|)}{\partial \boldsymbol{r}}d^3r'd^3p'$$
$$-\frac{\partial f(\boldsymbol{r},\boldsymbol{p})}{\partial \boldsymbol{p}}\cdot\int f(\boldsymbol{r}',\boldsymbol{p}')\frac{\partial\phi(|\boldsymbol{r}-\boldsymbol{r}'|)}{\partial \boldsymbol{r}}d^3r'd^3p'.$$

If the average single-particle distribution function does not vary appreciably over the range of interaction, the integrals in the first and last line of this expression vanish, and we finally obtain the expression

$$\frac{\partial f(\boldsymbol{r},\boldsymbol{p})}{\partial \boldsymbol{p}}\cdot\sum_{j=1}^{N}\frac{\partial\phi(|\boldsymbol{r}-\boldsymbol{r}_j|)}{\partial \boldsymbol{r}}$$

for the result of acting with $\mathcal{P}(f)$ on the expression in the last line of (7.61). This result corresponds to (7.62).

Exercise 167

Equation (7.70) follows from the symmetry of the δ function. Equation (7.71) can be established by considering the transformation $\boldsymbol{R}(t) \to -\boldsymbol{R}(t)$, corresponding to an exchange of the particles 1 and 2. Finally, (7.72) is a consequence of the time-reversal symmetry of the collision process.

Exercise 168

From (7.54), (7.73), and (7.77), we have

$$M\cdot\left.\frac{\delta E}{\delta x}\right|_{\boldsymbol{p}=0} = \frac{16}{k_{\mathrm{B}}m^2}\int|\boldsymbol{P}|f(\boldsymbol{r},\boldsymbol{p}')f(\boldsymbol{r},2\boldsymbol{P}-\boldsymbol{p}')\boldsymbol{p}'\cdot\bar{\boldsymbol{I}}_{\phi}(\boldsymbol{p}'-\boldsymbol{P},\boldsymbol{P})d^3Pd^3p'.$$

After introducing $\boldsymbol{p}'' = 2\boldsymbol{P}-\boldsymbol{p}'$, we obtain

$$M\cdot\frac{\delta E}{\delta x} = \frac{1}{k_{\mathrm{B}}m^2}\int f(\boldsymbol{p}')f(\boldsymbol{p}'')\boldsymbol{p}'\cdot\bar{\boldsymbol{I}}_{\phi}\left(\frac{\boldsymbol{p}'-\boldsymbol{p}''}{2},\frac{\boldsymbol{p}'+\boldsymbol{p}''}{2}\right)|\boldsymbol{p}'+\boldsymbol{p}''|d^3p'd^3p'',$$

where, for simplicity, we have suppressed the position argument of f and the condition $\boldsymbol{p} = 0$ in the notation. We now introduce the explicit expression (7.78) for hard spheres to obtain

$$M \cdot \frac{\delta E}{\delta x} = \frac{4d^2}{k_B m^2} \int\limits_{0 \le \boldsymbol{p}' \cdot \boldsymbol{p}''} f(\boldsymbol{p}')f(\boldsymbol{p}'')\boldsymbol{p}' \cdot \left(\frac{\boldsymbol{p}''\boldsymbol{p}''}{|\boldsymbol{p}''|^4} - \frac{\boldsymbol{p}'\boldsymbol{p}'}{|\boldsymbol{p}'|^4} \right) \cdot (\boldsymbol{p}' + \boldsymbol{p}'') d^3p' d^3p''.$$

By means of symmetry, this result can be simplified to

$$M \cdot \frac{\delta E}{\delta x} = \frac{4d^2}{k_B m^2} \int\limits_{0 \le \boldsymbol{p}' \cdot \boldsymbol{p}''} f(\boldsymbol{p}')f(\boldsymbol{p}'') \left[\frac{(\boldsymbol{p}' \cdot \boldsymbol{p}'')^2}{|\boldsymbol{p}''|^4} - 1 \right] d^3p' d^3p'',$$

so that we have an explicit expression in terms of moments of f. For an isotropic distribution depending only on the magnitude of the momentum, we obtain after carrying out the angular integrations

$$M \cdot \frac{\delta E}{\delta x} = \frac{32\pi^2 d^2}{3k_B m^2} \int\limits_0^\infty dp' \int\limits_0^\infty dp'' f(p')f(p'')(p'^4 - 3p'^2 p''^2).$$

For one-dimensional Gaussian distributions, the fourth moment is three times the square of the second moment, and the double integral hence vanishes. For other types of distributions, this is not true. In view of (7.75), the fact that, in general, the above integral does not vanish proves that M cannot be symmetric.

Exercise 169

After $f(\boldsymbol{r}, \boldsymbol{p}')f(\boldsymbol{r}, \boldsymbol{p}_t - \boldsymbol{p}')$ in (7.68) is replaced by $f(\boldsymbol{r}, (\boldsymbol{p}_t/2) - \boldsymbol{P})f(\boldsymbol{r}, (\boldsymbol{p}_t/2) + \boldsymbol{P})$, the only derivative with respect to $\boldsymbol{p}'$ is the divergence of $\boldsymbol{I}_\phi$, for which we have (7.74). We thus obtain

$$\begin{aligned} M &= \frac{\delta(\boldsymbol{r}' - \boldsymbol{r})}{k_B} \int v_{rel} f(\boldsymbol{r}, \frac{\boldsymbol{p}_t}{2} - \boldsymbol{P}) f(\boldsymbol{r}, \frac{\boldsymbol{p}_t}{2} + \boldsymbol{P}) \Big[\delta(\boldsymbol{p}' - \frac{\boldsymbol{p}_t}{2} + \boldsymbol{Q}) \\ &+ \delta(\boldsymbol{p}' - \frac{\boldsymbol{p}_t}{2} - \boldsymbol{Q}) - \delta(\boldsymbol{p}' - \frac{\boldsymbol{p}_t}{2} + \boldsymbol{P}) - \delta(\boldsymbol{p}' - \frac{\boldsymbol{p}_t}{2} - \boldsymbol{P}) \Big] \\ &\times \delta(\boldsymbol{p} - \frac{\boldsymbol{p}_t}{2} + \boldsymbol{Q}) d\sigma' d^3P d^3p_t, \end{aligned}$$

or the explicitly symmetric and positive semidefinite version

$$M = \frac{\delta(\boldsymbol{r}' - \boldsymbol{r})}{4k_B} \int d\sigma' d^3P d^3p_t v_{rel} f(\boldsymbol{r}, \frac{\boldsymbol{p}_t}{2} - \boldsymbol{P}) f(\boldsymbol{r}, \frac{\boldsymbol{p}_t}{2} + \boldsymbol{P})$$

$$\times \begin{pmatrix} \delta(\boldsymbol{p} - \frac{\boldsymbol{p}_t}{2} + \boldsymbol{Q}) \\ \delta(\boldsymbol{p} - \frac{\boldsymbol{p}_t}{2} - \boldsymbol{Q}) \\ \delta(\boldsymbol{p} - \frac{\boldsymbol{p}_t}{2} + \boldsymbol{P}) \\ \delta(\boldsymbol{p} - \frac{\boldsymbol{p}_t}{2} - \boldsymbol{P}) \end{pmatrix} \cdot \begin{pmatrix} 1 & 1 & -1 & -1 \\ 1 & 1 & -1 & -1 \\ -1 & -1 & 1 & 1 \\ -1 & -1 & 1 & 1 \end{pmatrix} \cdot \begin{pmatrix} \delta(\boldsymbol{p}' - \frac{\boldsymbol{p}_t}{2} + \boldsymbol{Q}) \\ \delta(\boldsymbol{p}' - \frac{\boldsymbol{p}_t}{2} - \boldsymbol{Q}) \\ \delta(\boldsymbol{p}' - \frac{\boldsymbol{p}_t}{2} + \boldsymbol{P}) \\ \delta(\boldsymbol{p}' - \frac{\boldsymbol{p}_t}{2} - \boldsymbol{P}) \end{pmatrix}.$$

In this equation, the symmetry of M in $\boldsymbol{p}$ and $\boldsymbol{p}'$ is manifest. The symmetric 4×4-matrix occurring in that formulation has the eigenvalues 4 [eigenvector $(1, 1, -1, -1)$] and 0 [eigenvectors $(1, 1, 1, 1)$, $(1, -1, 0, 0)$, $(0, 0, 1, -1)$] and is hence positive semidefinite.

Exercise 170

For the linearized entropy gradient (7.56), we obtain the following with (7.32) and (7.87):

$$\frac{\delta S[f]}{\delta f} = -k_B \ln \frac{\rho}{Nm\sqrt{2\pi m k_B T}^3} + k_B \frac{(\boldsymbol{p} - m\boldsymbol{v})^2}{2mk_B T} - k_B \hat{f}^{(1)}.$$

By means of the collisional invariants (particle number, momentum, kinetic energy), we realize that only the last term contributes to the irreversible dynamics and, with the result of Exercise 169, we obtain the following irreversible contribution to the Boltzmann equation:

$$\int d\sigma' d^3P d^3p_t v_{\text{rel}} f_{\rho,\boldsymbol{v},T}^{\text{Maxw}}(\boldsymbol{r}, \frac{\boldsymbol{p}_t}{2} - \boldsymbol{P}) f_{\rho,\boldsymbol{v},T}^{\text{Maxw}}(\boldsymbol{r}, \frac{\boldsymbol{p}_t}{2} + \boldsymbol{P}) \delta(\boldsymbol{p} - \frac{\boldsymbol{p}_t}{2} + \boldsymbol{P})$$

$$\times \left[\hat{f}^{(1)}(\boldsymbol{r}, \frac{\boldsymbol{p}_t}{2} + \boldsymbol{Q}) + \hat{f}^{(1)}(\boldsymbol{r}, \frac{\boldsymbol{p}_t}{2} - \boldsymbol{Q}) - \hat{f}^{(1)}(\boldsymbol{r}, \frac{\boldsymbol{p}_t}{2} + \boldsymbol{P}) - \hat{f}^{(1)}(\boldsymbol{r}, \frac{\boldsymbol{p}_t}{2} - \boldsymbol{P})\right] =$$

$$f_{\rho,\boldsymbol{v},T}^{\text{Maxw}}(\boldsymbol{r}, \boldsymbol{p}) \int d\sigma' d^3p' v_{\text{rel}} f_{\rho,\boldsymbol{v},T}^{\text{Maxw}}(\boldsymbol{r}, \boldsymbol{p}')$$

$$\times [\hat{f}^{(1)}(\boldsymbol{r}, \boldsymbol{q}) + \hat{f}^{(1)}(\boldsymbol{r}, \boldsymbol{q}') - \hat{f}^{(1)}(\boldsymbol{r}, \boldsymbol{p}) - \hat{f}^{(1)}(\boldsymbol{r}, \boldsymbol{p}')].$$

By ignoring the zeroth-order terms in the kinetic equation, which is certainly justified for an expansion around global equilibrium in the absence of external forces, and by equating the first-order terms, we finally obtain

$$\begin{aligned}\frac{\partial \hat{f}^{(1)}(\boldsymbol{r}, \boldsymbol{p})}{\partial t} &= -\frac{\boldsymbol{p}}{m} \cdot \frac{\partial}{\partial \boldsymbol{r}} \hat{f}^{(1)}(\boldsymbol{r}, \boldsymbol{p}) + \int d\sigma' d^3p' v_{\text{rel}} f_{\rho,0,T}^{\text{Maxw}}(\boldsymbol{r}, \boldsymbol{p}') \\ &\times \quad [\hat{f}^{(1)}(\boldsymbol{r}, \boldsymbol{q}) + \hat{f}^{(1)}(\boldsymbol{r}, \boldsymbol{q}') - \hat{f}^{(1)}(\boldsymbol{r}, \boldsymbol{p}) - \hat{f}^{(1)}(\boldsymbol{r}, \boldsymbol{p}')].\end{aligned}$$

This linearized Boltzmann equation has, for example, been discussed by Grad.[18]

Exercise 171

By means of the symmetry properties (7.20) and (7.21), the expression (7.88) can be rewritten as

$$M = \frac{1}{4} \frac{\delta(\boldsymbol{r}' - \boldsymbol{r})}{k_B} \int w(\boldsymbol{q}_1, \boldsymbol{q}_2 | \boldsymbol{p}_1, \boldsymbol{p}_2) \frac{f(\boldsymbol{q}_1) f(\boldsymbol{q}_2) - f(\boldsymbol{p}_1) f(\boldsymbol{p}_2)}{\ln[f(\boldsymbol{q}_1) f(\boldsymbol{q}_2)] - \ln[f(\boldsymbol{p}_1) f(\boldsymbol{p}_2)]}$$

[18] See p. 280–284 of Grad, Principles of the Kinetic Theory of Gases (1958).

$$\times \begin{pmatrix} \delta(\boldsymbol{p}_1 - \boldsymbol{p}) \\ \delta(\boldsymbol{p}_2 - \boldsymbol{p}) \\ \delta(\boldsymbol{q}_1 - \boldsymbol{p}) \\ \delta(\boldsymbol{q}_2 - \boldsymbol{p}) \end{pmatrix} \cdot \begin{pmatrix} 1 & 1 & -1 & -1 \\ 1 & 1 & -1 & -1 \\ -1 & -1 & 1 & 1 \\ -1 & -1 & 1 & 1 \end{pmatrix} \cdot \begin{pmatrix} \delta(\boldsymbol{p}_1 - \boldsymbol{p}') \\ \delta(\boldsymbol{p}_2 - \boldsymbol{p}') \\ \delta(\boldsymbol{q}_1 - \boldsymbol{p}') \\ \delta(\boldsymbol{q}_2 - \boldsymbol{p}') \end{pmatrix}$$
$$d^3p_1 d^3p_2 d^3q_1 d^3q_2 .$$

As in Exercise 169, the symmetry of M in $\boldsymbol{p}$ and $\boldsymbol{p}'$ is manifest, and the 4×4-matrix occurring in that formulation has the eigenvalues 4 [eigenvector $(1, 1, -1, -1)$] and 0 [eigenvectors $(1, 1, 1, 1)$, $(1, -1, 0, 0)$, $(0, 0, 1, -1)$] and is hence positive semidefinite; because the transition probability $w(\boldsymbol{q}_1, \boldsymbol{q}_2|\boldsymbol{p}_1, \boldsymbol{p}_2)$ is nonnegative and the logarithm is a monotonically increasing function, the positive semidefinite character of M is implied.

The degeneracy requirement (1.5) follows from the conservation of kinetic energy in the elastic collisions (the external potential energy is conserved because the collisions are assumed to take place in a single point in space).

From (7.56) and (7.88), we obtain

$$M \cdot \frac{\delta S}{\delta x} = \int \delta(\boldsymbol{p}_1 - \boldsymbol{p}) w(\boldsymbol{q}_1, \boldsymbol{q}_2|\boldsymbol{p}_1, \boldsymbol{p}_2)[f(\boldsymbol{q}_1)f(\boldsymbol{q}_2) - f(\boldsymbol{p}_1)f(\boldsymbol{p}_2)]$$
$$d^3p_1 d^3p_2 d^3q_1 d^3q_2 .$$

We replace the integrations d^3q_1 and d^3q_2 by integrations over the total momentum and the final relative momentum. In view of the fact that, in (7.89), one can replace $\delta(\boldsymbol{q}_1^2 + \boldsymbol{q}_2^2 - \boldsymbol{p}_1^2 - \boldsymbol{p}_2^2)$ by $\delta(2\boldsymbol{Q}^2 - 2\boldsymbol{P}^2)$, we recover the previous result (7.84) for the irreversible contribution to Boltzmann's equation by observing

$$d^3Q = Q^2 d\Omega' dQ = \frac{1}{4} Q d\Omega' d(2Q^2) = \frac{m}{8} v_{\rm rel} d\Omega' d(2Q^2).$$

Exercise 172

For $f(\boldsymbol{p}) = f^{\rm Maxw}_{\rho,\boldsymbol{v},T}(\boldsymbol{p})[1 + \hat{f}^{(1)}(\boldsymbol{p})]$, we can linearize the two terms containing the single-particle distribution function in (7.88) as follows:

$$f(\boldsymbol{q}_1)f(\boldsymbol{q}_2) - f(\boldsymbol{p}_1)f(\boldsymbol{p}_2) = f^{\rm Maxw}_{\rho,\boldsymbol{v},T}(\boldsymbol{p}_1) f^{\rm Maxw}_{\rho,\boldsymbol{v},T}(\boldsymbol{p}_2)$$
$$\times \; [\hat{f}^{(1)}(\boldsymbol{q}_1) + \hat{f}^{(1)}(\boldsymbol{q}_2) - \hat{f}^{(1)}(\boldsymbol{p}_1) - \hat{f}^{(1)}(\boldsymbol{p}_2)]$$

and

$$\ln[f(\boldsymbol{q}_1)f(\boldsymbol{q}_2)] - \ln[f(\boldsymbol{p}_1)f(\boldsymbol{p}_2)] = \hat{f}^{(1)}(\boldsymbol{q}_1) + \hat{f}^{(1)}(\boldsymbol{q}_2) - \hat{f}^{(1)}(\boldsymbol{p}_1) - \hat{f}^{(1)}(\boldsymbol{p}_2).$$

For the ratio of these two building blocks, we thus obtain $f^{\rm Maxw}_{\rho,\boldsymbol{v},T}(\boldsymbol{p}_1) f^{\rm Maxw}_{\rho,\boldsymbol{v},T}(\boldsymbol{p}_2)$, which is independent of $\hat{f}^{(1)}$, so that, after carrying out the $\boldsymbol{q}$ integrations as described in Exercise 171, the resulting friction matrix (7.88) agrees with the one found in Exercise 169.

Exercise 173
By using the chain rule and (7.32), (7.105), we obtain

$$\begin{aligned}\left(\frac{\partial}{\partial t}+\frac{\boldsymbol{p}}{m}\cdot\frac{\partial}{\partial \boldsymbol{r}}\right)f^{(0)} &= \frac{f^{(0)}}{\rho}\left(\frac{\partial}{\partial t}+\frac{\boldsymbol{p}}{m}\cdot\frac{\partial}{\partial \boldsymbol{r}}\right)\rho \\ &+ f^{(0)}\frac{\boldsymbol{p}-m\boldsymbol{v}}{k_{\mathrm{B}}T}\cdot\left(\frac{\partial}{\partial t}+\frac{\boldsymbol{p}}{m}\cdot\frac{\partial}{\partial \boldsymbol{r}}\right)\boldsymbol{v} \\ &- \frac{3}{2}\frac{f^{(0)}}{T}\left(\frac{\partial}{\partial t}+\frac{\boldsymbol{p}}{m}\cdot\frac{\partial}{\partial \boldsymbol{r}}\right)T \\ &+ f^{(0)}\frac{(\boldsymbol{p}-m\boldsymbol{v})^2}{2mk_{\mathrm{B}}T^2}\left(\frac{\partial}{\partial t}+\frac{\boldsymbol{p}}{m}\cdot\frac{\partial}{\partial \boldsymbol{r}}\right)T.\end{aligned}$$

By eliminating all the time derivatives by means of the Euler equations (7.96), (7.106), and (7.107) and rearranging terms, we obtain (7.109) after a lengthy calculation. In particular, the terms containing gradients of the mass density cancel.

Exercise 174
From Exercise 159 for $n_\phi = 5$ and (7.91), we obtain

$$\begin{aligned}\left[\frac{\partial f}{\partial t}\right]_{\mathrm{coll}}^{f_1 f_2} &= \sqrt{\frac{2\hat{\phi}}{m}}\int\frac{\hat{b}(\theta')}{\sin\theta'}\left|\frac{d\hat{b}(\theta')}{d\theta'}\right| \\ &\times [f_1(\boldsymbol{r},\boldsymbol{q})f_2(\boldsymbol{r},\boldsymbol{q}') - f_1(\boldsymbol{r},\boldsymbol{p})f_2(\boldsymbol{r},\boldsymbol{p}')]d\Omega' d^3p',\end{aligned}$$

where the relative velocity turns out to be irrelevant when the scattering process for the Maxwell gas is described in the center-of-mass frame.

Exercise 175
By inserting (7.109) and (7.110) into (7.108) and using the expression for the collision integral for the Maxwell gas found in Exercise 174, we obtain the following integral equations for $\bar{A}$ and $\bar{B}$:

$$\begin{aligned}\left(\frac{p^2}{2mk_{\mathrm{B}}T}-\frac{5}{2}\right)\boldsymbol{p} &= \sqrt{\frac{2\hat{\phi}}{m}}\int\frac{\hat{b}(\theta')}{\sin\theta'}\left|\frac{d\hat{b}(\theta')}{d\theta'}\right| f_{\rho,0,T}^{\mathrm{Maxw}}(\boldsymbol{p}') \\ &\times [\bar{A}(q)\boldsymbol{q}+\bar{A}(q')\boldsymbol{q}'-\bar{A}(p)\boldsymbol{p}-\bar{A}(p')\boldsymbol{p}']d\Omega' d^3p'\end{aligned} \quad \text{(F.47)}$$

and

$$\begin{aligned}\boldsymbol{p}\boldsymbol{p}-\frac{p^2}{3}\boldsymbol{1} &= \sqrt{\frac{2\hat{\phi}}{m}}\int\frac{\hat{b}(\theta')}{\sin\theta'}\left|\frac{d\hat{b}(\theta')}{d\theta'}\right| f_{\rho,0,T}^{\mathrm{Maxw}}(\boldsymbol{p}') \\ &\times \left[\bar{B}(q)\left(\boldsymbol{q}\boldsymbol{q}-\frac{q^2}{3}\boldsymbol{1}\right)+\bar{B}(q')\left(\boldsymbol{q}'\boldsymbol{q}'-\frac{q'^2}{3}\boldsymbol{1}\right)\right.\end{aligned}$$

$$-\bar{B}(p)\left(\boldsymbol{pp}-\frac{p^2}{3}\mathbf{1}\right)-\bar{B}(p')\left(\boldsymbol{p}'\boldsymbol{p}'-\frac{p'^2}{3}\mathbf{1}\right)\Bigg]\,d\Omega' d^3p'. \tag{F.48}$$

In simplifying the integrals, the symmetric and isotropic nature of $f^{\rm Maxw}_{\rho,0,T}(\boldsymbol{p}')$ is of great help. For example, the terms involving $\boldsymbol{p}'$ in the square brackets of (F.47) and (F.48) vanish immediately. More complicated are the terms involving $\boldsymbol{q}$ and $\boldsymbol{q}'$ because they depend on the fixed parameter $\boldsymbol{p}$ but also on all integration variables, $\boldsymbol{p}'$, and the solid angle. We here follow a strategy of simplifying the integrands by using the symmetric and isotropic nature of $f^{\rm Maxw}_{\rho,0,T}(\boldsymbol{p}')$ and the averaging implied by the integration over the azimuthal angle φ'. In order to do that, we describe the scattering process by the equations

$$\boldsymbol{q}=\frac{1+\cos\theta'}{2}\boldsymbol{p}+\frac{1-\cos\theta'}{2}\boldsymbol{p}'-\sin\theta'\boldsymbol{n}(\varphi'),$$

$$\boldsymbol{q}'=\frac{1-\cos\theta'}{2}\boldsymbol{p}+\frac{1+\cos\theta'}{2}\boldsymbol{p}'+\sin\theta'\boldsymbol{n}(\varphi'),$$

where $\boldsymbol{n}(\varphi')$ describes a full circle in the plane perpendicular to $\boldsymbol{p}'-\boldsymbol{p}$ when φ' goes from 0 to 2π. This implies that, upon integration over φ', all odd powers of $\boldsymbol{n}(\varphi')$ disappear (we need only the first and third powers) and that we can make the following replacement for second powers:

$$\boldsymbol{n}(\varphi')\boldsymbol{n}(\varphi')\to\frac{1}{8}\left[(\boldsymbol{p}'-\boldsymbol{p})^2\mathbf{1}-(\boldsymbol{p}'-\boldsymbol{p})(\boldsymbol{p}'-\boldsymbol{p})\right].$$

Once all occurrences of $\boldsymbol{n}(\varphi')$ are eliminated by these rules, one can simplify the dependence on $\boldsymbol{p}'$ by means of the properties of $f^{\rm Maxw}_{\rho,0,T}(\boldsymbol{p}')$. It should be noted that this strategy is possible only because no extra factors of $|\boldsymbol{p}'-\boldsymbol{p}|$ occur for the Maxwell gas (see Exercise 174).

We first establish the constant solution $\bar{B}$ of (F.48) given in (7.117). As an illustrative example, we consider the integrand $\boldsymbol{qq}$, and we neglect isotropic contributions because they are cancelled in the combination $\boldsymbol{qq}-(q^2/3)\mathbf{1}$. With the ideas for eliminating $\boldsymbol{n}(\varphi')$, we obtain the following equivalent replacement:

$$\begin{aligned}\boldsymbol{qq}\ \to\ &\left(\frac{1+\cos\theta'}{2}\boldsymbol{p}+\frac{1-\cos\theta'}{2}\boldsymbol{p}'\right)\left(\frac{1+\cos\theta'}{2}\boldsymbol{p}+\frac{1-\cos\theta'}{2}\boldsymbol{p}'\right)\\ &-\frac{(\sin\theta')^2}{8}(\boldsymbol{p}'-\boldsymbol{p})(\boldsymbol{p}'-\boldsymbol{p}).\end{aligned}$$

Because the first moments in $\boldsymbol{p}'$ vanish by symmetry and the second moments are isotropic, we immediately obtain

$$\boldsymbol{qq}\to\left(\frac{1+\cos\theta'}{2}\right)^2\boldsymbol{pp}-\frac{(\sin\theta')^2}{8}\boldsymbol{pp}=\left(\frac{1+\cos\theta'}{2}-\frac{3}{8}(\sin\theta')^2\right)\boldsymbol{pp}.$$

The corresponding result for $\boldsymbol{q}'\boldsymbol{q}'$ is obtained by changing the sign of $\sin\theta'$ and $\cos\theta'$, and we hence get

$$\boldsymbol{q}\boldsymbol{q} - \frac{q^2}{3}\mathbf{1} + \boldsymbol{q}'\boldsymbol{q}' - \frac{q'^2}{3}\mathbf{1} - \left(\boldsymbol{p}\boldsymbol{p} - \frac{p^2}{3}\mathbf{1}\right) \to -\frac{3}{4}(\sin\theta')^2 \left(\boldsymbol{p}\boldsymbol{p} - \frac{p^2}{3}\mathbf{1}\right).$$

With this equivalent replacement we realize that (7.117) is indeed a solution of (F.48), provided that

$$c^{\text{Maxw}} = \int_0^\pi (\sin\theta')^2 \hat{b}(\theta') \left|\frac{d\hat{b}(\theta')}{d\theta'}\right| d\theta'.$$

With the explicit result for $\theta(\hat{b})$ given in (F.46), this constant can be evaluated as the integral

$$c^{\text{Maxw}} = \int_0^\infty [\sin\theta(\hat{b})]^2 \hat{b}\, d\hat{b}.$$

A numerical Romberg integration based on the complete elliptic integral (6.11.19) and the subroutine `qromb` of the Numerical Recipes[19] gives the result $c^{\text{Maxw}} \approx 0.87239$.

We next turn to the solution $\bar{A}$ of (F.47). As an auxiliary result, which can be obtained by exactly the same procedure as established for modifying the integrand $\boldsymbol{q}\boldsymbol{q}$, we obtain the following after a rather lengthy calculation:

$$q^2\boldsymbol{q} \to \frac{1+\cos\theta'}{2} p^2\boldsymbol{p} - (\sin\theta')^2 \left(\frac{1}{4}p^2 - \frac{5}{12}p'^2\right)\boldsymbol{p}.$$

We thus obtain an equivalent replacement in which again $(\sin\theta')^2$ arises as a prefactor,

$$q^2\boldsymbol{q} + q'^2\boldsymbol{q}' - p^2\boldsymbol{p} \to -(\sin\theta')^2 \left(\frac{1}{2}p^2 - \frac{5}{6}p'^2\right)\boldsymbol{p}.$$

With the moments listed in the solution to Exercise 176, one thus realizes that a solution of (F.47) is given by the quadratic function

$$\bar{A}(p) = -\frac{1}{\pi}\frac{1}{c^{\text{Maxw}}}\sqrt{\frac{m}{2\hat{\phi}}}\frac{m}{\rho}\frac{p^2}{2mk_{\text{B}}T}.$$

However, this solution is not unique. In view of the momentum conservation in a collision, any constant could be added. The value of the constant is fixed by the auxiliary equation (7.111), and we finally obtain the unique result (7.116).

Exercise 176

The evaluation of the integrals (7.113) and (7.115) is based on the following Gaussian moments:

$$\frac{m}{\rho}\int f^{\text{Maxw}}_{\rho,0,T}(\boldsymbol{r},\boldsymbol{p})|\boldsymbol{p}|^2 d^3p = 3mk_{\text{B}}T,$$

[19] Press et al., *Numerical Recipes* (Cambridge, 1992).

$$\frac{m}{\rho}\int f^{\text{Maxw}}_{\rho,0,T}(\boldsymbol{r},\boldsymbol{p})|\boldsymbol{p}|^4 d^3p = 15(mk_{\text{B}}T)^2,$$

$$\frac{m}{\rho}\int f^{\text{Maxw}}_{\rho,0,T}(\boldsymbol{r},\boldsymbol{p})|\boldsymbol{p}|^6 d^3p = 105(mk_{\text{B}}T)^3.$$

Exercise 177

Compared with upper convected time derivatives, the signs of the terms $+\boldsymbol{\kappa}\cdot\mathring{\boldsymbol{c}}+\mathring{\boldsymbol{c}}\cdot\boldsymbol{\kappa}^T$ are wrong, whereas $\boldsymbol{\kappa}$ and $\boldsymbol{\kappa}^T$ would have to be exchanged in these terms to obtain lower convected derivatives. Any other choice than $\xi = 0$ or $\xi = 2$ would introduce unwanted factors $\boldsymbol{\kappa}$ and $\boldsymbol{\kappa}^T$ from both sides. Therefore, the time derivative shown in (7.140) violates the principle of material objectivity, as a consequence of inertial effects on the second moment tensor in velocity space.

Exercise 178

Note that the averages ρ and $\boldsymbol{v}$ involve f. A straightforward calculation of the functional derivatives (7.146) with the rules of Appendix C then gives

$$\hat{A}(\rho(\boldsymbol{r}),\boldsymbol{p}) = A(\rho(\boldsymbol{r}),\boldsymbol{p}) + m\int \frac{\partial A(\rho(\boldsymbol{r}),\boldsymbol{p}')}{\partial\rho} f(\boldsymbol{r},\boldsymbol{p}')d^3p'$$

and

$$\hat{A}(\boldsymbol{v}(\boldsymbol{r}),\boldsymbol{p}) = A(\boldsymbol{v}(\boldsymbol{r}),\boldsymbol{p}) + \frac{1}{\rho(\boldsymbol{r})}[\boldsymbol{p} - m\boldsymbol{v}(\boldsymbol{r})]\cdot\int \frac{\partial A(\boldsymbol{v}(\boldsymbol{r}),\boldsymbol{p}')}{\partial\boldsymbol{v}} f(\boldsymbol{r},\boldsymbol{p}')d^3p'.$$

Exercise 179

By applying the results of Exercise 178, we obtain

$$A = m \quad\Rightarrow\quad \hat{A} = m,$$

$$A = \frac{\boldsymbol{p}}{\rho} \quad\Rightarrow\quad \hat{A} = \frac{\boldsymbol{p} - m\boldsymbol{v}}{\rho},$$

$$A = \frac{(\boldsymbol{p} - m\boldsymbol{v})^2}{3\rho k_{\text{B}}} \quad\Rightarrow\quad \hat{A} = \frac{(\boldsymbol{p} - m\boldsymbol{v})^2 - 3mk_{\text{B}}T}{3\rho k_{\text{B}}},$$

and

$$A = \frac{(\boldsymbol{p} - m\boldsymbol{v})^2\boldsymbol{1} - 3(\boldsymbol{p} - m\boldsymbol{v})(\boldsymbol{p} - m\boldsymbol{v})}{3\rho k_{\text{B}}T} \quad\Rightarrow$$
$$\hat{A} = \frac{(\boldsymbol{p} - m\boldsymbol{v})^2(\boldsymbol{1} - \mathring{\boldsymbol{c}}) - 3(\boldsymbol{p} - m\boldsymbol{v})(\boldsymbol{p} - m\boldsymbol{v})}{3\rho k_{\text{B}}T}.$$

Exercise 180

According to Exercise 179, we need to use the third and fourth variables

$$\hat{A} = \frac{(\boldsymbol{p} - m\boldsymbol{v})^2 - 3mk_{\text{B}}T}{3\rho k_{\text{B}}}$$

and

$$\hat{B} = \frac{(\boldsymbol{p} - m\boldsymbol{v})^2(\mathbf{1} - \mathring{\boldsymbol{c}}) - 3(\boldsymbol{p} - m\boldsymbol{v})(\boldsymbol{p} - m\boldsymbol{v})}{3\rho k_{\mathrm{B}} T}$$

in (7.149). For example, with

$$\frac{\partial \hat{A}}{\partial \boldsymbol{p}} = \frac{2(\boldsymbol{p} - m\boldsymbol{v})}{3\rho k_{\mathrm{B}}},$$

the second contribution in (7.149) contains vanishing odd moments unless $\partial/\partial\boldsymbol{r}$ acts on the velocity in one of the factors $(\boldsymbol{p}-m\boldsymbol{v})$ of $\hat{B}$ so that nonvanishing even moments are produced. We thus obtain for this contribution to L_{34}

$$\frac{2m}{3\rho k_{\mathrm{B}}} \int f \frac{2(\boldsymbol{p} - m\boldsymbol{v})(\boldsymbol{p} - m\boldsymbol{v}) : \boldsymbol{\kappa}(\mathbf{1} - \mathring{\boldsymbol{c}}) - 3\boldsymbol{\kappa} \cdot (\boldsymbol{p} - m\boldsymbol{v})(\boldsymbol{p} - m\boldsymbol{v})}{3\rho k_{\mathrm{B}} T} d^3p$$
$$- \frac{2m}{3\rho k_{\mathrm{B}}} \int f \frac{3(\boldsymbol{p} - m\boldsymbol{v})(\boldsymbol{p} - m\boldsymbol{v}) \cdot \boldsymbol{\kappa}^T}{3\rho k_{\mathrm{B}} T} d^3p,$$

which, after evaluation of the second moments, gives

$$\frac{2m}{3\rho k_{\mathrm{B}}} \left[\frac{2}{3} (\mathbf{1} - \mathring{\boldsymbol{c}}) : \boldsymbol{\kappa}(\mathbf{1} - \mathring{\boldsymbol{c}}) - \boldsymbol{\kappa} \cdot (\mathbf{1} - \mathring{\boldsymbol{c}}) - (\mathbf{1} - \mathring{\boldsymbol{c}}) \cdot \boldsymbol{\kappa}^T \right].$$

For the first contribution in (7.149), we similarly obtain

$$-\frac{2m}{3\rho k_{\mathrm{B}}} \left[\frac{2}{3} (\mathbf{1} - \mathring{\boldsymbol{c}}) : \boldsymbol{\kappa}(\mathbf{1} - \mathring{\boldsymbol{c}}) - \boldsymbol{\kappa}^T \cdot (\mathbf{1} - \mathring{\boldsymbol{c}}) - (\mathbf{1} - \mathring{\boldsymbol{c}}) \cdot \boldsymbol{\kappa} \right],$$

and, by adding these two contributions, we find the result of (7.154).

Exercise 181

The degeneracy condition results in the following equation:

$$\frac{\partial}{\partial \boldsymbol{r}} \cdot \rho \left[\mathbf{1} + 2(\mathbf{1} - \mathring{\boldsymbol{c}}) \cdot \frac{\partial s(\mathring{\boldsymbol{c}})}{\partial \mathring{\boldsymbol{c}}} - (\mathbf{1} - \mathring{\boldsymbol{c}}) - \frac{2}{3} (\mathbf{1} - \mathring{\boldsymbol{c}})(\mathbf{1} - \mathring{\boldsymbol{c}}) : \frac{\partial s(\mathring{\boldsymbol{c}})}{\partial \mathring{\boldsymbol{c}}} \right] = 0.$$

Because this condition should hold for every possible choice of the mass density field ρ we conclude that the factor in square brackets must vanish, so that we obtain a linear equation for $\partial s(\mathring{\boldsymbol{c}})/\partial\mathring{\boldsymbol{c}}$. A particular solution of this equation is given by

$$\frac{\partial s(\mathring{\boldsymbol{c}})}{\partial \mathring{\boldsymbol{c}}} = -\frac{1}{2} (\mathbf{1} - \mathring{\boldsymbol{c}})^{-1},$$

which corresponds to the entropy (7.144). To find possible additional solutions, we consider the corresponding homogeneous equation

$$\frac{\partial s(\mathring{\boldsymbol{c}})}{\partial \mathring{\boldsymbol{c}}} = \frac{1}{3} (\mathbf{1} - \mathring{\boldsymbol{c}}) : \frac{\partial s(\mathring{\boldsymbol{c}})}{\partial \mathring{\boldsymbol{c}}} \mathbf{1}.$$

Thus, $\partial s(\mathring{\boldsymbol{c}})/\partial \mathring{\boldsymbol{c}}$ must be isotropic, and every isotropic tensor solves the homogeneous equation and can be added to the particular solution associated with (7.144). Such additional terms occur when we add a multiple of $\mathrm{tr}\mathring{\boldsymbol{c}}$ to $s(\mathring{\boldsymbol{c}})$ and are irrelevant because they should be fixed by performing the derivative $\partial s(\mathring{\boldsymbol{c}})/\partial \mathring{\boldsymbol{c}}$ under the constraint that $\mathring{\boldsymbol{c}}$ is traceless. We thus arrive at the conclusion that the functional form of $s(\mathring{\boldsymbol{c}})$ is uniquely fixed by the degeneracy requirement.

This conclusion is in accordance with the general experience that the entropy function determines the stress tensor. Because the stress tensor is known from (7.127), it should not be surprising that only a particular form of the entropy is consistent with it.

Exercise 182

By inserting (7.78) into (7.163), we find

$$\hat{I}(P') = 4d^2 \frac{1}{P'} \int_{P>P'} \frac{(\boldsymbol{P}\cdot\boldsymbol{P}')^2 \boldsymbol{P}'\cdot(\boldsymbol{P}-\boldsymbol{P}')(\boldsymbol{P}-\boldsymbol{P}')\cdot\boldsymbol{P}}{|\boldsymbol{P}-\boldsymbol{P}'|^4} d^3P.$$

If we use spherical coordinates with the z-axis in the direction of $\boldsymbol{P}'$, then we obtain

$$\hat{I}(P') = 8\pi d^2 P' \int_{P'}^{\infty} dP \int_{-1}^{1} dz \frac{z^2 P^4 (zPP' - P'^2)(P^2 - zPP')}{(P^2 + P'^2 - 2zPP')^2},$$

or, after use of the integration variable $u = P'/P$,

$$\hat{I}(P') = 8\pi d^2 P'^6 \int_0^1 du \int_{-1}^{1} dz \frac{z^2 (z-u)(1-zu)}{u^5(1+u^2-2zu)^2}.$$

Equation (7.164) implies

$$c^{\mathrm{hs}} = 4\pi \int_0^1 du \frac{1}{u^5} \int_{-1}^{1} dz z^2 (1-zu) \frac{\partial}{\partial u} \frac{1}{1+u^2-2zu}$$

or

$$c^{\mathrm{hs}} = 4\pi \left(\int_0^1 \frac{du}{u^5} \frac{\partial}{\partial u} \int_{-1}^{1} \frac{z^2 dz}{1+u^2-2zu} - \int_0^1 \frac{du}{u^4} \frac{\partial}{\partial u} \int_{-1}^{1} \frac{z^3 dz}{1+u^2-2zu} \right).$$

After the straightforward z-integrations,[20] we obtain

$$c^{\mathrm{hs}} = 2\pi \int_0^1 \left[\frac{1}{u^8} - \frac{2}{3u^6} + \frac{1}{u^4} - \frac{1}{2}\left(\frac{1}{u^9} - \frac{1}{u^7} - \frac{1}{u^5} + \frac{1}{u^3} \right) \ln \frac{1+u}{1-u} \right] du.$$

The u-integration would also be straightforward; however, the integrand is too singular near $u = 0$ ($\propto 1/u^4$) so that the integral diverges.

[20] See 2.111.3 of Gradshteyn & Ryzhik, *Tables* (Academic, 1980).

Exercise 183

From (7.145) and (7.175), we obtain

$$\begin{aligned}\left(M \cdot \frac{\delta S}{\delta x}\right)_{4\,ij} &= -\frac{k_{\rm B}\rho}{2m}(M_{44})_{ijkl}(\mathbf{1} - \mathring{\boldsymbol{c}})^{-1}_{kl} \\ &= -\frac{\rho}{m^4 k_{\rm B}^2 T^2}\int\int (\sin\theta')^2 d\sigma' P'^3 f_{\rm r}(\boldsymbol{P}')(P_i'\hat{\delta}_{jk} + P_j'\hat{\delta}_{ik}) \\ &\quad \times(\mathbf{1} - \mathring{\boldsymbol{c}})^{-1}_{kl} P_l' d^3P'.\end{aligned}$$

By means of (7.171), we obtain after an integration by parts

$$\left(M \cdot \frac{\delta S}{\delta x}\right)_{4\,ij} = \frac{-\rho}{2m^3 k_{\rm B} T}\int\int (\sin\theta')^2 d\sigma' P'^3 f_{\rm r}(\boldsymbol{P}')\frac{\partial}{\partial P_k'}(P_i'\hat{\delta}_{jk} + P_j'\hat{\delta}_{ik})d^3P',$$

which, after carrying out the differentiation, coincides with the nonlinear relaxation term (7.174).

Exercise 184

In the entropy expression (7.144), we can use the following quadratic expansion:

$$\ln\det(\mathbf{1} - \mathring{\boldsymbol{c}}) \approx -\frac{1}{2}\mathring{\boldsymbol{c}} : \mathring{\boldsymbol{c}}.$$

The degeneracy requirement (1.4) of GENERIC is not satisfied for a quadratic entropy or linearized entropy gradient.

Exercise 185

We found that all the reversible terms of the structured kinetic theory coincide naturally with the convective terms given by Grad, and we can thus focus on the relaxation term, that is, the last term in (7.131). The original relaxation term can be reproduced by evaluating the friction matrix at local equilibrium and by linearizing the entropy gradient, as we noted in (7.178). The resulting friction matrix has all the desirable properties (it is symmetric, positive semidefinite, and degenerate), but, according to Exercise 184, linearization of the entropy gradient violates the degeneracy requirement $L\delta S/\delta x = 0$. Exercise 181 showed that the degeneracy requirement actually forces us to use the entropy (7.144), which leads to the relaxation term

$$-\frac{1}{\tau}\left[(\mathbf{1} - \mathring{\boldsymbol{c}})^{-1} - \frac{1}{3}\mathbf{1}\mathrm{tr}(\mathbf{1} - \mathring{\boldsymbol{c}})^{-1}\right]$$

instead of $-(1/\tau)\mathring{\boldsymbol{c}}$ in (7.131).

Exercise 186

To benefit from the lengthy calculations in the solutions to Exercise 79 and 80, we write

$$\mathring{\boldsymbol{c}} = \bar{x}(\boldsymbol{\kappa} + \boldsymbol{\kappa}^T) + \bar{y}\boldsymbol{\kappa}\cdot\boldsymbol{\kappa}^T + \bar{z}\boldsymbol{\kappa}^T\cdot\boldsymbol{\kappa} - \frac{1}{3}(\bar{y} + \bar{z})\dot{\gamma}^2\mathbf{1},$$

where $\bar{x}$, $\bar{y}$, and $\bar{z}$ remain to be determined either from

$$-\boldsymbol{\kappa}\cdot\mathring{\boldsymbol{c}} - \mathring{\boldsymbol{c}}\cdot\boldsymbol{\kappa}^T + \boldsymbol{\kappa} + \boldsymbol{\kappa}^T + \frac{2}{3}(\mathbf{1} - \mathring{\boldsymbol{c}})\mathring{\boldsymbol{c}} : \boldsymbol{\kappa} = \frac{1}{\tau}\mathring{\boldsymbol{c}}$$

or from

$$-\boldsymbol{\kappa}\cdot\mathring{\boldsymbol{c}} - \mathring{\boldsymbol{c}}\cdot\boldsymbol{\kappa}^T + \boldsymbol{\kappa} + \boldsymbol{\kappa}^T + \frac{2}{3}(\mathbf{1} - \mathring{\boldsymbol{c}})\mathring{\boldsymbol{c}} : \boldsymbol{\kappa} = \frac{1}{\tau}\left[(\mathbf{1} - \mathring{\boldsymbol{c}})^{-1} - \frac{1}{3}\mathbf{1}\mathrm{tr}(\mathbf{1} - \mathring{\boldsymbol{c}})^{-1}\right]$$

for the original and the structured moment method, respectively. The right-hand side of the first of these equations is the linearized version of the right-hand side of the second equation. Once we have determined $\bar{x}$, $\bar{y}$, and $\bar{z}$, the pressure tensor expression (7.124) leads to the following expressions for the viscometric functions:

$$\eta = \frac{\rho}{m}k_{\rm B}T\bar{x},$$

$$\Psi_1 = \frac{\rho}{m}k_{\rm B}T(\bar{y} - \bar{z}),$$

$$\Psi_2 = \frac{\rho}{m}k_{\rm B}T\bar{z}.$$

For Grad's original method, we obtain (the isotropic terms cancel by construction)

$$\left(1 + \frac{1}{3}\bar{y}\dot{\gamma}^2 - \frac{2}{3}\bar{z}\dot{\gamma}^2\right)(\boldsymbol{\kappa} + \boldsymbol{\kappa}^T) - 2\bar{x}\boldsymbol{\kappa}\cdot\boldsymbol{\kappa}^T =$$

$$\left(\frac{1}{\tau} + \frac{2}{3}\bar{x}\dot{\gamma}^2\right)[\bar{x}(\boldsymbol{\kappa} + \boldsymbol{\kappa}^T) + \bar{y}\boldsymbol{\kappa}\cdot\boldsymbol{\kappa}^T + \bar{z}\boldsymbol{\kappa}^T\cdot\boldsymbol{\kappa}],$$

and, by equating coefficients, we find

$$\bar{z} = 0, \quad \left(1 + \frac{1}{3}\bar{y}\dot{\gamma}^2\right)\tau = \left(1 + \frac{2}{3}\tau\bar{x}\dot{\gamma}^2\right)\bar{x}, \quad -2\tau\bar{x} = \left(1 + \frac{2}{3}\tau\bar{x}\dot{\gamma}^2\right)\bar{y}.$$

These results coincide with the ones given in (7.138) and (7.139). The explicit solution of the cubic equation for $\bar{x}$ can be written as

$$\bar{x} = \frac{1}{2\dot{\gamma}^2\tau}\left(\lambda_1^{1/3} - \lambda_1^{-1/3}\right)^2$$

with

$$\lambda_1 = \sqrt{1 + \frac{9}{2}\dot{\gamma}^2\tau^2} + \sqrt{\frac{9}{2}}\,\dot{\gamma}\tau, \qquad \lambda_1^{-1} = \sqrt{1 + \frac{9}{2}\dot{\gamma}^2\tau^2} - \sqrt{\frac{9}{2}}\,\dot{\gamma}\tau.$$

For the structured moment method, we construct $(\mathbf{1} - \mathring{\boldsymbol{c}})^{-1}$ from the results of Exercise 80, with a sign change in $\bar{x}$, $\bar{y}$, and $\bar{z}$. We then obtain the equation

$$\left(1 + \frac{1}{3}\bar{y}\dot{\gamma}^2 - \frac{2}{3}\bar{z}\dot{\gamma}^2\right)(\boldsymbol{\kappa} + \boldsymbol{\kappa}^T) - 2\bar{x}\boldsymbol{\kappa}\cdot\boldsymbol{\kappa}^T =$$

$$\frac{2}{3}\bar{x}\dot{\gamma}^2[\bar{x}(\boldsymbol{\kappa}+\boldsymbol{\kappa}^T)+\bar{y}\boldsymbol{\kappa}\cdot\boldsymbol{\kappa}^T+\bar{z}\boldsymbol{\kappa}^T\cdot\boldsymbol{\kappa}]-\frac{1}{\dot{\gamma}^2\tau[1+(\bar{y}+\bar{z})\dot{\gamma}^2/3]}(\boldsymbol{\kappa}\cdot\boldsymbol{\kappa}^T+\boldsymbol{\kappa}^T\cdot\boldsymbol{\kappa})$$

$$+\frac{1}{\tau\lambda_2}\left[\bar{x}(\boldsymbol{\kappa}+\boldsymbol{\kappa}^T)+\left(1+\frac{1}{3}\bar{y}\dot{\gamma}^2-\frac{2}{3}\bar{z}\dot{\gamma}^2\right)\frac{\boldsymbol{\kappa}\cdot\boldsymbol{\kappa}^T}{\dot{\gamma}^2}\right.$$
$$\left.+\left(1-\frac{2}{3}\bar{y}\dot{\gamma}^2+\frac{1}{3}\bar{z}\dot{\gamma}^2\right)\frac{\boldsymbol{\kappa}^T\cdot\boldsymbol{\kappa}}{\dot{\gamma}^2}\right],$$

with

$$\lambda_2=\left(1+\frac{1}{3}\bar{y}\dot{\gamma}^2-\frac{2}{3}\bar{z}\dot{\gamma}^2\right)\left(1-\frac{2}{3}\bar{y}\dot{\gamma}^2+\frac{1}{3}\bar{z}\dot{\gamma}^2\right)-\bar{x}^2\dot{\gamma}^2.$$

By equating coefficients, we find

$$\left(1+\frac{1}{3}\bar{y}\dot{\gamma}^2-\frac{2}{3}\bar{z}\dot{\gamma}^2\right)\tau=\left(\frac{1}{\lambda_2}+\frac{2}{3}\tau\bar{x}\dot{\gamma}^2\right)\bar{x},$$

$$-2\tau\bar{x}=\left(\frac{1}{\lambda_2}+\frac{2}{3}\tau\bar{x}\dot{\gamma}^2\right)(\bar{y}-\bar{z}),$$

and

$$\frac{2}{3}\tau\bar{x}\dot{\gamma}^4\bar{z}=\frac{1}{1+(\bar{y}+\bar{z})\dot{\gamma}^2/3}-\frac{1}{\lambda_2}\left(1-\frac{2}{3}\bar{y}\dot{\gamma}^2+\frac{1}{3}\bar{z}\dot{\gamma}^2\right).$$

By forming the ratio of the first two of these equations, we obtain the following relationship between the viscometric functions predicted by the structured moment method:

$$-\frac{2\eta^2}{\Psi_1}=\frac{\rho}{m}k_BT+\frac{1}{3}(\Psi_1-\Psi_2)\dot{\gamma}^2.$$

For small shear rates, the viscometric functions are given by

$$\eta=\frac{\rho}{m}k_BT\tau\left(1-\frac{2}{3}\tau^2\dot{\gamma}^2\right),$$

$$\Psi_1=-2\tau^2\frac{\rho}{m}k_BT\left(1-\tau^2\dot{\gamma}^2\right),$$

$$\Psi_2=-\tau^2\frac{\rho}{m}k_BT\left(1-3\tau^2\dot{\gamma}^2\right).$$

The important new feature compared with (7.138) is the nonvanishing second normal-stress coefficient. Although this is a natural consequence of the full nonlinear form of the entropy gradient, one should not rely on the detailed prefactors because we have used the equilibrium friction matrix.

Exercise 187

The counterpart of (7.157) is

$$M_{AB}(\boldsymbol{r}) \quad = \quad \frac{2}{k_Bm}\int\hat{A}(\boldsymbol{r},\frac{\boldsymbol{p}_t}{2}+\boldsymbol{P})|\boldsymbol{P}|$$

$$
\begin{aligned}
&\times \quad \frac{f(\boldsymbol{r},\frac{\boldsymbol{p}_t}{2}+\boldsymbol{Q})f(\boldsymbol{r},\frac{\boldsymbol{p}_t}{2}-\boldsymbol{Q})-f(\boldsymbol{r},\frac{\boldsymbol{p}_t}{2}+\boldsymbol{P})f(\boldsymbol{r},\frac{\boldsymbol{p}_t}{2}-\boldsymbol{P})}{\ln[f(\boldsymbol{r},\frac{\boldsymbol{p}_t}{2}+\boldsymbol{Q})f(\boldsymbol{r},\frac{\boldsymbol{p}_t}{2}-\boldsymbol{Q})]-\ln[f(\boldsymbol{r},\frac{\boldsymbol{p}_t}{2}+\boldsymbol{P})f(\boldsymbol{r},\frac{\boldsymbol{p}_t}{2}-\boldsymbol{P})]} \\
&\times \quad [\hat{B}(\boldsymbol{r},\frac{\boldsymbol{p}_t}{2}+\boldsymbol{P})+\hat{B}(\boldsymbol{r},\frac{\boldsymbol{p}_t}{2}-\boldsymbol{P}) \\
&- \quad \hat{B}(\boldsymbol{r},\frac{\boldsymbol{p}_t}{2}+\boldsymbol{Q})-\hat{B}(\boldsymbol{r},\frac{\boldsymbol{p}_t}{2}-\boldsymbol{Q})]d\sigma' d^3P d^3p_t. \qquad \text{(F.49)}
\end{aligned}
$$

This expression is actually symmetric in $\hat{A}$ and $\hat{B}$ (see Exercise 171). From the combination of $\hat{B}$ with different arguments in (F.49) and from the collisional invariants, it follows that the first three rows, and hence the first three columns, of (7.161) must vanish. Only M_{44} remains to be calculated. For the fourth-rank tensor M_{44}, we obtain with the expressions of Exercise 179

$$
\begin{aligned}
M_{44} &= \frac{1}{(\rho k_B T)^2}\frac{2}{k_B m}\int |\boldsymbol{P}|(\boldsymbol{PP}-\boldsymbol{QQ}) \\
&\times \frac{f(\boldsymbol{r},\frac{\boldsymbol{p}_t}{2}+\boldsymbol{Q})f(\boldsymbol{r},\frac{\boldsymbol{p}_t}{2}-\boldsymbol{Q})-f(\boldsymbol{r},\frac{\boldsymbol{p}_t}{2}+\boldsymbol{P})f(\boldsymbol{r},\frac{\boldsymbol{p}_t}{2}-\boldsymbol{P})}{\ln[f(\boldsymbol{r},\frac{\boldsymbol{p}_t}{2}+\boldsymbol{Q})f(\boldsymbol{r},\frac{\boldsymbol{p}_t}{2}-\boldsymbol{Q})]-\ln[f(\boldsymbol{r},\frac{\boldsymbol{p}_t}{2}+\boldsymbol{P})f(\boldsymbol{r},\frac{\boldsymbol{p}_t}{2}-\boldsymbol{P})]} \\
&\times (\boldsymbol{PP}-\boldsymbol{QQ})d\sigma' d^3P d^3p_t. \qquad \text{(F.50)}
\end{aligned}
$$

When the products of two single-particle distribution functions are refactored into f_c and f_r [see (7.158) and (7.159)], the integration over $\boldsymbol{p}_t$ can be carried out and we obtain

$$
\begin{aligned}
M_{44} &= \frac{2}{m^3 k_B^3 T^2}\int |\boldsymbol{P}|(\boldsymbol{PP}-\boldsymbol{QQ})(\boldsymbol{PP}-\boldsymbol{QQ}) \\
&\times \frac{f_r(\boldsymbol{r},\boldsymbol{Q})-f_r(\boldsymbol{r},\boldsymbol{P})}{\ln f_r(\boldsymbol{r},\boldsymbol{Q})-\ln f_r(\boldsymbol{r},\boldsymbol{P})}d\sigma' d^3P. \qquad \text{(F.51)}
\end{aligned}
$$

Equation (F.51) is the counterpart of (7.168). By using (7.159) and exploiting symmetries, we obtain the alternative formula

$$
M_{44} = -\frac{4}{m^2 k_B^2 T}\int |\boldsymbol{P}|\frac{(\boldsymbol{PP}-\boldsymbol{QQ})(\boldsymbol{PP}-\boldsymbol{QQ})}{(\boldsymbol{PP}-\boldsymbol{QQ}):(\boldsymbol{1}-\mathring{\boldsymbol{c}})^{-1}} f_r(\boldsymbol{r},\boldsymbol{P})d\sigma' d^3P, \qquad \text{(F.52)}
$$

which implies

$$
M_{44} : \frac{\delta S}{\delta \mathring{\boldsymbol{c}}} = \frac{2\rho}{m^3 k_B T}\int |\boldsymbol{P}| \int (\boldsymbol{PP}-\boldsymbol{QQ})d\sigma' f_r(\boldsymbol{r},\boldsymbol{P})d^3P.
$$

Because this result coincides with (7.173), the structured moment method based on the atomistic and the phenomenological friction matrices lead to identical time-evolution equations.

Near equilibrium, where $f_r(\boldsymbol{Q})/f_r(\boldsymbol{P})$ is close to unity, we have

$$
\begin{aligned}
f_r(\boldsymbol{Q})-f_r(\boldsymbol{P}) &= f_r(\boldsymbol{P})\left[\frac{f_r(\boldsymbol{Q})}{f_r(\boldsymbol{P})}-1\right] \approx f_r(\boldsymbol{P})\ln\frac{f_r(\boldsymbol{Q})}{f_r(\boldsymbol{P})} \\
&= f_r(\boldsymbol{P})[\ln f_r(\boldsymbol{Q})-\ln f_r(\boldsymbol{P})],
\end{aligned}
$$

so that, from (F.51), we obtain

$$M_{44} = \frac{2}{m^3 k_B^3 T^2} \int P(\boldsymbol{PP} - \boldsymbol{QQ})(\boldsymbol{PP} - \boldsymbol{QQ}) f_r(P) d\sigma' d^3 P. \tag{F.53}$$

The friction matrix in (F.53) must be of the form (7.176), and $\bar{M}_{44} = (M_{44})_{lkkl}/10$ reproduces (7.177). At equilibrium, not only the time-evolution equations obtained from the atomistic and the phenomenological friction matrices for Boltzmann's equation coincide but also the friction matrices on the level of moments.

F.8 SIMULATIONS

Exercise 188

Equation (8.12) is obtained by differentiating the sum of a geometric series,

$$\sum_{n=0}^{\infty} \lambda^n = \frac{1}{1-\lambda},$$

which gives

$$\sum_{n=1}^{\infty} n\lambda^{n-1} = \frac{1}{(1-\lambda)^2}.$$

Exercise 189

From Figure 8.4, we find the following transition probabilities:

$$P = \begin{pmatrix} \frac{1}{2} & 0 & \frac{1}{2} & 0 & 0 \\ 0 & \frac{1}{4} & 0 & \frac{3}{4} & 0 \\ 0 & 0 & \frac{1}{3} & 0 & \frac{2}{3} \\ \frac{1}{4} & \frac{1}{2} & 0 & \frac{1}{4} & 0 \\ \frac{1}{3} & 0 & \frac{1}{3} & 0 & \frac{1}{3} \end{pmatrix}.$$

Because the matrix $1-P$ is singular, we regularize it by introducing a small parameter ϵ in the $(1,1)$ entry,

$$1 - P = \begin{pmatrix} \frac{1}{2}+\epsilon & 0 & -\frac{1}{2} & 0 & 0 \\ 0 & \frac{3}{4} & 0 & -\frac{3}{4} & 0 \\ 0 & 0 & \frac{2}{3} & 0 & -\frac{2}{3} \\ -\frac{1}{4} & -\frac{1}{2} & 0 & \frac{3}{4} & 0 \\ -\frac{1}{3} & 0 & -\frac{1}{3} & 0 & \frac{2}{3} \end{pmatrix}.$$

By inverting the regularized matrix $1-P$, we obtain the potential matrix R [according to (8.15)] in the form

$$R = \begin{pmatrix} \frac{1}{\epsilon} & 0 & \frac{3}{2\epsilon} & 0 & \frac{3}{2\epsilon} \\ \frac{1}{\epsilon} & 4 & \frac{3}{2\epsilon} & 4 & \frac{3}{2\epsilon} \\ \frac{1}{\epsilon} & 0 & \frac{3}{2\epsilon}+3 & 0 & \frac{3}{2\epsilon}+3 \\ \frac{1}{\epsilon} & \frac{8}{3} & \frac{3}{2\epsilon} & 4 & \frac{3}{2\epsilon} \\ \frac{1}{\epsilon} & 0 & \frac{3}{2\epsilon}+\frac{3}{2} & 0 & \frac{3}{2\epsilon}+3 \end{pmatrix},$$

where we now can take the limit $\epsilon \to 0$. Our final result is the potential matrix

$$R = \begin{pmatrix} \infty & 0 & \infty & 0 & \infty \\ \infty & 4 & \infty & 4 & \infty \\ \infty & 0 & \infty & 0 & \infty \\ \infty & \frac{8}{3} & \infty & 4 & \infty \\ \infty & 0 & \infty & 0 & \infty \end{pmatrix}.$$

The states "1," "3," and "5" are recurrent, whereas "2" and "4" are transient. This situation should be clear from Figure 8.4 because there is no possibility to return from "1" to "4." Once we go from "4" to "1," the Markov chain will only visit the three recurrent states "1," "3," and "5." The entries 0 and ∞ are thus immediately clear from the figure defining the Markov chain. The interesting information in R is about the expected number of visits of the transient states "2" and "4" when one starts at "2" or "4." The matrix F can be computed according to (8.12) and (8.13) from the (regularized) potential matrix. We find

$$F = \begin{pmatrix} 1 & 0 & 1 & 0 & 1 \\ 1 & \frac{3}{4} & 1 & 1 & 1 \\ 1 & 0 & 1 & 0 & 1 \\ 1 & \frac{2}{3} & 1 & \frac{3}{4} & 1 \\ 1 & 0 & 1 & 0 & 1 \end{pmatrix}.$$

In this matrix, the entries 0 and 1 are immediately clear from the figure defining the Markov chain. In particular, it is clear that the transient state "4" is reached with probability one if one starts at the transient state "2". The fractional probabilities are not immediately obvious, but they can be verified by elementary considerations. The irreducible Markov chain on the recurrent states is aperiodic because every state has a nonzero probability for going into itself in the next step.

For the purpose of illustration, we calculate the first two columns of F directly from (8.10). The linear systems for the first and second column vector are

$$\begin{pmatrix} 1 & 0 & -\frac{1}{2} & 0 & 0 \\ 0 & \frac{3}{4} & 0 & -\frac{3}{4} & 0 \\ 0 & 0 & \frac{2}{3} & 0 & -\frac{2}{3} \\ 0 & -\frac{1}{2} & 0 & \frac{3}{4} & 0 \\ 0 & 0 & -\frac{1}{3} & 0 & \frac{2}{3} \end{pmatrix} \cdot \begin{pmatrix} F_{11} \\ F_{21} \\ F_{31} \\ F_{41} \\ F_{51} \end{pmatrix} = \begin{pmatrix} \frac{1}{2} \\ 0 \\ 0 \\ \frac{1}{4} \\ \frac{1}{3} \end{pmatrix}$$

and

$$\begin{pmatrix} \frac{1}{2} & 0 & -\frac{1}{2} & 0 & 0 \\ 0 & 1 & 0 & -\frac{3}{4} & 0 \\ 0 & 0 & \frac{2}{3} & 0 & -\frac{2}{3} \\ -\frac{1}{4} & 0 & 0 & \frac{3}{4} & 0 \\ -\frac{1}{3} & 0 & -\frac{1}{3} & 0 & \frac{2}{3} \end{pmatrix} \cdot \begin{pmatrix} F_{12} \\ F_{22} \\ F_{32} \\ F_{42} \\ F_{52} \end{pmatrix} = \begin{pmatrix} 0 \\ \frac{1}{4} \\ 0 \\ \frac{1}{2} \\ 0 \end{pmatrix},$$

respectively. The equations for the first column vector have a unique solution that coincides with the previously given result for F. For the second column vector, we get a one-dimensional linear space of solutions,

$$F_{12} = F_{32} = F_{52}, \quad F_{22} = \frac{3}{4} + \frac{1}{4}F_{12}, \quad F_{42} = \frac{2}{3} + \frac{1}{3}F_{12}.$$

The correct solution can only be selected after observing that the state "2" cannot be reached from the states "1," "3," and "5" ($F_{12} = F_{32} = F_{52} = 0$).

Finally, we turn to the relative frequencies for visiting the recurrent states over a long period of time or to the limiting probability distribution. At this point, we are not prepared to find these relative frequencies yet, and this exercise is hence also meant as a motivation for the subsequent study of the limit distribution. In the subsequent development, it is shown that the limiting probability distribution can be found by solving the system of linear equations in (8.18). The solution is

$$\pi_1 = \frac{1}{4}, \quad \pi_3 = \frac{3}{8}, \quad \pi_5 = \frac{3}{8}.$$

Exercise 190

From (8.4), we obtain

$$1 - P = \tilde{p} \begin{pmatrix} 1 & -1 & 0 & 0 & \\ 0 & 1 & -1 & 0 & \cdots \\ 0 & 0 & 1 & -1 & \\ & & \vdots & & \ddots \end{pmatrix}.$$

By a direct matrix multiplication, we can verify that

$$R = \frac{1}{\tilde{p}} \begin{pmatrix} 1 & 1 & 1 & 1 & \\ 0 & 1 & 1 & 1 & \\ 0 & 0 & 1 & 1 & \cdots \\ 0 & 0 & 0 & 1 & \\ & & \vdots & & \ddots \end{pmatrix}$$

is the inverse of $1 - P$ and thus, according to (8.15), the potential matrix. Because all diagonal entries are finite ($\tilde{p} > 0$), all states are transient.

Exercise 191

Let j be a recurrent state with a consequent k. We thus have $R_{jj} = \infty$, and there exists an n such that $P^n_{jk} > 0$. From (8.14), we conclude

$$R_{jk} \geq \sum_{l=0}^{\infty} P^{l+n}_{jk} \geq \sum_{l=0}^{\infty} P^l_{jj} P^n_{jk} = R_{jj} P^n_{jk} = \infty.$$

Equation (8.12) then implies $F_{kk} = 1$, and (8.13) gives the desired result $R_{kk} = \infty$, which expresses the recurrent nature of the state k.

Exercise 192
The possible moves given in Figure 8.5 can be expressed in the stochastic matrix

$$P^{\rm mov} = \frac{1}{2}\begin{pmatrix} 0 & 0 & 1 & 1 \\ 0 & 0 & 1 & 1 \\ 1 & 1 & 0 & 0 \\ 1 & 1 & 0 & 0 \end{pmatrix},$$

where it has been assumed that flips of the left and right spins are offered with equal probability. When the construction (8.30) is performed for this symmetric stochastic matrix $P^{\rm mov}$, we find (8.32).

Exercise 193
For a flip of the left spin, we have

$$P_{\rm l}^{\rm mov} = \begin{pmatrix} 0 & 0 & 0 & 1 \\ 0 & 0 & 1 & 0 \\ 0 & 1 & 0 & 0 \\ 1 & 0 & 0 & 0 \end{pmatrix},$$

and the construction (8.30) yields

$$P_{\rm l} = \begin{pmatrix} 1-e^{-\Delta} & 0 & 0 & e^{-\Delta} \\ 0 & 1-e^{-\Delta} & e^{-\Delta} & 0 \\ 0 & 1 & 0 & 0 \\ 1 & 0 & 0 & 0 \end{pmatrix}.$$

We similarly find

$$P_{\rm r}^{\rm mov} = \begin{pmatrix} 0 & 0 & 1 & 0 \\ 0 & 0 & 0 & 1 \\ 1 & 0 & 0 & 0 \\ 0 & 1 & 0 & 0 \end{pmatrix}, \quad P_{\rm r} = \begin{pmatrix} 1-e^{-\Delta} & 0 & e^{-\Delta} & 0 \\ 0 & 1-e^{-\Delta} & 0 & e^{-\Delta} \\ 1 & 0 & 0 & 0 \\ 0 & 1 & 0 & 0 \end{pmatrix}.$$

For $P_{\rm l}$ we have $\{1,4\}$ and $\{2,3\}$ as classes of communicating states, whereas the classes for $P_{\rm r}$ are $\{1,3\}$ and $\{2,4\}$. By a straightforward matrix multiplication, we obtain

$$P = P_{\rm l} \cdot P_{\rm r} = \begin{pmatrix} (1-e^{-\Delta})^2 & e^{-\Delta} & e^{-\Delta}(1-e^{-\Delta}) & 0 \\ e^{-\Delta} & (1-e^{-\Delta})^2 & 0 & e^{-\Delta}(1-e^{-\Delta}) \\ 0 & 1-e^{-\Delta} & 0 & e^{-\Delta} \\ 1-e^{-\Delta} & 0 & e^{-\Delta} & 0 \end{pmatrix}.$$

For this stochastic matrix P, any state can be reached from any other state in one or two steps. The detailed balance condition is violated because, for example, P_{14} is zero (according to the Metropolis algorithm, if the left spin is flipped, the right one will also be flipped in order to bring back the energy to the lower level), whereas P_{41}

is nonzero. Because P_l and P_r were constructed by the Metropolis method, we have $\pi \cdot P_l = \pi \cdot P_r = \pi$ and thus

$$\pi \cdot P = \pi \cdot P_l \cdot P_r = \pi \cdot P_r = \pi.$$

Exercise 194

By means of the Fokker-Planck equation (8.57), we can evaluate

$$\frac{d\langle g(X_t)\rangle}{dt} = \int g(x) \frac{\partial f(x,t)}{\partial t} dx.$$

After several integrations by part, we obtain

$$\frac{d\langle g(X_t)\rangle}{dt} = \int f(x,t) \left[A(x,t) \cdot \frac{\partial g(x)}{\partial x} + \frac{1}{2} D(x,t) : \frac{\partial^2 g(x)}{\partial x \partial x} \right] dx,$$

which we can rewrite as

$$\frac{d\langle g(X_t)\rangle}{dt} = \left\langle A(X_t,t) \cdot \frac{\partial g(X_t)}{\partial x} + \frac{1}{2} D(X_t,t) : \frac{\partial^2 g(X_t)}{\partial x \partial x} \right\rangle. \tag{F.54}$$

From these equations it becomes clear that, for a "sufficiently smooth" function g, the second-order derivative should be postulated to be piecewise continuous.

To obtain the result (F.54) from the stochastic differential equation (8.58), we start with the second-order Taylor expansion

$$dg(X_t) = dX_t \cdot \frac{\partial g(X_t)}{\partial x} + \frac{1}{2} dX_t dX_t : \frac{\partial^2 g(X_t)}{\partial x \partial x}, \tag{F.55}$$

where the second-order term was kept because, in view of (8.62), it contributes to the first-order rate of change. We now insert (8.58) into (F.55), average the result, and systematically keep only terms of order dt, in order to arrive at

$$d\langle g(X_t)\rangle = \left\langle A(X_t,t) \cdot \frac{\partial g(X_t)}{\partial x} \right\rangle dt + \frac{1}{2} \left\langle D(X_t,t) : \frac{\partial^2 g(X_t)}{\partial x \partial x} \right\rangle dt,$$

in agreement with (F.54). Here, we have used the fact that, within the Itô discretization scheme (8.60), the increment dW_t is independent of X_t, vanishes on average, and has second moments given by (8.61).

As a general rule, one can always use the replacement [see (8.65)]

$$dW_t dW_t^T \to 1 dt,$$

which simplifies (F.55) to the first-order expression

$$\begin{aligned} dg(X_t) &= \frac{\partial g(X_t)}{\partial x} \cdot dX_t + \frac{1}{2} D(X_t,t) : \frac{\partial^2 g(X_t)}{\partial x \partial x} dt \\ &= \left[\frac{\partial g(X_t)}{\partial x} \cdot A(X_t,t) + \frac{1}{2} D(X_t,t) : \frac{\partial^2 g(X_t)}{\partial x \partial x} \right] dt \\ &\quad + \frac{\partial g(X_t)}{\partial x} \cdot B(X_t,t) \cdot dW_t. \end{aligned} \tag{F.56}$$

The transformation rule (F.56) for the stochastic differential (8.58) of a process X_t is known as the *Itô formula*. This unusual transformation rule involving second-order derivatives is the cornerstone of stochastic calculus.

Exercise 195

According to the integration scheme (8.60), we have

$$X_{t_{j+1}} = X_{t_j} + [a(t_j) + A(t_j) \cdot X_{t_j}](t_{j+1} - t_j) + B(t_j) \cdot (W_{t_{j+1}} - W_{t_j}),$$

that is, $X_{t_{j+1}}$ is a linear combination of the Gaussian random variables X_{t_j} and $W_{t_{j+1}} - W_{t_j}$, with a shift in the average determined by $a(t_j)$; therefore, $X_{t_{j+1}}$ is also Gaussian.

From (F.54) given in the solution to Exercise 194, we get

$$\frac{d\langle X_t \rangle}{dt} = \langle A(X_t, t) \rangle = a(t) + A(t) \cdot \langle X_t \rangle, \tag{F.57}$$

which is a deterministic linear differential equation for $\langle X_t \rangle$. By applying (F.54) to $g(x) = xx^T$, we further obtain

$$\frac{d\langle X_t X_t^T \rangle}{dt} = \langle A(X_t, t) X_t^T \rangle + \langle X_t A(X_t, t)^T \rangle + \langle D(X_t, t) \rangle .$$

After specializing to the linear stochastic differential equation (8.67), we by means of (F.57) obtain the following time-evolution equation,

$$\frac{dC_t}{dt} = A(t) \cdot C_t + C_t \cdot A(t)^T + D(t), \tag{F.58}$$

for the covariance matrix

$$C_t = \langle X_t X_t^T \rangle - \langle X_t \rangle \langle X_t \rangle^T .$$

Exercise 196

The Gaussian solution of Exercise 195 has the explicit form

$$f(x, t) = (2\pi)^{-n/2} (\det C_t)^{-1/2} \exp\left\{ -\frac{1}{2}(x - \langle X_t \rangle) \cdot C_t^{-1} \cdot (x - \langle X_t \rangle) \right\}.$$

By differentiation, we obtain

$$\begin{aligned}
\frac{\partial f(x,t)}{\partial t} &= \Bigg[-\frac{1}{2} \frac{1}{\det C_t} \frac{d}{dt} \det C_t + \frac{d\langle X_t \rangle}{dt} \cdot C_t^{-1} \cdot (x - \langle X_t \rangle) \\
&\quad -\frac{1}{2}(x - \langle X_t \rangle) \cdot \frac{dC_t^{-1}}{dt} \cdot (x - \langle X_t \rangle) \Bigg] f(x, t),
\end{aligned}$$

$$\frac{\partial f(x,t)}{\partial x} = -C_t^{-1} \cdot (x - \langle X_t \rangle) f(x, t),$$

$$\frac{\partial^2 f(x,t)}{\partial x \partial x} = [C_t^{-1} \cdot (x - \langle X_t \rangle)(x - \langle X_t \rangle) \cdot C_t^{-1} - C_t^{-1}] f(x,t),$$

and hence

$$\frac{\partial}{\partial x}[a(t) + A(t) \cdot x] f(x,t) = \left\{ \mathrm{tr} A(t) - [a(t) + A(t) \cdot x] \cdot C_t^{-1} \cdot (x - \langle X_t \rangle) \right\} \times f(x,t).$$

By inserting all these ingredients into the Fokker-Planck equation (8.57) corresponding to the linear stochastic differential equation (8.67) and equating all terms constant, linear, and quadratic in $x - \langle X_t \rangle$ in the prefactors of f, we obtain

$$\frac{1}{\det C_t} \frac{d}{dt} \det C_t = 2 \mathrm{tr} A(t) + D(t) : C_t^{-1}, \tag{F.59}$$

$$\frac{d \langle X_t \rangle}{dt} = a(t) + A(t) \cdot \langle X_t \rangle, \tag{F.60}$$

and

$$-\frac{dC_t^{-1}}{dt} = A(t)^T \cdot C_t^{-1} + C_t^{-1} \cdot A(t) + C_t^{-1} \cdot D(t) \cdot C_t^{-1}, \tag{F.61}$$

where the quadratic form yields an equation for the *symmetrized* coefficient matrix. Equation (F.60) is identical to (F.57), and (F.61) is equivalent to (F.58), as can be seen by multiplying (F.61) from both sides by C_t and using the standard rule for the derivative of the inverse of a matrix. Finally, (F.59) follows from (F.61) by double contraction with C_t, where the result of Exercise 108 for the derivative of a determinant has been used. Note that in this exercise we have found a rigorous solution of the Fokker-Planck equation for linear stochastic differential equations by means of purely deterministic methods.

Exercise 197

We rewrite the linear differential equation (8.67) as

$$\frac{dX_t}{dt} = A(t) \cdot X_t + a(t) + B(t) \cdot \frac{dW_t}{dt},$$

and first consider the homogeneous equation

$$\frac{dX_t}{dt} = A(t) \cdot X_t.$$

We introduce the particular matrix of solution vectors of this homogeneous equation

$$\Phi_{tt'} = \mathcal{T}\text{-}\exp \left\{ \int_{t'}^{t} A(t'') dt'' \right\},$$

where the time-ordering operation expressed by $\mathcal{T}$ is important for noncommuting matrices $A(t)$ at different times t (in a Taylor expansion, factors with larger time arguments need to be further to the left), and the initial condition $\Phi_{t't'}$ is given by the unit matrix. By employing the standard method of variation of constants to solve the inhomogeneous equation, we obtain

$$\begin{aligned} X_t &= \Phi_{tt_0} \cdot X_{t_0} + \int_{t_0}^{t} \Phi_{tt'} \cdot \left[a(t') + B(t') \cdot \frac{dW_{t'}}{dt'} \right] dt' \\ &= \Phi_{tt_0} \cdot X_{t_0} + \int_{t_0}^{t} \Phi_{tt'} \cdot a(t') dt' + \int_{t_0}^{t} \Phi_{tt'} \cdot B(t') \cdot dW_{t'} . \end{aligned}$$

The stochastic character of X_t results from a linear superposition of increments of the Wiener process (and a Gaussian initial condition) so that X_t is Gaussian at all times. By averaging X_t, we obtain

$$\langle X_t \rangle = \Phi_{tt_0} \cdot \langle X_{t_0} \rangle + \int_{t_0}^{t} \Phi_{tt'} \cdot a(t') dt', \tag{F.62}$$

which is the solution of (F.57). By averaging $X_t X_t^T$ and using (1.59), we further obtain

$$C_t = \Phi_{tt_0} \cdot C_{t_0} \cdot \Phi_{tt_0}^T + \int_{t_0}^{t} \Phi_{tt'} \cdot D(t') \cdot \Phi_{tt'}^T dt', \tag{F.63}$$

which is the solution of (F.58). We have thus shown that the procedure to solve deterministic linear differential equations leads to the correct result even in the stochastic case. This observation is related to the fact that, if B is independent of x, there is no difference between the Itô and Stratonovich integration schemes and the corresponding Fokker-Planck equations.

Exercise 198

From Exercise 194, we obtain

$$\frac{d \langle \boldsymbol{Q}_t \boldsymbol{Q}_t \rangle}{dt} = \boldsymbol{\kappa} \cdot \langle \boldsymbol{Q}_t \boldsymbol{Q}_t \rangle + \langle \boldsymbol{Q}_t \boldsymbol{Q}_t \rangle \cdot \boldsymbol{\kappa}^T - \frac{1}{\tau_0} \left(\left\langle \frac{\boldsymbol{Q}_t \boldsymbol{Q}_t}{1 - \boldsymbol{Q}_t^2 / b} \right\rangle - \mathbf{1} \right),$$

where, despite (8.71), we have introduced τ_0 for clarity. The FENE-P approximation is given by

$$\left\langle \frac{\boldsymbol{Q}_t \boldsymbol{Q}_t}{1 - \boldsymbol{Q}_t^2 / b} \right\rangle \to \frac{\langle \boldsymbol{Q}_t \boldsymbol{Q}_t \rangle}{1 - \langle \boldsymbol{Q}_t^2 \rangle / b}.$$

Because the conformation tensor $\boldsymbol{c}$ in Section 4.2.6 is normalized to the unit tensor at equilibrium, the equilibrium solution of the evolution equation for $\langle \boldsymbol{Q}_t \boldsymbol{Q}_t \rangle$ implies the proportionality

$$\langle \boldsymbol{Q}_t \boldsymbol{Q}_t \rangle = \frac{b}{b+3} \boldsymbol{c}$$

and thus

$$\frac{d\boldsymbol{c}}{dt} = \boldsymbol{\kappa} \cdot \boldsymbol{c} + \boldsymbol{c} \cdot \boldsymbol{\kappa}^T - \frac{b+3}{b} \frac{1}{\tau_0} \left(\frac{b}{b+3-\mathrm{tr}\boldsymbol{c}} \boldsymbol{c} - \mathbf{1} \right).$$

A comparison with (4.99) leads to the identification

$$\tau = \frac{b}{b+3} \tau_0.$$

With the present identifications, and up to an additive constant, (4.96) for the dimensionless FENE-P spring entropy contribution can be rewritten as

$$s^{(\phi)} = -\frac{\phi^{(S)}}{k_{\mathrm{B}}T} = \frac{1}{2} \frac{HQ_0^2}{k_{\mathrm{B}}T} \ln \left(1 - \frac{\langle \boldsymbol{Q}^2 \rangle}{Q_0^2} \right),$$

which should be compared with (8.68).

Exercise 199
We have

$$\begin{aligned} \left\langle \left\{ \int_0^\tau \left[1 - e^{-\zeta(\tau - t)/m} \right] dW_t \right\}^2 \right\rangle &= \int_0^\tau dt \int_0^\tau dt' \left[1 - e^{-\zeta(\tau - t)/m} \right] \\ &\quad \times \left[1 - e^{-\zeta(\tau - t')/m} \right] \left\langle \frac{dW_t}{dt} \frac{dW_{t'}}{dt'} \right\rangle \\ &= \int_0^\tau dt \left[1 - e^{-\zeta(\tau - t)/m} \right]^2 \\ &= \tau - \frac{2m}{\zeta} \left(1 - e^{-\zeta\tau/m} \right) + \frac{m}{2\zeta} \left(1 - e^{-2\zeta\tau/m} \right), \end{aligned}$$

where (1.59) has been used.

Exercise 200
From (8.91) and (8.93), we have

$$\begin{aligned} 2\tau \langle D_{\mathrm{HE}} \rangle &= \langle \dot{x}^{\mathrm{f}}(0)^2 \rangle \frac{m^2}{\zeta^2} \left(1 - e^{-\zeta\tau/m} \right)^2 \\ &\quad + \frac{2k_{\mathrm{B}}T}{\zeta} \left\langle \left\{ \int_0^\tau \left[1 - e^{-\zeta(\tau - t)/m} \right] dW_t \right\}^2 \right\rangle \\ &= 2\tau \frac{k_{\mathrm{B}}T}{\zeta} \left[1 - \frac{m}{\zeta\tau} \left(1 - e^{-\zeta\tau/m} \right) \right], \end{aligned}$$

where (8.89) and (8.98) have been used.

Exercise 201

From (8.90) and (8.93), we have

$$\begin{aligned}
\langle D_{\rm GK}^2\rangle &= \left\langle \dot{x}^{\rm f}(0)^2\left[\int_0^\tau \dot{x}^{\rm f}(t)dt\right]^2\right\rangle \\
&= \langle \dot{x}^{\rm f}(0)^4\rangle \frac{m^2}{\zeta^2}\left(1-e^{-\zeta\tau/m}\right)^2 \\
&\quad +\langle \dot{x}^{\rm f}(0)^2\rangle \frac{2k_{\rm B}T}{\zeta}\left\langle\left\{\int_0^\tau \left[1-e^{-\zeta(\tau-t)/m}\right]dW_t\right\}^2\right\rangle \\
&= \left(\frac{k_{\rm B}T}{\zeta}\right)^2\left[3\left(1-e^{-\zeta\tau/m}\right)^2\right. \\
&\quad \left. +2\frac{\zeta}{m}\left\langle\left\{\int_0^\tau \left[1-e^{-\zeta(\tau-t)/m}\right]dW_t\right\}^2\right\rangle\right] \\
&= 2\left(\frac{k_{\rm B}T}{\zeta}\right)^2\left(\frac{\zeta\tau}{m}-e^{-\zeta\tau/m}+e^{-2\zeta\tau/m}\right),
\end{aligned}$$

where, for the last step, (8.98) has been used. After subtracting $\langle D_{\rm GK}\rangle^2$ we obtain (8.96).

Equation (8.97) follows directly from the properties of Gaussian random variables, for which fourth moments are directly related to second moments,

$$\left\langle\left(\int_0^\tau \dot{x}^{\rm f}(t)dt\right)^4\right\rangle = 3\left\langle\left(\int_0^\tau \dot{x}^{\rm f}(t)dt\right)^2\right\rangle\left\langle\left(\int_0^\tau \dot{x}^{\rm f}(t)dt\right)^2\right\rangle.$$

Exercise 202

As a general definition of the conditional averages, we have

$$\langle A(z)\rangle_{\boldsymbol r} = \frac{\int \rho_x(z)A(z)\delta(\boldsymbol{r}-\boldsymbol{r}_j)d^{6N}z}{\int \rho_x(z)\delta(\boldsymbol{r}-\boldsymbol{r}_j)d^{6N}z},$$

where the choice of j should be clear from the context. By differentiating (8.110) with respect to $\boldsymbol{p}_j$, we obtain

$$\frac{\partial \rho_x}{\partial \boldsymbol{p}_j} = -\left\{\frac{1}{k_{\rm B}T}\left[\frac{\boldsymbol{p}_j}{m}-\boldsymbol{v}(\boldsymbol{r}_j)\right]+\lambda\frac{\partial \Pi(z)}{\partial \boldsymbol{p}_j}\right\}\rho_x. \tag{F.64}$$

By multiplying (F.64) with $\delta(\boldsymbol{r}-\boldsymbol{r}_j)$ and integrating over all phase space, we obtain

$$\left\langle \frac{\boldsymbol{p}_j}{m}-\boldsymbol{v}(\boldsymbol{r})+k_{\rm B}T\lambda\frac{\partial \Pi(z)}{\partial \boldsymbol{p}_j}\right\rangle_{\boldsymbol r} = 0.$$

With the first identity in (8.111) we thus arrive at (8.112). If we multiply (F.64) with $\boldsymbol{p}_j\delta(\boldsymbol{r}-\boldsymbol{r}_j)$ and integrate over all phase space, we find

$$3k_{\rm B}T = \left\langle \frac{\boldsymbol{p}_j^2}{m} \right\rangle_{\boldsymbol{r}} - \langle \boldsymbol{p}_j \rangle_{\boldsymbol{r}} \cdot \boldsymbol{v}(\boldsymbol{r}) + k_{\rm B}T \left\langle \lambda \boldsymbol{p}_j \cdot \frac{\partial \Pi(z)}{\partial \boldsymbol{p}_j} \right\rangle_{\boldsymbol{r}} .$$

With $\langle \boldsymbol{p}_j \rangle_{\boldsymbol{r}} = m\boldsymbol{v}(\boldsymbol{r})$, we obtain (8.113) after summing over j.

Exercise 203

From $d\boldsymbol{r}'_j/dt = \partial H'/\partial \boldsymbol{p}'_j$, we obtain by means of (8.115)

$$\frac{d\boldsymbol{r}'_j}{dt} = \frac{1}{q_{\rm L}q_{\rm T}}\frac{\partial H'}{\partial \boldsymbol{p}_j} = \frac{1}{q_{\rm L}}\left[\frac{\boldsymbol{p}_j}{m} - \boldsymbol{v}(\boldsymbol{r}_j) + k_{\rm B}T\lambda\frac{\partial \Pi(z)}{\partial \boldsymbol{p}_j}\right],$$

from which we get (8.117) after using $\boldsymbol{r}_j = q_{\rm L}\boldsymbol{r}'_j$. Similarly, $d\boldsymbol{p}'_j/dt = -\partial H'/\partial \boldsymbol{r}'_j$ implies

$$\begin{aligned} \frac{d\boldsymbol{p}'_j}{dt} &= -q_{\rm L}\frac{\partial H'}{\partial \boldsymbol{r}_j} \\ &= -q_{\rm L}q_{\rm T}\left[\frac{\partial \phi_{\rm tot}}{\partial \boldsymbol{r}_j} - [\boldsymbol{p}_j - m\boldsymbol{v}(\boldsymbol{r}_j)] \cdot \left[\frac{\partial}{\partial \boldsymbol{r}_j}\boldsymbol{v}(\boldsymbol{r}_j)\right]^T + k_{\rm B}T\lambda\frac{\partial \Pi(z)}{\partial \boldsymbol{r}_j}\right], \end{aligned}$$

and thus (8.118). From $dq_{\rm L,T}/dt = \partial H'/\partial p_{\rm L,T}$, we obtain (8.119) and (8.120). From $dp_{\rm L,T}/dt = -\partial H'/\partial q_{\rm L,T}$, we finally have

$$\frac{dp_{\rm L}}{dt} = -q_{\rm T}\frac{\partial \phi_{\rm L}(q_{\rm L})}{\partial q_{\rm L}} + \frac{1}{q_{\rm L}}\sum_{j=1}^{N}\left(\boldsymbol{p}_j \cdot \frac{\partial H'}{\partial \boldsymbol{p}_j} - \boldsymbol{r}_j \cdot \frac{\partial H'}{\partial \boldsymbol{r}_j}\right)$$

and

$$\frac{dp_{\rm T}}{dt} = -\frac{H'}{q_{\rm T}} + \frac{1}{q_{\rm T}}\sum_{j=1}^{N}\boldsymbol{p}_j \cdot \frac{\partial H'}{\partial \boldsymbol{p}_j} - \bar{f}k_{\rm B}T,$$

where, throughout the solution of this exercise, H' is to be considered as a function of the arguments occurring in (8.116). These last two equations lead to (8.121) and (8.122).

Exercise 204

Because the particle momenta relax rapidly compared with the slowest configurational modes, we do not need to consider the position dependence of the velocity field; that is, we can take $\boldsymbol{\lambda}_M(\boldsymbol{r})$ as constant (zero). In view of the stochastic thermostat in (8.134), we don't need a Nosé thermostat, and Andersen's barostat is irrelevant because we consider the toy model of noninteracting Rouse chains. The relevant part of (8.117), (8.118) can hence be expressed as a modification of the total potential,

$$\phi_{\rm tot} \rightarrow \phi_{\rm tot} + k_{\rm B}T\lambda\Pi.$$

For the specific situation of the Rouse model, we obtain the modified spring potential,

$$\frac{1}{2}H\sum_{j=1}^{N-1} \boldsymbol{Q}_j^2 + k_{\mathrm{B}}T\boldsymbol{\lambda}_c : \boldsymbol{Q}_1'\boldsymbol{Q}_1' = \frac{1}{2}H\left(\sum_{j=1}^{N-1} {\boldsymbol{Q}_j'}^2 + 2\boldsymbol{\lambda}_c : \boldsymbol{Q}_1'\boldsymbol{Q}_1'\right).$$

Therefore, all modes with $j > 1$ are unaffected, and, in the spring-force term for the slowest mode, we need to make the replacement

$$\boldsymbol{Q}_1' \to \boldsymbol{Q}_1' + (\boldsymbol{\lambda}_c + \boldsymbol{\lambda}_c^T)\cdot\boldsymbol{Q}_1',$$

which leads to (8.135). With (8.52), we then obtain

$$\frac{d\boldsymbol{Q}_1'}{dt} = \frac{1}{2}(\boldsymbol{\kappa} + \boldsymbol{c}\cdot\boldsymbol{\kappa}^T\cdot\boldsymbol{c}^{-1})\cdot\boldsymbol{Q}_1' - \frac{1}{2\tau_1}\boldsymbol{Q}_1' + \sqrt{\frac{1}{\tau_1}}\frac{d\boldsymbol{W}_1'}{dt}, \tag{F.65}$$

where we have made use of our knowledge that $\boldsymbol{\lambda}_c$ is symmetric so that $\boldsymbol{\lambda}_c$ and $\boldsymbol{\lambda}_c^T$ can be used interchangeably. Note that (F.65) is similar to (8.134), so that the multiplostat has introduced a flow term. If we write down the corresponding equation for the second moment tensor $\boldsymbol{c} = \langle\boldsymbol{Q}_1'\boldsymbol{Q}_1'\rangle$, the result is

$$\frac{d\boldsymbol{c}}{dt} = \boldsymbol{\kappa}\cdot\boldsymbol{c} + \boldsymbol{c}\cdot\boldsymbol{\kappa}^T - \frac{1}{\tau_1}(\boldsymbol{c} - \boldsymbol{1}),$$

which, according to (4.93), contains exactly the right flow term on the coarse-grained level of conformation tensors.

Exercise 205

Using (8.142) for the leap-frog scheme at two different times, we have

$$\boldsymbol{r}_j(t+\Delta t) - \boldsymbol{r}_j(t) = \boldsymbol{r}_j(t) - \boldsymbol{r}_j(t-\Delta t) + \Delta t[\boldsymbol{v}_j(t+\tfrac{1}{2}\Delta t) - \boldsymbol{v}_j(t-\tfrac{1}{2}\Delta t)],$$

and (8.141) then implies

$$\boldsymbol{r}_j(t+\Delta t) = 2\boldsymbol{r}_j(t) - \boldsymbol{r}_j(t-\Delta t) + \Delta t^2\frac{\boldsymbol{F}_j(t)}{m_j}.$$

Similarly, we obtain from (8.144)

$$\begin{aligned}\boldsymbol{r}_j(t+\Delta t) - \boldsymbol{r}_j(t) &= \boldsymbol{r}_j(t) - \boldsymbol{r}_j(t-\Delta t) + \Delta t[\boldsymbol{v}_j(t) - \boldsymbol{v}_j(t-\Delta t)] \\ &+ \frac{\Delta t^2}{2m_j}[\boldsymbol{F}_j(t) - \boldsymbol{F}_j(t-\Delta t)]\end{aligned}$$

for the velocity Verlet algorithm, so that (8.145) shifted by $-\Delta t$ again implies (8.138).

Exercise 206

Equation (8.110) reduces to the simpler form

$$\rho_x \propto \exp\left\{-\frac{\phi_{\mathrm{tot}}(\boldsymbol{r}_1,\ldots\boldsymbol{r}_N)}{k_{\mathrm{B}}T} - \boldsymbol{\lambda}_c' : \boldsymbol{\Pi}'(z)\right\}. \tag{F.66}$$

The kinetic energy is contained in the trace of $\boldsymbol{\Pi}'(z)$, and the barostat term is absent because, for a gas, we have chosen to consider a constant volume. Note that the temperature occurs only as a Lagrange multiplier for the potential energy in (F.66) but not in the velocity distribution governed by $\boldsymbol{\Pi}'$. The Lagrange multiplier $\boldsymbol{\lambda}_c'$ is identified by comparing (F.66) and (7.141),

$$\boldsymbol{\lambda}_c' = \frac{N}{2k_{\mathrm{B}}T}(\mathbf{1} - \mathring{\boldsymbol{c}})^{-1},$$

and by evaluating Gaussian integrals we further find

$$\langle \boldsymbol{\Pi}' \rangle_x = k_{\mathrm{B}}T(\mathbf{1} - \mathring{\boldsymbol{c}}).$$

The average $\langle \boldsymbol{\Pi}' \rangle_x$ hence indeed contains the information about T and $\mathring{\boldsymbol{c}}$, which can be separated in the well-known way by the trace operator (see Exercise 179).

Equations (8.125), (8.126), and (8.128) without the kinetic energy terms, and after neglecting the barostat, read

$$\frac{d\boldsymbol{r}_j}{dt} = k_{\mathrm{B}}T\frac{\partial[\boldsymbol{\lambda}_c' : \boldsymbol{\Pi}'(z)]}{\partial \boldsymbol{p}_j},$$

$$\frac{d\boldsymbol{p}_j}{dt} = -\frac{\partial \phi_{\mathrm{tot}}}{\partial \boldsymbol{r}_j} - k_{\mathrm{B}}T\frac{\partial[\boldsymbol{\lambda}_c' : \boldsymbol{\Pi}'(z)]}{\partial \boldsymbol{r}_j} - \frac{p_{\mathrm{T}}}{M_{\mathrm{T}}}\boldsymbol{p}_j,$$

and

$$\frac{dp_{\mathrm{T}}}{dt} = \sum_{j=1}^{N} k_{\mathrm{B}}T\boldsymbol{p}_j \cdot \frac{\partial[\boldsymbol{\lambda}_c' : \boldsymbol{\Pi}'(z)]}{\partial \boldsymbol{p}_j} - 3Nk_{\mathrm{B}}T,$$

respectively. With

$$k_{\mathrm{B}}T\frac{\partial}{\partial \boldsymbol{p}_j}\boldsymbol{\lambda}_c' : \boldsymbol{\Pi}'(z) = (\mathbf{1} - \mathring{\boldsymbol{c}})^{-1} \cdot \left(\frac{\boldsymbol{p}_j}{m} - \boldsymbol{\kappa} \cdot \boldsymbol{r}_j\right)$$

and

$$k_{\mathrm{B}}T\frac{\partial}{\partial \boldsymbol{r}_j}\boldsymbol{\lambda}_c' : \boldsymbol{\Pi}'(z) = -\boldsymbol{\kappa}^T \cdot (\mathbf{1} - \mathring{\boldsymbol{c}})^{-1} \cdot (\boldsymbol{p}_j - m\boldsymbol{\kappa} \cdot \boldsymbol{r}_j),$$

we recover (8.149)–(8.151).

Exercise 207

At equilibrium, M_{44} is of the form (7.176), so that we need only to estimate the scalar coefficient $\bar{M}_{44}$. After inserting (8.154) into (8.153), we obtain a double sum over particles, where only those pairs of particles contribute that interact during the time τ_{s}. Therefore, each particle contributes with the collision probability, τ_{s}/τ, where a collision leads to an almost complete change of momentum, so that we have

$$\bar{M}_{44} \sim \frac{V}{k_{\mathrm{B}}\tau_{\mathrm{s}}}\frac{1}{N^2}N\frac{\tau_{\mathrm{s}}}{\tau} = \frac{1}{nk_{\mathrm{B}}\tau},$$

except for prefactors. We then obtain for the viscosity [see (7.179)]

$$\eta = \frac{T}{\bar{M}_{44}} \sim nk_{\mathrm{B}}T\tau,$$

in agreement with (7.136).

F.9 APPENDICES

Exercise 208
The solution to this exercise is based on the explicit expression (B.18) for the exterior derivative of a differential form (B.17). It is convenient to start from

$$\omega = \omega_{jj_1\ldots j_k} d\xi^j \otimes d\xi^{j_1} \otimes \ldots \otimes d\xi^{j_k},$$

because, for forms of degree zero, the result is trivial and, for one-forms, we have already established (B.25). We then obtain

$$\begin{aligned} i_{\mathsf{v}} i_{\bar{\mathsf{v}}} d\omega &= \mathsf{v}^j \bar{\mathsf{v}}^n \left[\frac{\partial \omega_{jj_1\ldots j_k}}{\partial \xi^n} - \frac{\partial \omega_{nj_1\ldots j_k}}{\partial \xi^j} - \sum_{l=1}^{k} (-1)^l \frac{\partial \omega_{njj_1\ldots \not{j_l} \ldots j_k}}{\partial \xi^{j_l}} \right] \\ &\times \; d\xi^{j_1} \otimes \ldots \otimes d\xi^{j_k}, \end{aligned}$$

$$i_{\bar{\mathsf{v}}} d i_{\mathsf{v}} \omega = \bar{\mathsf{v}}^n \left[\frac{\partial (\mathsf{v}^j \omega_{jj_1\ldots j_k})}{\partial \xi^n} - \sum_{l=1}^{k} (-1)^l \frac{\partial (\mathsf{v}^j \omega_{njj_1\ldots \not{j_l} \ldots j_k})}{\partial \xi^{j_l}} \right] d\xi^{j_1} \otimes \ldots \otimes d\xi^{j_k},$$

and

$$d i_{\bar{\mathsf{v}}} i_{\mathsf{v}} \omega = \sum_{l=1}^{k} (-1)^l \frac{\partial (\mathsf{v}^j \bar{\mathsf{v}}^n \omega_{njj_1\ldots \not{j_l} \ldots j_k})}{\partial \xi^{j_l}} d\xi^{j_1} \otimes \ldots \otimes d\xi^{j_k}.$$

By adding up all the terms on the left-hand side of (B.26), we obtain after a number of cancellations:

$$\left(\mathsf{v}^n \frac{\partial \bar{\mathsf{v}}^j}{\partial \xi^n} - \bar{\mathsf{v}}^n \frac{\partial \mathsf{v}^j}{\partial \xi^n} \right) \omega_{jj_1\ldots j_k} d\xi^{j_1} \otimes \ldots \otimes d\xi^{j_k},$$

which, according to (B.10), is the right-hand side of (B.26).

Exercise 209
For one-forms ω, (B.31) can be rewritten as

$$i_{\mathsf{g}} d_{\bar{\mathsf{g}}} \omega - i_{\bar{\mathsf{g}}} d i_{\mathsf{g}} \omega = i_{[\mathsf{g},\bar{\mathsf{g}}]} \omega,$$

where (B.29) and (B.36) have been used. Further simplification of the notation yields

$$i_{\mathsf{g}} d_{\bar{\mathsf{g}}} \omega - d_{\bar{\mathsf{g}}} \omega(\mathsf{g}) = -\omega([\bar{\mathsf{g}}, \mathsf{g}]).$$

The definition (B.38) is chosen such that we have the product rule

$$d_{\bar{\mathsf{g}}}[\omega(\mathsf{g})] = [d_{\bar{\mathsf{g}}} \omega](\mathsf{g}) + \omega(d_{\bar{\mathsf{g}}} \mathsf{g}).$$

Exercise 210
The evaluation of the exterior derivative of the two-form Ω is based on (B.31), which,

together with (B.29) and (B.30), allows us to reduce the problem to differential forms of lower degree. In a first step, we obtain

$$\begin{aligned} d\Omega(\mathsf{g},\bar{\mathsf{g}},\bar{\bar{\mathsf{g}}}) &= i_{\bar{\bar{\mathsf{g}}}} i_{\bar{\mathsf{g}}} i_{\mathsf{g}} d\Omega = i_{\bar{\bar{\mathsf{g}}}} (i_{\mathsf{g}} d i_{\bar{\mathsf{g}}} - i_{\bar{\mathsf{g}}} d i_{\mathsf{g}} - d i_{\bar{\mathsf{g}}} i_{\mathsf{g}} - i_{[\mathsf{g},\bar{\mathsf{g}}]})\Omega \\ &= i_{\bar{\bar{\mathsf{g}}}} i_{\mathsf{g}} d i_{\bar{\mathsf{g}}}\Omega - i_{\bar{\bar{\mathsf{g}}}} i_{\bar{\mathsf{g}}} d i_{\mathsf{g}}\Omega - d_{\bar{\bar{\mathsf{g}}}}[\Omega(\mathsf{g},\bar{\mathsf{g}})] + \Omega(\bar{\bar{\mathsf{g}}},[\mathsf{g},\bar{\mathsf{g}}]). \end{aligned}$$

In the second step, we further simplify $i_{\bar{\bar{\mathsf{g}}}} i_{\mathsf{g}} d i_{\bar{\mathsf{g}}}\Omega$ and $i_{\bar{\bar{\mathsf{g}}}} i_{\bar{\mathsf{g}}} d i_{\mathsf{g}}\Omega$. We then obtain

$$i_{\bar{\bar{\mathsf{g}}}} i_{\mathsf{g}} d i_{\bar{\mathsf{g}}}\Omega = (i_{\mathsf{g}} d i_{\bar{\bar{\mathsf{g}}}} - i_{\bar{\bar{\mathsf{g}}}} d i_{\mathsf{g}} - i_{[\mathsf{g},\bar{\bar{\mathsf{g}}}]}) i_{\bar{\mathsf{g}}}\Omega = d_{\mathsf{g}}[\Omega(\bar{\mathsf{g}},\bar{\bar{\mathsf{g}}})] + d_{\bar{\bar{\mathsf{g}}}}[\Omega(\mathsf{g},\bar{\mathsf{g}})] + \Omega(\bar{\mathsf{g}},[\bar{\bar{\mathsf{g}}},\mathsf{g}]),$$

and, by exchanging g and $\bar{\mathsf{g}}$,

$$i_{\bar{\bar{\mathsf{g}}}} i_{\bar{\mathsf{g}}} d i_{\mathsf{g}}\Omega = d_{\bar{\mathsf{g}}}[\Omega(\mathsf{g},\bar{\bar{\mathsf{g}}})] - d_{\bar{\bar{\mathsf{g}}}}[\Omega(\mathsf{g},\bar{\mathsf{g}})] + \Omega(\mathsf{g},[\bar{\bar{\mathsf{g}}},\bar{\mathsf{g}}]).$$

By combining these results we arrive at (B.39).

Exercise 211
From the definition (B.36), we obtain

$$d_{\mathsf{g}}\Omega(\bar{\mathsf{g}},\bar{\bar{\mathsf{g}}}) = i_{\bar{\bar{\mathsf{g}}}} i_{\bar{\mathsf{g}}} i_{\mathsf{g}} d\Omega + i_{\bar{\bar{\mathsf{g}}}} i_{\bar{\mathsf{g}}} d i_{\mathsf{g}}\Omega.$$

Compared with the solution of Exercise 210, the terms corresponding to $i_{\bar{\bar{\mathsf{g}}}} i_{\bar{\mathsf{g}}} d i_{\mathsf{g}}\Omega$ need not be subtracted, so that we arrive at (B.40). In the standard setting of differential forms, the derivative of $\bar{\mathsf{v}}$ in the direction of v is given by the Jacobi-Lie bracket (B.10), implying

$$\left[\frac{\partial}{\partial \xi^j}, \mathsf{v}\right] = \frac{\partial \mathsf{v}^l}{\partial \xi^j}\frac{\partial}{\partial \xi^l}.$$

By inserting $\mathsf{g} = \mathsf{v}$, $\bar{\mathsf{g}} = \partial/\partial\xi^j$, and $\bar{\bar{\mathsf{g}}} = \partial/\partial\xi^k$ into (B.40), we obtain

$$(d_{\mathsf{v}}\Omega)_{jk} = \mathsf{v}^l \frac{\partial \Omega_{jk}}{\partial \xi^l} + \Omega\left(\frac{\partial}{\partial \xi^j}, \frac{\partial \mathsf{v}^l}{\partial \xi^k}\frac{\partial}{\partial \xi^l}\right) + \Omega\left(\frac{\partial \mathsf{v}^l}{\partial \xi^j}\frac{\partial}{\partial \xi^l}, \frac{\partial}{\partial \xi^k}\right),$$

which, in view of the linearity of differential forms, leads us to (B.41).

Exercise 212
By using (B.46) and the product rule (see Exercise 209), we obtain

$$d_{\mathsf{v}}[\Omega(\bar{\mathsf{v}},\bar{\bar{\mathsf{v}}})] = d_{\mathsf{v}}[\bar{\Omega}(\bar{\mathsf{v}})\cdot\bar{\bar{\mathsf{v}}})] = i_{\bar{\bar{\mathsf{v}}}} d_{\mathsf{v}}\bar{\Omega}(\bar{\mathsf{v}}) + \bar{\Omega}(\bar{\mathsf{v}})\cdot d_{\mathsf{v}}\bar{\bar{\mathsf{v}}} = i_{\bar{\bar{\mathsf{v}}}} d_{\mathsf{v}}\bar{\Omega}(\bar{\mathsf{v}}) - \Omega(\bar{\mathsf{v}},[\bar{\bar{\mathsf{v}}},\mathsf{v}]).$$

From this result and cyclic permutations, we can use (B.39) to show that $d\Omega(\mathsf{v},\bar{\mathsf{v}},\bar{\bar{\mathsf{v}}}) = 0$ translates into (B.47).

Exercise 213
Equation (B.71) follows immediately by realizing that the inner product operator i_{g} requires insertion of the vector field g. To establish (B.72), we evaluate $i_{\bar{\mathsf{g}}} d_{\bar{\bar{\mathsf{g}}}}\omega$ by means of (B.36) and (B.31),

$$i_{\bar{\mathsf{g}}} d_{\bar{\bar{\mathsf{g}}}}\omega = i_{\bar{\mathsf{g}}} i_{\bar{\bar{\mathsf{g}}}} d\omega + i_{\bar{\mathsf{g}}} d i_{\bar{\bar{\mathsf{g}}}}\omega = i_{[\bar{\mathsf{g}},\bar{\bar{\mathsf{g}}}]}\omega + i_{\bar{\bar{\mathsf{g}}}} d i_{\bar{\mathsf{g}}}\omega = \omega([\bar{\mathsf{g}},\bar{\bar{\mathsf{g}}}]) + d_{\bar{\bar{\mathsf{g}}}}\omega(\bar{\mathsf{g}}).$$

Equation (B.72) then follows by adding the cyclic permutations of this result.

Exercise 214

According to (B.74), for a given Poisson operator L, a set of base vectors for the corresponding Dirac structure can be obtained by choosing a complete set of base vectors of the space of differential forms (D_1) and extending them by the components in the space of vector fields (G) by means of L. In our example, we have $\mathsf{G} = \mathsf{D}_1 = \mathbb{R}^4$ and from the standard base vectors of $\mathsf{D}_1 = \mathbb{R}^4$ we obtain, with the L given in the solution to Exercise 3 (with $m = A_c = 1$), the following base vectors of $\mathsf{D} \subseteq \mathsf{G} \oplus \mathsf{D}_1$:

$$\begin{pmatrix} 0 \\ -1 \\ 0 \\ 0 \\ \hline 1 \\ 0 \\ 0 \\ 0 \end{pmatrix}, \quad \begin{pmatrix} 1 \\ 0 \\ -p_1 \\ p_2 \\ \hline 0 \\ 1 \\ 0 \\ 0 \end{pmatrix}, \quad \begin{pmatrix} 0 \\ p_1 \\ 0 \\ 0 \\ \hline 0 \\ 0 \\ 1 \\ 0 \end{pmatrix}, \quad \begin{pmatrix} 0 \\ -p_2 \\ 0 \\ 0 \\ \hline 0 \\ 0 \\ 0 \\ 1 \end{pmatrix},$$

where the upper and lower components corresponding to G and D_1, respectively, have been separated by a line. The elements of the four-dimensional Dirac structure for this problem can be written explicitly as linear combinations of the above base vectors; we thus have an example of an *image representation* of a Dirac structure. For any two vectors from the above list, including two identical ones, the antisymmetry (B.68) or (B.71) of Dirac structures implies the symbolically written identity

$$\text{upper} \cdot \text{lower} + \text{lower} \cdot \text{upper} = 0.$$

By exchanging the upper and lower components and using properly transposed vectors, we hence obtain the following matrix defining a linear mapping with the Dirac structure spanned by the above base vectors as a kernel,

$$\left(\begin{array}{cccc|cccc} 1 & 0 & 0 & 0 & 0 & -1 & 0 & 0 \\ 0 & 1 & 0 & 0 & 1 & 0 & -p_1 & p_2 \\ 0 & 0 & 1 & 0 & 0 & p_1 & 0 & 0 \\ 0 & 0 & 0 & 1 & 0 & -p_2 & 0 & 0 \end{array}\right),$$

in which the unit matrix and $-L = L^T$ can be recognized nicely. We thus have an example of a *kernel representation* of a Dirac structure. Through $p_1(q, E_1, E_2)$ and $p_2(q, E_1, E_2)$, the Dirac structure of this exercise depends on the point of the manifold.

Exercise 215

The more detailed transformation looks as follows,

$$d\bar{\xi}^j \wedge d\bar{\xi}_j^* \quad = \quad d\xi^j \wedge d\xi_j^* + \frac{\partial\bar{\xi}^j}{\partial\xi^k} \frac{\partial^2 \xi^i}{\partial\xi^l \partial\bar{\xi}^j} \xi_i^* \, d\xi^k \wedge d\xi^l$$

$$
\begin{aligned}
&= d\xi^j \wedge d\xi_j^* + \left[\frac{\partial}{\partial \xi^l}\left(\frac{\partial \bar{\xi}^j}{\partial \xi^k}\frac{\partial \xi^i}{\partial \bar{\xi}^j}\right) - \frac{\partial^2 \bar{\xi}^j}{\partial \xi^l \partial \xi^k}\frac{\partial \xi^i}{\partial \bar{\xi}^j}\right] \xi_i^* d\xi^k \wedge d\xi^l \\
&= d\xi^j \wedge d\xi_j^* - \frac{\partial^2 \bar{\xi}^j}{\partial \xi^l \partial \xi^k}\frac{\partial \xi^i}{\partial \bar{\xi}^j} \xi_i^* d\xi^k \wedge d\xi^l = d\xi^j \wedge d\xi_j^*,
\end{aligned}
$$

in view of the skew-symmetry of $d\xi^k \wedge d\xi^l$.

Exercise 216

With the use of (B.6), the condition (B.107) becomes

$$\frac{\partial \gamma_{kl}}{\partial \xi^i} = \Gamma_{ik}^m \gamma_{ml} + \Gamma_{il}^m \gamma_{km}.$$

By multiplying with γ^{ij} and summing over i, we then obtain

$$\gamma^{ij}\frac{\partial \gamma_{kl}}{\partial \xi^i} = \gamma^{ij}\Gamma_{ik}^m \gamma_{ml} + \gamma^{ij}\Gamma_{il}^m \gamma_{km}. \tag{F.67}$$

If we exchange i and k and also i and l before multiplying with γ^{ij}, then we obtain

$$\gamma^{ij}\frac{\partial \gamma_{il}}{\partial \xi^k} = \gamma^{ij}\Gamma_{ki}^m \gamma_{ml} + \Gamma_{kl}^j \tag{F.68}$$

and

$$\gamma^{ij}\frac{\partial \gamma_{ki}}{\partial \xi^l} = \Gamma_{lk}^j + \gamma^{ij}\Gamma_{li}^m \gamma_{km}. \tag{F.69}$$

After adding (F.68) and (F.69) and subtracting (F.67), we arrive at (B.108).

Exercise 217

We consider the small transformations

$$g_t(\xi) = \xi + t\boldsymbol{v}_\xi, \quad \bar{g}_{t'}(\xi) = \xi + t'\bar{\boldsymbol{v}}_\xi.$$

If we keep only those second-order terms involving the product tt', then we have

$$\bar{g}_{t'} \circ g_t(\xi) = \bar{g}_{t'}(\xi + t\boldsymbol{v}_\xi) = \xi + t\boldsymbol{v}_\xi + t'\bar{\boldsymbol{v}}_\xi + tt' v_\xi^l \frac{\partial \bar{\boldsymbol{v}}_\xi}{\partial \xi^l}$$

and thus

$$\bar{g}_{t'} \circ g_t - g_t \circ \bar{g}_{t'} = tt'[\boldsymbol{v}, \bar{\boldsymbol{v}}].$$

The remarks after (B.11) lead to the desired result.

Exercise 218

With $\bar{\xi} = f_1(\xi)$, the left-hand side of (B.128) for the transformation rule (B.134) is

$$\bar{F}\left(\frac{\partial f_1(\xi)}{\partial \xi^k} q^k, \bar{\xi}, f_2\right) = \frac{\partial f_2(\bar{\xi})}{\partial \bar{\xi}^j}\frac{\partial f_1^j(\xi)}{\partial \xi^k} q^k,$$

which, according to the chain rule, indeed coincides with the right-hand side of (B.128). For the transformation rule (B.135), the linearity of $\bar{F}(q,\xi,f)$ in q and the consistency of $\bar{F}(q,\xi,f)$ lead to

$$\begin{aligned} F(F(q,\xi,f_1),\bar{\xi},f_2) &\doteq \frac{\bar{F}(\bar{F}(q,\xi,f_1),\bar{\xi},f_2)}{|\bar{F}(\bar{F}(q,\xi,f_1),\bar{\xi},f_2)|} \\ &= \frac{\bar{F}(q,\xi,f_2\circ f_1)}{|\bar{F}(q,\xi,f_2\circ f_1)|} = F(q,\xi,f_2\circ f_1). \end{aligned}$$

We further have

$$\frac{\delta \bar{F}^l(q,\xi,f)}{\delta f^j(\xi')} = \delta^l_j q^k \frac{\partial \delta(\xi-\xi')}{\partial \xi^k},$$

implying

$$\bar{\mathcal{L}}^l_j(q,\xi) = \delta^l_j q^k \frac{\partial}{\partial \xi^k}$$

and

$$\frac{\delta F^l(q,\xi,f)}{\delta f^j(\xi')} = \left[\delta^l_j - \frac{\bar{F}^l(q,\xi,f)q^n}{|\bar{F}(q,\xi,f)|^2}\frac{\partial f^j(\xi)}{\partial \xi^n}\right]\frac{q^k}{|\bar{F}(q,\xi,f)|}\frac{\partial \delta(\xi-\xi')}{\partial \xi^k},$$

which leads to

$$\mathcal{L}^l_j(q,\xi) = \left(\delta^l_j - \frac{q^l q^j}{|q|^2}\right)\frac{q^k}{|q|}\frac{\partial}{\partial \xi^k}.$$

Note that, for the transformation rule (B.135), the condition (B.129) actually requires $|q| = 1$.

Exercise 219

By definition of the functional considered in this exercise, we have

$$\rho + \lambda\delta\rho \mapsto \rho(\boldsymbol{r}') + \lambda\delta\rho(\boldsymbol{r}') = \rho(\boldsymbol{r}') + \lambda\int \delta\rho(\boldsymbol{r})\delta(\boldsymbol{r}-\boldsymbol{r}')d^3r.$$

After differentiation with respect to λ, a comparison with the definition (C.1) gives the desired result.

Exercise 220

The first two contributions to $\partial\phi^{(S)}/\partial\boldsymbol{r}_k$ result from the term $j = k$ in the sum (C.15) and a further one for $j = k-1$,

$$\begin{aligned} \frac{\partial\phi^{(S)}}{\partial\boldsymbol{r}_k} &= \left.\frac{\partial f(\boldsymbol{r},\Delta\boldsymbol{r})}{\partial\boldsymbol{r}}\right|_{\boldsymbol{r}=\boldsymbol{r}_k,\Delta\boldsymbol{r}=\boldsymbol{r}_{k+1}-\boldsymbol{r}_k} \\ &\quad - \left\{\left.\frac{\partial f(\boldsymbol{r},\Delta\boldsymbol{r})}{\partial\Delta\boldsymbol{r}}\right|_{\boldsymbol{r}=\boldsymbol{r}_k,\Delta\boldsymbol{r}=\boldsymbol{r}_{k+1}-\boldsymbol{r}_k} - \left.\frac{\partial f(\boldsymbol{r},\Delta\boldsymbol{r})}{\partial\Delta\boldsymbol{r}}\right|_{\boldsymbol{r}=\boldsymbol{r}_{k-1},\Delta\boldsymbol{r}=\boldsymbol{r}_k-\boldsymbol{r}_{k-1}}\right\}, \end{aligned}$$

for $k = 2,\ldots,N-1$. This result should be compared with (C.14). In the field-theoretic formulation of polymer models, in particular, for the purpose of renormalization-group calculations, one actually uses the continuous formulation of (C.15).[21]

Exercise 221

By applying the formula (C.14), we obtain

$$\frac{\delta\Sigma(T)}{\delta T(\boldsymbol{r})} = \left.\frac{df}{dT}\right|_{T=T(\boldsymbol{r})} \frac{\partial T(\boldsymbol{r})}{\partial \boldsymbol{r}} \cdot \frac{\partial T(\boldsymbol{r})}{\partial \boldsymbol{r}} - 2\frac{\partial}{\partial \boldsymbol{r}} \cdot \left[f(T(\boldsymbol{r}))\frac{\partial T(\boldsymbol{r})}{\partial \boldsymbol{r}}\right].$$

Exercise 222

Introducing the functional

$$\delta^2\Sigma = \left.\frac{d^2}{d\lambda d\lambda'}\Sigma(T+\lambda\delta T+\lambda'\delta T')\right|_{\lambda=0,\lambda'=0},$$

which depends linearly on δT and $\delta T'$, an elementary calculation gives

$$\begin{aligned}\delta^2\Sigma &= \int\left[2f\frac{\partial\delta T}{\partial\boldsymbol{r}}\cdot\frac{\partial\delta T'}{\partial\boldsymbol{r}} + 2\frac{df}{dT}\frac{\partial T}{\partial\boldsymbol{r}}\cdot\left(\delta T\frac{\partial\delta T'}{\partial\boldsymbol{r}} + \delta T'\frac{\partial\delta T}{\partial\boldsymbol{r}}\right)\right.\\ &+ \left.\frac{d^2 f}{dT^2}\left(\frac{\partial T}{\partial\boldsymbol{r}}\right)^2\delta T\delta T'\right]d^3r.\end{aligned}$$

In view of the second-order derivative operator in (C.21), we rewrite this expression in the slightly more complicated form

$$\begin{aligned}\delta^2\Sigma &= \int\left\{2f^{1-2\alpha}\frac{\partial(f^\alpha\delta T)}{\partial\boldsymbol{r}}\cdot\frac{\partial(f^\alpha\delta T')}{\partial\boldsymbol{r}} + 2(1-\alpha)\frac{df}{dT}\frac{\partial T}{\partial\boldsymbol{r}}\cdot\frac{\partial(\delta T\delta T')}{\partial\boldsymbol{r}}\right.\\ &+ \left.\left[\frac{d^2 f}{dT^2} - 2\frac{\alpha^2}{f}\left(\frac{df}{dT}\right)^2\right]\left(\frac{\partial T}{\partial\boldsymbol{r}}\right)^2\delta T\delta T'\right\}d^3r.\end{aligned}$$

Assuming $\delta T = \delta T' = 0$ at the boundaries, we obtain after an integration by parts for the term involving $\partial(\delta T\delta T')/\partial\boldsymbol{r}$

$$\begin{aligned}\delta^2\Sigma &= \int\left\{2f^{1-2\alpha}\frac{\partial(f^\alpha\delta T)}{\partial\boldsymbol{r}}\cdot\frac{\partial(f^\alpha\delta T')}{\partial\boldsymbol{r}} - 2(1-\alpha)\frac{df}{dT}\frac{\partial^2 T}{\partial\boldsymbol{r}^2}\delta T\delta T'\right.\\ &+ \left.\left[-(1-2\alpha)\frac{d^2 f}{dT^2} - 2\frac{\alpha^2}{f}\left(\frac{df}{dT}\right)^2\right]\left(\frac{\partial T}{\partial\boldsymbol{r}}\right)^2\delta T\delta T'\right\}d^3r.\end{aligned}$$

At this point, we use the condition that the first functional derivative vanishes, which, according to Exercise 221, is

$$2f\frac{\partial^2 T}{\partial\boldsymbol{r}^2} = -\frac{df}{dT}\left(\frac{\partial T}{\partial\boldsymbol{r}}\right)^2,$$

[21] See, for example, Section 4.1 of Doi & Edwards, *Theory of Polymer Dynamics* (Clarendon, 1986), or Section XI.1.6 of de Gennes, *Scaling Concepts in Polymer Physics* (Cornell, 1979).

further leading to

$$\begin{aligned}\delta^2\Sigma &= \int \left\{ 2f^{1-2\alpha} \frac{\partial(f^\alpha \delta T)}{\partial \boldsymbol{r}} \cdot \frac{\partial(f^\alpha \delta T')}{\partial \boldsymbol{r}} \right. \\ &+ \left. \left[(1-\alpha-2\alpha^2)\frac{1}{f}\left(\frac{df}{dT}\right)^2 - (1-2\alpha)\frac{d^2 f}{dT^2} \right] \left(\frac{\partial T}{\partial \boldsymbol{r}}\right)^2 \delta T \delta T' \right\} d^3 r.\end{aligned} \tag{F.70}$$

According to the definition (C.17), this expression corresponds to the second functional derivative (C.21). Note that a particularly simple expression for the second functional derivative results for the choice $\alpha = 1/2$. However, to establish an inequality like (C.19), it may be more convenient to choose other values of α in (F.70) for $\delta T = \delta T'$. We then find that the stationary points of (C.16) are indeed minima.

References

Aharoni, S. M. On entanglements of flexible and rodlike polymers. *Macromolecules* 16 (1983) 1722–1728.

Akkermans, R. L. C., and W. J. Briels. A structure-based coarse-grained model for polymer melts. *J. Chem. Phys.* 114 (2001) 1020–1031.

Alder, B. J., and W. E. Alley. Generalized hydrodynamics. *Physics Today* 37/1 (1984) 56–63.

Allen, M. P., and D. J. Tildesley. *Computer Simulation of Liquids* (Clarendon, Oxford, 1987).

Andersen, H. C. Molecular dynamics simulations at constant pressure and/or temperature. *J. Chem. Phys.* 72 (1980) 2384–2393.

Attard, P. Stochastic molecular dynamics: a combined Monte Carlo and molecular dynamics technique for isothermal simulations. *J. Chem. Phys.* 116 (2002) 9616–9619.

Bao, D., J. Marsden, and R. Walton. The Hamiltonian structure of general relativistic perfect fluids. *Commun. Math. Phys.* 99 (1985) 319–345.

Barnes, H. A., J. F. Hutton, and K. Walters. *An Introduction to Rheology*. Rheology Series, Volume 3 (Elsevier, Amsterdam, 1989).

Benettin, G., and A. Giorgilli. On the Hamiltonian interpolation of near to the identity symplectic mappings with application to symplectic integration algorithms. *J. Stat. Phys.* 74 (1994) 1117–1143.

Beris, A. N. Bracket formulation as a source for the development of dynamic equations in continuum mechanics. *J. Non-Newtonian Fluid Mech.* 96 (2001) 119–136.

Beris, A. N., and B. J. Edwards. *The Thermodynamics of Flowing Systems* (Oxford University Press, New York, 1994).

Beris, A. N., and V. G. Mavrantzas. On the compatibility between various macroscopic formalisms for the concentration and flow of dilute polymer solutions. *J. Rheol.* 38 (1994) 1235–1250.

Binder, K., and D. W. Heermann. *Monte Carlo Simulation in Statistical Physics*. Springer Series in Solid-State Sciences, Volume 80 (Springer, Berlin, 1992), 2nd edition.

Bird, R. B., R. C. Armstrong, and O. Hassager. *Fluid Mechanics*, Volume 1 of *Dynamics of Polymeric Liquids* (Wiley, New York, 1987), 2nd edition.

Bird, R. B., C. F. Curtiss, R. C. Armstrong, and O. Hassager. *Kinetic Theory*, Volume 2 of *Dynamics of Polymeric Liquids* (Wiley, New York, 1987), 2nd edition.

Bird, R. B., and C. F. Curtiss. Fascinating polymeric liquids. *Physics Today* 37/1 (1984) 36–43.

Bird, R. B., H. H. Saab, and C. F. Curtiss. A kinetic theory for polymer melts. 3. Elongational flows. *J. Phys. Chem.* 86 (1982) 1102–1106.

Bird, R. B., H. H. Saab, and C. F. Curtiss. A kinetic theory for polymer melts. IV. Rheological properties for shear flows. *J. Chem. Phys.* 77 (1982) 4747–4757.

Bird, R. B., W. E. Stewart, and E. N. Lightfoot. *Transport Phenomena* (Wiley, New York, 2001), 2nd edition.

Bogolyubov, N. N. Problems of a dynamical theory in statistical physics. In J. de Boer and G. E. Uhlenbeck (Editors), *Collected Papers*, Volume I in Studies in Statistical Mechanics (North-Holland, Amsterdam, 1962) p. 1–118.

Bolder, H. Deformation of tensor fields described by time-dependent mappings. *Arch. Rational Mech. Anal.* 35 (1969) 321–341.

Boltzmann, L. Weitere Studien über das Wärmegleichgewicht unter Gasmolekülen. *Wien. Ber.* 66 (1872) 275–370.

Boltzmann, L. *Vorlesungen über Gastheorie, I. Theil* (Barth, Leipzig, 1896).

Boltzmann, L. *Lectures on Gas Theory* (Dover, New York, 1995).

Bond, S. D., B. J. Leimkuhler, and B. B. Laird. The Nosé-Poincaré method for constant temperature molecular dynamics. *J. Comput. Phys.* 151 (1999) 114–134.

Boon, J. P., and S. Yip. *Molecular Hydrodynamics* (Dover, New York, 1991).

Brenner, H. Navier-Stokes revisited (2005). Physica A, in press. A previous version of the manuscript, entitled "Inequality of Eulerian and Lagragian velocities in molecularly inhomogeneous fluids," was submitted to Phys. Rev. E in October 2002.

Bugaenko, N. N., A. N. Gorban, and I. V. Karlin. Universal expansion of three-particle distribution function. *Theor. Math. Phys.* 88 (1991) 977–985.

Burke, W. L. *Applied Differential Geometry* (Cambridge University Press, Cambridge, 1985).

Callen, H. B. *Thermodynamics and an Introduction to Thermostatistics* (Wiley, New York, 1985), 2nd edition.

Casas-Vázquez, J., and D. Jou. Temperature in non-equilibrium states: a review of open problems and current proposals. *Rep. Prog. Phys.* 66 (2003) 1937–2023.

Casimir, H. B. G. On Onsager's principle of microscopic reversibility. *Rev. Mod. Phys.* 17 (1945) 343–350.

Cattaneo, C. Sulla conduzione del calore. *Atti Sem. Mat. Fis. Univ. Modena* 3 (1948) 3–21.

Cercignani, C. *The Boltzmann Equation and Its Applications*. Applied Mathematical Sciences, Volume 67 (Springer, New York, 1988).

Clebsch, A. Ueber die Integration der hydrodynamischen Gleichungen. *J. Reine Angew. Math.* 56 (1859) 1–10.

Coleman, B. D., and W. Noll. The thermodynamics of elastic materials with heat conduction and viscosity. *Arch. Rational Mech. Anal.* 13 (1963) 167–178.

Coley, A. A., R. J. van den Hoogen, and R. Maartens. Qualitative viscous cosmology. *Phys. Rev. D* 54 (1996) 1393–1397.

Courant, T. J. Dirac manifolds. *Trans. Am. Math. Soc.* 319 (1990) 631–661.

Creutz, M. Monte Carlo study of quantized SU(2) gauge theory. *Phys. Rev. D* 21 (1980) 2308–2315.

Cross, M. C., and P. C. Hohenberg. Pattern formation outside of equilibrium. *Rev. Mod. Phys.* 65 (1993) 851–1112.

Curtiss, C. F., and R. B. Bird. A kinetic theory for polymer melts. I. The equation for the single-link distribution function. *J. Chem. Phys.* 74 (1981) 2016–2025.

Curtiss, C. F., and R. B. Bird. A kinetic theory for polymer melts. II. The stress tensor and the rheological equation of state. *J. Chem. Phys.* 74 (1981) 2026–2033.

Curtiss, C. F., and R. B. Bird. Diffusion-stress relations in polymer mixtures. *J. Chem. Phys.* 111 (1999) 10362–10370.

Da Prato, G., and J. Zabczyk. *Stochastic Equations in Infinite Dimensions*. Encyclopedia of Mathematics and Its Applications, Volume 45 (Cambridge University Press, Cambridge, 1992).

de Gennes, P. G. Reptation of a polymer chain in the presence of fixed obstacles. *J. Chem. Phys.* 55 (1971) 572–579.

de Gennes, P. G. *Scaling Concepts in Polymer Physics* (Cornell University Press, Ithaca, 1979).

de Groot, S. R. *Thermodynamics of Irreversible Processes* (North-Holland, Amsterdam, 1951).

de Groot, S. R., and P. Mazur. *Non-Equilibrium Thermodynamics* (Dover, New York, 1984), 2nd edition.

de Groot, S. R., W. A. van Leeuwen, and C. G. van Weert. *Relativistic Kinetic Theory* (North-Holland, Amsterdam, 1980).

de Pablo, J. J., and H. C. Öttinger. An atomistic approach to general equation for the nonequilibrium reversible-irreversible coupling. *J. Non-Newtonian Fluid Mech.* 96 (2001) 137–162.

Dealy, J. M. Official nomenclature for material functions describing the response of a viscoelastic fluid to various shearing and extensional deformations. *J. Rheol.* 39 (1995) 253–265.

Doi, M. A constitutive equation derived from the model of Doi and Edwards for concentrated polymer solutions and polymer melts. *J. Polym. Sci., Polym. Phys. Ed.* 18 (1980) 2055–2067.

Doi, M., and S. F. Edwards. Dynamics of concentrated polymer systems. Part 1. Brownian motion in the equilibrium state. *J. Chem. Soc. Faraday Trans. II* 74 (1978) 1789–1801.

Doi, M., and S. F. Edwards. Dynamics of concentrated polymer systems. Part 2. Molecular motion under flow. *J. Chem. Soc. Faraday Trans. II* 74 (1978) 1802–1817.

Doi, M., and S. F. Edwards. Dynamics of concentrated polymer systems. Part 3. The constitutive equation. *J. Chem. Soc. Faraday Trans. II* 74 (1978) 1818–1832.

Doi, M., and S. F. Edwards. Dynamics of concentrated polymer systems. Part 4. Rheological properties. *J. Chem. Soc. Faraday Trans. II* 75 (1979) 38–54.

Doi, M., and S. F. Edwards. *The Theory of Polymer Dynamics*. International Series of Monographs on Physics, Volume 73 (Clarendon, Oxford, 1986).

Doi, M., and T. Ohta. Dynamics and rheology of complex interfaces. I. *J. Chem. Phys.* 95 (1991) 1242–1248.

Dorfman, I. *Dirac Structures and Integrability of Nonlinear Evolution Equations*. Nonlinear Science: Theory and Applications (Wiley, Chichester, 1993).

Dressler, M., and B. J. Edwards. The influence of matrix viscoelasticity on the rheology of polymer blends. *Rheol. Acta* 43 (2004) 257–282.

Dressler, M., B. J. Edwards, and H. C. Öttinger. Macroscopic thermodynamics of flowing polymeric liquids. *Rheol. Acta* 38 (1999) 117–136.

Eckart, C. The thermodynamics of irreversible processes. III. Relativistic theory of the simple fluid. *Phys. Rev.* 58 (1940) 919–924.

Edwards, B. J. An analysis of single and double generator thermodynamic formalisms for the macroscopic description of complex fluids. *J. Non-Equilib. Thermodyn.* 23 (1998) 301–333.

Edwards, B. J., A. N. Beris, and V. G. Mavrantzas. A model with two coupled Maxwell modes. *J. Rheol.* 40 (1996) 917–942.

Edwards, B. J., A. N. Beris, and H. C. Öttinger. An analysis of single and double generator thermodynamic formalisms for complex fluids. II. The microscopic description. *J. Non-Equilib. Thermodyn.* 23 (1998) 334–350.

Edwards, B. J., H. C. Öttinger, and R. J. J. Jongschaap. On the relationships between thermodynamic formalisms for complex fluids. *J. Non-Equilib. Thermodyn.* 22 (1997) 356–373.

Ellero, M., P. Español, and E. G. Flekkøy. Thermodynamically consistent fluid particle model for viscoelastic flows. *Phys. Rev. E* 68 (2003) 041504, 1–19.

Español i Garrigós, J. Initial ensemble densities through the maximization of entropy. *Phys. Lett. A* 146 (1990) 21–24.

Español, P. Thermodynamics for a van der Waals fluid. *J. Chem. Phys.* 115 (2001) 5392–5403.

Español, P., and M. Revenga. Smoothed dissipative particle dynamics. *Phys. Rev. E* 67 (2003) 026705, 1–12.

Español, P., M. Serrano, and H. C. Öttinger. Thermodynamically admissible form for discrete hydrodynamics. *Phys. Rev. Lett.* 83 (1999) 4542–4545.

Español, P., and C. Thieulot. Microscopic derivation of hydrodynamic equations for phase-separating fluid mixtures. *J. Chem. Phys.* 118 (2003) 9109–9127.

Español, P., and F. Vázquez. Coarse graining from coarse-grained descriptions. *Phil. Trans. R. Soc. Lond. A* 360 (2002) 383–394.

Evans, D. J., and G. P. Morriss. *Statistical Mechanics of Nonequilibrium Liquids* (Academic Press, London, 1990).

Faller, R., and J. J. de Pablo. Density of states of a binary Lenard-Jones glass. *J. Chem. Phys.* 119 (2003) 4405–4408.

Faller, R., Q. Yan, and J. J. de Pablo. Multicanonical parallel tempering. *J. Chem. Phys.* 116 (2002) 5419–5423.

Fan, X. J., and R. B. Bird. A kinetic theory for polymer melts. VI. Calculation of additional material functions. *J. Non-Newtonian Fluid Mech.* 15 (1984) 341–373.

Fang, J., M. Kröger, and H. C. Öttinger. A thermodynamically admissible reptation model for fast flows of entangled polymers. II. Model predictions for shear and elongational flows. *J. Rheol.* 44 (2000) 1293–1317.

Feller, W. *An Introduction to Probability Theory and Its Applications*, Volume 1 (Wiley, New York, 1968), 3rd edition.

Ferry, J. D. *Viscoelastic Properties of Polymers* (Wiley, New York, 1980), 3rd edition.

Fetters, L. J., D. J. Lohse, and R. H. Colby. Chain dimensions and entanglement spacings. In J. E. Mark (Editor), *Physical Properties of Polymers Handbook*, AIP Series in Polymers and Complex Materials (AIP Press, New York, 1996) p. 335–340.

Fredrickson, G. H., V. Ganesan, and F. Drolet. Field-theoretic computer simulation methods for polymers and complex fluids. *Macromolecules* 35 (2002) 16–39.

Freed, K. F. *Renormalization Group Theory of Macromolecules* (Wiley, New York, 1987).

Frisch, U. *Turbulence* (Cambridge University Press, Cambridge, 1995).

Galea, T. M., and P. Attard. Constraint method for deriving nonequilibrium molecular dynamics equations of motion. *Phys. Rev. E* 66 (2002) 041207, 1–7.

Gallavotti, G. *Statistical Mechanics: A Short Treatise*. Texts and Monographs in Physics (Springer, Berlin, 1999).

García-Colín, L. S., and P. Goldstein. Onsager's reciprocity relations for the Soret and Dufour effects. *Rev. Mex. Fís.* 44 (1998) 420–426.

Gardiner, C. W. *Handbook of Stochastic Methods for Physics, Chemistry and the Natural Sciences*. Springer Series in Synergetics, Volume 13 (Springer, Berlin, 1990), 2nd edition.

Garrido, P. L., S. Goldstein, and J. L. Lebowitz. Boltzmann entropy for dense fluids not in local equilibrium. *Phys. Rev. Lett.* 92 (2004) 050602, 1–4.

Geigenmüller, U., U. M. Titulaer, and B. U. Felderhof. The approximate nature of the Onsager-Casimir reciprocal relations. *Physica A* 119 (1983) 53–66.

Gelfand, I. M., and S. V. Fomin. *Calculus of Variations* (Prentice-Hall, Englewood Cliffs, 1963).

Ginzburg, V. L., and L. D. Landau. On the theory of superconductivity. *Zhur. Eksp. Teor. Fiz.* 20 (1950) 1064–1082.

Goldstein, H. *Classical Mechanics* (Addison-Wesley, Reading, 1980), 2nd edition.

Gorban, A. N., I. Karlin, P. Ilg, and H. C. Öttinger. Corrections and enhancements of quasi-equilibrium states. *J. Non-Newtonian Fluid Mech.* 96 (2001) 203–219.

Gorban, A. N., I. Karlin, H. C. Öttinger, and L. L. Tatarinova. Ehrenfest's argument extended to a formalism of nonequilibrium thermodynamics. *Phys. Rev. E* 63 (2001) 066124, 1–6.

Gorban, A. N., and I. V. Karlin. Thermodynamic parameterization. *Physica A* 190 (1992) 393–404.

Gorban, A. N., and I. V. Karlin. General approach to constructing models of the Boltzmann equation. *Physica A* 206 (1994) 401–420.

Gorban, A. N., and I. V. Karlin. Method of invariant manifolds and regularization of acoustic spectra. *Transp. Theory Stat. Phys.* 23 (1994) 559–632.

Gorban, A. N., and I. V. Karlin. Method of invariant manifold for chemical kinetics. *Chem. Eng. Sci.* 58 (2003) 4751–4768.

Gorban, A. N., and I. V. Karlin. Geometry of irreversibility: the film of nonequilibrium states (2004). ArXiv:cond-mat/0308331. Submitted to Chem. Eng. Sci.

Grabert, H. *Projection Operator Techniques in Nonequilibrium Statistical Mechanics* (Springer, Berlin, 1982).

Grad, H. On the kinetic theory of rarefied gases. *Commun. Pure Appl. Math.* 2 (1949) 331–407.

Grad, H. Principles of the kinetic theory of gases. In S. Flügge (Editor), *Thermodynamics of Gases*, Volume XII in Encyclopedia of Physics (Springer, Berlin, 1958) p. 205–294.

Grad, H. Asymptotic theory of the Boltzmann equation. *Phys. Fluids* 6 (1963) 147–181.

Gradshteyn, I. S., and I. M. Ryzhik. *Table of Integrals, Series and Products* (Academic Press, San Diego, 1980), 4th edition.

Grmela, M. Bracket formulation of dissipative fluid mechanics equations. *Phys. Lett. A* 102 (1984) 355–358.

Grmela, M. Particle and bracket formulations of kinetic equations. In J. E. Marsden (Editor), *Fluids and Plasmas: Geometry and Dynamics*, Volume 28 in Contemporary Mathematics. American Mathematical Society (American Mathematical Society, Providence, Rhodes Island, 1984) p. 125–132.

Grmela, M., A. Aït-Kadi, and P. G. Lafleur. Suspensions of fibres in viscoelastic fluids: rheology. *J. Chem. Phys.* 109 (1998) 6973–6981.

Grmela, M., M. Bousmina, and J.-F. Palierne. On the rheology of immiscible blends. *Rheol. Acta* 40 (2001) 560–569.

Grmela, M., G. Lebon, and D. Lhuillier. A comparative study of the coupling of flow with non-Fickean thermodiffusion. Part II: GENERIC. *J. Non-Equilib. Thermodyn.* 28 (2003) 23–50.

Grmela, M., and H. C. Öttinger. Dynamics and thermodynamics of complex fluids. I. Development of a general formalism. *Phys. Rev. E* 56 (1997) 6620–6632.

Guo, G.-J., Y.-G. Zhang, and Y.-J. Zhao. Comment on "Computation of the viscosity of a liquid from time averages of stress fluctuations." *Phys. Rev. E* 67 (2003) 043101, 1–3.

Hairer, E. Backward analysis of numerical integrators and symplectic methods. *Ann. Numer. Math.* 1 (1994) 107–132.

Hamermesh, M. *Group Theory and its Application to Physical Problems* (Dover, New York, 1989).

Hansen, J. P., and I. R. McDonald. *Theory of Simple Liquids* (Academic Press, London, 1986), 2nd edition.

Harmandaris, V. A., V. G. Mavrantzas, and D. N. Theodorou. Atomistic molecular dynamics simulation of polydisperse linear polyethylene melts. *Macromolecules* 31 (1998) 7934–7943.

Harmandaris, V. A., V. G. Mavrantzas, and D. N. Theodorou. Atomistic molecular dynamics simulation of stress relaxation upon cessation of steady-state uniaxial elongational flow. *Macromolecules* 33 (2000) 8062–8076.

Harmandaris, V. A., V. G. Mavrantzas, D. N. Theodorou, M. Kröger, J. Ramírez, H. C. Öttinger, and D. Vlassopoulos. Crossover from the Rouse to the entangled polymer melt regime: signals from long, detailed atomistic molecular dynamics simulations, supported by rheological experiments. *Macromolecules* 36 (2003) 1376–1387.

Hauge, E. H., and A. Martin-Löf. Fluctuating hydrodynamics and Brownian motion. *J. Stat. Phys.* 7 (1973) 259–281.

Herrchen, M., and H. C. Öttinger. A detailed comparison of various FENE dumbbell models. *J. Non-Newtonian Fluid Mech.* 68 (1997) 17–42.

Hess, S., and D. J. Evans. Computation of the viscosity of a liquid from time averages of stress fluctuations. *Phys. Rev. E* 64 (2001) 011207, 1–6.

Hess, S., M. Kröger, and D. J. Evans. Crossover between short- and long-time behavior of stress fluctuations and viscoelasticity of liquids. *Phys. Rev. E* 67 (2003) 042201, 1–4.

Hilbert, D. Begründung der kinetischen Gastheorie. *Math. Ann.* 72 (1912) 562–577.

Hillery, M., R. F. O'Connell, M. O. Scully, and E. P. Wigner. Distribution functions in physics: fundamentals. *Phys. Rep.* 106 (1984) 121–167.

Hirschfelder, J. O., C. F. Curtiss, and R. B. Bird. *Molecular Theory of Gases and Liquids* (Wiley, New York, 1954).

Hiscock, W. A., and L. Lindblom. Stability and causality in dissipative relativistic fluids. *Ann. Phys. (N.Y.)* 151 (1983) 466–496.

Hiscock, W. A., and L. Lindblom. Generic instabilities in first-order dissipative relativistic fluid theories. *Phys. Rev. D* 31 (1985) 725–733.

Hiscock, W. A., and J. Salmonson. Dissipative Boltzmann-Robertson-Walker cosmologies. *Phys. Rev. D* 43 (1991) 3249–3258.

Honerkamp, J. *Stochastic Dynamical Systems* (VCH Publishers, New York, 1994).

Honerkamp, J., and J. Weese. A nonlinear regularization method for the calculation of relaxation spectra. *Rheol. Acta* 32 (1993) 65–73.

Hoover, W. G. Canonical dynamics: equilibrium phase-space distributions. *Phys. Rev. A* 31 (1995) 1695–1697.

Hulsen, M. A., A. P. G. van Heel, and B. H. A. A. van den Brule. Simulation of viscoelastic flows using Brownian configuration fields. *J. Non-Newtonian Fluid Mech.* 70 (1997) 79–101.

Hütter, M. Thermodynamically consistent incorporation of the Schneider rate equations into two-phase models. *Phys. Rev. E* 64 (2001) 11209, 1–11.

Hütter, M. GENERIC treatment of compressible two-phase flow: convection mechanism of scalar morphological variables. *J. Non-Equilib. Thermodyn.* 27 (2002) 349–365.

Hütter, M., I. V. Karlin, and H. C. Öttinger. Dynamic mean-field models from a nonequilibrium thermodynamics perspective. *Phys. Rev. E* 68 (2003) 016115, 1–9.

Hütter, M., and H. C. Öttinger. Fluctuation-dissipation theorem, kinetic stochastic integral, and efficient simulations. *J. Chem. Soc. Faraday Trans.* 94 (1998) 1403–1406.

Ianniruberto, G., and G. Marrucci. Stress tensor and stress-optical law in entangled polymers. *J. Non-Newtonian Fluid Mech.* 79 (1998) 225–234.

Ilg, P., I. V. Karlin, and H. C. Öttinger. Generating moment equations in the Doi model of liquid-crystalline polymers. *Phys. Rev. E* 60 (1999) 5783–5787.

Ilg, P., and H. C. Öttinger. Nonequilibrium relativistic thermodynamics in bulk viscous cosmology. *Phys. Rev. D* 61 (2000) 023510, 1–10.

Irving, J. H., and J. G. Kirkwood. The statistical mechanical theory of transport processes. IV. The equations of hydrodynamics. *J. Chem. Phys.* 18 (1950) 817–829.

Israel, W. Nonstationary irreversible thermodynamics: a causal relativistic theory. *Ann. Phys. (N.Y.)* 100 (1976) 310–331.

Israel, W., and J. M. Stewart. Transient relativistic thermodynamics and kinetic theory. *Ann. Phys. (N.Y.)* 118 (1979) 341–372.

Jaynes, E. T. Information theory and statistical mechanics. *Phys. Rev.* 106 (1957) 620–630.

Jaynes, E. T. Information theory and statistical mechanics. II. *Phys. Rev.* 108 (1957) 171–190.

Jongschaap, R. J. J. Microscopic modeling of the flow properties of polymers. *Rep. Prog. Phys.* 53 (1990) 1–55.

Jongschaap, R. J. J. The matrix model, a diven state variables approach to non-equilibrium thermodynamics. *J. Non-Newtonian Fluid Mech.* 96 (2001) 63–76.

Jongschaap, R. J. J., and H. C. Öttinger. Equilibrium thermodynamics—Callen's postulational approach. *J. Non-Newtonian Fluid Mech.* 96 (2001) 5–17.

Jongschaap, R. J. J., and H. C. Öttinger. The mathematical representation of driven thermodynamic systems. *J. Non-Newtonian Fluid Mech.* 120 (2004) 3–9.

Jou, D., and J. Casas-Vázquez. Extended irreversible thermodynamics and its relation with other continuum approaches. *J. Non-Newtonian Fluid Mech.* 96 (2001) 77–104.

Jou, D., J. Casas-Vázquez, and G. Lebon. *Extended Irreversible Thermodynamics* (Springer, Berlin, 1996), 2nd edition.

Jou, D., J. Casas-Vázquez, and G. Lebon. Extended irreversible thermodynamics revisited (1988-98). *Rep. Prog. Phys.* 62 (1998) 1035–1142.

Kadanoff, L. P. Turbulent heat flow: structures and scaling. *Physics Today* 54/8 (2001) 34–39.

Karlin, I. V. Exact summation of the Chapman-Enskog expansion from moment equations. *J. Phys. A: Math. Gen.* 33 (2000) 8037–8046.

Karlin, I. V., A. N. Gorban, G. Dukek, and T. F. Nonnenmacher. Dynamic corrections to moment approximations. *Phys. Rev. E* 57 (1998) 1668–1672.

Kaufman, A. N. Dissipative Hamiltonian systems: a unifying principle. *Phys. Lett. A* 100 (1984) 419–422.

Kaufman, A. N. Lorentz-covariant dissipative Lagrangian systems. *Phys. Lett. A* 109 (1985) 87–89.

Kawasaki, K., and J. D. Gunton. Critical dynamics. In C. A. Croxton (Editor), *Progress in Liquid Physics* (Wiley, Chichester, 1978) p. 175–211.

Kevrekidis, I. G., C. W. Gear, J. M. Hyman, P. G. Kevrekidis, O. Runborg, and C. Theodoropoulos. Equation-free, coarse-grained multiscale computation: enabling microscopic simulators to perform system-level analysis. *Comm. Math. Sci.* 1 (2003) 715–762.

Kipnis, C., and C. Landim. *Scaling Limits of Interacting Particle Systems*. Grundlehren der Mathematischen Wissenschaften, Volume 320 (Springer, Berlin, 1999).

Kirkwood, J. G. Statistical mechanics of fluid mixtures. *J. Chem. Phys.* 3 (1935) 300–313.

Kirkwood, J. G. *Selected Topics in Statistical Mechanics*. John Gamble Kirkwood, Collected Works (Gordon and Breach, New York, 1967).

Kneller, G. R., and K. Hinsen. Computing memory functions from molecular dynamics simulations. *J. Chem. Phys.* 115 (2001) 11097–11105.

Kogan, A. M. Derivation of Grad's type equations and study of their relaxation properties by the method of maximization of entropy. *J. Appl. Math. Mech.* 29 (1965) 130–142.

Kremer, K., and G. S. Grest. Dynamics of entangled linear polymer melts: a molecular dynamics simulation. *J. Chem. Phys.* 92 (1990) 5057–5086.

Kreuzer, H. J. *Nonequilibrium Thermodynamics and its Statistical Foundations* (Oxford University Press, Oxford, 1981).

Kröger, M., M. Hütter, and H. C. Öttinger. Symbolic test of the Jacobi identity for given generalized "Poisson" bracket. *Comput. Phys. Commun.* 137 (2001) 325–340.

Kröger, M., and H. C. Öttinger. Beyond-equilibrium molecular dynamics of a rarefied gas subjected to shear flow. *J. Non-Newtonian Fluid Mech.* 120 (2004) 175–187.

Kröger, M., J. Ramírez, and H. C. Öttinger. Projection from an atomistic chain contour to its primitive path. *Polymer* 43 (2002) 477–487.

Kubo, R., M. Toda, and N. Hashitsume. *Nonequilibrium Statistical Mechanics*, Volume II of *Statistical Physics* (Springer, Berlin, 1991), 2nd edition.

Kuiken, G. D. C. *Thermodynamics of Irreversible Processes*. Wiley Tutorial Series in Theoretical Chemistry (Wiley, Chichester, 1994).

Lallemand, P., D. d'Humières, L.-S. Luo, and R. Rubinstein. Theory of the lattice Boltzmann method: three-dimensional model for linear viscoelastic fluids. *Phys. Rev. E* 67 (2003) 021203, 1–19.

Lamb, W., I. Murdoch, and J. Stewart. On an operator identity central to projection operator methodology. *Physica A* 298 (2001) 121–139.

Landau, D. P., and K. Binder. *A Guide to Monte Carlo Simulations in Statistical Physics* (Cambridge University Press, Cambridge, 2000).

Landau, L. D., and E. M. Lifshitz. *Fluid Mechanics*, Volume 6 of *Course of Theoretical Physics* (Pergamon, Oxford, 1987), 2nd edition.

Landau, L. D., and E. M. Lifshitz. *Statistical Physics, Part 2*, Volume 9 of *Course of Theoretical Physics* (Pergamon, Oxford, 1992), 2nd edition.

Lanford III, O. E. Time evolution of large classical systems. In J. Moser (Editor), *Dynamical Systems, Theory and Applications*, Volume 38 in Lecture Notes in Physics. Batelle Seattle 1974 Rencontres (Springer, Berlin, 1975) p. 1–111.

Laso, M., and H. C. Öttinger. Calculation of viscoelastic flow using molecular models: the CONNFFESSIT approach. *J. Non-Newtonian Fluid Mech.* 47 (1993) 1–20.

Lebon, G., M. Grmela, and D. Lhuillier. A comparative study of the coupling of flow with non-Fickean thermodiffusion. Part I: extended irreversible thermodynamics. *J. Non-Equilib. Thermodyn.* 28 (2003) 1–22.

Leimkuhler, B. A separated form of Nosé dynamics for constant temperature and pressure simulation. *Comput. Phys. Commun.* 148 (2002) 206–213.

Leimkuhler, B. J., and C. R. Sweet. The canonical ensemble via symplectic integrators using Nosé and Nosé-Poincaré chains. *J. Chem. Phys.* 121 (2004) 108–116.

Lewis, R. M. A unifying principle in statistical mechanics. *J. Math. Phys.* 8 (1967) 1448–1459.

Lhuillier, D., M. Grmela, and G. Lebon. A comparative study of the coupling of flow with non-Fickean thermodiffusion. Part III: internal variables. *J. Non-Equilib. Thermodyn.* 28 (2003) 51–68.

Lodge, A. S. *Body Tensor Fields in Continuum Mechanics* (Academic Press, New York, 1974).

Maartens, R., and V. Méndez. Nonlinear bulk viscosity and inflation. *Phys. Rev. D* 55 (1997) 1937–1942.

Macdonald (Editor), J. R. *Impedance Spectroscopy: Emphasizing Solid Materials and Systems* (Wiley, New York, 1987).

Marinari, E., and G. Parisi. Simulated tempering–a new Monte Carlo scheme. *Europhys. Lett.* 19 (1992) 451–458.

Marrucci, G., and G. Ianniruberto. Open problems in tube models for concentrated polymers. *J. Non-Newtonian Fluid Mech.* 82 (1999) 275–286.

Marsden, J. E. *Lectures on Mechanics* (Cambridge University Press, Cambridge, 1992).

Marsden, J. E., R. Montgomery, P. J. Morrison, and W. B. Thompson. Covariant Poisson brackets for classical fields. *Ann. Phys. (N.Y.)* 169 (1986) 29–47.

Marsden, J. E., and T. S. Ratiu. *Introduction to Mechanics and Symmetry*. Texts in Applied Mathematics, Volume 17 (Springer, New York, 1999), 2nd edition.

Marsden, J. E., and M. West. Discrete mechanics and variational integrators. *Acta Numerica* 10 (2001) 357–514.

Martyna, G. J., D. J. Tobias, and M. L. Klein. Constant pressure molecular dynamics algorithms. *J. Chem. Phys.* 101 (1994) 4177–4189.

Martyna, G. J., M. E. Tuckerman, D. J. Tobias, and M. L. Klein. Explicit reversible integrators for extended systems dynamics. *Mol. Phys.* 87 (1996) 1117–1157.

Maugin, G. A. *The Thermomechanics of Nonlinear Irreversible Behaviors. An Introduction*. Series on Nonlinear Science, Series A, Volume 27 (World Scientific, Singapore, 1999).

Mavrantzas, V. G., T. D. Boone, E. Zervopoulou, and D. N. Theodorou. End-bridging Monte Carlo: a fast algorithm for atomistic simulation of condensed phases of long polymer chains. *Macromolecules* 32 (1999) 5072–5096.

Mavrantzas, V. G., and H. C. Öttinger. Atomistic Monte Carlo simulations of polymer melt elasticity: their nonequilibrium thermodynamics GENERIC formulation in a generalized canonical ensemble. *Macromolecules* 35 (2002) 960–975.

Mavrantzas, V. G., and D. N. Theodorou. Atomistic simulation of polymer melt elasticity: calculation of the free energy of an oriented polymer melt. *Macromolecules* 31 (1998) 6310–6332.

Maxwell, J. C. *Theory of Heat* (Dover, New York, 2001).

McLeish, T. C. B., and R. G. Larson. Molecular constitutive equations for a class of branched polymers: the pom-pom polymer. *J. Rheol.* 42 (1998) 81–110.

Meirovitch, H. A new method for simulation of real chains: scanning future steps. *J. Phys. A: Math. Gen.* 15 (1982) L735–L741.

Meirovitch, H. Method for estimating the entropy of macromolecules with computer simulation. Chains with excluded volume. *Macromolecules* 16 (1983) 249–252.

Merriam-Webster's Collegiate Dictionary (Merriam-Webster, Springfield, 1993).

Metropolis, N., A. W. Rosenbluth, M. N. Rosenbluth, A. H. Teller, and E. Teller. Equation of state calculations by fast computing machines. *J. Chem. Phys.* 21 (1953) 1087–1092.

Miller, D. G. Thermodynamics of irreversible processes: the experimental verification of the Onsager reciprocal relations. *Chem. Rev.* 60 (1960) 15–37.

Misner, C. W., K. S. Thorne, and J. A. Wheeler. *Gravitation* (Freeman, New York, 1973).

Mori, H. A continued-fraction representation of the time-correlation functions. *Prog. Theor. Phys.* 34 (1965) 399–416.

Mori, H. Transport, collective motion, and Brownian motion. *Prog. Theor. Phys.* 33 (1965) 423–455.

Mori, H., I. Oppenheim, and J. Ross. Some topics in quantum statistics: the Wigner function and transport theory. In J. de Boer and G. E. Uhlenbeck (Editors), *Collected Papers*, Volume I in Studies in Statistical Mechanics (North-Holland, Amsterdam, 1962) p. 213–298.

Morrison, P. J. Bracket formulation for irreversible classical fields. *Phys. Lett. A* 100 (1984) 423–427.

Morrison, P. J. Hamiltonian description of the ideal fluid. *Rev. Mod. Phys.* 70 (1998) 467–521.

Moser, J. Lectures on Hamiltonian systems. *Mem. Am. Math. Soc.* 81 (1968) 1–60.

Mouritsen, O. G. *Computer Studies of Phase Transitions and Critical Phenomena*. Springer Series in Computational Physics (Springer, Berlin, 1984).

Müller, I. Zum Paradoxon der Wärmeleitungstheorie. *Z. Physik* 198 (1967) 329–344.

Muschik, W., S. Gümbel, M. Kröger, and H. C. Öttinger. A simple example for comparing GENERIC with rational non-equilibrium thermodynamics. *Physica A* 285 (2000) 448–466.

Muschik, W., C. Papenfuss, and H. Ehrentraut. *Concepts of Continuum Thermodynamics* (Kielce University of Technology, Kielce, 1996).

Nelson, E. *Dynamical Theories of Brownian Motion* (Princeton University Press, Princeton, 1967).

Nosé, S. A molecular dynamics method for simulations in the canonical ensemble. *Mol. Phys.* 52 (1984) 255–268.

Nosé, S. Constant temperature molecular dynamics methods. *Prog. Theor. Phys.* Suppl. 103 (1991) 1–46.

Onsager, L. Reciprocal relations in irreversible processes. I. *Phys. Rev.* 37 (1931) 405–426.

Onsager, L. Reciprocal relations in irreversible processes. II. *Phys. Rev.* 38 (1931) 2265–2279.

Öttinger, H. C. Computer simulation of three-dimensional multiple-chain systems: scaling laws and virial coefficients. *Macromolecules* 18 (1985) 93–98.

Öttinger, H. C. Renormalization-group calculation of excluded-volume effects on the viscometric functions for dilute polymer solutions. *Phys. Rev. A* 40 (1989) 2664–2671.

Öttinger, H. C. Renormalization of various quantities for dilute polymer solutions undergoing shear flow. *Phys. Rev. A* 41 (1990) 4413–4420.

Öttinger, H. C. Series expansions for shear-flow properties from the Doi-Edwards model. *J. Non-Newtonian Fluid Mech.* 37 (1990) 265–279.

Öttinger, H. C. A class of multiparticle constitutive equations. *J. Rheol.* 35 (1991) 1275–1301.

Öttinger, H. C. Incorporation of polymer diffusivity and migration into constitutive equations. *Rheol. Acta* 31 (1992) 14–21.

Öttinger, H. C. Modified reptation model. *Phys. Rev. E* 50 (1994) 4891–4895.

Öttinger, H. C. *Stochastic Processes in Polymeric Fluids: Tools and Examples for Developing Simulation Algorithms* (Springer, Berlin, 1996).

Öttinger, H. C. GENERIC formulation of Boltzmann's kinetic equation. *J. Non-Equilib. Thermodyn.* 22 (1997) 386–391.

Öttinger, H. C. General projection operator formalism for the dynamics and thermodynamics of complex fluids. *Phys. Rev. E* 57 (1998) 1416–1420.

Öttinger, H. C. Relativistic and nonrelativistic description of fluids with anisotropic heat conduction. *Physica A* 254 (1998) 433–450.

Öttinger, H. C. Thermodynamically admissible equations for causal dissipative cosmology, galaxy formation, and transport processes in a gravitational collapse. *Phys. Rev. D* 60 (1999) 103507, 1–9.

Öttinger, H. C. A thermodynamically admissible reptation model for fast flows of entangled polymers. *J. Rheol.* 43 (1999) 1461–1493.

Öttinger, H. C. Derivation of two-generator framework of nonequilibrium thermodynamics for quantum systems. *Phys. Rev. E* 62 (2000) 4720–4724.

Öttinger, H. C. Thermodynamically admissible reptation models with anisotropic tube cross sections and convective constraint release. *J. Non-Newtonian Fluid Mech.* 89 (2000) 165–185.

Öttinger, H. C. Thermodynamic admissibility of the pompon model for branched polymers. *Rheol. Acta* 40 (2001) 317–321.

Öttinger, H. C. GENERIC projection-operator derivation of Boltzmann's kinetic equation. *J. Non-Equilib. Thermodyn.* 27 (2002) 105–119.

Öttinger, H. C. Modeling complex fluids with a tensor and a scalar as structural variables. *Rev. Mex. Fís.* 48 (2002) Supl. 1, 220–229.

Öttinger, H. C. Coarse-graining of wormlike polymer chains for substantiating reptation. *J. Non-Newtonian Fluid Mech.* 120 (2004) 207–213.

Öttinger, H. C., and A. N. Beris. Thermodynamically consistent reptation model without independent alignment. *J. Chem. Phys.* 110 (1999) 6593–6596.

Öttinger, H. C., and M. Grmela. Dynamics and thermodynamics of complex fluids. II. Illustrations of a general formalism. *Phys. Rev. E* 56 (1997) 6633–6655.

Öttinger, H. C., and Y. Rabin. Renormalization-group calculation of viscometric functions based on conventional polymer kinetic theory. *J. Non-Newtonian Fluid Mech.* 33 (1989) 53–93.

Öttinger, H. C., B. H. A. A. van den Brule, and M. A. Hulsen. Brownian configuration fields and variance reduced CONNFFESSIT. *J. Non-Newtonian Fluid Mech.* 70 (1997) 255–261.

Parrinello, M., and A. Rahman. Crystal structure and pair potentials: a molecular dynamics study. *Phys. Rev. Lett.* 45 (1980) 1196–1199.

Parrinello, M., and A. Rahman. Polymorphic transitions in single crystals: a new molecular dynamics method. *J. Appl. Phys.* 52 (1981) 7182–7190.

Parrinello, M., and A. Rahman. Strain fluctuations and elastic constants. *J. Chem. Phys.* 76 (1982) 2662–2666.

Pasquali, M., and L. E. Scriven. Theoretical modeling of microstructured liquids: a simple thermodynamic approach. *J. Non-Newtonian Fluid Mech.* 120 (2004) 101–135.

Peter, C., C. Oostenbrink, A. van Dorp, and W. F. van Gunsteren. Estimating entropies from molecular dynamics simulations. *J. Chem. Phys.* 120 (2004) 2652–2661.

Press, W. H., S. A. Teukolsky, W. T. Vetterling, and B. P. Flannery. *Numerical Recipes in FORTRAN* (Cambridge University Press, Cambridge, 1992), 2nd edition.

Prigogine, I. *Introduction to Thermodynamics of Irreversible Processes* (Wiley, New York, 1967), 3rd edition.

Reichl, L. E. *A Modern Course in Statistical Physics* (University of Texas Press, Austin, 1980).

Robertson, B. Equations of motion in nonequilibrium statistical mechanics. *Phys. Rev.* 144 (1966) 151–161.

Ryckaert, J. P., and G. Ciccotti. Introduction of Andersen's demon in the molecular dynamics of systems with constraints. *J. Chem. Phys.* 78 (1983) 7368–7374.

Saab, H. H., R. B. Bird, and C. F. Curtiss. A kinetic theory for polymer melts. V. Experimental comparisons for shear-flow rheological properties. *J. Chem. Phys.* 77 (1982) 4758–4766.

Sanz-Serna, J. M., and M. P. Calvo. *Numerical Hamiltonian Problems*. Applied Mathematics and Mathematical Computation, Volume 7 (Chapman & Hall, London, 1994).

Sarti, G. C., and G. Marrucci. Thermomechanics of dilute polymer solutions: multiple bead-spring model. *Chem. Eng. Sci.* 28 (1973) 1053–1059.

Schäfer, L. *Excluded Volume Effects in Polymer Solutions* (Springer, Berlin, 1999).

Schwarzl, F. R. *Polymermechanik: Struktur und mechanisches Verhalten von Polymeren* (Springer, Berlin, 1990).

Serrano, M., and P. Español. Thermodynamically consistent mesoscopic fluid particle model. *Phys. Rev. E* 64 (2001) 046115, 1–18.

Struchtrup, H., and M. Torrilhon. Regularization of Grad's 13 moment equations: derivation and linear analysis. *Phys. Fluids* 15 (2003) 2668–2680.

Tesi, M. C., E. J. Janse van Rensburg, E. Orlandini, and S. G. Whittington. Monte Carlo study of the interacting self-avoiding walk model in three dimensions. *J. Stat. Phys.* 82 (1996) 155–181.

Theodorou, D. N., T. D. Boone, L. R. Dodd, and K. F. Mansfield. Stress tensor in model polymer systems with periodic boundaries. *Makromol. Chem., Theory Simul.* 2 (1993) 191–238.

Toda, M., R. Kubo, and N. Saitô. *Equilibrium Statistical Mechanics*, Volume I of *Statistical Physics* (Springer, Berlin, 1992), 2nd edition.

Tribus, M., and R. B. Evans. The probability foundations of thermodynamics. *Appl. Mech. Rev.* 16 (1963) 765–769.

Truesdell, C., and W. Noll. *The Non-Linear Field Theories of Mechanics* (Springer, Berlin, 1992), 2nd edition.

Tschoegl, N. W. *The Phenomenological Theory of Linear Viscoelastic Behavior* (Springer, Berlin, 1989).

Udey, N., and W. Israel. General relativistic radiative transfer: the 14-moment approximation. *Mon. Not. R. Astron. Soc.* 199 (1982) 1137–1147.

van der Schaft, A. J., and B. M. Maschke. On the Hamiltonian formulation of nonholonomic mechanical systems. *Rep. Math. Phys.* 34 (1994) 225–233.

van der Schaft, A. J., and B. M. Maschke. Hamiltonian formulation of distributed-parameter systems with boundary energy flow. *J. Geom. Phys.* 42 (2002) 166–194.

van Meerveld, J. Note on thermodynamic consistency of the integral pom-pom model. *J. Non-Newtonian Fluid Mech.* 108 (2002) 291–299.

Velasco, R. M., F. J. Uribe, and L. S. García-Colín. Inconsistencies in moment methods. *Phys. Rev. E* 66 (2002) 032103, 1–4.

Wagner, A. J. Derivation of a non-objective Oldroyd model from the Boltzmann equation (2001). ArXiv:cond-mat/0105067.

Wagner, N. J. The Smoluchowski equation for colloidal suspensions developed and analyzed through the GENERIC formalism. *J. Non-Newtonian Fluid Mech.* 96 (2001) 177–201.

Wagner, N. J., H. C. Öttinger, and B. J. Edwards. Generalized Doi-Ohta model for multiphase flow developed via GENERIC. *AIChE J.* 45 (1999) 1169–1181.

Wang, F., and D. P. Landau. Determining the density of states for classical statistical models: a random walk algorithm to produce a flat histogram. *Phys. Rev. E* 64 (2001) 056101, 1–16.

Wang, F., and D. P. Landau. Efficient, multiple-range random walk algorithm to calculate the density of states. *Phys. Rev. Lett.* 86 (2001) 2050–2053.

Wapperom, P., and M. A. Hulsen. Thermodynamics of viscoelastic fluids: the temperature equation. *J. Rheol.* 42 (1998) 999–1019.

Wehrl, A. General properties of entropy. *Rev. Mod. Phys.* 50 (1978) 221–260.

Wei, J. Irreversible thermodynamics in engineering. *Ind. Eng. Chem.* 58 (1966) 55–60.

Weinberg, S. *Gravitation and Cosmology, Principles and Applications of the General Theory of Relativity* (Wiley, New York, 1972).

Wilchinsky, A. V., S. H. Faria, and K. Hutter. Thermodynamic modelling of polymer solutions with molecular form birefringence. *J. Non-Newtonian Fluid Mech.* 99 (2001) 57–77.

Winter, H. H. Analysis of dynamic mechanical data: inversion into a relaxation time spectrum and consistency check. *J. Non-Newtonian Fluid Mech.* 68 (1997) 225–239.

Wodarzik, U. F. On relativistic heat conduction. *Phys. Lett. A* 91 (1982) 279–281.

Woods, L. C. *The Thermodynamics of Fluid Systems.* Oxford Engineering Science Series (Clarendon, Oxford, 1975).

Zimdahl, W. Bulk viscous cosmology. *Phys. Rev. D* 53 (1996) 5483–5493.

Zimdahl, W. Cosmological particle production and generalized thermodynamic equilibrium. *Phys. Rev. D* 57 (1998) 2245–2254.

Zimm, B. H., G. M. Roe, and L. F. Epstein. Solution of a characteristic value problem from the theory of chain molecules. *J. Chem. Phys.* 24 (1956) 279–280.

Zwanzig, R. Memory effects in irreversible thermodynamics. *Phys. Rev.* 124 (1961) 983–992.

Zwanzig, R. Hydrodynamic fluctuations and Stokes' law friction. *J. Res. Natl. Bur. Std. (U.S.)* 68B (1964) 143–145.

Zwanzig, R. *Nonequilibrium Statistical Mechanics* (Oxford University Press, New York, 2001).

Author Index

Subject Index